DENSE Z-PINCHES

THIRD INTERNATIONAL CONFERENCE

AIP CONFERENCE PROCEEDINGS 299

DENSE Z-PINCHES

THIRD INTERNATIONAL CONFERENCE

LONDON, UNITED KINGDOM 1993

EDITORS:

MALCOLM HAINES
ANDREW KNIGHT
IMPERIAL COLLEGE

AIP PRESS

American Institute of Physics New York

L.C. Catalog Card No. 93-74569
ISBN 1-56396-297-7
DOE CONF-9304126

Printed in the United States of America.

Contents

STABILITY

THEORY AND SIMULATIONS

X-RAY SOURCES AND X-RAY LASERS

PLASMA FOCUS

IMPLOSIONS

FIBER PINCHES

SPECTROSCOPY

X-PINCHES, VACUUM SPARKS, AND OTHER PINCHES

A WIDER VIEW

PREFACE

This book is the Proceedings of the Third International Conference on Dense Z-Pinches which was held at Imperial College, London, UK from April 19–23, 1993.

The first conference in this series was held in Alexandria, Virginia in March 1984, and was attended by 34 scientists who were mainly interested in the application of the Z-pinch to fusion. The second conference, held in Laguna Beach, California in April 1989, attracted 107 attendees, and the subject was expanded to include the Z-pinch as an x-ray source and other applications. Only 38 of the participants were from outside the U.S., and there was some apprehension about holding the third conference in Europe. However, there was an increased attendance of 117, despite the cutbacks in some of the programs, of whom 25 came from the U.S.

A special effort was made to raise money to provide travel grants for those, especially from Russia and other Eastern European countries, who would otherwise be unable to attend. We are very grateful to our generous sponsors who made this possible; these are AWE Aldermaston, The Royal Society, E.O.A.R.D., Kentech, Ltd., Maxwell Laboratories, and Imperial College.

The subject of Z-pinches is of growing importance, and the conference coincided with the grand opening ceremony of the new terawatt class generator MAGPIE (mega-ampere generator for plasma implosion experiments) for the dense Z-pinch project at Imperial College. At this ceremony Professor S. I. Braginsky and Dr. R. S. Pease FRS, who independently developed the theory of the critical current at which bremsstrahlung losses balance the Joule heating in a confined Z-pinch, met each other for the first time. Together with Mr. J. C. Martin, the inventor of pulse power, they gave some encouraging speeches to those in this field, especially to the team that designed and built MAGPIE.

At the conference the subject of greatest interest was stability. It is now well established that theory based on ideal MHD is of limited applicability, and there were several papers describing stabilizing effects due to collapsing multilayers, or if a plasma surrounds a metal fiber, or when microchannel preionization is employed. Of particular interest was a full kinetic treatment of large ion Larmor radius effects on stability. Even if instabilities grow, there is experimental evidence that the plasma column can be well confined and can even benefit from ion viscous heating. The subjects of neutron anisotropy, ion and electron beam formation, micropinches (hot spots) made progress with several more advanced diagnostics being employed. The implosion of shells, wire arrays, and gas puffs, principally for x-ray source development leads to a comparison of detailed radiative transport codes with experimental data. Recent changes in the former Soviet Union have led to a relaxation of secrecy of much of their Z-pinch and focus work, and indeed two papers were presented of international collaboration employing Angara-5 at Troitzk.

As Chairman of the Local Organizing Committee I would like to pay tribute to Michael Coppins and Bucker Dangor for their work on the scientific program, and most especially to Andrew Knight who made all the administrative arrangements, not least the splendid banquet with music in Skinners' Hall.

Malcolm Haines

Third International Conference on Dense Z-pinches
April 19–23, 1993, Imperial College, London, UK

List of Delegates
(Delegates are listed by country, under their current affiliations)

Argentina

INFID, Facultad de Ciencias Exactas Y Naturales
Pab. 1, Ciudad Universitaria
1428 Buenos Aires, Argentina — Dr. H. Bruzzone

Chile

Facultad de Fisica
Pontifica Universidad Catolica de Chile
Casilla 306
Santiago 22, Chile — Dr. H. Chuaqui

Czech Republic

Czech Acad. Sci.
Institute of Plasma Physics, P. O. Box 17
182 11 Prague 8, Czech Republic — Dr. A. Krejčí, Dr. J. Rauš

Department of Physics
Electrotechnical Faculty, CTU
Technická 2
166 27 Prague 6, Czech Republic — Dr. Prof. P. Kubeš

France

CEA, CAM/MR
BP 12
91680 Bruyères-le-Châtel, France — Dr. D. Friart

CEA, CESTA
BP 2, Route des Gargails
33114 Le Barp, France — Dr. P. Romary, L. Voisin

L.P.M.I.
Ecole Polytechnique
91128 Palaiseau Cédex
France — S. Attelan, Dr. H. Doucet, Dr. N. S. Edison, S. Sube, Dr. B. Etlicher, G. Dontevieux, F. Douieb, C. Rouillé, P. Zehnter

CEA
31-33, Rue de la Fédération
75752 Paris Cédex 15
France — Dr. A. Bernard

Gremi-Université d'Orléans 45067 Orléans Cédex 2, France	Dr. C. Fleurier
Laboratoire des Plasmas Denses CNRS, Université P. et M. Curie Tour 12, Etage 5, 4 Place Jussieu F-75252 Paris Cédex 05, France	Dr. J. Larour Dr. M. Skowronek
Germany	
RWTH Aachen, Steinbachstr. 15 D 5100 Aachen, Germany	Dr. R. Lebert D. Rothweiler
Inst. f. Experimentalphysik Heinrich-Heine-Universität Universitätsstr. 1 W-4000 Düsseldorf, Germany	Prof. Dr. G. Decker Dr. W. Kies P. Röwekamp F. Schmitz S. Stein
Inst. f. Angewandte Physik Heinrich-Heine-Universität (as above)	Prof. Dr. P. Thielemann B. Nensel
Institut f. Experimentalphysik V Ruhr-Universität Bochum Universitätsstr. 150 W-4630 Bochum 1, Germany	Prof. Dr. H.-J. Kunze Dr. F. B. Rosmej
Universität Erlangen, Physikalisches Institut I Erwin-Rommel-Str. 1 W-8520 Erlangen, Germany	P. J. Hagen T. C. Wagner
Universität Stuttgart, Inst. f. Plasmaforschung Pfaffenwaldring 31 D-7000 Stuttgart 80, Germany	Dr. H. Schmidt
Israel	
Racah Institute of Physics Hebrew University of Jerusalem GIVAT-RAM, 91904 Jerusalem, Israel	D. Stutman
Weizman Institute of Science 76100 Rehovot, Israel	Prof. Y. Maron
Italy	
Università di Ferrara, via del Paradiso 12 44100 Ferrara, Italy	Prof. J. G. Linhart

Japan	
Dept. of Electronic Eng., Gunma University Kiryu, Gunma 376, Japan	Dr. K. Hirano
Atomic Energy Research Institute Nihon University 1-8 Kanda-Surugadai, Chiyoda-ku 101 Tokyo, Japan	Dr. T. Miyamoto K. Takasugi
Malaysia	
Plasma Research Laboratory, Physics Department University of Malaya 59100 Kuala Lumpur, Malaysia	Dr. C. S. Wong
Mexico	
Instituto de Ciencias Nucleares, UNAM Apdo. Postal 70-543 Circuito Exterior de C.U. 04510 México D. F., Mexico	Dr. J. J. E. Herrera
Poland	
Institute of Plasma Physics and Laser Microfusion P.O. Box 49, 00-908 Warsaw, Poland	Dr. S. Denus Dr. L. Karpinski
Dept. of Thermonuclear Research Soltan Institute for Nuclear Studies 05-400 Swierk n. Warsaw, Poland	Prof. M. Sadowski
Romania	
Institute of Atomic Physics IFTAR, Lab. 22, P.O. Box MG-7 Bucharest R-76900, Romania	Dr. N.-B. Mandache Dr. I. Tiseanu
Faculty of Physics, University of Bucharest P.O. Box MG 11, Magurele Bucharest R-76900, Romania	C. Dumitrescu-Zoita
Russia	
High Current Electronics Institute Academichesky Ave. 4 634055 Tomsk, Russia	Prof. R. B. Baksht Dr. S. A. Sorokin
Institute of Spectroscopy Russian Academy of Science Troitzk, 142092 Moscow Region, Russia	Prof. K. N. Koshelev Dr. Yu. V. Sidelnikov
Institute of Technical Physics, Dept. 3	

P.O. Box 245, 454070 Chelyabinsk-70, Russia — Dr. O. V. Diyankov

Institute of Theoretical and Experimental Physics
B. Cheremushkinskaya 25
117259 Moscow, Russia
— Dr. P. V. Sasorov, Dr. A. V. Gerusov

I. V. Kurchatov Institute
Kurchatov Square 1
123182 Moscow, Russia
— Dr. L. Rudakov, Dr. E. O. Baronova, Dr. A. B. Kukushkin, Dr. V. V. Vikhrev

P. N. Lebedev Physical Institute
Russian Academy of Sciences
Leninsky Prospect 53
117924 Moscow, Russia
— Prof. V. A. Gribkov

Scientific Research Institute of Technical Glass
Krzhizhanovsky Street 29
117218 Moscow, Russia
— Dr. V. L. Kantsyrev, Dr. A. S. Shlyaptseva

TRINITI, Troitzk Branch of Kurchatov
Institute of Atomic Energy
Troitzk, Moscow Region, Russia
— Prof. V. P. Smirnov, Dr. S. L. Nedoseev

Sweden

Alfvén Laboratory, Royal Institute of Technology
100 44 Stockholm, Sweden
— Dr. J. Scheffel

Group of Theoretical Electronics
Uppsala University
Villavägen 4, 75121 Uppsala, Sweden
— Dr. A. B. Bud'ko

United Kingdom

AWE PLC
Aldermaston
Berkshire RG7 4PR, UK
— Dr. P. Davis, Dr. M. Goodman, Dr. K. Baird, Dr. D. Forster, Dr. P. C. Thompson

EOARD
223/231 Old Marylebone Road
London NW1 5TH, UK
— Dr. K. Hackett

Fraser–Nash Consultancy
Shelsley House, Randalls Way
Leatherhead, Surrey KT22 7TX, UK
— Dr. C. Christou

Kentech Instruments, Ltd., 10 Badgers Wood
Farnham Common, Bucks SL2 3HH, UK
— Dr. A. Dymoke-Bradshaw

Plasma Physics Group
Blackett Laboratory
Imperial College
— R. Aliaga-Rossel, Dr. J. Bayley, Prof. M. G. Haines, Dr. T. D. Arber, F. Beg

Prince Consort Road London SW7 2BZ, UK	Dr. A. R. Bell Dr. P. Choi A. E. Dangor P. Jaitly Dr. I. H. Mitchell	Dr. J. P. Chittenden Dr. M. Coppins Dr. A. Folkierski P. Lee Choon Keat J. Worley
Department of Engineering Science, Oxford University Parks Road, Oxford OX1 3PJ, UK		Dr. N. Graneau
Pease Partners, West Ilsley Newbury, Berks RG16 0AW, UK		Dr. R. S. Pease
United States		
Berkeley Research Associates, Inc. P.O. Box 852, Springfield, VA 22150		Dr. N. R. Pereira
Laboratory of Plasma Studies, Cornell University 369 Upson Hall, Ithaca, NY 14853		D. H. Kalantar
Defense Nuclear Agency 6801 Telegraph Road Alexandria, VA 22310		Dr. K. Ware
F.E.A.T. Corporation, Inc., 391-B Chipeta Way Salt Lake City, UT 84108		F. G. Jaeger
HY-Tech Research Corporation, 104 Centre Court Radford, VA 24141		Dr. E. J. Yadlowsky
Lawrence Livermore National Laboratory P.O. Box 808, L-637 Livermore, CA 94551		Dr. C. W. Hartman
Logicon/RDA, 2100 Washington Blvd. Arlington, VA 22204-5706		Dr. I. Vitkovitsky
Los Alamos National Laboratory P-1, MS E-526 Los Alamos, NM 87545	J. C. Cochrane Dr. J. V. Parker	Dr. J. S. Shlachter W. Matuska Dr. D. L. Peterson P. T. Sheehey
Maxwell Laboratories, Inc. 9244 Balboa Avenue San Diego, CA 92123		Dr. E. Waisman
Naval Research Laboratory Code 6773 4555 Overlook Avenue SW Washington DC 20375-5000		Dr. J. Davis Dr. A. E. Robson Dr. K. G. Whitney Dr. F. C. Young
Physics International Company 2700 Merced Street		

San Leandro, CA 94577	Dr. C. Deeney
Sandia National Laboratories Department 1273, P.O. Box 5800 Albuquerque, NM 87185	Dr. T. J. Nash Dr. R. B. Spielman
Physics Department University of California, Irvine Irvine, CA 92717-4575	G. G. Peterson
Institute of Geophysics & Planetary Physics University of California, Los Angeles Los Angeles, CA 90024-1567	Dr. S. I. Braginsky
University of Illinois 214 NEL, 103 S. Goodwin Avenue Urbana, IL 61801	Prof. G. H. Miley
Laboratory for Plasma Research University of Maryland College Park, MD 20742	Dr. A. L. Velikovich

STABILITY

IMPROVED STABILITY OF GAS-EMBEDDED Z-PINCHES USING A MICRO-CHANNEL PREIONISATION

John Benage,
P-1, Los Alamos National Laboratory, Los Alamos, NM 87545, USA

Maurice Skowronek and Paul Roméas
Laboratoire des Plasmas Denses, Tour 12 E5, Univ. P. & M. Curie
F-75252 Paris cedex 05 (France).

ABSTRACT

A micro-channel using corona effect at atmospheric pressure, either in hydrogen or in air, is created prior to the triggering of the main discharge. The channel diameter has been measured to have about 10 µm diameter, with a cw intensity of the order of 0.1 mA. The main discharge is powered by a Marx generator delivering a current pulse of 240 kA during 200 ns at the end of a water line having 1 Ω impedance. Four pictures of the pinch are taken at each shot delayed in time, each from the other, from 10 ns to 260 ns.

The discharge is seen to remain stable against usual instabilities during more than 500 ns. When a m = 1 instability appears sometime, its relative amplitude seem to remain constant as a function of time, at a very low level.

The expansion of the channel takes place at very high velocities. In air, the expansion is slowered at a time corresponding to the maximum intensity.

The influence of the micro-channel characteristics has been clearly checked : when it is not well-established, the discharge becomes unstable very early, even with the use of the pointed electrodes.

INTRODUCTION

Sausage instabilities (m = 0) or helix instabilities (m = 1) are well-known to perturb very rapidly, in a few tens of nanoseconds, the plasma column produced in a Z-Pinch, leading to its destruction. Recently, theoretical predictions and some experimental observations lead to proposals of new configurations avoiding these catastrophic evolutions[1,2]. Moreover, the current distribution as a function of the radius may present, in some cases, an inversion called "inverse skin effect" which has for consequence to enhance the pinch current intensity[3]. Experiments conducted at NRL[4] and at LANL[5] with fibers have shown some trend to an anomalous stability. One of the proposal for improving the pinch stability has been the use of a preionisation in a microchannel ($\Phi = 10\ \mu m$). The main current is injected through this channel and the produced pinch is observed to be stable[6]. Experiments conducted at Santiago de Chile have confirmed these observations[7,8]. By means of interferometric methods, the electron density profiles have been determined. These profiles are strongly peaked on pinch axis, showing two different evolutions: the axial zone goes inward due to the pinch, while the outer zone expands rapidly outward due to the accretion.

EXPERIMENTAL SET-UP

The preionisation channel is produced using corona effect between tungsten needles, separated by about 1 cm. The tip diameter is about 0.5 μm. Two copper rings centered on each needle equalize the potentials in view to stabilize the micro-channel. The current intensity flowing through it is about 100 μA. The experiment is conducted either in air at atmospheric pressure or in hydrogen at a pressure $p = 1.25$ atm. The main generator which is connected by means of a spark-gap to the channel, is composed by a Marx generator followed by a water line whose impedance is 1 Ω. The current intensity is measured by means of a Rogowski coil. The current derivative is recorded together with the voltage signal obtained through a capacitive voltage divider. The sweep speed is set to 100 ns/div.

RESULTS

On figure 1, a typical oscillogram shows the current derivative di/dt as given by a Rogowski coil and the voltage by a capacitor divider. The current derivative di/dt reaches the maximum value of 1.4×10^{12} A/s in about 30 ns. The maximum current intensity of 240 kA is reached after 220 ns. The instant that will be chosen as the origin of times is easily determined.

The pictures of the discharge are taken by triggering four light intensifiers and the trigger pulses are recorded on separate oscillographs allowing for each picture a precise determination of the picture timing in a 10ns picture duration. A set of four images is taken at each shot by means of four microchannel plate intensifiers triggered at different instants of the discharge. The delay time has been varied from 5 ns to 270 ns, the exposure time being less than 10 ns. The optical magnification has the value 2.5 for a first series of experiments and 0.6 for another.

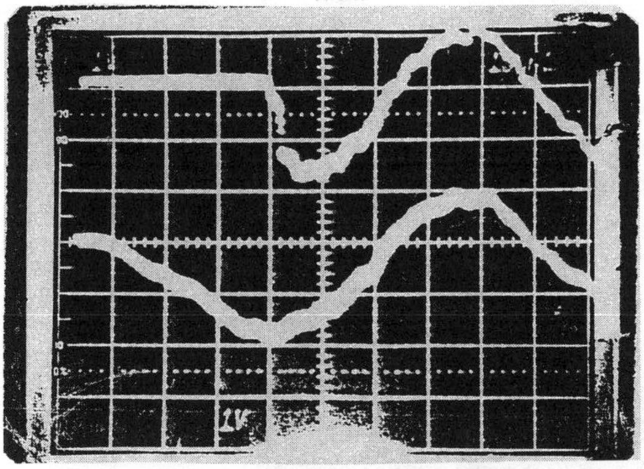

Fig.1. Oscillograms.
upper trace : $di/dt = 6.7x10^{11}$ A/s/div ;
lower trace : voltage ; time :100 ns/div.

t = 65 ns

t = 175 ns

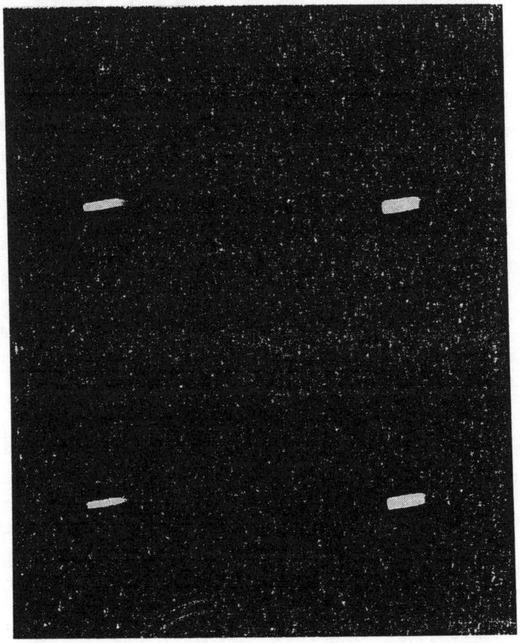

t = 130 ns

t = 150 ns

Fig. 2. Frame images in air. $\Delta t = 10$ ns.

On figure 2, a set of 4 pictures is displayed, concerning a discharge in air, at atmospheric pressure taken at times : $t_1 = 65$ ns,

t_2 = 130 ns, t_3 = 150 ns and t_4 = 175 ns. The original magnification was 0.6. The discharges appear to be stable. In air, in some cases, slight deformations are seen and may be related to a misalignment between the needles. In air, the discharge pictures have a sharp limit and do not show any structure. This is due to the high opacity of the discharge. The electron density may be very high, of the order of 10^{20} cm^{-3}.

t = 5 ns

t = 170 ns

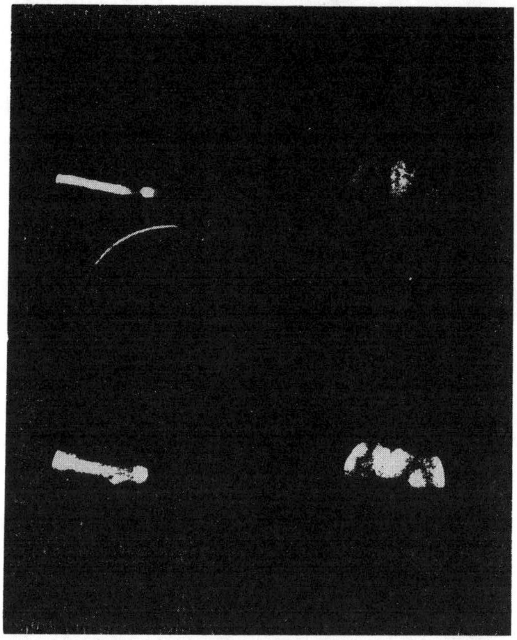

t = 70 ns

t = 110 ns

Fig.3. Frame images in air : p = 1 atm; Δt = 10 ns.

On figure 3, a set of 4 pictures is displayed, concerning a discharge in air, at p = 1 atm, taken at : t_1 = 5 ns, t_2 = 70 ns, t_3 = 110 ns and t_4 = 170 ns. The original magnification was G = 2.5.

On figure 4, a set of 4 pictures is displayed, concerning a discharge in hydrogen, at pressure p = 1.25 atm, taken at times : t_1 = 100 ns, t_2 = 175 ns, t_3 = 220 ns and t_4 = 270 ns. The original magnification was 0.6. In hydrogen, some helix is observed, with a thread of the screw varying from 5 mm to 8 mm; however, the discharge expands cylindrically from the initial shape without any further growth of the instability. In this case, the outer plasma is not opaque and the structure may be seen: the axial zone appears to be more luminous indicating that the current density and the electron density is maximum on axis.

t = 100 ns t = 270 ns

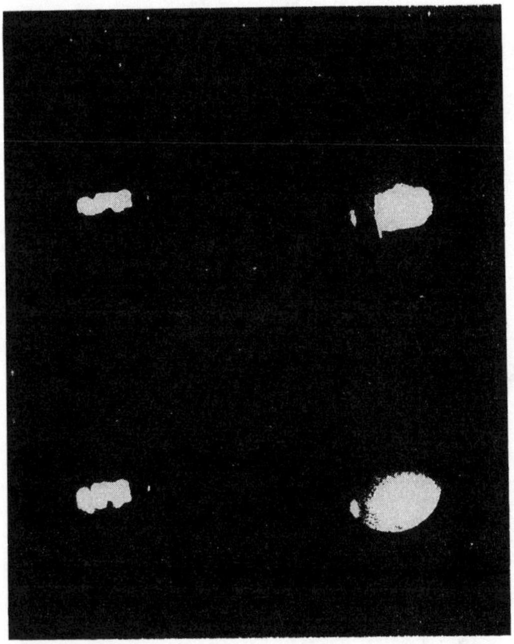

t = 175 ns t = 220 ns

Fig. 4. Frame images in H_2 : p = 1.25 atm; Δt = 10 ns.

On some pictures, especially in hydrogen, luminous overintensities are visible which may be related to "hot-spots" described by numerous authors in the X-ray domain. On figures **5**, the evolution of the discharge radius is displayed.

Figure **5 a** displays the expansion of a discharge in air. At about 100 ns, the discharge radius reaches 2 mm. The initial expansion velocity exceeds 20 km/s in air. After 150 ns, the expansion velocity becomes slower at a time which is near of the current maximum and due to high values of the magnetic field. In air, at 240 ns, the radius is 3.5 mm, the magnetic induction at the surface is B = 14 T and the magnetic pressure : P_{mag} = 750 atm.

Figure **5 b** displays the expansion of a discharge in hydrogen. At about 100 ns, the discharge radius reaches 3 mm. The initial expansion velocity exceeds 30 km/s. At t = 240 ns, the radius is 8 mm, the magnetic induction at the surface is B = 6 T. The magnetic pressure is : P_{mag} = 143 atm. These values of B and of P_{mag} should be even greater, if we consider that the current goes through the axial zone. Choosing an effective radius two times smaller lead to B x 2 and P_{mag} x 4.

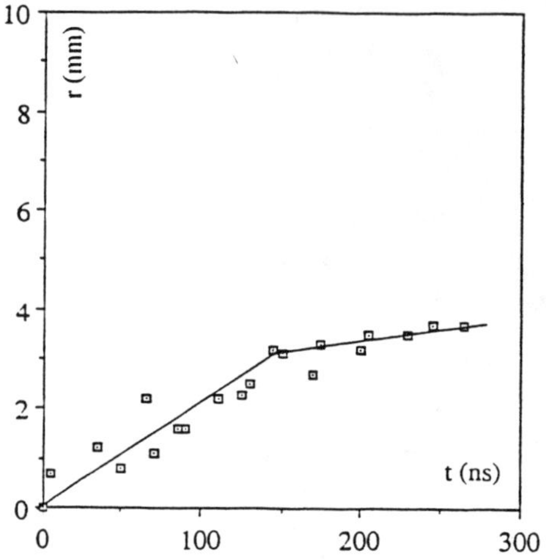

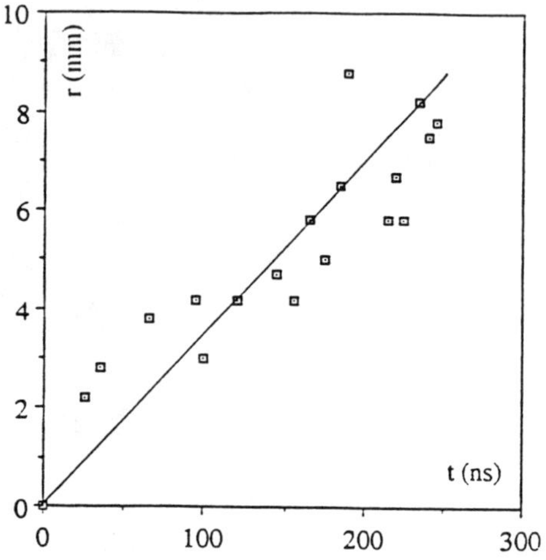

Fig.5a.Discharge radius in air   Fig.5b.Discharge radius in hydrogen

If the expansion is ascribed to a shock propagation, the associated Mach number would be M 67 in air and M 22 in H_2. It is more likely to ascribe the expansion of the outer part to the accretion phenomenon. The accretion may introduce, in the plasma column, important quantities of heated and ionised material due to the very intense emitted radiation at the beginning of the discharge. The following table summarize the caracteristics of the different experiments :

Table I : Expansion velocities of the discharge radius.

	t < 150 ns	t > 150 ns.
Air p = 1 atm	v = 20 km/s (M60)	v = 3 km/s (M10)
H_2 p = 1.25 atm	v = 35 km/s (M25)	

CONCLUSION

Gas-embedded Z-pinches triggered from a ionised micro-channel either in air or in hydrogen have shown an interesting stability. When an helical instability appears, it does not grow : the expansion of the discharge column continues cylindrically from the initial perturbed shape.

The initial state of the micro-channel is important for the stability of the pinch, in particular a good alignement of the needles is necessary.

The fact that, with the same very sharp electrodes, the stability occurs only in a case when the initial micro-channel is well established, seem to prove that the vaporisation of that electrode and the consequent cooling of the plasma is not the fundamental reason for the quenching of the instabilities in gas-embedded Z-Pinches.

REFERENCES

[1]M.G. Haines, Phys. and Controlled Fusion, **31**, p.759 (1989).

[2]M. Skowronek et al., 2^{d} Intern. Conf. on High Density Pinches, Laguna Beach, USA, p.491 (1989).

[3]E. P. Boggasch, J. Christiansen, K. Frank, R. Tkotz and H. Riege., IEEE Transactions on Plasma Science, **19**, p.866 (1991).

[4]J.D. Sethian, A.E. Robson, K.A. Gerber and A.W. Da Silva, Phys. Rev. Lett., **59**, p.892 (1987).

[5]J. Hammel, D.W. Scudder, Proc. 14th Eur. Conf. Controlled Fusion and Plasma Physics, Madrid, (E.P.S.) p.450 (1987).

[6]M. Skowronek, P. Roméas et Vu-Thien Gia, Rev. Phys. Appl.**12**, p.1723 (1977).

[7]L. Soto, H. Chuaqui, M. Favre, E. Wyndham, M. Skowronek, M. Coppins, P. Jaitly and J. Scheffel, Latin-American Workshop in Plasma Physics, 20-31 july 1992.

[8]H. Chuaqui, L. Soto, M. Favre, E.S. Wyndham and M. Skowronek, 3rd Int. Conf. on High Density Z-Pinch, London 20-23 april 1993.

ACKNOWLEDGEMENTS

This work has been supported by a CNRS-NSF agreement and funding. The LANL support is also acknowledged.

DYNAMICS AND STABILITY OF DENSE Z-PINCHES

N.A. Bobrova, V.V. Neudachin, T.L. Razinkova, P.V. Sasorov
Institute of Theoretical and Experimental Physics, Moscow, 117259, Russia

ABSTRACT

This is a review of theoretical works in the field of dense Z-pinch physics performed by ITEP plasma group. The problems under discussion are. One-dimensional dense Z-pinch dynamics is investigated for a wide range of parameters. A transition between classical (compressional) and dissipative regimes of Z-pinch dynamics is followed up. Analytical expressions, determining the type of Z-pinch dynamics and temporal evolution of hot rarefied corona of the dissipative Z-pinches are obtained. Heterogeneous equilibrium states of dense radiative Z-pinches, when the corona is hot and rarefied, while pinch core is very dense and "cold", are shown to exist. The problem of instabilities of dense Z-pinch equilibrium states is investigated, taking into account relevant dissipative processes. On the basis of the results obtained, we consider the general problem of an enhanced stability of dense Z-pinch.

INTRODUCTION

Following the experiments of Refs.[1,2] and the suggestion in Refs.[3,4], there has been increased interest in the problem of the dynamics of Z-pinches produced by the electrical explosion of condensed filaments and, in particular, frozen deuterium fibers. In the experiments of Refs.[1,2] a current with a strength 200 – 600 kA was passed through an initially frozen deuterium fibers 10 – 100 μm in diameter for $\sim$ 100 ns. Passing $\sim$ 10 MA current through a frozen DT column 0.5 – 1 cm in diameter for $\sim$ 100 ns has been suggested in Ref.[3] to accomplish ignition of the thermonuclear fuel in the sausage neck of Z-pinches, resulting in excess of released energy over the applied energy. One-dimensional MHD calculations were carried out in Ref.[5] and it was shown that under the experimental conditions of Refs.[1,2], at least in the initial part of the current pulse, the pinch is heterogeneous. The core of the pinch remains dense and cold, with a temperature $T \sim 0.1 - 1$ eV, whereas the corona of the pinch, through which almost all of the current flows, is tenuous and hot with $T \sim 100 - 200$ eV. The central part of the pinch gradually "evaporates" or ablates due to the heat flux from the hot corona.

A similar situation occurs in the explosion of thin metallic wires (see, e.g., Refs.[6,7]). The corresponding one-dimensional MHD simulation was made about ten years ago in Ref.[8]. In a plasma of strongly radiating elements, the heat flux from the corona into the cold core may be balanced by soft x-ray emission from the transition region between the corona and the dense core. We have constructed such equilibrium heterogeneous states in Ref.[9].

Weakly and strongly radiating dense Z-pinches thus have much in common. They are additionally related by the fact that heterogeneous pinches possess increased stability [2,6,7]. From all appearances, the cold dense core of a heterogeneous pinch plays a stabilizing role. The mode $m = 0$, for example, can be stabilized in the nonlinear stage as a result of filling in the neck of the plasma pinch, which evaporates more strongly from the surface of the core in this case.

This may serve to explain the two-dimensional numerical simulation of Ref.[10], in which the increased stability of axisymmetric Z-pinches was demonstrated. Many articles were devoted to the linear theory of the dense Z-pinch instabilities. See for example Refs.[11–13]. Our results [14,15] are summarized in these proceedings [22]. Only a general outlook on the problem of the dense Z-pinch stability is submitted in this brief review. It coincides to some extend with the opinion mentioned in some other works [10,16].

One can thus see that the dynamics of dense Z-pinches differs from that of classical or compressional ones (see Refs.[17,18]). Dissipative processes (Ohmic heating, heat conduction, and emission) play a major role in dense Z-pinch dynamics. Our results [18], concerning the relationship and mutual transition between the classical and dissipative[5] regimes of Z-pinch dynamics, are reviewed below.

We shall discuss also the process of a turbulent expansion [20] of the dense Z-pinch corona revealed in the experiments on Angara-5 installation [21].

THE TRANSITION BETWEEN THE DISSIPATIVE AND THE CLASSICAL REGIMES OF Z-PINCH DYNAMICS [19]

Let us consider a long axially and azimuthally uniform cylindrical plasma column which is confined by the interaction of the axial current, passing through the plasma column with the azimuthal magnetic field. The behavior of this system is described by the magnetohydrodynamic model. We take into account the difference between ion and electron temperatures and all dissipative processes in the electron gas.

The "cold-start" initial conditions are a solid, cryogenic fiber of the radius R_0 and density ρ_0. The initial temperature is very low, so that it's value does not influence on the basic results. We suppose that the electrical breakdown occurs on the fiber surface and that load impedance is much less than source impedance. Then the time dependence of the electric current through the pinch can be treated as given, $I(t) = I_{m\ \sin}\pi t/2t_m$, and is not affected by the physical processes in the pinch.

The series of computations of dense deuterium Z-pinch dynamics for different I_m, R_0 and ρ_0 were performed. In the entire range of parameters studied in this work, at least at the initial stage of the discharge, the pinch is heterogeneous. The pinch behavior is similar to that of classical pinches, when the corona thickness is small in comparison with the initial fiber radius. Classical pinches are characterized by noticeable compression of the pinch and plasma heating behind the converged shock wave front. In other case we have the situation, related to the experiments with frozen fibers discussed in the Ref.[5]. In this regime the dissipative processes play the main role. Therefore this regime will be called the dissipative one. Shown in Fig.1 are typical radial profiles for this case. The large difference in the temperature and the density between the corona and the core at nearly constant pressure is noticeable. As the result there is the heat transfer from the corona towards the core. The heated outer fiber layers expand radially, i.e., ablate gradually.

The analysis of computational results demonstrates that the overheating instability of the plasma in the limit of the low thermal conductivity is the physical reason of the high temperature corona formation. For the plasma temperature which is in accordance with the Bennet relation and for the initial value of the pinch density comparable with the density of solid state the plasma is really

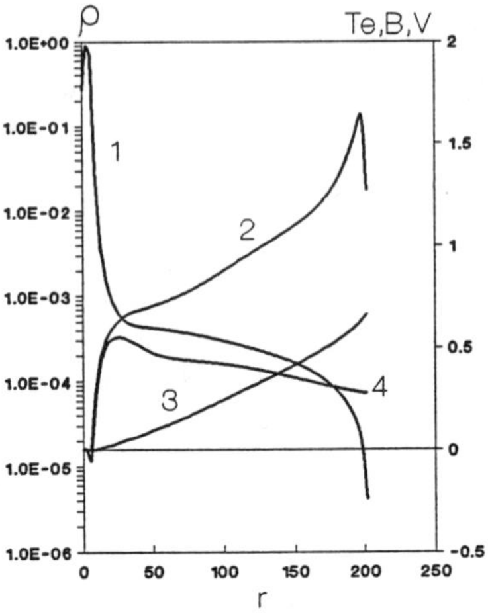

Figure 1: Radial profiles at 35 ns: curve 1, mass density; 2, electron temperature; 3, magnetic field: 4, radial velocity. Radius is in μm, ρ is in g/cm^3, T_e is in 100 eV, v is in 10^6 cm/s, B is in MG. $\rho_0 = 0.18$ g/cm^3, $R_0 = 15$ μm, $I_m = 250$ kA, $t_m = 200$ ns.

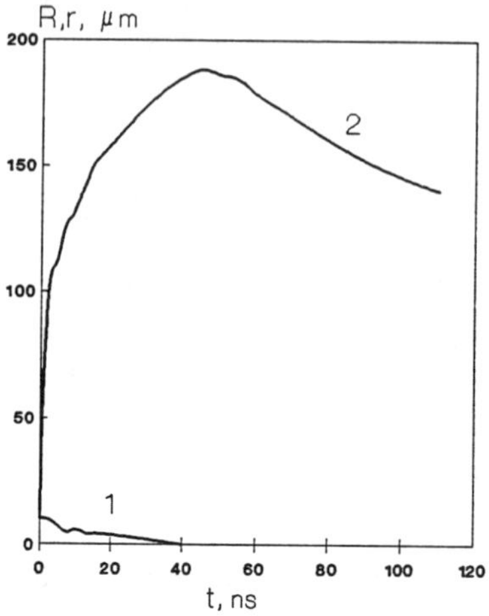

Figure 2: Temporal history of core radius (curve 1) and external radius (curve 2) for the case $\rho_0 = 0.09$ g/cm^3, $R_0 = 10$ μm, $I_m = 250$ kA, $t_m = 200$ ns.

always unmagnetized. Hence the value of a temperature conductivity, which is $\chi \approx \kappa_\perp / n_e$, is less than a magnetic viscosity, $\nu_m = c^2/4\pi\sigma_\perp$, where $\kappa_\perp$ and $\sigma_\perp$ are the thermal and electrical conductivity in perpendicular direction with respect to magnetic field. This is the reason for the overheat instability. In reality the overheated unstable plasma is located at the fiber boundary, because the breakdown occurs on the boundary surface. The growth of the instability is restricted by thermal conductivity. Hence heterogeneous pinch structure created by the electrical explosion of the solid-deuterium fiber, heterogeneous equilibrium states in radiative pinches (discussed below) have the similar physical reason.

The pinch expansion process has two stages. The first is the result of strong and rapid Ohmic heating of the pinch corona (when magnetic Reynolds number $R_m \ll 1$) accompanied by almost free supersonic plasma expansion into vacuum ($\beta = 4\pi p/B^2 \gg 1$). Noticeable part of Ohmic heat is transported by thermal conduction radially inward to the cold fiber and is spent on the core ablation. In our model this stage continues only few nanoseconds. During the second stage of the expansion the pinch is near equilibrium from the mechanical point of view. The heat from the corona is conducted radially inward to the cold, dense core. The core ablates. During this stage $\beta \sim 1, R_m > 1$. A simple analytical description of this stage of pinch expansion can be obtained. We shall consider that mechanical equilibrium exists and external corona radius is much more than the core one. Using Ohm's law, the Bennet relation, energy conservation law and equating to the order of magnitude the heat flow due to thermal conductivity and the power of Ohmic heating, we have [19] the following time dependences of the basic pinch corona parameters in the case $\dot{I} = dI/dt = const$:

$$\begin{aligned} R_{ex} &= 100\,\dot{I}^{-3/7}t^{2/7}\ \mu\text{m}\,, && (1)\\ T_e &= 20\,\dot{I}^{4/7}t^{2/7}\ \text{eV}\,, && (2)\\ M_c &= 5\times10^{-4}\,\dot{I}^{10/7}t^{12/7}\ \mu\text{g/cm}\,, && (3)\\ E^*_{z\,ex} &= 2\,\dot{I}\ \text{kV/cm}\,, && (4) \end{aligned}$$

where $\dot{I}$ is in TA/s, t is in ns. Here M_c is the corona line mass per length, $E_{z\,ex}$ is the axial electric field on the pinch surface and R_{ex} is the external pinch radius. The values of the coefficients in the expressions (1)-(4) are selected according to the results of the numerical simulation. In this case the time to fiber ablation if it is less than t_m is given approximately by the expression:

$$t_{ab} = 90\,\dot{I}^{-5/6}M^{7/12}\ \text{ns}\,, \qquad (5)$$

where the fiber mass per length, M, is in μg/cm.

The effects of a finite electron Larmor radius and especially Nernst effect on the pinch corona dynamics are of the great interest. Taking into account the Nernst effect, the expressions for a dissipative part of the axial component of the electric field, E^*_z and for radial component of the thermal flux must have the following form:

$$\begin{aligned} E^*_z &= j/\sigma_\perp - NB\partial T_e/\partial r\quad, && (6)\\ q_e &= -\kappa_\perp\partial T_e/\partial r + NT_eBj && (7) \end{aligned}$$

Thus we take into account all dissipative processes in the electron component of plasma for the geometry considered. We used the Braginskii formalism for all transport coefficients. See also Ref.[19]. To study this problem we compare the results of the calculations, taking into account both effects (curve 1 at the Fig.3) with the results, presented at the Fig.3 by curves 2 and 3. The curve 2 corresponds to the case $N = 0, x_e \neq 0$, where $x_e = \omega_{Be}/\nu_{ei}$ is an electron magnetization parameter. The curve 3 corresponds to the case $x_e = 0,\ N = 0$. From this comparison we have the following conclusions. In all these versions the pinch plasma is nonuniform and can be divided in hot, tenuous corona and cold dense core. However for $N = 0, x_e \neq 0$, the strong temperature increase at the periphery of the corona is observed. Both effects ($N \neq 0, x_e \neq 0$) lead to the increase of the external pinch radius. We see that the Nernst effect must be taken into account if we want to take into account the effects of electron magnetization(i.e. the effect of finite electron Larmor radius).

It is interesting to search the transition between dissipative and classical regimes, when external or initial paramaters of the pinch are changed. It is obvious that the role of the dissipative processes diminishes, when the magnetic Reynolds number R_m increases. We have four parameters, $\dot{I}$, R_0, ρ_0, t_m, ($\dot{I}_m = \pi I_m/2t_m$) characterizing the pinch behavior. We shall assume that the time of the first contraction $t_p \leq t_m$. Then we have only three parameters $\dot{I}_m$, R_0 and ρ_0 . It is well known that nonradiating Z-pinch dynamics is characterized only by two dimensionless parameters. That is why it is convenient to show our results on the plane $\dot{I}$, R_0, when ρ_0 is fixed. We choose $\rho_0 = 0.09 g/cm^3$.

The Fig.4 shows $R-t$ diagram of pinch dynamics obtained by our simulation for different values of $\dot{I}$ and R_0. The values of $\dot{I}_m$ and R_0 are changed concurrently, i.e. $\dot{I}_m \propto R_0$, so that the temperature, defined by the Bennet relation, was constant. The Fig.4 gives us gradual transition from the dissipative regime ($I_m = 250$ kA) to the classical regime ($I_m = 4$ MA), when $\dot{I}_m$ increases.

It is necessary to note, that the behavior of classical nonradiating Z-pinches is characterized by two main features. First the thickness of an outer layer, carrying almost all the current is much less than the radius. Secondly the plasma temperature behind the shock wave front is so high, that there is a strong skin effect with respect to the electric field. The skin-layer thickness is much less than the distance between the skin layer and the shock wave front. From our numerical simulations the conclusion follows. When $\dot{I}_m < \dot{I}_{cr1}(\rho_0, R_0)$ the pinch is dissipative. When $\dot{I}_m > \dot{I}_{cr2}(\rho_0, R_0)$ the pinch is classical. When $\dot{I}_{cr1} < \dot{I}_m < \dot{I}_{cr2}$ the plasma flow pattern is similar to that of classical pinch, but there is no skin effect. The electric field strength on the pinch axis is the same order of magnitude as the electric field out of the pinch. At the same time the electric current is carried by thin, hot, high conductive layer on the pinch surface. $\dot{I}_{cr1}$ is defined by the condition, that at the moment of first compression of the pinch core the total radius is equal to it's initial value. $\dot{I}_{cr2}$ is defined by the following condition. The electric field strength on the pinch axes is hundred times less than $E_z^*(R_{ex})$ at the moment, when the shock wave reaches the axis. $\dot{I} = \dot{I}_{cr1}(\rho_0, R_0)$ and $\dot{I} = \dot{I}_{cr2}(\rho_0, R_0)$, as functions of the pinch mass per length are shown in Fig.5. So, we have

$$\dot{I}_{cr1} = 2.4 \cdot 10^{10} (\rho_0/\rho_s)^{1/8} R_0^{-5/4} \text{ A/s} , \quad (8)$$

$$\dot{I}_{cr2} = 3.4 \cdot 10^{13} (\rho_0/\rho_s)^{1/2} R_0^{-1/2} \text{ A/s} . \quad (9)$$

Here R_0 is in cm. The dependences (8), (9) were obtained analytically, taking into account expressions (1)-(4). The values of the coefficients in (8), (9) were selected according to the numerical simulation. Let us note, that for the experiments in Ref.[2] $\dot{I}_m < \dot{I}_{cr1}$, for the experiments in Ref.[21], $\dot{I}_{cr1} < \dot{I} < \dot{I}_{cr2}$ and for the parameters proposed in Ref.[3], $\dot{I}_m \geq \dot{I}_{cr2}$.

EQUILIBRIUM STATES OF RADIATIVE Z-PINCHES[9]

The MHD-equilibria of plasma and magnetic field for dense Z-pinch was considered. Plasma and magnetic field are supposed to be in equilibrium with respect to dynamical, thermal and diffusion processes. We take into account the heating due to finite plasma resistivity, the cooling due to radiation of transparent plasma and the heat transfer due to electron thermal conductivity. We will demonstrate that the heterogeneous equilibrium states of radiative Z-pinches can occur. In such states the difference between the temperature in the inner and in the outer regions can be significant.

Equilibrium states are characterized by two dimensionless parameters $\varepsilon = \chi/\nu_m$ and $\theta = T/T_*$. The first parameter characterizes the role of thermal conductivity respect to Ohmic heating (the average Ohmic heating equals to bulk

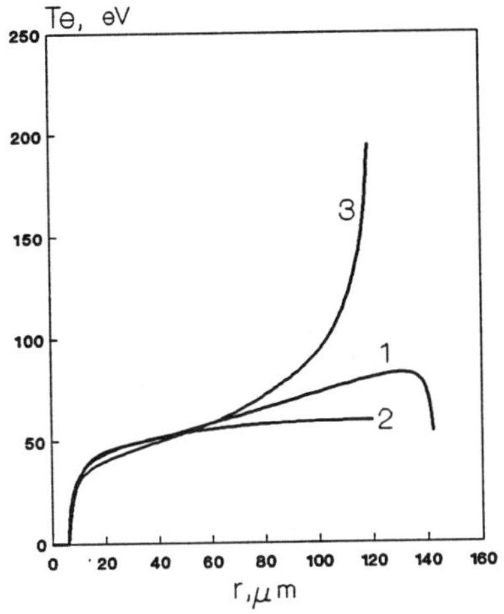

Figure 3: Radial profiles of electron temperature at $t = 10$ ns: curve 1, the effect of finite electron Larmor radius and the Nernst effect are taken into account; curve 2, both effects are not taken into account; curve 3, only the effect of finite electron Larmor radius is taken into account. $\rho_0 = 0.09$ g/cm^3, $R_0 = 10$ μm, $I_m = 250$ kA, $t_m = 200$ ns.

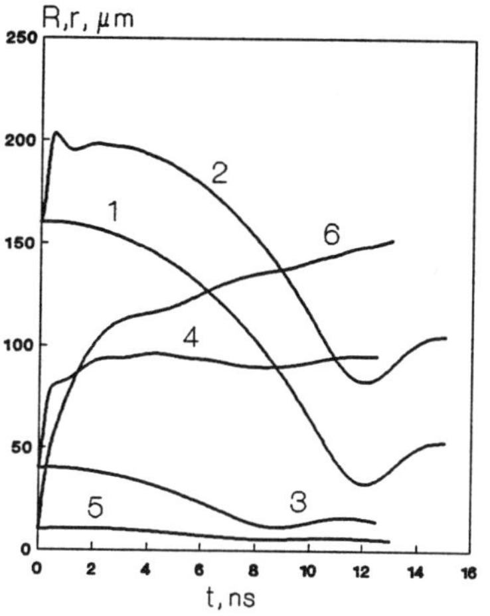

Figure 4: $R - t$ diagram for $\rho_0 = 0.09$ g/cm, $t_m = 200$ ns: curve 1, core radius; curve 2, external radius for $I_m = 4$ MA, $R_0 = 160$ μm; curve 3, core radius; curve 4, external radius for $I_m = 1$ MA, $R_0 = 40$ μm; curve 5, core radius; curve 6, external radius for $I_m = 250$ kA, $R_0 = 10$ μm.

radiative cooling rate), and the second one is the ratio of pinch temperature and some typical value of temperature T_*, which is defined by the dependence of radiated power per unit volume, Q on T, which is rather complicated. Here χ is the coefficient of temperature conductivity. Thermal conductivity smoothens the temperature along the pinch, if $\varepsilon > 1$. That is why, electric current density is nearly uniform. Equalizing Ohmic heating IE_z and radiative cooling $2\pi \int_0^{R_{ex}} Q\, r dr$, one determines relations between R_{ex}, I, and M.

Heterogeneous equilibrium states of radiative Z-pinches occur in low thermal conductivity limit, when $\varepsilon \ll 1$. In this limit the equation of thermal transport is

$$E_z^2 = Q/\sigma_\perp = \text{const} . \tag{10}$$

The type of the equilibrium is determined by the dependence of the product of plasma electric resistivity and radiative cooling rate per unit volume on the temperature for fixed plasma pressure. For the conditions, those are typical for the dense pinch this quantity, i.e. $(Q/\sigma_\perp)_{p=const}$ has a maximum at some temperature T_*.

In the central part of the pinch, where pressure is high, the solution of the equation (10) can exist. However, near the pinch boundary, where the pressure vanishes $p \to 0$, the equation (10) has no solution. It means, that in the

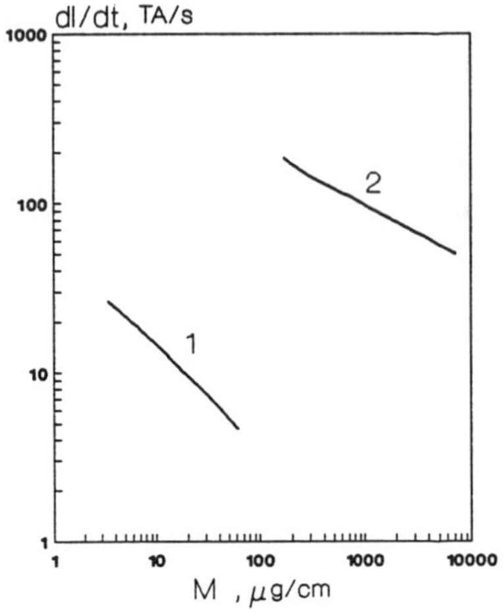

Figure 5: $\dot{I}_{cr1}$ (curve 1) and $\dot{I}_{cr2}$ (curve2) as the function of total pinch mass per length for $\rho = 0.09$ g/cm^3

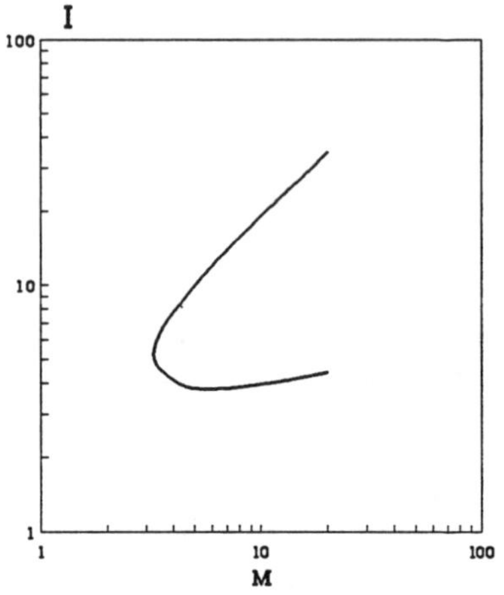

Figure 6: Region on the plane $(M,\ I)$, where heterogeneous equilibrium states of copper pinch exist. A unit for I is 80 kA and for M is 1.8×10^{-6} g/cm.

outer region radiative cooling rate per unit volume cannot compensate heating due to plasma resistivity. This outer part of the pinch gets hot until the effects of thermal transfer become essential. Thus the equilibrium states do not exist if thermal conductivity vanishes, and in outer region of the pinch it is always necessary to take into account thermal conductivity effects. Heterogeneous equilibrium states of radiative Z-pinches occur in low thermal conductivity limit ($\varepsilon < \varepsilon_{cr}$), and when $T \approx T_*$. Radial temperature distribution in these states is very nonuniform. Several hundreds of such equilibrium states have been calculated [9] for the pinch plasmas consisting of the elements with large Z. The Z is the number of the element. For such elements T_* is of the order of 200 eV. The region in the plane $(M,\ I)$, where the heterogeneous equilibrium states of copper pinches exist, is shown on Fig. 6. Examples of correspondent radial structure of the heterogeneous pinches can be seen in Ref.[9] and on Fig.1(b) of Ref.[22] in this volume.

MHD INSTABILITIES OF DENSE RADIATIVE Z-PINCHES[14,15]

Our special poster report [22] devoted to this problem was submitted on DZP Conference. That is why we will present here only some preliminary remarks and general conclusions.

We investigated instabilities of static equilibrium states of the pinches. They are supposed to be in equilibrium with respect to mechanical processes as well as to both thermal and diffusion processes. The instability problem under consideration is a well defined problem from the mathematical point of view on the contrary to the instability problem of the pinches with temporal evolution. Thus we are forced to take into account radiative cooling of the pinch to balance the Ohmic heating. We consider here the equilibrium states which were under discussion in the previous section of this paper. Besides Ohmic heating, magnetic diffusion and radiative cooling, a heat transport due to electron thermal conductivity is taken into account when temporal evolution of small two- and three-dimensional disturbances is investigated.

We have shown that dissipative Z-pinches with high thermal conductivity and consequently with almost constant temperature are unstable under MHD perturbations. The instability growth rate does not differ greatly from those in ideal MHD case and, in particular, does not almost depend upon the magnetic Reynolds number, R_m.

Heterogeneous Z-pinches (with a low average thermal conductivity) for R_m-$<\sim 1$, however, exhibit enhanced stability under large scale perturbation with $m = 0$. At the same time the pinch corona is strongly unstable under small-scale and small-amplitude perturbations. In the case of heterogeneous pinches with high R_m, growing large scale disturbances are also localized in the corona and do not affect the dense, cold core.

We believe that these results do not strongly depend on the radiation power, while the radiation can, of course, greatly affect the evolution of the pinch as a whole. The results obtained require modifying the part of the Haines diagram [23] that concerns dissipative pinches. We also note that for the almost collisionless pinches with a small mass per length one must take into account the anomalous processes that determine the resistance of Z-pinches[24].

GENERAL OUTLOOK ON STABILITY PROBLEM FOR DZP[14,15]

The dense Z-pinches, formed by electrically triggered explosion of frozen deuterium fibers and metallic wires, exhibit enhanced stability. See for example Refs.[2,6,7,16]. It seems to be related with heterogeneous structure of such pinches and with the existence of the cold core. The core makes dense Z-pinches relatively stable as a whole, while the corona is unstable. The analogous opinion concerning the DZP stability is mentioned and considered in Refs.[10,14–16,25]. In the case of relatively small deuterium pinches the lifetime of the cold core is limited due to ablation. When the core is ablated completely the dense Z-pinche becomes strongly unstable. It does not inspire optimism about magnetic confinement of plasma in dense Z-pinches.

TURBULENT CORONA OF DENSE Z-PINCHES[20]

We see that the corona of heterogeneous Z-pinch is unstable. It leads to excitation of MHD-turbulence in the corona and to origin of anomalous transport processes such as: magnetic field diffusion, thermal conduction and Ohmic heating, which may be more effective than usual ones. But when one-dimensional approximation gives that thickness of the corona is more than the core radius, then the MHD-turbulence does not lead to qualitative change of the pinch dynamics. It can be seen from the works [10,25] which simulated numerically two-dimensional dynamics of such Z-pinches. Nevertheless the rate of the core ablation is changed. In the case when the corona is thin in one-dimensional numerical simulations (i.e. when $\dot{I} > \dot{I}_{cr1}$), the MHD-instabilities of the corona can lead to drastic change of the pinch dynamics [20]. The enhanced transport processes evoke corona expansion. A thickness of the corona becomes more than the pinch core through the time [20]

$$t_1 \simeq 1.5\ I^{-0.2} R_0^{1.2}\ \text{ns}\ , \tag{11}$$

where I is the current in MA. Through the time

$$t_2 \simeq 500\ I^{1/15} R_0^{19/15}\ \text{ns} \tag{12}$$

after discharge beginning, the pressure of the corona becomes of the order of the magnetic pressure and the current can be recaptured by the extended hot and tenuous corona. This process can prevent the compression of the pinch and it corresponds to the resuls of the experiments [21].

This work was supported, in part, by the Soros Foundation Grant awarded by the American Physical Society.

REFERENCES

1. J. E. Hammel & P. W. Scudder, in Proc. 14th Eur. Con. on Contr. Fus. and Pl. Phys. 1987, F. Euglman and J. L. Alvares (eds.) (EPS, Geneva, 1987) 450.
2. J. D. Sethian et al., Phys. Rev. Lett. 59, 892 (1987).
3. V. V. Yan'kov, Sov. J. Plasma Phys. 17, 260 (1991); Preprint No4218 Inst. At. Energy., Moscow, (1985).
4. V. V. Vikhrev, et al., Sov. Phys. Dokl. 30, 492 (1985).
5. I. R. Lindemuth, et al., Phys Rev. Lett. 62, 264 (1989).
6. I. K. Aivazov, et al., JETP Lett. 41, 135 (1985).
7. A. E. Aranchuk, et al., Sov. J. Plasma Phys. 12, 765 (1986).
8. R. B. Baksht, et al., Sov. J. Plasma Phys. 9, 706 (1983).
9. N. A. Bobrova, et al., Sov. J. Plasma Phys. 14, 617 (1988).
10. I. R. Lindemuth, Phys. Rev. Lett. 65, 179 (1990).
11. M. Lampe,Phys. Fluids **B3**, 1521 (1991)
12. I. D. Culverwell, et al., Phys. Fluids. **B2**, 129 (1990).
13. F. L. Cochran, et al., ibid. page 123.
14. V. V. Neudachin, et al., Nucl. Fusion 31, 1053 (1991).
15. V .V. Neudachin, et al., J. Moscow Phys. Soc. 2, 23 (1992).
16. E. S. Figura, et al., Phys. Fluids **B3**, 2835 (1991)
17. M. A. Leontovich, et al., At. Energy 3, 81 (1956).
18. V. F. Dyachenko, et al., in: Revs. of Plasma Phys. 5 (M. A. Leontovich ed.) (Consultant Bureau, N.Y. 1970).
19. N. A. Bobrova, et al., Sov J. Plasma Phys. 18, 269 (1992).
20. P. V. Sasorov, Sov. J. Plasma Phys. 17, 874 (1991).
21. A. V. Batjunin, et al., in: Proc. of 8th Int. Conf. on High-Power Beams, (1990, B. N. Breizman and B. A. Knyazev (eds.)) Vol. 2,(World Sci., Singapore, 1991).
22. V. V. Neudachin, et al. in these proceedings
23. M. A. Haines, et al., Phys Rev. Lett. 66, 1462 (1991).
24. P. V. Sasorov, Sov. J. Plasma Phys. 18, 138 (1992).
25. P. Sheehey, et al., Phys. Fluids **B4**, 3698 (1992).

ANALYTICAL SOLUTIONS FOR THE GROWTH RATES OF LOCALIZED PRESSURE-DRIVEN MODES IN A SCREW PINCH

A.B.Bud'ko and M.A.Liberman

Department of Technology, Uppsala University,
Box 534, S-751 21, Uppsala, Sweden

ABSTRACT

The growth rates spectrum of the ideal pressure-driven localized modes in a general screw pinch configuration is obtained in an explicit analytical form under the condition of the Suydam criterion violation. The suppression effect of the magnetic shear near the singular surface is exponential and strongly depends upon relative values of two small parameters: $\mu = m/((kr_s)^2+m^2)$ and $\varepsilon = (P'-P'_s)/P'_s$, where P'_s is the Suydam marginally stable pressure gradient.

INTRODUCTION

Magnetohydrodynamic (MHD) plasma instabilities associated with finite kinetic pressure gradient and localized in the vicinity of singular surface can be essentially divided into two types: ideal modes and resistive g-modes. The subject of the present report is dynamics of the ideal pressure driven localized modes (IPDLM) which can develop under the condition of the Suydam (or Mercier, for toroidal configurations) criteria violation, and is strongly influenced by magnetic shear. Results obtained for a planar current-carrying layer[1] and in the tokamak ordering[2,3] demonstrated exponential stabilization effect of sheared magnetic field near a singular surface which can considerably increase growth time of these modes as compared to the characteristic MHD time scale.

Dynamics of these modes becomes significant for the toroidal and straight pinch configurations with sheared magnetic field, relatively high beta values and confinement times long compared to the MHD time scale. In the toroidal pinch experiments, such as reversed-field pinch (RFP), the present analysis is related to a partially relaxed state established after the turbulent transition processes. As concerning straight pinch systems with axial magnetic field, the theory developed is applicable to gas-puff pinches[4] and Extrap configuration[5] which have demonstrated much longer plasma confinement times as compared to the earlier pinch experiments.

EIGENVALUES SPECTRA

We consider ideal MHD perturbations in a steady-state cylindrically symmetric screw pinch. The usual procedure of decomposition of small MHD perturbations into Fourier components and algebraic manipulations with the full linearized system of the MHD equations results in the second-order equation for the radial component of a particle displacement ξ_r [6], which comprises (together with the corresponding boundary conditions) the boundary value problem for the growth rate σ for given poloidal and axial wavenumbers m and k.

The further analysis is based on two essential assumptions. First, we require the growth rate to be small as compared to characteristic Alfven frequency, that excludes considerable deviation of the unperturbed pressure gradient P' from the marginal Suydam value P'_s. Second, we assume eigenmodes to be localized in the vicinity of the singular surface $r=r_s$ defined as $f(r_s) = \vec{k}\cdot\vec{B}\ (r_s) = 0$ where perturbations tend to minimize bending of the magnetic field lines. Obviously, both of these assumptions have to be justified from the final solution of the problem.

Decomposing all the variables into a Taylor's series in $\Delta r=r-r_s$ near the singular surface and retaining only the essential first and second-order terms, one can obtain the equation for the localized eigenmodes in the form

$$\frac{d}{dy}\left(\left(y^2+1\right)\frac{d\xi_r}{dy}\right)-\left(\frac{C_1}{y^2+1+C_4}+C_2\Gamma^2y^2+C_3\Gamma y-\frac{\varepsilon+1}{4}\right)\xi_r=0, \qquad (1)$$

where the following parameters and variables are introduced

$$y=\frac{k_0}{\Gamma}\Delta r\ ,\ \Gamma=\frac{k_0R_0}{f_s}\sqrt{\mu_0\rho}\ \sigma,\ f_s=R_0\frac{df}{dr}(r=r_s),\ B^2=B_\vartheta^2+B_z^2\ ,\ k_0^2=k^2+\frac{m^2}{r^2}\ ,$$

$$C_1=\left(\frac{2kR_0B_\vartheta}{r_sf_s}\right)^2(r=r_s)\ ,\ C_2=1+\frac{k^2r_s^2-m^2}{k_0^4r_s^4}\ ,\ C_3=-\frac{4k^2mB_\vartheta}{r_s^2k_0^3f_s}(r=r_s) \qquad (2)$$

$$C_4=\frac{B^2}{\gamma P}(r=r_s)\ ,\ \varepsilon=\frac{P'-P'_s}{P'_s}(r=r_s)\ ,\ P'_s=-\frac{f_s^2}{8k^2r_s}$$

Here μ_0 is permeability of free space, γ is adiabatic index.

Note, that contrary to the slab model[1], in the cylindrical geometry eigenmode generally is nonsymmetrical with respect to the singular surface, excepting for the limit $\min[B_z(r_s),B_\vartheta(r_s)]/B(r_s) << 1$ realized in tokamak and in Z-pinches with a weak axial field ($B_z << B_\vartheta$).

In order to obtain analytically eigenvalues spectrum of Eq.(1) it is necessary to get rid of the first derivative of the dependent variable and thus reduce it to the one-dimensional Schrödinger equation with the aid of some special change of variables. The most convinient (but obviously not unique) way implies the transformation of the dependent and independent variables in the form[3]

$$y = \mathrm{sh}(z) \ , \ \xi_r = \frac{1}{\sqrt{\mathrm{ch}(z)}}\Psi(z) \ , \tag{3}$$

which results in the following equation for the displacement function $\Psi(z)$

$$\frac{d^2}{dz^2}\Psi + \left(\frac{\varepsilon}{4} - \frac{C_1}{\mathrm{ch}^2(z)+C_4} - C_2\Gamma^2\mathrm{sh}^2(z) - C_3\Gamma\mathrm{sh}(z) - \frac{1}{4\mathrm{ch}^2(z)}\right)\Psi = 0. \tag{4}$$

Equation (4) is formally identical to the stationary Schrödinger equation for a particle with the energy $E=\varepsilon/4$ in the one-dimensional potential

$$U(z) = \frac{C_1}{\mathrm{ch}^2(z)+C_4} + C_2\Gamma^2\mathrm{sh}^2(z) + C_3\Gamma\mathrm{sh}(z) + \frac{1}{4\mathrm{ch}^2(z)} \ , \tag{5}$$

which shape is specified by the unperturbed profiles and the parameter Γ. The eigenvalues spectrum Γ_n(n=0,1,2...) is determined by the condition of localized states existence with the given energy $\varepsilon/4$. The result for the dimensionless growth rate $\Gamma=\Gamma_0$ (the maximum eigenvalue) can be expressed in the following form (for the details see Appendix)

$$\Gamma = \frac{4\varepsilon}{\Phi+1}\sqrt{\frac{\Phi^3}{(4C_1+1)C_2}}\ exp\left\{-2 - \frac{\Phi-1}{2\sqrt{\Phi}}\arccos\left(\frac{\Phi-1}{\Phi+1}\right) - \frac{\pi}{\sqrt{\varepsilon}}\right\} \ , \tag{6}$$

where Φ, C_1, C_2, C_3 and ε are defined by Eqs.(2),(A4). Although expressed in an explicit analytical form, Eq.(6) can hardly be observed in a general case, and it would be useful to consider its further simplifications in the limiting cases.

The first case corresponds to the limit $C_3^2 << \varepsilon C_2$ when the asymmetry of the potential (5) is ignorable that can be expressed as $\mu = m(k_0 r_s)^{-2} << \varepsilon << 1$, and is related to the purely axial modes ($m << kr_s$) in Z-pinches with a weak axial field and the modes localized near the field-reversal surface in RFP as well as the tokamak flute-type modes. In this case $\Phi \approx 1$ and Eq.(6) yields

$$\Gamma = \frac{2}{e^2}\frac{\varepsilon}{\sqrt{(4C_1+1)C_2}}\, exp\left\{-\frac{\pi}{\sqrt{\varepsilon}}\right\} . \tag{6a}$$

The opposite limit $\varepsilon C_2 << C_3^2$ is associated with the pressure gradient very close to the Suydam value (the strong asymmetry in the potential (5)) and implies $\varepsilon << \mu << 1$. In this limit we obtain $\Phi \approx \varepsilon/(2C_3)^2 << 1$ and

$$\Gamma = \frac{\varepsilon^{3/2}}{2e^2|C_3^3|\sqrt{4C_1+1}}\, exp\left\{-\frac{\pi}{\sqrt{\varepsilon}}\left(1-|C_3|\right)\right\} . \tag{6b}$$

IPDLM GROWTH RATES IN THE RFP PARTIALLY RELAXED STATE

Now we shall apply these results to analyze the pressure-driven modes in partially relaxed states of RFP discharges. We must emphasize, that although being obtained in the cylindrical limit the results are directly applicable to the RFP configurations as the toroidal corrections in this case are of order of squared inverse aspect ratio[7] and can be neglected in the first approximation.

The general feature common in all the toroidal pinch experiments, in particular RFP, is that after an initially highly turbulent phase the plasma settles into a quiescent state with reduced fluctuation activity and profiles essentially independent of the previous history of the discharge. The profiles in the relaxed state of RFP as described by Taylor's Bessel function model (BFM)[8] correspond to the minimum energy state with zero pressure gradient and are stable both to ideal and resistive modes. Experiments done with relatively high beta values show, however, that the measured toroidal magnetic field has a small reversed flux, as compared to the BFM, and the radius of the toroidal field reversal is larger, $\mu = \mu_0 \vec{J}\cdot\vec{B}/B^2$ falls to a small value near the wall and the the normalized pinch current $\Theta = B_\vartheta(R_0)/\langle B_z\rangle$ does not saturate, as predicted by the BFM.

Following Ortolani et.al.[9] we shall consider the model of the partially relaxed state with the postulated μ profile in the form

$$\mu = \frac{2\Theta_0}{R_0}\left(1 - \left(\frac{r}{R_0}\right)^{\alpha}\right) \tag{7}$$

with two free parameters α and Θ_0 that reasonably fits the experimentally measured μ profiles. The magnetic field components are determined from the pressure balance condition where in the context of the present analysis we take the pressure gradient beyond the Suydam stability limit

$$P' = P_S'(1+\varepsilon)\ ,\ \varepsilon > 0\ , \tag{8}$$

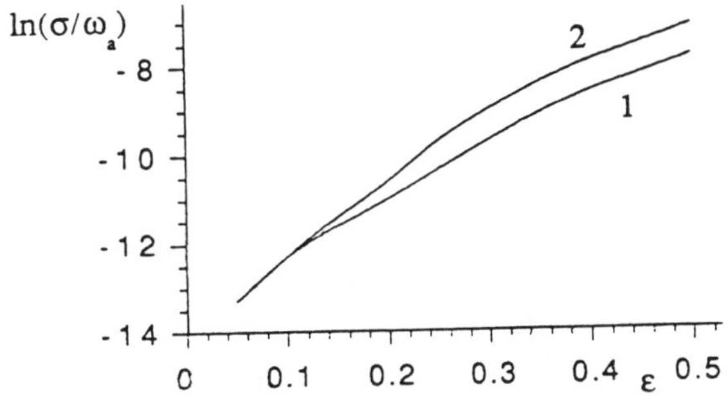

Fig.1 The IPDLM growth rates in the RFP partially relaxed state (α = 4).

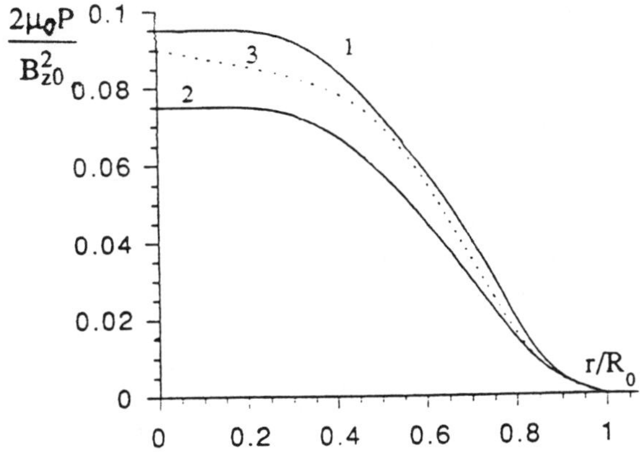

Fig.2 Pressure profiles in the RFP partially relaxed state (Θ = 1.77, α = 3).

but keeping in mind that the perturbations must not grow significantly for the energy confinement time scale, that is the growth rates have to be limited by the inverse energy confinement time. In Fig.1 the growth rates of ILPDM are plotted as a function of the normalized pressure gradient deviation from the Suydam value. Although the present analysis is valid for an arbitrary (but sufficiently smooth) profile ε(r), for the sake of simplicity we took ε = const in the numerical calculations. The profiles shape was calculated for each ε value. According to Eqs:(2), (6) with the profiles fixed, the growth rate σ is a function of the spatial coordinate r_s and the poloidal wavenumber m. The plotted growth rate represents the maximum over r_s and m value normalized on the characteristic Alfven freguency $\omega_a = B_z(0)/(\mu_0\rho)^{1/2}$. Curves 1, 2 correspond to the regimes with small ($\Theta \approx 1.6$, $F \approx -0.06$ for curve 1) and considerable ($\Theta \approx 2.4$, $F \approx -0.8$ for curve 2) reversed flux, where $F = B_z(R_0)/\langle B_z \rangle$.

In Fig.2 the pressure profiles are plotted calculated for $\varepsilon = 0.3$ (curve 1) and $\varepsilon = 0$ (curve 2). Free parameters were taken as following: $\alpha = 3$, $\Theta = 1.77$, $F = -0.25$ that corresponds to the parameters of the Extrap T1 RFP discharge reported recently[10]. Although the quantitative comparison with the experimental results (dashed curve 3) is rather difficult because of large experimental errors one can conclude that the present model with nonzero but sufficiently small IPDLM growth rates ($\sigma/\omega_a \leq 10^{-4}$ that corresponds to $0 < \varepsilon < 0.4$) provides reasonable agreement with the experimental profile.

And finally we considered the beta value achievable in the RFP discharges. To avoid a considerable growth of the unstable modes for the energy confinement time scale τ_E we chose $\sigma/\omega_a = 5 \cdot 10^{-4}$ as the upper limit for the growth rates ($\tau_E = 0.1 \div 1$ms, $\omega_a^{-1} = 0.1 \div 1\mu$s for typical RFP experiments). The analysis carried out demonstrates that development of the IPDLM does not prevent an extension of the achievable beta range by approximately 30% of its marginal stable value both for steep and smooth μ profiles. We must emphasize, however, that since this discussion is related only to the IPDLM, it represents the upper limit for beta and exceeds the current experimental values ($\beta_\vartheta \leq 15\%$, $\beta \leq 10\%$) which are limited by the current driven as well as tearing and resistive g modes[7].

CONCLUSION

In the present report we have investigated pressure driven modes of a screw pinch localized near a singular surface under the condition of the Suydam criterion violation. With the special change of variables the second-order equation for the eigenmodes was reduced to the one-dimensional Schrödinger equation with the shape of potential determined by the eigenvalues of the problem.

The expressions for spectrum and maximum eigenvalue - the growth rate were obtained in the explicit analytical form and revealed the exponential suppression effect of the magnetic shear near the singular surface (Eqs.(6),(6a),(6b)). This result is qualitatively similar to the spectra obtained in the planar layer approximation and for the flute-type perturbations in the tokamak[2,3] and generalizes it for the case $\varepsilon \leq \mu \ll 1$.

APPENDIX

We shall obtain the spectrum explicitly with the aid of the Bohr-Zommerfeld quantization rule[11], which yields an asymptotically exact result for short-wavelengths modes and is sufficiently accurate in a general case of nonlocalized eigenfunctions. We shall assume here $C_3 < 0$. In the case of the positive values of C_3 which is realized in the outer layers of RFP similar analysis obviously results in the same spectrum (A5) and expression (6) for the growth rate.

In our case the Bohr-Zommerfeld formula takes the following form

$$\int_{z_1}^{z_2} \sqrt{\frac{\varepsilon}{4} - U(z)}\, dz = (2n+1)\frac{\pi}{2} , \tag{A1}$$

where U(z) is expressed by Eq.(5). The turning points $z_{1,2}$ are determined from the condition $U(z_{1,2}) = \varepsilon/4$ and with the assumptions

$$\varepsilon \ll 1 , \; \Gamma_n \ll \frac{\varepsilon^{3/2}}{4|C_3|\sqrt{4C_1+1}} \tag{A2}$$

(the latter will be justified by the final result) one can obtain

$$sh(z_1) = \sqrt{\frac{4C_1+1}{\varepsilon}} , \quad sh(z_2) = \frac{\sqrt{C_3^2+\varepsilon C_2} - C_3}{2\Gamma_n C_2} .$$

By introducing a new variable $t = \Big(sh(z) - sh(z_1)\Big)\Big(sh(z_2) - sh(z_1)\Big)^{-1}$ we can reduce Eq.(A1) to

$$\sqrt{\frac{\varepsilon}{\Phi}} \int_0^1 \frac{\sqrt{t(1-t)(t+\Phi)(t+2\delta)}}{(t+\delta)^2}\, dt = (2n+1)\frac{\pi}{2} , \tag{A3}$$

where

$$\delta \equiv \frac{sh(z_1)}{sh(z_2)} \ , \ \Phi - \delta \equiv \frac{\sqrt{1 + \varepsilon C_2/C_3^2} \ - 1}{\sqrt{1 + \varepsilon C_2/C_3^2} \ + 1} \approx \Phi \ . \tag{A4}$$

Integration by parts of Eq.(A3) with accuracy of logarithmic and zeroth-order terms in δ yields

$$\sqrt{\varepsilon} \left\{ \ln\left(\frac{8\Phi}{(\Phi+1)\delta} \right) - 2 - \frac{\Phi-1}{2\sqrt{\Phi}} \arccos\left(\frac{\Phi-1}{\Phi+1} \right) \right\} = (2n + 1)\pi \ , \tag{A5}$$

which justifies the assumptions (A2) and directly results in the dispersion relation (6).

REFERENCES

1. H.P.Furth, J.Killeen, M.N.Rozenbluth, Phys. Fluids 6, 459 (1963).
2. A.B.Mikhailovskii, Theory of Plasma Instabilities (Consultants Bureau, N.Y., 1974) vol.2, ch.9.
3. O.K.Cheremnykh, S.M.Revenchuk, Plasma Phys. Contr. Fusion 34, 55 (1992).
4. F.S.Felber, M.M.Malley, F.J.Wessel, et.al., Phys. Fluids 31, 2053 (1988).
5. J.R.Drake, Plasma Phys. Contr. Fusion 26, 387 (1984).
6. K.Hain, R.Lust, Z.Naturforsch. Teil A 13, 936 (1958).
7. R.Paccagnella, A.Bondeson, H.Lütjens, Nuclear Fusion 31, 1899 (1991).
8. J.B.Taylor, Phys. Rev. Lett. 33, 1139 (1974).
9. S.Ortolani, in 20 Years of Plasma Physics (World Scientific Publishing, Philadelphia, PA, 1985), p.75.
10. P.Nordlund, S.Mazur, J.R.Drake, in Proceedings of the Joint International Conference on Plasma Physics, Part 1, p.599. The European Physical Society, Innsbruck,.1992.
11. L.D.Landau, E.M.Lifshitz, Quantum Mechanics (Pergamon Press, 1976), p.170

PARAMETER SPACE COMPARISON WITH UNIVERSAL DIAGRAM FOR Z-PINCH STABILITY

H. Chuaqui, L. Soto, M. Favre, E. S. Wyndham
Pontificia Universidad Católica de Chile,
Casilla 306, Santiago 22, Chile

M. Skowronek
Université Pierre et Marie Curie
T12, E5, 75230 Paris Cedex 05, France

ABSTRACT

A study of the stability of a gas embedded Z-pinch discharge is presented. The experiments have been carried out using pulse power generators capable of delivering dI/dt in excess of 10^{12} A·s^{-1}. The parameter space on the Universal Diagram of Z-pinch stability is explored by changing initial conditions (background gas pressure, preionization scheme: laser or microdischarge initiation and with or without preheating), as well as parameter change due to the evolution of each particular discharge. For every initial condition a trajectory is obtained in the Universal Diagram. Diagnostics used are Rogowskii loop and multiframe holographic interferometry, from which, current, line density and pinch radius are obtained. Results obtained are discussed in relationship with their position in the Universal Diagram. The applicability of the diagram is confirmed. Further work to explore different areas of the parameter space is envisaged.

INTRODUCTION

In recent times a considerable effort has gone into theoretical studies of Z-pinch stability. In general it is assumed that the pinch discharge has reached conditions under which the stability is analysed. However, in order to achieve the specified conditions the pinch has to evolve from a neutral gas or fibre. In so doing different plasma regimes can be encountered. The recent realization that for a Z-pinch in which the Bennett relation holds at all times, the regime at which the pinch is can be specified only by I^4a and N, where I is the pinch current, a the radius and N the line density, has made possible the construction of a universal diagram for regimes of Z-pinch stability[1] (UDZS). As the parameters required to locate a particular discharge in this diagram is small, the evolution of a particular Z-pinch discharge can be studied in terms of its stability properties. For each discharge a trajectory on the UDZS can be obtained and with it an assessment of the different regions of stability explored on the diagram given certain initial conditions.

EXPERIMENTAL DATA

The experimental results used for the present study were obtained at Pontificia Unversidad Católica de Chile using GEPOPU[2], a pulse power generator delivering 300kV maximum, 1.5Ω, 120ns pulse length, dI/dt > 10^{12} $A{\cdot}s^{-1}$, and the gas-embedded Z-pinch experiment at Imperial College[3]. For the experiments done in Santiago two different preionization schemes were used, namely a focused laser pulse from a Nd-YAG, 10ns, 1.06μm, 300-500mJ and a microdischarge obtained from a pair of 1.5mm diameter, 0.5μm radius tip tungsten needle electrodes, operated with preionizing discharge of 20μA [4]. In both cases Hydrogen gas was used at pressures 1/6, 1/3 and 1atm. In order to determine the position in the UDZS all the information required is current, pinch radius and line density. Diagnostics used to obtain these quantities are current measurement using a single turn Rogowskii loop and multiframe holographic interferometry from which radius and line density are obtained. A sequence of up to eight frames per shot, 10ns between frames is obtained, providing information over the whole time of interest[5]. Figure 1 shows

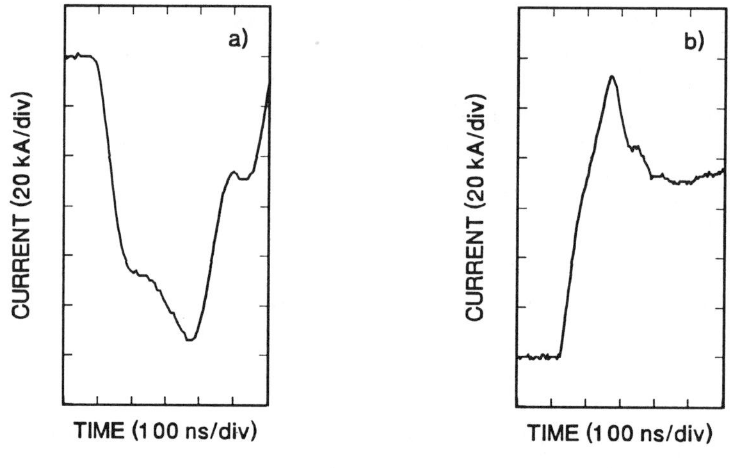

Fig. 1 Current traces for discharges at 1/3atm hydrogen: a) laser initiated; b) microdischarge initiated discharge.

current traces for a typical laser and microdischarge initiated pinch discharges. The difference in the current shape can be ascribed to differences in inductance between the two electrode configurations. Figure 2 shows interferogram and shadowgrams sequences obtained for both initial conditions. The experiment at Imperial College, London uses a 3Ω, 50ns, 120kV generator in which the line gap is shorted by a 200Ω resistor to a 10ns transfer section, such that about 1/7th of the charging voltage is available for preheating the laser initiated plasma column, driving up to 5kA of preheat current. Current measurements and single shot holographic interferometry are used to obtain the relevant information. For both experiments the pinch radius increases, the

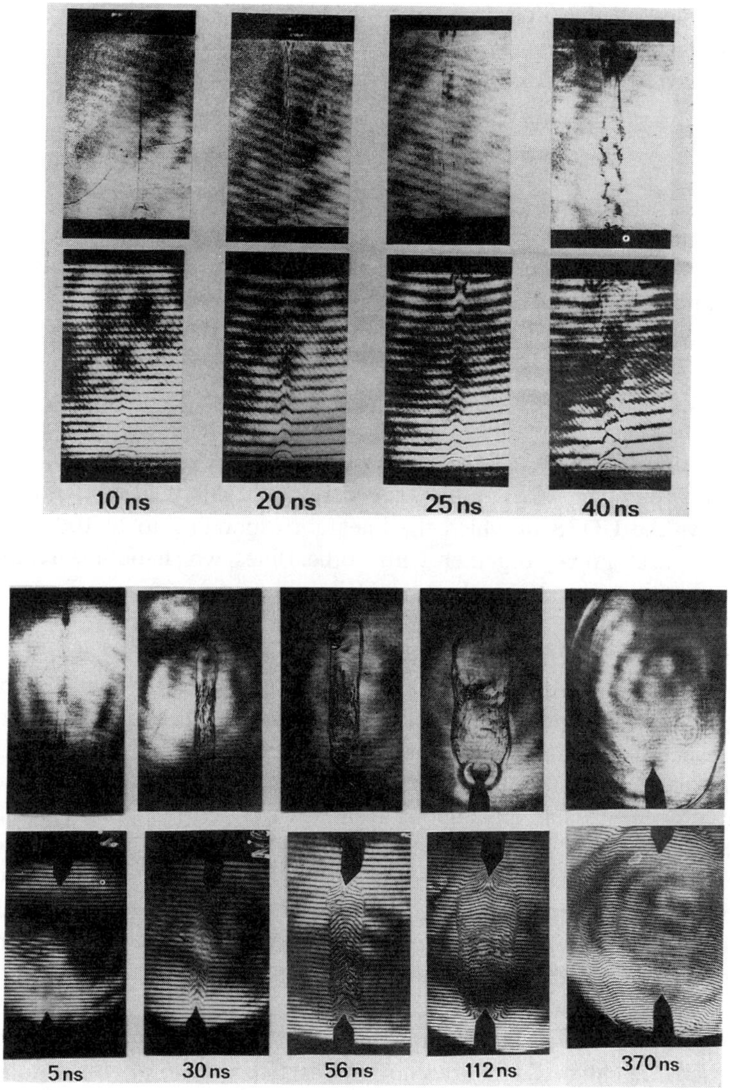

Fig. 2 Simultaneous shadowgram and interferogram sequences for laser initiated (top) and microdischarge initiated Z-pinch discharges in hydrogen at $^1/_3$atm. These are composite sequences of single shot simultaneous shadowgram and interferogram. Overall length of discharge in both cases is 10mm.

rate being smaller on the preheated case. In nearly all cases the discharge column evolves into a helix trough an m=1 instability. No m=0 instabilities are observed, as has been predicted theoretically[6].

PARAMETER SPACE

The parameter range explored with the different experimental conditions spans two orders of magnitude in line density and over five orders in I^4a. It is interesting to note that most results lie in the area of the UDZS in which transitions between different regimes occur. Resistive effects play a significant role in stabilizing pinch discharges, as has been shown [7-11]. The most relevant parameter appears to be the Lundquist number S. For values of S below about 100 the pinch should be stable. Another important stabilizing effect is due to finite Larmor-radius effects (FLR). For the Z-pinch it has been shown that there is a critical value of Larmor-radius to pinch radius ratio above which there is absolute stability[12]. Above $a_i/a=0.15$ FLR effects should stabilize the pinch. FLR effects are only applicable if $\Omega_i\tau_i$ is greater than 1. Figure 3 shows the UDZS in which the lines corresponding to S=100, $\Omega_i\tau_i=1$ and $a_i/a=0.1$ and 1 are shown, together with other lines which outline regions where different descriptions are valid, such as the Hall fluid model, viscous effects or electron anisotropy effects.

The different experimental results are shown in Fig. 3 for a range of initial conditions. Preheating does make a significant difference to the overall behaviour of the discharge. Initial line density is similar for the two pressures considered, and as the current increases the radius grows at a small rate. Trajectories on the UDZS are nearly vertical suggesting that the discharge evolves towards less resistive conditions. The discharge goes unstable after about 40ns, with the corresponding points coming close to the resistive behaviour boundary. For a situation in which there is no preheating the overall picture is quite different. In all cases the path followed in the UDZS is initially approximately parallel to the resistive instability boundary suggesting that the pinch maintains a fairly constant Lundquist number. It is interesting to compare discharges with the same background pressure, but with different preionization scheme. For $^1/_3$atm the laser initiated discharge starts at lower number density and goes unstable at later times, roughly at the time it crosses the S=100 line. The microdischarge initiated case does not go unstable in spite of crossing the resistive boundary. However, this happens for Larmor-radius of around a tenth of the pinch radius. It is therefore likely that FLR stabilization effects do play a role under these conditions. For a 1atm pressure in the microdischarge case the trajectory nearly follows the S=100 line. On this case some discharges are stable, whereas others are not. On average about 30% are stable. It would appear that this is a marginally stable situation. This is to be compared to the laser initiated discharge at 1atm, which is always unstable but our laser pulse is not short enough to freeze interference fringes. Preliminary analysis of results obtained for the lowest filling pressure used, namely $^1/_6$atm the discharge is unstable from early on. Line density is in the range $5\cdot10^{16}$-

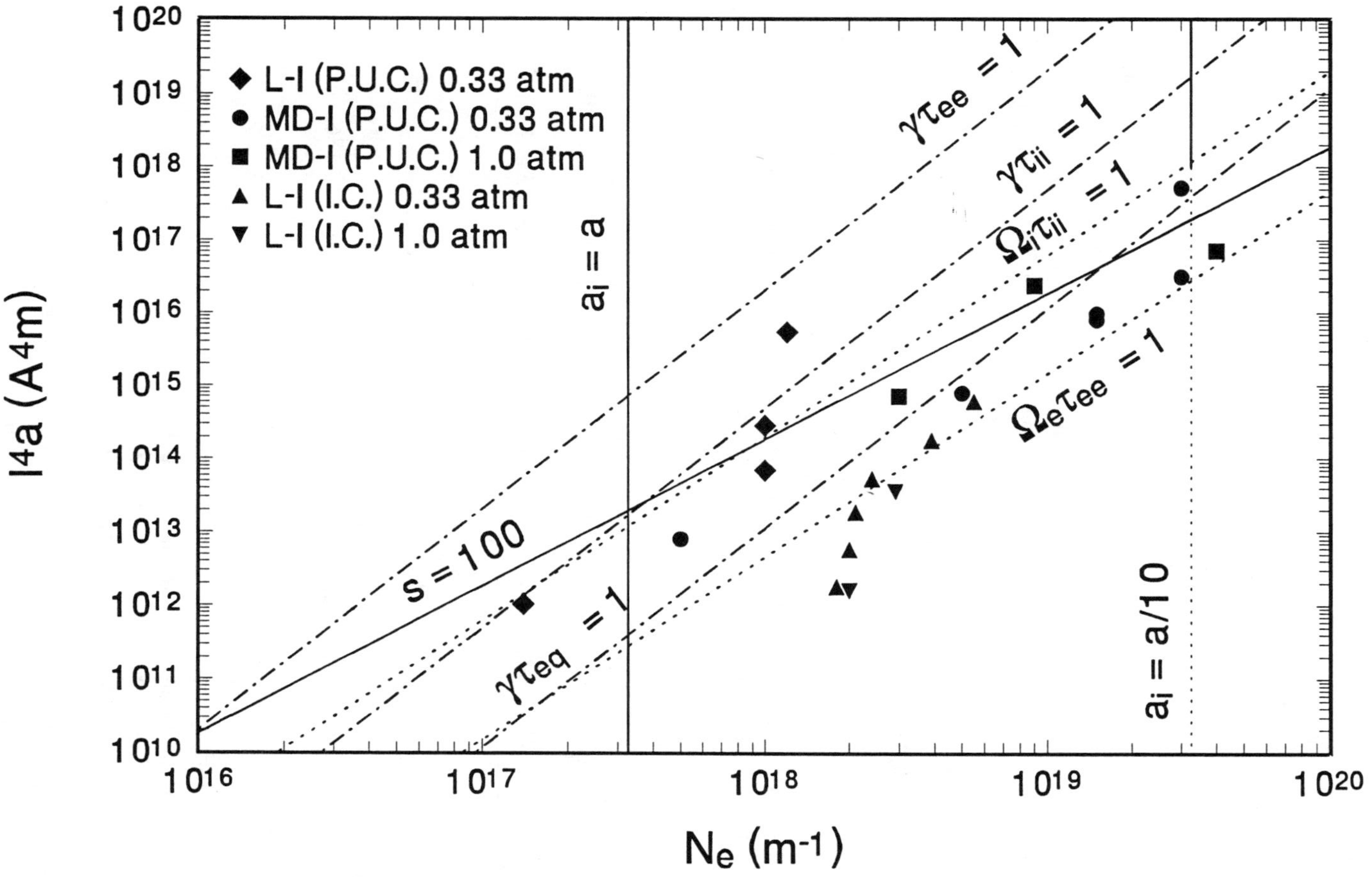

Fig 3: UDZPS for hydrogen. L-I: laser-initiated, MD-I: micro-discharge initiated.
Time intervals: ◆ 10-40 ns, ● 10-110 ns, ■ 20-110 ns, ▲ 6-40 ns, ▼ 5-15 ns.

10^{18} m^{-1} with radius and current such that the points are located slightly above the resistive instability threshold line.

DISCUSSION

The present range of gas embedded Z-pinch experiments described in terms of its behaviour related to the UDZS cover an important fraction of parameter space. Basically all the resistive regime is covered, even with some incursion into other regions. Results obtained suggest that the UDZS is accurate in describing the stability of Z-pinch discharges for the parameters investigated. It is noteworhty that this work has been carried out with small pulse power generators and does provide an important parameter base for larger experiments. With these results it is possible to assess the importance of different preionization or preheating techniques using generators which by virtue of their small size have a high repetition rate and are very versatile. The present work highlights the fact that an essential consideration in planning a Z-pinch experiment is the road towards the desired operation conditions. From the results obtained it would appear that a suitable trajectory to cross from the resistive regime to the FLR stabilized can be obtained under suitably controlled conditions. Further experimental work is certainly required to move into the unexplored areas of the UDZS, in particular in areas of interest for thermonuclear research.

ACKNOWLEDGEMENTS

We would like to thank professor M. G. Haines of Imperial College for stimulating conversations on these results. We thank FONDECYT, CONICYT, Fundación ANDES and CNRS for their various financial contributions to the research programme. L. Soto is a scholarship student of Fundación ANDES. We thank The British Council for their support under their Academic Link scheme.

REFERENCES

1. M. G. Haines and M. Coppins, Phys. Rev. Lett. **66**, 1462(1991)
2. M. Favre, Proc. Small Plasma Physics Experiments II, pp. 170-186, World Scientific(1989).
3. P. Choi, M. Coppins, A. E. Dangor, and M. B. Favre, Nuclear Fusion **28**, 1771(1988).
4. M. Skowronek, P. Romeas, J. Larour and A. E. Dangor, Second International Conference on Dense Z-pinches, Laguna Beach, 491(1989).
5. L. Soto et al, to be submitted for publication.
6. M. Coppins and J. Scheffel, Second International Conference on Dense Z-pinches, Laguna Beach, 211(1989).
7. M. G. Haines, Proc. Phys. Soc. **76**, 249(1960).
8. I. D. Culverwell and M. Coppins, Phys. Fluids B **2**, 129(1990).
9. F. L. Cochran and A. E. Robson, Phys. Fluids B **2**, 123(1990).

10. I. D. Culverwell, M. Coppins and M. G. Haines, Second International Conference on Dense Z-pinches, Laguna Beach, 246(1989).
11. P. M. Cox, Plasma Phys. Controlled Fusion **32**, 553(1990).
12. H. O. Åkerstedt, Phys. Scr. **37**, 117(1988).

THE KELVIN-HELMHOLTZ INSTABILITY IN A STRATIFIED PLASMA

R. Galvão
Instituto de Fisica, Universidade de São Paulo, São Paulo, SP, Brasil,

A. G. González† and J. Gratton†
INFIP-Lab. Física del Plasma, Facultad de Ciencias Exactas y Naturales, Universidad de Buenos Aires, Pabellón 1, Ciudad Universitaria, 1428 Buenos Aires, Argentina.

ABSTRACT

The hydromagnetic Kelvin-Helmholtz instability has been extensively studied in simple plasma configurations, but more complex situations relevant for fusion research involving the combined role of effective gravity due to acceleration, velocity shear, density and magnetic field stratification has not yet been investigated for compressible plasmas. We consider an isothermal equilibrium configuration with stratified density and magnetic field in a compressible plasma. The dispersion relation can be reduced to a highly complicated algebraic equation that involves many parameters. Various cases are examined. Analytical conditions for stability are given for a stratified plasma with no density jump. When there is a density discontinuity, the conditions for the existence of unstable modes are presented. The limiting case without gravity is examined.

INTRODUCTION AND GENERAL EQUATIONS

The Kelvin-Helmholtz instability (KHI) occurs in fluids and plasmas whenever there is a shear flow across the interface between two different media. It has been extensively studied in simple plasma configurations[1], but more complex configurations relevant for fusion research and astrophysics have not been very much explored. In particular, the combined role of effective gravity due to acceleration, velocity shear, density and magnetic field stratification has not yet been investigated in the compressible MHD regime. Recent experimental studies on the turbulence at the plasma edge in tokamak discharges suggest that the KHI may be an important mechanism during the so-called L-H transition[2]. These configurations have stratified density and gradients of velocity perpendicular to the curvature radius of the plasma torus. The effect of the magnetic field curvature is equivalent to a gravity in the direction of the radius, if we are only interested in small wavelengths. KHI may affect the non-linear evolution of Rayleigh-Taylor instability present in many accelerated, high density plasmas as in Plasma Focus or Liner experiments. In these complex problems, depending on many parameters, it is appropriate to start with special cases. We develop a theory of the compressible hydromagnetic KHI in the presence of gravity, considering an isothermal equilibrium with stratified density and magnetic field and velocity shear, extending our previous work on the Rayleigh-Taylor instabilities[3]. We consider a slab geometry where the density and all the other equilibrium quantities depend only on the coordinate in the direction of the effective gravity. Our purpose is to obtain general conditions for stability and we are not, therefore, trying to describe in all detail a real configuration, but to model it. We study the unstable modes that appear in a stratified plasma with a discontinuous jump in the equilibrium flow.

We assume ideal compressible MHD and consider a slab geometry in which a plasma in isothermal equilibrium is stratified along the y axis. The equivalent gravity is

†Member of the Consejo Nacional de Investigaciones Científicas y Técnicas, Argentina.

$\boldsymbol{g} = -g\boldsymbol{e}_y$, $\rho(y)$ is the density, $p(y)$ the pressure, $\boldsymbol{B} = B_x(y)\boldsymbol{e}_x + B_z(y)\boldsymbol{e}_z$ the magnetic field, $\boldsymbol{u} = u_x(y)\boldsymbol{e}_x + u_z(y)\boldsymbol{e}_z$ is the unperturbed flow velocity, $C_A = B/\sqrt{4\pi\rho}$, $C_S = \sqrt{\gamma p/\rho}$. The unperturbed equilibrium condition is

$$\frac{d}{dy}\left(p + \frac{B^2}{8\pi}\right) = -\rho g. \tag{1}$$

The y component (ζ) of the displacement of linear adiabatic perturbations to this state should satisfy the equation[4]

$$\frac{d}{dy}\left[\mathcal{H}\left(\frac{1}{\mathcal{M}} - 1\right)\frac{d\zeta}{dy}\right] + k^2\zeta\left[\mathcal{H} - g\frac{d\rho}{dy} - g\frac{d}{dy}\frac{\mathcal{H}}{\mathcal{M}\overline{\omega}^2} - g^2\frac{k^2\mathcal{H}}{\mathcal{M}\overline{\omega}^4}\right] = 0. \tag{2}$$

We consider normal modes $\delta Q = q(y)e^{i(k.x-\omega t)}$, $\boldsymbol{k} = (k_x, 0, k_z)$, $\overline{\omega}(y) = \omega - \boldsymbol{k}\cdot\boldsymbol{u}(y)$; $\psi(y)$ is the angle between $\boldsymbol{B}$ and $\boldsymbol{k}$, and

$$\mathcal{H} = \rho(C_A^2 k^2 \cos^2\psi - \overline{\omega}^2)\ , \qquad \mathcal{M} = 1 - (C_A^2 + C_S^2)(k/\overline{\omega})^2 + C_A^2 C_S^2 \cos^2\psi (k/\overline{\omega})^4 \tag{3}$$

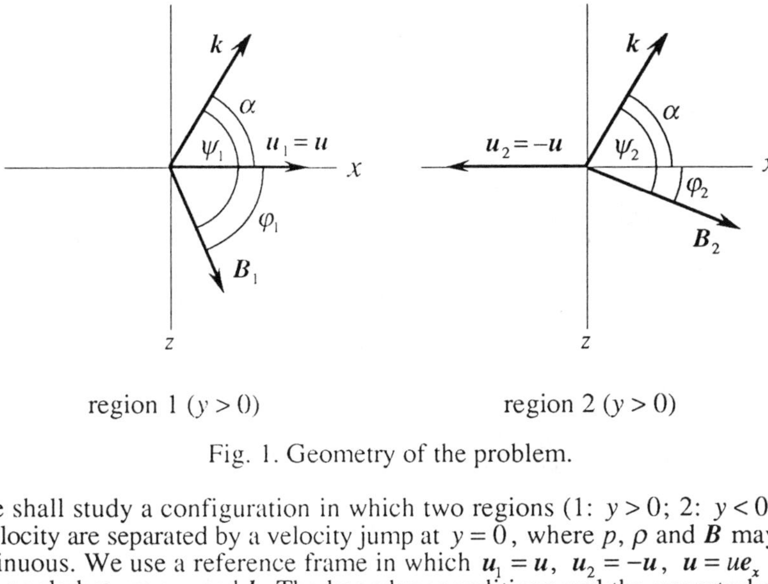

region 1 ($y > 0$) region 2 ($y > 0$)

Fig. 1. Geometry of the problem.

We shall study a configuration in which two regions (1: $y > 0$; 2: $y < 0$) of uniform velocity are separated by a velocity jump at $y = 0$, where p, ρ and $\boldsymbol{B}$ may also be discontinuous. We use a reference frame in which $\boldsymbol{u}_1 = \boldsymbol{u}$, $\boldsymbol{u}_2 = -\boldsymbol{u}$, $\boldsymbol{u} = u\boldsymbol{e}_x$ (Fig. 1); α is the angle between $\boldsymbol{u}$ and $\boldsymbol{k}$. The boundary conditions and the unperturbed equilibrium require[5] ($\{\mathcal{A}\} \equiv \mathcal{A}(y = 0+) - \mathcal{A}(y = 0-)$):

$$\{\zeta\} = 0, \quad \left\{\mathcal{H}\left(\frac{1}{\mathcal{M}} - 1\right)\frac{d\zeta}{dy} - k^2 g\zeta\left[\rho + \frac{\mathcal{H}}{\mathcal{M}\overline{\omega}^2}\right]\right\} = 0, \quad \left\{p + \frac{B^2}{8\pi}\right\} = 0. \tag{4}$$

We assume that in each region $p_i = p_{0,i}e^{-2q_i y}$, $\rho_i = \rho_{0,i}e^{-2q_i y}$, $B_i = B_{0,i}e^{-q_i y}$ (isothermal unperturbed profiles with effective height q_i, constant direction of $\boldsymbol{B}$, but B varies so that the plasma β is constant, $i = 1, 2$ for regions 1, 2); then the general solutions of (2) are $\zeta_i = \zeta_{-,i}\exp[q_i(1-\Gamma_i)y] + \zeta_{+,i}\exp[q_i(1+\Gamma_i)y]$ with $\mathrm{Re}(\Gamma_i) > 0$, or $\mathrm{Im}(\Gamma_i) > 0$ if $\mathrm{Re}(\Gamma_i) = 0$, $\zeta_{-,i}$, $\zeta_{+,i}$ constants and:

$$\Gamma_i = \sqrt{R_i - K_i T_i}\ , \tag{5}$$

$$R_i = 1 + \frac{4G_i[V_{A,i}^2 - (F_i^2 - V_{A,i}^2)\bar{V}_i^2] + 4G_i^2(\bar{V}_i^2 - V_{A,i}^2)}{F_i^2(\bar{V}_i^2 - V_{A,i}^2)(\bar{V}_i^2 - V_{B,i}^2)}\ ,\quad T_i = \frac{(\bar{V}_i^2 - V_{+,i}^2)(\bar{V}_i^2 - V_{-,i}^2)}{F_i^2(\bar{V}_i^2 - V_{B,i}^2)} \tag{6}$$

with $G_i = g/2q_iC_{S,i}^2 = \gamma^{-1} + M_{A,i}^2/2$, $\bar{V}_i = \bar{\omega}_i/kC_{S,i}$, $K_i = (k/q_i)^2$, $M_{A,i} = C_{A,i}/C_{S,i}$, and $F_i = \sqrt{1+M_{A,i}^2}$, $V_{A,i} = M_{A,i}\cos\psi_i$, $V_{B,i} = M_{A,i}\cos\psi_i/F_i$, $V_{\pm,i}^2 = (F_i^2 \pm \sqrt{F_i^4 - 4V_{A,i}^2})/2$. We define $M_e = C_{S,2}/C_{S,1}$, $f = \rho_{0,1}/\rho_{0,2}$, $\bar{V}_1 = V - U$, $\bar{V}_2 = (V+U)/M_e$, $V = \omega/k_TC_{S,1}$, $U = u\cos\alpha/C_{S,1}$; then $K_2 = K_1 f^2$ (equilibrium requires $f = M_e^2 G_2/G_1$).

Using the boundary conditions across the interface we obtain:

$$\begin{bmatrix}\zeta_{-,1}\\ \zeta_{+,1}\end{bmatrix} = \frac{1}{2fP_1\Gamma_1}\begin{bmatrix}\mathcal{E}_2 & \mathcal{E}_3\\ \mathcal{E}_1 & \mathcal{T}\end{bmatrix}\begin{bmatrix}\zeta_{-,2}\\ \zeta_{+,2}\end{bmatrix}, \tag{7}$$

with

$$\mathcal{T} = fP_1\Gamma_1 + P_2\Gamma_2 - fS_1 + S_2,\quad \mathcal{E}_1 = fP_1\Gamma_1 - P_2\Gamma_2 - fS_1 + S_2,$$
$$\mathcal{E}_2 = fP_1\Gamma_1 + P_2\Gamma_2 + fS_1 - S_2,\quad \mathcal{E}_3 = fP_1\Gamma_1 - P_2\Gamma_2 + fS_1 - S_2, \tag{8}$$

where

$$P_i = \frac{F_i^2(V_{A,i}^2 - \bar{V}_i^2)(\bar{V}_i^2 - V_{B,i}^2)}{2G_i(V_i^2 - V_{+,i}^2)(V_i^2 - V_{-,i}^2)},\quad S_i = P_i + \frac{(F_i^2 - V_{A,i}^2)\bar{V}_i^2 - V_{A,i}^2}{(V_i^2 - V_{+,i}^2)(V_i^2 - V_{-,i}^2)} \tag{9}$$

The matrix equation (7) describes a variety of problems concerning localized eigenmodes, and reflection and transmission of propagating internal waves.

For the unstable modes and the stable surface oscillations, $\mathrm{Re}(\Gamma_{1,2}) \neq 0$, and we must have $\zeta_{+,1} = \zeta_{-,2} = 0$ since both regions extend to infinity; then we must require

$$\mathcal{T} = 0, \tag{10}$$

so that (10) is the dispersion relation for eigenmodes localized around the interface. On the other hand internal propagating modes have $\mathrm{Re}(\Gamma_{1,2}) = 0$ so there is no restriction on $\zeta_{-,i}$, $\zeta_{+,i}$. In this case the condition (10), or

$$\mathcal{E}_1 = 0,\ \text{or}\ \mathcal{E}_2 = 0,\ \text{or}\ \mathcal{E}_3 = 0, \tag{11}$$

are the dispersion relations for radiation (poles of the reflection coefficient), or for total transmission of waves (zeros of the reflection coefficient), or for their time-reversed processes, according to the sign of the y component of the group velocity in each region. For conciseness we shall not enter into these details.

A polynomial equation associated to $\mathcal{T} = 0$ can be obtained by squaring twice (10) to get rid of the square roots. It is given by :

$$\Pi(K) = AK^2 + BK + C = 0, \tag{12}$$

with $K = K_1$ and $A = f^4W_-^2$, $B = f^2(QW_+ + X_-W_-)$, $C = Q(Q - 2X_+) + X_-^2$, $Q = (fS_1 - S_2)^2$,

$W_{\pm} = T_2P_2^2 \pm T_1P_1^2$, $X_{\pm} = f^2R_1P_1^2 \pm R_2P_2^2$.

The roots of Π are continuous functions of the parameters, that have poles at $A = 0$, but are otherwise finite; they encompass not only the solutions of (10), but also those of the three conditions (11), so that in looking for localized eigenmodes care must be taken to discard the spurious solutions (the roots of Π that satisfy any of the (11)). We call "true branch" ($\mathcal{T}$) the set of the solutions of (10) and "spurious branches" ($\mathcal{E}_1$, $\mathcal{E}_2$, $\mathcal{E}_3$) those belonging to (11), respectively. As a function of the parameters of the problem $\mathcal{T}$, $\mathcal{E}_1$, $\mathcal{E}_2$, $\mathcal{E}_3$ consist of several disjointed continuous manifolds. These pieces touch each other along boundaries where two branches coincide. Consider a root belonging to a certain piece of $\mathcal{T}$: as the parameters are varied, this root will move along $\mathcal{T}$ and may eventually cross its boundary, where $\mathcal{T}$ and a spurious branch coincide. At this point the nature of the root changes, becoming spurious. It can be shown that the branches can coincide only for $\Gamma_{1,2} = 0, \infty$, $K \to \infty$ or $Q = 0$ (see Table I).

Table I: coincidences of branches of the associated polynomial.

If:	The coincidences are:
$\Gamma_1 = 0, \infty$	$\mathcal{T} \leftrightarrow \mathcal{E}_3$ and $\mathcal{E}_1 \leftrightarrow \mathcal{E}_2$
$\Gamma_2 = 0, \infty$	$\mathcal{T} \leftrightarrow \mathcal{E}_1$ and $\mathcal{E}_2 \leftrightarrow \mathcal{E}_3$
$K \to \infty$	$\mathcal{T} \leftrightarrow \mathcal{E}_2$ and $\mathcal{E}_1 \leftrightarrow \mathcal{E}_3$
$Q = 0$	$\mathcal{T} \leftrightarrow \mathcal{E}_2$ and $\mathcal{E}_1 \leftrightarrow \mathcal{E}_3$

The problem of finding the localized eigenmodes is straightforward, but the analysis of the general case is exceedingly cumbersome as 8 independent parameters are involved ($M_{A,1}$, $M_{A,2}$, $\varphi_1 = \psi_1 - \alpha$, $\varphi_2 = \psi_2 - \alpha$, f, $U_0 = u / C_{S,1} = U / \cos\alpha$ that characterize the configuration, and α, K that define the perturbation mode). In addition Π leads to a polynomial equation of a very high degree in V that must be solved numerically. For this reasons we shall only consider here some special cases.

CASE WITH GRAVITY AND NO DENSITY JUMP

Let us suppose that both plasmas have the same β and there is neither a jump of the density ($f = 1$) nor of $\boldsymbol{B}$ ($\varphi_1 = \varphi_2$, $M_{A,1} = M_{A,2}$), but $\boldsymbol{g} \neq 0$ so that there is stratification ($q_1 = q_2 \neq 0$). There are no Rayleigh-Taylor unstable modes, but the internal gravity modes may be unstable[3] if $M_A < M_{AC} \equiv \sqrt{2(1 - 1/\gamma)}$. For simplicity we shall restrict our analysis to flute modes ($\psi_1 = \psi_2 = \pi / 2$, $V_{A,1} = V_{A,2} = 0$). Now Π can be written as a fourth degree polynomial in V^2:

$$aV^8 + 4bV^6 + 6cV^4 + 4dV^2 + e = 0, \tag{13}$$

with $a = K^2$, $b = -K^2(F^2 + U^2)$, $c = K[2F^4 + 4G(G - F^2) + K(2F^4 + 2F^2U^2 + 3U^4)]/3$, $d = K\{[4G(G-F^2) + KU^4](F^2 - U^2) + 2F^4U^2(K-1)\}$, $e = [4G(G-F^2) + KU^2(2F^2 - U^2)]^2 + 4KU^4(F^2 - 2G)^2$.

We define $\bar{H} = ac - b^2$, $\bar{I} = ae - 4bd + 3c^2$, $\bar{J} = ace + 2bcd - ad^2 - eb^2 - c^3$. Then

$$\begin{aligned} \Delta = \bar{I}^3 - 27\bar{J}^2 = 2^{16}F^4K^6(F^2 - 2G)^4(F^2G - G^2 - KF^2U^2)^2 \times \\ \{[F^2 - K(F^2 + 4U^2)]^2 + 4K(F^2 - 2G)^2\} \geq 0, \end{aligned} \tag{14}$$

and the roots of Π are all complex, or all real[6]. To be real, it is necessary that

$$\overline{H}=16K^3[F^4+(F^2-2G)^2-KF^2(F^2+4U^2)]<0 \quad (15)$$

and

$$a^2\overline{I}-12\overline{H}^2=4K^6F^2(F^2-2G)^2[2K(F^2+2U^2)-F^2]<0 \quad (16)$$

Since (15) and (16) cannot be satisfied simultaneously, the roots are always complex except for $\Delta=0$ when two roots become real and coincident. This happens for $K=K_c$, where

$$K_c = G(F^2-G)/F^2U^2 \ . \quad (17)$$

The critical value K_c exists always ($F^2-G=(M_A^2+M_{AC}^2)/2>0$) and marks a boundary between $\mathcal{T}$ and a spurious branch that gives the limit of the existence of the KHI, since for $K>K_c$ there is always an overstable mode. Then there is a large k region where linear KHI exists and in which Im(ω) increases without bounds with k. For $K<K_c$ there are no modes of the Fourier type. We conclude that there are no linearly stable configurations in this case, since for sufficiently large k the KHI will grow fast enough in the characteristic time of an experiment.

CASE WITH A DENSITY JUMP

If there is a density discontinuity at the interface unstable Rayleigh-Taylor modes can appear. We shall consider, as before, $M_{A,1}=M_{A,2}$. The resulting polynomial in V is of high degree (12 for flute modes) so that only numerical solutions are available. Here we shall give the conditions of existence and stability of the modes in a compact, but implicit form. The boundaries of the regions where the modes exist are determined by the conditions of Table I; it is found that in this case the relevant $\mathcal{T}\leftrightarrow\mathcal{E}_i$ contacts are:

(i) $\Gamma_1=0$ .From $\Pi=0$ one obtains $W_{-}(T_1Q^2-R_2T_1P_2^2+f^2R_1T_2P_2^2)=0$.

(ii) $\Gamma_2=0$. The resulting condition is $W_{-}(T_2Q^2+R_2T_1P_1^2-f^2R_1T_2P_1^2)=0$.

Another critical value comes from $Q=0$, in which case two complex conjugate roots coalesce yielding a double real root.

For flute modes there is always a small interval of parameters in the neighborhood of the critical values given by (28-29) where the $\mathcal{T}$ roots are real and the modes are purely oscillatory. Out of this interval the roots are complex, and the modes are overstable. The systematic analysis of this case will be the object of a future paper.

LIMITING CASE WITHOUT GRAVITY

If $g\to 0$ the plasma has no stratification in each region. This is mathematically equivalent to $K\to\infty$, $g\neq 0$; then Π reduces to $A=0$, so that $\mathcal{E}_2$, $\mathcal{T}$ and $\mathcal{E}_1$, $\mathcal{E}_3$ coalesce to a $\mathcal{T}$ and an $\mathcal{E}$ branch. If in addition ρ and $\boldsymbol{B}$ are continuous one finds

$$\Pi=F^2r^2-F^4rs-V_A^2[(2+F^2)V_A^2-2F^4]r-V_A^4(1-2F^2-V_A^2)s+F^2s^2+2V_A^6, \quad (18)$$

with $r=(V^2-U^2)^2$, $s=2(V^2+U^2)$. An equation equivalent to (18) was obtained in ref.[7], where a detailed study was made for the CGL model and $\boldsymbol{B}$ parallel to $\boldsymbol{u}$. Here we discuss the MHD stability domains for a general orientation of $\boldsymbol{B}$ and $\boldsymbol{u}$.

In Fig. 2 we show two typical stability diagrams, as a function of ψ and U, and for fixed M_A. Due to the symmetry of (18) only $U>0$, $0<\psi<\pi/2$ need to be considered. The following critical lines are relevant:

(i) The lines corresponding to $\Gamma_{1,2}=0,\infty$. are $U_r=(V_+ + V_-)/2$, $U_{r\pm}=|V_A\pm V_B|/2$; the

line $U_{r\pm}$ marks a $\mathcal{T}\leftrightarrow\mathcal{E}$ contact where stable surface modes appear; U_r marks a transition from a radiation mode to a surface mode.
(ii) The U corresponding to the $V=0$ roots of $\mathcal{T}$(marginal stability) are solutions of:

$$F^2U^8 - 2F^4U^6 - V_A^2\left(2+F^2\right)\left(V_A^2 - 2F^4\right)U^4 - 2V_A^4\left(1-2F^2-V_A^2\right)U^2 + 2V_A^6 = 0. \quad (19)$$

The (19) has always two real roots ($U=V_A, U=U_{i+}$). For $M_A>1$, no other real roots exists. For $M_A<1$ and ψ small another pair of real roots defines the line U_{i-}.
(iii) The lines $U_{d\pm}$ ($U_{d-}<U_{d+}$) correspond to the double roots of $\mathcal{T}$; they are obtained solving $\Pi=0$, $\partial\Pi/\partial V=0$ for U, and discarding the roots not belonging to $\mathcal{T}$. It is found that $U_{d+}\le V_A$; the equal sign holds for $\psi=0,\pi/2$; and also for an intermediate value $\psi=\psi_c$, if $M_A<1$. On $U_{d\pm}$ two real roots coalesce and become complex (marginal stability), or two pure imaginary roots coalesce and become complex (transition from monotonic instability to overstability).

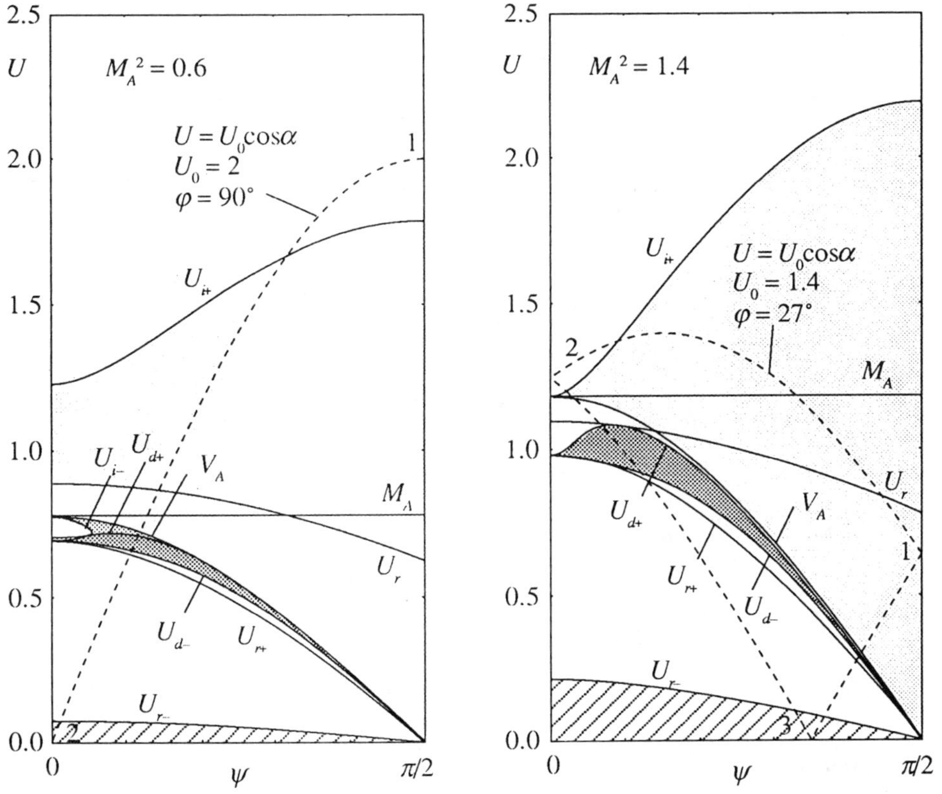

Fig. 2. Stability diagrams for: (a) $M_A=\sqrt{3/5}$; (b) $M_A=\sqrt{7/5}$. Light gray and medium gray areas correspond to one and two monotonically unstable modes respectively, the dark gray regions to overstable modes. In the hatched areas no modes are possible.

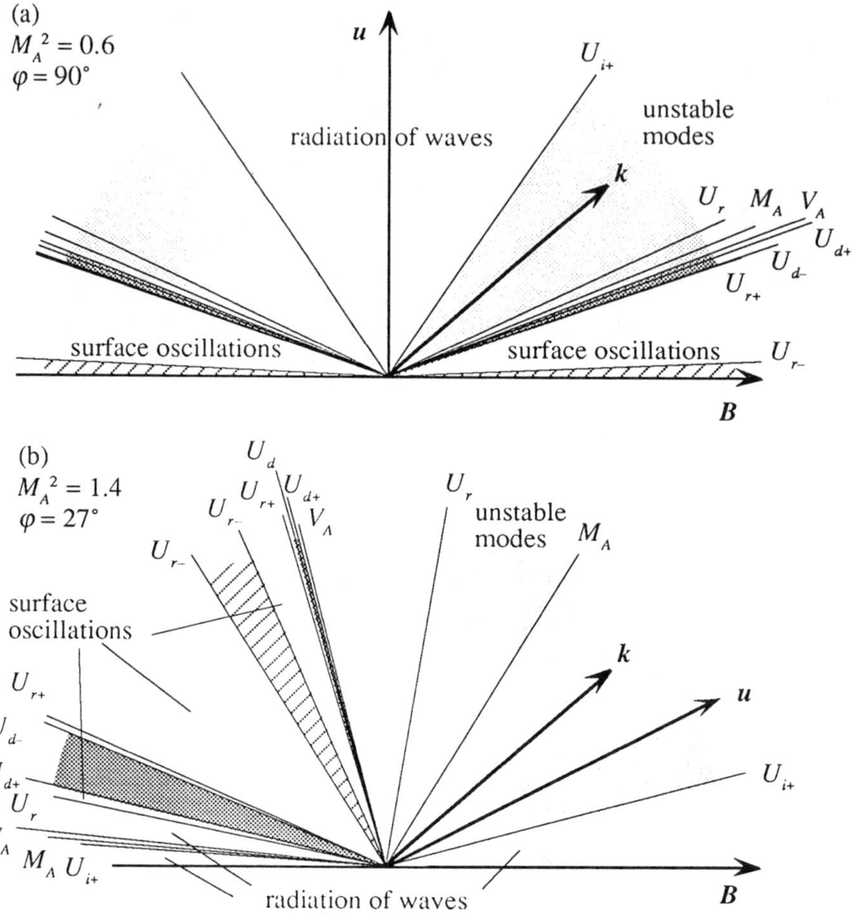

Fig. 3. Stability of configurations: (a) $M_A = \sqrt{3/5}$, $U_0 = 2$, $\varphi = \pi/2$; (b) $M_A = \sqrt{7/5}$, $U_0 = 7/5$, $\varphi = 27°$.

Two types of instabilities are shown as gray areas in Fig. 2:

(a) Monotonic instabilities (light gray) appear for $M_A < 1$, in the intervals $V_A < U < U_{i+}$ for $\psi > \psi_c$ and $U_{d+} < U < U_{i+}$ for $\psi < \psi_c$. In the last case there is a second monotonic instability in the interval $U_{d+} < U < V_A$, excluding the region to the left of U_{i-} (medium gray, Fig. 2a). The Im(V) of this mode is smaller than that of the first mode. For $M_A > 1$ there is a single unstable mode for $V_A < U < U_{i+}$ (Fig. 2b).

(b) Overstability exists for any M_A, in the interval $U_{d-} < U < U_{d+}$ (dark gray).

The stability diagram helps to analyze the modes of any configuration characterized by fixed M_A, U_0 and φ. Any case of interest can be easily analyzed. This can be shown by means of a couple of examples that show the variety of situations that can arise. The dashed line 1-2 in Fig. 2a represents $U = U_0 \cos \alpha$ for $M_A = \sqrt{3/5}$, $U_0 = 2$ and $\varphi = \pi/2$. In Fig. 3a we show the domains of stability of the modes as a function of ψ (only $0 < \psi < \pi$ is considered due to symmetry). The flute modes (point 1) are stable (since Landau's stability criterion[8] $U_0 > \sqrt{2}F$ is satisfied): there are two modes, both of the radiation type. As ψ (and correspondingly U) decreases, the phase velocity (V) of the radiation modes decreases. At $U = U_{i+}$ the slower one becomes marginally unstable and for smaller ψ there is a monotonically unstable mode and the remaining radiation mode. At $U = U_r$ the latter transforms into a stable surface oscillation, but as the line V_A is crossed it also becomes marginally unstable and a second monotonic instability appears. When the line U_{d+} is reached both instabilities merge giving an overstable and a damped mode. When U_{d-} is crossed these modes merge again to become two stable surface oscillations. For still smaller ψ the slower stable surface oscillation disappears at the $\mathcal{T}\leftrightarrow\mathcal{E}$ transition U_{r+}, but the other stable surface mode survives until the $\mathcal{T}\leftrightarrow\mathcal{E}$ transition U_{r-}. For $U < U_{r-}$ there is a ψ interval (hatched in Fig. 3a) where no modes exist, for α close to $\pi/2$.

Likewise, in Fig. 2b $M_A = \sqrt{7/5}$ and the line 1-2-3 represents $U = U_0 \cos\alpha$ for $U_0 = 7/5$, $\varphi = 27°$. Following this line one finds the transition points shown in Fig. 3b that limit the domains of stability as a function of ψ. Comparing Figs. 3a and 3b it can be seen that the domains are distorted due to the obliquity of $\boldsymbol{B}$ and $\boldsymbol{u}$, but the hatched region where no modes exist is again centered around $\alpha = \pi/2$

FINAL REMARKS

The general dispersion relation for the ideal compressible MHD modes localized near a velocity discontinuity between two plasmas in isothermal equilibrium in an effective gravity field has been derived. For continuous $\boldsymbol{B}$, ρ criteria for stability have been obtained for flute modes. Implicit stability criteria are given when ρ has a jump. The limiting case without gravity ($\boldsymbol{B}$ and ρ continuous) is studied for any relative orientation of $\boldsymbol{B}$, $\boldsymbol{u}$ and $\boldsymbol{k}$. The results of this last case indicate that the stability of a given configuration cannot be guessed on the basis of the analysis of particular orientations of $\boldsymbol{k}$ (say flute or parallel modes). The present work is a first attempt to achieve a systematic understanding of a very involved problem, and may be a useful starting point in studies of more realistic configurations as those present in experiments.

ACKNOWLEDGMENTS

This work started during a visit to the ICTP, Trieste, Italy. We are glad to acknowledge the encouragement of Prof. S. Mahajan and the hospitality and support then received.

REFERENCES

1. See J. A. Fejer, Phys. Fluids 7, 499 (1964), A. K. Sen, Phys. Fluids 7, 1293 (1964), D. J. Southwood, Planet. Space Sci. 16, 587 (1968).
2. See R. J. Groebner, K. H. Burrell, and N. P. Seraydarian, Phys. Rev Lett. 64, 3015 (1990), K. Ida, S. Hidekuma, Y. Miura, T. Fujita, M. Mori, K. Hoshino, N. Susuki, and T. Yamuchi, Phys. Rev. Lett. 65, 1364 (1990), A. R. Field, G. Fussmannn, J. V. Hofmann, Nucl. Fusion 32, 1191 (1992).
3. A. G. González, J. Gratton, and F. T. Gratton, in *Dense Z-Pinches* (edited by N. Pereira, J. Davies, and N. Rostoker), AIP Conf. Proc. 195, 280 (1989).
4. J. Gratton, F. T. Gratton, and A. G. González, Plasma Phys. Controlled Fusion 31, 435 (1988).
5. A. G. González, and J. Gratton, Solar Phys. 134, 211 (1991).
6. W. S. Burnside and A. W. Panton, The Theory of Equations, Vol. I (Dover, New York, 1960).
7. S. Duhau, and J. Gratton, J. Plasma Phys. 13, 451 (1975).
8. L. Landau, Compt. Rend. Acad. Sci. URSS 44, 139 (1944).

SNOWPLOW MECHANISM AND STABILITY OF IMPLODING MULTICASCADE LINER SYSTEMS

S. M. Gol'berg
Branch of I. V. Kurchatov Institute of Atomic Energy, Troitsk 142092, Russia

A. L. Velikovich
Laboratory for Plasma Research, University of Maryland, College Park, MD 20742

ABSTRACT

Rayleigh-Taylor (RT) instability developing in a layer of matter accelerated by magnetic pressure is shown to be suppressed if the accelerated layer scoops unperturbed matter, entraining it into motion. This stabilizing mechanism is effective for multicascade systems like explosive generators of high pulsed magnetic field or multiple (nested) gas-puff Z-pinches, where unperturbed elements are successively involved into accelerated motion. Linear stability analysis of one-dimensional solutions of the piston problem demonstrates that perturbation of the given wavelength λ does not grow appreciably until the thickness of the accelerated layer $L(t)$ exceeds λ. Before that, if acceleration is increased rapidly enough, amplitudes of the long-wavelength perturbations remain almost constant and an oscillatory eigenmode localized near its front surface. If acceleration is increased not too rapidly, stays constant, or is decreased, then the long- wavelength perturbations with $\lambda > 2L(t)$ are damped. For the case when acceleration of the first shell is reversed in collision, short-wavelength perturbations are stabilized. The optimal mass ratio is estimated for a double gas-puff Z-pinch: The inner shell must be ~ 3 times heavier than the outer one.

BASICS OF SNOWPLOW STABILIZATION

The snowplow stabilization has received almost no theoretical attention until recently (see [1]), though its efficiency has been verified experimentally in many cases. This is illustrated in Fig. 1. The well-known sausage instability of a steady plasma column with axial current [Fig. 1(a)] has been shown to be similar (in a sense) to the RT instability. An annular plasma liner imploded by an axial current [Fig. 1(b)] exhibits a real RT instability: Plasma is accelerated toward the axis, and in its own frame of reference it is supported by the pressure of massless azimuthal magnetic field against inertial forces. This instability limits the compression ratio of both metallic and gaseous liners on the level of about $R_i/R_f \approx 5-7$.

Consider now implosion of a solid gaseous column [Fig. 1(c)], with a shock wave S propagating to the axis ahead of the accelerated current sheath. The difference between Figs. 1(b) and 1(c) may appear inessential: In both cases, the plasma boundary is accelerated toward the axis, the conditions for the RT instability to develop being evidently satisfied. But the dynamic plasma configurations like that shown in Fig. 1(c), turn out to be remarkably stable: For experiments with plasma focus devices 1000-fold radial compression ratios are quite typical: $R_i/R_f \approx 10^3$ ($R_i \sim 10^2$ cm, $R_f \sim 1$ mm). [2]

To explain this, both types of acceleration are compared in Fig. 2 for planar geometry. Figure 2(a) corresponds to standard conditions of the RT instability development: A plane layer of constant mass is accelerated in vacuum by the pressure of magnetic field $\mathbf{B}$ (which is the same as a massless fluid if $\mathbf{kB} = 0, \mathbf{k}$ being the wavevector of the perturbation). Pressure p is constant over the whole front surface of the layer, in particular, it is the same in points A_1 and A_2, where perturbed velocities are different. In the case of Fig. 2 (b), the layer acts as a snowplow: It is moved into a half-space filled by unperturbed matter, which is entrained into motion by a shock wave S propagating ahead of the accelerated surface. The boundary of the perturbed motion is now represented by the shock front, hence, the pressure is greater where the perturbed velocity is higher: $p(A_1) > p(A_2)$. As the flow behind the shock wave is subsonic, a local pressure perturbation generates sonic waves,[3] which decrease pressure in A_1 and increase it in A_2. In other words, in the case of Fig. 2(a), a RT unstable eigenmode localized near the accelerated surface of the layer and an oscillatory eigenmode localized near its front surface are fully independent and uncoupled. On the contrary, in the case of Fig. 2(b) the RT mode is strongly coupled to the wave running along the shock front, and the stabilization is due to this very coupling.

SNOWPLOW STABILIZATION OF A SHOCK-PISTON FLOW

The snowplow stabilization mechanism starts to act at the linear stage of the RT instability development. The most straightforward way of studying it is a linear stability analysis of a plane shock-piston flow [Fig. 2 (b)], where the magnetic piston surface, being accelerated by a massless fluid, is RT unstable. This has been done in Ref. 1, and some of the results are presented below.

The magnetic piston starts moving at $t = 0$ according to the law

$$z(t) \equiv L(t) = g_0 t^m/m, \quad m > 1, \tag{1}$$

into a half-space $z > 0$ filled by a cold nonmoving gas. The flow of the shock-heated gas (plasma) ahead of the piston turns out to be self-similar,[4] and stability analysis is done for the self-similar solutions of Ref. 4. A noninertial frame of reference is used, where the piston is at rest, and the acceleration of the inertial force acting on the plasma is $g(t) = (m-1) \times g_0 t^{m-2}$, it is directed toward negative z, time-dependent, and spatially uniform.

We introduce a dimensionless time variable $\tau = kL(t) = kg_0 t^m/m$ and dimensionless displacement of the accelerated surface $\xi = k\delta z_0$. The displacement normalized to its initial value at $\tau = 0$ for the fastest-growing perturbation eigenmodes is shown in Fig. 3 for several values of m. For comparison, the dashed lines present the results of a similar calculation carried out for a constant mass of incompressible fluid supported by the pressure of a massless fluid against gravity whose time-dependent acceleration is the same as above.

The dashed lines show conventional exponential growth (for constant acceleration, $m = 2$, this is $\exp[\sqrt{g_0 k t}]$). The solid lines illustrate the stabilizing action of the snowplow mechanism for long-wavelength perturbations (small τ). In particular, for $m = 2$, expansion of the exponent $\exp[\sqrt{g_0 k t}]$ in power series includes terms of order of t, t^2, and t^3, all with positive coefficients, whereas the expansion for the corresponding $m = 2$ solid line starts from the term of order of t^4, the coefficient being negative!

The latter means that perturbations whose wavelengths are sufficiently large compared to the thickness $L(t)$ of the gas layer scooped by the piston, are damped. Appreciable change of initial perturbation starts not at $\tau > 1$, but rather at $\tau > \pi$, i. e., when $L(t)$ exceeds at least half of the perturbation wavelength λ.

This is easily explained in terms of wave coupling mentioned above. Indeed, the eigenfunction representing RT unstable modes decay as $\exp(-kz)$ with increasing distance z from the unstable surface. On the other hand, the actual thickness $\Delta L(t)$ of the accelerated layer is several times less than $L(t)$ due to both shock and adiabatic compression. Initial perturbation generates a standing wave, so that displacement of the accelerated surface tends to oscillate. However, the sign of displacement does not change more than once: With increasing τ, and $k\Delta L(t)$ exceeding unity, the unstable surface is being decoupled from the stabilizing shock front, the RT instability gains importance. In the long run ($k\Delta L(t) \gg 1$) the standard behavior is reproduced.

SNOWPLOW STABILIZATION IN A COLLISION

The shock-piston flow described above represents a limiting case of snowplow stabilization. Separate shells to be successively involved into accelerated motion are replaced by a continuum (with number of shells tending to infinity, and mass of each tending to zero, their product held constant). This case, being analytically tractable,[1] does not appear to be optimal for stabilization. In effect, with actual collision, the snowplow mechanism is expected to be even more effective, stabilizing the most rapidly growing short-wavelength perturbations. This has been confirmed in recent experiments with multiple gas-puff Z-pinch implosions. [5,7]

Indeed, consider the case when an impacting shell is decelerated due to shock pressure produced in the impact. Basically, this is the same Fig. 2(b), with the direction of acceleration **a** being reversed. In addition, in the dense (shock-compressed) region of Fig. 2(b), a contact interface C, separating the impacting shell from the compressed region of impacted shell, would appear.

With vector **a** reversed, density gradient and effective gravity near the magnetic piston surface have the same direction. The perturbation eigenmodes localized near this surface are no more unstable; instead, they are running surface

(Rayleigh) waves with dispersion relation $\omega = \sqrt{|a| k}$. This certainly means a stabilization, a self-healing: Local perturbations, developed on the piston surface in the course of RT unstable acceleration of the first shell, will no longer be eigenfunctions of the flow. They will be spread over the whole surface by the running waves, in other words, averaged, smoothed out. Of course, the contact interface C may now exhibit both Rayleigh-Taylor and Richtmyer-Meshkov instabilities. However, we limit ourselves to the case when we do not mind internal mixing of the two shells (for instance, if both of them are layers of the same gas), provided that the accelerated layer enclosed between the two stable surfaces, the shock and the piston, is smooth as a whole.

We come to the following stabilization criterion: the deceleration time t_d should be sufficient for spreading the perturbations over the surface. Since gravity acceleration $g = |a|$ depends on time, we can present it in the form

$$\int_0^{t_d} [g(t)k]^{\frac{1}{2}} dt > 1, \tag{2}$$

Of course, unity in the right-hand side of Eq. (2) is quite arbitrary; possibly, some other constant of order unity, like π or 2π would be more appropriate there. However, the results that follow are not sensitive to the value of this constant.

Let us now estimate the left-hand side of Eq. (2) for a double gas-puff configuration (Fig. 4) imploded by a constant current I_m switched on at $t = 0$. We use a thin-shell approximation, supposing δ_1, $\delta_2 \ll R_1, R_2$, and therefore neglecting variation of magnetic pressure in the course of collision.

Before collision, the first shell is accelerated inward and implodes from $R = R_1$ to $R = R_2$. Its equation of motion is

$$\mu_1 \frac{du}{dt} = \frac{I_m^2}{c^2 R}, \tag{3}$$

where $u = -dR/dt, \mu_1$ is the mass of the first shell per unit length. Equation (3) is easily integrated. At the moment of collision

$$u = u_1 = \frac{I_m}{c} \left[\frac{1}{\mu_1} \ln \frac{R_1}{R_2} \right]^{\frac{1}{2}}, \tag{4}$$

Collision generates a strong shock wave in the second shell. The mass velocity behind the shock wave in our approximation equals u. Consequently, velocity of the shock front is $D = \frac{\gamma+1}{2} u$, and the pressure behind the shock front is $p = \frac{\gamma+1}{2} \rho_2 u^2$, ρ_2 being the unperturbed density of the second shell. Equation of motion of the first shell now becomes

$$\mu_1 \frac{du}{dt} = \frac{I_m^2}{c^2 R_2} - 2\pi R_2 \frac{\gamma+1}{2} \rho_2 u^2 = \frac{\gamma+1}{2} \frac{\mu_2}{\delta_2} (u_2^2 - u^2), \tag{5}$$

where

$$u_2 = \frac{I_m}{c}\left[\frac{2}{\gamma+1}\frac{\delta_2}{R_2}\frac{1}{\mu_2}\right]^{1/2}$$

(recall that we neglect variation of R in collision). It would be convenient to use here dimensionless velocity $V = u/u_2$ and dimensionless time $\tau = (\gamma+1)\mu_2 u_2 t/(2\mu_1\delta_2)$. We see that the sign of acceleration is reversed in (5) in comparison to (3) if the initial value

$$V_1 \equiv \frac{u_1}{u_2} = \left[\frac{2}{\gamma+1}\frac{\mu_2}{\mu_1}\frac{R_2}{\delta_2}\ln\left(\frac{R_1}{R_2}\right)\right]^{1/2} > 1. \tag{6}$$

Since we suppose $\delta_2 \ll R_2$, the inequality (6) is easily satisfied, provided that the mass ratio μ_2/μ_1 is not too low.

Integrating the equation of motion (5), we obtain

$$V = \frac{(V_1+1)e^{\tau} + V_1 - 1}{(V_1+1)e^{\tau} - V_1 + 1}, \tag{7}$$

Deceleration lasts at least as long as the shock wave propagates through the second shell, that is, until the moment t_d determined from $\int_0^{t_d} D(t)dt = \delta_2$, or, which is the same,

$$\int_0^{\tau_d} V(\tau)d\tau = \frac{2\mu_2}{\mu_1}. \tag{8}$$

With the aid of Eq. (7) this could be expressed as

$$\tau_d = 2\ln\frac{1}{V_1+1}\left[\left(\exp(2\mu_2/\mu_1) + V_1^2 - 1\right)^{1/2} + \exp(\mu_2/\mu_1)\right]. \tag{9}$$

Let us substitute $g = \mid du/dt \mid$ from Eq. (5) into (2), and perform integration with time dependence $V(\tau)$ given by Eq. (7), and the upper limit of integration - by Eq. (9).

The result for the range of stabilized wavelengths $0 < \lambda < \lambda_m$, could be presented in the form

$$\frac{\lambda_m}{\delta_2} = \frac{4\pi}{\gamma+1}\frac{\mu_1}{\mu_2}\left[\ln\frac{F_-(\mu_2/\mu_1, V_1)F_+(0,V_1)}{F_+(\mu_2/\mu_1, V_1)F_-(0,V_1)}\right]^2, \tag{10}$$

where

$$F_\pm(a,b) = [e^{2a} + b^2 - 1]^{\frac{1}{2}} + e^a \pm (b^2-1)^{\frac{1}{2}}.$$

For low and high values of the mass ratio μ_2/μ_1 the right-hand side of Eq. (10) is proportional to μ_2/μ_1 and μ_1/μ_2, respectively. Consequently, there is some optimal value of μ_2/μ_1, which corresponds to the maximum value of λ_m for given ratios R_1/R_2, δ_2/R_2. However, the value of λ_m decreases slowly with increased mass ratio, so that the maximum is not sharp; for a broad range of mass ratios near the optimal value the snowplow stabilization is effective, see Fig. 5. E. g., for $2.5 < \mu_2/\mu_1 < 3, \lambda_m$ is in most cases within 15% of its maximum value.

The optimal mass ratio μ_2/μ_1 is not sensitive to the parameters of the problem, and for most cases is between 3.2 and 3.8, being slightly greater for higher compression ratios R_2/R_1 and for smaller values of δ_2/R_2. The maximum value of λ_m grows with increasing R_2/R_1, though not fast: a 2.5-fold increase of the latter value, from 4 to 10, increases λ_m at most for 20%. Optimal choice of the relative thickness of the second cascade, δ_2/R_2, however, depends on the perturbation wavelength range we consider the most dangerous, and seek to stabilize. If they are of order of the shell thickness (as is the case for ablatively accelerated foils), we must increase the ratio λ_m/δ_2. For this, higher density of the second cascade would be better, though not very much so: a 4-fold decrease in δ_2/R_2 from 0.4 to 0.1 results in a 30% increase in λ_m. On the other hand, if the most dangerous wavelengths are of order of the final radius R_2 (this is quite possible in a Z-pinch geometry), then we are interested in increasing λ_m/R_2. In this case, the looser the second cascade is, the better, because λ_m/R_2 is roughly proportional to δ_2/R_2.

CONCLUSION

We have shown that collision between the outer (accelerated) and inner (stationary) cascades of a multiple gas-puff Z pinch is capable of self-healing the perturbations developed in the course of acceleration of the outer cascade. In the course of collision, velocity of the impacting shell decreases. However, it continues to implode, so that the magnetic energy is still being transformed into kinetic energy of the imploding plasma. In the end, one can expect producing a reasonably uniform combined shell, whose inward velocity is smaller than that of the incident first shell, but still appreciable. The gain is evident: on the one hand, though one can easily produce a thin initial liner, making the total current flow through it would be much less easy. On the other hand, the outer liner, having imploded to the radius of the inner one, would be severely distorted by RT instability, and therefore unlikely to produce uniform hot plasma in further implosion. With a double shell, a tighter plasma column is produced near the axis at the final stagnation stage. [5–7] There is also some experimental evidence confirming better performance of the double gas-puff systems with a heavier inner cascade.[7]

For a single cascade, the mass per unit length needed to match implosion dynamics to the pulsed power driver is known to be determined by the initial radius, current pulse amplitude and duration. With multiple shells, we have more room for optimization. Though detailed discussion of this issue is beyond the scope of the present paper, it should be mentioned that a heavier second cascade does not necessarily require accordingly longer current pulse. Indeed, the combined shell whose inertia is greater, is closer to the axis, and has some starting inward velocity. The thin-shell model predicts the implosion time in a double gas-puff system to increase by factor of

$$1 + \left(\frac{\mu_1 + \mu_2}{\mu_1} \right)^{1/2} \frac{R_2}{R_1} \tag{11}$$

compared to implosion of a single first cascade. For $\mu_2/\mu_1 = 3$, and R_2/R_1 varying from 0.1 to 0.2, we obtain from Eq. (11) a 20% to 40% increase in implosion time, compared to a two-fold increase estimated for a single shell of mass $\mu_1 + \mu_2$ per unit length.

The snowplow stabilization is readily combined with the stabilization by magnetic shear.[8,9] Magnetic field could be used to stabilize implosions of successive cascades separately or as a "magnetic cushion", to transmit pressure between the cascades, making at the same time the collision softer. [10,11] Combined method of optimization allows greater flexibility for optimization, making it possible, for instance, to decrease the mass of the inner shell, thereby increasing energy per ion available for thermalization. Each of those methods applied separately is capable to stabilize implosions up to compression ratio of ~30. Therefore, combining both methods to produce carefully optimized implosions, one can reasonably expect achieving compression ratio ~ 100. (The latter estimate seems to be confirmed by recent experimental results.[10,11]). This level of stability appears to be quite sufficient for the most interesting applications of pulsed power systems operating in a nanosecond range: namely, for creating an x-ray laser (in the range of I_m between 10 and 20 MA), and for realizing breakeven conditions in a DT fusion plasma (with I_m between 20 and 30 MA).[12]

REFERENCES

1. S. M. Gol'berg and A. L. Velikovich, Phys. Fluids B **5**, 1164 (1993).
2. N. V. Filippov, T. A. Filippova, and V. P. Vinogradov, Nucl. Fusion Suppl. **2**, 577 (1962).
3. G. B. Whitham, *Linear and Nonlinear Waves* (Wiley, New York, 1974).
4. N. A. Krasheninnikova, Izv. Akad. Nauk S.S.S.R., Otd. Tekh. Nauk, No. 8, 22 (1955).
5. R. B. Baksht, A. V. Luchinskii, and A. V. Fedyunin, Sov. Phys.-Tech. Phys. **62**, no. 8 (1992)
6. R. B. Spielman, T. Nash, M. Krishnan, Bull. Am. Phys. Soc. **37**, 1578 (1992).
7. R. B. Baksht, A. V. Fedyunin, I. M. Datsko, Oreshkin V. I., Third International Conference on Dense Z-pinches, Imperial College, London, UK, 19-23 April 1993, Conference Programme, p. P4.
8. F. S. Felber, M. M. Malley, F. J. Wessel et al., Phys. Fluids **31**, 2053 (1988).
9. A. B. Bud'ko, M. A. Liberman, A. L. Velikovich, and F. S. Felber, Phys. Fluids B **2**, 1159 (1990).

10. S. A. Sorokin, S. A. Chaikovsky, Preprint, Tomsk Sci. Center, Acad. Sci. U.S.S.R. , Siberian Branch, no. 12 (1992).
11. S. A. Sorokin, S. A. Chaikovsky, op. cit. in Ref. 7, p. P63.
12. S. M. Gol'berg, M. A. Liberman, and A. L. Velikovich, Plasma Phys. and Contr. Fusion **32**, 319 (1990).

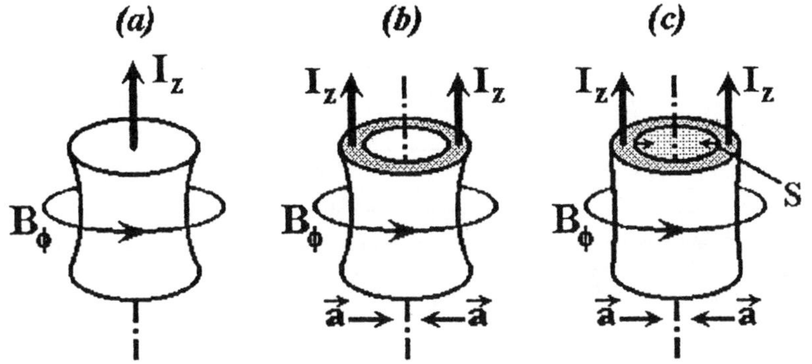

Fig. 1. Plasma confinement and acceleration in cylindrical geometry.

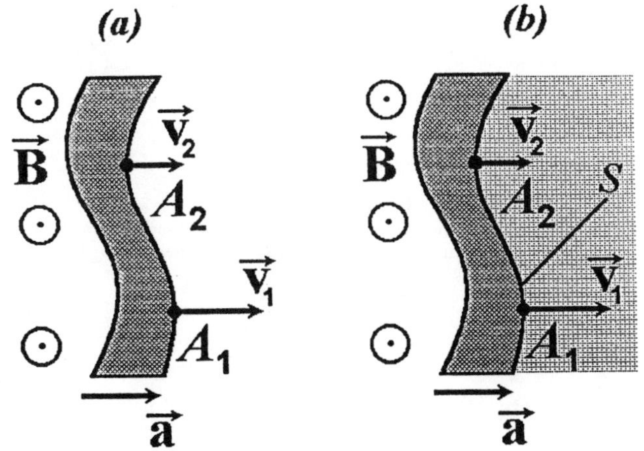

Fig. 2. Acceleration of a plasma layer by a magnetic piston: (a) constant mass of the accelerated plasma; (b) snowplow.

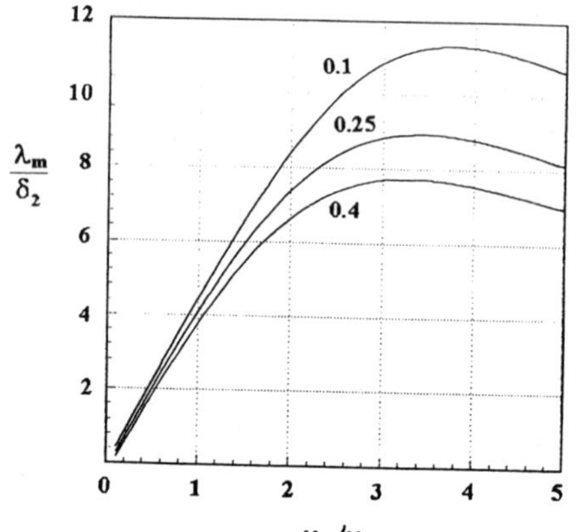

Fig. 5.

Range of stabilized wavelengths versus mass ratio for $R_2/R_1 = 7.5$ and three values of δ_2/R_2.

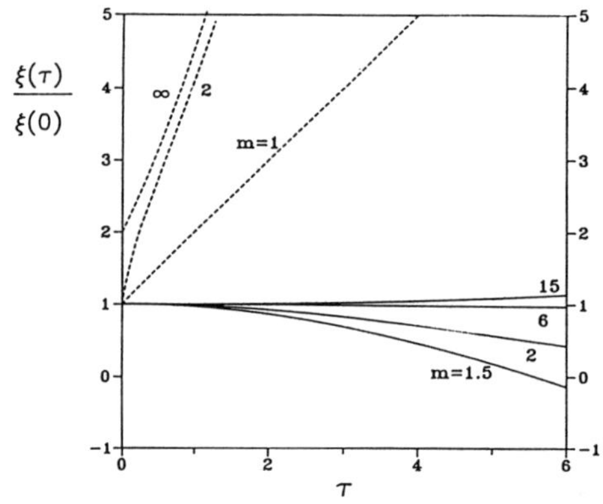

Fig. 3.

Displacement of the accelerated surface versus dimensionless time τ for several values of m: snowplow (solid lines) and constant mass (dashed lines).

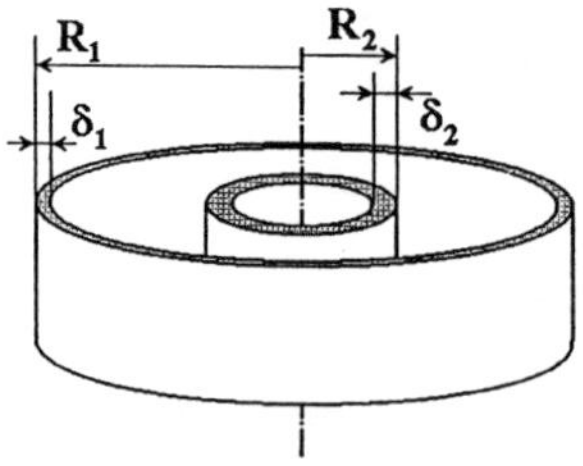

Fig. 4. Scheme of a double gas-puff configuration.

MHD STABILITY OF SELF-SIMILAR SOLUTIONS TO A DYNAMIC Z-PINCH

P. Jaitly and M. Coppins
Blackett Laboratory, Imperial College, London SW7 2BZ, U.K.

ABSTRACT

A general class of self-similar, time dependent, non-stationary z-pinch equilibria is examined in detail. It is shown that there are several zones of parameter space, corresponding to various types of solution. One-dimensional simulations reveal that the self-similar solutions act as attractors. The instaneous linear MHD growth rates for the m=0 and m=1 instabilities are found to be almost the same for all the solutions. Thus there is no especially favourable subset of the equilibria to aim for experimentally.

1. INTRODUCTION

To achieve an equilibrium z-pinch which remains stationary while the plasma is undergoing ohmic heating requires a current rising as $I \propto t^{1/3}$ (in the absence of radiation).[1,2] Generators used for z-pinch experiments do not deliver a current of exactly this form. Radial expansion or contraction is therefore an inherent feature of experimental pinches. The fact that z-pinches are, in general, dynamic rather than static may be important for stability. Here we describe a general class of self-similar, non-stationary z-pinch equilibria, characterized by $I \propto t^{\alpha}$. Some features of these solutions for different α are described. Using one-dimensional (1-D) simulations we show that z-pinches spontaneously relax to self-similar equilibria. The effect of profile changes (corresponding to changes in α) on stability are studied.

By "equilibria" we mean states in which pressure balance is maintained even though variables are time dependent and, in general, non-stationary. We make the following assumptions: (i) $Z = 1$, (ii) $n_i = n_e = n$, (iii) $T_i = T_e = T$, (iv) $\overline{\Omega_{ci}\tau_{ii}} > 1$, (v) $\Gamma = 5/3$ (Γ is the ratio of specific heats), (vi) $ln\Lambda$ is constant (when a numerical value is required we set it to 10), (vii) $v_\theta = v_z = 0$, (viii) inertia is neglected, (ix) radiation is neglected. The time dependent z-pinch equilibrium equations are:

Pressure Balance
$$\frac{\partial P}{\partial r} = -jB, \tag{1}$$

Ohm's Law
$$E_z = \eta j - vB, \tag{2}$$

Faraday's Law
$$\frac{\partial B}{\partial t} = \frac{\partial E_z}{\partial r}, \tag{3}$$

Heat Flow
$$q = -k_B \chi \frac{\partial T}{\partial r}, \tag{4}$$

Energy Equation
$$\frac{3}{2} n^{5/3} \frac{d}{dt}\left(\frac{P}{n^{5/3}}\right) = \eta j^2 - \frac{1}{r}\frac{\partial}{\partial r}(rq), \tag{5}$$

Continuity Equation
$$\frac{dn}{dt} = -\frac{n}{r}\frac{\partial}{\partial r}(rv), \tag{6}$$

Amperes Law
$$j = \frac{1}{\mu_0 r}\frac{\partial}{\partial r}(rB), \tag{7}$$

Ideal Gas Law
$$P = 2nk_B T, \tag{8}$$

where k_B is Boltzmann's constant, and v is the plasma radial velocity. Assumption (iv) allows us to use the magnetized plasma form of resistivity and thermal conductivity, i.e.,

$$\eta = \frac{\sqrt{m_e} e^2 ln\Lambda}{6\sqrt{2}\pi^{3/2}\epsilon_0^2} (k_B T)^{-3/2}, \tag{9}$$

$$\chi = \frac{\sqrt{m_i} e^2 ln\Lambda}{6\pi^{3/2}\epsilon_0^2} n^2 B^{-2} (k_B T)^{-1/2}. \tag{10}$$

2. SELF-SIMILAR SOLUTIONS

We now seek *self-similar* solutions to (1)–(10). These are separable solutions in which the profile shapes as functions of some similarity variable, ξ, are preserved, but the scale changes with time. Self-similar time dependent z-pinch equilibria were first explored by Braginskii and Shafranov.[1] Here we use a modified form of Glasser's[3] general formulation.

The solutions impose:

$$I \propto \left(1+\gamma\frac{t}{t_0}\right)^{\alpha} \tag{11}$$

where, in our formulation, $\gamma = \pm 1$, α is a free parameter, and t_0, which is essentially a resistive diffusion time, is determined as part of the solution (21).

The equations can be reduced to a set of coupled first order ordinary differential equations in $\xi(r,t) = r/a(t)$ [where a is a length scale which, like t_0, is determined as part of the solution, (21)] for a set of normalized profile functions $\{u_1 \ldots u_5\}$. These have the following physical interpretation: u_1 corresponds to P; u_2 to rB (i.e., the current flowing within radius r); u_3 to ηj (i.e., the resistive part of the electric field); u_4 to T; and u_5 to rq. The equations are:

$$\frac{du_1}{d\xi} = -\xi^{-1}u_2u_3u_4^{3/2}, \tag{12}$$

$$\frac{du_2}{d\xi} = \xi u_3 u_4^{3/2}, \tag{13}$$

$$\frac{du_3}{d\xi} = \gamma\alpha\frac{u_2}{\xi}, \tag{14}$$

$$\frac{du_4}{d\xi} = -\xi^{-3}u_1^{-2}u_2^2u_4^{5/2}u_5, \tag{15}$$

$$\frac{du_5}{d\xi} = \frac{\xi}{\lambda}\left[u_3^2u_4^{3/2} - \gamma u_1\right], \tag{16}$$

where $\lambda = \sqrt{m_i/8m_e}$. The boundary conditions are $u_1(0) = u_4(0) = 1$ and $u_2(0) = u_5(0) = 0$. We now define another free parameter[4], Ω, which sets the u_3 boundary condition as follows $u_3(0) = \Omega^{-1/2}$.

Equations (12)–(16) can now be integrated from $\xi = 0$ out to some upper limit $\xi = \xi^*$ to obtain the profile functions $\{u_1 \ldots u_5\}$. To find physical variables we adopt the following procedure: (i) choose values of $\gamma(= \pm 1)$, α, Ω, ξ^*; (ii) integrate (12)–(16)

over the range $0 \leq \xi \leq \xi^*$ to obtain $\{u_1 ... u_5\}$; (iii) specify r_0 (the initial pinch radius), I_0 (the initial current), and N (the pinch line density); (iv) evaluate the following constants:

$$a_0 = r_0/\xi^*, \tag{17}$$

$$B_0 = \frac{\mu_0 I_0}{2\pi a_0 u_2(\xi^*)}, \tag{18}$$

$$T_0 = \frac{\pi a_0^2 B_0^2}{N\mu_0} \int_0^{\xi^*} \xi \frac{u_1}{u_4}\, d\xi, \tag{19}$$

$$t_0 = \frac{6\sqrt{2}\pi^{3/2}\epsilon_0^2}{\sqrt{m_e}\mu_0^2\, ln\Lambda}\, a_0^2 T_0^{3/2}, \tag{20}$$

(v) calculate physical variables from the following equations:

$$a(t) = a_0 \tau^{\frac{1}{2}(1-3\alpha)}, \tag{21}$$

$$P(r,t) = \frac{B_0^2}{\mu_0} u_1(\xi)\tau^{5\alpha-1}, \tag{22}$$

$$B(r,t) = B_0 \frac{u_2(\xi)}{\xi} \tau^{\frac{1}{2}(5\alpha-1)}, \tag{23}$$

$$j(r,t) = \frac{B_0}{\mu_0 a_0} \frac{1}{\xi} \frac{du_2}{d\xi} \tau^{4\alpha-1}, \tag{24}$$

$$n(r,t) = \frac{B_0^2}{2T_0\mu_0} \frac{u_1(\xi)}{u_4(\xi)} \tau^{3\alpha-1}, \tag{25}$$

$$E(r,t) = \frac{a_0 B_0}{t_0} \left[u_3(\xi) + \gamma \frac{(3\alpha-1)}{2} u_2(\xi) \right] \tau^{\alpha-1}, \tag{26}$$

where

$$\tau \equiv 1 + \gamma \frac{t}{t_0}. \tag{27}$$

By varying the free parameters we can generate a wide range of equilibria. γ and α set the form of the current time history; ξ^* sets the position of the edge [in practice this is defined as the position of the first zero or the first minimum in $n(r)$]; and Ω sets the form of the heat flow at the edge.

There are two basic types of solution: (i) *gas-embedded*, in which the density is finite and the heat flow is outward at the pinch edge, and (ii) *thermally isolated*, in which the density is zero and the heat flow is inward at the pinch edge. In the latter case the heat flow arises from a skin current flowing on the plasma surface. Only gas-embedded solutions are found for $\gamma = -1$.

Figure 1 shows (for $\gamma = +1$) that (α, Ω) parameter space can be partitioned into zones corresponding to the various types of solutions. We reject those in the region labelled "unphysical" because they represent gas-embedded pinches with an inward heat flow from the surrounding gas.

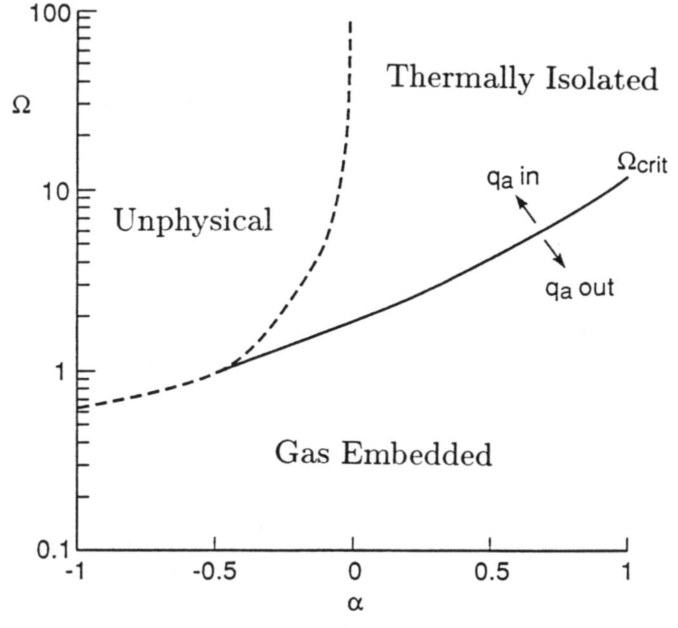

Fig. 1 (α, Ω) *parameter space for* $\gamma = +1$.

For every α greater than $-1/2$, in the $\gamma = +1$ case, there is a unique critical Ω for which $n_{edge} = 0$ and $q_{edge} = 0$. These Ω_{crit} solutions probably provide the best representation of fibre pinches *in vacuo*. In this case, therefore, a well behaved solution exists for a linearly rising current ($\gamma = +1$, $\alpha = +1$), but *not* for a linearly falling current ($\gamma = -1$, $\alpha = +1$). In the NRL fibre pinch experiment[5] the current waveform was approximately triangular. The plasma remained quiescent during the current rise but the start of the current fall was strongly correlated with the onset of a high level of MHD activity. This may have been triggered by the absence of a suitable equilibrium.

3. GENERATION AND STABILITY OF THE SELF-SIMILAR EQUILIBRIA

For many complex physical systems self-similar solutions act as attractors.[6] Previous computational studies[7,8] have shown that this property is a feature of time dependent z-pinch equilibria in the $\alpha = 1/3$ case. We have found that it also applies to other values of α.

We use a 1-D Lagrangian fluid code[9] to simulate the evolution of the z-pinch. This essentially integrates (1)–(8), with the important addition of inertia in the equation of motion (1). The code imposes zero heat flow at the plasma surface. Thus we would expect the Ω_{crit} solutions to arise, and this is indeed found over the whole range of α for which they exist (i.e., $\alpha > -1/2$).

Figure 2 shows three successive snapshots of the (normalized) density profile from such a simulation for $\alpha = 1$ (i.e., a linear current rise). In each case the broken line is the self-similar profile, and the solid line is the code result. The dimensionless time is defined as $t' \equiv t/t_0$. The spontaneous relaxation to the self-similar equilibrium is clearly seen.

Turning to the subject of stability, we notice that the problem falls into two parts; (i) the effect of equilibrium profile changes, and (ii) the dynamic effect of equilibrium motion. Here we consider only (i). This is a necessary first step in carrying out the full calculation. The neglect of equilibrium motion is also strictly justified in the long time limit for $\alpha > -1/3$.

We use a 1-D shooting code to solve the ideal MHD linear eigenvalue equation. This assumes a stationary, steady state equilibrium, for which we use the instantaneous self-similar equilibrium profiles. For any α we choose the Ω_{crit} solution.

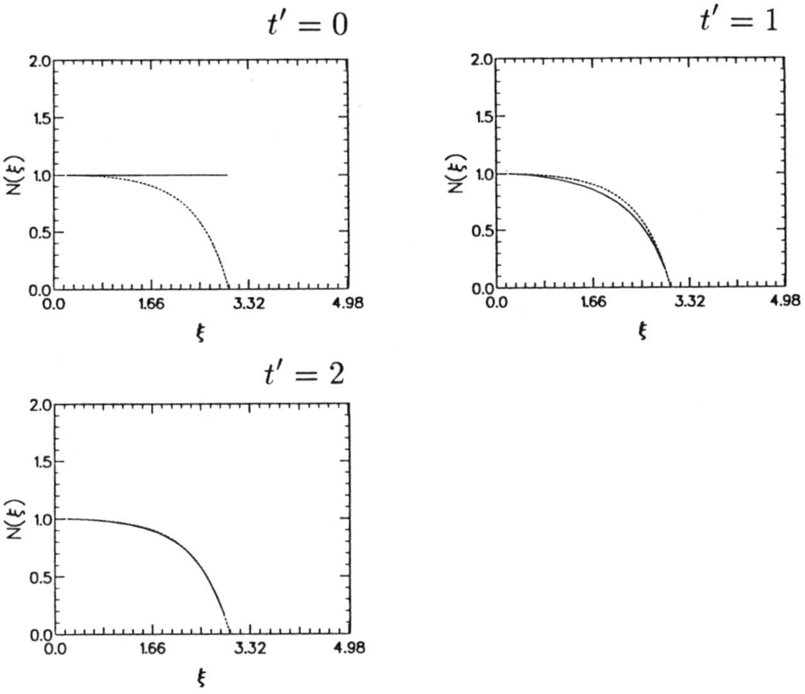

Fig. 2 Relaxation to self-similar density profile for $\gamma = +1$ and $\alpha = 1$.

We find that for the $m = 0$ and the $m = 1$ modes, the ideal MHD growth rate is almost independent of α. Previous work[10] has shown that the growth rate is also almost independent of Ω for fixed α. Therefore we cannot significantly improve stability by tailoring the current waveform to achieve a favoured profile shape.

REFERENCES

1. S.I. Braginskii and R.D. Shafranov, in *Plasma Physics and Problem of Controlled Thermonuclear Reactions*, edited by M.A. Leontovich (London, Pergamon, 1959), p. 39.
2. M.G. Haines, Proc. Phys. Soc. **76**, 249 (1960).
3. A.H. Glasser, J. Comp. Phys. **85**, 159 (1989).
4. M. Coppins, I.D. Culverwell, and M.G. Haines, Phys. Fluids **31**, 2688 (1988).

5. J.D. Sethian, A.E. Robson, K.A. Gerber, and A.W. DeSilva, Phys. Rev. Lett. **59**, 892 (1987).
6. G.I. Barenblatt, *Similarity, Self-similarity, and Intermediate Asymptotics* (New York, Consultants Bureau, 1979).
7. P. Rosenau, R.A. Nebel, and H.R. Lewis, Phys. Fluids B **1**, 1233 (1989).
8. M. Coppins, J.P. Chittenden, and I.D. Culverwell, J. Phys. D: Appl. Phys. **25**, 178 (1992).
9. The code was originally written by A.R. Bell, see: I.D. Culverwell, M. Coppins, M.G. Haines, A.R. Bell, and G.J. Rickard, Plasma Phys. and Contr. Fusion **31**, 387 (1989).
10. I.D. Culverwell and M. Coppins, Plasma Phys. and Contr. Fusion **41**, 1443 (1989).

INSTABILITY HEATING OF THE HDZP

R.H. Lovberg
University of California San Diego, La Jolla, CA 92093

R.A. Riley, and J.S. Shlachter
Los Alamos National Laboratory, Los Alamos, NM 87545

ABSTRACT

We present a model of dense Z-Pinch heating. For pinches of sufficiently small diameter and high current, direct ion heating by **m=0** instabilities becomes the principal channel for power input. This process is particularly important in the present generation of dense micro-pinches (e.g., **HDZP-II**) where instability growth times are much smaller than current risetimes, and a typical pinch diameter is several orders smaller than that of the chamber. Under these conditions, **m=0** formation is not disruptive: the large $\mathbf{E_z}$ field reconnects the instability cusps externally, after which the ingested magnetic flux decays into turbulent kinetic energy of the plasma. The continuous process is analogous to boiling of a heated fluid.

A simple analysis shows that an equivalent resistance

$$R_t = \frac{\ell}{4\sqrt{Nm_i}}\left(\frac{\mu_0}{\pi}\right)^{3/2}\frac{I}{r}$$

appears in the driving circuit, where **I** is the pinch current, **N** is the line density, $\boldsymbol{\ell}$ is the pinch length, $\mathbf{m_i}$ is the ion mass, and **r** is the pinch radius. A corresponding heating term has been added to the ion energy equation in a **0-D**, self-similar simulation, which had been written previously to estimate fusion yields and radial expansion of $\mathbf{D_2}$ fiber pinches. The simulation results agree well with the experimental results from **HDZP-II**, where the assumption of only joule heating produced gross disagreement. Turbulent ion heating should be the dominant process in any simple pinch carrying meg-ampere current and having submillimeter radius.

INTRODUCTION

Initial experiments with **HDZP-II** have given some unexpected results. Among them are:

1. Expansion of the pinch appears to be extremely rapid. Shadowgrams and interferograms show the column expanded to approximately **1 mm** diameter in **20** to **30 ns** after the beginning of current flow, where theory that assumes joule heating and Spitzer resistivity would predict

only slight expansion. **Fig. 1** is an interferogram of the pinch taken **20 ns** after the onset of current. At this time, the pinch diameter appears to be slightly greater than **1 mm**.

2. Neutron yield is two orders of magnitude smaller than would have been expected for the parameters of the experiment.
3. Intense **m=0** instability appears from the earliest times at which images have been obtained; it does not appear to disrupt the pinch. However, **m=1** instability has not been observed.

In this paper, we make an estimate of the effect of direct plasma heating by **m=0** instability growth and show that the observed expansion and radiation behavior can plausibly arise from it.

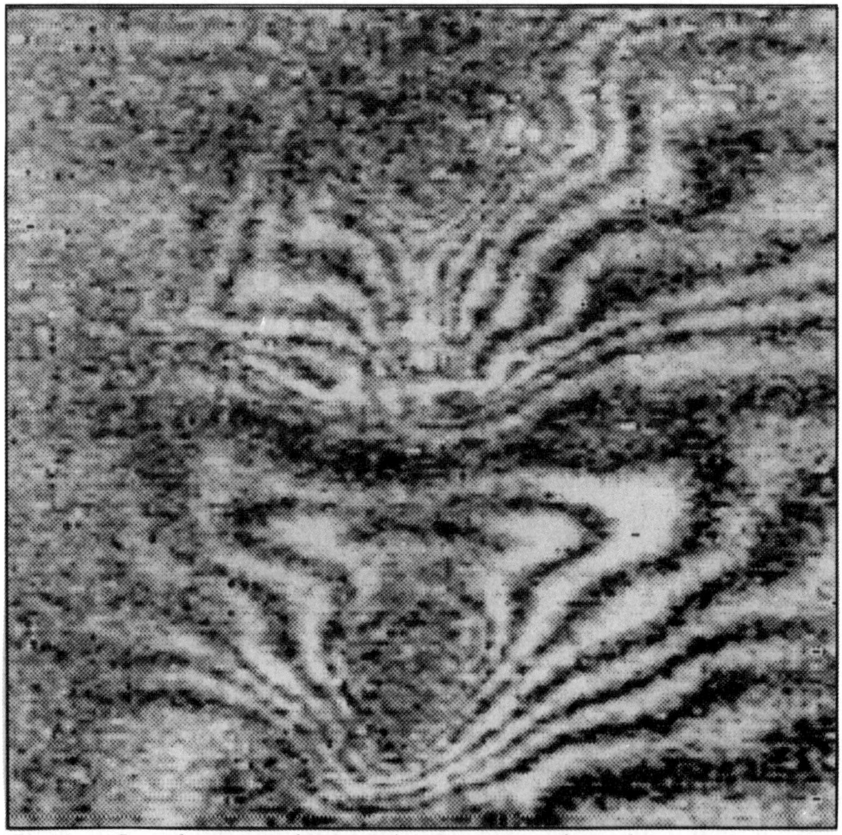

Fig. 1. Interferogram (**6x6 mm**) taken **20 ns** into the current.

INSTABILITY HEATING, QUALITATIVE

It is assumed for this discussion that there is no **m=1** (kink) instability of the plasma column. This is in accord with observation, although we offer no explanation for its absence.

When **m=0** deformations occur, the plasma acquires a macroscopic fluid-kinetic energy associated with deformation of its original cylindrical shape. In the nonlinear limit of the growth of the **m=0** mode, circular cusps of plasma are ejected radially away from the pinch (**Fig. 2**). In early experiments with gas pinches, these cusps generally reached the discharge tube wall, so that full development of instability growth usually resulted in disruption and termination of the discharge. In the **HDZP-II**, radially ejected plasma is still very far from the wall, so no disruption occurs. Moreover, the applied electric field in the **HDZP-II** remains strong for hundreds of instability growth times, so that one should expect an external reconnection of the cusps, with the result that cells of $\mathbf{B}_\theta$ are isolated from the main circuit, and the main current is diverted to the region outside of the cusps. The total energy contained within the larger pinch now consists of ion and electron thermal energy, turbulent kinetic energy of ions, and isolated cells of magnetic energy.

A detached torus of flux within the plasma has no equilibrium state short of a collapse of its inner interface with the plasma to zero radius. In reaching this limit, it will have given up most of its magnetic energy to work against the plasma, and as a result the ions receive most of this energy.

This process then continually repeats itself, amounting to a virtual "boiling" of the pinch.

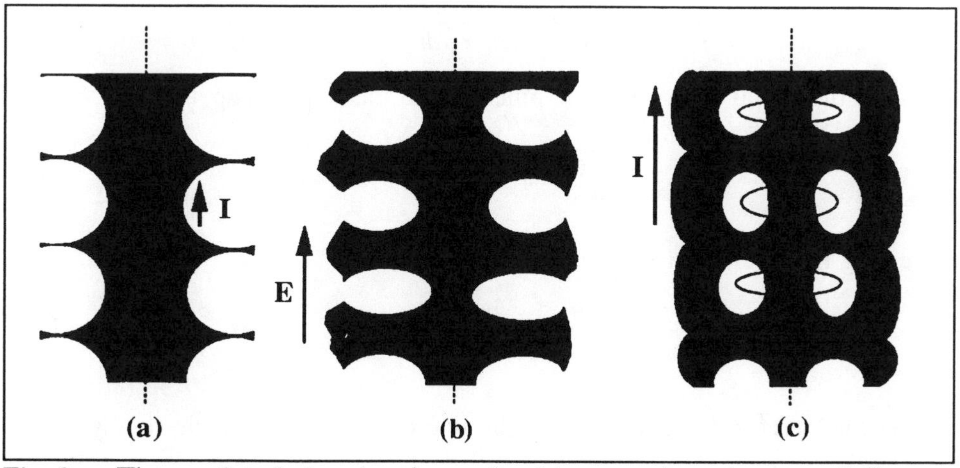

Fig. 2. The **m=0** turbulent heating cycle: **(a)** cusps eject plasma radially, **(b)** expanding cusps reconnect current, and **(c)** pinch current flows outside, trapped cells of **B**.

INSTABILITY HEATING, QUANTITATIVE

A convenient way to estimate energy transfer by an MHD instability is to describe its geometry as a time-varying inductance, particularly if the plasma current is externally imposed. The total power input to an inductive circuit for which resistance is negligible is

$$P = IV = I\frac{d}{dt}(LI) = IL\frac{dI}{dt} + I^2\frac{d}{dt}L \quad . \tag{1}$$

A slight rearrangement separates the terms functionally:

$$P = \frac{d}{dt}\left(\frac{1}{2}LI^2\right) + \frac{I^2}{2}\frac{dL}{dt} \quad . \tag{2}$$

The second term represents the power input to gross kinetic energy of the deformable conductor, while the first term is the rate of increase of magnetic energy.

The **m=0** instability raises the inductance of sections of the pinch through a shrinking of the inner coaxial conductor (plasma) diameter. The inductance of such a pinch section with radius **r**, length $\boldsymbol{\ell}$, and wall radius **b** is

$$L = \frac{\mu_0 \ell}{2\pi}\ln\left(\frac{b}{r}\right) \quad , \tag{3}$$

so that the rate of inductance change associated with pinch radius change is

$$\frac{dL}{dt} = -\frac{\mu_0 \ell}{2\pi r}\frac{dr}{dt} \quad . \tag{4}$$

Equating the radial velocity of the pinch to the local Alfvén speed,

$$\frac{dr}{dt} = v_a = \frac{B_\theta}{\sqrt{\mu_0 \rho}} \quad , \tag{5}$$

expressing $\mathbf{B}_\theta$ in terms of current and radius,

$$B_\theta = \frac{\mu_0 I}{2\pi r} \quad , \tag{6}$$

and setting $\boldsymbol{\rho\pi r^2 = Nm_i}$, where **N** is the line density of the pinch, finally yields

$$\frac{dL}{dt} = \frac{\ell}{4\sqrt{Nm_i}}\left(\frac{\mu_0}{\pi}\right)^{3/2}\frac{I}{r} \quad . \tag{7}$$

This rate of inductance increase does not imply that the overall circuit inductance is continually increasing; multiplied by the current, **Eq. (7)** represents the rate at which magnetic flux is being detached from the circuit and ingested by the turbulent pinch. Multiplied by $\mathbf{I^2/2}$, it gives the power input to macroscopic plasma motion, as mentioned earlier. This turbulence decays directly into ion thermal energy.

The inductance decrease associated with the outward moving cusps is small compared to the increase from the necks, if the motion is assumed to be approximately volume-conserving. The rate of inductance decrease associated with the overall expansion of the pinch is negligible, since the expansion goes at about 2% of the Alfvén speed.

The rate of loss of magnetic energy from the circuit through ingestion into the pinch is contained in the first right-hand term of **Eq.(2)**. Since, in the unstable motion, the relative growth rate of inductance is far greater than that of the current, it is reasonable to approximate **I** as constant over an instability growth time. In this case, the two right terms in **Eq.(2)** become identical, so the total power into ions is

$$P_i = I^2 \frac{dL}{dt} \tag{8}$$

PINCH SIMULATION

This heating term has been incorporated into a **0-D** pinch simulation, which was modified from one written previously to study the effects of alpha-particle retention at large burn rates. This code, **BURN4**, assumes self-similarity of the plasma profile during radius change. It incorporates joule heating, bremsstrahlung radiation, energy exchanges among various particle species, and gas dynamics that, as is usual under the self-similar assumption, assume the instantaneous transmission of pressure changes through the pinch. It also calculates **D-D**, or **D-T**, reaction rates. The relevant differential equations are given in **Appendix A**.

The new heating term is added to the ion energy equation. An arbitrary guess must be made of the fraction of the whole pinch length undergoing unstable radial motion at any instant. Based on analysis of a number of images showing **m=0** activity, we provisionally set this fraction to **0.1**.

SIMULATION RESULTS

For comparison, runs were made with and without the instability heating term. **Fig. 3** is a plot of radius, ion and electron temperatures, and current as a function of time, with the sinusoidal current waveform adjusted to the same risetime and peak current as those in the initial set of **HDZP-II** data taken through September, 1990. The turbulent heating is turned off. From an initial radius of **15 μm**, the pinch undergoes a rapid initial expansion and is subsequently recompressed to somewhat less than **100 μm**. Ion and electron temperatures remain nearly identical throughout the current rise and reach slightly over **2 keV**. **Fig. 4** shows calculated neutron output during the same interval. Without turbulent heating, the total yield up to **100 ns** is **7×10^{11}**, which is two orders more than what has been observed experimentally.

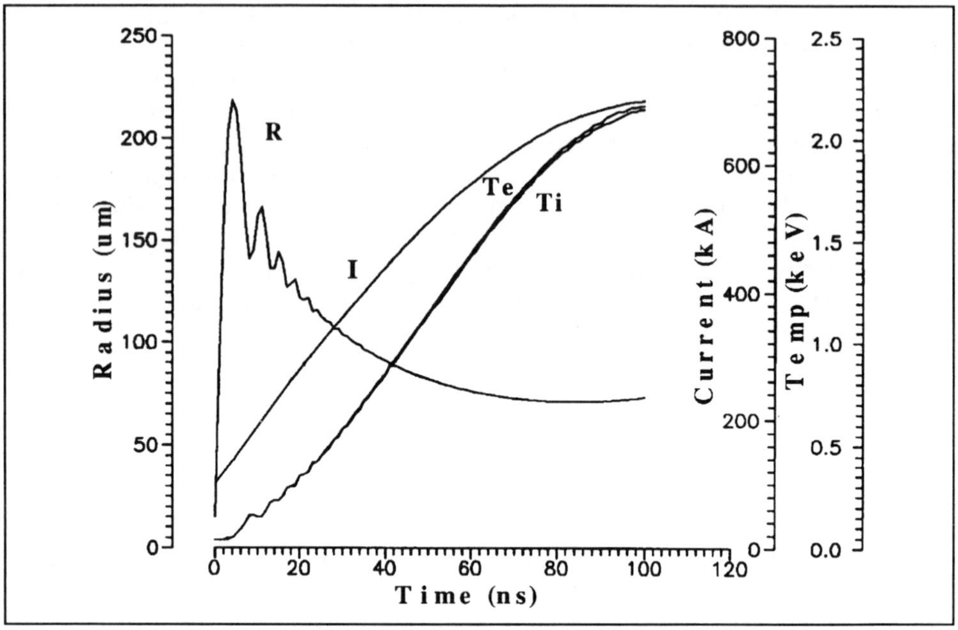

Fig. 3. Joule-heated **HDZP-II 0-D** simulation results.

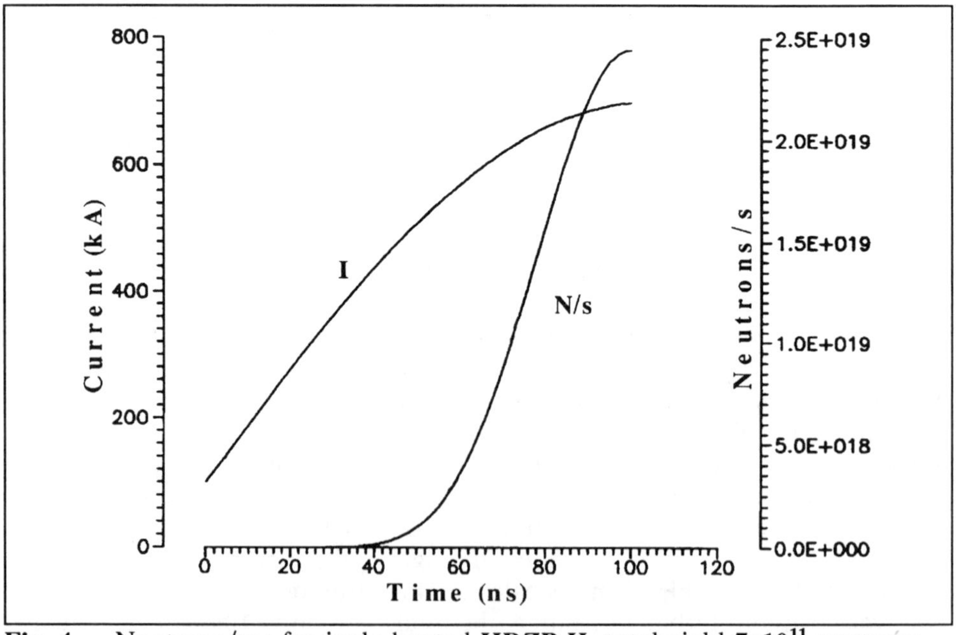

Fig. 4. Neutrons/sec for joule-heated **HDZP-II**, total yield **7x10^{11}** neutrons.

In contrast, results of the simulation with turbulent heating turned on are remarkably consistent with observation. **Fig. 5** shows the radius increasing to nearly **3 mm** during the current rise. At **30 ns**, it has reached a radius of **500 μm**, which is its approximate value in the interferogram of **Fig. 1**. Ion temperature is nearly **4 keV** at **100 ns**, while the electrons remain fairly cold, not exceeding **700 eV**. The very rapid plasma expansion decouples ions and electrons, and the joule heating of electrons can barely supply the temperature lost to adiabatic decompression.

The neutron yield for this case is **8 x 10^{9}**, which is in the range of yields obtained experimentally.

The low electron temperature can explain some difficulty in obtaining unambiguous soft x-ray signals at low-**keV** energies.

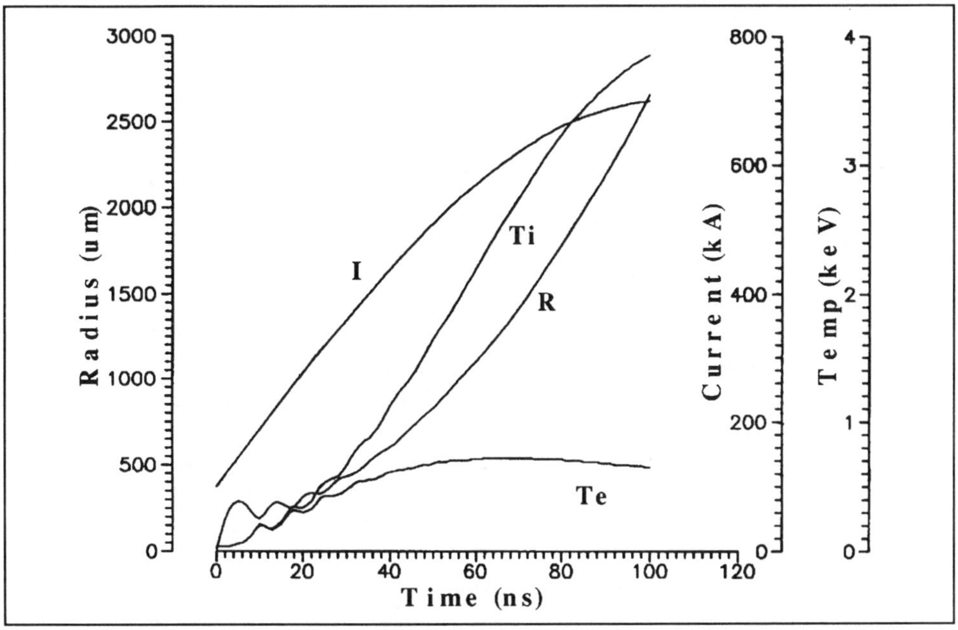

Fig. 5. The **m=0** turbulent-heated **HDZP-II 0-D** simulation results.

IMPURITY ADMIXTURE

One experiment in the series that generated the data quoted above involved loading the **D_2** fiber with approximately 10% of neon. The motive was to produce significant early-time radiation loss that would inhibit the sudden initial expansion, and also to add inertia to the pinch. The neutron yield under these conditions was expected to decrease by a very large factor, if not to disappear altogether.

Contrary to expectation, the neutron output from this shot (**#193**) remained at the level of the pure D_2 shots. This result cannot be understood if one supposes that ion heating is through transfer from joule-heated electrons. However, it follows immediately from a turbulent-heating model, in that ions receive their energy directly from fluid kinetic energy, and are insensitive to electron temperature. T_e only needs to be high enough to keep the magnetic Reynolds number large, so that **B**-field energy goes mostly into fluid motion rather than resistive dissipation.

DISCUSSION

The occurrence of this kind of turbulent heating ought to be a universal property of simple (unstabilized) pinches. However, for several reasons one might not have expected it to be a dominant process for pinches formed by the traditional breaking down of a tenuous gas:

1. The explosive radial expansion resulting from turbulent heating would drive a "classical" gas pinch into its chamber walls almost immediately, since the wall/pinch radius ratio for these devices is usually not greater than about ten. It has been observed that such discharges usually disrupt shortly after instabilities appear. The **HDZP-II** is unique in being so far removed from the chamber wall ($> 10^4$ pinch radii) that no plasma-wall encounter can occur; turbulent heating can proceed throughout the current rise.
2. At the very low densities and much larger geometrical scale of gas pinches, the time required for turbulent plasma motion to decay into ion thermal energy is probably not much less than the lifetime of the pinch. As a consequence, the process suggested above may not have time in which to achieve a steady state.
3. The effective turbulent resistance, dL/dt, varies as I/r. In the **HDZP-II**, this is probably three orders larger than in any previous pinch, so that here, uniquely in pinch research, turbulent heating is the dominant process.

The analysis presented above is very approximate in most of its details. However, one may be confident about some of its features. The radial expulsion of plasma by $m=0$ instabilities, with external circuit reconnection, and isolation and subsequent decay of cells of flux and turbulence, is a very likely mechanism. The expression for this effective power input to ions is subject to large uncertainties in its constant multipliers, but the scaling with current and radius is probably correct. The crude simulation employed here would ideally be followed by more sophisticated efforts; this may not be practical, however, because the next step should be a **2-D** code that would simulate instabilities in detail, and the cost and time needed to follow the system through hundreds of growth times and over hundreds of original pinch radii could be prohibitive.

To the extent that this model is correct, one may predict that rapid turbulent heating will be prominent in all of the currently planned **HDZP** experiments. In cases where the original fiber is significantly larger than the present **30 μm**, thus raising both **r** and **N**, and peak current smaller, the effect may be somewhat mitigated.

APPENDIX A: DIFFERENTIAL EQUATIONS

The set of coupled **ODE**'s for the $\mathbf{D_2}$ fiber simulation are for seven variables: ion density $\mathbf{n_i}$ (equal to the electron density $\mathbf{n_e}$), their corresponding energy densities $\mathbf{U_i}$, $\mathbf{U_e}$, the pinch radius **r**, radial velocity **v**, integrated neutron yield $\mathbf{N_n}$, and the total pinch current **I**. The deuterium of the pinch is assumed to be fully ionized at the beginning of the calculation, and to have a temperature of **10 eV**; the starting current is then set at a value that produces dynamic equilibrium at the initial pinch radius.

SI units are used, with the exception that the temperatures are in **keV**:

$$\frac{dn_i}{dt} = -\frac{2v}{r}n_i \qquad (expansion) \tag{9}$$

$$\frac{dU_e}{dt} = \frac{3000\,e\,n_e}{2}\frac{(T_i - T_e)}{\tau_{ei}} - \frac{2v\gamma}{r}U_e - \mathring{U}_{Brems} + \mathring{U}_{JouleHeat}$$

$$\mathring{U}_{Brems} = 5.3\times 10^{-37} n_e^2 \sqrt{T_e} \qquad \mathring{U}_{JouleHeat} = 3.3\times 10^{-10}\frac{I^2 \ln\Lambda}{r^4 T_e^{3/2}} \tag{10}$$

$$\frac{dU_i}{dt} = \frac{3000\,e\,n_i}{2}\frac{(T_e - T_i)}{\tau_{ei}} - \frac{2v\gamma}{r}U_i + \mathring{U}_{TurbHeat}$$

$$\mathring{U}_{TurbHeat} = 1.0\times 10^{-12}\frac{I^3}{r^3\sqrt{m}} \tag{11}$$

$$\frac{dr}{dt} = v \tag{12}$$

$$\frac{dv}{dt} = \frac{2\pi r}{m}(P_{int} - P_{ext}) = \frac{2\pi r}{m}\left(\frac{2}{3}\sum U - \frac{\mu_0 I^2}{8\pi^2 r^2}\right) \tag{13}$$

$$\frac{dN_n}{dt} = \pi r^2 \ell_p n_i^2 \overline{\sigma v} \tag{14}$$

$$\frac{dI}{dt} = \frac{2\pi I_{max}}{\tau_I}\cos\frac{2\pi t}{\tau_I} \tag{15}$$

where ℓ_p is the pinch length, **5 cm** in the present experiment, and **m** in **Eqs. (11)** and **(13)** is the total mass/meter.

The **D-D** reaction rate and particle energy exchange time constant in the above equations are:

$$\overline{\sigma v}_{DD} = \frac{2.3\times10^{-20}}{T_i^{2/3}}\exp\left(-\frac{18.76}{T_i^{1/3}}\right) \quad \frac{m^3}{s} ,$$

$$\tau_{ei} = 2.5\times10^{19}\frac{T_3^{3/2}}{n_e \ln\Lambda} \quad .$$

In addition, the temperatures and energy densities are related as

$$T_{e,i} = 4.16\times10^{15}\frac{U_{e,i}}{n_{e,i}} \quad .$$

It is assumed that $\gamma = 5/3$.

Additional simplifying assumptions are made in **Eqs (9)-(15)** are:

- The plasma is assumed to retain a uniform distribution over **r**; the expansion terms in the particle and energy density equations reflect this assumption (**0-D** self-similarity).
- In **Eq. (13)**, the pinch mass is assumed to reside at the outer boundary.
- Since it is assumed in this calculation that the total burn fraction is too small to produce significant pressure from charged reaction products, equations for their number and energy conservation are not included.

Finally, a small viscosity term was included in **Eq. (13)** to damp oscillations in the solution near **t=0**. It appears to have had no significant effect on later progress of the calculation. Some residual oscillation is still apparent.

MHD INSTABILITY OF DENSE DISSIPATIVE Z-PINCHES

V.V. Neudachin, P.V. Sasorov
Institute of Theoretical and Experimental Physics, Moscow, 117259, RUSSIA

ABSTRACT

The MHD instability of dissipative Z-pinches has been investigated with magnetic field diffusion, Ohmic heating, thermal conductivity and radiative losses taking into account. We pay main attention in this paper to stability of heterogeneous Z-pinches and to investigation, how thermal conductivity influences on instability of pinches with almost constant temperature.

INTRODUCTION

In several experiments enhanced stability of dense Z-pinches has been revealed. It relates both to the pinches, formed by electrical explosion of the metallic wires [1,2], and to the pinches, formed by the electrical explosions of condensed deuterium fibers [3]. This property of dense Z-pinches may be important for their applications. It is the reason for that a number of works [4–9,15] appeared at last time, which investigate the instabilities of Z-pinches with dissipative processes accounted for, which are important just for dense Z-pinches.

The complicated radial structure of dense Z-pinches was revealed in the theoretical works [10–12], and in some experiments [1,2]. Thus Z-pinches have a heterogeneous structure under certain conditions. In this case Z-pinch has a dense cold core and hot rarefied corona through which almost the entire current flows.

The present paper briefly reviews the latest works in instabilities of dissipative Z-pinches devoted, among other things, to removing several contradictions found in [4–9,15], to which end additional calculations have been carried out. In this work we investigate the MHD stability of Z-pinches with respect to small three dimensional perturbations and with the next processes, accounted for: diffusion of the magnetic field, Ohmic heating, thermal conductivity and volume radiative losses. Investigation of stability of static equilibrium states is the problem, well defined in the mathematical sense. So we limit ourselves by the consideration of instability only of equilibrium states of Z-pinches. The dissipative processes mentioned above are allowed for in a self-consistent manner both in the equilibrium problem and in the problem of instability of a Z-pinch.

We'll pay main attention in this paper to stability of heterogeneous Z-pinches and studies how the effect of thermal conductivity influences on instability of pinches with almost constant temperature.

FORMULATION OF PROBLEM AND METHOD OF ITS SOLUTION

For the investigation of equilibrium states of Z-pinches and their stability we use the following equations of MHD:

$$\rho\partial\mathbf{v}/\partial t + \rho(\mathbf{v},\nabla)\mathbf{v} = -\nabla p - \mathbf{H}\times curl\mathbf{H}, \tag{1}$$

$$\partial\rho/\partial t + div(\rho\mathbf{v}) = 0, \tag{2}$$

$$\partial\mathbf{H}/\partial t = curl(\mathbf{v}\times\mathbf{H}) - \nu curl(\frac{curl\mathbf{H}}{T^s}), \tag{3}$$

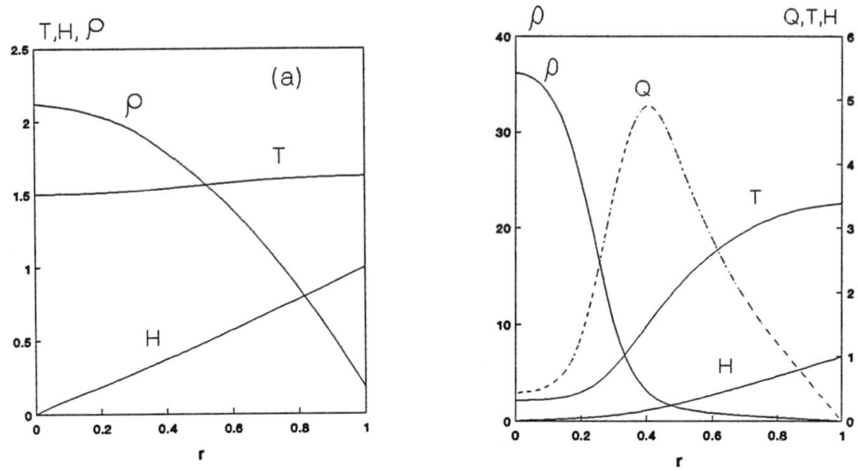

Figure 1: (a) Radial structure of considered equilibrium state of type D ($\delta = 1$, $q = 2$, $f = T^{1/2}$, $s = 3/2$, $b = 1/2$, $h = 1$). (b)H_δ-type equilibrium state of Z-pinch, which is investigated on instability.

$$\rho\,\partial\epsilon/\partial t + \rho(\mathbf{v},\nabla)\epsilon + p\,div\mathbf{v} = \nu\{\frac{(curl\mathbf{H})^2}{T^s} + \delta div(T^b\nabla T) - \rho^q f(T)\}. \quad (4)$$

Notations are usual. Note only, that ϵ is the specific internal energy of the plasma per unit mass, $\nu\rho^q f(T)$ is the volume radiative capacity of the plasma, ν/T^s is the specific electrical resistance of the plasma, and $\nu\delta T^b$ is the heat conductivity coefficient of the plasma. Parameters ν, δ, s, b, q are some positive constants. We use certain special system of units in eq.(1-4), which permits to show clearly the dimensionless parameters of problem [9].

Parameters ν and δ are connected closely with the dimensionless parameters of problem: magnetic Reynolds number and Peclet number:

$$R_m = B^{1/2}T(1)^{s+h/2}/\nu, \quad (5)$$

$$Pe = R_m BT(1)^{h-1-b-s}/\delta \quad (6)$$

The equilibrium states of Z-pinch, corresponding to eq.(1–4), were taken from works [11,13]. To investigate the stability of equilibrium states with respect to the small disturbances we solved numerically the linear initial value problem. This problem corresponds to system of equations (1–4), which is linearized in the vicinity of the equilibrium state of the eq.(1–4). The boundary conditions are also linearized. Such approach to stability problem permits us to find the unstable mode with the maximum growth rate for fixed m and k , where $m = 0, 1, 2...$ and $k > 0$ are the azimuthal and axial wave numbers of perturbations, which are proportional to $e^{i(m\varphi+kz)}$. Thus we considered briefly the formulation of problem and method of its solution,Refs.[9,17] contain more details.

EQUILIBRIUM STATES [13,11]

We investigate stability of equilibrium states of two types: D and H_δ. Equilibrium state is determined, if two parameters are given, parameter δ, determining heat conductivity, and full current I for instance. Note, that parameter ν falls out from the equations for equilibrium states.

The character of equilibrium depends mainly on δ. For large $\delta(\delta \geq 0.2 \div 0.3)$ the role of heat conductivity are also large in comparison with Ohmic heating and radiative losses. That is why the pinch is nearly isothermal [11,13]. Such type of equilibrium we denote as type D (Fig. 1a). For small $\delta(\delta \leq 0.2)$ the role of heat conductivity is small, and the equilibrium state is heterogeneous state [11] with the cold dense core and hot, rarefied corona. High temperature in the corona leads to compensation of smallness of δ and to relatively large local role of heat conductivity in corona. Such equilibrium states we will denote as H_δ-type (Fig. 1b).

INSTABILITY OF ISOTHERMAL Z-PINCH

Cochran and Robson [4] and Coppins and Culverwell [5] revealed the stabilizing influence of resistivity of plasma and/or nonstationarity of Z-pinch on sausage instability. This fact may open good possibilities for interpretation of the experiments [3], where enhanced stability of dense deuterium Z-pinches was found. Unperturbed pinch in [4,5] is isothermal and in this sense it corresponds to our type D equilibrium states.

Main results of [4,5] relates to nonradiating, nonstationary Z-pinches, but in [4] the equilibrium Pease-Braginskii Z-pinch also was considered, and stabilization effect of resistivity on sausage instability was obtained. It is easy to see, that the contradiction exists between the results of [4,5] and the results of works [6,7,9] (and [8], after detail analysis). All this works may be considered as almost independent. Our results [9] were obtained before comprehensive information concerning [4,5] was published.

We had undertaren a number of detailed calculations, before we discovered the cause of contradiction between the results of [4] and [9] (for example).

For the simplification of the problem Z-pinch was considered in [4] (and in [5]) as isothermal in radial direction, but the heat transfer along z-direction was neglected. One can consider that the radial coefficient of thermal conductivity in their calculations was very large, but the same in Z-direction was neglected. We should note, that sufficiently great value of radial heat conductivity is quite necessary in formulation of the problem, corresponding to [4] and [5]. In nonstationary problem, which is considered in [4,5], the quasi-equilibrium states of type D can't exist, if the heat conductivity is small, see [12,14]. (We already said about it above, in fact). One would think, that stability of Z-pinch can not depend on fact, that heat conductivity in Z-direction have not been taken into account. However, that is not so.

Our direct calculations (with $q = 2, f = \sqrt{T}, S = 3/2, b = 5/2, h = 1, \delta = 1$) show, that we reproduce the results of [4], if the heat conductivity in Z-direction is turned off, and the results of [9], if heat conductivity is described completely, (see Fig.2). The latter results are in agreement with works [6–8].

The stabilizing effect, which takes place, when the heat conductivity in Z-direction is turned off, may be explained qualitatively. When neck of sausage instability contracts, plasma in neck of z-pinch heats, but axial thermal flux is

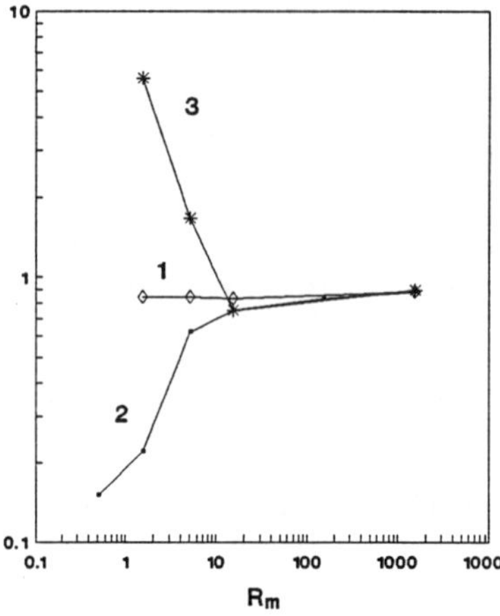

Figure 2: The growth rates of MHD instability ($m = 0, k = 1$) of equilibrium pinch of type D (see Fig. 1a). Their dependences upon $R_m(\propto \nu^{-1})$ for different methods of heat conductivity description in linearized equations : (1) complete description ($\delta = 1$), (2) heat conductivity in Z-direction is absent, (3) full absence of heat conductivity ($\delta = 0$).

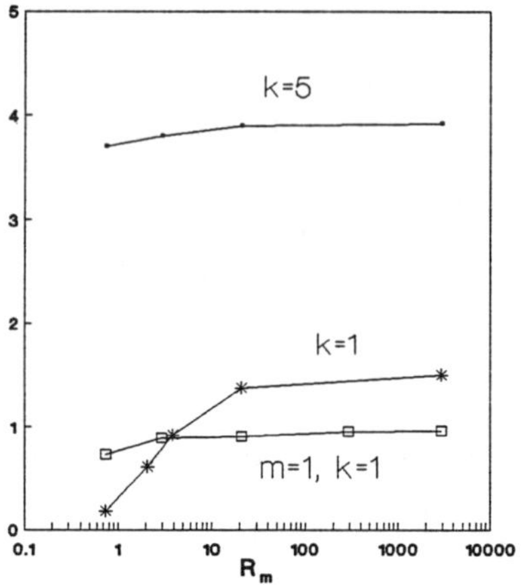

Figure 3: Dependence of growth rate of MHD instability with $m = 0, k = 1$ and $k = 5$,and $m = 1, k = 1$ on R_m for heterogeneous equilibrium Z-pinch (Fig. 1b).

absent. So plasma temperature and pressure in neck rise, and contraction of neck may be stopped.

If the description of heat conductivity is full, the growth rate of MHD instability for equilibrium states of isothermal Z-pinches weakly depends on resistivity ν (or on R_m) for $k \sim 1$ (Fig.2). We should note, that if we turn off artificially the heat conductivity on the whole, i.e. assume $\delta = 0$, then we discover strong thermal instability of radiating Z-pinch for $R_m \leq 1$ [9,16], (see Fig.2).

INSTABILITIES OF THE HETEROGENEOUS EQUILIBRIUM STATES OF Z-PINCHES

Let us consider now the equilibrium Z-pinches of H_δ type. We present, for example, the results for the equilibrium with $\delta = 0.03$; $I = 1$; $q = 1$; $h = 7/4$; $s = 3/4$; $b = 7/4$; $L = 1$; $f(T) = T^{5/2}\exp(-5/4\ln^2 T)$. The radial structure of this equilibrium Z-pinch is shown at Fig. 1b. The dependence of growth rate on ν are given in Fig.3, which has been calculated for modes with $m = 0, k = 1$ and $k = 5$.

It is seen, that the strong suppression of instability with $k = 1$ takes place for $R_m \leq 1$. Such behavior of the growth rate can be explained in the following manner. Arising and the development of sausage instability lead to the

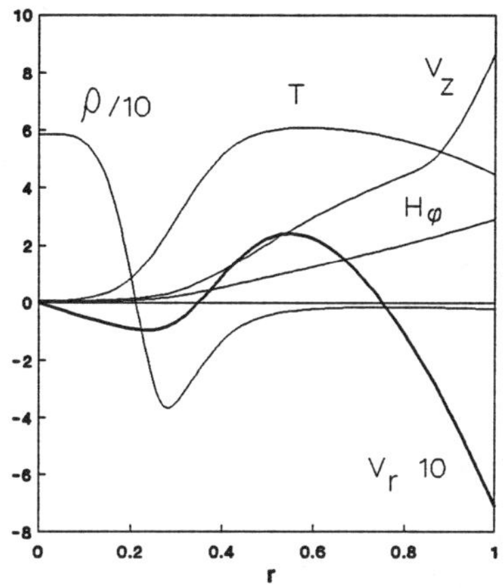

Figure 4: Eigenfunctions of the mode $m = 0$, $k = 1$ for H_δ equilibrium state (Fig. 1b), $\nu = 1.4$ ($R_m = 2.06$).

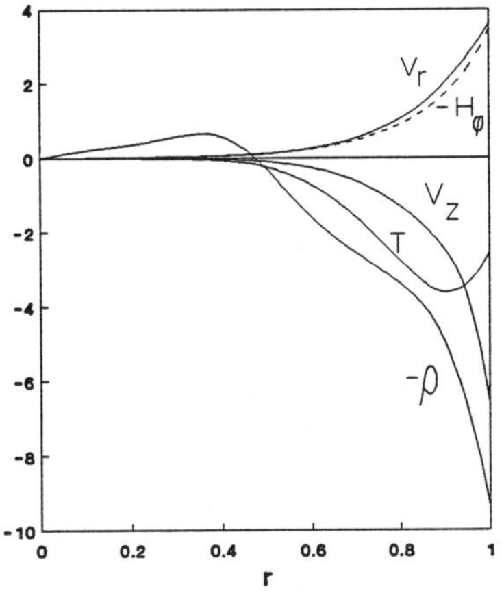

Figure 5: Eigenfunctions of the mode $m = 0$, $k = 5$, $\nu = 0.14$ ($R_m = 20.6$) for H_δ equilibrium state .

increasing of Ohmic heating in the narrow part of Z-pinch. The heating begins to exceed the radiative losses, the neck of pinch is heated and the "evaporation" or ablation of the external layers of core occurs because of the heat conductivity.

This results in filling of neck of Z-pinch by the ablating plasma, and, as consequence, in stabilizing effect. We may come to such conclusion looking at the V_r and other eigenfunctions of unstable mode with great resistivity, ν (see Fig.4).

For mode with shorter wave length, $k = 5$, the instability does not affects the central part of the pinch, therefore the increasing of ν does not lead to stabilization.

Thus, growth rates of large scale MHD modes ($m = 0$) decreased strongly decreasing for heterogeneous Z-pinches with

$$R_m \leq 1 \tag{7}$$

At the same time for $R_m \leq 1$ the instabilities, which doesn't affect the core of pinch can develop in the corona (Fig.5). For kink mode $m = 1, k = 1$ our calculations give only weak reducing of growth rate, when resistivity ν is increased (Fig.3). The supplement information about instability of the heterogeneous Z-pinches is contained in [9].

CONCLUSIONS

Thus we have shown, that the dissipative isothermal pinches with great heat conductivity are unstable with respect to MHD perturbations. The growth rates of instability differ only weakly from the same for the ideal MHD case.

Heterogeneous Z-pinches (with little average heat conductivity) for $R_m \leq 1$ demonstrate the enhanced stability with respect to large scale perturbations of small amplitude. In that time, however, the corona of pinch may be strongly unstable with respect to small scale perturbations, which are localized in corona.

REFERENCES

1. I. K. Aivazov, et al., JETP Lett. 41, 111 (1985).
2. L. E. Aranchuk, et al., Sov. J. Plasma Phys. 12, 765 (1986).
3. J. D. Sethian, et al., Phys. Rev. Lett. 59, 892 (1987).
4. F. L. Cochran, et al., Phys Fluids **B2**, 123 (1990).
5. I. D. Culverwell,et al., Phys Fluids **B2**, 129 (1990).
6. M. Lampe,Phys. Fluids **B3**, 1521 (1991)
7. P. M. Cox, Plasma Phys. and Contr. Fusion, 32, 553 (1990).
8. Ir. R. Lindemuth, Phys. Rev. Lett. 65, 179 (1990).
9. V. V. Neudachin,et al., Nucl. Fus. 31, 1053 (1991).
10. R. B. Bakst, et al., Sov. J. Plasma Phys. 14, 706 (1988).
11. N. A. Bobrova, et al., Sov. J. Plasma Phys. 14, 617 (1988).
12. Ir. R. Lindemuth,et al., Phys. Rev. Lett. 62, 264 (1989).
13. N. A. Bobrova, et al., Sov. J. Plasma Phys. 13, 53 (1987).
14. N. A. Bobrova, et al., Sov. J. Plasma Phys. 18, 269 (1992).
15. M. G. Haines, et al., Phys. Rev. Lett. 66, 1462 (1991).
16. V. S. Imshennik,et al., Sov. J. Plasma Phys. 14, 393 (1988).
17. V. V. Neudachin, et al., J. Moscow Phys. Soc. 2, 23 (1992).

LINEAR STABILITY OF THE LARGE LARMOR RADIUS Z-PINCH

Jan Scheffel

Alfvén Laboratory, Royal Institute of Technology, S-100 44 Stockholm, Sweden.

Tony Arber, Michael Coppins

Blackett Laboratory, Imperial College, London SW7 2BZ, UK.

ABSTRACT

For the first time, calculations of large Larmor radius (LLR) effects on the linear stability of realistic Z-pinch equilibria have been performed. The fixed boundary m=0 instability of the pure Z-pinch (no external magnetic field) is considered (free-boundary and m=1 modes are presently under study). We use the Vlasov-Fluid model, where ions are treated fully kinetically and electrons are modelled as a cold, massless fluid. Stability is found to be remarkably sensitive to equilibrium profiles. A flat current equilibrium is increasingly stabilized (smaller growth rate) by LLR effects as the normalized average Larmor radius ε is increased to about 0.1. Complete stabilization cannot be obtained. For larger values of ε the growth rate increases to reach above the small Larmor radius value when $\varepsilon \approx 0.3$. The Bennett equilibrium, however, is increasingly destabilized as ε increases.

BACKGROUND

Historically, the experimental evolution of the m=0 mode has been the major reason for abandoning the Z-pinch as a magnetic fusion candidate. Its prediction from simple fluid models[1] has further strengthened its position as an uncurable artefact of nature. The mode has a simple intuitive interpretation, arising as a collective $\mathbf{E}\times\mathbf{B}$ drift where the electric field $\mathbf{E}$ originates from the separation of ions and electrons due to the centrifugal ('gravitational') drift. Early papers on the finite Larmor radius effect[2,3] indicated a possible remedy for this instability, caused by the differentiated ion and electron drifts in inhomogeneous $\mathbf{E}$ fields. The theory was, however, based on the small Larmor radius limit in the fluid approach, excluding important effects such as those of resonant ion interaction. Hence, until this day, no theory has convincingly determined whether a satisfying stabilization of realistic profiles can be achieved by LLR effects. This paper presents the answer for the linear regime with fully kinetic ions in a circular cylinder Z-pinch.

Depending on the particular type of Z-pinch, other mechanisms than LLR effects can salvage *m*=0 stability. We are here only concerned with collisionless stability; a fusion Z-pinch features collision parameters $\omega_{ci}\tau_{ii} \geq 10^3$ and Lundquist numbers $S \geq 10^5$. It has been shown[4] that presence of a surrounding gas (Extrap, gas-embedded pinch) can aid in satisfying the fluid condition for stability, which requires the boundary pressure to be of the order 10-50% of the peak pressure. Further, visco-resistive effects can stabilize the boundary region of a gas-enclosed plasma[5]. In 2-D cross-section geometry, magnetic X-points due to the presence of a magnetic octupole field (as in Extrap) diminishes the fraction of boundary pressure required for stability. Dynamic effects have experimentally shown to be of importance for *m*=0 stability of cryogenic fibre pinches[6,7]. Finally there remains the so far unexplored possibilities of saturation at a low level or mode coupling, in the non-linear regime.

RESULTS

In Fig.1 growth rates are shown for a constant current density equilibrium (defined later), as function of Larmor radius and axial wavenumber. The parameter $\varepsilon = (\int dr/r_L)^{-1}$ is the averaged Larmor radius, normalized to the pinch radius a, and k is the axial wavenumber. Growth rates γ are normalized to the thermal transit time a/v_{th}, with v_{th} being the thermal ion speed.

Obviously there is good agreement between the two different schemes employed; an initial-value and a variational method. For reference, values for the Chew-Goldberger-Low (CGL) ε=0 limit are also given.

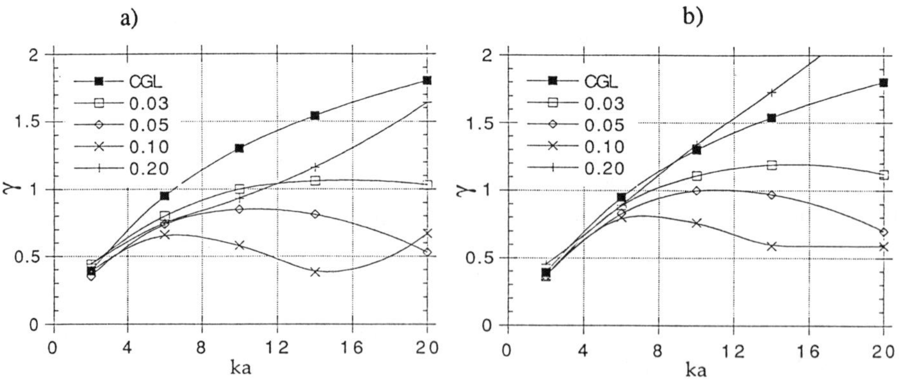

Fig.1. Vlasov-fluid stability of constant current density equilibrium. Normalized growth rate vs wavenumber ka for different values of normalized average Larmor radius ε. a): Initial value method, b): variational method.

The existence of a minimum in the growth rate vs ε is evident from Fig.2. The position of the minimum is localized in the range 0.08<ε<0.16 for $2 \leq ka \leq 20$.

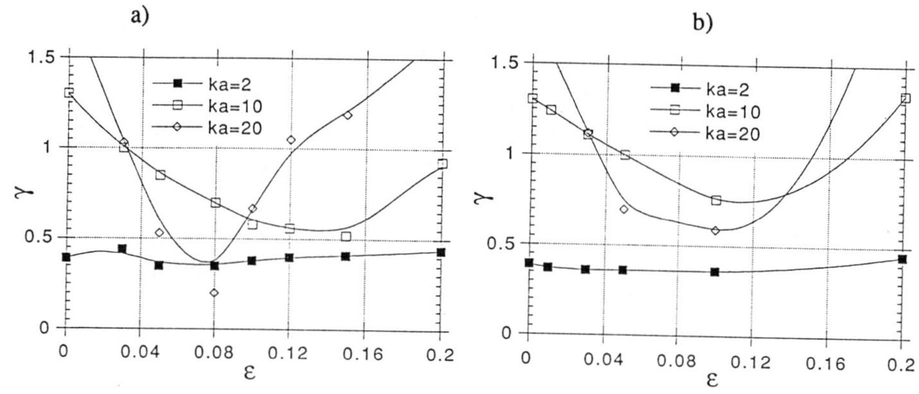

Fig.2. Growth rates as function of ε for different axial wavelengths. a): Initial value method, b): variational method.

In Fig.3 growth rates for the Bennett equilibrium are shown. For all values of ε and ka LLR effects are *destabilizing*. No significant minimum in the growth rate for any values of ε can be found. We conclude that LLR effects are highly sensitive to the type of equilibrium. The Vlasov-fluid stability of yet another equilibrium, the skin current pinch, has indeed been determined earlier[8]. The growth rate was found to saturate at the value $0.5\sqrt{\pi} \approx 0.89$ for $ka \geq 5$ (ε cannot be defined for this equilibrium).

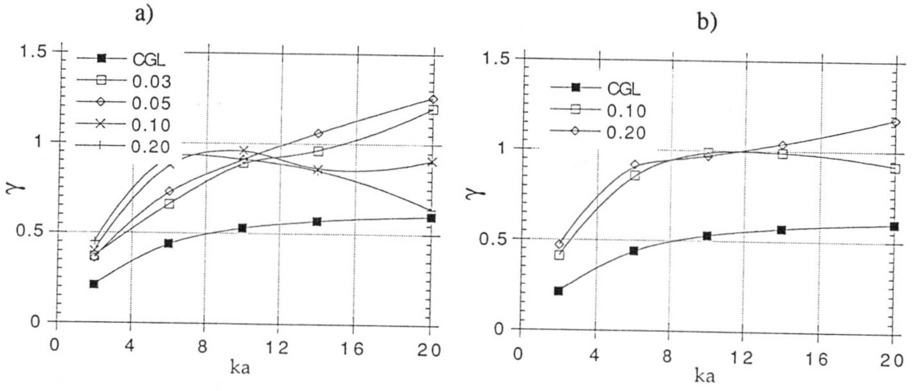

Fig.3. Stability of Bennett equilibrium.

We will now go somewhat into the details of our stability models, describe the equilibria used and discuss the results.

THE VLASOV-FLUID MODEL

The model was proposed by Freidberg[9] in an effort to determine the difference between ideal MHD stability thresholds and growth rates to those of a kinetic plasma model. The enormous complexity of the full Boltzmann equation for ions and electrons is alleviated through assuming that ion dynamics is collision-free on time scales considered, and that electron dynamics is fluid-like across the magnetic field. Also MHD frequencies are assumed; electrons are moving along field lines free enough to short out any parallel electric fields that would arise, but collide frequently enough to prevent low-frequency instabilities driven by resonant electrons. The resulting, linearized, equations can be written

$$\frac{\partial f_1}{\partial t} + \mathbf{v}\cdot\nabla f_1 + \frac{e}{m}(\mathbf{E}_0 + \mathbf{v}\times\mathbf{B}_0)\cdot\nabla_v f_1 \equiv \frac{df_1}{dt'} = \frac{ef_0}{k_B T_{i0}}\,\mathbf{v}\cdot\mathbf{E}_1 \tag{1}$$

$$\frac{\partial \mathbf{B}_1}{\partial t} = -\,\nabla\times\mathbf{E}_1 \tag{2}$$

$$\mathbf{E}_1 = -\mathbf{u}_{e0}\times\mathbf{B}_1 - \mathbf{u}_{e1}\times\mathbf{B}_0 \tag{3}$$

$$\mathbf{u}_{e1} = \mathbf{u}_{i1} - \frac{n_1}{n_0}\mathbf{u}_{e0} - \frac{1}{\mu_0 e n_0}\nabla\times\mathbf{B}_1 \tag{4}$$

where standard notations have been used, 'e' and 'i' denote electron and ion variables respectively. Moments are taken of the ion Vlasov equation (1) to obtain

$$n_1 = \frac{1}{n_0}\int f_1 d^3v \tag{5}$$

$$\mathbf{u}_{i1} = \frac{1}{n_0}\int f_1 \mathbf{v}\, d^3v\ . \tag{6}$$

Elegant comparison theorems have been derived. Seyler and Lewis[10] showed that, assuming the equilibrium distribution function $f_0'(v) < 0$, ideal MHD and the Vlasov fluid model have the *same* stability thresholds. FLR or LLR effects may decrease the growth rate significantly, but resonant ions maintain the stability limit.

Curiously, the case ε=0 for *m*=0 modes of a Z-pinch represents a singular exception; here growth rates are given by the Chew-Goldberger-Low (CGL) equations, and *not* by ideal MHD. The CGL model is derived for the limit $\varepsilon \ll \omega_{th}/\omega \ll (ka\varepsilon)^{-1}$, hence when $\varepsilon \rightarrow 0$[4]. Åkerstedt[11] has shown that, neglecting FLR and resonant contributions, the CGL stability condition for *m*=0[12] is obtained exactly from the Vlasov-fluid model.

Physically, the reason for this is that the *m*=0 mode is associated with the sound wave, being sensitive to the form of the energy (pressure transport) equation. The CGL model gives the correct anisotropic pressure equations in the collisionless, zero heat flow limit.

The condition for zero heat flow is exactly fulfilled by *m*=0 modes of the small Larmor radius Z-pinch, where $\mathbf{B}\cdot\nabla \equiv 0$. Ideal MHD, being derived as a collision-dominated model, has a poor energy equation. Most modes are, however, associated with the shear Alfvén wave which is unsensitive to the form of the energy equation. Apart from the energy equation, the equations of ideal MHD coincide with the (collisionless) Vlasov-fluid equations in the $\varepsilon \to 0$ limit.

Vain attempts have been made to include LLR effects into fluid models. For small, but finite, ε a set of collisionless one-fluid equations can be derived from the Maxwell-Vlasov equations:

$$\frac{\partial n}{\partial t} + \nabla\cdot(n\mathbf{u}) = 0 \quad (7)$$

$$mn\frac{d\mathbf{u}}{dt} = \mathbf{j}\times\mathbf{B} - \nabla\cdot P - \nabla\cdot(\pi_i+\pi_e) \quad (8)$$

$$\frac{d}{dt}\left(\frac{p_\perp}{nB}\right) = \text{heat flow terms} \quad (9)$$

$$\frac{d}{dt}\left(\frac{p_{//}B^2}{n^3}\right) = \text{heat flow terms}$$

where $\mathbf{u} \approx \mathbf{u}_i$ and π is the off-diagonal pressure tensor. Other relations, required to close the set of equations, are given by Eqs. (2-4). It is here assumed that T_e=0 in Ohm's law, Eq.(3), causing the ∇p_e term to vanish, and that the Nernst term is negligible. This has no bearing on the following argument.

For large ε Eqs.(7-9) no longer can be regarded as fluid equations. This is because macroscopic quantities like $\mathbf{u}_i$ are obtained by taking moments over particles moving in orbits too large to preserve localised regions in phase space. The pressure tensor π e.g. cannot be obtained without use of f_1, thus information outside the fluid model is required. A self-consistent LLR formulation of a fluid model is impossible. Finally, fluid equations exclude the important velocity space effect of resonant ions; a parallel is the disappearance of Landau damping in fluid models of plasma oscillations.

TWO METHODS

Two radically different methods have been employed to solve the linearized Vlasov-fluid stability problem defined by Eqs[1-6]; an initial-value solution and a variational method.

Both methods are presently focused on fixed boundary *m*=0 modes in circularly cylindrical geometry, for which the boundary conditions are:

$$r=0: \quad u_{e1r} = B_1 = E_{1r} = E_{1z} = 0 \quad (10)$$

$$r=a: \quad u_{e1r} = B_1 = E_{1z} = 0.$$

A Maxwellian ion equilibrium distribution function $f_0(v)$ is assumed in both cases. We consider the behaviour of a single normal mode, i.e. all perturbed variables are taken to have a dependence $\propto e^{i(m\theta+kz)}$.

In the initial value approach time derivatives are retained, rather than eliminating them by using the exponential time dependence $e^{i\omega t}$ of the normal mode. Thus ω does not appear in the equations. A random perturbation is applied and its time evolution is followed. After some growth times the fastest growing mode will dominate, and the solution converges to exponential growth with a characteristic growth rate and with a well defined eigenfunction structure.

The code FIGARO embodies the numerical implementation of the method. Individual orbits of an ensemble of typically $1\text{-}2\cdot 10^4$ particles are integrated in their equilibrium fields. Each particle, which represents a large number of neighboring particles in phase space, carries with it values of f_0 (constant of motion) and f_1. At any time step, values of f_1 are determined from integrating Eq.(1) and used for obtaining the moments (5) and (6). The perturbed magnetic field $\mathbf{B}_1$ is obtained from the perturbed electric field $\mathbf{E}_1$, defined at the previous time step, using Eq.(2). The perturbed electron macroscopic velocity $\mathbf{u}_{e1}$ is found from Eq.(4) and inserted into Eq.(3) to give $\mathbf{E}_1$ at the present time step. This value is now used to advance f_1 and so on.

Macroscopic variables are specified in a set of (typically 50) 1-D radial cells, and the moment taking involves averaging over particles in the cell. For cell j the calculation is performed as follows:

$$\int f_1 Q\, d^3v \rightarrow \frac{n_{0j}}{K_j}\sum_{k=1}^{K_j}\frac{f_{1k}}{f_{0k}}Q_k \ , \qquad (11)$$

where Q is any particle variable, k is the label for individual particles in cell j, and K_j is the total number of particles in the cell.

The second method is based on the energy relation derived from the Vlasov-fluid eigenvalue equation which, although being non-Hermitian, is still variational in Z-pinch geometry. Eigenfunctions derived from a CGL shooting code have been substituted into the Vlasov-fluid energy relation. The complex eigenvalue which satisfies the resulting equation is then determined to a higher order of accuracy than the initial 'guess' eigenfunction. A method which uses a finite number of a complete set of test functions is also employed. The trajectory integrals which appear in Vlasov-fluid theory have been Fourier analysed and reduced to integrals over a gyro period. This technique has previously been used to study rotating theta pinches[13]. For numerical reasons the energy relation is rewritten in a form in which the kinetic and fluid parts are more clearly separated[14]. No FLR expansions are used and the resulting energy relation is still analytically exact for all orbits.

EQUILIBRIA

The Vlasov-fluid model assumes that the equilibrium distribution $f_0 = f_0(H)$ only, where $H=mv_{th}^2/2 + e\phi_0$ is the particle energy, with ϕ_0 being the equilibrium electrostatic potential. We therefore choose equilibria in which the ion pressure is isotropic, the ion temperature is uniform, and the macroscopic velocity is zero.

The constant current density equilibrium has a magnetic field $B_0(x) \sim x$, where $x \equiv r/a$, and a density profile given by $n_0(x) \sim 1+\alpha-x^2$. The case $\alpha=0$ would correspond to a singular eigenfunction[4]; we have here chosen the value $\alpha=0.1$. The Bennett equilibrium is characterized by a uniform electron velocity. In this case $B_0(x) \sim x/(1+\beta^2x^2)$ and $n_0(x) \sim (1+\beta^2x^2)^{-2}$, with β a constant (here chosen to be 3.0). The current density profile is centrally peaked for this case.

CONCLUSION

Large Larmor radius (LLR) effects on the linear stability of realistic Z-pinch equilibria are studied. Using two radically different methods, the Vlasov-Fluid model is applied to fixed boundary *m*=0 modes of the pure Z-pinch (no external magnetic field). A basic result is that stability depends sensitively on the equilibrium profiles. Growth rates of a flat current equilibrium are decreased by LLR effects up to a factor of four as the normalized average Larmor radius ε is increased to about 0.1. Complete stabilization is not obtained for any values of ε or ka. For $\varepsilon>0.1$ the growth rate increases to reach above the small Larmor radius value when $\varepsilon=0.2$-0.4. Growth rates of a Bennett equilibrium are destabilized by LLR effects for all values of ε and ka.

Since the ion line density N is related to ε through $N=4\cdot10^{17}/\varepsilon^2\ m^{-1}$, the minimum growth rate for a constant current density equilibrium is obtained when $N\approx 4\cdot10^{19}/m^{-1}$.

REFERENCES

[1] B.B. Kadomtsev, in Reviews of Plasma Physics, edited by M.A. Leontovich, Consultants Bureau, New York, Vol II, 1966.

[2] B. Lehnert, Phys. Fluids **4**(1961)525, 847 and 1053.

[3] M.N. Rosenbluth, N.A. Krall, N. Rostoker, Nuc. Fus. Suppl., part I (1962)143.

[4] J. Scheffel and M. Coppins, Nuc. Fus. **33**(1993)101.

[5] B. Lehnert, Phys. Scr. **16**(1977)147.

[6] J.D. Sethian, A.E. Robson, K.A. Gerber and A.W. DeSilva, Phys. Rev. Lett. **59**(1987)892 and 1790.

[7] J.E. Hammel and D.W. Scudder, Proc. 14th European Conf. on Controlled Fusion and Plasma Physics, Madrid, 1987, Part II, 450.

[8] T.D. Arber and M. Coppins, Phys. Fluids B **1**(1989)2289.

[9] J.P. Freidberg, Phys. Fluids **15**(1972)1102.

[10] C.E. Seyler and H.R. Lewis, J. Plasma Physics **27**(1982)37.

[11] H.O. Åkerstedt, J. Plasma Physics **41**(1989)45.

[12] M. Coppins, Phys. Fluids **B1**(1989)591.

[13] C.E. Seyler, Phys. Fluids **22**(1979)2324.

[14] C.E. Seyler and D.C. Barnes, Phys. Fluids **24**(1981)1989.

IMPLODING LINER STABILIZATION EXPERIMENTS ON THE SNOP-3 GENERATOR

S. A. Sorokin, S. A. Chaikovsky
High Current Electronics Institute, Tomsk, Russia

ABSTRACT

Some methods of the plasma liner stabilization in addition to entraining of an initial axial magnetic field were used in the experiments on the SNOP-3 generator ($I_m \simeq 1.1$ MA, $\tau \simeq 100$ ns). Helical current return rods were used to stabilize the liner (annular gas puff) at the beginning of the implosion. A 50-fold stable radial compression of the krypton liner was observed using a time-integrated pinhole camera.

Double shell liners with an initial magnetic field are attractive from the viewpoint of stable high-radius-ratio liner implosions and efficient energy transfer from the generator to the plasma pinch of mass $m < m_1$ (m_1 is the liner mass required to implode the liner at the peak of the current pulse).

Model 0-dimensional calculations of the motion dynamics of the shells and experiments on double shell liner implosions were perfomed. The radius of a stable compressed argon inner shell was measured to be 50-60 μm (at an initial radius of 5 mm).

The maximum magnetic field $B_m \simeq 6000$ T was estimated using the measured radius of the compressed nitrogen inner shell.

The effect of liner stabilization on the X-ray yield and the possible future experiments using the stable liner implosions are discussed.

INTRODUCTION

It is well known that plasma liner implosions are subject to MHD instabilities. The imploding liner breaks up when implosion approaches the axis and, as the result, the liner radial compression ratio is restricted. An initial axial magnetic field B_0 has a stabilizing effect on the liner implosion. The results of the experiments [1,2] have shown that the initial magnetic field to maximum current ratio should be $B_0/I_m \simeq 1\text{-}1.5$ T/MA to stabilize the implosion of the liner of an initial radius $r_0 \simeq 1$cm. As a lower initial magnetic field corresponds to a higher compressed magnetic field, then the initial magnetic field B_0[T] = (1.0-1.5)I_m[MA] is optimum to generate the highest magnetic fields. A stable radial compression ratio of the liner inner boundary $r_0/r_f \simeq 22$ (at B_0= 2.0 T) was observed using a time-integrated pin-hole camera [2]. If the magnetic flux inside the liner was conserved during these implosions, then the maximum of compressed magnetic field was $B_m \simeq 1000$ T.

The axial magnetic field B_z substantially decreases the growth rates of instabilities, if $B_z \simeq B_\varphi$, where B_φ is the azimuthal magnetic field. For $B_0/I_m \simeq 2$ T/MA and $r_0 \simeq 1$ cm the condition $B_z \simeq B_\varphi$ is fulfilled for the liner radius $r \simeq r_0/10$. Therefore, the liner is more unstable at the beginning of the implosion, when $r \simeq (0.3-1)r_0$ and $B_z \ll B_\varphi$.

Stable liner implosions at $B_0/I_m < 1.0-1.5$ T/MA can be obtained using some methods of liner stabilization in addition to the entraining of the initial magnetic field. Helical current return rods can be used to stabilize the liner at the beginning of the implosion. In this case, the driving magnetic field has two components $B_\varphi = \mu_0 I/2\pi r$ and $B_z = \mu_0 I \cdot tg\psi/2\pi R$. (Here ψ is the angle between the direction of the current return rods and the z-axis, R-the current return sructure radius). As the liner implodes, the B_z to B_φ ratio changes and perturbations with different wave numbers $\vec{k}$ (which do not bend the magnetic field lines) have maximum growth rates at different times during the implosion.

Double shell liners with an initial magnetic field are attractive from the viewpoint of stable high-radius-ratio liner implosions and efficient energy transfer from the generator to the plasma pinch of mass $m < m_1$ (m_1 is the liner mass required to implode the liner at the peak of the current pulse).

Together with an outer hollow plasma shell of mean radius r_{10} and mass per unit length m_1, the double shell liner has an inner (annular or solid) plasma shell of radius r_{20} and mass per unit length m_2. The outer shell is accelerated by the magnetic field of the generator current and transfer its kinetic energy to the inner shell through the axial magnetic field compressed between the shells.

If the plasma conductivity is high enough and the magnetic Reynolds number (the ratio of the diffusion time of the magnetic field out of the plasma to the implosion time) is $R_m \gg 1$, then a 0-dimensional model can be used to calculate the dynamics of motion of the shells. In this model an electrical circiut equation and two equations of motion of an infinitely thin and an infinitely conducting shells are solved numerically.

The results of the calculations for the implosion of a double shell liner with $m_1 = 10$ μg, $m_2 = 5$ μg, $r_{10} = 13$ mm, $r_{20} = 6$ mm, with the initial magnetic field inside the inner shell $B_{20} = 0.7$ T, and the initial field between the shells $B_{10} = 1.2$ T are shown in Fig. 1.

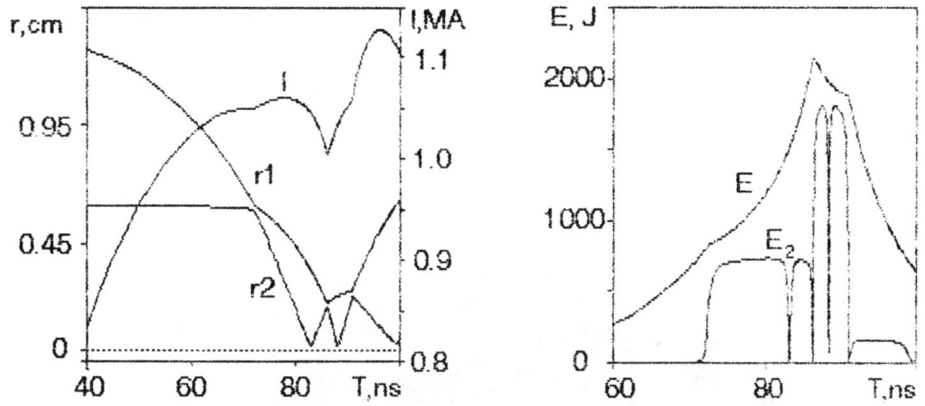

Fig. 1. The results of 0-dimensional calculations of the double shell liner. Here I is the current through the liner, r_1 and r_2-the radii of the outer and the inner shells, respectively, E - the energy coupled to the liner, E_2- the kinetic energy of the inner shell.

One can see that once the shells collide for the second time, 84% of the liner energy

$$E = 0.5 \int \dot{L} I^2 dt$$

are transferred to the inner shell, which implodes to $r_{2f} = 0.067$ mm, that corresponds to the inner shell radial compression ratio $r_{20}/r_{2f} \simeq 90$ and the maximum magnetic field $B_m \simeq 5.5 \cdot 10^3$ T.

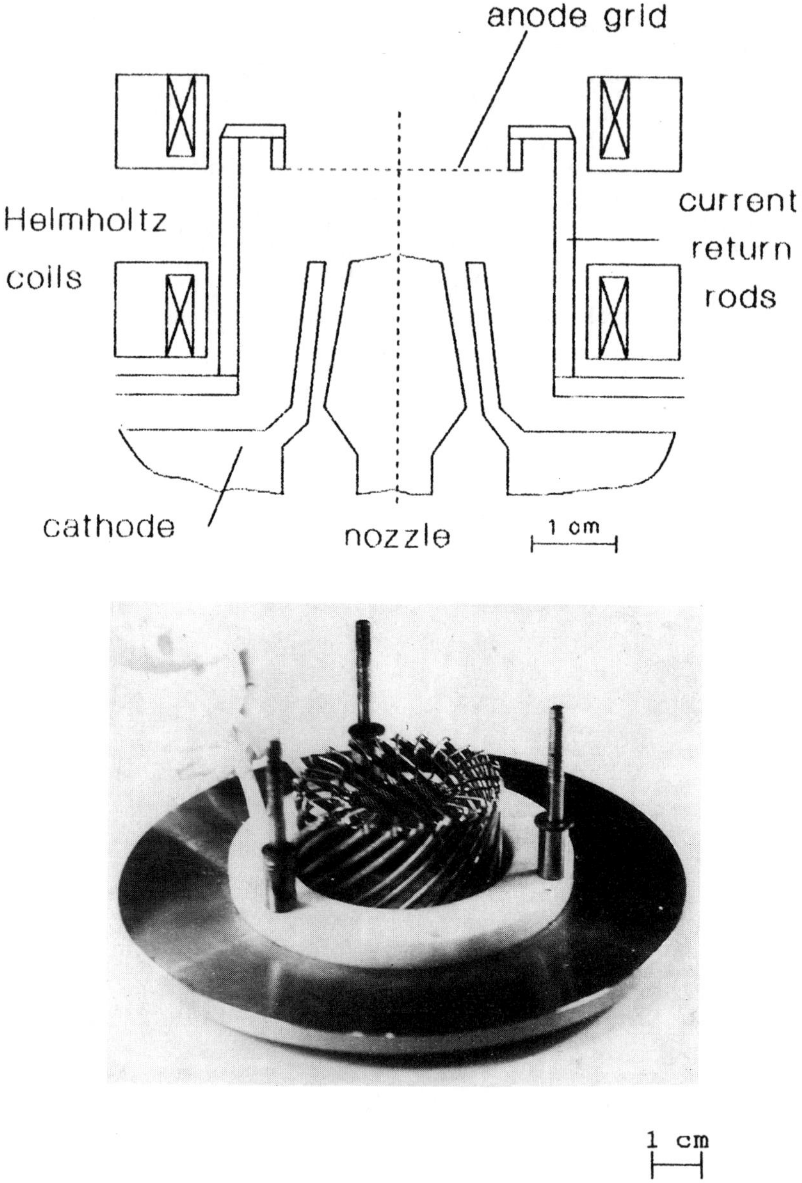

Fig.2. A schematic of the load region and a photograph of the helical current return structure.

EXPERIMENTS

The SNOP-3 generator ($I_m \simeq 1.1$ MA, $\tau \simeq 100$ ns) was used to perform experiments on stabilization of imploding plasma liners by helical current return rods and on implosion of double shell liners.

Magnetic probes, Rogowski coils, X-ray diodes, pinhole cameras, a visible light streak camera and time-integrating bolometer were included for diagnostics.

A schematic of the load region and a photograph of the current return structure (24 helical rods) are shown in Fig. 2. The initial magnetic field was produced by Helmholtz coils. A hollow gas liner was injected into a 1.2-1.5 cm long load region using a fast electrodynamic gas valve coupled to a supersonic nozzle.

The X-ray pinhole photographs of the krypton pinch with different B_0 and ψ are shown in Fig. 3. The time-integrated pinhole camera had a 30 μm pinhole and viewed the pinch through a 8 μm aluminium filter. One can see a 50-fold stable radial compression in shot with B_0= 0.86 T and $\psi = 45^\circ$. Only a slight m=1 perturbation is observed.

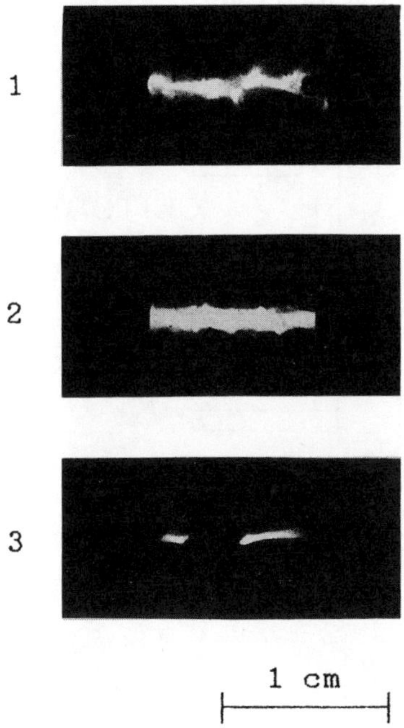

Fig.3 X-ray pinhole pictures for krypton implosions with: 1) ψ = 0, B_0=0 2) ψ = 0, B_0=0.86 T, 3) $\psi = 45^\circ$, B_0=0.86 T.

A schematic of the load region for double shell liner implosion experiments is shown in Fig. 4. Copper coils were used to reduce the initial magnetic field in the region of the inner shell ($B_{20} \simeq B_{10}/2$). An additional nozzle formed a solid or hollow gas column on the liner axis. In these experiments the time-integrated pinhole camera had 3 pinholes: 1) 45 μm-diam pinhole with a 3 μm mylar and 0.2 μm aluminum filter, 2) 45 μm-diam pinhole with a 8 μm aluminum filter, and

3) 10-15 μm-diam pinhole with a 8 μm aluminum filter. X-ray pinhole photographs of the argon pinch are shown in Fig. 5. In shot with $\psi = 45°$ some regions of the pinch are shadowed by the current return rods . The inner shell is stable for $\psi = 45°$ and $B_{10} = 1.4$ T. The diameter of the compressed inner shell was measured to be $2r_{2f} \simeq 100-120$ μm (from the 10-15 μm pinhole). Analysis of the photographs has shown that the X-ray yield of the outer shell was substantionaly lower, when the double shell liner structure was used. The outer shell was only visible on a photograph through a 3 μm mylar and 0.2 μm aluminum filter.

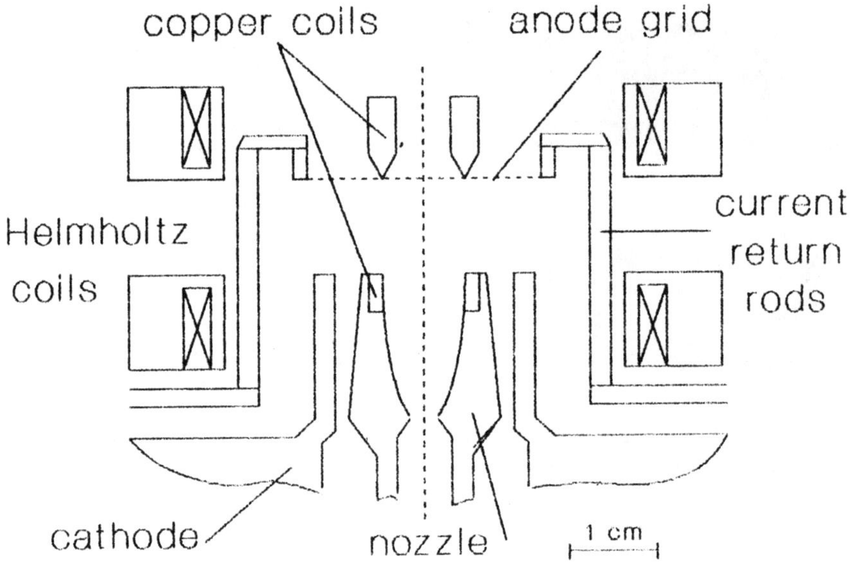

Fig. 4. A schematic of the load region for double shell liner experiments.

Argon implosions for $B_{10} = 1.4$ T gave total X-ray yield $\simeq 2$ kJ both with and without the inner shell. The total X-ray yields were measured with an unfiltered time-integrating bolometer.

The following experiments were performed to prove that the sort of material of the outer shell has a minor effect on the inner shell compression and the X-ray yield. In these experiments a capillary discharge in Al_2O_3 was used to inject the inner shell [3]. The experiments have shown that the time-integrated pinhole photographs, the total X-ray yield, and the aluminum K-shell yield did not noticeable vary, when different gases (argon, deuterium, nitrogen) were used to form the outer liner shell.

Three aluminum cathode XRDs filtered with 12.5 μm mylar, 15 μm mylar, and 17.5 μm mylar were used to measure the aluminum K-shell yield. XRD's signal ratio corresponds to the filter transmissions ratio in the 1.5-2 keV spectral range, that confirmed that the radiation was from the aluminum K-shell.

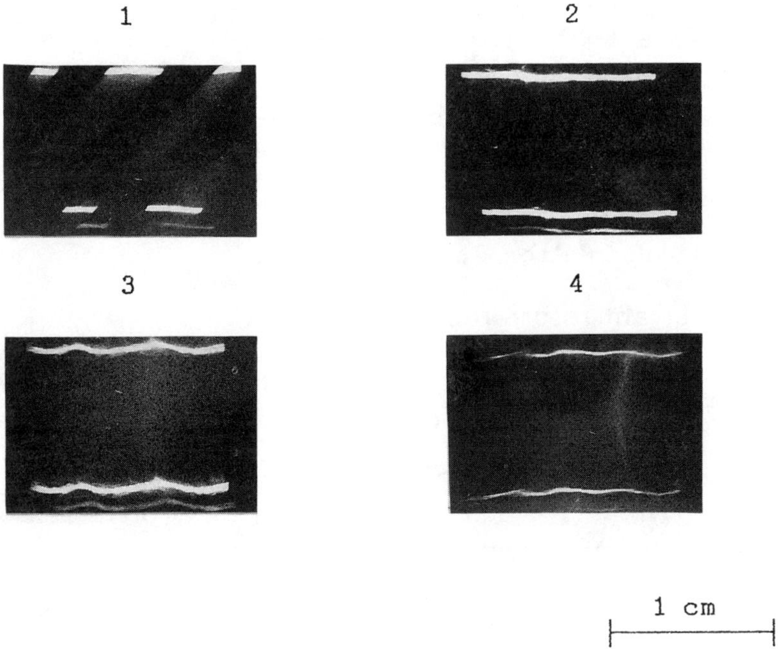

Fig.5. X-ray pinhole photographs for argon double shell liner implosions with: 1) r_{20}= 5 mm, B_{10}= 1.4 T, ψ = 45°. 2) r_{20}= 5 mm, B_{10}= 1.4 T, ψ = 0. 3) r_{20}=5 mm, B_{10}= 0.8 T, ψ = 0. 4) r_{20}= 2.5 mm, B_{10}= 0.8 T, ψ = 0.

The time averaged electron temperature T_e was measured with spatial resolution along the z-axis by the absorption method [4] using the free-bound and free-free continuum spectrum and assuming the Maxwell velocity distribution of electrons. The UFSH-S X-ray film was used to detect X-ray radiation. The spatial resolution along the z-axis was provided by a slit mounted between the film and the pinch perpendicular to the z-axis.

Mylar, aluminum, titanium and teflon were used as filters. The measured electron temperature of a nitrogen liner was uniform along z-axis and amounted to 0.8-1 keV in various shots.

The stagnation nonsimultaneity ("zippering") time Δt was estimated using two pin-diodes (SPPD11-04) with the time resolution 1.5 ns. The pin-diodes were filtered to look above 3 keV in shots with argon liner. The first pin-diode detected the radiation from the whole liner length, the second one from a part of the liner length Δz = 1 mm. The pulse width of the first pin-diode was $\simeq$ 3.5-ns FWHM. The pulse width of the second pin-diode was $\simeq$ 2-ns FWHM, i.e. of the order of pin-diode time resolution. It allowed us to estimate the "zippering" time $\Delta t \leqslant 3$ ns.

A visible light streak camera with the slit parallel to the z-axis was used to observe the stagnation nonsimultaneity along the pinch axis [5]. Fig. 6b shows, that the nonsimultaneity passing the radius r=2 mm by the outer shell doesn't exceed 2 ns with the nozzle design shown in Fig. 6a .

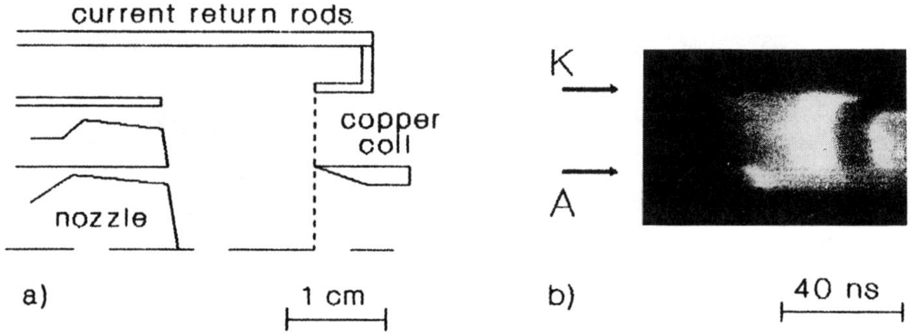

Fig 6. (a) A nozzle design, (b) a streak photograph of the argon liner. The slit of the streak camera was parallel to z-axis and viewed the liner at the radius r=2 mm.

The free-free and free-bound continuum X-ray radiation power of the nitrogen liner in the 4-5 keV region of the spectrum, measured with pin-diodes with a 80 μm titanium filter , agrees with that calculated for a nitrogen plasma column with a diameter of 200 μm, mass of 4 μg and electron temperature of 1 keV. This fact allows one to suppose that the 200 μm diameter pinch observed using a pinhole camera contained the whole mass of the inner liner shell.

DISCUSSION and CONCLUSION

Using some methods of stabilization of the liner implosion, such as a helical driving magnetic field and a double shell liner structure in addition to the entraining of the initial magnetic field, a stable up to 100-fold radial compressions of the inner shell of a double shell liner were attained.

Stable liner implosions at a relatively low initial magnetic field make it possible to increase essentially the value of the generated magnetic field. Estimation of the magnetic Reynolds number R_m for a compressed nitrogen inner shell performed for T_e= 1 keV gives

$$R_m = \tau_d / \tau_i \simeq 20 ,$$

where τ_d is the diffusion time of the magnetic field out of the plasma cylinder of radius r_{2f} and Spitzer's conductivity and $\tau_i = 2r_{2f}/v_{2f}$-the inertial confinement time of the inner shell. At the beginning of the inner shell implosion ($r_{20} \simeq 6$ mm, $T_e \simeq 10$ eV) R_m is estimated to be $\geq$ $\geq$10. Since the inner shell plasma $R_m \gg 1$ the magnetic flux is conserved during implosion. Thus, the maximum magnetic field (at B_{20}=0.6 T) can be estimated as

$$B_{2f} = B_{20}(r_{20}/r_{2f})^2 \simeq 6000 \text{ T}.$$

The inner shell plasma can be used for x-ray lasing experiments. Instabilities are the main obstacles to achieve X-ray lasing in z-pinch plasma. Instabilities 1) decrease the effective lasant length and the maximum electron density, 2) result in refractive index gradients, which induce beam losses, 3) result in the differential axial plasma velocity ($v > 10^7$ cm/s) and, hence, in the differential Doppler shift, which reduce the gain.

Most of these problems are solved in the double shell liner scheme with an initial magnetic field. The inner shell is a straight and homogeneous plasma column at the final stage of implosion. As the inner shell is compressed by the axial magnetic field, it is stable to axial perturbations . The differential Doppler shift caused by the differential radial velocity mitigates the opacity effect of reducing gain. The hollow inner shell has a waveguiding effect on the X-ray laser beam propagation and can reduce the deleterious effects of refraction.

A gain of 10 cm^{-1} can be achieved for iron plasma optimum parameters in collisional excitation scheme [6,7].

Preliminary experiments have been perfomed to obtain a stable pinch of iron plasma. A capillary discharge in Fe_2O_3 was used to form the inner shell, while the outer shell was formed with a gas nozzle. A stable pinch of iron and oxygen plasma with a diameter 150-200 μm was observed (Fig. 7).

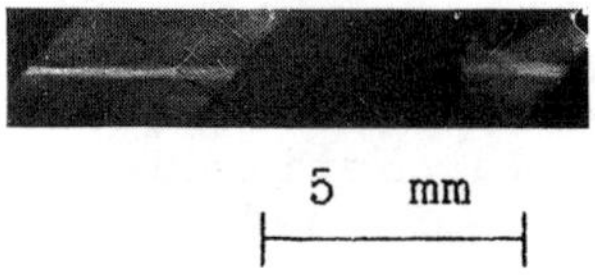

Fig. 7. A pinhole photograph of the iron bearing pinch

Besides for the megagauss magnetic field generation and the x-ray lasing, the results of the liner implosion stabilization experiments can be used in research on:

1) plasma compression by a hollow liner [8],
2) ion acceleration and neutron production [9,10], and
3) increasing the K-shell radiation yield.

REFERENCES

1. F.S. Felber, F.J. Wessel, N.C. Wild et al., J. Appl. Phys. **64**, 3831 (1988).
2. S.A. Sorokin, S.A. Chaikovsky, in Megagauss Fields and Pulsed Power Systems, (Nova Science Publ. NY 1990), p.719.
3. F.S. Young, S.J. Stephanakis, V.E. Scherrer et al., Appl. Phys. Lett. **50**, 1053 (1987).
4. F.C. Jahoda, E.M. Little, W.E. Quinn et al., Phys. Rev. **119**, 843 (1960).
5. S.A. Sorokin, A.V. Khachaturian, S.A. Chaikovsky, Fizika Plasmy **17**, 1453 (1991).
6. U. Feldman and J.F. Seely, J. Appl. Phys. **56**, 2475 (1984).
7. V.I. Oreshkin and V.V. Loskutov, VINITI, N1713-B92 (1992).
8. S.M. Golberg, M.A. Liberman and A.L. Velikovich, Proc. Second Int. Conf. on Dense Z-pinches, Laguna Beach, 1989, p.345.
9. Ia.P. Terletskii, Sov. Phys. JETP **32**, 310 (1957).
10. S.A. Sorokin and S.A. Chaikovsky, Preprint N 12, HCEI, Tomsk (1992).

OBSERVATION OF A M=0 INSTABILITY AND ITS TIME EVOLUTION IN A NEON GAS PUFF

P.Zehnter[(a)], B. Etlicher, A.S. Chuvatin [(b)]

Laboratoire de Physique des Milieux Ionisés, Laboratoire du CNRS, Ecole Polytechnique, 91128 Palaiseau, France.

D. Friart, M.Darrigol

CEA, Service CEM, BP 12, 91680 Bruyères-le-Châtel, France

L.Voisin, J.C Couturaud, J. Ribolzi

CEA, CESTA, Service PE, B.P 2, 33114 Le Barp, France

ABSTRACT

In this paper, we will describe the time evolution of a m=0 instability which was recorded on AGLAE, a 1 TW generator, during the collapse of a neon gas puff. We observe a correlation between images produced by a visible streak camera with slit parallel to the axis of the pinch and that of a temporally resolved pinhole array. The pinhole images are recorded by a four strip X-Ray MCP camera (Slix) which uses three different filters showing the temporal evolution of a simple spectrum. We also use a two channel time integrated pinhole camera. The behavior of the instability will be correlated with a simple theoretical analysis of the phenomenon. A brief description of the experiment and the diagnostic arrangement will be given. Then detailed analysis of experimental data and their correlation with a snowplow MHD model will be given.

INTRODUCTION

The hot spot regime in Z-pinches is a well known phenomena which has been described earlier in many publications,[1-4] essentially through the analysis of time integrated pinhole X-ray pictures. The analysis of the X-ray pictures is usually correlated with MHD instability models without any possible time evolution of either hot spots or plasma jets. We will describe the evidence of plasma jets and their time evolution which usual time-integrated signature is the flares around the neck formation. The time evolution of such an instability was first recorded in plasma focus experiments[5,6] and correlated with an accelerated electron beam[7]. The aim of our experiment is to analyze the time behavior of the m=0 instability from its beginning to its end and the macroscopic effect on the plasma along the axis.

(a) Also at : CEA, Service CEM, BP 12, 91680 Bruyères-le-Châtel, France

(b) Permanent address I.V. Kurchatov Institute of Atomic Energy - Moscow, 123182 - Russia

GENERAL APPARATUS

For the present experiment, the AGLAE generator[8] is used to drive a Neon Z-pinch load at an electrical impulsion of 1 TW. The AGLAE pulsed power generator consists of a 170 kJ, 2 MV Marx bank feeding a four segment coaxial water-transmission line with varying impedance. The lines are calculated to maximize energy transfer into a 0.56 Ω diode. The high-voltage pulse is applied directly to the cathode via a conical self magnetically insulated vacuum-feed transition. The line voltage and current are measured upstream of the load region by a resistive probe and B_θ-probe, respectively. The total vacuum feed inductance is 25 nH and the current delivered into the pinch reaches a value of 1.3 MA in 100 ns.

The Neon gas puff is produced with a supersonic annular Mach 3 design nozzle[8]. The mean diameter of the nozzle is 1.1 cm with a shell thickness of 1.4 cm. The gas is stored in a 3 bar plenum pressure which is opened by means of fast electromagnetic vane. The total mass injected is approximatively 50 μg. The anode is composed of fine wire mesh of stainless steel to permit uninterrupted gas flow. The maximum Neon K-shell yield reaches 2 kJ, measured with a Molybdenum photocathode X-ray diode. The half width of the X-ray pulse is 10 ns.

Fig. 1 gives a schematic view of the detectors arrangement around the pinch. The time evolution of the pinch compression and the instability formation is recorded through two different cameras.

- The first one is an optical streak camera which allows the time evolution analysis (1 ns of time resolution) and the spatial evolution of the plasma along the z axis. The slit of the streak is parallel to the axis of the pinch, (Fig.1) to follow the formation of the instability. In this case, we record the axial time behavior of the Z-pinch formation. The spatial resolution of this detector is 100 μm and the plasma history is followed during 160 ns.

- The second detector is a four stripline X-ray Microchannel Plate with a 25 Ω gold photocathode (thickness 500 Å). Those strip can be gated with a 1 kV electrical pulse with an adjustable FWHM between 1 and 12 ns. On each strip, the plasma is imaged through three 50 μm diameter pinholes filtered with different filters (3 μm Al, 5 μm Al, 1 μm Cu). This system gives temporal, energy and spatial resolution in two dimensions. The spatial resolution is ~ 100 μm and the time resolution is chosen to be 8 ns, which is compatible with the generator jitter. The spectral transmission of the filters is presented on Fig.2.

Triggering of the Slix and the visible streak is done to fit with final compression which occurs near maximum current. We also use a two, time integrated, 150 μm diameter pinhole camera with two different filters (20 μm Al and 12 μm Ti). This camera is installed 90° prior to the the streak camera and the Slix. The integration time of the pinhole and the Slix images are correlated to an array of 10 fast filtered PIN diodes with the same energy spectrum.

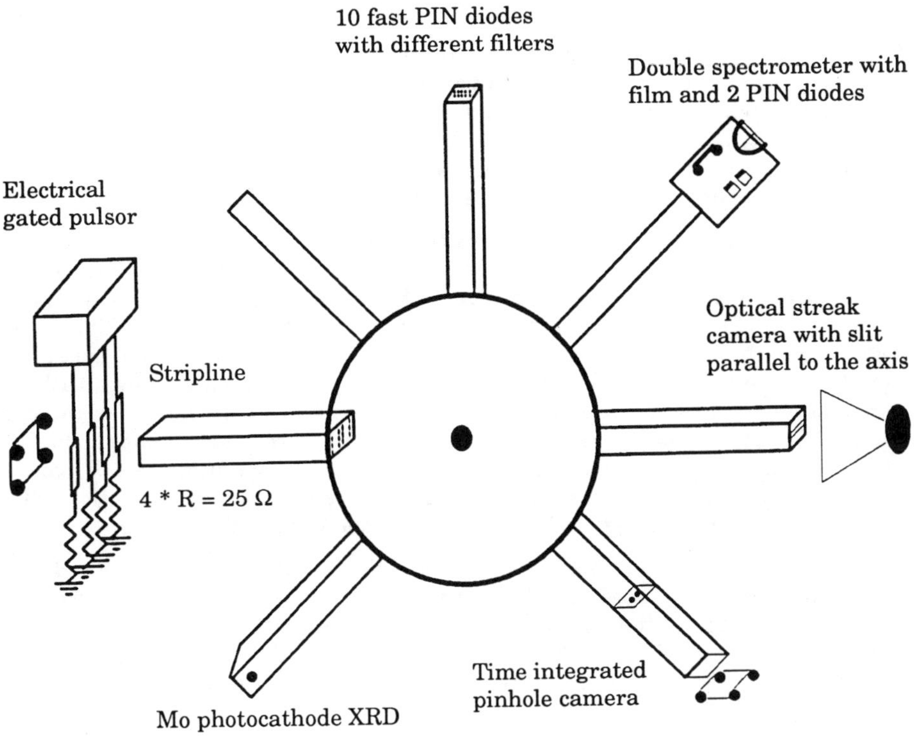

Fig.1 General view of the experimental setup with location of the diagnostics

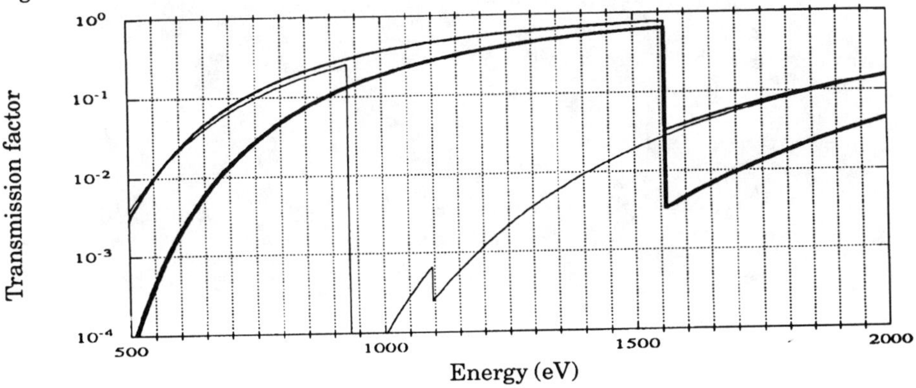

Fig.2 Transmission factor of the Slix filters — *Al 5μm,* — *Al 3μm,* — *Cu 1μm*

EXPERIMENTAL RESULTS

In this section, we will describe the time and spatial evolution of the m=0 instability which appears during compression of the Neon puff. Fig.3 shows the nine pinhole Slix results. The fourth strip was not activated due to shadow of the mounting system. The exposure time was 8 ns, with 8 ns interframe which means 24 ns continuous recording of the phenomenon. Fig.4 shows the visible streak camera results and Fig.5 is the time integrated pinhole pictures.

The first stripline, recorded at t_0, shows the beginning of a hot spot formation as indicated on Fig.3a. Locally, this spot formation introduces a small inhomogeneity in the density or temperature profile. The perturbation evolve rapidly by amplification of the magnetic field pressure and turn into a m=0 instability. At the beginning of the collapse, density and temperature increase and the hot spot begin to radiate. The local internal energy of the plasma perturbation increases with the compression velocity and introduces a large discontinuity with the mean internal energy of the surrounding column. Such a local pressure discontinuity induces a mass flow along the axis to insure the density flux conservation law. From the second strip (t_0 + 8 ns) and third one (t_0 + 16 ns), we observed this z-motion of the plasma which is time and spatial correlated with the visible streak picture.

The axial velocity of the plasma jets generated by the instability is measured from the Slix picture to be between 20-30 cm/μs in the cathode direction and 5-10 cm/μs to the anode. The better temporal resolution of the streak camera gives a more precise evolution of the jets. We measured a velocity of 25 cm/μs during the first 3 ns. After that, because of the collision regime and the ponderomotive strength due to a ionic

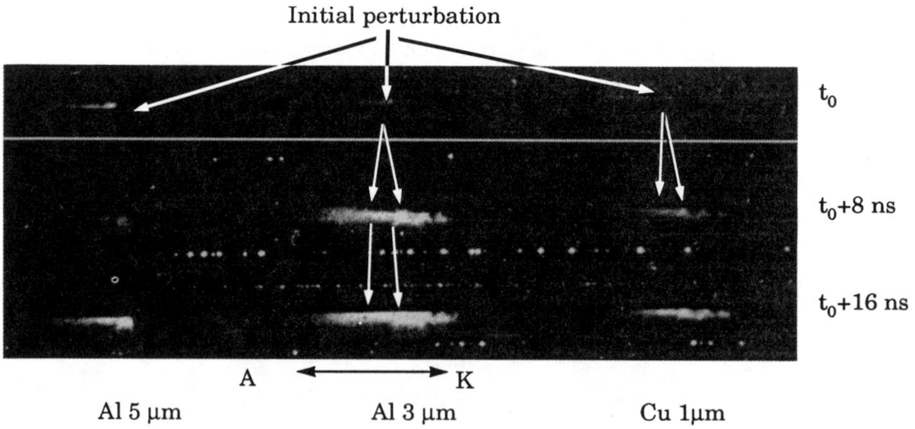

Fig.3a *Slix results. Distance between hot spots 0.5< d <0.8 (cm)*

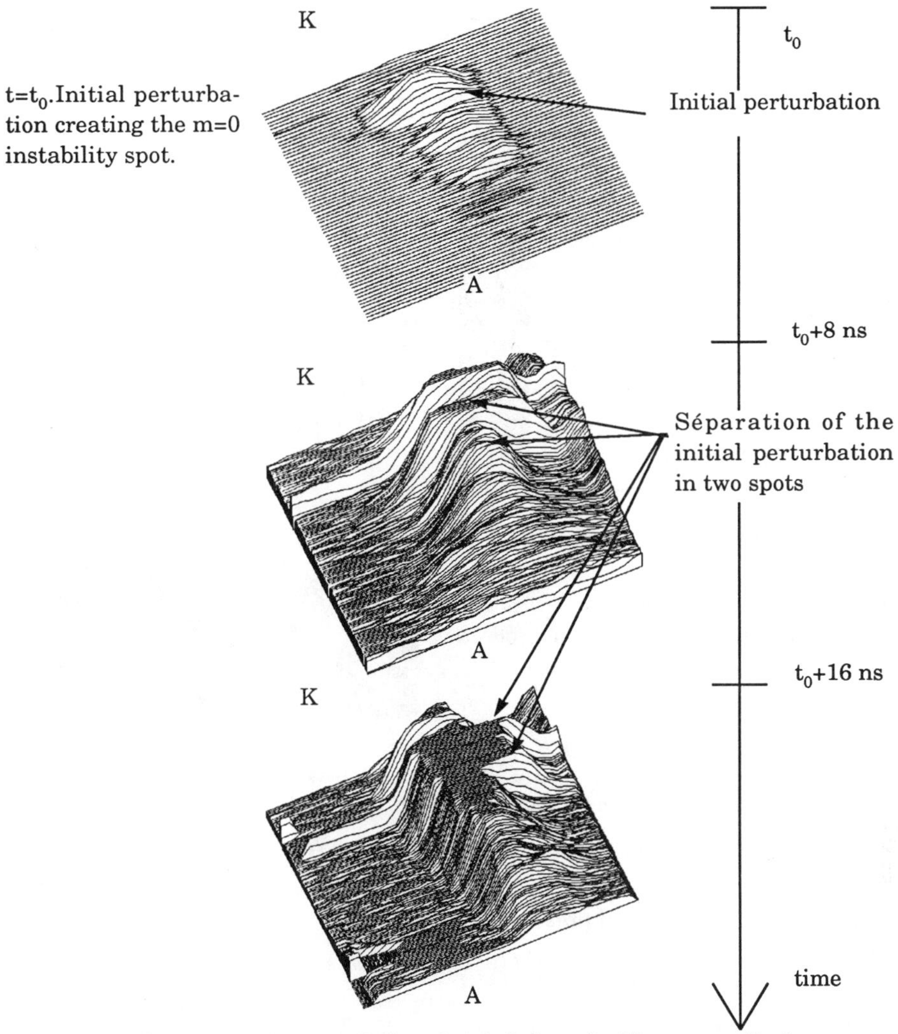

Fig.3b Digital analysis of the 3 μm Al filtered pinhole from the Slix camera. We observe separation of the initial spot into two spots moving with different speeds along the z axis.

density gradient, the velocity decreases to 10 cm/μs. One of the important results is that the radiated yield from those two spots is much higher than the yield from the initial "hot spot", due to the density increases induced by the longitudinal shock propagation. The velocity of the two propagating shocks are different as shown on Fig.4. The discrepancy in the z-velocity can be easily related to the electric field and the ponderomotive force.

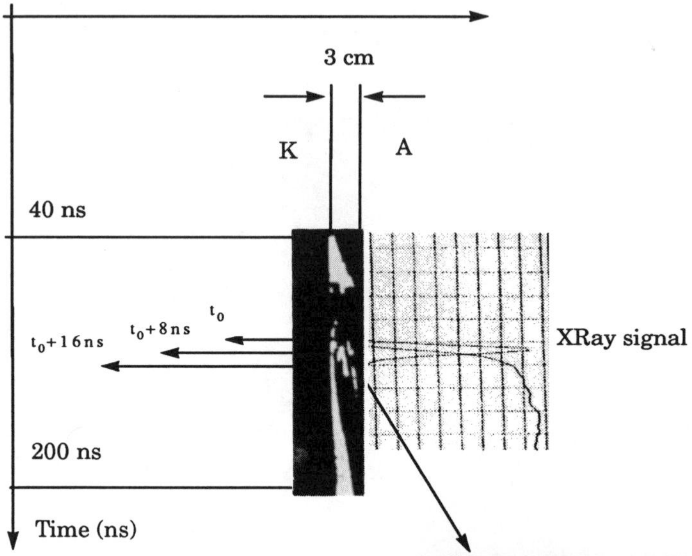

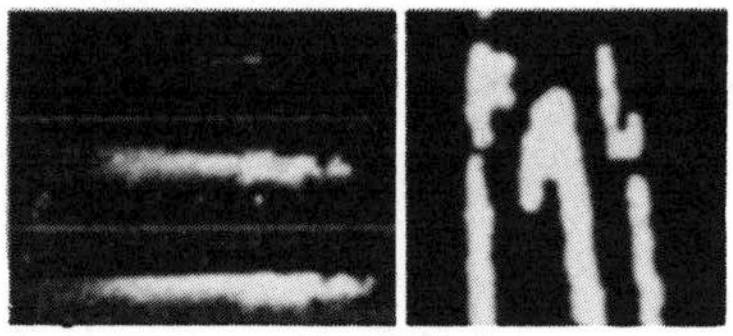

Fig.4 Visible streak results compare in time with a XRD signal. We observed the hot spot separation Distance d between hot spots 0.5 < d < 0.8 cm

If we consider an adiabatic gas, $\gamma = 1 + 2/n$, where n is the number of degrees of freedom, we can calculate an ionic temperature from the velocity of the moving spots

$$kT_i / e = m_i \, C_s^2 / (\gamma Z e) \approx 1.2 \text{ keV}.$$

From a simple 0-D MHD model with an incompressible fluid plasma, the axial velocity is related to the radial pressure perturbation

$$V_z = k P^* / (\rho_0 \, \omega) \approx 10 \text{ cm/}\mu\text{s}$$

where : ρ_0 is the density
: k is then wavelength vector
: ω is the frequency of the growth rate of the instability
: Vz is the axial speed of the pinch

In our case, $\rho_0 = 10^{20}$ cm^{-3}, $\omega = 10^8$ s^{-1}, k = 65 cm^{-1}, $V_z = 10^7$cm/s

The order of magnitude of the perturbed pressure is 1kBar< P*<10 kBar (the magnetic pressure at the equilibrium state is 1Mbar) for the 50 μg of Neon.

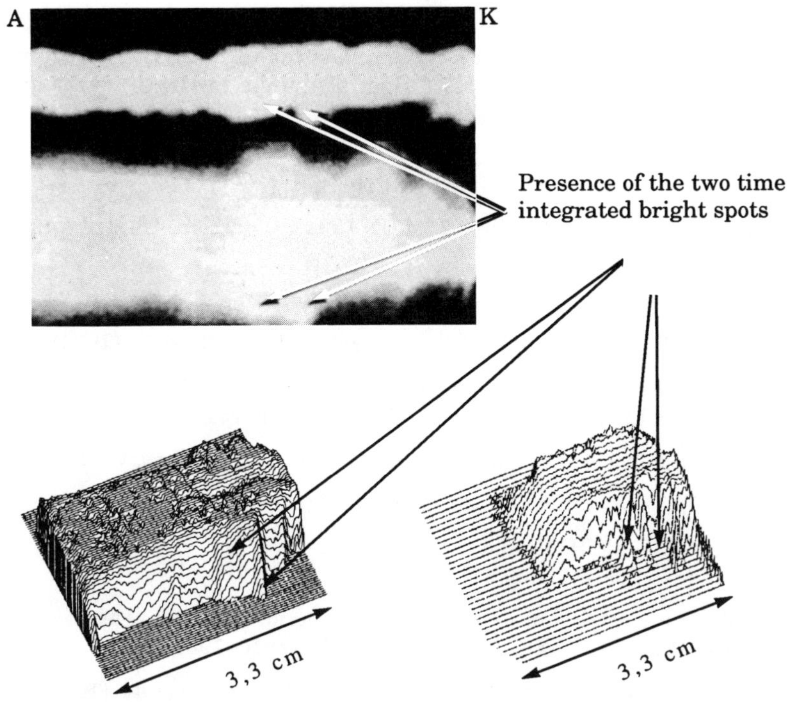

Fig.5 Time integrated pinholes. Top image and right digitalized curve 12 µm Ti filter, bottom image and left digitalized curve 20 µm Al.The distance d between bright spots correlates with Fig.4 (0.5 < d < 0.8 cm)

CONCLUSIONS

We demonstrate the m=0 formation in a standard gas puff and its evolution into two bright spots. It is shown that the radial collapse induces an axial flow which result in a shock propagation along the z-axis. Density and temperature of the perturbation increase[9] resulting in an increase of the X-ray yield. We, also, demonstrate that the radial compression induces the axial flow of matter resulting in a shock formation along the z-axis. Time integrated pinholes show the final two spots spatially correlated with both visible streak and Slix images. The time integrated radiation comes from the shock which first heat up the plasma to temperature in excess of 1 keV and secondly, reduce its velocity due to axial increased density gradient. The theoretical analysis of the m=0 instability results in an axial flow which fit the experimental data.

ACKNOWLEDGEMENTS

We would like to acknowledge Pr. V. Smirnov for its remarks on the physical process and very fruitful discussions. Special thanks to Dr. C. Nazet and J.-M. Angles which give us the opportunity to complete this experiment and provide the technical support.

REFERENCES

1 J.L Tuck, Wash 115, p.121 Proc 2 nd UN Con. On peaceful Uses of Atomic Energy **32** (3), 4.

2 MD Kruskal and M.Schwarzchild, Proc.Roy Soc.London.A No **223**, 348, (1954).

3 F.J Wessel, B. Etlicher, and P. Choi, Phys. Rev. Lett. **69**, 3181, (1992).

4 N.S. Edison, A.S Chuvatin, B. Etlicher, P. Zehnter, S. Attelan, C. Rouillé. 3rd Int. Conf. on Dense Z-Pinches, Londres, 20-23 Avril 1993.

5 A. Bernard, A. Jolas, J. Launspach, JP. Watteau, Plasma Phys, Vol.**15**, 1022 (1973)

6 A. Jolas, Thèse à l'université de Paris Sud, Orsay, (1981).

7 P. Choi, C.S Wong, H. Herold, Laser and particles Beams, vol **7**, part 4, 763, (1989)

8 R Bailly-Salins, C. Bruno, J. Chevallier, J. David, J. Delvaux, M. Picod, L .Voisin, Proc. Beams'88, July 4-8, Karsruhe, 1210, 1988

9 V.V. Vikhrev, V.V. Ivanov, G.A. Rozanova, Soviet.J. Phys. **15**(1), 44, (1989)

THEORY AND SIMULATIONS

Micro-turbulence in the fibre Z-pinch.

J.P. Chittenden,

The Blackett Laboratory, Imperial College, London, UK.

Abstract

Experimentally the radial extent of fibre Z-pinch plasmas is typically found to be much greater than predicted by simple MHD theory. Usually these differences are attributed to the presence of large scale MHD instabilities. Here an alternative explanation is offered in terms of micro-instability behaviour. Since fibre Z-pinches are surrounded by vacuum, the plasma density is zero at the pinch surface. If the current density remains finite near the pinch surface, the electron drift speed will exceed the local ion sound speed and micro-instabilities will result. These micro-instabilities are self-limiting in that they give rise to anomalous resistivity which heats the ions and reduces the local current density thereby reducing the ratio of the electron drift velocity to the ion sound speed. The lower hybrid drift instability is found to be particularly important in the pinch formation phase where the line density of charged particles is very small. Inclusion of micro-turbulent resistivity in 1-D simulations of the ionisation stage of fibre Z-pinch experiments shows the formation of a high temperature, low density, low current density coronal plasma which expands out to large radius. This corona persists throughout the duration of the experiment and has consequences for the overall degree of plasma confinement attainable in fibre Z-pinches.

1. Introduction

The degree of plasma confinement in hydrogenic fibre z-pinch experiments has to date been disappointing[1,2] and certainly less than predicted by, for example, 1-D simulations of the ablation phase[3]. In particular there is a general lack of plasma confinement for low line density pinches and contrary to simple theory, in experiments with a higher $^{dI}/_{dt}$ the degree of confinement is worse[4]. This lack of confinement is usually attributed to conversion of magnetic energy into thermal energy by the turbulence caused by macroscopic MHD instabilities[5,6]. In this paper an alternative explanation is given in terms of the turbulence caused by micro-instabilities.

Micro-instabilities are an inevitable consequence of the vacuum-plasma boundary which such a dominant feature of fibre z-pinches. If the current density at the plasma surface remains finite as the density falls to zero, then the electron drift velocity $v_d = {}^{j}/_{ne}$ becomes infinite. This obviously unphysical situation is ignored in the majority of theoretical treatments of fibre z-pinches. As usual the theoretical anomaly can be overcome simply by including more physics in the model. When v_d exceeds the local ion sound speed c_s, then micro-instabilities are exited. As we shall see in section 3, the main macroscopic consequence of these micro-instabilities is high resistivity in the surface plasma which tends to reduce the ratio v_d / c_s. The plasma therefore attempts to converge to an equilibrium where $v_d \leq c_s$ and in which micro-instabilities are self limiting.

In section 2 we investigate how microscopic turbulence and in particular the lower hybrid drift instability can affect the macroscopic behaviour of z-pinch plasmas by increasing the electron-ion collision frequency. In section 3 we investigate the effect of this anomalous collision frequency upon quasi-stationary z-pinch equilibria using a two temperature, 1-D, Lagrangian, resistive MHD code. In section 4 we simulate conditions more applicable to experiment, starting with a cold unionised hydrogen fibre and using non-equilibrium ionisation dynamics in the 1-D code in order to model the ablation phase of experiment and to study the affects of micro-turbulence during this phase.

2. Micro-Instabilities in Z-pinch plasmas

No detailed investigation of micro-instabilities in z-pinch plasmas exists in the literature. However micro-instabilities were studied extensively by the θ pinch community in the 1970s and more recently included in theoretical investigations of RFPs and the Earth's magneto-tail. Since the θ pinch and z pinch both have simple orthogonal field arrangements, we assume that over the small scale length of micro-instabilities, the two plasma configurations are indistinguishable and that θ pinch micro-instability theory is directly translatable to the z-pinch. This has been the assumption of other authors[7,8].

A survey of θ pinch papers[9,10,11] indicates that the micro-instability with the most readily surpassable threshold and highest growth rate for fibre z-pinch plasmas is the lower hybrid drift (LHD) instability which is driven by the cross field current. For a 200 μm diameter hydrogen z-pinch carrying 100 kA, this instability grows at approximately the lower hybrid frequency $\omega_{LH} = \omega_{pi}/\sqrt{1+\omega_{pe}/\omega_{ce}} \approx \sqrt{\omega_{ce}\omega_{ci}} = 8.2\text{x}10^{11}s^{-1}$. Therefore the instability grows over peco-second time scales and will in fact saturate over much faster time scales than the dynamical evolution of the plasma. Furthermore the saturation amplitude for the lower hybrid waves in the above plasma is far smaller than any plasma structures. Therefore the only way in which the micro-instability affects the macroscopic plasma properties is through an anomalously high electron-ion collision frequency (ν_{an}) as the electrons are now scattered by both individual ions and a saturated wave structure in the ion fluid. Thus the lower hybrid drift instability modifies the macroscopic plasma equations wherever υ_{ei} appears, i.e. in all electron transport coefficients (including $\beta_\wedge$ terms) and the electron-ion equilibration rate. However it is the resultant high resistivity that has the dominant effect.

The very rapid growth and saturation of the lower hybrid drift instability means that the non-linear interaction between the background plasma and the growth of the instability can be ignored thereby greatly simplifying the treatment of its macroscopic consequences. From the θ pinch literature[9,10,11] the modification to the electron-ion collision frequency due to saturated lower hybrid drift waves is $\upsilon_{ei} \rightarrow \upsilon_{ei} + \upsilon_{an}$ where;

$$\upsilon_{an} = \omega_{LH}\left\{\frac{v_d}{c_s}\right\}^2 \frac{\varepsilon_F}{\frac{1}{2}n_e m_e v_d^2} \qquad \{1\}$$

where ε_F is the energy density in the fluctuating electric field of the wave. The ratio $\varepsilon_F/\frac{1}{2}n_e m_e v_d^2$ is essentially the efficiency with which energy is converted from the kinetic energy of the drift electrons into the energy of the instability. For a saturated wave it is safe to adopt the maximum value of ε_F which is thought to lie between an upper bound given by Fowler's thermodynamic method,

$$\varepsilon_F = \frac{1}{2} n_e m_e v_d^2 \qquad \{2a\}$$

and a lower bound given by equating $\frac{1}{2} n_e m_e v_d^2$ with the total wave energy (the field energy plus the kinetic energy of the oscillating particles) giving;

$$\varepsilon_F = \frac{1}{2}\left(1 + \frac{\omega_{pe}^2}{\omega_{ce}^2}\right)\frac{1}{2} n_e m_e v_d^2 \qquad \{2b\}$$

Numerical studies of the growth and saturation of the lower hybrid drift instability[9,10,12] give estimates of ε_F which lie between these two bounds but which are dependent on β, T_e/T_i etc. and may vary considerably. However as we shall see in the next section in quasi-stationary pinches, the plasma will converge to an approximately isothermal equilibrium where v_d/c_s is self-limiting and less than or equal to unity. The convergence of the ratio v_d/c_s dominates the level of anomalous resistivity and the choice of ε_F less relevant. In the non-equilibrium case, described in section 4, the plasma does not relax to isothermal equilibrium consequently we can only place upper and lower bounds on the effect of micro-turbulence on the plasma's behaviour.

In the absence of more accurate data we shall make use of the estimate[12] for ε_F which is 0.01 m_e/m_i times equation 2b, which is based on numerical simulation of the instability. Ensuring that this value does not exceed the Fowler upper bound gives us;

$$\varepsilon_F = \frac{1}{2} n_e m_e v_d^2 \min\left[1, 0.01\frac{m_i}{m_e}\frac{1}{2}\left(1 + \frac{\omega_{pe}^2}{\omega_{ce}^2}\right)\right] \qquad \{2c\}$$

3. Equilibrium Modelling

If we simulate a pinch in which the total current rises as $I \propto t^{1/3}$, the pinch remains approximately stationary and will converge to self-similar solutions which act as numerical attractors. This procedure has been used to verify analytically obtained self-similar profiles for stationary equilibria for both high and low ion magnetisation[13]and to explore the effects of the Nernst and Ettinghausen terms upon such equilibria[14]

For the purposes of this paper we will make use of a two temperature, 1-D, Lagrangian, resistive MHD code described in more detail in reference 14. The Nernst and Ettinghausen effects are included since these have been found to be necessary in two temperature simulations of fibre pinches in order to eliminate thermal instabilities near the plasma surface and are important in determining v_d/c_s[14]. The Epperlein-Haines[15] expressions for the scaling of electron thermal conductivity and resistivity with magnetisation are used. Unfortunately since these expressions are derived for simple electron-ion collisions and not for the case of scattering from drift waves, therefore their direct implementation may give rise to inaccuracies. However since the dominant effect of micro-turbulence is enhanced resistivity and the magnetisation parameter changes the resistivity only by a factor of ~2, the error in using the Epperlein-Haines scaling will be insignificant compared to the effects of the micro-instabilities.

The equations of the model are identical to those described in the next section but with

$^{dn_e}/_{dt} = 0$. The current is assumed to evolve as $I = 10^7\ t^{1/3}$ and a line density of $N=10^{20}$ is used. In the absence of the effects of micro-instabilities, the results of reference 12 are obtained with the Nernst and Ettinghausen effects combining to eliminate the electro-thermal instability near the surface and producing an almost isothermal pinch of uniform current density.

Figure 1 shows the results for the same conditions with the electron-ion collision frequency modified according to equations 1 and 2c.

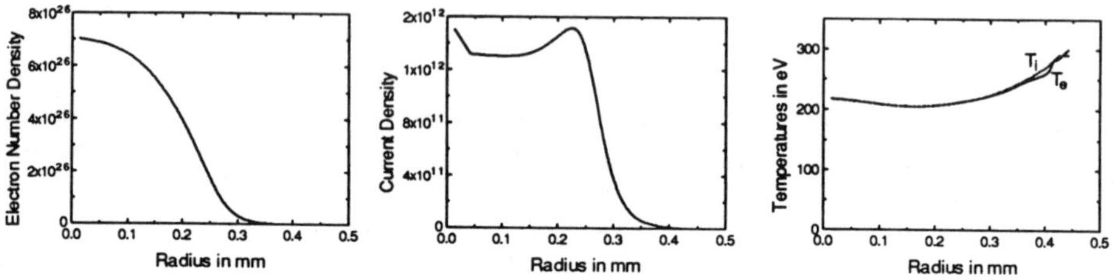

Figure 1, plasma profiles from quasi-stationary simulation including ν_{an}.

The most obvious feature of these new equilibrium profiles is the presence of a low density, low current density coronal plasma. Surprisingly, there are no obvious signs of dramatic Ohmic heating due to anomalously high resistivity. This is because in equilibrium the magnetic flux is rapidly diffused through a highly resistive surface region and consequently the local current density and Ohmic heating rates are reduced. In fact the plasma converges to an almost isothermal state. Note that T_e and T_i remain equal near the surface, this is because for LHD waves the Ohmic heating due to ν_{an} is shared equally between ions and electrons (see next section). The high thermal conductivity of the ions ensures that they are isothermal whereas $\beta_\wedge$ terms ensure that the electrons are isothermal and since the temperatures of the two species are coupled in the high density core, they remain equal throughout the pinch.

Also note that the current density now falls to zero at the surface of the pinch and the surface drift velocity of the electrons is no longer infinite.

The ratio $^{v}d/c_s$ drives the micro-instability which increases the resistivity which in turn changes the plasma profiles in order to limit $^{v}d/c_s$. Eventually an equilibrium is obtained where $^{v}d/c_s \leq 1$.

Therefore the profiles obtained are qualitatively similar to those obtained by Robson[7], who made the assumptions of uniform temperature and vd/cs $\leq$ 1 in order to generate profiles of density and current density analytically. It is no surprise that the outer regions of the plasma profiles in figure 1 resemble the Bennett profiles[16] for which constant v_d is the determining factor.

If the minimum value for the fluctuation energy of the lower hybrid drift waves given by equation 2b is used, the plasma profiles obtained are qualitatively similar to those in figure 2 with a marginally smaller coronal layer. This is because in equilibrium, the level of anomalous resistivity is simply that required to limit $v_d \leq c_s$.

4. Simulations of Fibre Ablation

In actual experiments, the plasma profiles described in the previous section may well be representative of the plasma structure after the pinch has been fully ionised. However as we shall see in this section, the duration of the fibre ablation phase can be a significant fraction of the experimental duration. Since the effect of micro-instabilities upon the transport is increasingly important at lower line densities[7], when the line density of ionised particles is zero (as in an unionised fibre), we expect equation 1 to dominate the plasma transport. The effects of equation 1 can lead to plasma conditions far from the equilibrium described in the previous section and to a substantial reduction in plasma confinement during the ablation phase.

Simple scaling calculations suggest that the early plasma consists of three Maxwellian fluids whose individual temperatures are strongly dependent on the rates of ionisation, recombination and equilibration. Furthermore, the plasma cannot be considered to be in Saha equilibrium since the time scale to relax to such an equilibrium is considerably longer than the time scale over which the internal energy of a species changes.

The ablation phase has previously been modelled[3] using a 3 species (electrons, ions and neutrals) and two temperature (electrons and ions) version of the 1-D code described in the previous section. The collisional ionisation and three body recombination rate equations are added to the resistive MHD equations in order to model ionisation levels which are far from Saha equilibrium.

The equations used in the model are as follows;

Continuity Equation

$$\frac{\mathrm{dr}}{\mathrm{dt}}+\frac{\partial \mathrm{v_r}}{\partial \mathrm{r}}=0 \qquad \{3\}$$

Momentum Equation

$$\rho\frac{dv_r}{dt}+j_z B_\theta+\frac{\partial}{\partial r}(p_e+p_i)=0 \qquad \{4\}$$

Electron energy equation

$$\rho_e^{\gamma/\gamma-1}\frac{d}{dt}\frac{p_e}{\rho_e^\gamma}=\eta_\perp j_z^2-\frac{1}{r}\frac{\partial}{\partial r}(rq_e)+\frac{n_e k}{\gamma-1}(T_i-T_e)3\upsilon_{ei}\frac{m_e}{m_i} \qquad \{5\}$$

$$+\frac{n_e k}{\gamma-1}(T_n-T_e)3\upsilon_{en}\frac{m_e}{m_n}-\frac{kT_e}{\gamma-1}\frac{dn_e}{dt}-E_{ion}\frac{dn_e}{dt}$$

Ion energy equation

$$\rho_i^{\gamma/\gamma-1}\frac{d}{dt}\frac{p_i}{\rho_i^\gamma}=-\frac{1}{r}\frac{\partial}{\partial r}(rq_i)+\frac{n_i k}{\gamma-1}(T_e-T_i)3\upsilon_{ei}\frac{m_e}{m_i} \qquad \{6\}$$

Electron heat flow

$$q_e=-\mathrm{K}_e\frac{\partial(kT_e)}{\partial r}-\frac{\beta_\wedge j_z kT_e}{e} \qquad \{7\}$$

Ion heat flow

$$q_i=-\mathrm{K}_i\frac{\partial(kT_i)}{\partial r} \qquad \{8\}$$

Ohm's Law

$$E_z + v_r B_\theta = \eta_\perp j_z + \frac{\beta_\wedge}{e}\frac{\partial (kT_e)}{\partial r} \qquad \{9\}$$

Collisional ionization and three body recombination

$$\frac{dn_e}{dt} = n_e n_n \langle \sigma_{ion} v_e \rangle - n_i n_e^2 \frac{\langle \sigma_{ion} v_e \rangle}{K(T_e)} \qquad \{10\}$$

Non degenerate Saha equation

$$K(T_e) = \frac{2(2\pi m_e kT_e)}{h^3} e^{-E_{ion}/kT_e} \qquad \{11\}$$

here the terms in $\beta_\wedge$ are the Ettinghausen heat flow and the Nernst term in Ohm's law. The perpendicular resistivity is a function of a total collision frequency which is the sum of the classical electron-ion contribution, the anomalous collision frequency due to the lower hybrid drift instability (equation 1) and the electron-neutral collision frequency taken from Ovcharenko[16]. Note that the Ohmic heating due to the anomalous part of the collision frequency is shared equally between the electron and ion energy equations. This is because particle scattering from a LHD wave, deposits equal energies into both species once the wave becomes saturated[10]. The last term in equation 5 is the energy required to liberate an electron. The ionisation energy E_{ion} is calculated including the contributions due to continuum lowering[17]. The Maxwellian averaged ionisation cross-section $<\sigma_{ion} v_e>$ is taken from numerical fits to experimental results[18]. The penultimate term in equation 5 is the energies required to thermalise aliberated electron. Terms in T_α-T_β in equations 5 and 6 are equilibration terms between electrons and ions and between electrons and neutrals. The ion and neutral equilibration is assumed to be instantaneous, i.e. $T_i = T_n$.

The plasma simulation is started with a cold unionised fibre at solid density (line density $N=10^{20}$ ions m^{-1}) and a voltage pulse similar to that expected from the MAGPIE generator[19] is applied. Breakdown occurs over the surface of the pinch forming a low density, high temperature, fully ionised coronal plasma. The ablation phase is characterised by the propagation of an ablation wave from this corona to the dense, cold, unionised core.

In the absence of the effects of micro-turbulence, the pinch becomes fully ionised after some 30 ns during which time the radius plasma has expanded to ~70μm radius.

With the effects of micro-turbulence included by way of equation 1, we find that in order to resolve the extremely large gradients in plasma parameters that occur during simulation, it is necessary to operate with a large number of computational cells (~200) and to adopt very fine meshing near the surface such that the mass represented by the surface cell is 10^4 times less than that represented by the cell on axis.

We are now faced with the dilemma that the concept of pinch radius is somewhat meaningless since for simulations including lower hybrid drift effects the plasma density near the pinch surface is over eight orders of magnitude less than that on axis. Experimentally such a low density part of the corona would be invisible in comparison to the adjacent core plasma. Therefore it is more useful to consider radii containing fractions of the total mass of the pinch which are less than 100%.

Figure 2 shows the radii within which lie 90%, 95%, 99% and 99.9% of the total mass of the pinch for simulation in which the best estimate for the fluctuation energy of the lower hybrid drift waves has been assumed (equation 2c). During the first ~70ns of the discharge, the majority of the plasma is apparently confined within 100-200 μm radius. However during this period the pinch is less than 20% ionised. Complete ablation of the fibre does not occur until 122ns through the discharge. The accelerated rate of ablation after 70ns is accompanied by expansion of the plasma column to ~2 mm radius.

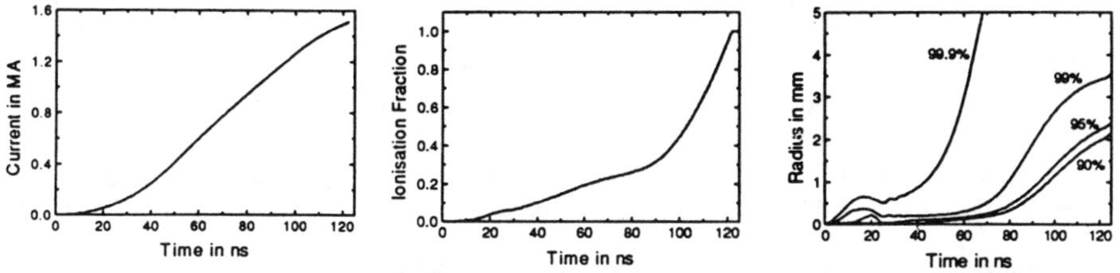

Figure 2 current, ionisation fraction and radii containing fixed fraction of the total mass versus time up to complete ablation, for υ_{an} given by equation 2c.

Figure 3 shows the first 600 μm of plasma profiles after 25ns and is representative of the internal structure of the pinch throughout ablation.

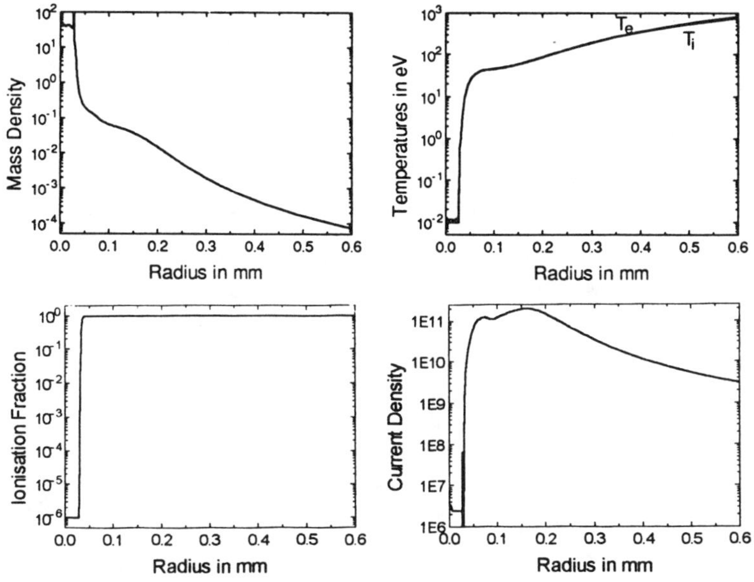

Figure 3 Plasma profiles 25ns into ablation phase using υ_{an} given by equation 2c.

The pinch consists of three distinct regions. There is a cold, high density, unionised core; an intermediate density, intermediate temperature fully ionised region in which the majority of the current flows and a very low density, high temperature, fully ionised, coronal plasma.

Ablation is occurring due to thermal conduction down the large temperature gradient at the core surface. Continuum lowering assists the ablation rate by reducing the ionisation energy in the core material which is being compressed to substantially above solid density in reaction to the blow-off of the corona.

As in the equilibrium case, the high resistance near the surface causes rapid magnetic flux diffusion through this region and the surface current density is low. However, the reduction in current density is not sufficiently rapid to prevent the effects of micro-turbulence from superheating the coronal plasma. It is unclear whether the high temperature of such a small fraction of material would be observable experimentally.

If instead of using the estimate for ε_F given in equation 2c, we adopt the minimum value for ν_{an} given by equation 2b, we obtain qualitatively similar results.

Figure 4 shows the pinch evolution from a simulation where the driving voltage was trebled. The time to complete ablation is now reduced to 57ns. However the experimentally observed degree of plasma confinement may well be less than in the lower voltage case since the expansion of the pinch due to complete ablation occurs at an earlier time.

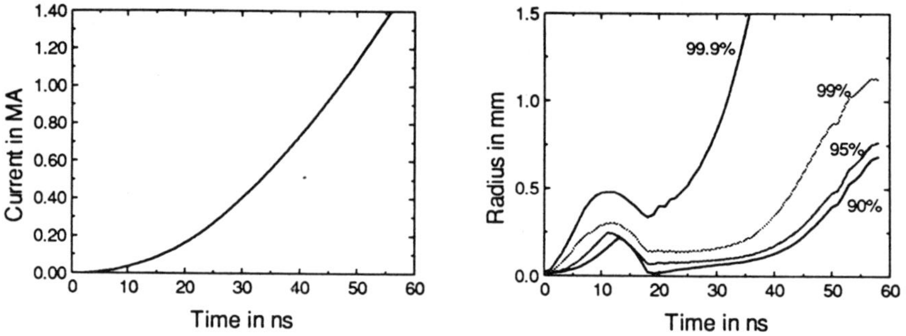

Figure 4, pinch dynamics with higher voltage and ν_{an} given by equation 2c.

In figure 5, a lower line density pinch of $N=10^{19}$ has been simulated with the lower applied voltage. With a much lower mass of material to ionise, the time to complete ablation is reduced to 37ns and there is little evidence of rapid expansion associated with complete ablation.

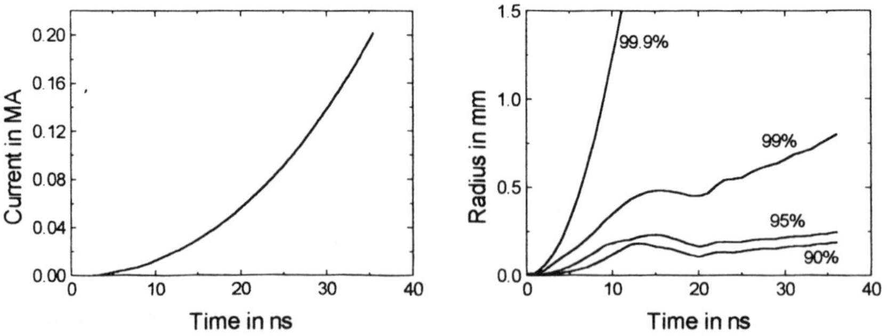

Figure 5, pinch dynamics for $N=10^{19}$ and ν_{an} given by equation 2c.

It must be stressed that the results of this section are only preliminary. The ablation of the fibre material is governed by the rate of thermal conduction from corona to core. The thermal conductivity coefficients used here are an base on Braginskii's values with an ad hoc modification to include the effect of the neutral fluid. The inclusion of a more comprehensive set of transport coefficients for partially plasma is a obvious requirement. Also the effect of fast particles on the ionization of the core material has not been addressed in this work. Finally the mechanism of the lower hybrid drift instability (as well as other micro-instabilities) and in particular the fluctuation level at saturation requires a proper investigation.

6. References

{1} D.W. Scudder, J.S. Schlacter, P.R. Forman, R.A. Riley and R.H. Lovberg, in these proceedings.

{2} J.D. Sethian, A.E. Robson, K.A. Gerber and A.W. DeSilva, Physics of Alternative Magnetic Confinement Schemes, Varenna, p511 (1990)

{3} W. Kies, et. al. J. Appl. Phys. **70** p7261 (1991)

{4} J.P. Chittenden, Bull. Am. Phys. Soc. **35** p1961 (1990)

{5} R.H. Lovberg, R.A. Riley and J.S. Schlacter, in these proceedings.

{6} I.R. Lindemuth, Phys. Rev. Lett. **65** p179 (1990)

{7} A.E. Robson, Phys. Fluids **B3** p1461 (1991)

{8} V.V. Vikhrev and V.M. Korzhavin, Sov. J. Plasma Phys. **4** p411 (1978)

{9} R.C. Davidson and N.A. Krall, Nuclear Fusion **17** p1313 (1977)

{10} P.C. Liewer and R.C. Davidson, Nuclear Fusion **17** p85 (1977)

{11} R.C. Davidson and N.T. Gladd, Phys. Fluids **18** p1327 (1975)

{12} J.F. Drake, P.N. Guzdar, A.B. Hassam and J.D. Huba, Phys. Fluids **27** p1148 (1984)

{11} M. Coppins, J.P. Chittenden and I.D. Culverwell, J. Phys. D **25** p178 (1992)

{12} J.P. Chittenden and M.G. Haines, to be published in J. Phys. D (1993)

{13} E.M. Epperlein and M.G. Haines, Phys. Fluids **29** p1029 (1986)

{15} W.H. Bennett Phys. Rev. **45** p890 (1934)

{16} V.A. Ovcharenko, High Temp. **8** p1012 (1970)

{17} R.M. More, Lawrence Livermore Report, UCRL-84991 (1981)

{18} J. Killeen, G. Gibson and S.A. Colgate, Phys. Fluids **3** p387 (1960)

{19} I.H. Mitchell et al, in these proceedings.

RADIATION HYDRODYNAMICS OF GAS-PUFF Z-PINCH PLASMAS

J. DAVIS, J. GIULIANI, M. MULBRANDON
PLASMA PHYSICS DIVISION
NAVAL RESEARCH LABORATORY
WASHINGTON, DC 20375

and

F. L. COCHRAN
BERKELEY RESEARCH ASSOCIATES
SPRINGFIELD, VA. 22151

ABSTRACT

Non-LTE radiation hydrodynamic numerical simulations in 1-D and 2-D are performed for multi-terawatt driven argon and krypton gas puff loads. The influence of enhanced transport coefficients on the plasmas' implosion dynamics and the effects it has on the radiation yield and spectral distribution are discussed. Also presented are simulations for the performance of a krypton gas puff driven by a class of future generators.

INTRODUCTION

Over the years there has been sustained enthusiastic interest in and fascination with Z-pinch plasmas. Whether the interest is in radiation source development, fusion plasmas, or basic research there exits an extensive bibliography of literature dedicated to describing and understanding the behavior and performance of Z-pinch plasmas. Recently, the development and application of multi-terawatt generators to drive cylindrically symmetric annular material loads, particularly puff gas loads, has made it possible to attain hotter plasmas of moderate Z-material loads leading to significant increases in the soft x-ray production accompanied with a corresponding shift to shorter wavelengths. In this paper we investigate and summarize the behavior of a SATURN- and super- class multi-terawatt generator driving argon and krypton gas puff plasmas. The simulation employs a non-LTE radiation magnetohydrodynamics model to evaluate and determine the radiative properties and performance of the pinched plasmas for several combinations of enhanced transport coefficients. The motivation for exploring variations of the transport coefficients is based on the lack of agreement between the complete suite of experimental data (including x-ray pulse lengths, radiative yield, temperature, density, and pinch radius of the x-ray emitting region.) and 1-D numerical simulations using classical transport coefficients

In this study the focus will be on the behavior and magnitude of the x-ray power radiated for several mass loadings self-consistently driven by a circuit representative of the SATURN and future generators. The results presented here are based on a series of 1-D and 2-D benchmark simulations both for classical and enhanced transport and resistivity coefficients, respectively. The numerical simulations are compared with experimental radiation data from imploded argon and krypton gas puffs. Finally, a limited number of numerical simulations have been performed in order to provide estimates of the radiation production of krypton gas puffs as a function of load current for future generators.

MODEL

The equations describing the model include the equations of hydrodynamics, Maxwell's equations, and atomic rate equations representing the population of the atomic levels as well as the charge states.

$$d\rho/dt=-\rho\nabla\bullet u \qquad [1]$$

$$\rho du/dt=-\nabla(P_e+P_i+Q_a) + JXB/c\ , \qquad [2]$$

$$d\varepsilon_{e/dt} + P_e dV/dt = -V\nabla\bullet q_e + V\eta J^2 + P_{rad} + V\dot{C}_{ei}(T_i - T_e)\ , \qquad [3]$$

$$d\varepsilon_i/dt + (P_i + Q_a)dV/dt = -V\nabla\bullet q_i + V\dot{C}_{ei}(T_e - T_i) \qquad [4]$$

To these equations are added Maxwell's equations, viz.,

$$\nabla X B = 4\pi J/c, \qquad [5]$$

$$\nabla X E = -\partial B/c\partial t\ , \qquad [6]$$

along with Ohm's law,

$$E = \eta J - u X B/c. \qquad [7]$$

In eqs. [1] - [7], $P_{e,i}$ are the material pressures for electrons and ions, $\varepsilon_{e,i}$ are the specific energies, ρ is the density, V is the inverse of the density, $q_{e,i}$ are the heat fluxes, $\dot{C}_{ei}$ is the electron-ion energy exchange term. P_{rad} is the radiative power, and ηJ^2 is the ohmic heating. In eqs. (3) and (4), Q_a is an artificial viscosity used for numerical stability in regions of strong compression. The remaining symbols have their conventional meaning.

The atomic physics model contains all the ground states and an extensive manifold of energetically and diagnostically important excited states distributed throughout the various ionization stages. There are a sufficient number of argon K- and L-shell spectral lines carried in the calculation to describe adequately the energetics and total radiative emissions. The krypton K-shell is more than sufficient, however the L-shell, still under improvement, provides only a crude estimate of the plasmas' radiative performance. Radiation transport and opacity are based on the local approximation and is employed to evaluate the line opacity within each computational zone. The spectral lines are represented as Voigt profiles and include broadening contributions from natural, Doppler random and directed, and Stark broadening. Equation Of State tables are constructed from this model and used in the dynamic calculations. Table lookups for P_{rad}, T_e, and <Z> based on internal energy and density are employed. The radiation spectra are produced in a post-process calculation using the detailed atomic physics model and restart data from the dynamic simulations. For the 2D simulations the computational model is a cylindrically symmetric annular R-Z geometry and employs a Lagrangian push followed by an Eulerian remap phase. The simulation uses a voltage waveform to drive a circuit model characteristic of the generator and load to calculate a self-consistent current.

RESULTS

Both the argon and krypton gas puff loads are 2 cms. long with a linear mass density distribution with center radius of 1.25 cm, a radius of 0.4 cm and infinite Mach number nozzle (i.e., no divergence) in the 1-D simulations. The nozzle radius is 0.4 cm. for all the runs shown except as noted in Tables I and II below, where the nozzle radius is 0.2 cm. For the typical voltage pulse shown in Fig.1 a self-consistently calculated current pulse is determined from a circuit model representing the SATURN generator characteristics, including the different argon mass loadings. The circuit load couplings for a variety of mass loadings, of which Fig.1 is representative, exhibits the current evolution. The simulations indicate that the peak current into the load varies as a function of mass and, to a lesser extent, opacity. For most of the simulations the peak current reached about 9 MA and then decreased as a result of the increasing dL/dt of the load. The power radiated is given in terawatts. For the 350 μgm/cm load, using the local opacity model, the plasma pinches to a minimum radius at about 90 ns which coincides with the peak in the K-shell x-ray pulse. In this 2-D simulation, the nozzle expansion rate was 10° and there was no tilt angle for the nozzle. The peak K-shell power radiated for this case is slightly less than 3 terawatts. The K- and L-shell x-ray power spectrum (in units of ergs/sec/KeV) for this simulation is shown on Figs. 2 and 3 at 90 and 95 ns, respectively, which is at and approximately just after peak compression, respectively.

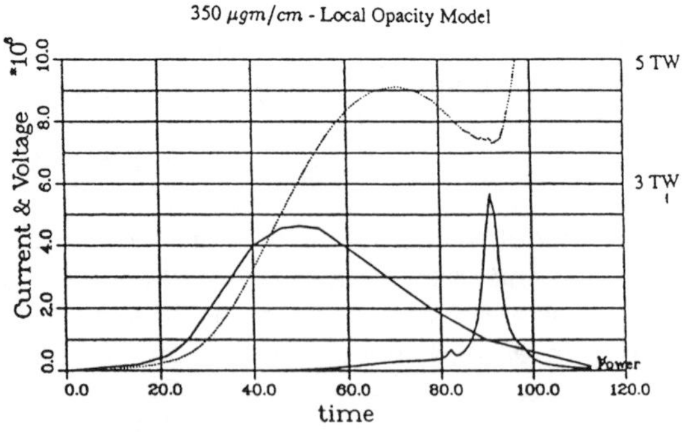

Fig. 1. Current, Voltage, and Radiation waveforms vs. time.

The radiative yield is clearly dominated by the L-shell line spectra at both times indicating that the implosion dynamics was insufficient to heat the bulk plasma into the excited K-shell manifold. In the 5 ns elapsed time duration between the two figures the plasma has significantly cooled by radiation. The major reduction occurs in the free-bound continuum beyond the K-shell lines where the difference is at least two orders of magnitude. Additional free-bound reductions of factors of two to three occurs between the L- and K-shell lines, and factors of two reduction occur below about 20 eV. The intensity of the line spectra is also reduced and redistributed reflecting the lower temperatures resulting from the cooling effects from the burst of x-rays at peak compression. For the assumed initial conditions this scenario is repeated for a number of different mass loadings in the range from 200 to 600 μgms/cm.

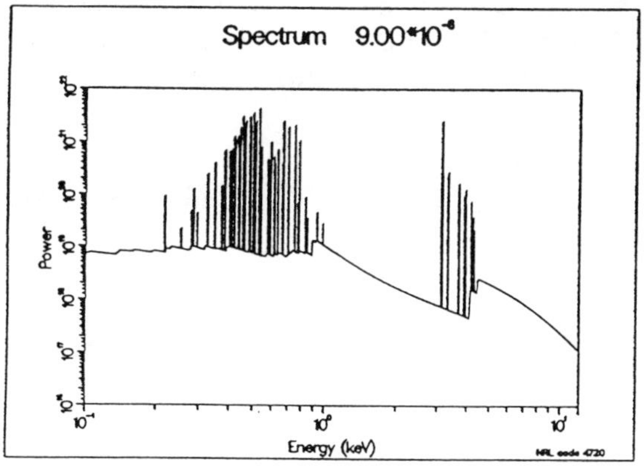

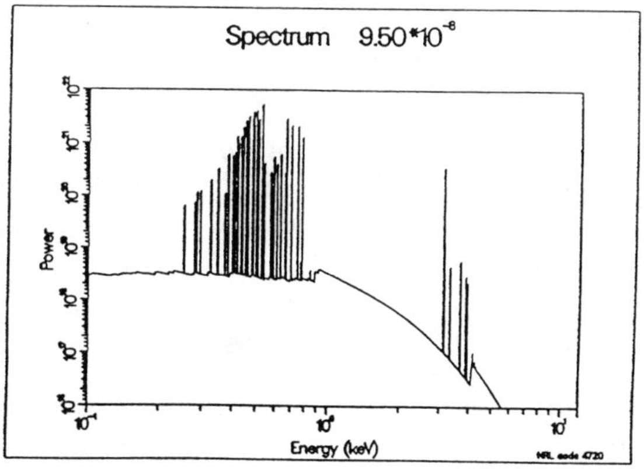

Figures 2 and 3. Radiated Power vs. photon energy for argon.

A measure of the argon K-shell yield as a function of linear mass density is shown in Fig. 4 using classical transport coefficients for a radius of 1.25 cm. All of these 2-D simulations had an inward 10° nozzle tilt angle. For comparison two additional simulations were performed for a mass of 300 µgm/cm: 100 times Spitzer and an optically thin model. The 200 µgm/cm case resulted in overheating producing an inefficient K-shell line radiator. At 300 µgm/cm the enhanced resistivity calculation is closer to the experimental result than the classical or optically thin result. The "thin" result allows the L-shell emitted radiation to cool the plasma as a volume emitter. The competition between plasma heating and L-shell cooling makes it difficult, if not impossible, under these conditions to attain temperatures for optimum K-shell emission. The trend to higher mass loadings shows increased K-shell emission at this radius. However, the slope of the curve levels off suggesting that the yield will not significantly increase. In fact, this behavior is corroborated for this radius for masses greater than 500 µgm/cm indicating that the K-shell yield has already peaked. A 2-D simulation with a inward nozzle tilt angle of 10° and radius of 0.2 cm, 2 cms long, 2T, 100 times classical resistivity and no other enhancements is shown in Fig. 5.

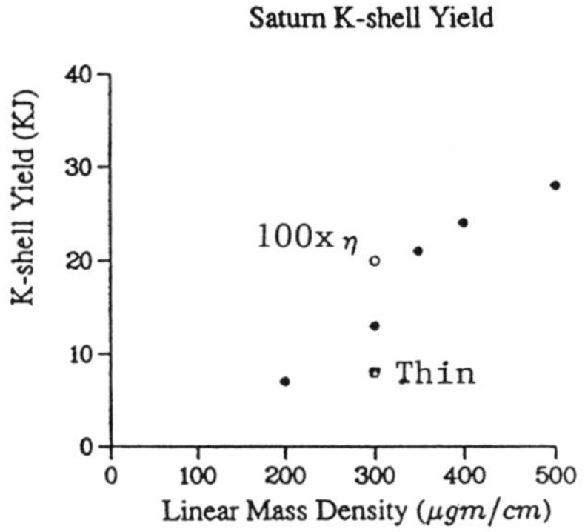

Fig. 4. K-Shell Yield vs. Mass Density.

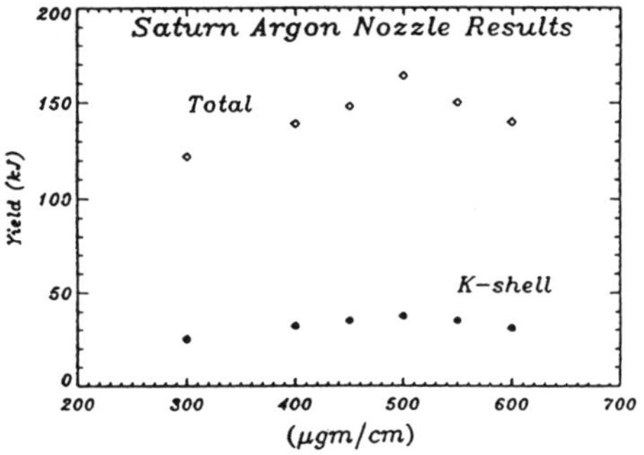

Fig. 5. Radiation Yield vs. Mass Density.

It is becoming fashionable to asses the figure of merit with which any given implosion efficiently produces K-shell radiation. Such a parameter has come to be known as η_r. The subscript r denotes the fact that in 2-D we consider only the radial kinetic energy. It is defined by the expression[1] $K_i=\eta_r E_{min}$ where K_i is the maximum implosion kinetic energy per ion and E_{min} is the sum of the ionization energies and the ion and electron thermal energies required to reach the heliumlike stage of the relevant element of atomic number Z. A scaling law for E_{min} is given by $E_{min} = 1.012Z^{3.662}$ eV/ion. The temperature used to determine the thermal energies necessary to provide high K-shell occupancy is discussed in Ref.(2). The values of peak η_r as a function of linear mass density is shown in Fig 6. For the lower mass cases there is a poorer x-ray conversion efficiency due to overheating and reduced K-shell occupation. For a fixed radius and mass the closer η_r is to unity the more likely the K-shell yield is maximized. Finally, for this series of simulations we present in Fig. 7 the time difference between the peak η_r and the peak K-shell power. This result provides a measure of when the K-shell radiation appears on the hydrodynamic time scale, i.e., a kind of thermalization time.

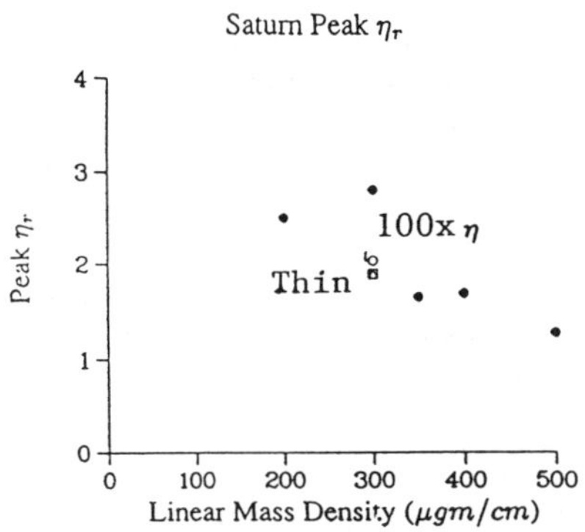

Fig. 6. Figure of Merit vs. Mass Density.

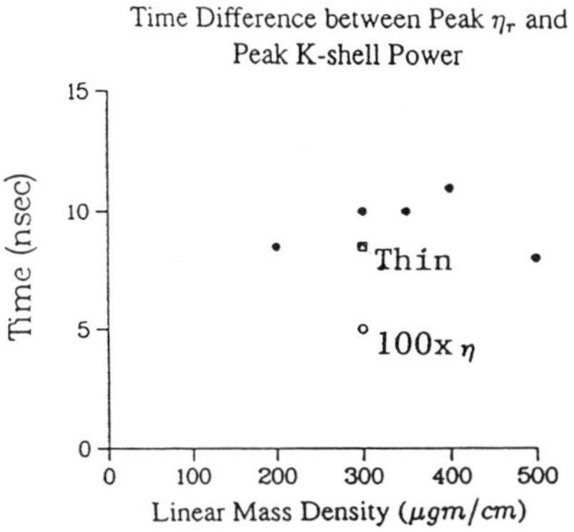

Fig. 7. Time Difference of η_r vs. Mass Density.

Table I presents a number of 1-D argon simulations for several combinations of transport coefficients and mass loadings. Under the various headings are labeled identifying information about the simulation. For instance, under 'Run' is listed the run number, mass in μgms/cm, and AR for argon. The majority of simulations are two temperature except where noted. In the column labeled 'art. Heat Conduction', on means that there is an enhanced thermal conduction component to help remove the energy generated by the artificial viscosity. This helps alleviate hollow implosions which occur as a result of overheating the central region of the plasma. The term ion (or in some cases el) heating refers to which plasma component is heated with the artificial viscosity. The columns labeled 'Transp.' and 'Res.' 'Mult.' represent the scaling factor by which the classical thermal conduction and resistivity are multiplied by. The Yield is in Kilojoules and the 'Comments' provide information on the element (in this case AR), KSH refers to K-shell, and the number of radial zones contained in the numerical simulation. The first entry represents a one temperature simulation for a 300 μgms/cm using classical transport coefficients. In comparison with Run 1136, which is identical except it represents a two temperature model, the K-shell yield results are very different. Under these circumstances, the two temperature simulation predicts a much cooler implosion and cannot significantly populate into excited K-shell line radiation. The principal reason for this behavior is that the ions are very hot and confined to a few zones and at implosion are unable to equilibrate on a radiation time scale. This problem is commonly encountered in and plagues numerical modeling. The un-optimized experimental argon K-shell yields on SATURN varied between about 40 to 55 Kjoules. From Table I it can be seen that several entries are within the experimental range. The simulations with 100 times classical resistivity significantly raises the joule heating in the plasma and accounts for much of the increased radiation yields. The last entry in the Table Run 20101 is for a nozzle radius of 0.2 cm and in comparison with Run 11348 for 0.4 cm radius shows essentially no difference in the K-shell yield. However, in 2-D there is significant differences in the K-shell yield on the order of 25% due to the more compact plasma that is produced with the smaller nozzle radius which inhibits the "zippering" effect. A 2-D simulation was made for a 10° nozzle tilt angle with an enhanced thermal conductivity of 30 and an enhanced resistivity of 10 for a 300 μgms/cm mass loading. The resolution for the numerics is a 121x31 grid. The K-shell yield is 19.2 Kjoules. This is to be compared and contrasted with Run 11337. The difference can be attributed to the limited number of zones which is unable to resolve gradients adequately in the 2-D simulation. This is corrected by increasing the number of zones as well as the computational time!

Table II presents similar data for krypton. Since there is a paucity of experimental K-shell yields for the SATURN simulator at present current levels and the krypton plasma is a prolific L-shell radiator the L-shell yields are presented. Conclusions similar to the argon results apply to krypton. For example, Run 11353 represents an optically thin simulation, i.e. a volume emitter. The L-shell is generally opaque, particularly in some of the stronger lines, and generally does not radiate as a volume emitter. However, in comparison with Run 11351 which includes opacity, the yield is not that different suggesting that the density is low enough that the optical depths of the stronger lines do not significantly affect the yield, i.e., the plasma is effectively thin. In both the argon and krypton cases where a comparison is made between a one and two temperature simulation, all else equal, there is an indication that the two temperature simulation contains hot ions confined to a few numerical zones that are unable to equilibrate fast enough to affect the radiation. This result points out the danger of unconditionally accepting the predictions of computer simulations. Clearly, each simulation must be assessed on its merits and on how well it predicts and agrees (whenever possible) with the suite of experimental data.

Run	2T	Art. Heat Conduction	Transp. Mult.	Res. Mult.	Yield (kJ)	Comments
11302 300 AR	1T	on,ion heating	1.0	1.0	62	AR-KSH 250 zones
11308 300 AR	2T	on,el. heating	1.0	1.0	23	AR-KSH 250 zones
11314 300 AR	2T	off,el. heating	1.0	1.0	13	AR-KSH 250 zones
11326 300 AR	2T	on,ion heating	1.0	1.0	11	AR-KSH 250 zones
11337 300 AR	2T	on,ion heating	30.	10.	30	AR-KSH 250 zones
11348 300 AR	2T	on,ion heating	30.	30.	42	AR-KSH 250 zones
11346 300 AR	2T	on,ion heating	30.	100.	53	AR-KSH 250 zones
11340 300 AR	2T	on,ion heating	1.0	10.	25	AR-KSH 250 zones
11337 300 AR	2T	on,ion heating	30.	10.	30	AR-KSH 250 zones
11338 300 AR	2T	on,ion heating	40.	10.	30	AR-KSH 250 zones
11339 300 AR	2T	on,ion heating	50.	10.	30	AR-KSH 250 zones
12901 300 AR	2T	on,ion heating	30.	10.	39	AR-KSH 500 zones *
11352 400 AR	2T	on,ion heating	30.	30.	55	AR-KSH 250 zones
11354 400 AR	2T	on,ion heating	30.	100.	72	AR-KSH 250 zones
11362 500 AR	2T	on,ion heating	30.	10.	59	AR-KSH 250 zones
11358 500 AR	2T	on,ion heating	30.	30.	59	AR-KSH 250 zones
20101 300 AR	2T	on,ion heating	30.	30.	44	AR-KSH 250 zones moz=0.2†

Table I
Argon

Run	2T	Art. Heat Conduction	Transp. Mult.	Res. Mult.	Yield (kJ)	Comments
11301 300 KR	1T	on,ion heating	1.0	1.0	239	KR-LSH 250 zones
11307 300 KR	2T	on,el. heating	1.0	1.0	21	KR-LSH 250 zones
11313 300 KR	2T	off,el. heating	1.0	1.0	12	KR-LSH 250 zones
11323 300 KR	2T	on,ion heating	1.0	1.0	10	KR-LSH 250 zones
11342 300 KR	2T	on,ion heating	30.	10.	11	KR-LSH 250 zones
11347 300 KR	2T	on,ion heating	30.	30.	21	KR-LSH 250 zones
11345 300 KR	2T	on,ion heating	30.	100.	49	KR-LSH 250 zones
12902 300 KR	2T	on,ion heating	30.	10.	14	KR-LSH 500 zones
11351 400 KR	2T	on,ion heating	30.	30.	25	KR-LSH 250 zones
11353 400 KR	2T	on,ion heating	30.	30.	23	KR-LSH-thin 250 zones
11355 400 KR	2T	on,ion heating	30.	100.	55	KR-LSH 250 zones
11361 500 KR	2T	on,ion heating	30.	10.	8	KR-LSH 250 zones
11357 500 KR	2T	on,ion heating	30.	30.	22	KR-LSH 250 zones
20102 300 KR	2T	on,ion heating	30.	30.	20	KR-LSH 250 zones moz=0.2

Table II
Krypton

Finally, a series of simulations were performed for a class of future generators driving a krypton gas puff plasma with an initial radius of 10 cm and a thickness 25% of the radius. The mass distribution is Gaussian. The peak current was put into a short circuit load so the actual current into the load is somewhat less. Note that the loads are considerably more massive than those discussed above. Normalized K-shell and total radiative yields as a function of mass are shown in Fig. 8 for several currents. The lower set of curves predicts the K-shell yield and the upper set the total yield. Figure 9 shows the variation of the normalized K-shell yield as a function of load mass and initial puff radius for a 60 MA generator. The results are self-explanatory. At the larger radii, i.e., 10 cm and above, mhd simulations shows the plasma becoming unstable and breaking up during the run-in. Just how far out the load can be placed and still be a viable x-ray radiation source is currently under investigation.

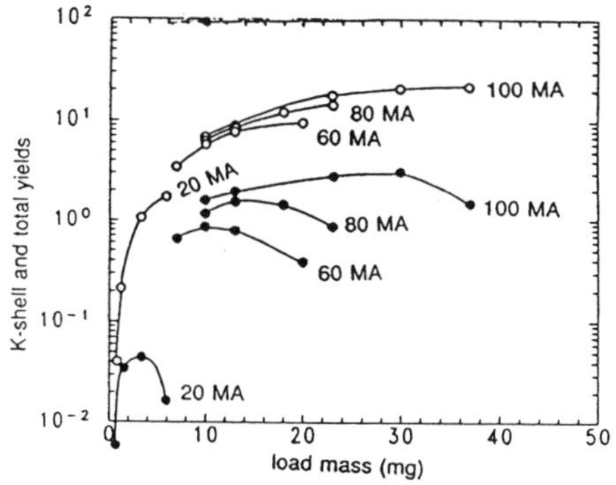

Fig. 8. Radiation Yield vs. Mass For Krypton.

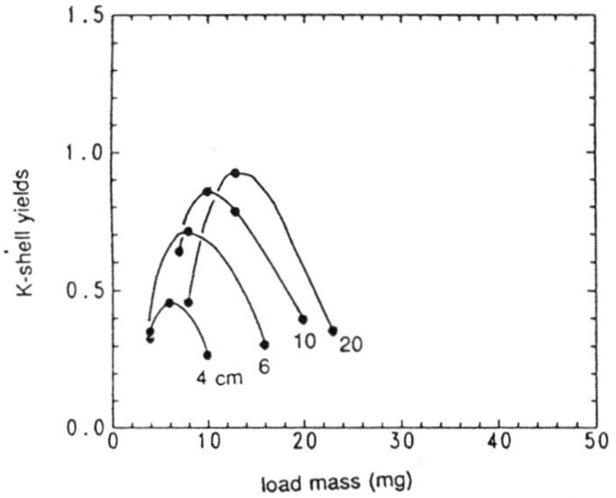

Fig. 9. K-Shell Yield vs. Mass as function of radius.

SUMMARY AND CONCLUSIONS

The radiative performance of argon and krypton gas puff plasmas driven by terawatt generators was investigated for a variety of conditions with a non-LTE radiation MHD model in 1-D and 2-D. Numerical simulations were performed for the SATURN simulator for several mass loadings and transport coefficients with different multiplicative factors to assess their influence on the plasmas' radiative performance. The results indicate the importance and sensitivity of the radiative performance to the choice of transport coefficients. This all points to the need to return to fundamental physics issues involving energy transport, laminar versus turbulent flow, etc., if we are to fully understand how these plasmas evolve through the cold start to the final burst of x-rays. There are questions raised by how well the numerics capture the physics. Great strides have been made and are continuing in the numerical simulation arena but for our part we look more for systematic trends in the results rather than absolute numbers. Care must always be exercised and in the final analysis it is how well the results agree with and predict experiments that counts.

ACKNOWLEDGEMENT

The authors extend their thanks to their colleagues in the Radiation Hydrodynamics Branch for helpful discussions during the course of this investigation.

REFERENCES

1. C. Deeney, T. Nash, R. Prasad, L. Warren, et. al. Phys. Rev. A 44, 6762 (1991).
2. K. G. Whitney , J. W. Thornhill, J. P. Apruzese and J. Davis, J. Appl. Phys. 67, 1725 (1990).

METHOD FOR NUMERIC SOLVING OF 2.5D MHD EQUATIONS IN MOVING COORDINATE SYSTEM

O.V. Diyankov and S.A. Terekhoff.
Russian Institute for Technical Physics
454070 RUSSIA, Chelyabinsk-70, P.O.Box 245. E-mail sta@ch70.chel.su

ABSTRACT

This paper is devoted to the describing of numeric method for solving 2.5D MHD equations. The method is based on splitting the original system onto two systems : hyperbolic and diffusive. Explicit difference scheme for the first system is of TVD type, and implicit scheme for the second one has been got through the approximation of energy function. The results of some illustrative calculations are shown.

INTRODUCTION

The investigation of dynamics of axis-symmetric implosive systems, such as Z-pinches, liners, exploding wires, remains an actual problem of modern physics, besides of half-century way. The actuality is caused most of all by the necessity of the laboratory soft X-ray sources developing and continuing search for possible applications to the fusion problem.

The computer modelling plays the increasing role in these investigations. It stays the main instrument for theoretic analysis. Essential understanding of the variety of phenomena in high temperature plasmas has been achieved in Russian Institute of Technical Physics (RITP, Chelyabinsk-70) using 1D ZARYA code[1, 2] .

The progress in development of adequate to considering phenomena multi-dimensional codes is restrained not only by cumberness of physics models (caused by the necessity of taking into account, besides the gas dynamics, the non-equilibrium photon and particle transport, magnetic field diffusion, anisotropy of kinetic coefficients in magnetic field and non-equilibrium energy and pulse redistribution between plasma components), but mathematic peculiarities of solving partial differential equations.

During the motion of considering cumulative systems the arbitrary two-dimensional flow of compressible gas (plasma) occurs. When the magnetic field is present, this flow permits nonconflicting 2.5 dimensional description, when all vectors have three components and all variables depend on two spatial variables and time. The flow is accompanied by multiple interaction of shocks and rarefation waves, which representation in numeric solution has great influence on the whole dynamics. While perturbations grow, large deformations of flow appear, which breaks the work of facility. The correct numeric solution up to the moments of loosing of plasma continuity is essential for treating of stability and life time of plasma formations. The appearing diffusive processes are nonlinear, so the accurate description of their fronts propagation is needed.

This paper is devoted to the developing of numeric methods, satisfying above demands.

2.5D MHD MODEL

Our model is based on well-known MHD equations [4]. While its developing we used the results of the papers [2], [3].

The equations, describing plasma flows are :

$$\frac{\partial \rho}{\partial t} + div(\rho \cdot \mathbf{u}) = 0 \tag{1}$$

$$\rho \frac{\partial \mathbf{u}}{\partial t} + \rho \cdot (\mathbf{u} \cdot \mathbf{grad})\mathbf{u} + \mathbf{grad} p = \frac{1}{4\pi} \cdot [\mathbf{B} \times \mathbf{rot} \mathbf{B}] \tag{2}$$

$$\rho \frac{\partial \varepsilon_e}{\partial t} + p_e \cdot div \mathbf{u} + div \mathbf{q}_e = -Q^{ei} + Q^{fe} + Q^e_{\text{ext}} + + \frac{1}{4\pi\sigma}(\|\mathbf{J}^{|}\|^2 + \|\mathbf{J}^{\perp}\|^2) \tag{3}$$

$$\rho \frac{\partial \varepsilon_i}{\partial t} + p_i \cdot div \mathbf{u} + div \mathbf{q}_i = Q^{ei} + Q^i_{\text{ext}} \tag{4}$$

$$\rho \frac{\partial \varepsilon_f}{\partial t} + p_f \cdot div \mathbf{u} + div \mathbf{q}_f = -Q^{fe} + Q^f_{\text{ext}} \tag{5}$$

$$\frac{\partial \mathbf{B}}{\partial t} = -c \cdot \mathbf{rot} \mathbf{E} \tag{6}$$

$$\mathbf{E} = \frac{1}{\sigma^{|}} \mathbf{J}^{|} + \frac{1}{\sigma^{\perp}} \mathbf{J}^{\perp} - \frac{1}{c} \cdot [\mathbf{u} \times (\mathbf{rot} \mathbf{B})] - \frac{1}{N_e e} \cdot (\mathbf{grad} p_e - \mathbf{R}_T - \frac{1}{c} \cdot [\mathbf{J} \times \mathbf{B}]) \tag{7}$$

$$\mathbf{J} = \frac{\mathbf{c}}{4 \cdot \pi} \cdot \mathbf{rot} \mathbf{B} \tag{8}$$

$$\mathbf{q}_e = \kappa_e^| \cdot \mathbf{grad}^| T_e + \kappa_e^\perp \cdot \mathbf{grad}^\perp T_e \tag{9}$$

$$\mathbf{q}_i = \kappa_i^| \cdot \mathbf{grad}^| T_i + \kappa_i^\perp \cdot \mathbf{grad}^\perp T_i \tag{10}$$

$$\mathbf{q}_f = -\frac{l_f c}{3} \cdot \mathbf{grad}\ U_f \cdot \left(1 + \frac{2}{3} l_f U_f \cdot \|\mathbf{grad} U_f\|\right)^{-1} \tag{11}$$

Here symbols $^|$ and $^\perp$ denote vector components along magnetic field and perpendicular to it responsibly, Q^{fe} , Q^{ei} - interchanging terms, Q_{ext} - sources, other designations correspond to [3]. The equations of state and formule for coefficients of heat conductivity and conductivity are added to the system (1)–(11).

In this paper we describe the first stage of realization of the model : numeric method for solving one-temperature 2.5D MHD equations with isotropic coefficients.

Equations (1)–(11) are written in vectors in Cartesian coordinate system. For getting of 2.5D MHD model one should write these equations in arbitrary Euler coordinate system and reduce the dependence on the third spatial coordinate. The last action may be fulfilled not for arbitrary coordinate system, but only for that one, in which metric tensor doesn't depend on the third coordinate. The authors could find only two such systems : cylinder system, all variables depend on x and z; and planar system, variables depend on x and y.

Equations of 2.5D MHD model, which have been got, were written in moving coordinate system. To get these equations in divergent form, we have written them with respect to Cartesian vector components.

The system of equations, which has appeared as the result of above operations, is rather complicate, so we used the widely spread method : slitting onto physics processes. We have got two systems : one of hyperbolic type and another of parabolic type. First one is expanded gas dynamics system of equations, and the second one describes diffusion processes : heat and magnetic field diffusion.

NUMERIC METHOD FOR SOLVING HYPERBOLIC SYSTEM

For hyperbolic system we used method, appearing in many papers, describing multidimensional gas dynamics codes (see for example [5, 6, 7]).

Its essence is in following procedure. Equations are discretized with respect to time and at each time step the equations in lagrange coordinate system are solved. In this system, which is the special case of moving coordinate system, the equations have especially simple form. Then the equations for coordinate system movement are solved (that is equivalent to mesh reconstruction), and all values are recalculated. This method has shown its effectiveness for gas dynamics problems.

We have suggested explicit difference scheme for the equations in lagrange coordinate system of first order accuracy. This scheme has been got, using well-known technique of A.Harten (TVD - scheme). The main advantage of this scheme, as we think, is the using in it viscosity, which is minimal essential for monotonicity of numeric solution. That's why the dissipation of discontinuities is also minimal. The donor-acceptor scheme is used at the second stage.

Now we'll discuss briefly the algorithms of transporting the coordinate system. Of course we have two simplest cases : Euler (stationary) and lagrange (moving with mass velocity) coordinate systems.

Besides, for some tasks, in which cumulative motion occurs, it is usefull to have moving along one direction coordinate system. In this coordinate system sites move along corresponding vector, generally different for each site.

We also have the coordinate system, in which the position of coordinate site at the next moment is determined using the information about the movement of neighbour sites.

And in addition we use the most general algorithm - the algorithm of minimization of some function of quality of coordinate system. This function decreases when coordinate system streams to orthogonal and the length of each of two local basis vectors stream one to another.

This set of algorithms allows to fulfil numeric modelling of plasma flows with rather complex geometry.

NUMERIC METHOD FOR SOLVING OF DIFFUSION SYSTEM OF EQUATIONS

The system of equations of diffusive type, got after splitting, consists of two equations of diffusion : for heat diffusion and the diffusion of the third component of magnetic field; and system of two equations for diffusion of remaining two components of magnetic field. All these equations coupled through the coefficients. So the simple iteraions is used to resolve this dependance.

The equations for heat diffusion and for the third component of magnetic field are solved independently. The implicit difference scheme for numeric solving of these equations is used. This scheme was suggested by Kershaw in the paper [9]. For resolving of obtained linear system we use ICCG method [10].

For the rest system we suggested implicit scheme, which is got from the approximation of magnetic energy. For the unique equation it coincides with the scheme [9]. The linear system is solved by the block ICCG method.

REALIZATION IN CODE AND ILLUSTRATIONS

The described approach was implemented in code MAG for VAX computer, consisting of more than 50000 lines. The language of realization is VMS PASCAL. We perform te set of one- and multi- dimensional methodic calculations.

The results of numeric modelling of thin liner, imploding by the pressure of magnetic field, appearing due to the current through the liner, are shown below. The problem corresponds to the experiments and caculations, performed in US Air Force Weapons Laboratory (Kirtland) [11]. The aluminium liner has initial radius of $5cm$ and thickness of $68\mu gm/cm^2$. The current on load has pick value of $7MA$ with the risetime of $250nsec$. In our calculations we considered only pondermotive effect of magnetic field on liner through the external magnetic pressure and isotropic electron heat diffusion. Liner was treated as ideal conductor.

In fig.1 the results of 1D calculations of gas dynamics stage of symmetric liner compressing are shown. Thick lines show the inner and outer fronts of liner. The numeric and experimental results from [11] are also shown here. Observed distinguishes may be explained by our simplifications : only kinematics and heat transfer are taken into account.

2D calculations were performed to illustrate the influence of 2D disturbances. In the first calculation in their initial state the liner of $50\mu gm/cm^2$ thickness has $\pm$ 2% uniform distributed disturbances in density. In fig.2 the form of liner for the time $t = 570nsec$ from the beginning of the pulse is shown. The following evolution of the flow is accompanied by the intensive radiation, which has not been taken into account in our numeric experiments. In the second calculation liner slides along stiff walls, converging while approaching to the axis. At the axis the distance between walls equals to $1.5cm$, and at the initial radius - to $2cm$. While sliding the perturba-

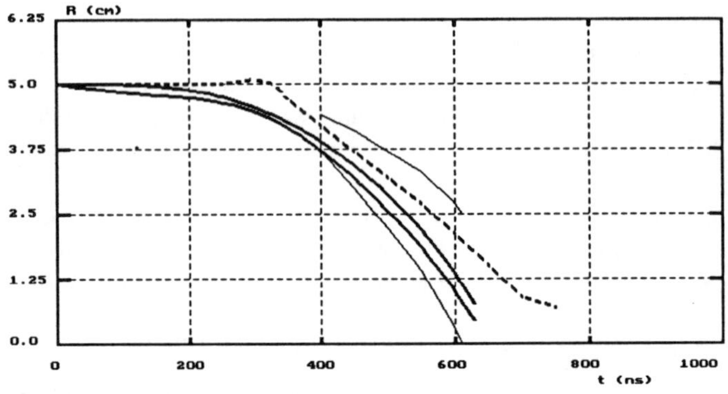

Figure 1: The dynamics of imploding liner. Thick lines - MAG 1D, thin lines - calculations [11], dashed lines - electric radius from the experiment [11]

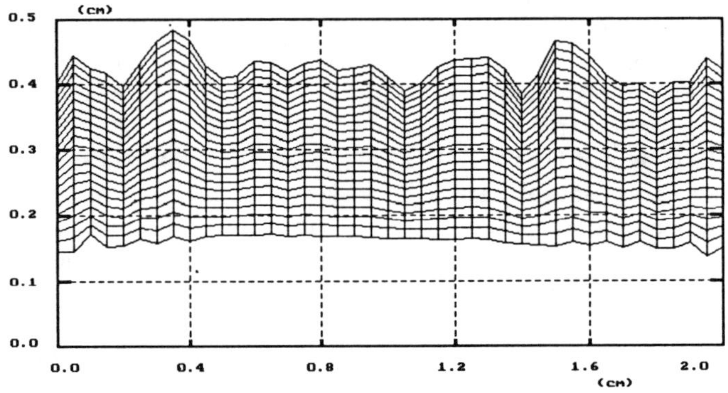

Figure 2: The form of liner with initial disturbances of ± 2% in density at the moment t = 570 nsec.

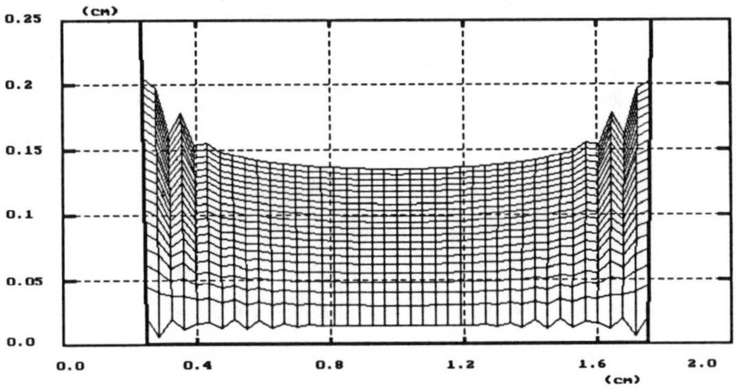

Figure 3: The form of liner sliding along converging walls at t = 590 nsec.

tions appear near the walls, which lead to the form of liner at the moment $t = 590nsec$, as it is shown in the fig.3.

CONCLUSION

The described approach to the development of numeric method for solving 2.5D MHD equations has been realized in MAG code. The progress of MAG assumes both füther development of mathematic methods and computer implementation of whole described three-temperature model. The main practical aspect is broadening of MAG code possibilities in describing problems with complex geometry.

Authors are grateful to V.Lykov and now late A.Zuev for multiple discussions, consultations and attention to the work.

REFERENCES

[1] A.I. Zuev, Rus. J. of Comp. Mat. and Mat. Phys., v.32, 82 (1992).

[2] E.N. Avrorin, V.A. Lykov. Computations of Targets for ICF by ZARYA code. Proc. Int. Conf. LASER'90, San-Diego USA, 811 (1990).

[3] S.I. Braginskii, Sov. JETP, V.6, 358 (1958).

[4] L.D. Landau, E.M. Lifshitz, Electrodynamics (Nauka, Moscow, 1982).

[5] C.W. Hirt, A.A. Amsden, J.L. Cook, J. Comp. Phys. V.14, N3 (1974).

[6] S.K. Godunov, A.V. Zabrodin et al., Numeric Solving of Multi-Dimensional Gas Dynamics Problems, (Nauka, Moscow, 1976), in Russian.

[7] N. Eskov, V. Volkov et al, Report to Russian/US Weapons Laboratories Introductory Technical Exchange in Computational and Computer Science. (Livermore, 1992).

[8] A. Harten; J. Comp. Phys., V.49, N 3., 357-393 (1983).

[9] D.S. Kershaw, J. Comp. Phys., V.39, N 2., 375-395 (1981)

[10] D.S. Kershaw, J. Comp. Phys., V.26, N 1., 43-65 (1978).

[11] W.L. Baker, J.H. Degnan, R.E. Reinovsky, Proc. 3-rd Int. Conf MEGAGAUSS-3, Novosibirsk, 39-49 (1984).

NUMERICAL SIMULATION OF A HOLLOW Z-PINCH DISCHARGE IN A MULTICHARGED GAS

A.V.Gerusov
Institute for Theoretical and Experimental Physics
B.Cheremushkinskaya, 25, Moscow, 117259, Russia

ABSTRACT

One dimensional magnetohydrodynamic code that takes into account a resonance line reabsorbtion and time-dependence of average ionization level is presented. It is based on Braginskii type MHD equations. Reabsorbtion of resonance lines is described by Biberman-Holstein method. Distribution on ionization stages is calculated in quasi-stedy-state approximation. The worked out code allows to study a discharge plasma consisted from $K-, L-, M-, N$-shell ions. With using of developed code the possibility of $[Ne]$-like ions creation in hollow Z pinch that implodes on a target placed on an axis have been investigated. The current amplitude is 0.5 MA, the timerise is 200 ns. It is assumed that a target is very massive and is described by undisturbed, nonconductive cylinder. The series of computations was carried out for Kr and Fe filling plasma. For typical variant the detailed consideration of implosion is presented.

INTRODUCTION

It is well known that today Z pinch discharge in a multicharged gas is important object of investigations in many laboratories of the world [1,2]. For its theoretical study in this paper one dimensional magnetohydrodynamic (MHD) code that takes into account a resonance line reabsorbtion and time-dependence of average ionization level is presented. It is based on Braginskii tipe MHD equations. Reabsorbtion of resonance lines is described by Biberman-Holstein method. Distribution on ionization stages is calculated in quasi-stedy-state approximation. The worked out code allows to study a discharge plasma consisted from $K-$, $L-$, $M-$, $N-$shell ions.

With using of developed code the possibility of $[Ne]$-like ions creation in hollow Z pinch that implodes on a target placed on an axis have been investigated. The current amplitude is 0.5 MA, the timerise is 200 ns. It is assumed that a target is very massive and described by undisturbed, nonconductive cylinder with radius $0.05r_0$, r_0 — is the initial radius of pinch column.

These calculations have been initiated by Semenov and Shliaptsev (Lebedev Physical Institute, Moscow) to examine the possibility of X–ray lasing in target by resonant photoexcitation from imploded plasma shell. Population inversions are supposed to obtain on $3p - 3s$ transition in Ne-like ions. Necessary condition for this is the appearing of Ne-like ions in imploded shell.

The series of computations was carried out for Kr and Fe filling plasma with various values of r_0 and ρ_{L0}, (ρ_{L0} — the line density of the pinch): $\rho_0 \sim 1.5 - 2$

cm, $\rho_{L0} \sim (2.6-24)\ \mu g/cm$. Computations show that $[Ne]$-like ions never appear in Kr plasma and appear in Fe plasma for wide range of initial conditions.

There are a few papers devoted to describing of Z pinch dynamics in a multicharged gas. Therefore it is interesting to investigate the distribution of plasma parameters at various phases of Z pinch compression. The detailed considiration is carried out for typical variant of Z pinch discharge in Fe: $r_0 = 1.5$ cm, $\rho_{L0} = 5.84\ \mu g/cm$, $I_0 = 0.5$ MA.

I.EQUATIONS DESCRIBING Z PINCH IN A MULTICHARGE GAS

a) MHD equations

If PF operates in a multicharged gas the essential part of energy may be lost on radiative cooling and ionization. Hence, MHD equations describing PF discharge in a multicharged gas must take into account these effects.

In this paper the next one-dimensional Braginskii type MHD equations are used:

$$
\begin{aligned}
&\frac{d\rho}{dt} + \rho \mathrm{div}\bar{V} = 0, \quad \rho\frac{d\bar{V}}{dt} = -\nabla(P_e + p_i) + 1/c[\bar{j}\bar{H}] - \mathrm{Div}\check{\pi}, \\
&\rho\frac{d\varepsilon_e}{dt} + P_e \mathrm{div}\bar{V} = -\mathrm{div}(\bar{q}_e) + \frac{c^2}{16\pi^2\sigma}(\mathrm{rot}\bar{H})^2 - Q_{ex} - Q_r, \\
&\rho\frac{d\varepsilon_i}{dt} + P_i \mathrm{div}\bar{V} = -\mathrm{div}(\bar{q}_i) - \mathrm{div}(\check{\pi}\bar{V}) + \bar{V}\mathrm{Div}\check{\pi} + Q_{ex}, \qquad (1)\\
&\frac{\partial \bar{H}}{dt} - \mathrm{rot}[\bar{V}\bar{H}] = -c\mathrm{rot}(\bar{j}/\sigma), \quad \bar{j} = \frac{c}{4\pi}\mathrm{rot}\bar{H}, \quad \rho = Am_pN, \\
&P_e = nkzT_e, \quad \varepsilon_e = \frac{3/2zkT_e + U(z)}{M}, \quad P_i = nkT_i, \quad \varepsilon_i = \frac{3/2kT_i}{M}, \\
&\frac{d}{dt} = \frac{\partial}{\partial t} + \bar{V}\nabla, \quad \bar{H} = (0, H_\varphi, 0), \quad \bar{V} = (V, 0, 0), \\
&\frac{\partial}{\partial\varphi} = \frac{\partial}{\partial Z} \equiv 0
\end{aligned}
$$

They are written in common notations. The ion viscosity tensor $\check{\pi}$, the electron and ion heat flux q_e, q_i, the conductivity σ and the exchange rate of energy Q_{ex} are calculated on Braginskii-Spitzer type formulas. Their form for mixture of deuterium and a multicharged gas are given in [3]. The average ionization level z and radiative cooling power Q_r [4] are determined from equations described in the next items.

b) Ionization dynamics equations

Hydrodynamic time scales of Z pinches at a collapse phase are about $10^{-9} - 10^{-8}$ s. In this case, density populations of excited states can be calculated under steady-state assumption. Then the density of excited level n of k ion can be determined through the densities of ground levels N_{kn_0}, N_{k+1,n_0} from equation:

$$
\begin{aligned}
&-N_{kn}\left(\sum_{n'<n}\left(C_{knn'}+\frac{(\theta A)_{knn'}}{N_e}\right)+\sum_{n'>n}C_{knn'}+C^i_{kn}\right)+\\
&+\sum_{n'>n}N_{kn'}\left(C_{kn'n}+\frac{(\theta A)_{kn'n}}{N_e}\right)+\\
&+\sum_{n'<n}N_{kn'}C_{kn'n}=-\left(N_{kn_0}C_{kn_0n}+N_{k+1,n0}\beta_{kn}\right) \qquad (2)
\end{aligned}
$$

In rate equation (2) the next processes have been taken into account: excitation
$(C_{knn'}, n < n')$ and *de*-excitation $(C_{knn'}, n < n')$ by electron collisions, ionization by electron collisions C^i_{kn}, spontaneous radiative transition $A_{knn'}$, radiative β^R_{kn}, dielectronic β^D_{kn} and three-body recombination β^T_{kn}: $\beta_{kn} = \beta^R_{kn} + \beta^D_{kn} + \beta^T_{kn}$, resonance line reabsorbtion $\theta_{knn'}$.

Let us denote by α_{kl} the ratio of the population density N_{kl} to the total ion density N: $\alpha_{kl} = \frac{N_{kl}}{N}$. Then rate equations for the populations of the ground levels α_{kn_0} can be written as:

$$
\frac{d\alpha_{kn_0}}{dt} = -\alpha_{kn_0}N_e(C^i_k+\beta_{k-1})+\alpha_{k+1,n_0}N_e\beta_k+\alpha_{k-1,n_0}N_eC^i_{k-1} \qquad (3)
$$

where

$$
\begin{aligned}
C^i_k &= C^i_{kn_0}+\sum_{n>n_0}C^i_{kn}\alpha^c_{kn}\\
&\qquad\qquad\qquad\qquad\qquad\qquad (4)\\
\beta_k &= \beta_{kn_0}+\sum_{n>n_0}\beta_{kn}-C^i_{kn}\beta^c_{kn}
\end{aligned}
$$

In equations (4) $\alpha^c_{kn}, \beta^c_{kn}$ denotes the solutions of (2) when its right side equals to $-C_{kn_0n}, -\beta_{kn}$ respectevely. The equations system (3) is completed by equation

$$
\sum \alpha_{kn_0} = 1 \qquad (5)
$$

which is true when

$$
\sum_{n>n_0}\alpha_{kn} \gg \alpha_{kn_0}
$$

The rate equation for average ionization level $y = \sum y_k\alpha_{kn_0}$ can be deduced from (3):

$$
\frac{dz}{dt} = \sum N_e(\alpha_{kn_0}C^i_k - \alpha_{k+1,n_0}\beta_k)
$$

c) Level structures

The number of equations (2) is equal to the number of all excited levels of ions of all ionization stages. Numerical calculation of line reabsorbtion by Biberman--Holstein method requieres that population densities of all ion levels would be determined at each moment at each spatial point of MHD descrete mesh. Hence level structures of ions have to be enough simple. In this paper special attention is devoted to simplicity of level structure.

The quasilevel structure which can be applied to plasma consisted from $K-, L-, M-$, and $N-$ electronic shell ions is shown in Fig.1. In dense plasmas the ground quasilevel ($\Delta n = 0$) unites all levels of all electronic configurations obtained in result of $\Delta n = 0$ transitions from the ground electronic configuration.

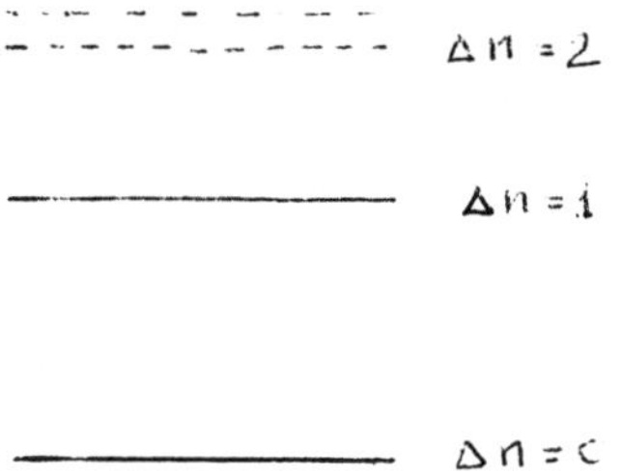

Fig. 1

For example, the ground quasilevel of [N]-like ion unites levels of the next electronic configurations:

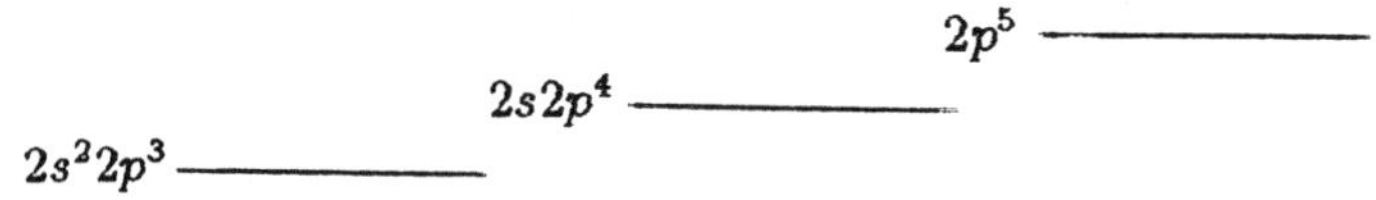

Here, $2s^22p^3$ is the ground electronic configuration.

In low density (coronal) plasmas the ground quasilevel unites levels only of the ground electronic configuration.

The excited quasilevel ($\Delta n \geq 1$) unites states with the same n of excited electron and residual electronic configurations which are included in the ground quasilevel of the next ionization stage. Within each quasilevel the sublevel populations are statistically distributed.

Quasilevel with $\Delta n \geq 1$) are considered in hydrogenlike approximation. Energies and oscillator strengths of $\Delta n = 0$ transitions are taken from [5].

d) Resonance line reabsorbtion

The resonance lines reabsorbtion is very important effect for determing of Z pinch plasma parameters at the compression phase of a Z pinch because of radiative cooling of a multicharged plasma can be very important in an energy balance. The best method for describing of line reabsorbtion in a Z pinch geometry is the method developed in [6]. However it need very much computer time especially in using mutually with MHD equations. In this paper the more simple method is used — the method by Biberman-Holstein [7]. This method is based on inducing of escape factor θ for each reabsorbed line. Its value have been determined for a slab layer and for a center of a cylinder or a sphere in cases of Doppler, Voigt and Lorentz line profiles [7]. In this paper I suppose that imploding hollow Z pinch shell in the most degree corresponds to a slab layer and Voigt profile. Then the escape factor can be written as [7]:

$$\theta_{knn_0}(r) = \frac{1}{2+3\sqrt{\tau_r/a}} + \frac{1}{2+3\sqrt{(\tau_b-\tau_r)/a}} \qquad (6)$$

where $a = \Delta V_{st}/\Delta V_D$, τ_b, τ_r — is the optical depthes in the center of $n \to n_0$ line of k ion calculated for Doppler broadening in the points with the coordinates r and r_b.

For transitions between excited levels escape factors equal to 1: $\theta_{kn'n} = 1, n', n > n_0$ in equations (2).

e) Radiative cooling power

Determing values of level populations from equations (2) the radiative cooling power Q_r (1) can be calculated from relations:

$$Q_r = QN_eN$$
$$Q = Q_L + Q_R + Q_B$$
$$Q_L = Q_{L1} + Q_{L2} + Q_{L3}$$
$$Q_{L1} = \sum_k \alpha_{kn_0} \Delta E_{12k} \frac{C_{12k}A_{21k}\theta_{21k}}{N_eC_{21k} + A_{21k}\theta_{21k}}$$
$$Q_{L2} = \sum_k \sum_{n\geq n_0} \sum_{n'>n} \alpha_{kn'}(E_{kn'} - E_{kn})A_{n'nk}\theta_{n'nk}$$
$$Q_{L3} = \sum \alpha_{k+1,n_0} \sum_j \beta^D_{knj}(E_{k+1,j} - E_{k+1,n_0}) \cdot \frac{\Gamma_{knj}\theta_{k+1,jn_0}}{\Gamma_{knj} + N_eC^i_{kn}}$$

Q_{L1} is the line cooling rate resulted from $\Delta n = 0$ transitions from the lowest level. Q_{L2} is the line cooling rate resulted from $\Delta n \geq 1$ transitions. Q_{L3} is the line cooling rate resulted from transitions following by dielectronic recombination. ΔE_{12k} is the energy of $\Delta n = 0$ transition of k ion, A_{21k} — its decay probability, C_{12k}, C_{21k} — its rates of excitation and de-excitation by electronic collisions, θ_{21k} — its escape factor. β_{knj} is the dielectronic recombination rate in tenuous plasma into the level n of k ion accompanied by the excitation $k+1n_0 \to k+1, j$, Γ_{knj} — the autoionization rate of the inverse process, θ_{k+1,jn_0} — the escape factor of $j \to n_0$ transition of $k+1$ ion.

f) Initial and boundary conditions

This paper is devoted to examining of hollow Z pinch discharge that implodes on a target placed on an axis of symmetry. It is assumed that a target is very massive and is described by undisturbed, nonconductive cylinder with radius $0.05r_0$, r_0 is the initial external radius of pinch column. The internal radius of initial gas jet is equal to $0.85r_0$. Then the initial conditions can be written as:

$$t = 0: \quad v = 0, \quad H = 0,$$
$$0.05r_0 \leq r \leq 0.85r_0: \quad \rho = 0 \quad T_i = T_e = 0,$$
$$0.85r_0 \leq r \leq r_0: \quad \rho = \frac{\rho_0}{1 - 0.85^2}, \quad \rho_0 = \frac{\rho_{L_0}}{\pi r_0^2}, \quad T_i = T_e = 1eV$$

where m_{L_0} is the linear mass density of Z pinch.

The boundary conditions are traditional:

$$r = 0.05r_0: \quad v = 0, \quad H = 0, \quad \frac{\partial T_e}{\partial r} = \frac{\partial T_i}{\partial r} = 0,$$
$$r = r_b: \quad p = 0, \quad H = \frac{2I}{r_b}, \quad \frac{\partial T_e}{\partial} = \frac{\partial T_i}{\partial r} = 0$$

Here I is measured in $CGSM$ unit ($1CGSM = 0.1A$).

It follows from experiment (Lebedev Physical Institute) that the current waveform without load can be expressed by relationships:

$$\varphi(t) = t/t_s I_0, \quad t \lesssim t_s,$$
$$\varphi(t) = I_0, \quad t \gtrsim t_s,$$

Here I_0 is the current amplitude, t_s is the time of switch opening. Because of the capacitor discharge time is much greater than t_s, the pinch inductance can be taking into account on the base of magnetic flux conservation. Then the current through the pinch may be written as:

$$I = \frac{\varphi(t) L_0}{L_0 + 2h_p \ln \frac{r_0}{r_b}}$$

h_p is the length of pinch column, L_0 is the external inductance ($h_p = 2$ cm, $L_0 = 40$ cm).

II. DYNAMICS OF Z PINCH IN A MULTICHARGED GAS

The calculations on above model pursued two main goals. At first, as it is mentioned in Introduction it is necessary to determine a possibility of Ne-like ions appearing for $I_0 = 0.5$ MA, $t_s = 200$ns. The installation with such parameters was suggested at Lebedev Physical Institute to construct photopumped X-ray laser on $3p - 3s$ transition of $[Ne]$-line ion. Secondly, it is interesting to investigate distributions of magnetohydrodynamics values in Z-pinch shell because of information on this question published elsewhere is very scanty.

Two series of calculations have been carried out. At the first series the filling material is krypton (Kr), the initial radius is equal to 2 cm, and the line density is changed in the range 3 – 24 mg/cm. At the second series the filling material is ferrous (Fe), the initial radius is equal to 1.5 cm, the line density is changed in the range 4 – 11 mg/cm.

a) Z pinch in Kr

The calculations show that for $r_0 = 2$ cm Ne-like ions Kr never appear for all questioned range of ρ_{L_0}. The character value of plasma temperature

$$T_0 = \frac{T_0^2}{2\pi N_0 r_0^2 k_B z_0}$$

$\pi N_0 A m_p = \rho_{L_0}$, with help of relation for character value of time

$$t_0 = \frac{r_0^2 \sqrt{2\pi N_0 A m_p}}{I_0}$$

can be written as

$$T_0 = \frac{A m_p}{k_B z_0} \left(\frac{r_0}{t_0}\right)^2 \qquad (7)$$

The character value of average ionization level z_0 increases with increasing of T_0. The value of t_0 has to be in accordance with t_s. Then it follows from (7) that maximum values of T_0, z_0 will achieve for the largest value of r_0. Hense for $r_0 < 2$ cm Ne-like ions of Kr will not also appear.

b) Z pinch in Fe

Because of in Kr [Ne]-like ions never appear it is very interesting to investigate the possibility of their appearance in another filling material with smaller nuclear charge. For this purpose Fe have been chosen. Computations shows that [Ne]-like ions of Fe appear in the wide range of initial values. Particularly for $r_0 = 1.5$ cm they appear for $4 \leq \rho_{L_0} \leq 11$ mg/cm.

Let us consider in detail the magnetohydrodynamic values distribution for typical variant $\rho_{L_0} = 5.87$ mg/cm, $\rho_0 = 1.5$ cm, $I_0 = 0.5$ MA. They are shown in Fig. 2–7. The profiles of plasma density are shown in Fig.2, of electron temperature — in Fig.3, of ion temperature —in Fig.4, of average ionization level — in Fig.5, of velocity — in Fig.6, of magnetic field in Fig.7. The time moments A – 205.7 ns, B – 206.3 ns, C – 206.9 ns correspond to imploding of the Z pinch shell to an axis, the moment D – 207.6 ns is the moment of the first stop of the external pinch boundary (the maximum compression), the moment E is the moment of the second stop of the external pinch boundary after following expansion.

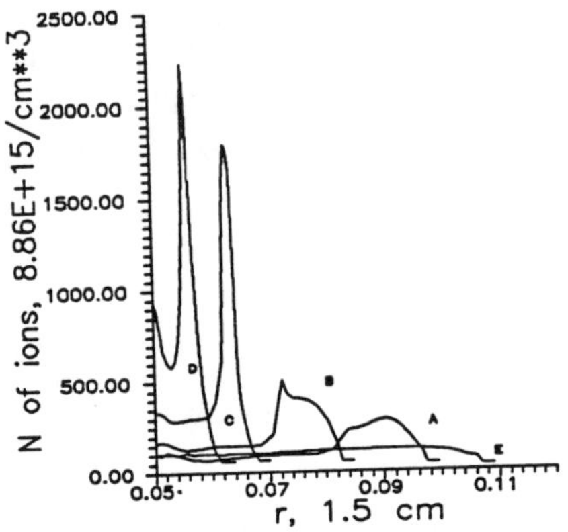

Fig. 2

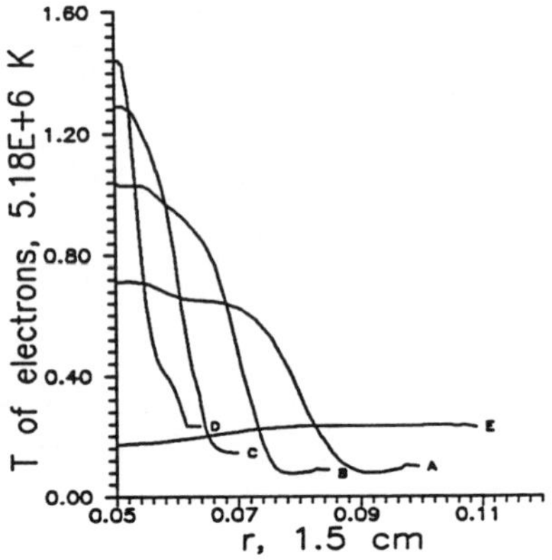

Fig. 3

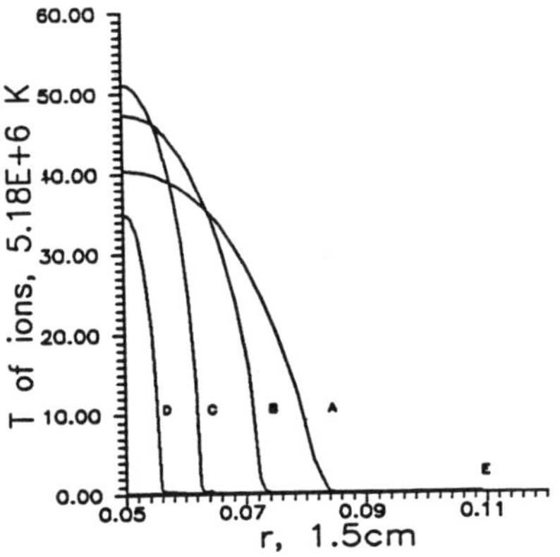

Fig. 4

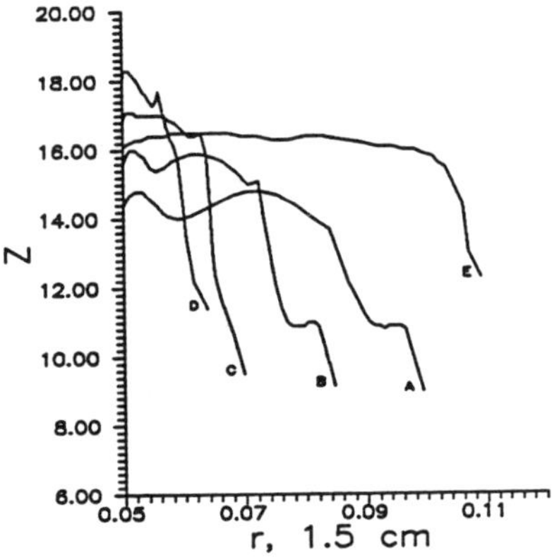

Fig. 5

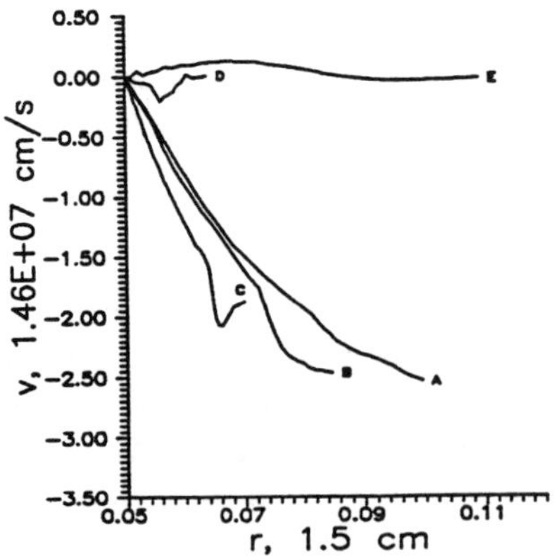

Fig. 6

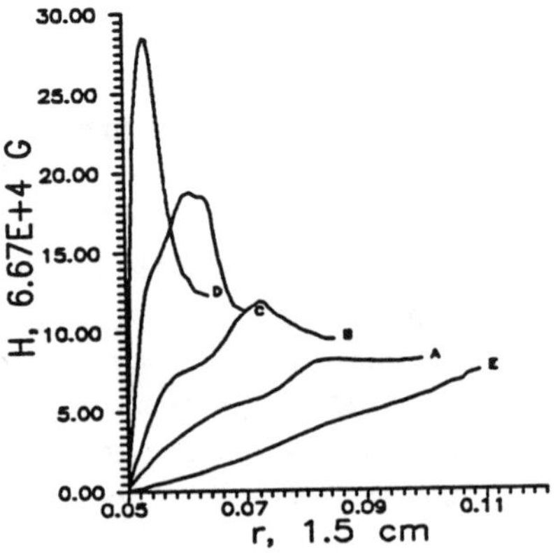

Fig. 7

As follows from Fig.2-7 the plasma parameters distributions are strongly nonuniform. On the compresses phase ($A-D$) the plasma density in the internal region is smaller than in the external one, but the temperature is distributed in the opposite manner. It is connected with two reasons, the first one is the sharpening of shock wave in approaching to an axis. The second reason is connected with the fact that denser plasma more strongly emits and it leads to further decreasing of the plasma temperature in dense region. The main characteristic of the plasma parameter distribution on imploding and compression phases is the creation of tenuous hot plasma in the internal region of the Z pinch shell: $T_e \sim 0.3-0.6$ KeV, $T_i \sim 5 \div 25$ KeV, $z \sim 14-18$, $N \sim (1-5) \cdot 10^{19}$ cm^{-3}, and dense, enough cold plasma in the external region: $T_e \simeq T_i \sim 20-30eV, z \sim 10-12, N \sim (0.2-2) \cdot 10^{19} cm^{-3}$. The mass of the internal region approximately equals to a helf of all pinch mass.

In the hotter plasma region the average ionization level z achives 16 between the moments B and C and does not decrease up to the moment E in spite of strong cooling and rarefining of plasma. In the external region z achieves 16 on the expansion phase and does not also decrease after that. z slightly exceeds 16 practically in the whole plasma and at the moment of the second pinch compression in spite of the energy of the pinch is approximately 4 times smaller than at the moment of the first compression. The effect of the z fixing slighty over 16 is connected with two reasons. At first, there is a rapid changing in ionization potential in transition from [Na]-like to [Ne]-like ion, so that the decreasing of Te by three times leads to decreasing of z only from $\simeq 17$ to $\simeq 16$.

Secondly, the reabsorbtion of resonance lines of [Ne]-like ions is much larger than the reabsorbtion of resonance lines of ions with a lot of valent electrons. Hence, the radiative cooling rate of plasma consisted mainly from [Ne]-like ions is weakend and it promotes to fixing of z slightly over 16.

It is interesting to note the distribution of magnetic field (Fig.7). During compression it achives the maximum in an interior point of the pinch shell rather than on the external boundary of Z pinch. It is connected with the fact that the characteristic time of magnetic field diffusion is much greater than the compression time. On the expansion phase the profile of H becomes monotonic as in traditional deuterium Z pinches.

Calculations show that ratio of radiation flux from pinch to flux of blackbody with average temperature of electrons are about $\sim 10^{-4} - 10^{-3}$, and distribution on ionization stages are much closer to coronal than to statistical.

REFERENCES

1. N.R.Pereira, J.Davis, J.Appl.Phys., 64, R1, (1988).

2. Proc. 2nd Int. Conference on High Density Z pinches (at Laguna Beach, CA, April 26-28) ed. N.R.Pereira, J.Davis and N.Postoker (New York: American Institute of Physics, 1989).

3. A.V.Gerusov, V.S.Imshennik, Sov. J. Plasma Phys., 11, 332 (1985).

4. A.V.Gerusov, Preprint N 171. Institute of Theoretical and Experimental Physics, (Moscow, CCRI Atominform, 1989)

5. D.E.Post, R.V.Jensen, C.B.Tarfer, et al., Atomic Data and Nuclear Data Tables, 20, 397, (1977).

6. J.P.Aprusese, J.Davis, D.Duston, K.G.Whitney, JQSRT, 23, 479, (1980).

7. L.M.Biberman, V.S.Vorob'ev, I.T.Jacubov, Kinetics of nonequilibrium lowtemperature plasma (Moscow, Nauka, 1982).

ONE-DIMENSIONAL MODELING OF DOUBLE GAS-PUFF IMPLOSION WITH ANOMALOUS RESISTIVITY CONSIDERATION.

I.V.Glazyring, N.G.Karlykhanov, A.A.Kondrat'ev, V.G.Nikolaev, and M.S.Timakova.

National Institute of Technical Physics, P.O.Box 245, Chelyabinsk-70, RUSSIA, E-mail: nio3@ch70.chel.SU.

The implosion dynamics of argon plasma is investigated using 1D two- temperature radiation MHD code ZARYA. A detailed configuration atomic model has been employed to calculate the line emission. The model includes effects of line opacity, Doppler effect due to ions motions and ionic state variation. The state populations are calculated using a set of time- dependent atomic rate equations for electron collisional excitation and de-excitation, dielectronic and radiative recombination, photoexcitation and others processes. The ionization state of the plasma is calculated self-consistently with line and continuum radiation transport. Perfect gas equations of state taking into account the energy of ionization are used for electrons and ions pressure and internal energy calculations.

Anomalous resistivity mechanism phenomenologically considered is low- hybrid drift instability model. Microturbulence effects are included by increasing the resistivity coefficients. Finally, a comparison is made with calculations using Braginskii transport coefficients. We find that numerical results taking into account anomalous collisions are more closed to the experimental data.

1. INTRODUCTION

In recent years a number of research groups have studied the effect of radiation on the evolution of a Z-pinch. Because the imploding Z-pinches are efficient X-ray sources, they have important applications in microlithography and microscopy[1]. Layered, hollow cylindrical plasma implosions have been investigated as an efficient mean of the radiation pumping of X-ray lasant plasma[2]. During the last time there have been many theoretical investigations concerned with describing the dynamics of double gas-puffs.

An accurate theoretical description of MHD processes taking into account radiation transfer and ionization dynamics is necessary for the optimization of pulsed-power system together with liner. To date, there are several codes studied such systems from a variety points of view[2-7]. Z-pinch experiments show that gas-puffs have produced longer radiation pulse width, lower ion densities at peak implosion, higher resistivity heating over classical 1D MHD results. These differences are probably connected with: 1) 2D hydromagnetic instabilities 2) microturbulence effects due to current-drive microinstabilities. When considering the parameters of Z-pinch, the effects of anomalous collisions are normally considered minimal. However, in low-density pinch plasmas, anomalous collisions due to plasma microinstabilities may exceed Columb collisions by order of magnitude. In our 1D consideration we discuss the role of anomalous resistivity. Enhanced resistivity typically results from ion-acoustic or lower-hybrid drift (LHD) microturbulence. The anomalous transport properties in high-temperature plasmas with applications to Θ-pinch, associated with various microinstabilities driven by cross-field currents have been investigated in review paper[8]. The increased resistivity resulting from the LHD instability has been studied by A.E.Robson[9]. The kinetic model has been developed to estimate the anomalous resistivity in a Z-pinch system in paper[10]. A simple and widely used model for LHD resistivity is included in our MHD Code.

2. ZARYA MODEL.

All calculations have been carried out using an extended version of the one-dimensional two-temperature Lagrangian code ZARYA[12]. Previously this model was used to study radiation effects in laser-fusion target experiments. We have modified the code to include some effects for application in numerical simulation of Z-pinch.

ZARYA includes the basic hydrodynamic equations of continuity, momentum and energy, Ohm's law and Maxwell equations. In conventional version of the code, differ from described one in Chapter 3, the relevant Braginskii transport coefficients[13] are used. The generator current is specified as a function of time and sets the outside boundary conditions on the magnetic field according $B_0=2I/rc$. Perfect gas equations of state taking into account the energy of ionization are used.

The state populations are calculated using a set of time-dependent atomic rate equations[14] for electron collisional ionization, excitation and de-excitation, three-body recombination, dielectronic recombination, photoionization, photoexcitation, radiative recombination and spontaneous emission. The ionization state of the plasma is calculated self-consistently with line and continuum radiation transport. The argon model contains all the ground states from the neutral to the fully ionized state. We consider 13 emission lines (6 principal resonance lines), each selected for their energetic contribution and diagnostic usefulness. They are performed from H-, He-, Li-like ionization states.

To include anomalous collisions model[8] for lower hybrid drift instability has been added:

$$\nu_{LH} = (\pi/2)^{1/2} (1+\omega_{pe}^2/\omega_{ce}^2)\cdot\omega_{LH}(u_d/u_i)^2 \frac{E_F}{n\cdot m_e\cdot u_d^2/2},$$

$$\omega_{LH} = \frac{\omega_{pi}^2}{1+\omega_{pe}^2/\omega_{ce}^2},$$

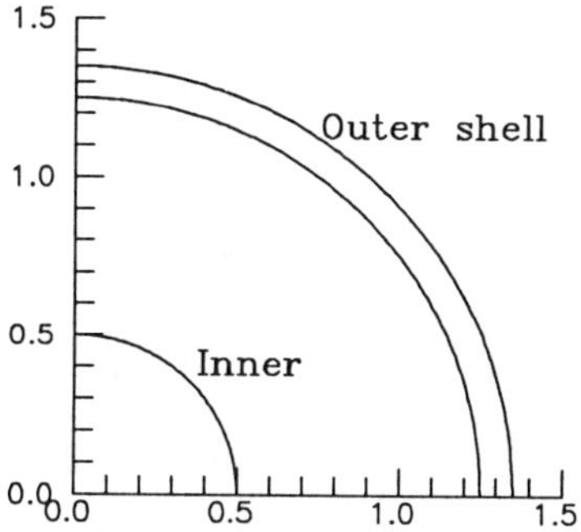

Fig.1. Double gas–puff configuration

$$E_F = \alpha \cdot n \cdot m_e u_d^2/2, \quad \alpha = 1 / 2(1+\omega_{pe}^2/\omega_{ce}^2).$$

where ω_{ce} - electron cyclotron frequency, ω_{pi} - ion plasma frequency, u_d - drift velocity, u_i - thermal ion velocity.

3. NUMERICAL RESULTS.

All simulations were performed for argon double gas-puff with parameters are chosen close to experiments being conducted at the High-Current Electronics Institute (Tomsk). Total mass of plasma is chosen according 0-dimensional model[15]

$$m\dot{r}_0^2 = 10^{-3}\ I_0^2\ t_f^2$$

where I_0(A) - current amplitude, t_f(s) - current rise time,

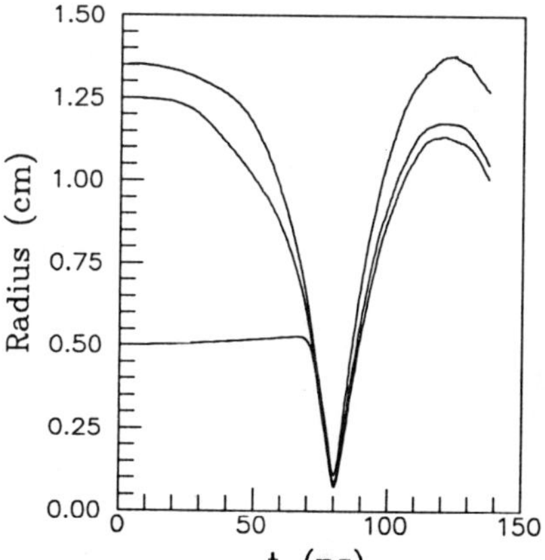

Fig.2a. Boundary radii vs. time. Classical Conductivity

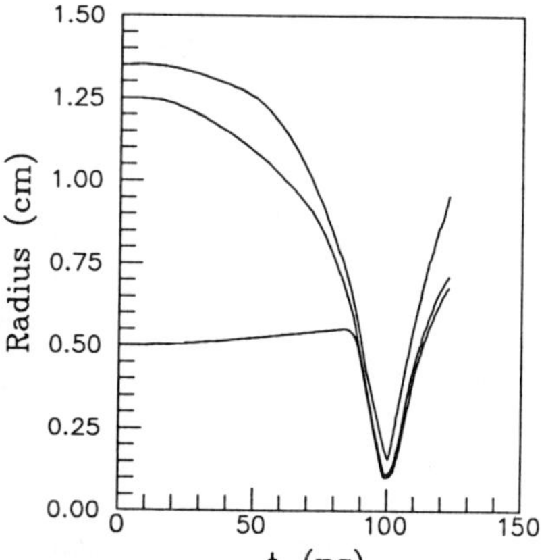

Fig.2b. Boundary radii vs. time Anomalous collisions

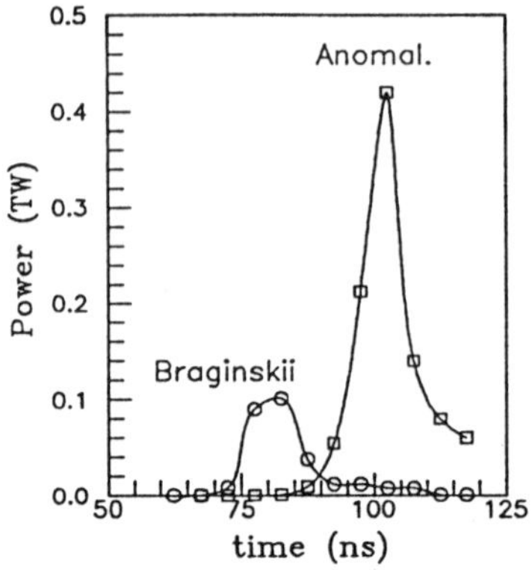

Fig.3. Radiation Power vs. time

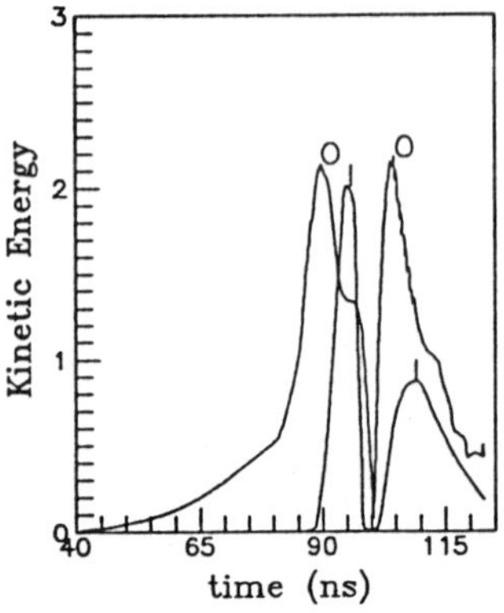

Fig.4. Kinetic energy history

m(g/cm) - load mass and r_0 (cm) - initial radius of outer shell. Equation gives the mass loading of 40 μg/cm so density of $2.6*10^{-5}$ g/cm (equal in outer and inner shells) and radius of inner column 0.5 cm, average radius of outer shell of 1.3 cm and thickness of 0.1 cm were chosen. The density of the background plasma between shells was set for numerical convenience and was taken $1.3*10^{-6}$ g/cm. The initial configuration is shown in Fig.1. The temperature was taken to be about 3 eV initially. The current driving the plasma dynamics was a sine wave with an amplitude of 1MA and a quarter of period of 100 ns.

Fig.2 shows plasma shells radii versus time calculated by ZARYA for two cases: Braginskii transport coefficients (a) and

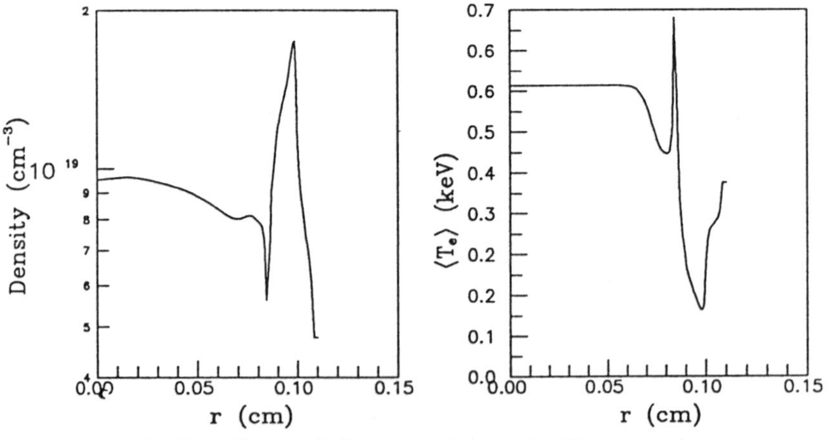

Fig.5. Density and Temperature profiles at stagnation

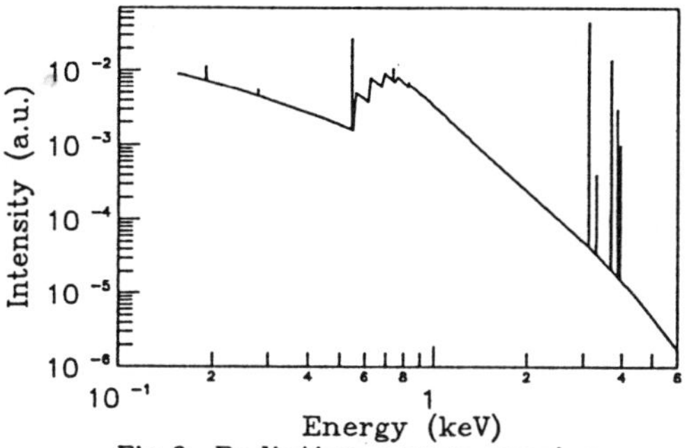

Fig.6. Radiation power spectrum at stagnation

anomalous coefficients (b). In the second case the liner radius reaches minimum value at the time which corresponds to maximum current time t_f, while first graph gives the minimum plasma radius time leads t_f.

Radiation power for anomalous (a) and classical (b) cases is shown in fig. 3. The total radiation output reaches 16 percent of the kinetic energy or about 5 kJ. For the first nanoseconds, the outer shell is moving radially inward. Skin depth in the classical case is smaller than in anomalous one and therefore the shock wave is stronger. As a result the outer shell stays thicker in case (b) at stagnation that one in the case (a). This shock, propagating through background plasma, heats ion component. The forward edge of the shell is heated to a few tens of eV in both cases. By the 70 ns (first case) and 80 ns (second case) the shock wave has reaches inner column. So that the case of enhanced resistivity is more adequate, further results will be given for the case. Thermal condition creates a nearly isothermal regime. In Fig. 4 the dependence kinetic energy versus time is shown. At the stagnation the density quickly reaches 10^{19} cm^{-3} region, the peak electron temperature exceeds. 600 eV (see Fig. 5). This region of very hot, dense plasma is radiating in the K-and L-shell. The emission spectrum at 100 ns for anomalous case is shown in Fig. 6. A half of total radiative power is carried by the H-like and He-like resonance lines.

4. CONCLUSIONS

The evolution of argon double gas-puff plasma has been studied from the context of anomalous resistivity addition to conventional 1D MHD code. Anomalous resistivity mechanism phenomenologically considered is low-hybrid drift instability model. Finally, a comparison is made with calculations using Braginskii transport coefficients. We find that numerical results taking into account anomalous collisions are more closed to the experimental data. Numerical results show lower ion densities at peak implosion, higher resistivity heating and relatively longer radiation pulse width over classical 1D MHD results.

ASKNOWLEGMENTS

The authors would like to thank Dr. V.I.Afonin for his initiative of investigating Z-pinch at Institute of Technical Physics, Professor A.D.Zoubov, for useful discussions of plasma implosion physics, Dr. V.E.Chernyakov for his interest in this investigation and for his initial work on the code, V.I.Ostashev for helpful assistance regarding line spectra of argon and to acknowledge the valuable comments made by Dr. V.A.Lykov. We would like to particulaly thank to Dr. R.B.Bakst, Dr. V.V.Loskutov and Dr. A.V.Luchinskii for stimulating discussions of the physics of double gas-puff implosion and liner radiation.

REFERENCES

1. N.R.Pereira and J.Davis, J.Appl.Phys. 64 (3), R1 (1988).
2. F.L.Cochran and J.Davis, Phys.Pluids B 2 (6), 1238 (1990).
3. V.V.Loskutov and V.I.Oreshkin, Pis'ma Zh. Eksp. Teor. Fiz, 52 (12), 1245 (1990).
4. Yu.A.Mankelevich, A.T.Rahimov, N.V.Suetin, S.S.Filippov, Fiz. plasmi, 16 (6), 664 (1990).
5. J.P.Chittenden, M.G.Haines, Phys.Pluids B, 2(8), 1889 (1990).
6. T.W.Hussey, M.K.Matzen, E.J.McGuire and H.E.Dalhed, J.Appl. Phys. 66(9), 4112 (1989).
7. S.V.Zakharov, A.G.Lisitsyn, 9th Int. Conf. on High-Particle Beams, BEAMS-92, Washington, DC, May 25-29, 1992, Abstracts, PH-17C, p.384.
8. R.C.Davidson and N.A.Krall, Nuclear Fusion 17(6), 1313 (1977).
9. A.E.Robson, Phys.Pluids B 3(6), 1461 (1991).
10. P.E.Pulsifer, K.G.Whitney, and J.W.Thornhill, 9th Int. Conf. on High-Power Part. Beams, BEAMS'92, Washington, DC, May 25-29, 1992, Abstracts, p.397.
11. R.E.Stewart, D.D.Deitrich, P.O.Egan, R.J.Forter, and R.J.Dukart, J.Appl.Phys. 61, 126 (1987).
12. E.N.Avrorin, A.I.Zuev, N.G.Karlykhanov et.al., Pis'ma Zh. Eksp. Teor. Fiz. 32, 457 (1980).
13. S.I.Braginskii, in Reviews of Plasma Physics (Consultants Bureau, New York, 1965), vol.1, p.205.
14. V.Yu.Politov, A.V.Lykov,M.K.Shinkarev, Int. Symp. on Laser-plasma Interactions, China, Abstracts, O-12, p.23.
15. R.B.Baksht, I.M.Datsko, A.V.Fedunin, Zh. Teh. Fiz., 3, 37 (1991).

ANALYTIC METHODS FOR RADIATIVE-COLLISIONAL PROCESSES IN PLASMAS WITH MULTIPLY CHARGED IONS

V.I.Kogan, A.B.Kukushkin, V.S.Lisitsa
Russian Research Center "Kurchatov Institute"
123182 Moscow Russia

ABSTRACT

The main results of a new, essentially classical method[1] for the analytic description of inelastic atomic processes, both radiative and collisional, caused by electrons of low and moderate energies, which are typical for most plasmas including hot dense plasmas with multiply charged ions (MCI), are presented. The results are to be used as building blocks in the numerical modeling of high-complexity plasmas.

1. INTRODUCTION

Numerical modeling of high-complexity plasmas with essential role of radiation processes faces the nessesity of a universal description for the elementary radiative and radiative-collisional processes in as simple as possible form. The present paper presents the results of a method[1] which deals with radiative and radiative-collisional processes, caused by electrons of low and moderate energies which are typical for most plasmas including hot dense plasmas with multiply charged ions (MCI).

The essence of the method[1], designated as "Kramers Electrodynamics" (KrED) and substantiated quantum-mechanically, is that the spectral dependence of the probability of the excitation of a quantum oscillator of frequency ω by an electron moving (quasi)classically in an attractive central potential, appears to retain its classical structure (Fourier-transform of the respective electron classical trajectory), with increasing ω and, hence, energy exchange $|\Delta E|=\hbar\omega$, up to very large values of the inelasticity parameter $\hbar\omega/E$ (E is the electron initial energy). This holds for the excitation of both electromagnetic field oscillator (i.e., emission of a real photon) and "atomic" oscillator (i.e., excitation of an atom by the Fermi equivalent photon "emitted" by the exciting electron).

The method[1] allows a substantial universalization of description for radiative-collisional electron-atomic processes, both elementary and kinetic ones. The method is applied[1] to the treatment of the following processes: Bremsstrahlung radiation of the electrons on many-electron atoms and ions; photorecombination and line radiation; radiative cascade between Rydberg atomic states;

polarization-induced radiation for free-free, free-bound and bound-bound transitions, both resonant and non-resonant ones; multiphoton processes.

In present paper we give an outline of the method[1] of classical description for the spectra, emitted by quasiclassically moving particle in the field of the MCI (this pertains to purely radiative processes, i.e. Bremsstrahlung, photorecombination), and its extension to the emission of equivalent photons (this pertains to collisional and radiative-collisional processes). The success of the methods[1] is illustrated with some analytic results which are of interest to the problems of radiation hydrodynamics of plasmas with the MCI.

2. QUALITATIVE DESCRIPTION OF THE CLASSICAL CHARACTER OF THE METHOD

Inelastic transitions, both radiative and (dipole) radiationless, induced by electron-atomic collisions may be approximately described classically for low and moderate electron energies E (i.e. in the domain of their quasiclassical motion for the entire class of (attractive) potentials $|U(r\to 0)| \approx r^{-\nu}$, $\nu<2$) irrespectively of the degree of inelasticity $\hbar\omega/E$.

The fact is that those transitions may be approximately represented in the form of the excitation of a quantum oscillator ω ("real" radiative or "equivalent" atomic one) by a classical current, the latter being formed by the attractive force center. The corresponding (mean) energy exchange is known to be a classical one. Hence, the classical describability of a (dipole) inelastic transitions depends only on the approximate non-distortion of the electron trajectory (more exactly, only its "radiating" section) for the transition under consideration.

The rate of energy exchange between classical current and oscillator contains an integral over the structure

$$\cos\left(\int_{r_0}^{r} [\omega \pm \omega_{rot}(r')] \frac{dr'}{\sqrt{\frac{2}{m}[E+|U(r')|-M^2/(2mr'^2)]}}\right)$$

where r_0 is the turning point of the electron trajectory, $\omega_{rot}(r)=M/mr^2$ is the local angular velocity of its rotation. Therefore, inelastic transition ΔE is caused, roughly speaking, by electron trajectory which has the (maximum) angular velocity of revolution ("rotation")

$$\omega_{rot}(r_0) \sim \omega \equiv \Delta E/\hbar \qquad (1)$$

The exact classical relation between the trajectory parameters at r_0 and the angular momentum M is (E^{kin}_{max} - maximum electron kinetic energy)

$$\omega_{rot}(r_0)/E^{kin}_{max}(r_0) = 2/M \qquad (2)$$

Multiplying eq.(2) by $\hbar$ and allowing for Eq.(1), one has

$$\frac{\hbar\omega}{E^{kin}_{max}(r\ )} \sim \frac{\hbar}{M} \sim \frac{1}{\ell} \qquad (3)$$

Thus, in domain of quasiclassical motion, $\ell \gg 1$, the actual, "dynamical" recoil of the electron in the radiation emission act is automatically small. This is, as can be easily proved, equivalent to the approximate conservation of the entire "radiating" section of electron trajectory in the transition act irrespectively of the value of "kinematic" inelasticity $h\omega/E$.

The condition for the existence of the wide domain of quasiclassical motion, $\ell \gg 1$, is the inequality

$$\tilde{\ell} \equiv \frac{mva}{\hbar} = \frac{a}{\lambda} \gg 1,$$

λ is the de Broglie electron wave length at infinity, "a", its scattering amplitude for an angle ≈ 1, defined by the equality $|U(a)|=E=\frac{mv^2}{2}$. For the class of potentials under consideration (see above), the width of the domain

$$1 \ll \ell \ll \tilde{\ell} \qquad (4)$$

(classical and strongly curved electron trajectories) increases with decreasing v (e.g. for the Coulomb case $\tilde{\ell} = Ze^2/\hbar v$).

The interval (4) of angular momenta l is in one-to-one correspondence with the interval of rotation velocities $\omega_{rot}(r_0)$ (and hence, due to eq.(1), with the interval of radiation frequencies ω both for real and Fermi equivalent photons)

$$\tilde{\omega} \ll \omega \ll \omega^{*} \qquad (5)$$

where $\tilde{\omega} \equiv v/a$, $\omega^{*} = \omega_{rot}(r_0)\Big|_{l=1}$. (For the Coulomb case $\tilde{\omega} = mv^3/Ze^2$, $\omega^{*} = Z^2me^4/\hbar^3$, i.e. the Bohr frequency).

The domain (5) is just the domain of "Kramers Electrodynamics"[1]. For this domain the "radiating" section of the electron trajectory around r_0 lies entirely within the (classically accessible just for the electron) space region r

$$|U(r)| \approx |U(r_0)| \gg E\ , \qquad (6)$$

wherein the electron is strongly accelerated (cf.eq.(3)) i.e. its energy constant E (and, hence, the "kinematic" inelasticity $\hbar\omega/E$) is eliminated ("switched-off") from the radiation emission mechanism.

3. BREMSSTRAHLUNG RADIATION (BR) ON MANY-ELECTRON ATOMS AND IONS

The analytic description for the Bremsstrahlung radiation (BR) of quasiclassical electrons on many-electron atoms and ions is shown[2] to be described within the framework of classical method. Analytic simplification of general expression for the classical BR spectrum of an electron in the field of many-electron atom (or ion) described by the Thomas-Fermi (TF) model,

$$U(r)= -\frac{Ze^2}{r}\,\chi\left(\frac{r}{a_{TF}},\ q_i\right);\quad a_{TF}=\frac{b\hbar^2}{me^2Z^{1/3}};\quad q_i=\frac{Z_i}{Z},\qquad (7)$$

gives the following final result[3] for the Gaunt-factor :

$$g(\omega,\varepsilon,q_i) = \frac{\sqrt{6}}{\pi}\, q_i^{\mu}\ \ln\left\{\exp\left[\frac{\pi}{\sqrt{6}}\, q_i^{-\mu}\ \max\left(q_i^2, g_{at}(\Omega,\varepsilon)\right)\right] + \right.$$

$$\left. + \left[4\varepsilon^{3/2}/\gamma\Omega q_i\right]^{q_i^{2-\mu}/\sqrt{2}}\right\},\qquad \mu= 1/2\ (1- \ln\sqrt{\varepsilon}),\qquad (8)$$

Here Z and Z_i are atomic number and ionic charge, respectively, $b = 0.885$, $\chi \equiv \chi(x,q_i)$ is the TF screening function, the dimensionless variables

$$\Omega = \omega\left(\frac{ma_{TF}^3}{2Ze^2}\right)^{1/2} = \left(\frac{b^3}{2}\right)^{1/2}\frac{\hbar^3}{me^4}\,\frac{\omega}{Z} = 21,6\ \frac{\hbar\omega(\mathrm{keV})}{Z}\qquad (9)$$

$$\varepsilon = \frac{Ea_{TF}}{Ze^2} = \frac{b\hbar^2}{me^4}\,\frac{E}{Z^{4/3}} = 32,6\ \frac{E(\mathrm{keV})}{Z^{4/3}}\qquad (10)$$

are, in fact, of purely classical nature, since the $\hbar$ entering them is simply a parameter of a fixed potential.

The function g_{at} in Eq.(8) is the Gaunt-factor[2,3] for the BR-spectrum on the neutral atom:

$$g_{at}(\Omega,\varepsilon)=\begin{cases} g_0(\varepsilon)+\left(\Omega/\Omega^*(\varepsilon)\right)^{\beta(\varepsilon)}\left[g_2(\varepsilon)-g_0(\varepsilon)\right], & \Omega<\Omega^*(\varepsilon), \\ g_{rot}(\Omega,\varepsilon), & \Omega>\Omega^*(\varepsilon) \end{cases}\qquad (11)$$

where

$$\Omega^*(\varepsilon) = 2.4\ \sqrt{\varepsilon}\qquad (12)$$

$$\beta(\varepsilon) = 1 + 0.006\ \varepsilon^{-1} = 1 + 10^{-3}\ (\Omega^*/\varepsilon)^2\qquad (13)$$

Here the functions $g_0 \equiv g(\omega=0)$ and g_2 also admit analytic presentation, e.g. for $0.01 \le \varepsilon \le$ few units,

$$g_2(\varepsilon) \equiv g_{rot}\left(\Omega^*(\varepsilon), \varepsilon\right) \qquad (14)$$

$$g_{rot}(\Omega,\varepsilon) = 3\,(\chi - x\chi')^2 \left[2 + \frac{\chi - x\chi'}{\chi + \varepsilon x}\right]^{-1} \qquad (15)$$

where $\chi' \equiv \partial\chi(x,q_i)/\partial x$, and $x \equiv x\,(\Omega,\ \varepsilon)$ is the root of

$$\frac{\chi + \varepsilon x}{x^3} = \Omega^2 \qquad (16)$$

For practical use the following table may be of interest.

Table 1. Universal classical functions $g_o(\varepsilon)$ and $g_2(\varepsilon)$.

i	g_o	g_2	i	g_o	g_2
0	0.537(-2)	0.220	7	0.260	1.16
1	0.700(-2)	0.450	8	0.360	1.16
2	0.152(-1)	0.751	9	0.478	1.16
3	0.330(-1)	1.05	10	0.613	1.16
4	0.651(-1)	1.15	11	0.762	1.16
5	0.113	1.16	12	0.924	1.16
6	0.177	1.16	13	1.10	1.16

Here, the values of the argument are $\varepsilon = 10^{\alpha(i)}$ with $\alpha(i)= -2 + i/5$, and $a(b) \equiv a\ 10^b$.

Classical description for the BR on neutral atom, formula (11), and its angular-distribution counterparts[2], are in good agreement with the results of quantum numerical calculations[4-6] and experimental data[7].

Formula (8) describes the results of yet cumbersome exact classical calculation with accuracy not worse than ~ 35%, the maximum inaccuracy being comitted within a sufficiently small (obviously, intermediate) domain of Ω at small values of ε and q_i (i.e. for small values of the Gaunt-factor). Formula (8) differs from the quantum numerical results[8] in their quasiclassical region, $0.1 \le \varepsilon \le 1$, by $\le$ 30% .

The criterion of electron quasiclassical motion varies from the well-known condition $Z_i e^2/\hbar v_e \gg 1$ for the BR in the Coulomb field of the fully stripped ion to the condition of small[2] dimensionless energy ε, Eq.(10), for the neutral TF-atom (for detail see[1,2]).

Thus, the approximate analytic description (8) for the BR spectra appears to be properly substantiated, practically tractable and sufficiently accurate to be used, e.g., in the calculations of radiation kinetics in a plasmas in a wide range of parameters.

4. METHOD OF EQUIVALENT PHOTONS FOR RADIATIVE-COLLISIONAL PROCESSES

The radiative-collisional processes appear to be a wide domain for the application of the method[1].

The applicability of the KrED is demonstrated[1] for the description of collisional (e.g.excitation of ions by electron impact) and radiative-collisional (e.g. dielectronic recombination (DR) and polarizational radiation (PlR)) processes in plasmas. The most natural domain for application the method[1] is the physics of multicharged ions (MCI) . The description of these processes is obtained within the following framework:

(i) According to Fermi concept[9] of equivalent photons (EPHs) the electromagnetic field produced by an external particle (e.g. electron) at the MCI location may be interpreted as a flux of equivalent photons incident on the MCI .It may be shown that such a description is applicable provided the dipole approximation for the interaction between bound electron of the MCI and incident electron of plasma is valid. The latter approximation treats universally all the processes of energy loss by incident electron (either due to radiation emission during collision with ion or due to inelastic radiationless collision with ion) as the processes of the emission of (respectively ,real or equivalent) photons, the probability of both processes being determined by the dipole matrix element for the corresponding inelastic (radiative or radiationless) transition of incident electron.

(ii) The spectral intensity distribution of the EPHs can be decsribed on the basis of classical radiation theory. In this case the intensity of the EPH flux is determined simply by the Fourier-transforms of electron classical trajectory.

For the Fermi method to be applicable to the processes involving the MCI, the effective distances r_{eff} which are responsible for the main contribution to the cross-section of inelastic collision of incident electron with the MCI should be much greater than the characteristic size of bound electron orbit in the MCI.This condition is satisfied especially well just for the MCI. This may be illustrated for the process of excitation the $\Delta n=0$ transitions. The electron orbit size r^{bound}_{eff} is of the order 1/Z (in atomic units), transition energy ΔE for $\Delta n=0$ transitions in MCI is typically of the order Z, and the values of r^{inc}_{eff} for the corresponding cross section can be estimates as

$$r^{incident}_{eff} \sim r_{\omega=\Delta E/\hbar} \sim (Z/\Delta E^2)^{1/3} \sim Z^{-1/3} \gg Z^{-1} \equiv r^{bound}_{eff} \quad (17)$$

The approach (i),(ii) makes it possible to treat a number radiative-collisional processes (see Table 2).
- the excitation of the ion by electron impact as the absorption of the EPHs by this ion ;
- the same excitation with subsequent re-emission of real photon as the resonance fluorescence of the EPHs;
- the dielectronic recombination as the resonance fluorescence of the EPHs, which results in the recombination of incident electron onto the ion.

The essential advantage of this method comes from the application of available results for purely radiative processes to the description of radiationless both collisional and radiative- collisional processes.

The processes discussed above are of resonant character with respect to the absorption of the EPHs by the ion. Except for these processes, there are non-resonant processes as well, for which the intermediate state of a two-step "absorption-reemission" process is not real and consequently is not obeying the energy conservation law (this state is formed by the process of virtual excitation with the energy $E\neq\hbar\omega_0$, where ω_0 is the frequency of resonant transition). These non-resonant processes known as polarization radiation[10] can be treated as the non-resonant scattering of the EPHs by the ion and are determined by dynamical polarizability $\alpha(\omega)$ of the ion for non-resonant frequencies.

Table 2. Interrelation between radiative and radiative-collisional processes

Type of Transition	Processes with Real Photons	Processes with Equivalent Photons	
		line core	line wing
free-free	Bremsstrahlung radiation	Impact excitation	Polarizational Bremsstrahlung radiation
free-bound	Photorecombination radiation	Dielectronic recombination	Polarizational recombination
bound-bound	Line radiation	Dielectronic bound-bound transitions	Polarizational bound-bound transitions

5. NEW PROCESS OF ATOMIC STATE RELAXATION: POLARIZATION-INDUCED BOUND-BOUND TRANSITIONS

The concept of the EPHs allows[11] to obtain a universal relationship between the rates of emission the real photons (I_{real}) and equivalent ones (I_{EPH}), i.e.

between second and fourth columns of the Table 2:

$$\frac{I_{EPH}}{I_{real}} = \left[\frac{m_e \omega^2 \alpha(\omega)}{Z_C e^2}\right]^2 \equiv R(\omega) \qquad (18)$$

where Z_C is the charge of the ion (or ion core, for the lower row of the Table 2).

Table 2 indicates a new process of polarizational (or better to say, polarization-induced) bound-bound transitions. Equation (18) allows to estimate the rate of this process for the case of highly excited ion. Here, $\alpha(\omega) \sim f/2\omega_0(\omega-\omega_0)$, where f is the oscillator strength of closest radiative transition in the core of frequency ω_0, we have (n is the principal quantum number of excited electron)

$$R(\omega) \cong (\omega_0 f/ 2Z_c \Delta\omega)^2 \cong (\omega_0 f/2)^2 (n/Z_c)^6 \qquad (19)$$

A typical value of the factor $\omega_0 f/2$ for transitions without change of principal quantum number in the core of Li- , Be-like and more complex ions varies within the range 0.1 - 1.0 a.u. (e.g., it equals approximately to 0.2 for transitions $2s^2$-2s2p in ions N^{+3}, O^{+4}, Ne^{+6}, etc). Therefore, for the states with $n \geq 2Z_c$, the polarization-induced bound-bound transitions dominate the radiative ones with the same frequency.

The importance of polarization-induced bound-bound transitions is illustrated here for the ion Ne IV. The level which corresponds to the double-excited state $2s2p^3 3l$ is located in the region of Rydberg spectrum of one-electron excited states $2s^2 2p^2 n'l$, namely, between levels $n' = 9$ and $n' = 10$. In this case $\Delta\omega/\omega \sim 10^{-2}$, and, finally, $R \geq 30$. It follows that the population of the level $2s2p^3 3l$ by transitions from those upper-lying levels which are close to double-excited level $2s2p^3 3l$ (in this case, these are the $2s^2 2p^2 10l$ and $2s^2 2p^2 9l$ levels) will be produced effectively by the polarizational transition (i.e. transitions via a virtual double-excited state, for level $2s^2 2p^2 10l$ indicated in Fig.1) rather than by a direct radiative transition. This fact, in turn, can influence the kinetics of the $2s2p^3 3l$ level population, especially in case when this level is populated predominantly by transitions from highly excited states (in the case considered, from the level $2s^2 2p^2 9l$ and higher levels). The latter situation is typical for, e.g., recombining non-equilibrium plasmas with the MCI.

REFERENCES

1. V.I.Kogan, A.B.Kukushkin, V.S.Lisitsa, Phys.Reports, 213, 1 (1992).
2. V.I.Kogan, A.B.Kukushkin, Sov. Phys. JETP, 60, 665 (1984).
3. V.V.Ivanov, A.B.Kukushkin, and V.I.Kogan, Sov.J.Plasma Phys. 15, 892 (1989).
4. C.M.Lee , L. Kissel, R.H.Pratt and H.K.Tseng, Phys. Rev. 13A, 1714 (1976).
5. V.P.Zhdanov, Sov. J. Plasma Phys. 4, 71 (1978).
6. H.K.Tseng, R.H.Pratt and C.M.Lee, Phys.Rev. 19A, 187 (1979).
7. M. Semaan and C.Quarles, Phys. Rev. 26A, 3152 (1982).
8. C.M.Lee, R.H.Pratt and H.K.Tseng, Phys. Rev. 16A 2169 (1977).
9. E. Fermi, Zs. f.Physik. 29 315 (1924).
10. Polarization Bremsstrahlung, Eds. V.N.Tsytovich and I.M.Oiringel (Plenum Press, London, 1992).
11. A.B.Kukushkin, V.S.Lisitsa, In Ref.10, Chapter 11, p.261.

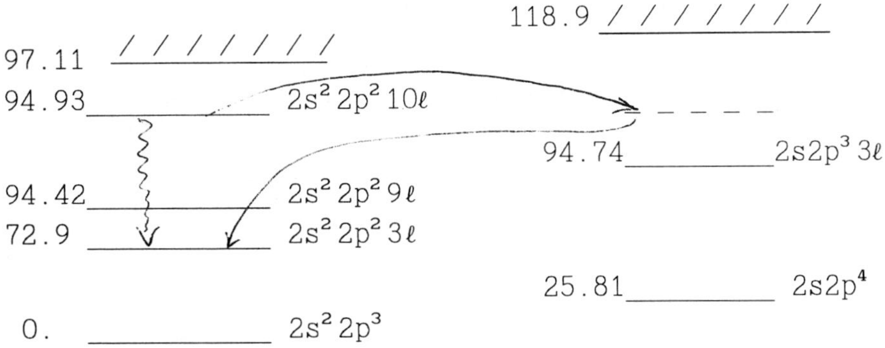

Fig.1 The scheme of ion Ne IV levels corresponding to one- and two-electron excitation (excitation energies in keV are indicated). The radiative and polarization-induced channels of decay of the $2s^2 2p^2 10l$ state are shown.

A QUALITATIVE MODEL FOR THE ENHANCED-RATE PROPAGATION OF MAGNETIC FIELD ALONG THE ANODE

A. B. Kukushkin
Russian Research Center "Kurchatov Institute"
123182 Moscow Russia

ABSTRACT

A qualitative model is proposed for the enhanced-rate propagation of the magnetic field along the anode. For infinite inertia of ions and homogeneous electron density, the scaling laws are obtained for the characteristic dimensions of magnetic field front propagating along the anode.

1. INTRODUCTION

The dynamics of the enhanced-rate propagation of the magnetic field along the anode is of increasing interest to the problems of transporting and accumulating the energy in Z-pinches, plasma switches etc. The present paper proposes a simple model for this phenomenon, which is based on (i) hydrodynamic character of the transfer of magnetic field by rotH-induced current, and (ii) braking the magnetic field flow perpendicular to the anode.

The model is in agreement with the results of 2D numerical modeling[1] and 1D analytic description[2].

2. SCALING LAWS FOR THE FRONT OF MAGNETIC FIELD PROPAGATION

Let us consider the dynamics of the magnetic field in the medium with inhomogeneous electron density and infinite inertia of ions for the following boundary and initial conditions: the anode lies in the region $z > 0$ and at time $t = 0$ magnetic field $H \equiv H_y$ of the value H_0 occupies the region $\{x < 0, z < 0\}$.

The dynamics of the magnetic field is described by equation (see, e.g.,[3])

$$\frac{\partial H}{\partial t} - \mathrm{rot}[\mathbf{u},\mathbf{H}] = -\,\mathrm{rot}\,(D_\sigma \mathrm{rot}\mathbf{H}), \tag{1}$$

$$\mathbf{u} = -\frac{c}{4\pi e n_e}\mathrm{rot}\mathbf{H} \tag{2}$$

where u is the electron current velocity, $D_\sigma = c^2/4\pi\sigma$, and σ is the Spitzer electric conductivity. The second term in the left-hand side of Eq. (1) describes the transfer of magnetic field by rotH-induced current.

Far from the anode the propagation of magnetic field in x-direction is determined by the finite σ. Here, the time dependence of characteristic electron current velocity scales as

$$u_z \equiv u_\perp \simeq \frac{j}{e n_e} \simeq \frac{c}{4\pi e n_e}\frac{H_0}{\Delta x_{\mathrm{dif}}(t)}, \tag{3}$$

where $\Delta x_{\mathrm{dif}} = (2D_\sigma t)^{1/2}$.

The transfer of magnetic field with velocity (3) leads to braking the magnetic field at the anode for $x>0$ due to gradient of electron density. Accumulation of magnetic field near the anode gives the gradient along x-direction (i.e. perpendicularly to the anode), which produces $j_z=(c/4\pi)\text{rot}H = (c/4\pi)\partial H_y/\partial x$. This current transfers magnetic field along the anode. The hydrodynamic character of left-hand side of Eq. (1) results finally in the fact that the flow of magnetic field parallelly to its initial boundary $x=0$ simply turns its velocity vector to z-direction, i.e. parallelly to the anode,

$$(u_{\|}(t))_{\text{near}} = (u_{\perp}(t))_{\text{far}} \tag{4}$$

with the distribution of magnetic field within the transition region between undisturbed value H_0 and zero value being formed exclusively by diffusion regime described by the right-hand side of Eq. (1), $\Delta z_{\text{eff}}(t) = \Delta x_{\text{dif}}(t)$.

Equation (4) enables us to find the rate of propagation along the anode:

$$\Delta x_{\text{eff}}(t) \cong \int_0^t u_{\|} dt \cong \frac{c}{4\pi e n_e} \frac{H_0\sqrt{2t}}{\sqrt{D_\sigma}} \cong \sqrt{2D_{\text{eff}}\, t} \tag{5}$$

with

$$D_{\text{eff}} = \left(\frac{cH_0}{4\pi e n_e}\right)^{1/2} \frac{1}{D_\sigma} \propto (\omega_H \tau_{ei})^2 D_\sigma \tag{6}$$

where $\omega_H = eH_0/mc$, and τ_{ei} is the characteristic time for electron-ion pair collisions.

Equation (6) exactly coincides with the results of 1D analytic model[2] and 2D numerical modeling[1].

The value of the layer width, $\Delta z_{\text{eff}}(t) = \Delta x_{\text{dif}}(t)$, satisfy the boundary condition at the anode surface, $E_z=0$, which gives $j_x \equiv j_{\|} = (1/\omega_H \tau_{ei}) j_{\perp}$, $(j_{\perp} \equiv j_z)$, and appears to be the condition of closing the current $j_{\|}$ at the anode:

$$\Delta z_{\text{eff}} \sim \omega_H \tau_{ei} \Delta x_{\text{eff}} \sim \Delta x_{\text{dif}} \cong (2D_\sigma t)^{1/2} = \Delta x_{\text{dif}}(t). \tag{7}$$

The above qualitative model for the enhanced-rate propagation of the magnetic field along the anode may be also applied to the case of finite inertia of ions and inhomogeneous media.

The author is grateful to V. V. Vikhrev and O. Z. Zabajdullin for helpful discussions.

REFERENCES

1. V. V. Vikhrev, O. Z. Zabajdullin, Fizika Plazmy (Sov. J. Plasma Phys.) (to be published).
2. A. V. Gordeev, A. V. Grechikha, and Ja. L. Kalda, Fizika Plazmy (Sov. J. Plasma Phys.) **16**, 95 (1990).
3. V. V. Vikhrev, S. I. Braginskii, In: Reviews of Plasma Physics, Ed. M. A. Leontovich (Consultants Bureau, N.Y., 1986), v. 10, p. 425.

MHD SIMULATION OF DEUTERIUM-FIBER-INITIATED Z-PINCHES WITH TWO-FLUID EFFECTS

Peter Sheehey
Department of Physics, University of California
Los Angeles, California 90024 USA

Irvin R. Lindemuth
Los Alamos National Laboratory
Los Alamos, New Mexico 87545 USA

ABSTRACT

Two-dimensional "cold-start" resistive MHD computations of formation and evolution of deuterium-fiber-initiated Z-pinches [1,2] have been extended to include separate ion and electron energy equations and finite-Larmor-radius ordered terms[3,4]. In the Ohm's Law (magnetic field evolution) equation, Hall and diamagnetic pressure terms have been added, and corresponding terms have been added to the energy equations. Comparison is made of the results of these computations with previous computations and with experiment[5].

INTRODUCTION

Deuterium-fiber-initiated Z-pinch experiments, with current peaks up to about 600 kA, reported very long-lived, compact plasmas showing little indication of disruption by m=0 "sausage" or m=1 "kink" instabilities[6–8]. Second-generation machines[5,9,10] have been designed to reach the Pease-Braginskii current (about 1.4 MA for deuterium)[11–13], in the hope that fusion conditions could be approached, if the earlier observed "anomalous stability" were to hold. Discharges at greater than half Pease-Braginskii current (700-900 kA)[5,9], however, have shown stronger indication of expansion and m=0 instability growth. We have done extensive computational modeling[1,2] of low- and high-current deuterium-fiber-initiated experiments on the Los Alamos machines HDZP-I[7,8] and -II[5,8]. Here we present some results of the extension of our computational model to include: 1) separate ion and electron energy equations; 2) Hall and diamagnetic pressure terms in the magnetic field evolution equation.

First, a brief discussion of the scope of the computational problem: For a significant fraction of its lifetime, a fiber-initiated Z-pinch plasma meets classical requirements for description as a magnetohydrodynamic (MHD) fluid (e.g., ion-ion collision time $\ll$ ion thermal transit time)[2,14]. Furthermore, the consistent (but so far unexplained) observation that three-dimensional behavior (e.g., growth of m=1 "kink" instabilities) is virtually absent in such experiments (diagnostic images are highly symmetric about the axis)[5–9] gives us some confidence in the results of MHD simulation in only two dimensions. This is fortunate, because the inclusion of vital experimental and physics details discussed below would at present make full 3-d simulation prohibitively expensive.

Linear ideal MHD stability theory for a Z-pinch plasma in general predicts instability to "sausage" (m=0) and "kink" (m=1) modes[15]. However, the growth rate of such instabilities is dependent on radial pressure profiles of the plasma; indeed, "Kadomtsev" profiles exist which are m=0 stable. Any actual experiment is likely to move through several non-ideal regimes (e.g., resistive MHD, Hall MHD), as density, temperature, etc., vary during the discharge; nonlinear effects, as well, are likely to be encountered.

Therefore, it is highly desirable to simulate such experiments starting from time zero (zero current, frozen fiber) if possible, in order for realistic plasma profiles to form and develop linearly/nonlinearly, as they will. Energy terms such as thermal conduction, Joule heating and radiation are clearly going to be important. The plasma column must be free to develop as if in vacuum, without the influence of an unrealistically confining or insufficiently resolved grid.

METHOD

The computations reported here are an extension of previous one-[16] and two-dimensional[1,2] deuterium-fiber-initiated Z-pinch modeling. We have used an alternating-direction-implicit numerical method, utilizing Newton-Raphson-like iteration to deal with nonlinear quantities, to solve the two-dimensional (r,z) MHD equations for mass density, specific internal energy, azimuthal magnetic field, and perpendicular velocity $(v_r,\ v_z)$[17]:

$$\frac{\partial\rho}{\partial t} + \nabla\cdot(\rho\vec{v}) = 0$$

$$\frac{\partial(\rho\vec{v})}{\partial t} + \nabla\cdot(\rho\vec{v}\vec{v}) + \nabla p - \vec{J}\times\vec{B} = 0$$

$$\frac{\partial(\rho\epsilon)}{\partial t} + \nabla\cdot(\rho\vec{v}\epsilon) + p\,\nabla\cdot\vec{v} - \nabla\cdot(\kappa_\perp\nabla T) - \eta\vec{J}^2 + Q_{rad} = 0$$

$$\frac{\partial\vec{B}}{\partial t} + \nabla\times(-\vec{v}\times\vec{B} + \frac{\eta}{\mu_0}\nabla\times\vec{B}) = 0$$

where ρ is mass density, $\vec{v}$ is velocity, $\vec{B}$ is magnetic field, $\vec{J}(=\nabla\times\vec{B}/\mu_0)$ is electrical current density, ϵ is specific internal energy, p is pressure, T is temperature, Q_{rad} is radiative energy loss, η is electrical resistivity, and $\kappa_\perp$ is (perpendicular) thermal conductivity. To obtain the equation of state (specific energy and pressure), the ionization level, the radiative energy loss, and the resistivity, we use the Los Alamos SESAME[18] tabulated atomic data base computer library. Thermal conductivity follows Braginskii[19].

For two-temperature runs, the above total energy equation is replaced by separate ion and electron energy equations:

$$\frac{\partial(\rho\epsilon_e)}{\partial t} + \nabla\cdot(\rho\vec{v}\epsilon_e) + p_e\,\nabla\cdot\vec{v} - \nabla\cdot(\kappa_{\perp e}\nabla T_e) - \eta\vec{J}^2 + Q_{rad} + Q_{ei} = 0$$

$$\frac{\partial(\rho\epsilon_i)}{\partial t} + \nabla\cdot(\rho\vec{v}\epsilon_i) + p_i\,\nabla\cdot\vec{v} - \nabla\cdot(\kappa_{\perp i}\nabla T_i) - Q_{shock} - Q_{ei} = 0$$

where ϵ_α, p_α, T_α, and $\kappa_{\perp\alpha}$ refer to the appropriate ion or electron quantities, Q_{ei} is the electron-ion energy equilibration term, and Q_{shock} is an ion shock-heating term. Because of problems encountered with the SESAME electron and ion equations of state at very low temperatures, it was necessary to use ideal gas equations of state; in one-temperature runs, this was found to have no major effect on the results, compared to SESAME one-temperature runs. Resistivity, radiation loss, and ionization level were still taken from SESAME.

Hall MHD runs replace the above magnetic field evolution equation with:

$$\frac{\partial \vec{B}}{\partial t} + \nabla \times (-\vec{v} \times \vec{B} + \frac{\eta}{\mu_0} \nabla \times \vec{B} + \frac{1}{en_e}(\vec{J} \times \vec{B} - \nabla p_e)) = 0$$

where n_e is electron number density and e is the charge magnitude of the electron. It was found that the Hall $(\nabla \times (\vec{J} \times \vec{B}/n_e))$ term acted as a strong nonlinear magnetic field convection operator, destabilizing the usual time- and space-centered alternating-direction-implicit algorithm. A stable adaptation of the algorithm was developed by treating the effective convective velocity (proportional to $\vec{J}/n_e$) explicitly and using a donor-cell (but still time-centered) treatment of the $\vec{B}$ convected. This two-dimensional Hall MHD program was successfully tested on the Kingsep-Mokhov-Chukbar magnetic penetration problem discussed by Mason[20]. To preserve total energy conservation, a term complementary to the diamagnetic pressure term was added to the total energy equation: $p_e \nabla \cdot (-\vec{J}/en_e)$. Hall MHD runs to date have not yet been done with the two-temperature model, although we expect to do this shortly.

Actual experimental current vs. time values provide the boundary condition for magnetic field at the outer radial wall. "Cold-start" initial conditions are a solid, cryogenic deuterium fiber, surrounded to about twice the fiber radius by a low density, "warm" halo plasma (density $10^{-3}\times$solid, temperature 1 eV), which provides an initial current conduction path. Our computed results are insensitive to the details of this halo plasma after a short-lived (10 nsec) transient, because of the small mass involved relative to the fiber-generated plasma. The surrounding vacuum is simulated by a cold, very low density region extending out to a zero-temperature, electrically insulating wall. The early fiber-ablation stage of the discharge necessitates relatively fine radial grid spacing, but because this stage can be followed by an explosive expansion of the heating plasma, the radial grid is checked at each timestep, and adjusted so that the outer boundary is always at least 150% of the radius within which 95% of the total axial current is contained.

Radial grids of ~100 points, more finely spaced near the axis to better resolve the fiber/plasma column, cover a radius as small as 1 mm, but ultimately as large as several cm, if rapid expansion is followed. In the original, single-temperature MHD runs, axial sections from 2 cm down to 0.25 mm, covered with uniformly spaced axial grids of 31 to 62 points, were used, respectively capable of resolving the largest (X-ray "beads") and smallest (shadowgram "spicules") features observed in any of the experiments. Although the smallest, most finely resolved grids did show fine-scale instability growth starting earlier than the larger grids, saturation of shorter wavelengths resulted in larger grids ultimately showing faster expansion. Hence the timing of instability development and expansion varied by as much as 20 nsec for different grid sizes, but this is comparable to experimental timing uncertainties (e.g., the relation between driving voltage, current, and diagnostic images in time). Extended model runs to date have only been done with 1 mm and 1 cm, 31-point axial grids; more complete coverage of potential instability scales is planned.

RESULTS

Our single-temperature MHD simulations of the Los Alamos experiments HDZP-I (250 kA peak at 125 nsec)[1,2,7] and HDZP-II (750 kA peak at 100 nsec)[2,5,8] showed early development of m=0 instabilities in the plasma corona, which carries most of the current as the fiber ablates. These instabilities, and associated expansion of the plasma column in terms of mass or current, were not well reflected in the shadowgram images of the early experiment (Fig. 1(a-d)). The later, higher-current experiment HDZP-II reached complete ionization of the fiber much more quickly, at which point full nonlinear instability development led to intense nonuniform heating and rapid expansion of the plasma column (Fig. 2(a-d)). The agreement between simulation-generated diagnostic images (shadowgrams and interferograms), and actual experimental data, is strong. The variations we have modeled to date in fiber thickness, current ramp, and plasma initiation show some differences in timing of the instability-driven expansion, but expansion soon after complete ionization always appears.

As temperatures rise and density drops (e.g., in narrow m=0 "necks"), the appropriateness of the collisional fluid model (requiring ion-ion collision time $\ll$ ion thermal transit time)[2,14] breaks down. Even before this, however, is lost the more stringent requirement for ion-electron energy equilibration:

$$\frac{\left(\frac{m_i}{m_e}\right)^{\frac{1}{2}} v_{th}\tau_{ii}}{a} = \frac{\left(\frac{m_i}{m_e}\right)^{\frac{1}{2}} \tau_{ii}}{\tau_{th}} \ll 1.$$

Simulation results with two energy equations were fundamentally similar to those with only one temperature. An interesting detail was noted: Early in the discharge (and always in 1-d simulations), electron temperature, driven by ohmic heating, remained above or equal to ion temperature. When the 2-d runs began to show drastic nonlinear instability development (such as in Fig. 2), shock heating of ions associated with steep velocity gradients became dominant, causing ion temperatures to significantly exceed electron temperatures, both peak (Fig. 3(a)) and average (Fig 3(b)). Hence an instability heating mechanism may be further contributing to plasma column expansion.

The Hall and diamagnetic pressure terms are usually ordered out of the fluid model on the basis of small relative Larmor radius[14]. Haines and others[3,4] have noted that these terms may indeed not be small in the case of the Z-pinch, with its field null on axis; as well, there may be regions in the low-density coronal plasmas generally seen in our simulations (particularly in areas of high electron density gradient) where these terms will be important. In our simulations with these terms, we do see small-wavelength instabilities appear at the edge of the plasma corona, before any instabilities were seen in the simple MHD runs (Fig. 4(a-b)). However, the ultimate pattern of instability-driven heating and expansion is still seen in the Hall runs done to date. There is a strong convection of magnetic field in the axial (electron current) direction, leading to a visible depletion of field at the cathode and build-up at the anode. Although something like this might actually occur at the true experimental axial boundaries, it is probably unrealistic at computational "mirror" boundaries only 1 mm apart. We may implement periodic axial boundary conditions, in order to allow a more realistic field flow through the computational axial section taken, in addition to doing a variety of axial section lengths (computing an actual 5-cm axial section could be done, but would require a coarse axial grid which would potentially suppress small-wavelength instabilities).

There are other fluid terms of finite-Larmor-radius order: the transverse thermoelectric (Nernst and Ettinghausen) effects, an energy convection by (electron) current term, and terms in the viscous stress tensor. We hope to implement the first two of these soon, although as we found with the Hall term, new physics may entail new numerical difficulties of unknown severity. Within the collisional fluid regime treated here, we argue that viscous stress will be negligible relative to pressure gradient effects[14]. Such an argument may not hold in the "collisionless MHD" regime, in which experiments proposed by Haines and others[10,21] are purported to dwell. A further extension of the present work would be into the "collisionless MHD" regime, perhaps following an experimental plasma as it passes from collisional to collisionless.

A fiber-initiated Z-pinch plasma may provide an appropriately heated and magnetized plasma for implosion inside a heavy liner to fusion conditions, in a "magnetized target fusion" scheme. In addition, implosion of a hollow, initially solid deuterium cylinder by a fiber-Z-pinch-style fast current ramp may be a route to fusion that avoids the instability problems seen in the fiber Z-pinches simulated here. We are employing the computational tools developed for the work reported here in the evaluation of these related concepts.

CONCLUSIONS

Detailed two-dimensional MHD simulations of deuterium-fiber Z-pinches have shown good agreement to the Los Alamos experiments HDZP-I and HDZP-II. Late in low-current and early in higher-current experiments, when the fiber has become fully ionized, m=0 instabilities develop rapidly, and drive intense nonuniform heating and rapid expansion of the plasma column, dropping densities orders of magnitude below the high densities desired for fusion conditions. We believe the "cold-start" approach taken here is vital to the realistic simulation of such experiments, due to the close tie between plasma profiles and stability. Inclusion of the finite-Larmor-radius effects and two-temperature model presented here does not change these basic conclusions. If an experiment is to reach potentially stabilizing conditions in a "collisionless MHD" or other regime, it would appear desirable to avoid the dangerously unstable collisional MHD area explored in our simulations. Fiber-initiated Z-pinches may have potential in magnetized target fusion and related concepts.

ACKNOWLEDGMENTS

We would like to acknowledge the assistance of the HDZP-I/II experimental group: R. Lovberg, D. Scudder, J. Shlachter, et al; and stimulating discussions with N. Bobrova, J. Chittenden, P. Choi, M. Coppins, J. Dawson, H. Etlicher, M. Haines, M. Liberman, R. Mason, A. Robson, P. Sasorov, and J. Sethian.

REFERENCES

1. I. R. Lindemuth, Phys. Rev. Lett. **65**, 179 (1990).
2. P. Sheehey, J. Hammel, I. Lindemuth, R. Lovberg, R. Riley Jr., D. Scudder, J. Shlachter, Phys. Fluids B **4**, 3698 (1992).
3. M. G. Haines, J. Phys. D **11**, 1709 (1978).
4. M. Coppins, D. J. Bond, and M. G. Haines, Phys. Fluids **27**, 2886 (1984).
5. D. W. Scudder, J. S. Shlachter, J. E. Hammel, F. Venneri, R. Chrien, R. Lovberg, R. Riley, in Physics of Alternative Magnetic Confinement Schemes, Proceedings of the Workshop Held at Varenna, Italy, 1990, ed. by S. Ortolani and E. Sindoni (SIF, Bologna, 1991), p. 519.
6. J. D. Sethian, A. E. Robson, K. A. Gerber, and A. W. DeSilva, Phys. Rev. Lett. **59**, 892 (1987); **59**, 1790(E) (1987).
7. J. Hammel and D. Scudder, in Proceedings of the Fourteenth European Conference on Controlled Fusion and Plasma Physics, Madrid, Spain, 87 F. Engelmann and J. L. Alvarez Rivas (EPS, Petit-Lancy, Switzerland, 1987), p. 450.
8. J. E. Hammel, in Dense Z-Pinches, ed. by N. R. Pereira, J. Davis, and N. Rostoker (American Institute of Physics, New York, 1989), p. 303.
9. J. Sethian, A. Robson, K. Gerber, and A. DeSilva, in Physics of Alternative Magnetic Confinement Schemes (Ref. 5), p. 511.
10. M. Haines, in Physics of Alternative Magnetic Confinement Schemes (Ref. 5), p. 277.
11. R. S. Pease, Proc. Phys. Soc. B **70**, 11 (1957).
12. S. I. Braginskii, Sov. Phys. JETP **6**, 494 (1958).
13. N. R. Pereira, Phys. Fluids B **2**, 677 (1990).
14. J. Freidberg, Ideal Magnetohydrodynamics (Plenum Press,New York,1987) Ch. 2.
15. B. B. Kadomtsev, in Reviews of Plasma Physics, edited by M. A. Leontovich (Consultants Bureau, New York, 1966), Vol. 2, p. 165.
16. I. Lindemuth, G. McCall and R. Nebel, Phys. Rev. Lett. **62**, 264 (1989).
17. I. R. Lindemuth, UC Lawrence Livermore Laboratory Report UCRL-52492 (1979).
18. Los Alamos National Laboratory Report LA-10160-MS, K. S. Holian (1984)
19. S. I. Braginskii, in Reviews of Plasma Physics, edited by M. A. Leontovich

(Consultants Bureau, New York, 1965), Vol. 1, p. 205.
20. R. Mason, P. Auer, R. Sudan, B. Oliver, C. Seyler, and J. Greenly, Phys. Fluids B **5**,(4)(1993).
21. M. Haines and M. Coppins, Phys. Rev. Lett. **66**, 1462 (1991).

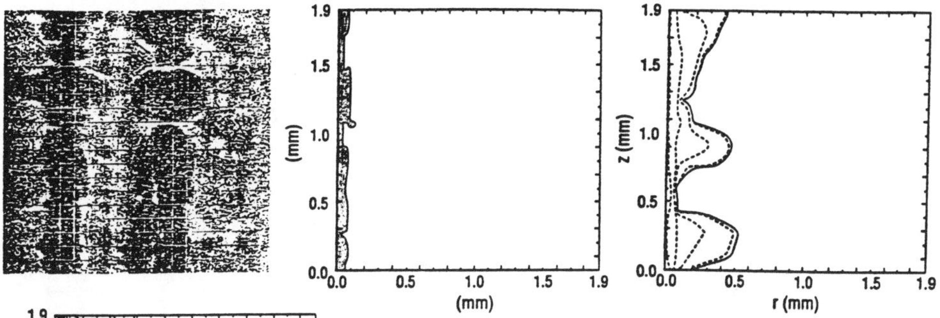

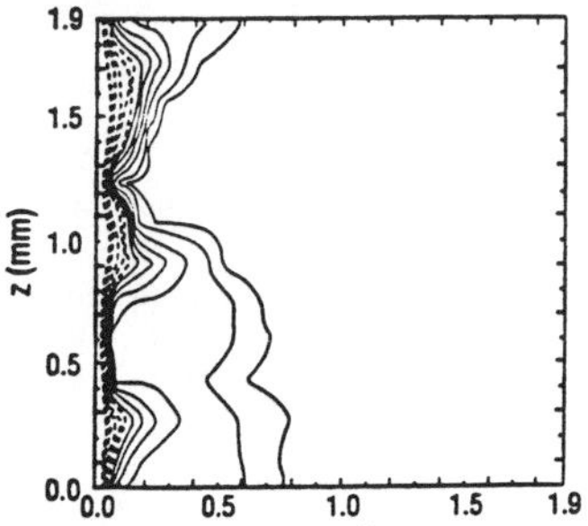

Fig. 1 (a) Experimental shadowgrams, HDZP-I: left-hand image, ~30 nsec (~50 kA); right-hand image, ~40 nsec (~65 kA); each grid block is 0.1 mm square; (b) Simulation shadowgram from section of same size as 1(a), HDZP-I, 30 nsec (50 kA); (c) Corresponding simulation density contours, 30 nsec (50 kA); rightmost contour contains 95% of the total mass; (d) Simulation axial current contours, 30 nsec (50 kA); rightmost contour contains 95% of the total axial current.

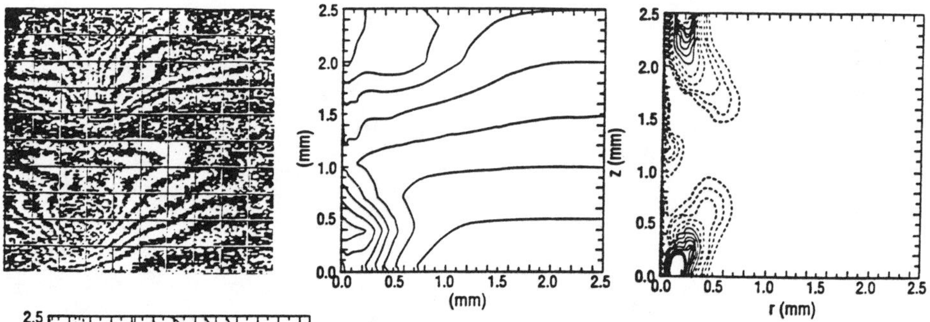

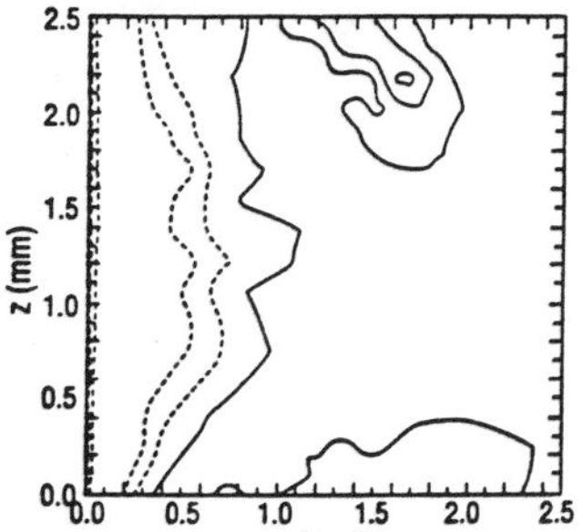

Fig. 2 (a) Experimental interferogram, HDZP-II, ~20 nsec (~200 kA); each grid block is 0.25 mm square; (b) Simulation interferogram, HDZP-II, 32 nsec (230 kA); (c) Simulation density contours, HDZP-I, 50 nsec (85 kA); (d) Simulation axial current contours, HDZP-I, 50 nsec (85 kA).

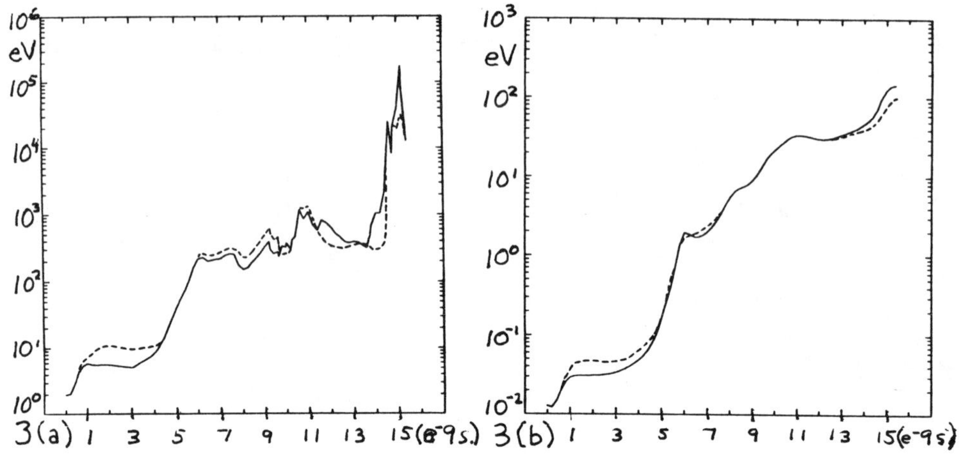

Fig. 3 (a) HDZP-II simulation maximum ion temperature (eV) vs. time (sec)–solid line; maximum electron temperature vs. time–dotted line; (b) Average (mass-weighted) ion temperature (eV) vs. time (sec)–solid line; average electron temperature vs. time–dotted line.

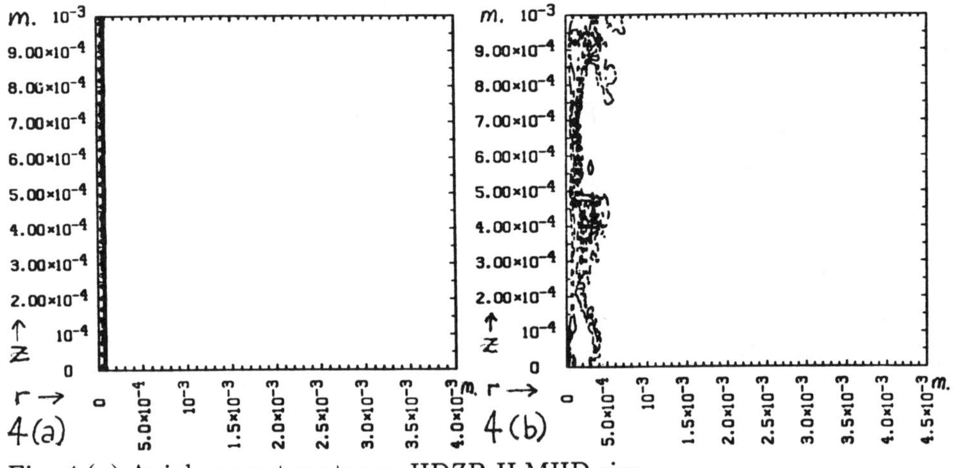

Fig. 4 (a) Axial current contours, HDZP-II MHD simulation, ~8 nsec (~30 kA); (b) Axial current contours, HDZP-II Hall MHD simulation, ~8 nsec (~30 kA).

Modeling of the enhanced rate of the magnetic field propagation along the anode in Z-pinch.

V.V. Vikhrev, O.Z. Zabaidullin

Russian Research Centre "Kurchatov Institute"

Institute of Nuclear Fusion

Moscow, Russia

Studies of the plasma Z - pinch dynamics in the frame of ideal MHD-equations predicts the profile of the moving plasma sheath like on figure 1. Really, that profile is observed in dense plasma discharge, for instance in Marshall's gun. However, the decreasing of initial gas density below the critical value of order $10^{16} - 10^{17}\ cm^{-3}$ lead to a sharp qualitatively changing of the sheath profile (fig.2). Experimentally, this changing was discovered in [1-3] and named as a slipping of plasma sheath along the anode. The theoretical analysis of this phenomenon developed in several directions, which are presented [1-9]. But, the more fruitful approach was the considering of influence of the Hall effect to magnetic field dynamics. With treating of Hall effect as a transferring of the magnetic field with current velocity by electrons along the current line. Firstly, the influence of Hall effect to magnetic field dynamics was shown on paper [4]. The modeling [4] of the ideal MHD-equations with account of Hall effect in boundary conditions showed the enhanced rate propagation of the magnetic field near the anode. The result of that magnetic field dynamics is a moving out of plasma from anode under the force of magnetic pressure $\nabla(H^2/8\pi)$. The moving out plasma from the anode produce the plasma sheath parallel to the electrode. But in paper [4] the problem of breaking of the magnetic field "frozenness" into a plasma was unsufficiently discussed. For the solving of that problem we additionally carried out the two-dimensional modeling of magnetic field dynamics in frame of electron magnetohydrodynamics and two - dimensional modeling of plasma Z - pinch dynamics with overall account of Hall effect. The results of such mode lings present in first and second part correspondingly.

1.The problem about penetration of the magnetic field into a plasma along the boundary of two media.

It was considered the plasma medium occupying all (x,y) space and composed from two region, distinguished by line y=0. The plasma parameters in each region are uniform: at y the electron concentration is n and plasma conductivity is s and at y0 are n and s correspondingly. Magnetic field have the only z - component and at initial moment occupy the left part at x and equal to H . At x0 magnetic field is absent. The orientation of the magnetic field was taken such for which the electron move in the direction of y-axis. The equation for the evolution of the magnetic field in frame of electron magneto hydrodynamics without of gradient of the electron pressure used as follow

$$\frac{\partial H}{\partial t} = rot\,[u, H] - rot(\ D\,rot\,H) \qquad (1)$$

Where u is current velocity: $u=-c/(4\pi qn)\,rotH$, D is coefficient of the magnetic field diffusion $c/(4\pi\sigma)$ and q is a value of the elemental charge. In our case of plane coordinates (x,y) the eq.(1) be the following

$$\frac{\partial H}{\partial t}=\frac{\partial}{\partial x}\left(D\frac{\partial H}{\partial x}\right)+\frac{\partial}{\partial y}\left(D\frac{\partial H}{\partial y}\right)+\frac{cH}{4\pi q}\frac{\partial H}{\partial y}\frac{\partial}{\partial x}\left(\frac{1}{n}\right)-\frac{cH}{4\pi q}\frac{\partial H}{\partial x}\frac{\partial}{\partial y}\left(\frac{1}{n}\right) \quad (2)$$

For our task $\partial n/\partial x=0$ in all space and $\partial n/\partial y$ is not equal zero only at line $y=0$. So, the changing of the magnetic field due to Hall effect occur only at the boundary of two media $y=0$. It means that in all space (x,y) the evolution of the magnetic field have diffusional character and only at boundary $y=0$ is the additional apportionment of the magnetic field due to Hall effect. Magnetic field apportionmented at line $y=0$ diffusionally expanded in the surroundings

$$\frac{\partial}{\partial y}\left(\frac{1}{\sigma}\frac{\partial H}{\partial y}\right)=\frac{H}{qn}\frac{\partial H}{\partial x}\frac{\partial}{\partial y}\left(\frac{1}{n}\right) \quad (3)$$

It knows, that the transferring of the magnetic field with current is a consequence of the presence of hall electric field $E = 1/(4\pi qn)\,[\mathrm{rot}H,H]$ in a plasma. In crossing fields E and H the transferring of the magnetic field is in the direction of electron move. The value of the magnetic energy flow in the direction perpendicular to x - axis defined by Pouting vector

$$S=\frac{c}{4\pi}\iint\,[E_{hall}+E_{joul},H]\,dx\,dz \quad (4)$$

where the ohmic component of the electric field $E_{joul}=\frac{c}{4\pi\sigma}\mathrm{rot}\,H$. At distance $h_1=\left(\frac{c^2 t}{4\pi\sigma_1}\right)^{1/2}$ and $h_2=\left(\frac{c^2 t}{4\pi\sigma_2}\right)^{1/2}$ from the boundary line $y=0$ the diffusional process is negligible in comparison with the transferring of the magnetic field due to Hal! effect. Then, the value S at distance yh from line $y=0$

$$S_1=\frac{z_0 H_0^3 c}{48\pi^2 q n_1}, \quad (5)$$

where z_0 is a size of the considering plasma region along z - axis. And at distance yh

$$S_2=\frac{z_0 H_0^3 c}{48\pi^2 q n_2} \quad (6)$$

The modeling of the eq.(2) was performed for different values of $\sigma_1, n_1, n_2, \sigma_2$. As a result the computer - plotted graphics of the magnetic field distribution in case of $\sigma_2=10\sigma_1$, $n_2>n_1$, $(\omega\tau)_1=\sigma_1 H/qcn=10$ are presented on fig.3a-3c. Gradation of grey colors corresponds to changing of the magnetic field by a factor 1.5. The fig.3a-3c show that the penetration of the magnetic field in volume of media have diffusional character. And only near the boundary we see the enhanced penetration of the magnetic field. Dependence of the way length L, which front of magnetic field come along the boundary to time t have diffusional character: $L=t^{1/2}$. Now, let us qualitatively consider the dynamics of the magnetic field (2). From the modeling of eq.(2) we obtain that the magnetic field due to Hall effect appears on boundary ($y=0$) and further expended diffusionally in up and down (fig.4). The width of the penetration region in each media is $h_1=\left(\frac{c^2 t}{4\pi\sigma_1}\right)^{1/2}$ and $h_2=\left(\frac{c^2 t}{4\pi\sigma_2}\right)^{1/2}$, correspondingly. The flow of the magnetic field energy on the boundary of the considered media defined by subtraction of (5) and (6)

$$S_1 - S_2 = \frac{Z_0 c H_0^3}{48 \pi^2 q} \left(\frac{1}{n_1} - \frac{1}{n_2} \right) \tag{7}$$

The flow $(S_1 - S_2)$ at time t make the region of magnetic field of length L. The calculations show that the penetration region have parabolic shape boundary (fig.4). The energy of the magnetic field in the penetration region is $2/3\ (h_1 + h_2) L_0 H_0^3 / 8\pi$. With account of eq.(7) we have the following equation for length

$$\frac{(h_1 + h_2) L H_0^3}{12 \pi} = \frac{Z_0 c H_0^3}{48 \pi^2 q} \left(\frac{1}{n_1} - \frac{1}{n_2} \right) t. \tag{8}$$

Substitution the value for the sum $(h_1 + h_2)$ into (8) give the final result

$$L = \frac{H_0 t^{1/2} (n_2 - n_1)(\sigma_1 \sigma_2)^{1/2}}{2 q \pi^{1/2} n_1 n_2 (\sigma_1^{1/2} + \sigma_2^{1/2})} . \tag{9}$$

The velocity of penetration process

$$V = \partial L / \partial t = \frac{H_0 t^{1/2} (n_2 - n_1)(\sigma_1 \sigma_2)^{1/2}}{4 q \pi^{1/2} n_1 n_2 (\sigma_1^{1/2} + \sigma_2^{1/2})} . \tag{10}$$

The penetration (9) have diffusion coefficient

$$D = L^2 / t = \frac{H_0^2 (n_2 - n_1)^2 (\sigma_1 \sigma_2)}{4 q^2 \pi (n_1 n_2)^2 (\sigma_1^{1/2} + \sigma_2^{1/2})^2}$$

In case of well conductive second medium like a metal ($\sigma_1 > \sigma_2$, $n_2 > n_1$) the formulas (9,10) will have the following shape

$$L_a = (\omega \tau)_{pl} (c^2 t / 4 \pi \sigma_{pl})^{1/2}, \tag{12}$$

$$V_a = (\omega \tau)_{pl} (c^2 t / 16 \pi \sigma_{pl})^{1/2}, \tag{13}$$

Where σ_{pl} is plasma conductivity and $(\omega \tau)_{pl}$ is a magnetization value for electron. In case of $\sigma_1 > \sigma_2$, $n_2 > n_1$ the diffusion coefficient (11) coincide with the result of [10].

2.The modeling of the Z-pinch plasma dynamics near the electrodes.

Z-pinch plasma dynamics was considered in frame of two - fluid two-dimensional MHD-equations with overall account of Hall effect on the assumption of the axisymmetric discharge development. Assuming the equity of ion and electron temperature [3,4] the MHD - equations used for plasma density n, strength of magnetic field H, mass velocity V, current density j and plasma temperature T:

$$\frac{\partial n}{\partial t}+div(nV)=0\ , \tag{14}$$

$$\frac{\partial H}{\partial t}=rot\,[V,H]-rot\,[\frac{c}{4\pi qn}rot H,H]-rot\,[\frac{c^2}{4\pi\sigma}rot\,H]\ , \tag{15}$$

$$\frac{\partial(nV)}{\partial t}+div(nVV)=-\frac{2}{m_i}\nabla(nT)+\frac{1}{4\pi m_i}[rot\,H\,,H]\ , \tag{16}$$

$$j=\frac{c}{4\pi}rot\,H\ , \tag{17}$$

$$\frac{3}{2}\left(\frac{\partial(nT)}{\partial t}+div(nT)\right)=-nTdiv\,V+j^2/\sigma\ , \tag{18}$$

where σ is plasma conductivity, m_i is ion mass. Simulation of the eq.(14-18) was carried out for plasma system, having the following initial distribution. The plasma column of radius R_0 is surrounded by magnetic field $H=H_0 R_0/r$ for $r\geq R$. The plasma density n_0 and temperature T_0 are uniforn in plasma column. Plasma is bounded by electrodes: lower is anode. In this paper we use numerical modeling to study the dependence of plasma dynamics from the parameters: magnetization electron value $(\omega\tau)_0=\frac{1.23\,10^{11}\,I_0\,T_0^{3/2}}{R_0\,n_0}$, exchange parameter $\Pi_i^{-1/2}=\left(\frac{m_i c^2}{4\pi n_0 q^2 R_0^2}\right)^{1/2}$ and Reynolds magnetic value $R_m=\Pi_i^{-1/2}/(\omega\tau)_0$, where I_0 is current value. The calculations were performed in the region $0\leq z\leq Z_m, 0\leq r\leq R$, where $R=1.2R_0$ and d Z_m is a distance between anode and cathode. The boundary conditions at r=R was the following: $W_z=0, W_r=0, n=n_{ext}=10^{-2}n_0; H=2I_0/cr$, where n is a plasma density outside the main plasma column. For r=0 we take the natural conditions of cylindrical symmetry: $\partial W_z/\partial r=0$; $W_r=0$; $H=0$; $\partial n/\partial r=0, \partial T/\partial r=0$. And at the electrodes surface z=0 and z=Z_m we set the conditions of a sticking plasma to electrodes and the absent of the plasma flow into its: $W_r=0, W_z=0$. As a boundary conditions for magnetic field for $z=0$ and $z=Z_m$ we used the eq.(17) itself. In which we put zero flows of magnetic field into electrodes due to effects of Hall $\frac{H}{n}\frac{\partial(rH)}{r\partial r}=0$ and of diffusion $\frac{1}{T^{3/2}}\frac{\partial H}{\partial z}=0$. This is assumes infinite as conductivity (T= ∞) and electron density (n= ∞) of the electrodes.

Dynamics of the plasma Z-pinch discharge near the electrodes was studied for various values of $\Pi_i^{-1/2}$ without diffusion, that means $(\omega\tau)_0=\infty$.

Figures 5a-5c present the computer-produced plots of the plasma density n(r,z) at the left part and magnetic field H(r,z) at the right part at the moment when magnetic field reaches the axis of the system. The grey-levels of the distributions in Fig.5 differs by a factor 2. The results of computation show that the front of the magnetic field overtakes plasma sheath for $\Pi_i^{-1/2}$ > 0.065. This condition appears to be a criterion for the penetration of the magnetic field into a nonperturbed plasma near the anode which, in turn, is a reason for the slipping of the current sheath along the anode surface. The radial component of the current gives rise to a "magnetic snow plough" pushing the plasma away from the anode. As the plasma moves away from the anode the sheath mass in the region near the anode decreases and the radial velocity increases here. For instance, on fig.5a the time of coming of the plasma sheath to z - axis is t_0 and on fig.5c this time is $0.6t_0$. That behavior of Z-pinch plasma near the anode was experimentally discovered in [1,3]. Numerically, from our computation we obtained the condition for such this phenomenon to occur:

1) $(\omega\tau)_0 > 1$,

2) $n_0 R_0^3 < z_i^{-2} M/M_H \, 5 \, 10^{18} \, cm^{-3}$,

where M_HM is effective (i.e. averaged over ion species) mass of atoms(ions) in plasma sheath, M is hydrogen mass atom and Z_i is effective charge of plasma ions. Fig.5 show the breaking of the ideal-MHD symmetry of the magnetic field dynamics near the electrodes, which is caused by the transfer of magnetic field by electrons with current velocity along the current line from cathode to anode. This effect was found numerically in [2].

The nontrivial fact of our computation is a finite velocity of magnetic field propagation along the anode to the contrary of infinite velocity penetration (10,13) in case of $(\omega\tau)_o = \infty$. The infinite velocity (10) was derived with account of the only diffusional mechanism of going out the magnetic field from the anode. But, our more full two - fluid MHD modeling of eq. (14-18) have two different mechanism of going out magnetic from anode: diffusional and hydrodynamical, that means the moving magnetic field with "snow ploughing" plasma. And the second - hydrodynamical mechanism does not disappear in case of $(\omega\tau)_o = \infty$.

Comparison of experimental data and numerical results is presented in Fig.6 and Fig.7. In Fig.6 the distribution of plasma density at the moment of Z-pinch column formation near the anode is shown (here the anode is on the bottom, cathode is absent). This picture was obtained by laser interferometry method in [3]. In Fig.7, plasma Z-pinch column is shown for the entire electrode gap (anode is on the bottom, cathode is on the top). Fig.7 presents the photograph from [11] of Z-pinch in scattered visible light. The experimental and numerical results are in qualitative agreement.

Conclusions.

In our work we considered the nature of the magnetic field penetration along the anode into a plasma. It was shown that in the frame of two - dimensional MHD - equations with overall account of Hall effect the phenomenon of the plasma sheath slipping along the anode have a threshold character. We found the conditions for which this phenomenon to occur:

1) $(\omega\tau)_0 > 1$,

2) $n_0 R_0^3 < z_i^{-2} M/M_H \, 5 \, 10^{18} \, cm^{-3}$,

where M_H is effective (i.e. averaged over ion species) mass of atoms(ions) in plasma sheath, M is hydrogen mass atom and Z_i is effective charge of plasma ions. The velocity of the magnetic field propagation along the anode is finite in case of $(\omega\tau)_o = \infty$.

References.

1. S.V.Bazdenkov, K.G.Gureev, N.V.Filippov, T.I.Filippova //Sov. Jour. Letts. in JETF 1973 .N.18 .v.3 .p. 199.

2. A.I.Morozov, L.S. Soloviev // Reviews of plasma physics. Ed. M.A.Leontovich. 1974.N.8 p. 94.

3. A.R.Terentiev, Ph.D. Thesis "Influence of Hall effect on plasma dynamics near the anode", Kurchatov Institute, Moscow 1987.

4. Vikhrev V.V., Gureev K.G.// Nuclear Fusion. 1977,v.17.p.291.

5. Mason R.J.,Jones M.E.,Grossman J.M.,Ottinger P.F.// Phys. Rev. Lett. 1988.V61.p.1835.

6. S.I. Braginskii, V.V. Vikhrev // Reviews of plasma physics. Ed. M.A.Leontovich. 1980. N.10 p. 243.

7. A.V.Gerusov, V.S. Imshennik // Sov. Jour. of Plasma Phys.1985. T.11.p.568.

8. P.V. Sasorov // Sov. Jour. of Plasma Phys.1990. N.16.p.1236.

9.S.F.Garanin, V.I.Malyshev // Sov. Jour. of Plasma Phys. 1990. N.16. p. 1218.

10. A.V. Gordeev, A.V. Grechiha, I.L. Kalda// Sov. Jour. of Plasma Phys. 1990.T.16. p.95.

11. I.F.Kvartshava, N.G.Reshetnyak, E.F.Hautiev // Plasma accelerators. Ed .L.A.Artsimovich, Moscow 1973 .p. 207.

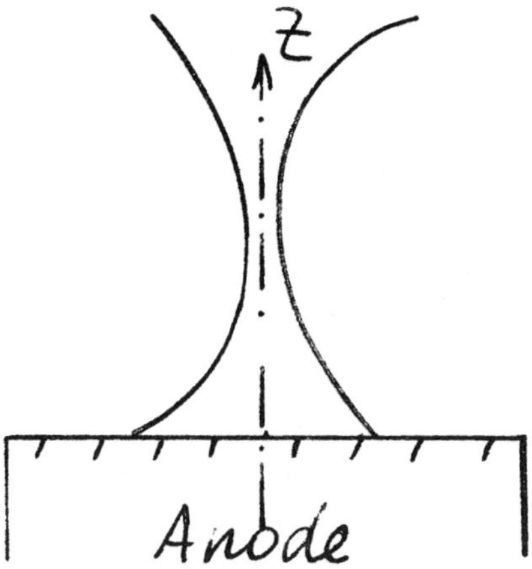

Fig.1. Profile of plasma sheath for dense z-pinch discharge.

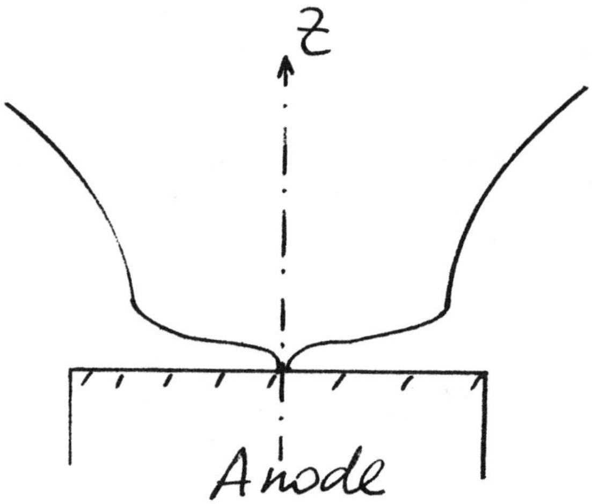

Fig.2. Profile of plasma sheath for z-pinch discharge with initial gas density less than 10^{17} sm^{-3}.

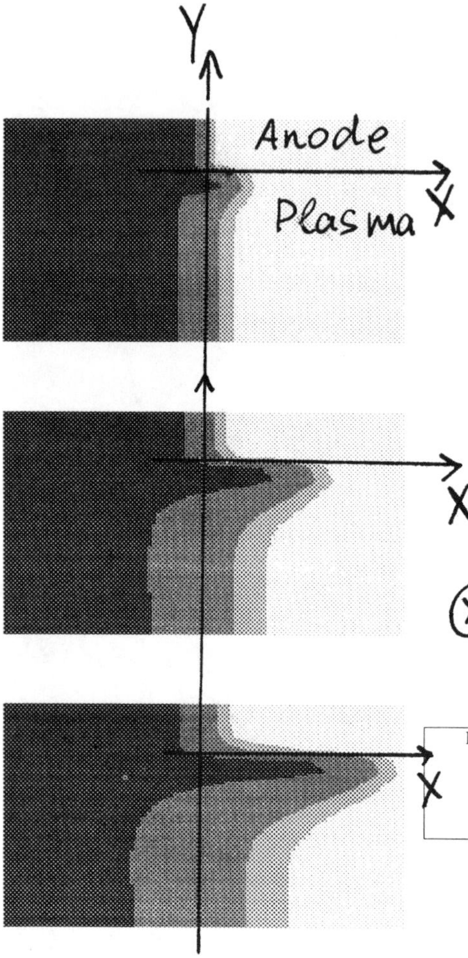

Fig.3(a,b,c). Distribution of the magnetic field during it penetration into a plasma along the anode.

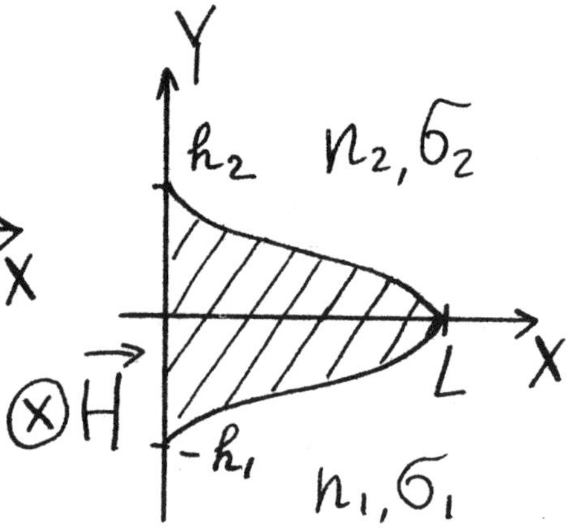

Fig.4. Penetration region of the magnetic field near the boundary of two different media.

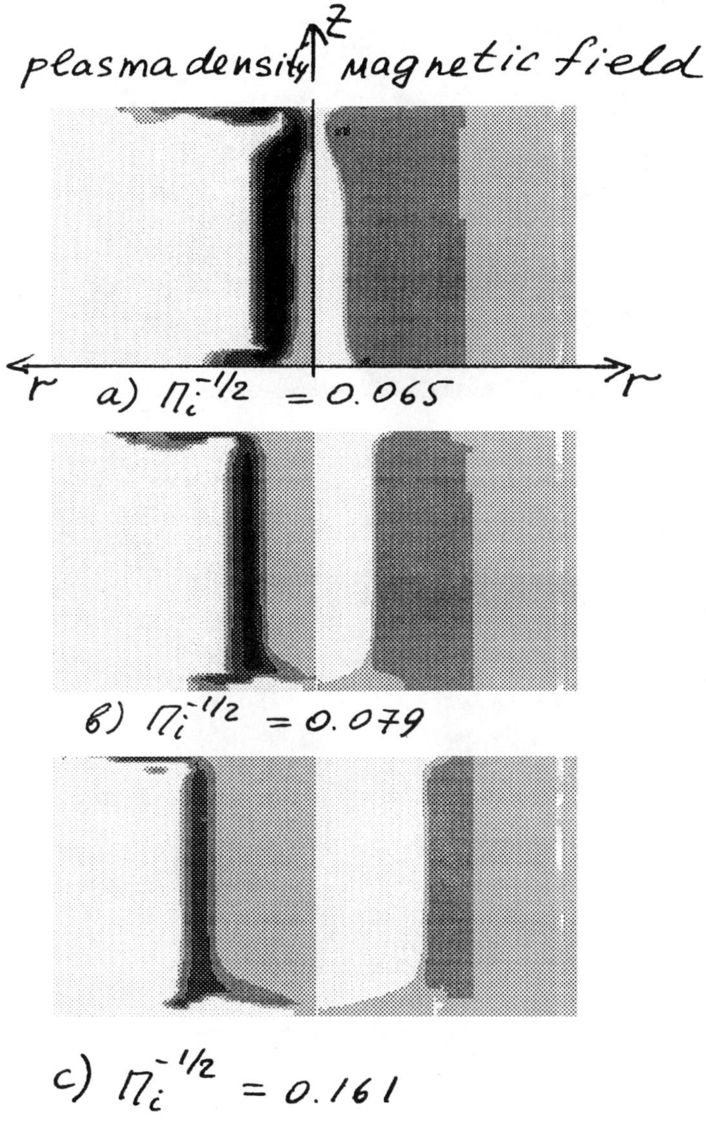

Fig.5. Computer-plotted distributions of plasma density n(r,z) at left part and magnetic field H(r,z) at right part. The upper boundary is cathode and lower is anode.

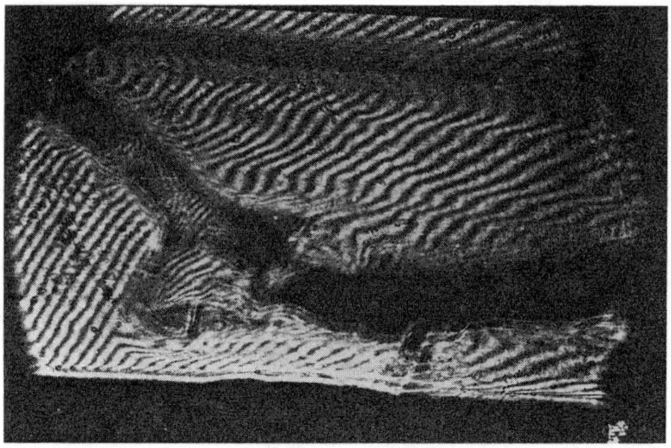

Fig.6. The photograph of the profile of plasma sheath at the moment of z-pinch column formation made by a laser interferometry method. Anode is lower boundary.

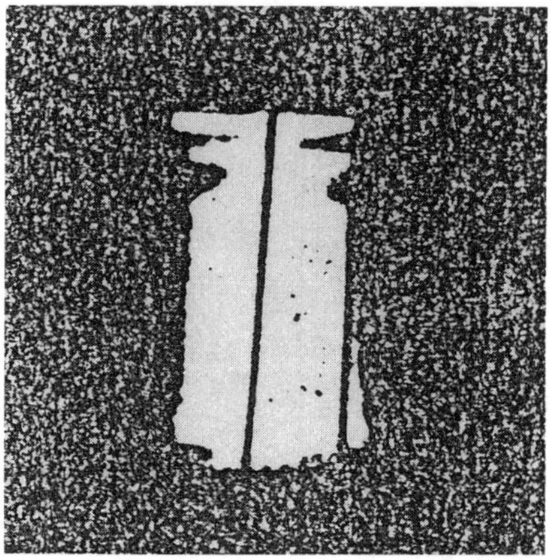

Fig.7. The photograph of plasma z-pinch column in a gap between anode (down) and cathode (up). The photo was produced in visible light.

Two-dimensional model of thermonuclear combustion wave initiation in Z-pinch.

V.V. Vikhrev, G.A. Rozanova
Russian Research Centre "Kurchatov Institute"
Institute of Nuclear Fusion
Moscow, Russia

At present it is known that a high temperature plasma is produced at the necks of the z-pinch [1]. Its emergence is related with the fact that Bennet's condition - radial quasiequilibrium - is satiafied in the process of compression in each cross-section of the pinch

$$NT = I^2/4c^2 \qquad (1)$$

where I is the current, T is the temperature, N is the amount of ions in the pinch cross-section. The plasma outflow from the neck zone reduces the amount of particles in the cross section and thus results in an increase in the temperature there up to the fusion magnitudes.

The presence of energy losses by radiation from the pinch plasma results in a noticeable rise in the parameter ρr in the high temperature plasma of the neck zone, where ρ=mn is the plasma density (m is the ions mass, n is the ion density), r is the characteristic transversal size of the plasma-occupied zone, e.g., radius. Indeed, taking (1) into account, one has

$$\rho r = mnr = mNr / \pi r^2 = f(T)\, I^2 / r. \qquad (2)$$

At the currents higher than 10 MA in the pinch, ρr of high temperature plasma at the necks, due to compression because of the energy loss by radiation, can attain the magnetude of the order of o,5 g/sm^2 In case of a DT-mixture, such a plasma initiates the burn wave which results in propagation of a high temperature plasma zone along the pinch.

The device in which the fusion wave burn initiation and propagation along the z-pinch can be is follows. The energy storage with the power of about 10 Mj, effective voltage of about 10^5 V and with the discharge current of about 10^7 A, should serve as an energy sourse. The loading is shown in Fig.1, being a pinch with the solid rod of DT-mixture, 1-5 mm in diameter. This rod is surrounded by a medium (solid at the initial instant of time) having the density 0,01 - 0,05 g/sm^3 10-20 sm in radius. This medium is nessesary for the condition of attaing the boundary of the rod by current sheath at the moment of maximum current.

In Fig.2 the DT-rod, which has already passed into a plasma state and the MHD-instability is being developed there, is schematically depicted. The simulation has been done for this dense high temperature zone, where the plasma temperature is T_0=2 keVat the initial instant of time. At this temperature and at the effective plasma charge $Z_{eff} = 1$, the fusion heat release in α-particles is accurately compensated by volumetric energy losses by radiation, due to bremsstrahlung. It means that if an inhomogeneity resulting in the emergence of a neck had not been developed in this zone, the plasma column would be in the initial state for the infinitely-long time and all the processes described bellow occur due to the development of perturbation only.

To satisfy the condition of fusion burn wave propagation along the pinch it iss nessesary that the fusion fuel be compressed by a magnetic field so that ρr>A in it, where A is the constant determined by the nature of fusion fuel and by the conditions of compression.

The results of calculating the neck at $\rho r = 0.23$ g/sm^2 are given in Fig.3. At high ρr values the fusion burn wave initiation starts at the earlier stage of the neck development. At $\rho r < 0.23$ g/sm^2 the fusion burn wave is damped.

Since the magnetic field pressure in the perturbed zone is somewhat greater than the plasma pressure in the column, the perturbed zone is expanded that results in the emergence of a cavern in the plasma column. The cavern is developed due to the entry of a magnetic field into it from the rarefied peripheral plasma (See, Fig.3a). In the near-axis zone the plasma column compacting and a gradual increase in its temperature occur. When the temperature > 5 keV is attained, an intense fusion heat release resulting in the emergence of a burn wave propogating along the pinch axis starts in a dense plasma. In this case, a dense plasma configuretion around the axis disappears beyond the propagating wave front. the process of wave propagation is accompanied by an esssential plasma temperature rise in the near-axis zone (See, Fig.3c). In Fig.s 4,5 one can see the graphs of temporal plasma temperature and density dependences in the produced burn wave.

For the parameter under consideration, $\rho r = 0.23$ g/sm^2, and for the plasma temperature at the initial instant of time, $T_0 = 2$ keV, the initial plasma column radius is r_0(sm) $= 4{,}6 \cdot 10^{-18} I^2$ (A), and the initial plasma density in the column ρ_0(g/sm^3) $= 0{,}5 \cdot 10^{17} I^2$ (A). At the pinch currents ~10 MA the plasma column parameters at which the fusion burn wave emergence is possible due to the MHD-instability development are $\rho_0 = 0{,}5 \cdot 10^3$ (g/sm^3) and $r_0 = 5 \cdot 10^{-3}$ mm, respectively. In the more powerful discharges (with the currents ~ 100 MA and higher) the requirements to these parameters are essentially reduced, and the fusion burn wave propagation along the pinch is already possible at a lower initial plasma density in it.

Calculations corresponding to the different ratio between the plasma energy loss by radiation and the fusion heat release in the plasma have also been done. When the fusion heat release at the initial instant of time considerably exceeds the energy loss by radiation, a rapid rise in the plasma temperature and the plasma column expansion occur. It is related with the fact that the fusion heat release power, with the temperature rise, rises faster than the radiated energy power from the optically-transparent plasma. When the fusion heat release is low at the initial instant of time in comparison with the plasma energy loss by radiation, a radiation pinch compression mode resulting in the production of a plasma neck zone with essentially-higher parameters is produced. This, in its turn, increases the parameter ρr and thus alleviates the process of fussion burn wave emergence and propagation along the pinch.

Thus the realized studies support the idea that an inhomogeneity emergin as a result of MHD-instability development in the pinch is not a barrier to the fusion burn wave emergence and propagation; on the contrary, it results in the initiation of this wave; the emergence and propagation of a fusion burn wave is essentially alleviated, when the plasma energy loss by radiation exceeds the fusion heat release in the plasma at the initial instant of time.

References.

1. V.V.Vikhrev, V.V. Ivanov, G.A. Rozanova. Nuclear Fusion, Vol.33, No.2 (1993), p.311.

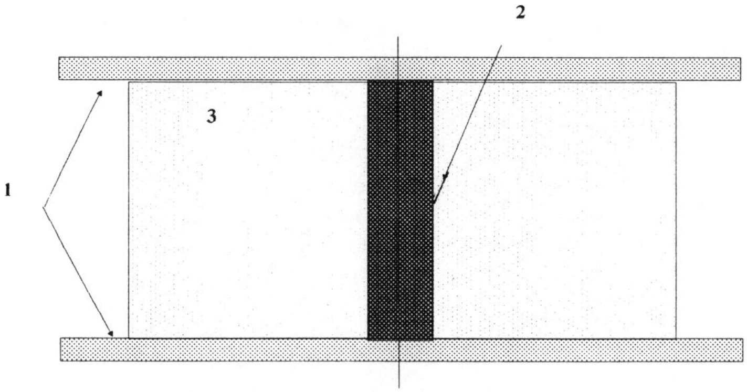

Fig.1. Scheme of the pinch. 1 - electrodes; 2 - solid rod fmade of frozen DT-mixture with density ~1 g/sm^3; 3 -solid medium with density ~0,01 g/sm^3

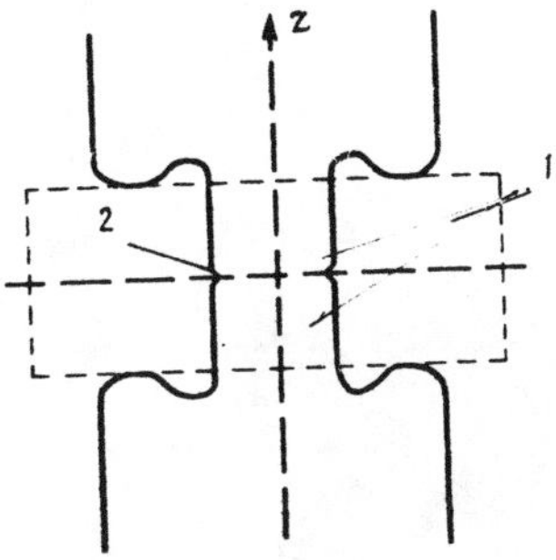

Fig.2. Scheme of DT-plasma column with MHD instability;
1 - high density plasma with initial temperature 2 keV;
2 - initial disturbance of the boundary

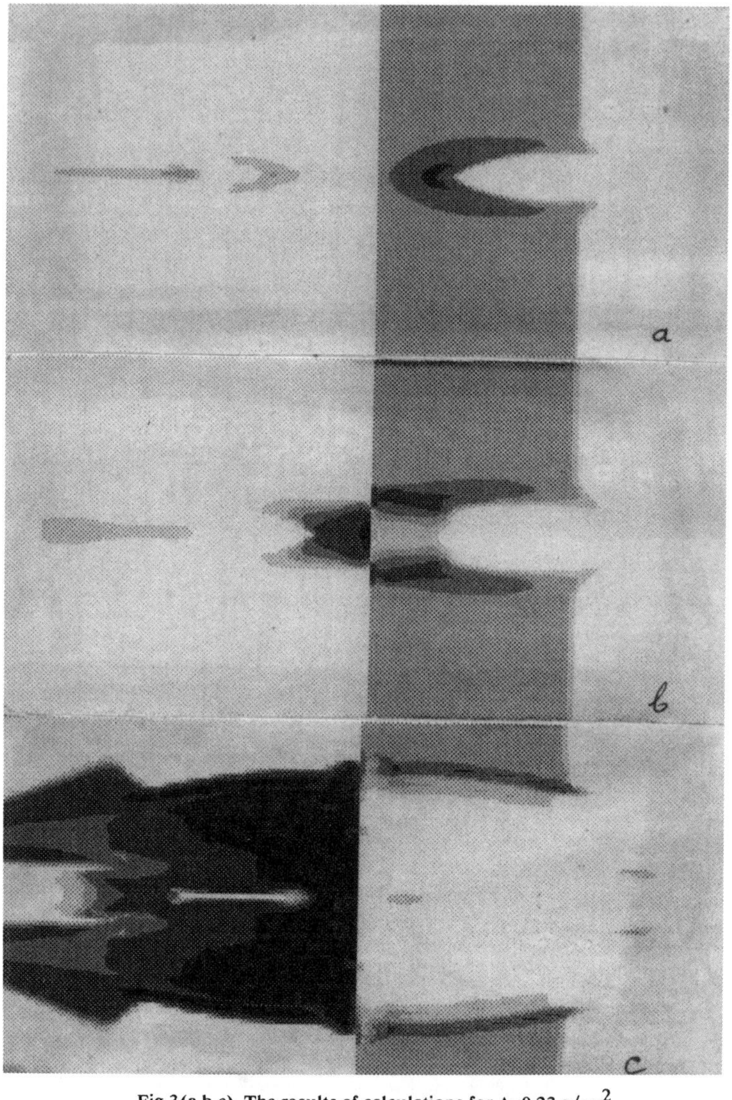

Fig.3(a,b,c). The results of calculations for A=0,23 g/sm^2
Left side - temperature distribution in plasma column ;
Right side - density distribution in plasma column.

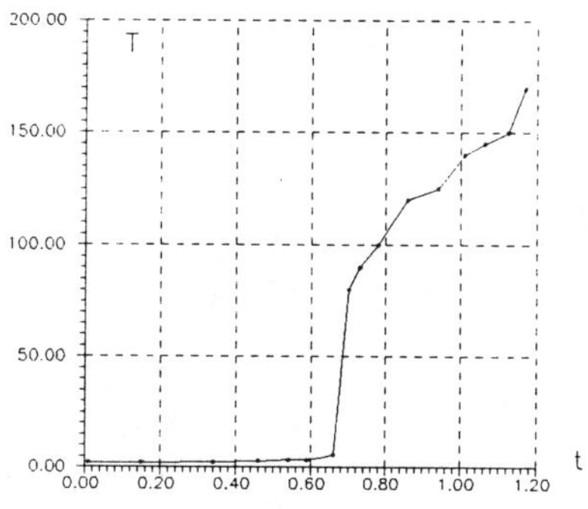

Fig.4.Graph of temporal plasma temperature dependence in the burn wave.

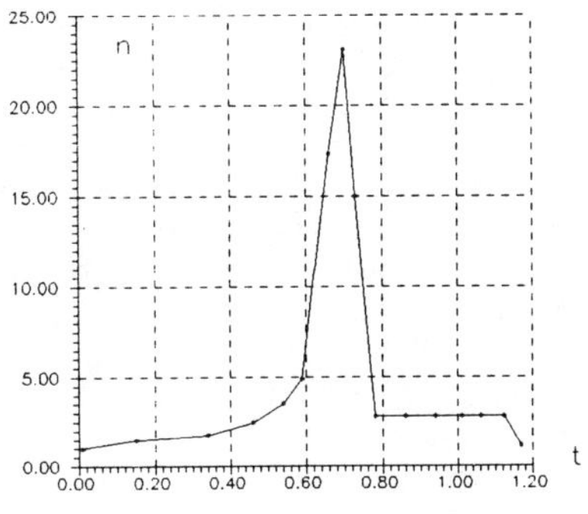

Fig.5.Graph of temporal plasma density dependence in the burn wave.

TWO-DIMENSIONAL NUMERICAL SIMULATION OF AN IMPLODING DOUBLE GAS-PUFF PLASMA.

A.D.Zoubov, G.A.Adamkevich, I.V.Glazyring, and A.A.Kondrat'ev.

National Institute of Technical Physics, Chelyabinsk-70,
P.O.Box 245, RUSSIA. E-mail: nio3@ch70.chel.SU

A computation analysis is performed to investigate hydromagnetic instabilities in double gas-puff plasma. The focus of this research is on the Rayleigh-Taylor instabilities suppression by means of an axial magnetic field. The scheme will led to more stable implosion results. The dynamic evolution of longwave perturbations produced by gas injection, which can led to "zippering" effect is also studied. The computations are performed using the 2D half-implicit single fluid magnetohydrodynamics (MHD) code TIGR-M. It is based on Euler- Lagrange description of motion with special velocity vector decomposition in movable coordinate basis.

1. INTRODUCTION

During the past several years, Z-pinches and gas-puffs have become an important means to provide pump radiation for X-ray laser physics experiments, to produce X-rays for material studies, to form high temperature and density plasmas, ultrahigh magnetic fields and for nuclear fusion (see review paper[1]). Since the early days, it has been generally accepted that simple Z-pinches (compressed plasma column) are violently unstable to both the sausage and hydromagnetic Rayleigh-Taylor (RT) modes[2-4]. These instabilities represent the main obstacles for using the pulse plasmas because they can destroy the symmetry of the imploding shell, thus degrade radiation production. To date, a number of research groups have studied the instability development. Since the early papers it is known that sausage modes of instability can be suppressed by axial magnetic field[5]. But the classical pinch instability sausage modes grow slower than the hydromagnetic RT instability when the shell radius is large compared to the wavelength[6]. The analysis by A.B.Bud'ko et.al.[3] shows that a relatively weak axial magnetic field present in the beginning of implosion produces a "window of stability", i.e., a domain in the space of the parameters of the plasma motion for which the RT instability modes are completely suppressed. Experimental investigations at High-Current Electronics Institute (Tomsk) have also demonstrated[7] that the largest degree of radial compression and suppression of instabilities can be achieved with the use of axial magnetic field. In this paper we investigate the dynamics of argon double gas-puff. The focus of this study is on the dynamic evolution and treatment of suppression effect in detail rather than the determination of an optimal design configuration. The effects are investigated: evolution of disturbances caused by instabilities at the plasma-magnetic field interface, the longwave perturbations produced by a axial injection for gas-puff implosions. These long perturbations can lead to "zippering" effect[8], i.e. axial nonuniformity at stagnation. The numerical calculations were carried out using the two-dimensional half-implicit single fluid Euler-Lagrange MHD code TIGR-M[9].

2. TWO-DIMENSIONAL MHD CODE TIGR-M.

The TIGR-M simulation model is code for numerical simulation of two-dimensional flows of heat-conducting gases in complex systems[9]. The model has been modified to include effects of fluids moving in magnetic and other fields. Euler-Lagrange description of motion with special velocity vector decomposition in movable coordinate basis is taken as a principle of this technique.

The whole system of MHD equations which takes into account heat conduction and is solved according to the technique TIGR-M has the form[10]:

$$\frac{d\rho}{dt} + \rho \cdot \mathrm{div}\mathbf{u} = 0; \tag{1}$$

$$\frac{d\mathbf{u}}{dt} + \frac{1}{\rho}\cdot \mathrm{grad}P = - \frac{1}{4\pi\rho}\cdot [\mathbf{H}\cdot \mathrm{rot}\mathbf{H}]; \tag{2}$$

$$\frac{\partial \mathbf{H}}{dt} = \mathrm{rot}[\mathbf{u}\cdot\mathbf{H}] - \mathrm{rot}(\chi\cdot \mathrm{rot}\mathbf{H}) ; \tag{3}$$

$$\mathrm{div}\mathbf{H} = 0 ; \tag{4}$$

$$\frac{dE}{dt} + \frac{P}{\rho}\cdot \mathrm{div}\mathbf{u} = \frac{1}{\rho}\,\mathrm{div}(\varkappa\cdot \mathrm{grad}T) + \frac{\chi}{4\pi\rho}(\mathrm{rot}\mathbf{H})^2, \tag{5}$$

where equations of states for matter and kinetic coefficients have been determined by relations: $P = P(\rho,T)$, $E = E(\rho,T)$, $\varkappa = \varkappa(\rho,T)$, $\sigma = \sigma(\rho,T)$, $\chi = c^2/4\pi\sigma$, and ρ – density of matter, E – specific internal energy, P – pressure, **u** – velocity, T – temperature, $\varkappa$ – coefficient of heat conduction, σ – coefficient of electrical conduction, **H** – magnetic field strength.

In the numerical technique TIGR-M, a mixed Euler-Lagrange way of gas motion description is applied. One coordinate lines family coinciding with interface is Lagrange one, the other is Euler one. This way allows to follow interfaces and to compute overflowing of matter in layers.

Euler's family of coordinate lines represents a set of straight lines (rays) not intersecting in the region of solution.

Mixtures are computed according to a concentration method.

Equations of continuum motion are written in this movable curvilinear coordinates with consideration of the fact that the corresponding contravariant component of velocity vector remains continuous on the boundaries coinciding with Lagrange family of coordinate lines. Such expansion allows not to separate interfaces especially while performing computations.

Equations of discontinuity (1), motion (2), magnetic field (3) and energy (5) are solved by the method of splitting according to directions[11]: angular (i.e. Euler direction) and radial (i.e. Lagrange one). Implicit algebraic equations thus obtained (after linearization of non-linear terms) are solved by run.

In overwhelming number of applications it is possible to confine oneself to two particular cases of magnetic fields configuration:

- two components of velocity and two components of magnetic field lying in the same plane as the velocity vector does;
- two (or three) components of velocity and one magnetic intensity vector component which is normal to velocity vector.

For the first case the representation of field via vector potential appears to be advantageous because in this case it is sufficient to solve the equation for a single non-zero component of vector potential for determination of the field.

For the whole series of practically interesting MHD problems the condition for stability of separate runs[10] appears to be too much strict; however this condition can be avoided in the so-called combined runs method [12], in which in the first stage, equations of dynamics, kinematics and electrodynamics are solved by Newton's iterations but in the second stage equation of energy is solved and dissipative processes are taken into account.

In the technique TIGR-M, equation of heat conduction is solved by the scheme of splitting according to directions[11]. The difference equation is written on a nine-dotted pattern. Equations for vector potential in the case of longitudinal magnetic field (the poloidal one for axial symmetry) and equation for field strength in the case of "transversal" (azimuthal) magnetic field have the form of a two-dimensional diffusion equation, therefore it is possible to apply the known difference methods for solution of such equations.

Some computations have been performed for two-dimensional problems of a rather common class of motions with uniform deformation which are characterized by linear relationship between velocity and coordinates. Such exact solutions of the MHD-equations system appear to be highly useful for theoretical analysis of multidimensional motions and represent a rather non-trivial class of solutions used for testing numerical simulation programs of continuum multidimensional flows.

Detailed investigation of such class of MHD-flows has been performed in work[13] by V.A.Simonenko and one of the authors of the given work.

Among multiple applications of the TIGR-M it is necessary to note the following:
- development of axially asymmetric perturbations in A.I.Pavlovskii's cascade magnetocumulative generator;
- astrophysical problem on gravitational compression of a rotating homogeneous gas cloud;
- 2D effects in laser-fusion target experiments and other applied problems.

At present the TIGR-M is included in the more general complex of computer programs (the method of splitting according to geometry has been applied) with allowance for many other physical processes in addition to possibility to take into account transfer of energy in three-temperature approximation.

The assumption that the calculations of double gas-puff implosion will be performed in the one-temperature approach allows to use of the code for simulation during first stage of implosion. Two-temperature 1D code ZARYA predicts that the one-temperature approximation is adequate till outer shell reaches inner core.

3. RT INSTABILITY GROWTH RATE WITHOUT EXTERNAL MAGNETIC FIELD.

RT instability could be severe enough to disrupt gas-puff during acceleration of plasma. This instability, its nonlinear saturation and wall instabilities have been studied mainly in connection with the relevant experiments[6,8,14,15]. RT modes are found to be fastest growing and most dangerous. RT instability generally refers to the well-known instability of a "heavy fluid", which is usually assumed to be incompressible, supported by a "light fluid" in a gravitational field. It occurs in accelerating fluid systems. A magnetic field may be regarded as a "light fluid". Classically the RT instability is considered as the linear growth of a small amplitude perturbation. We will study its initial nonlinear feature.

The liner configuration has been chosen to be modeled is representative of experiments being conducted at Tomsk[16]. The current driving the plasma dynamics was a sine wave with an amplitude of 1 MA and a quarter of period of 100 ns. All simulations were performed for argon double gas-puff with parameters: mass loading of 40 $\mu g/cm$, density of 27 $\mu g/cm^3$ (equal in outer and inner shells), radius of inner core of 0.5 cm, average radius of outer shell of 1.35 cm and thickness of 0.1 cm. The temperature of plasma was taken to be about 3 eV initially. The density of the background plasma between shells was set for numerical convenience and was taken ten times less than in shell's one.

To investigate the development of disturbances a series of MHD calculations were performed with different wavelengths and different initial perturbations. Note that all these calculations were carried out using sinusoidal current pulse.

The initial small sinusoidal displacement of outer shell is described by

$$r(z) = r_0 + A\cdot\cos(2\pi kz),$$

where k is the wave number of the perturbation, A is the amplitude, r_0 is initial radius. On account of the periodicity of the cosine function, we can take the range of z from 0 to 1/2k instead of wave length $Z_0=1/k$.

The history of initial perturbation for A/r_0 = 3% and k=2 (two wavelengths on the full length of liner) are shown in Figure 1.

At the first stage the outer shell accelerated to the axis. Although it has being significantly distorted during implosion, the inner core remains to be uniform to about 100 ns. Because the model is non-radiative, the time of stagnation exceeds the time of maximum current. Initial stage of growth of RT instability is a linear, exponential growth which transitions to a nonlinear phase. Linear stage is limited the bubble rise when there is no more mass of thin shell to retard the growth of the bubble. By 100 ns outer shell reaches the inner core. The last picture at 117 ns corresponding approximately to stagnation shows the maximum perturbations. The boundary (left and right) radii versus time are shown in Figure 2.

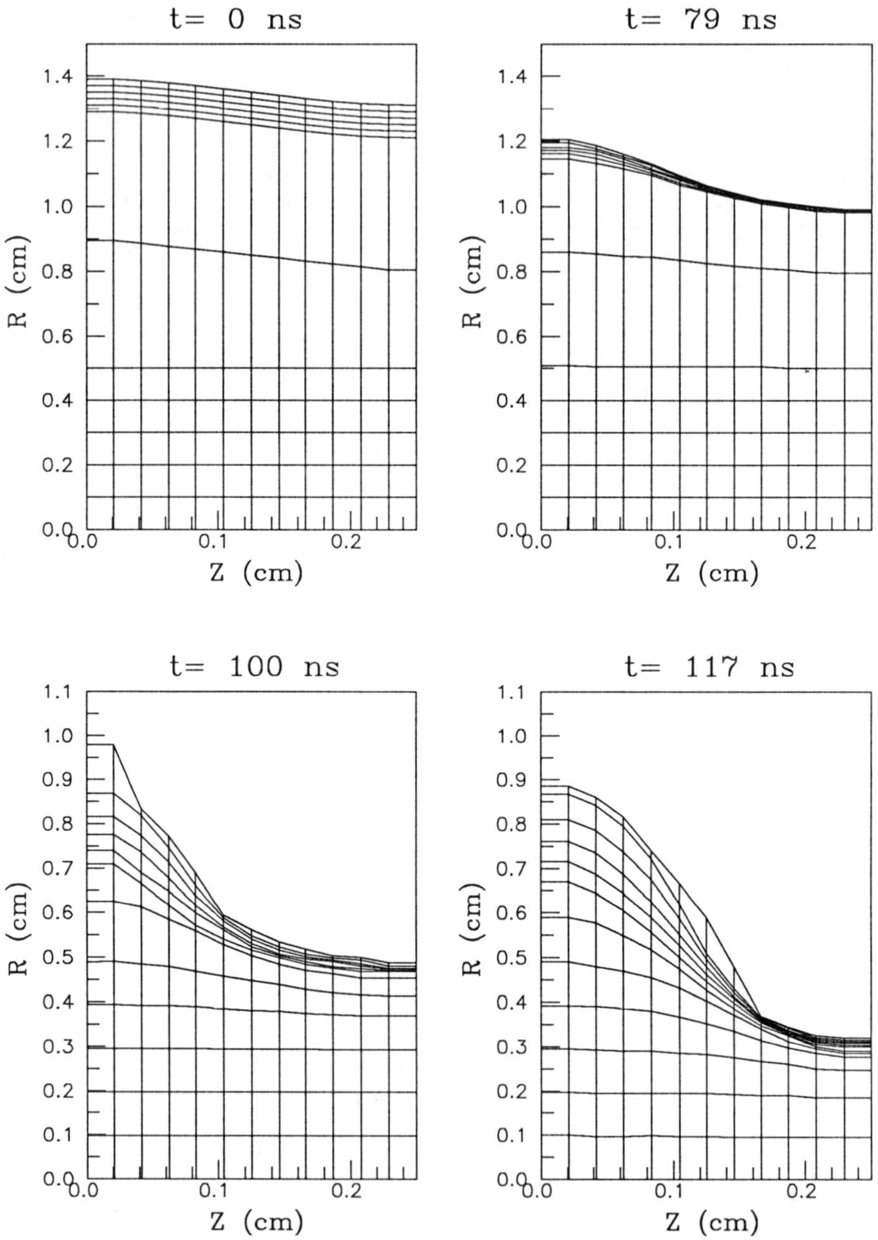

Fig.1. TIGR grid for Rayleigh–Taylor instability simulation in double gas–puff.

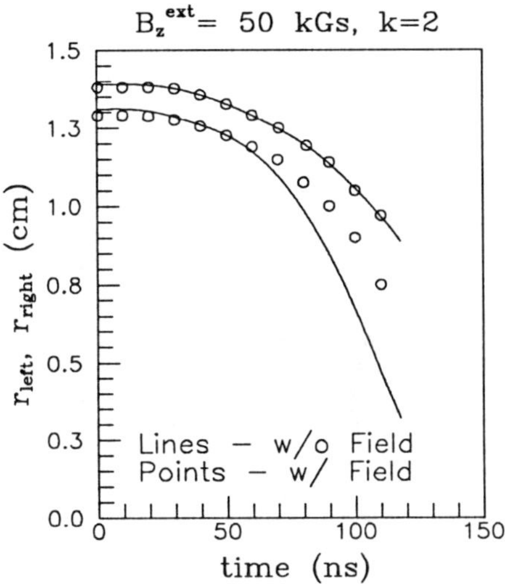

Fig.2. Boundary radii vs. time

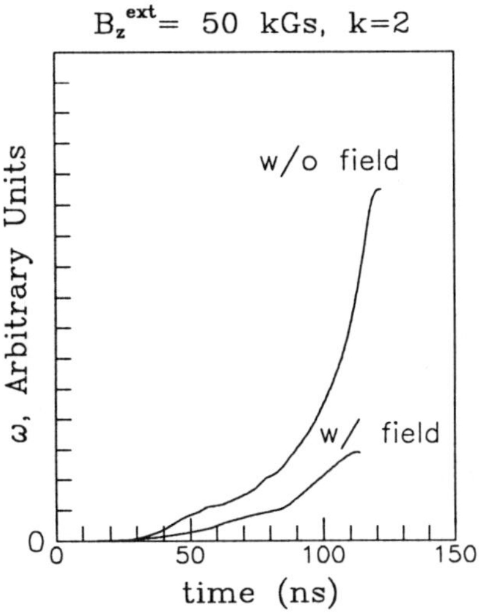

Fig.3. Growth rate vs. time

We may to propose two different ways to estimate the nonlinear growth of perturbations in numerical simulations:

a) integral ($\Omega(t)$ - growth rate),

$$\Delta r^{n} = \Delta r^{0} \exp(\Omega_n t_n) \ ;$$

b) local ($\omega(t)$ - growth rate),

$$\Delta r^{n} = \Delta r^{n-1} \exp(\omega_n \ \Delta t_n) \ ,$$

where $\Delta r^n = r_{max}(t_n) - r_m(t_n)$, r_{max} - radius of "bubble", r_m = $r(z_0/2)$ - middle radius of the perturbation, z_0= 1/2k.

The connection between Ω_n and ω_n is given by the equation

$$\Omega_n = \frac{1}{t_n} \sum_{k=1}^{n} \omega_k \ \Delta t_k.$$

We shell consider the local growth rate ω_n in our numerical study because the rate is more responsible to development of perturbation versus time.

Figure 3 shows the growth rate coefficient ω versus time for k=2.

Calculations performed for different wave number of disturbances are shown growth rate increasing for greater k that is in accordance with linear theory of RT instability.

Computational results indicate that RT instability plays a significant role in implosion process.

4. CALCULATIONS USING AXIAL MAGNETIC FIELD.

A number of calculations were carried out using previously imposed axial magnetic field. Because of the high magnitude of plasma conductivity the field penetration into the body of plasma is

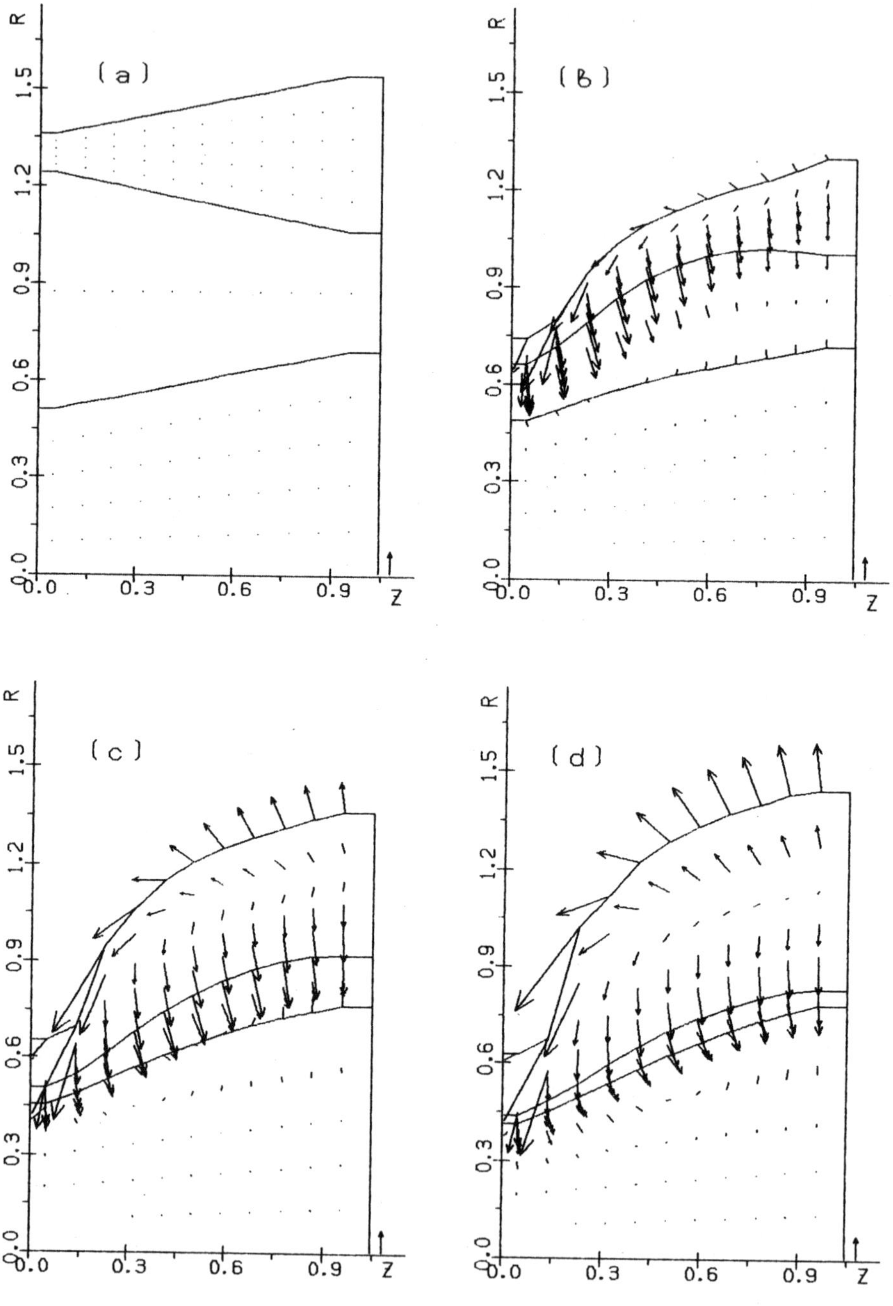

Fig.4. Velocity fields for "zippering" calculation.

small. For usual parameters: t=30 ns, T=100 eV the skin depth is about 0.03 cm. The same value is predicted by numerical simulation. Thus we assume that the magnetic pressure can be calculated by $H^2/8\pi$ applying to the external bound of plasma body. According to the recommendation of the paper[3], we used H^{ext} = 50 kGs. Boundary radii and growth rate dependence are shown in Fig.2, 3. Comparison of the curves shows that magnetic field effectively suppresses instability.

5. "ZIPPERING" EFFECT MODELLING.

"Zippering" effect is a serious obstacle to obtain a uniform implosion. Some methods for the zipper time reducing have been suggested in the way of using of inwardly titled gas puff nozzle. To study this effect numerically, two dimensional simulations were made for the case of initial spatial gas density distribution. In Fig.4 the velocity fields for different times: (a) - 0 ns, (b) - 100 ns, (c) - 113 ns, (d) - 117 ns are shown. Calculations give the zipper time of about 15 ns, which is in good agreement with the experimentally measured one. It is interesting note that the angle of slope to axis of liners surfaces at the time t=120 ns is about two times of the angle of initial inner plasma core configuration.

6. CONCLUSION.

Disturbances during implosion of double gas-puff have been studied by numerical simulation. Results show that the most dangerous modes are the small wavelength. These instabilities can be suppressed by using of axial magnetic field. The dynamic evolution of longwave perturbations produced by gas injection is also studied. More detailed calculations are now in progress to use more adequate model for three component magnetic field calculations and to extend obtained results over a wider range.

ACKNOWLEDGMENTS

The authors would like to thank Dr. V.I.Afonin for his initiative of investigating Z-pinch at NITPh, A.N.Shushlebin for his work on the code development, and V.A.Rot'ko for his help in numerical simulations. We are also grateful to Dr. R.B.Bakst, Dr. S.A.Sorokin from HCEI for stimulating discussions of the physics of double gas-puff implosion and axial magnetic field using.

REFERENCES

1. N.R.Pereira and J.Davis, J.Appl.Phys. 64(3), R1 (1988).
2. E.G.Harris, Phys.Fluids 5(9), 1057 (1962).
3. A.B.Bud'ko, M.A.Liberman, A.L.Velikovich, and F.S.Felber, Phys.Fluids B 2(6), 1159 (1990).
4. M.Lampe, Phys.Fluids B 3(7), 1521 (1991).
5. G.Bateman, MHD Instabilities (MIT Press, Cambridge, MA, 1978).
6. N.F.Roderick, and T.W.Hussey, Second Int. Conf. on Dense Z-pinches, Laguna Beach, 1989, p.157.
7. S.A.Sorokin and S.A.Chaikovsky, Proc. of 5th Int. Conf. on Megagauss Magnetic Field and Pulsed Power Systems, MG-5, Ed. by

V.M.Titov and G.A.Shvetsov, p.719.
8. T.W.Hussey, M.K.Matzen, and N.F.Roderick, J.Appl.Phys., 59, 2677 (1986).
9. A.Yu.Bisyarin, V.M.Gribov, A.D.Zubov, N.A.Pervinenko, V.E.Neuvazhaev, F.D.Frolov, VANT, Ser.: Metodiki i Programmy Chisl. Resheniya Zadach Matem. Fisiki", 3 (17), 34 (1984) (In Russian).
10. A.A.Samarskii,Yu.P. Popov, Difference methods for solution of gas dynamics problems (Nauka, Moskva, 1980), (In Russian).
11. N.N.Yanenko, Method of fractional steps for solution of multidimensional mathematical physics problem (Nauka, Novosibirsk, Sib. otd., 1967), (In Russian).
12. V.A.Gasilov, V.Ya.Karpov, A.Krukovskii, Preprint IPM Ak. Nauk SSSR, No.54, (1984). (In Russian).
13. A.D.Zubov, V.A.Simonenko, VANT, ser.: Teoreticheskaya i prikladnaya fizika, 2 (45), 45 (1987). (In Russian).
14. F.L.Cochran and J.Davis, Phys.Fluids B, 2 (6), 1238 (1990).
15. I.R.Lindemuth, Proc. of 5th Int. Conf. on Megagauss Magnetic Field and Pulsed Power Systems, MG-5, Ed. by V.M.Titov and G.A.Shvetsov, p.719.
16. S.A.Sorokin and S.A.Chaikovsky, Preprint No.12, Tomsk Scientific Center, Siberian Division of RAN, 1992.

X-RAY SOURCES AND X-RAY LASERS

DEMONSTRATION OF THE X-PINCH AS A LABORATORY SCALE SOFT X-RAY SOURCE

C Christou, Imperial College, London SW7†

ABSTRACT

The X-pinch plasma is generated by passing a high electric current through a crossed wire load. X-pinches investigated in this work are driven by two current generators: a 4kJ, 30kV capacitor bank with a quarter-period of 1.2μs and maximum current of 320kA; and a 4kJ, 360kV generator with a maximum current of 100kA and a 90ns rise-time. Plasma light and X-ray emission are monitored, and pinch dynamics are studied using schlieren photography.

The X-pinch is observed to be axially asymmetric about the central crossing point and a plasma column is seen to form along the bisector of the angle between the two twisted wires.

X-ray emission is dependent on the plasma atomic number, *Z*, with substantial emission from the anode side of the X-pinch for a high *Z* load, but a single intense source with low *Z*. Hard X-rays are always generated at the anode. Emission characteristics are compared with z-pinches driven by the same generators.

X-pinch X-ray yield in the 3 to 12Å range is measured to be up to 1J per shot, and analysis of soft X-ray continuum emission indicates the presence of a 1 to 2keV plasma in the aluminium X-pinch. This point-like, intensely emitting plasma has potential applications in X-ray microscopy and lithography.

THE DISCHARGE CIRCUITS

Schematic diagrams of the two discharge circuits are shown in figures 1 and 2. The "long pulse" circuit consists of nine 1μF capacitors mounted in parallel on a copper sheet which is isolated from the evacuated discharge chamber by a nitrogen-pressurised spark gap. The capacitor bank is generally charged to 30kV. In the "short pulse" circuit, shown in figure 2, a 360kV Marx bank is discharged into a water filled coaxial transmission line, causing an SF_6 filled gap to break down, thus delivering around 100kA to the load. The quarter-period time for the long pulse circuit is 1.2μs whereas the 10% to 90% rise time of the short pulse circuit is 90ns.

Previous X-pinch work[1,2] has been carried out using discharge circuits similar to that of figure 2; this work is the first study of X-pinches driven by a low voltage capacitor bank with a relatively long current rise time.

†Present address: Frazer-Nash Consultancy, Leatherhead, Surrey KT22 7TX.

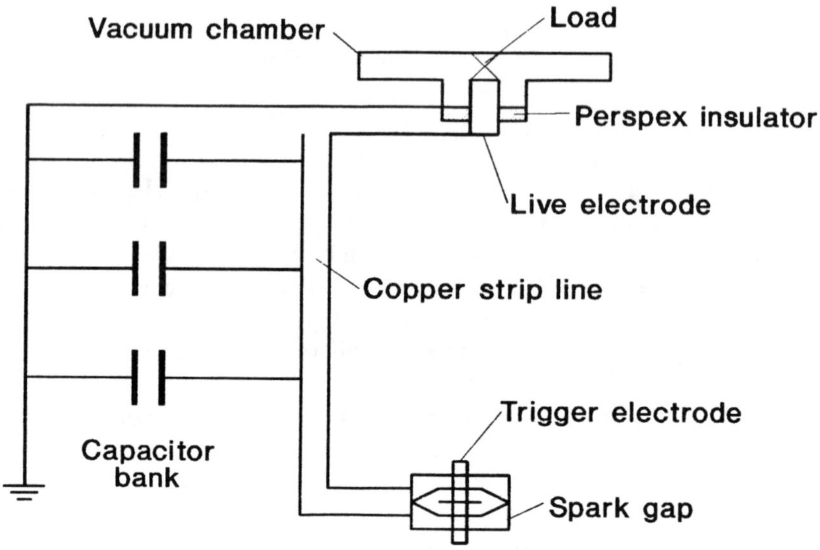

Figure 1: Schematic diagram of long pulse discharge circuit.

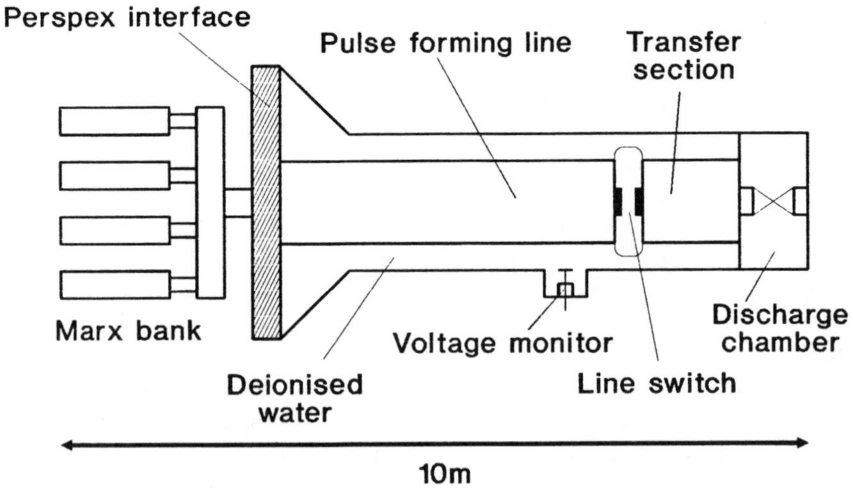

Figure 2: Schematic diagram of transmission line.

OPTICAL PICTURES

Figure 3 shows an X-pinch formed from two 18μm tungsten wires and driven by the long pulse capacitor bank. The picture is simply a long exposure image taken with a heavily filtered open shutter camera. This figure clearly shows the existence of a new column of emitting material passing along the anode-cathode axis through the cross point of the wires. Many similar images were obtained for a great variety of materials. A 50ns optical framing camera was used to obtain figures 4 and 5, driven by the long pulse and short pulse bank respectively. The aluminium X-pinch of figure 4 shows the asymmetric behaviour of the X-pinch apparent at times later than the initiation of X-ray emission. The cathode side of the discharge retains the cross wire structure and shows the new limb leading from the cross point, but the anode side has degenerated into a poorly defined plasma filament and cloud. The tungsten X-pinch in figure 5 shows the same asymmetry, but with a less well defined axial filament.

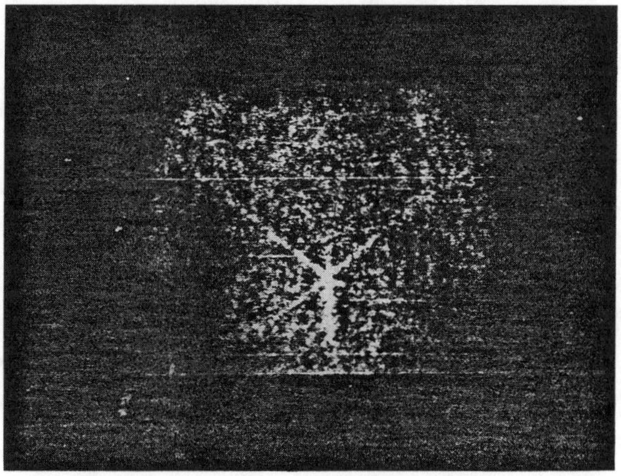

Figure 3: 18μm tungsten X-pinch.

SOFT X-RAY PINHOLE PHOTOGRAPHY

The asymmetry of the X-pinch is most obvious in the soft X-ray emission from the discharge. Figure 6 shows a 13μm molybdenum X-pinch driven by the long pulse generator. The image is obtained using a 100μm pinhole and a 10μm beryllium and 2.5μm mylar filter. It can be seen in this figure that soft X-rays are emitted from the cross point of the wires and from the limbs of the cross on the anode side of the cross point. The emission pattern is dependent on the atomic number of the load, as can be seen from figure 7, a 200μm pinhole picture of an aluminium X-pinch driven by the long pulse bank and filtered by beryllium and mylar as before. Here, the emission is almost entirely localised to the crossing point of the wires. Previous analysis[3] suggests that the intense emission along the limbs of the anode side of the wire is due to the acceleration of electron beams at the cross point. The number of free electrons generated at the cross point increases with atomic number and so the limb emission from low atomic number X-pinches is lower than that in high atomic number X-pinches.

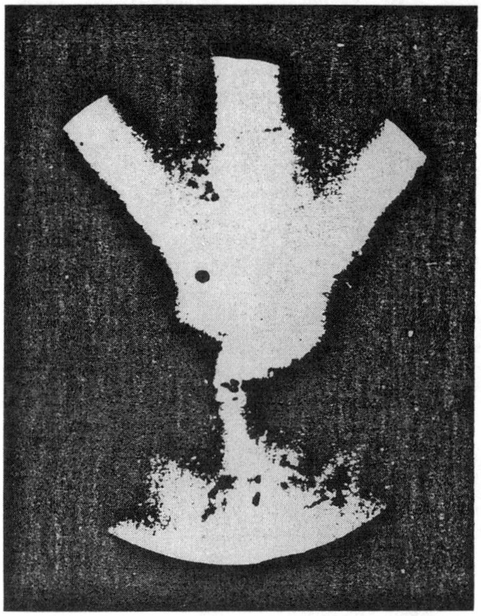

Figure 4: 50ns optical frame of aluminium X-pinch - long pulse.

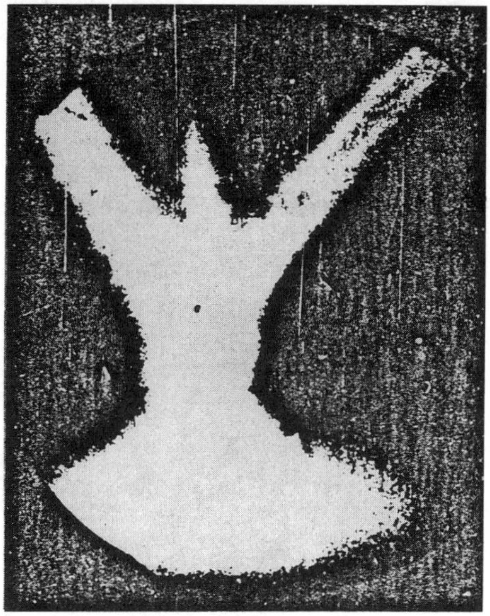

Figure 5: 50ns optical frame of tungsten X-pinch - short pulse.

Geometrical interpretation of the backlighting of wire arrays[4] shows that in low atomic number discharges driven by the long pulse capacitor bank, the diameter of the emitting region at the cross point of the X-pinch can be as little as 50μm.

Short pulse X-pinch discharges show similar characteristics of soft X-ray emission as the long pulse but with a much more intense anode emission.

SCHLIEREN PHOTOGRAPHY

Schlieren photographs of long pulse X-pinches and z-pinches were obtained using a rhodamine 6G dye cell which was pumped by a nitrogen laser. The pulse length was 3ns. Figure 8 is a schlieren photograph of a 25μm Cu/Ni alloy X-pinch. This figure shows the axial plasma column extending from the centre out to the cathode and towards the anode. Schlieren photographs taken at different times

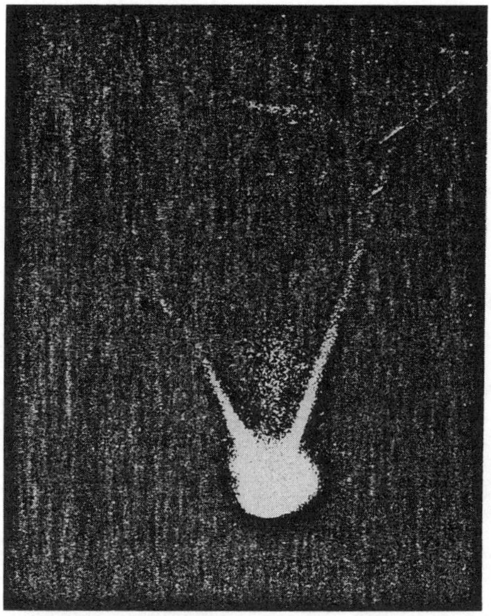

Figure 6: Soft X-ray pinhole photograph of 13μm molybdenum X-pinch .

Figure 7: Soft X-ray pinhole photograph of 15μm aluminium X-pinch.

in the discharge show that this middle column grows from the cross point out towards the electrodes, and also show that the plasma structure on the anode side of the cross point is dispersed before the plasma on the cathode side[3], as shown in figures 4 and 5.

Figure 9 is a schlieren photograph of a long-pulse z-pinch formed from a 122μm copper wire. This figure shows the regular spikes characteristic of an m=0 or Rayleigh-Taylor instability. It is interesting to note the relative lack of these instabilities in the X-pinches carrying comparable currents; this is thought to be due to the fact that there are two current paths in the X-pinch, resulting in a lower current density in each limb.

X-RAY CALORIMETRY

The total soft X-ray yield in the 1 to 20Å region of interest for X-ray lithography and microscopy was measured to be of the order of 100mJ to 1J per shot for the long pulse X-pinch[5]. Increasing the diameter of the original load wires acts to increase the total yield of the discharge up to a maximum beyond which the X-pinch does not form. There is a significant increase in X-pinch yield as the atomic number of the load wires is increased, but this radiation is emitted from a much larger region of the discharge.

Figure 8: Schlieren photograph of 25μm copper/nickel alloy X-pinch.

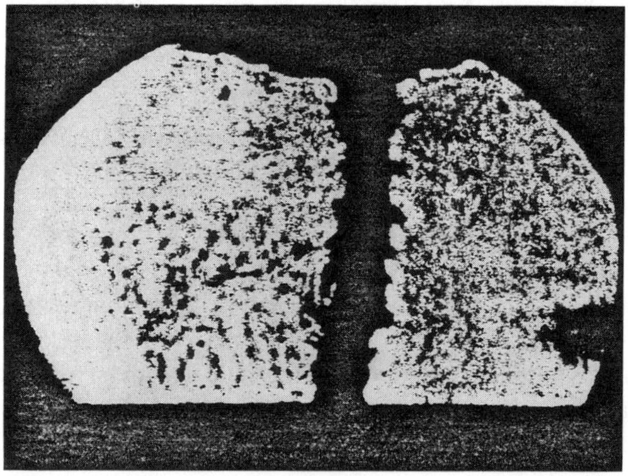

Figure 9: Schlieren photograph of 122μm copper z-pinch.

Z-pinch soft X-ray yield increases with atomic number of the load but the absolute soft X-ray yield from an exploding wire z-pinch driven by the long pulse generator is smaller than that from the X-pinch by a factor of approximately 50 and is observed to come from a number of hot spots randomly located at points along the axis of the discharge[5].

X-pinches driven by the short pulse capacitor bank emit a similar flux of soft X-radiation as long pulse X-pinches carrying comparable currents, and the yield is seen to increase with atomic number as with the long pulse X-pinches. The main difference in the X-ray emission from the two banks is the emission of much more intense hard X-radiation from the short pulse discharges. This is due to the much higher operating voltage of the short pulse generator.

A rough determination of the temperature of an aluminium X-pinch was made by a two filter method. Two PIN diodes were exposed to the discharge, one filtered with a 10μm copper foil and one with a 25μm titanium foil. Comparison of the ratio of the signals to the ratio expected from a radiation model calculated to include bremsstrahlung and recombination radiation[6] allows a temperature to be determined. Figures 10 and 11 show the theoretical and experimental signal ratios. According to this model, the X-pinch temperature rises to a maximum of between 1 and 2keV during the soft X-ray emission phase. The two filter technique neglects high energy electron beams and radiation from plasma impurities, and so this plasma electron temperature should be taken as an order of magnitude estimate at best.

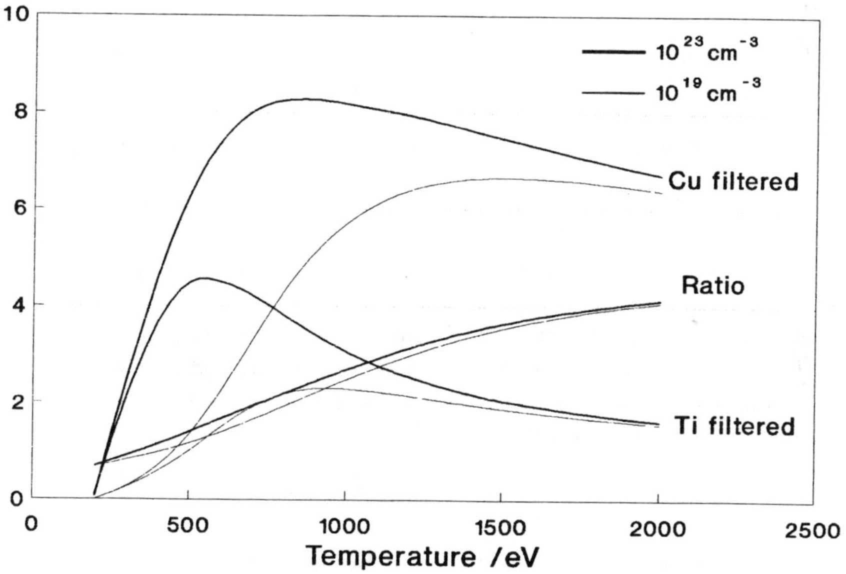

Figure 10: Theoretical prediction of PIN diode signal ratios.

SUMMARY

Work carried out in this study has shown that X-pinches and exploding wire z-pinches driven by a low power long pulse bank behave in a similar way to those driven by high power pulsed power devices. A new plasma column is formed between the limbs of the original wire cross and substantial soft X-ray emission is evident in both cases. Fast imaging of the X-pinch and X-ray pinhole photography show the gross asymmetries of the X-pinch discharge. The localisation of the emitting region of a plasma discharge to a single point allows the X-pinch to be used as a point source of radiation in a way that the z-pinch cannot.

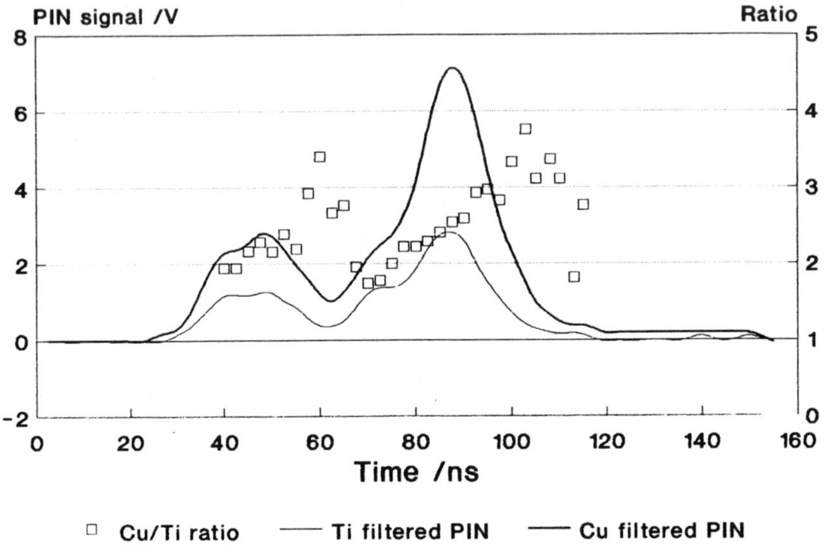

Figure 10: Experimental measurement of PIN diode signal ratios.

The soft X-ray yield from the long pulse and short pulse generators was of a similar intensity and, when driven by the long pulse circuit, the X-pinch source was found to be relatively free from hard X-ray emission. It would therefore appear that for practical applications, such as microscopy and lithography, the long pulse circuit is prefeıable to the short pulse circuit because it is cheaper, more mechanically robust and gives a "cleaner" soft X-ray spectrum.

REFERENCES

1 S M Zakharov, G V Ivanenkov, A A Kolomenskii, S A Pikuz & A I Samokhin, Sov. Tech. Phys. Lett., **8**, p155 (1982).

2 N Qui, D A Hammer, D H Kalantar & G D Rondeau, J. Quant. Spectrosc. Transfer, **44(5/6)**, p519 (1990).

3 P Choi, C Christou & R Aliaga, Proc. SPIE - Int. Soc. Opt. Eng. (USA),**1552**, p270 (1991).

4 C Christou & P Choi, Bull. APS, **36(9)**, p2413 (1991).

5 C Christou & P Choi, Proc SPIE - Int. Soc. Opt. Eng. (USA), **1552**, p278 (1991).

6 B V Robouch & J P Rager, J. Appl. Phys., **44(4)**, p1527 (1973).

ENHANCEMENT OF X-RAY PRODUCTION IN Z-PINCH PLASMAS USING MAGNETIC FIELDS

N.S. Edison, B. Etlicher, S. Attelan, C. Rouillé
Laboratoire PMI, Ecole Polytechnique, 91128 Palaiseau, France

A.S. Chuvatin
I.V.Kurchatov Institute of Atomic Energy, 123182 Moscow, Russia

R. Aliaga
Blackett Laboratory, Imperial College, London SW7 2BZ, UK

ABSTRACT

We are investigating the effects of an axial magnetic field to stabilize an aluminum vapor z-pinch. An aluminum plasma jet is created from an exploding foil in a DC magnetic field ($B_{z0} \leq 300$ G). The applied field is small compared to the azimuthal field, $B_{z0} << B_\theta$, and is intended to reduce the growth of instabilities during the compression phase. The pinch is driven by a 2 Ω, 0.1 TW generator (250 kA in 80 ns). Additionally, a micron sized wire may be placed on the pinch axis leading to the plasma-on-wire (POW) configuration. Qualitatively, increasing the axial magnetic field improves the pinch with the m=1 instabilities becoming negligible for fields higher than 150 G. We find that the externally applied fields can enhance x-ray production up to a critical field. Above this critical field x-ray emission decreases even though the pulse length of the radiation may still be increasing. As the applied field increases, the period of x-ray emission increases with the harder spectrum affected the least. The x-ray yield peaks for the POW and Al jet alone configurations at 150 G and 50 G respectively. Diagnostics include filtered PIN x-ray diodes, time-resolved schlieren photography, and time-integrated multiple filtered pinholes. We will present the results comparing the POW and aluminum jet configurations described above.

INTRODUCTION

Stabilization of z-pinches has long been a topic of interest in physics. Applications such as x-ray lasers or fusion depend upon the ability to consistently create a stable and uniform medium. Previous studies[1,2] show that targets placed at the center of an implosion can behave stably, for example, the POW configuration.[3] Much of our work revolves around improving the stability of the imploding shell. We are currently working with or have in the past worked with current shaping,[4,5] internally generated,[6] and externally applied magnetic fields.[7]

One method of reducing the growth of instabilities is to allow a magnetic field to penetrate into the implosion shell parallel to the pinch axis before the initiation of compression. For fast implosion times the magnetic field

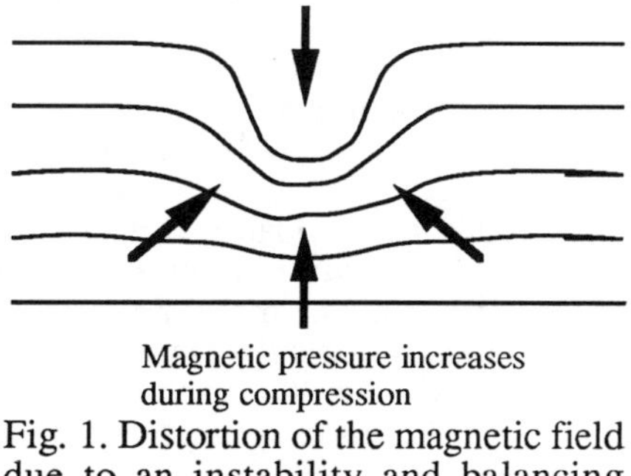

Fig. 1. Distortion of the magnetic field due to an instability and balancing magnetic pressure.

cannot diffuse out of the plasma and, in effect, would be frozen into the plasma. This axial magnetic field would stiffen the plasma along the pinch axis. Since the magnetic field is inside the plasma, the plasma sees a magnetic pressure increasing as the square of the local compression radius. Instabilities would be required to expend energy in order to distort the magnetic field (see figure 1), thus, reducing the kinetic energy of the ions and the growth rate of the instability.

Here we have purposely chosen the applied axial field, B_{z0}, to be small compared to the azimuthal field generated by the pinch, B_θ. Since $B_{z0} << B_\theta$, the work done in compressing the applied magnetic field is negligible compared to the discharge energy of the compression. The average kinetic energy per ion is independent of the applied field. Thus, the final compression diameter should be unchanged but with a more uniform distribution of density and temperature along the pinch axis.

EXPERIMENT

We have conducted this series of experiments on GAEL, a 2 Ω, 0.1 TW pulse-line generator (250 kA in 80 ns).[8] The load consists of a 25 μm diameter copper wire immersed in an aluminum plasma. The aluminum plasma is created by exploding a 5 μm foil and injecting the resulting plasma through a hole in the anode. Figure 2 shows a cross section of the load region and the aluminum plasma injection nozzle. The mass in the aluminum plasma is controlled by adjusting the time delay between the firing of the generator and exploding the foil.[9] A pair of Helmholtz coils provide a DC magnetic field up to 300 G parallel to the pinch axis.

Figure 3 shows the arrangement of diagnostics around the load region.

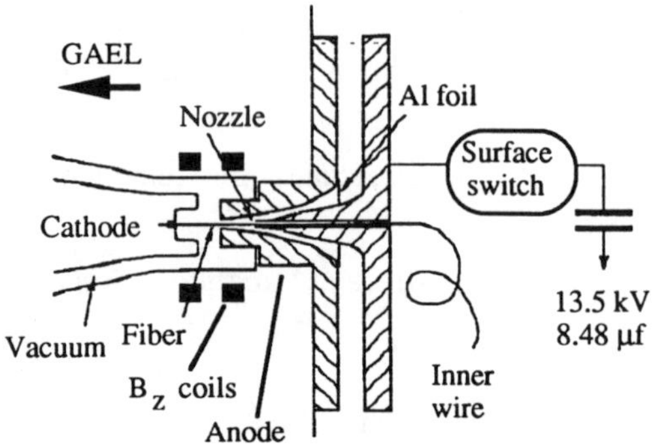

Fig. 2. Load region on GAEL.

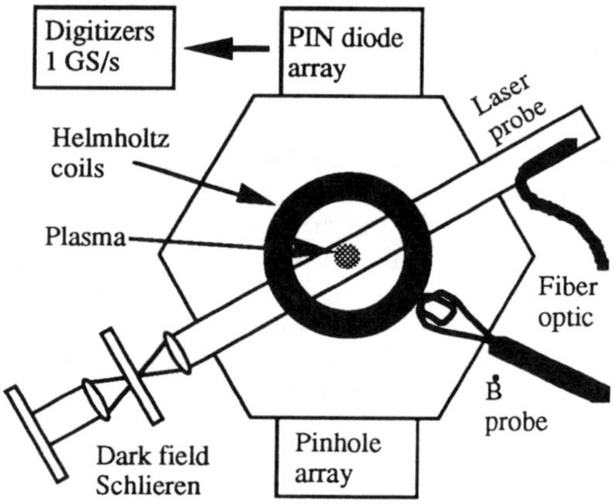

Fig. 3. Arrangement of diagnostics used on GAEL.

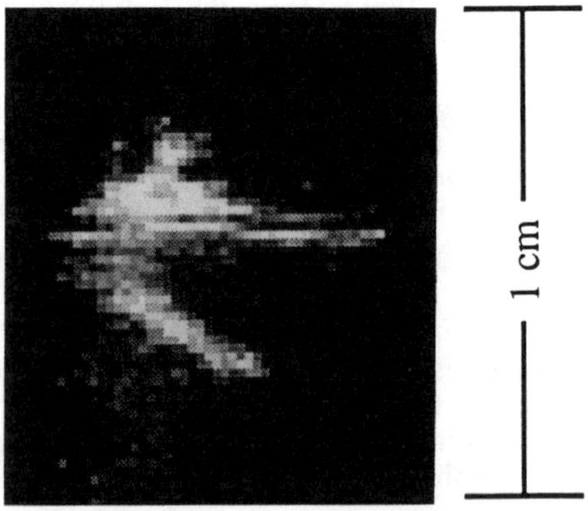

Fig. 4. Schlieren image of the POW configuration prior to the arrival of the main current pulse.

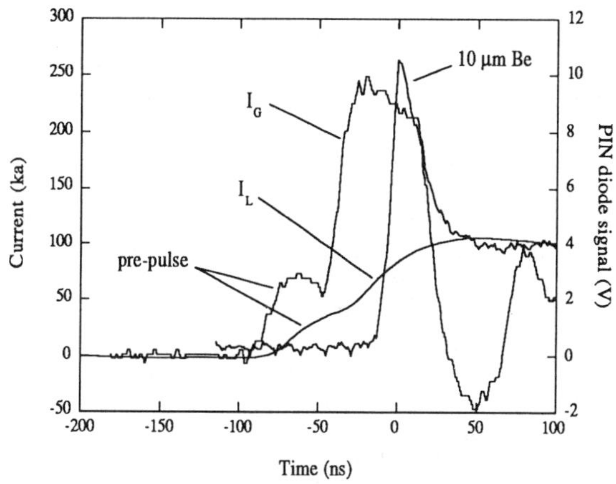

Fig. 5. Timing for generator current, I_G, load current, I_L, and a diode signal.

The view in figure 3 is along the load with the generator located behind the load region. The XRD array consists of ten filtered p-i-n diodes. Several of the filters have been chosen as Ross matched pairs in order to determine temperature from the bound-free and bound-bound spectrum. All XRD signals as well as the other electrical signals are recorded on digitizers driven by a single timebase at 500 MS/s or 1 GS/s. A fiber optic picks off part of the laser probe beam in order to determine timing of the schlieren photography.

We image the schlieren dark field with a 200 µm diameter wire parallel to the pinch axis to block the focal spot of the laser. The laser operates at 5896 Å with a 5 ns pulse width. Figure 4 is an example of a schlieren image showing the POW configuration just before the main current pulse arrives. One can see the fiber at the center of a conical structure formed by the plasma jet. The divergence of the jet implies a velocity near Mach 1.9.

dB/dt probes monitor the currents in the generator and in the load. The load current is integrated numerically and is assumed to be reliable only during the first 120 ns of the main discharge while the generator current is integrated electronically. Figure 5 shows the generator current, load current, and an XRD channel from a typical shot. The pinhole array consists of nine pinholes utilizing various pinhole diameters and filters to resolve a time integrated image of the pinch.

RESULTS

Figure 6 shows the load current for various applied magnetic fields. Both load configurations exhibit the same current characteristics which suggests that the current flows only in the outer plasma shell. Since the dB/dt probe is not reliable late in the discharge, we cannot see when the fiber has been ionized and begins to conduct. The current pulses begin with a 50 ns prepulse followed a main pulse lasting about 60 ns. The differences of the ratios of the amplitudes of the prepulse compared to the main pulse is due to differences in the final inductance of the load. For low magnetic fields, instabilities increase the load inductance which manifests itself as increased

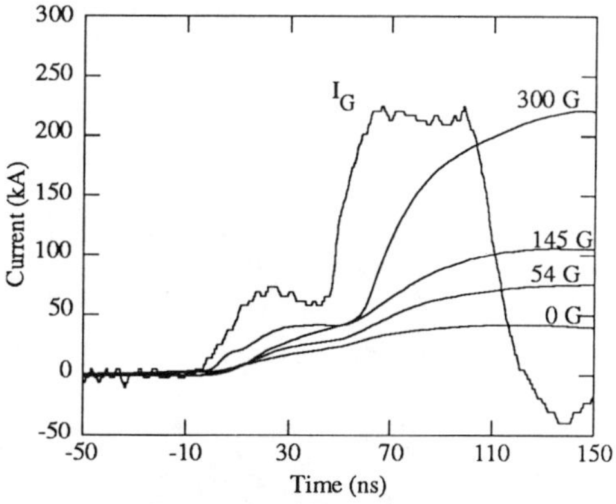

Fig. 6. Generator current, I_G, and load current for various applied magnetic fields.

current losses. As the magnetic field increases, the higher order instabilities (m≥1) are preferentially reduced, and the load current nearly matches that of the generator. The load current data is very reproducible with shot-to-shot variation only a few percent. Correlation of the load current, pinhole images, and x-ray diode signals gives an overall shot-to-shot variation to within 20% for identical experimental configurations.

Figures 7 and 8 show the results of the schlieren photography. The time of each image is determined with respect to the peak in the x-ray signals as given by the p-i-n diodes. The aluminum jet alone appears in figure 7. 15 ns before the maximum compression, we see zippering of the shell plasma from the cathode to about the center of the liner. This is probably due to the initial conical shape of the aluminum jet. The POW configuration data is shown in figure 8. In these images the Cu wire has been vaporized and is stable at the center of the implosion. Before and up to the time of peak compression, the imploding jet in both experimental configurations displays a large amount of structure. However, after the peak of the implosion (≈10 ns) the plasma is cylindrically symmetric and much more homogeneous along the pinch axis.

The late time schlieren images in figure 8 clearly show the remnants of the Cu wire on the pinch axis several nanoseconds after the peak of the implosion. Even though the entire wire has been vaporized and ionized, the original geometry remains essentially intact throughout the implosion. Measurements of the size of the core remnants reveal that the fiber has expanded to a diameter of about 200 μm. This gives a volume expansion factor around 100 which implies that the density has dropped to approximately $10^{21}/cm^3$ from solid density. One should note that the drop in density is observed well after the peak compression, so that during the peak of the implosion we would expect the density on axis to be even higher.

Figures 9 and 10 show three images from the pinhole camera array for four different applied magnetic fields. We have chosen typical data from a soft filter (10 μm Be), a medium filter (5 μm Ni), and a hard filter (10 μm Ti). Figure 9 shows the aluminum plasma jet alone. No signals are seen for the Ti filter and the signals for the Ni filter for fields above 140 G are very weak, but they have been included for contrast with the POW images.

For the aluminum jet configuration, increasing the applied field reduces the growth of instabilities, most notably m=1 instabilities. At magnetic fields less than 60 G, m=1 instabilities are readily apparent. However, the magnitude of the 'kinks' become smaller as the field increases from 0 G to 54 G. Above fields of 140 G, m=0 instabilities dominate the implosion, and the m=1 instabilities have apparently disappeared. In addition, the homogeneity of the aluminum plasma at the peak of the implosion tends to increase with the axial field. Even though the m=0 instabilities

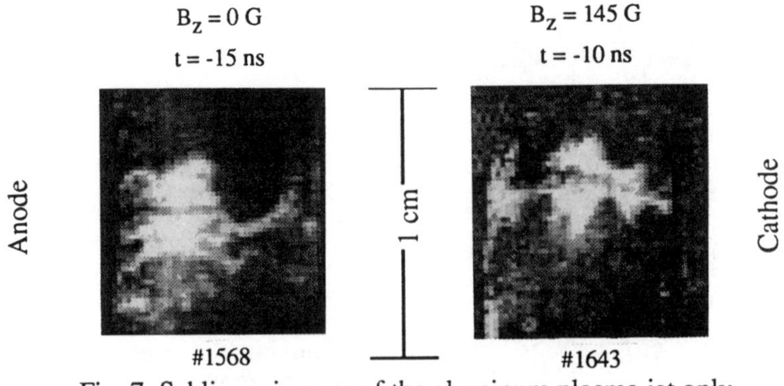

Fig. 7. Schlieren images of the aluminum plasma jet only.

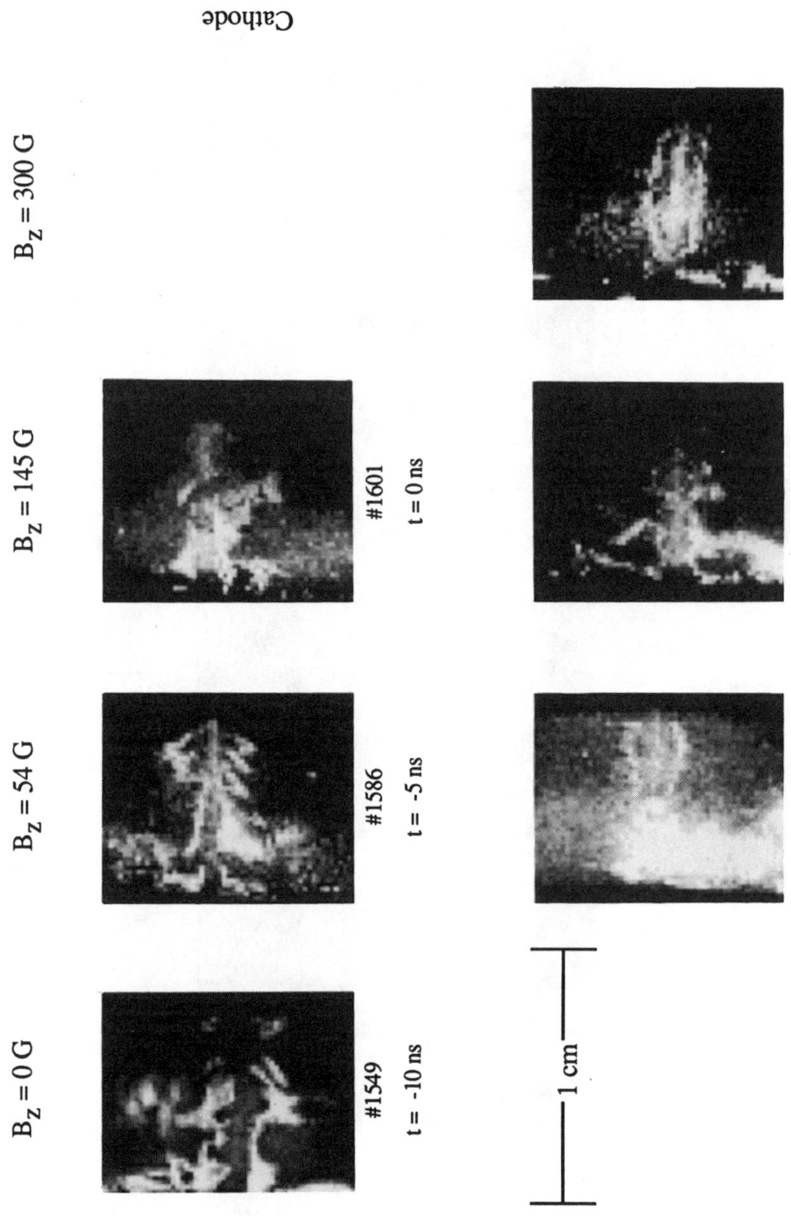

Fig. 8. Schlieren images for the POW configuration using various magnetic field strengths.

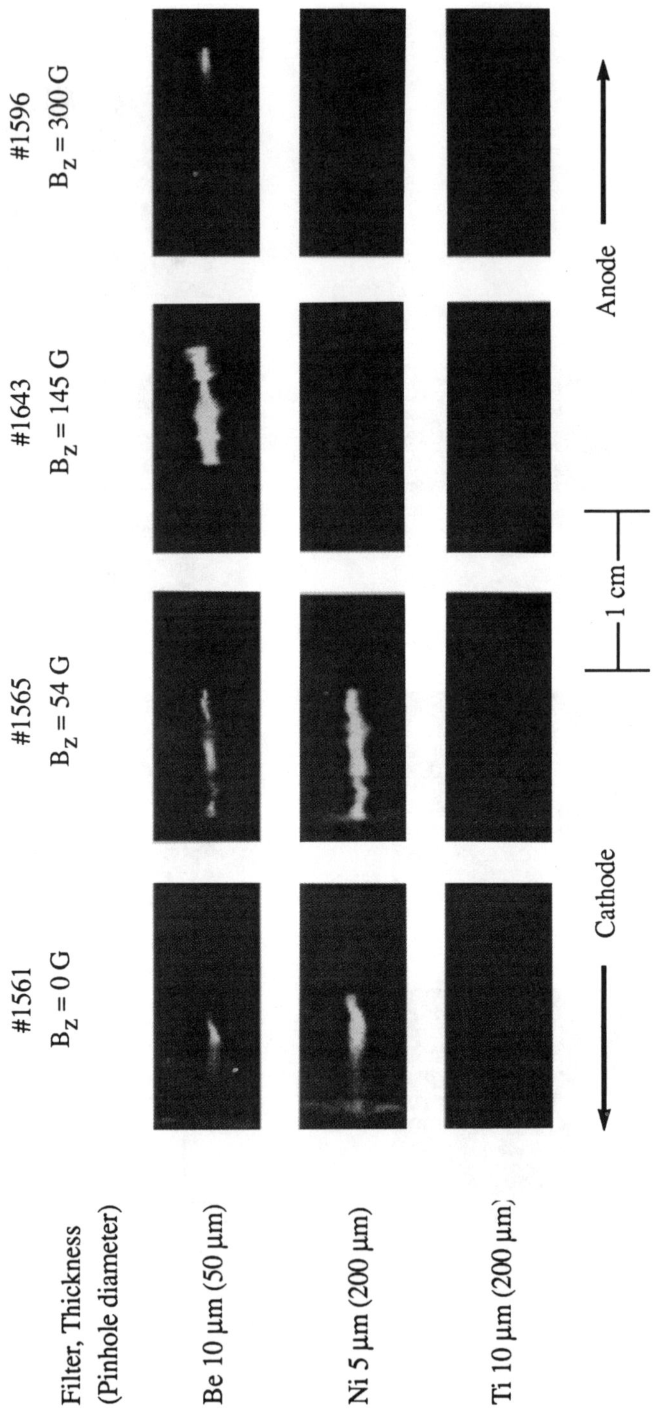

Fig. 9. Pinhole images for the aluminum jet only using various magnetic field strengths.

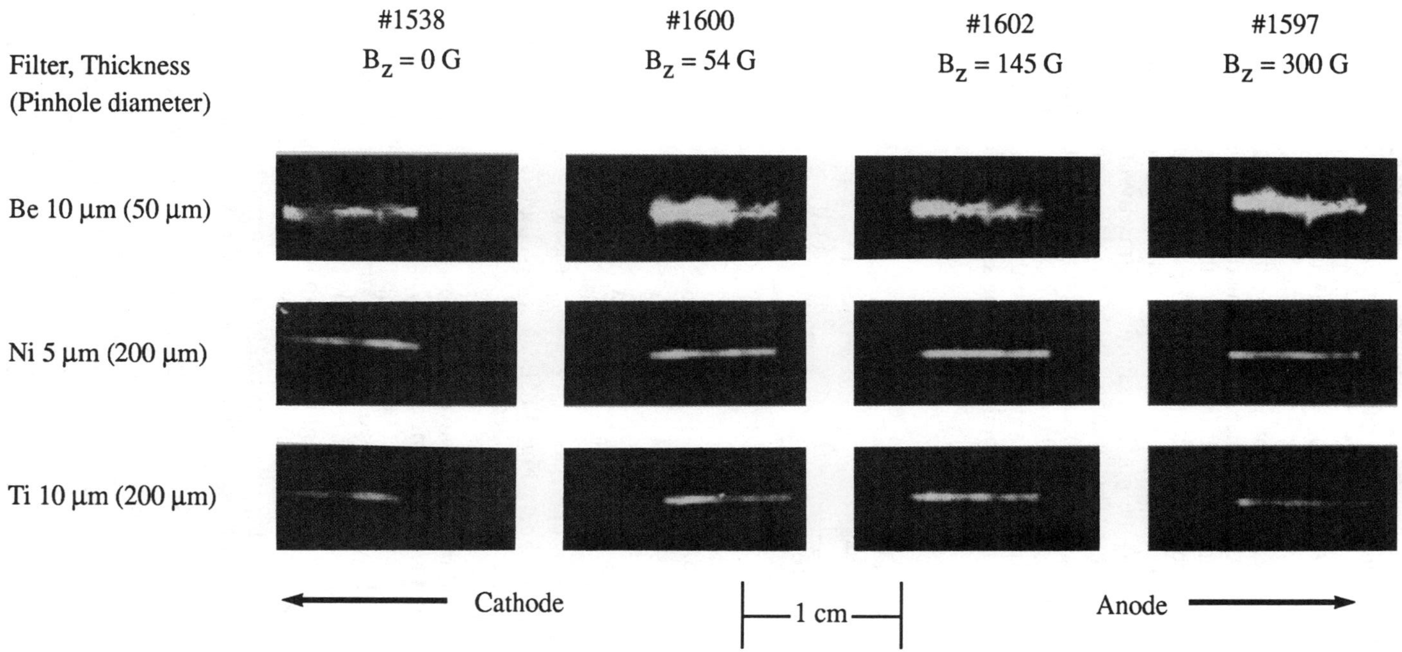

Fig. 10. Pinhole images for the POW configuration using various magnetic field strengths.

continue to be present, their spatial frequency and amplitudes have been reduced. For a field of 300 G, both m=0 and m=1 instabilities have been suppressed giving a plasma liner with radiating regions of homogeneity greater than 3 mm along the pinch axis. However, the compression at 300 G is not very good due to compression of the magnetic field.

Radiation from the aluminum jet is seen only on the Be and Ni channels which gives a maximum energy of 5 keV. As the magnetic field is increased the proportion of soft to hard radiation increases dramatically. This indicates that the temperature is lower and probably not capable of producing a significant amount of radiation above 2 keV. The increase in the soft radiation on the Be channel compared to the decrease in the peak x-ray signals (11) suggests that the pinch is emitting primarily in the UV and XUV spectrum for long periods of time. This is supported by the data in figures 12 and 13 which show that the radiation emission increases with the increasing magnetic field.

Figure 10 shows several of the pinhole images for the POW configuration. The introduction of the fiber markedly changes the emission from the pinch. For the POW configuration, data consistently appears on all pinhole channels. The hardest filter that we used is 5 μm Au (not shown here) and emission from the wire is clearly present. The Ni and Ti channels show emission from the wire only whereas on the softer channels both the copper wire and the aluminum plasma are present. The images from the Ti and Au channels require radiation of at least 6 keV. This strongly implies that we are exciting the K-shell of copper.

The pinhole images show that the wire maintains its shape during peak emission from the pinch. Data from a 20 μm pinhole with a 12 μm aluminum filter reveals that the wire has not expanded by more than a factor of six during the compression. This result agrees with the schlieren results showing a slightly larger expansion later in the evolution of the pinch. The beryllium filtered images, also, indicate that the wire has remained intact. However, these images are complicated by the presence of emission from the aluminum plasma. The other images exhibit penumbral blurring due to pinhole diameters several times larger than that of the wire.

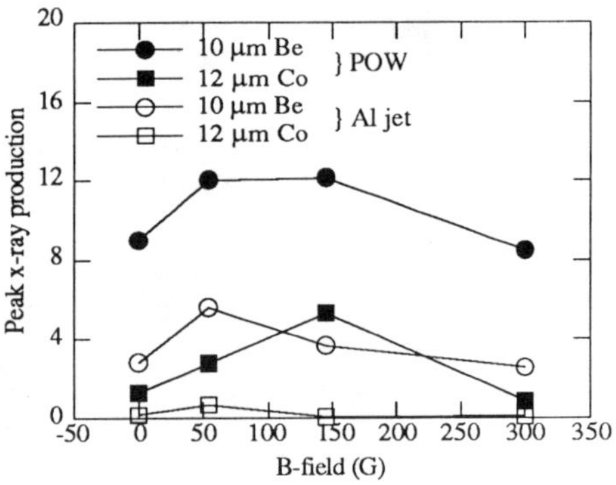

Fig. 11. Peak x-ray signals for different applied magnetic fields.

Another interesting difference between the two load configurations is the absence of emission from the aluminum plasma on the Ni channels for the POW configuration. This suggests that the fiber is absorbing a significant amount of energy from the imploding shell. If this transfer of energy is sufficient to vaporize and ionize the fiber, then we might expect the resistance of the fiber material to become much less than that of the imploding plasma. This provides a mechanism by which the current is switched from the shell plasma and into the fiber at the peak of the compression. This is very desirable since it eases constraints on the rise time of the current pulse, especially, with respect to the prepulse associated with capacitive pulse-forming generators.

One can correlate the x-ray diode data with the pinhole data to get an idea of the temporal history of the implosions. Additionally, since the x-ray signal widths depend on the hardness of the filters, one could, in principle, derive a temporal resolution based on the pinhole images alone. Temporal information from the p-i-n diodes can be seen in figures 12 and 13. In Figure 12 typical POW data temporally resolved is presented for fields of 0 G and 145 G, and the peaks of the signals have been normalized to unity in order to compare the pulse characteristics of all channels. In the absence of an applied field, all x-ray pulses tend to follow nearly identical temporal histories. For the aluminum jet alone, the softer x-ray channels see more radiation from the prepulse than the hard x-ray channels. As the magnetic field increases, the x-ray pulse widths become larger with the lower energy x-radiation pulses the most affected.

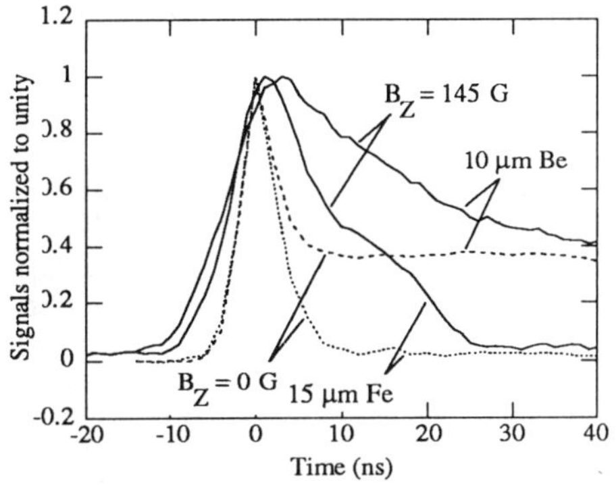

Fig. 12. Temporal evolution of the x-ray emission for two magnetic field strengths.

Figure 13 displays pulse widths taken from data such as that in figure 12 for various magnetic field strengths. The pulse widths in figure 13 show that the aluminum jet alone and POW configurations behave differently. Pulse widths for the POW configuration peak between 50 G and 150 G, whereas the aluminum jet widths tend to increase with the applied field. However, the diode signals of both configurations look very similar with respect to their rise times. This indicates that radiation in the POW configuration is primarily due to the presence of a wire which begins to radiate just before the peak of the shell plasma implosion. Note that the hard x-radiation for some of the aluminum jet implosions are very weak resulting in a large measurement error especially above fields of 100 G.

Another difference between the aluminum jet and POW configurations is in the magnetic field necessary to optimize peak x-ray emission. Figure 11 shows that the peak x-ray signals are maximum for different applied fields. Peak x-ray emission is observed on all p-i-n diode channels near 50 G for the aluminum jet alone. For the POW configuration the x-ray signals are always larger than for the jet alone, and x-ray emission is maximum at about 150 G with the hard channels being affected the most. If one estimates the x-ray yield by taking the product of the pulse widths with the peak signals, we see results similar to those of figure 11 even though the pulse widths are, also, increasing. The POW configuration consistently produces brighter x-ray emission with a maximum for 150 G while the aluminum jet alone is weakly optimized at 50 G. Our theoretical investigation of the pinch process indicates that the magnetic field necessary to optimize a pinch depends on the pinch mass and scales as the square root of the mass independent of the field direction.

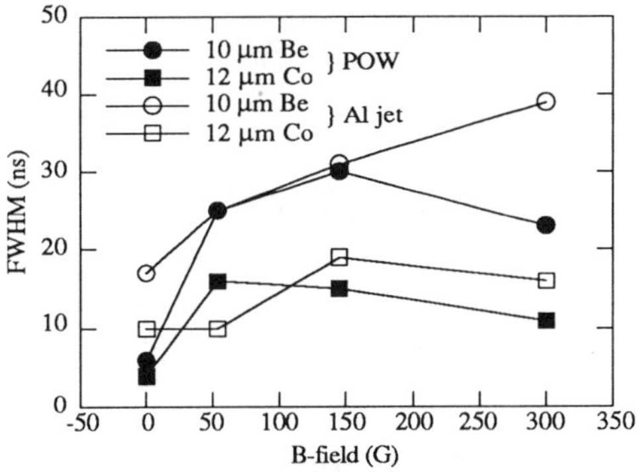

Fig. 13. X-ray diode signal widths for different applied magnetic fields.

Temperature information is also provided by the p-i-n diode array. Recall that the PIN diode signals scale almost identically in the absence of a magnetic field (see figure 12). This implies that the temperature is nearly constant throughout the implosion and that the radiating volume changes in time. The same effect is seen during compression phase for non-zero magnetic fields, but the difference in x-ray pulse widths show that the plasma is cooling immediately after the peak of the compression, *i.e.*, a higher proportion of soft to hard radiation.

CONCLUSIONS

Comparison of the behavior of the aluminum plasma jet with and without the copper fiber suggest that the dynamics of the implosion is independent of the presence of the wire. This is indicated by both the schlieren and the pinhole data (figures 7-10). We have observed that the current rise in the load does not depend on the presence of the copper wire, at least prior to the time of maximum compression. This is a very desirable feature for future investigation since optimization and development of the plasma jet can be done separately from the POW configuration. However, we have observed that the magnetic field required to optimize x-ray yield is different for the two configurations. Our theoretical investigation of the implosion suggests that the the field necessary to optimize the pinch is related to the pinch mass.

The application of an axial magnetic field enhances the stability of the pinch resulting in an increase of x-ray production. Cylindrical symmetry of the aluminum plasma pinch is enhanced in both load configurations as the axial field increases. This is due to the absence of m=1 instabilities at high field strengths. Even though m=0 instabilities remain, their growth has been reduced and axial homogeneity improved. The final compression appears to be independent of the magnetic field except at fields around 300 G. Above 200 G the assumption that $B_{z0} << B_\theta$, begins to break down.

The POW configuration consistently produces harder radiation as evidenced by the pinhole images. In addition, the lack of emission from the aluminum plasma in the POW configuration suggests that kinetic energy is being transferred to the wire at the peak of the implosion. A cool outer shell would have a much higher resistance than the hot core providing a mechanism for switching current into the fiber. Furthermore, a dense plasma ($n_e \approx 10^{21}/cm^3$) in the original shape of the wire remains at the core of the implosion after the the main current pulse has begun to subside. Densities of this magnitude in conjunction with the axial uniformity provided by the POW configuration have applications for x-ray lasers and fusion research.

Development of staged loads such as the POW configuration and improvement of the stability of imploding shells are creating exciting new avenues in z-pinch research. We foresee many interesting and unexpected of phenomena arising out of this work.

ACKNOWLEDGEMENTS

The authors thank and R. Spielman for providing code used to analyze the x-ray data. This work is supported by ETCA/CEG under contract #420/115/01.

REFERENCES

[1]R. B. Spielman, M. K. Matzen, M. A. Palmer, P. B. Rand, T. W. Hussey, and D. H. McDaniel, Appl. Phys. Lett. 47, 229 (1985).

[2]H. U. Rahman, P. Ney, F. J. Wessel, A. Fisher, and N. Rostoker, in *Dense Z-Pinches*, edited by N. Pereira, J. Davis, and N. Rostoker, AIP Conf. Proc. No. 195 (AIP, New York, 1989), p. 351.

[3]F. J. Wessel, B. Etlicher, and P. Choi, Phys. Rev. Lett. 69, 3181 (1992).

[4]A.S. Chuvatin, L. Véron, N.S. Edison, B. Etlicher, C. Rouillé, J.P. Stephan, H. Lamain and P. Auvray, APS Bulletin 37, 1565 (1992).

[5]B. Etlicher, L. Véron, S. Attelan, C. Rouillé, and F.J. Wessel, 9th International Conference on High-Power Particle Beams, Program Abstracts, p. 388 (1992).

[6]S. Attelan, B. Etlicher, C. Rouillé, F.J. Wessel, and P. Zehnter, APS Bulletin 36, 2398 (1991).

[7]F.J. Wessel, B. Etlicher, N.S. Edison, A.S. Chuvatin, L. Véron, C. Rouillé, and S. Attelan, APS Bulletin 37, 1578 (1992).

[8]J. Delvaux, H. Lamain, C. Rouille, H. J. Doucet, J. M. Buzzi, M. Gazaix, and B. Etlicher, in *Proceedings of the Fourth International Topical Conference on High Power Electron and Ion Beam Research and Technology* (Ecole Polytechnique, Palaiseau, France, 1981), Vol. 2, p. 775.

[9]P. Audebert, H. Lamain, B. Dufour, C. Rouille, B. Etlicher, L. Voisin, and P. Romary, in *Proceedings of the Eighth International Conference on High Power Particle Beams* (World Scientific, Singapore, 1990), Vol. 1, p. 422.

A VUV Recombination Radiation Experiment in a Fast Dynamic z-Pinch

P. Hagen, T. Wagner, M. Stetter, J. Christiansen, R. Tkotz

Physikalisches Institut I, University of Erlangen, 8520 Erlangen, Germany

H. Langhoff, A. Mehling

Physikalisches Institut, University of Würzburg, Germany

Abstract

A z-pinch type experiment capable of generating hot, dense, short-lived, high aspect ratio plasmas in a variety of gases is described. Acetylene is particularly interesting for x-ray laser applicatons. First studies by means of optical spectroscopy promise temperatures and densities indicating a highly ionized carbon plasma. Studies of expansion velocity point at possibilities to maximize cooling in the laser plasma.

Introduction

Small-scale recombination pumped x-ray lasers have been the focus of current research. For example capillary discharges have been proposed to be feasible for a table-top x-ray laser [1,2] as an alternative for large-scale laser driven systems. However these schemes are limited to single shot experiments or low repetition rates in most cases.

In this contribution we propose the possibility of developping a soft x-ray recombination laser based on a plasma produced by a steep shock front. This shock front is driven by a fast z-pinch discharge in a spatially homogeneous gas filled discharge tube. Due to a very effective preionisation it is possible to produce very uniform shock waves with high velocities leading to a plasma cylinder with a high length to diameter ratios (up to 80) on the axis. In this system most of the stored energy is used to accelerate the gas in the shock wave. This kinetic energy thermalizes when the shock front reaches the axis of the discharge tube and is therefore directly used to heat the laser plasma.

Experimental setup

The schematic setup with the diagnostic tools is shown in figure 1. The energy is stored in a 3.5 Ω water filled coaxial pulse forming line producing a rectangular current pulse of 100 ns duration and up to 50 kA amplitude. The discharge tube has typically an inner diameter of 14 mm and a length of 40 mm. The tube is terminated by a plane anode (high voltage side) and a hollow cathode to enable an axial optical access. A conical z-pinch behind the hollow cathode produces a homogeneous preionisation of the discharge volume.

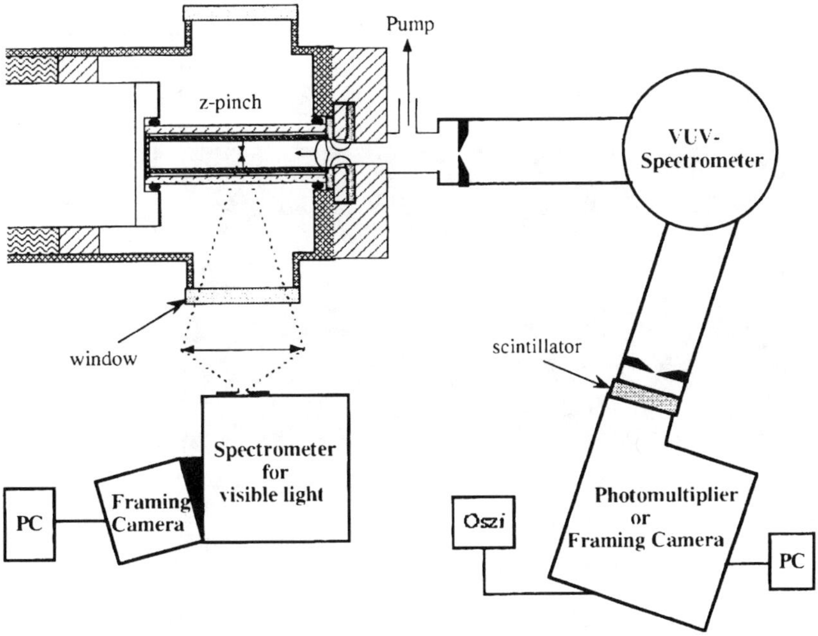

Fig. 1: Experimental setup

Features of the pinch

Because of high contraction velocity (~ 10^5 m/s, M»1) of the pinch, a shockwave detaches itself from the imploding piston. The width of the shockwave controls the radial plasma dimensions and the shock velocity is correlated to the plasma temperature after thermalizing on the axis. Estimates are given in [3] on the importance of the various heating processes.

Ohmic heating contributes only 10%, shock wave heating does 20%, but most of the heating is effected by transforming directed kinetic energy into thermal energy. If density gets too low the shock wave smears out due to the increased mean free path for neutrals penetrating the contracting shock wave. Too high densities increases the width of the shock wave and energy density becomes too low.

In the described geometry we got optimal results in regard of homogenity and stability of the shock produced plasma cylinder around mass densities of ~0.5 μg/cm^3 in the neutral gas filling. The corresponding width of the shockfront was determined from Schlieren measurements to be less than 250 μm [5]. Therefore the hot plasma core on the axis should be about 500 μm. Framing pictures with 10 ns exposure time indicate that the diameter is well below 1 mm, the spatial resolution due to the motional blurr (fig. 2).

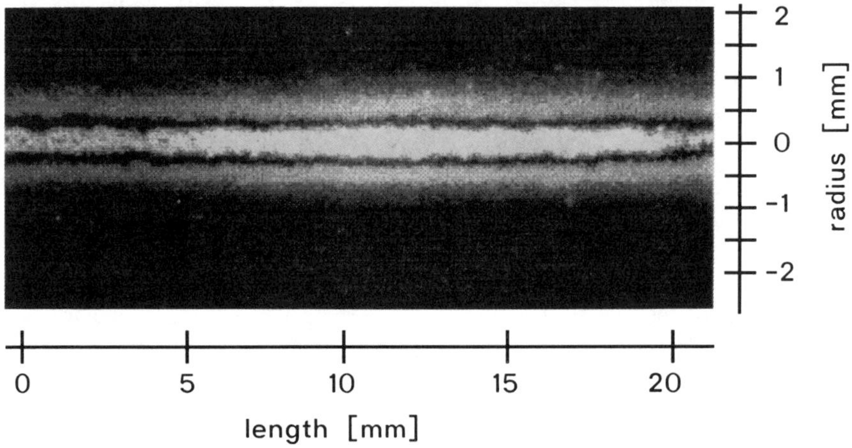

Fig. 2: Framing picture with 10 ns exposure time of a schock wave produced plasma in Helium, 300 Pa, current 38 kA

Plasma parameters in helium

During the formation of the plasma cylinder intense VUV-radiation can be seen on axis. A comparison of the time dependence of light output and discharge current is shown in figure 3. Electron density of the shock wave heated plasma could be measured temporally and spatially resolved by means of optical spectroscopy [5]. From end-on VUV-spectroscopic measurements electron temperature on the axis was determined in collaboration with University Würzburg (fig.4).

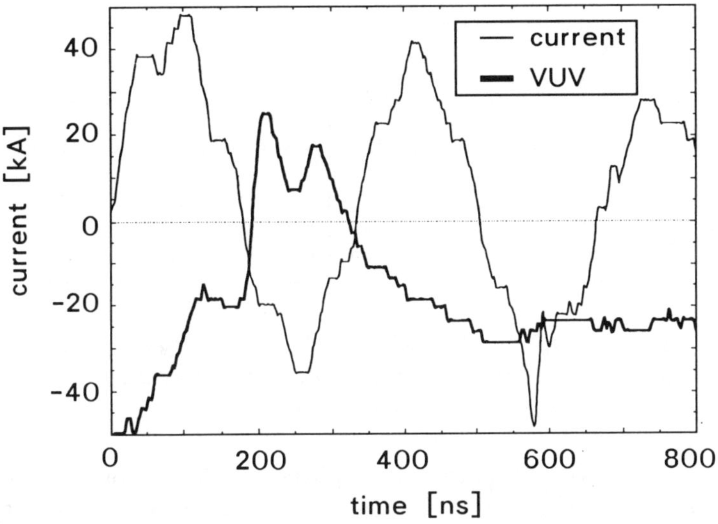

Fig. 3: Discharge current and VUV-radiation (150-180 nm) in a Helium plasma

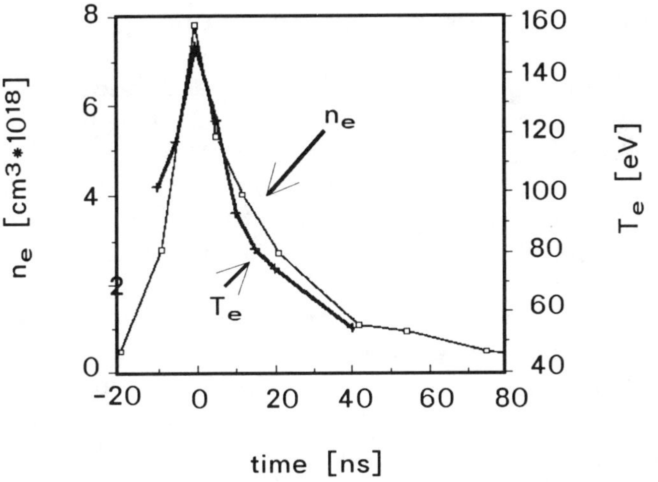

Fig. 4: Electron density n_e and temperature T_e of a 300 Pa helium plasma on axis, current 40 kA.

Taking this plasma data one can look for possible candidates for a short wavelength laser pumped by electron-collisional recombination. The data in table 1 were calculated for the Balmer-α-transition of a hydrogenic ion [6].

	He	Li	Be	B	C
nuclear charge Z	2	3	4	5	6
wavelength [nm]	164	72.9	41	26.2	18.2
max. n_e [cm^{-3}]	$3.9 \cdot 10^{16}$	$6.6 \cdot 10^{17}$	$4.9 \cdot 10^{18}$	$2.3 \cdot 10^{19}$	$8.4 \cdot 10^{19}$
max. T_e [eV]	5	12	22	34	49
gain(F~0.3) [cm^{-1}]	0.0008	0.02	0.2	1.2	5

Tab. 1: Requirements for a short wavelength laser pumped by electron-collisional recombination calculated for the Balmer-α transition of a hydrogenic ion.

In the case of Helium the plasma does not meet the requirements for laser action, as collisional mixing of adjacent levels due to the high electron density destroys any inversion.

First studies of an Acetylene plasma

The measured plasma temperatures and densities encourage to move on to higher z-gases, especially to carbon (see tab. 1). Providing that the parameters are comparable, the conditions for the onset of inversion could be fulfilled in a carbon plasma. Time integrated spectroscopic measurements in acetylene are plotted in figure 5. Recombination lines of C_{IV} indicate a substential amount of C_V.

Investigations of the Acetylene plasma parameters had to resort to optical spectroscopy in the visible as no VUV-diagnostics were available. In this case opacity cannot be neglected in the dense plasma core. This is affirmed by the absence of C_{IV} emission. Without exact knowledge of the radial electron density distribution it is difficult to predict the degree of opacity. Rough estimates can be made [7] assuming similar density as for Helium. The expected absorption of emission from the core region is affirmed by the observation of self-reversed lines some tens of nanoseconds after pinching.

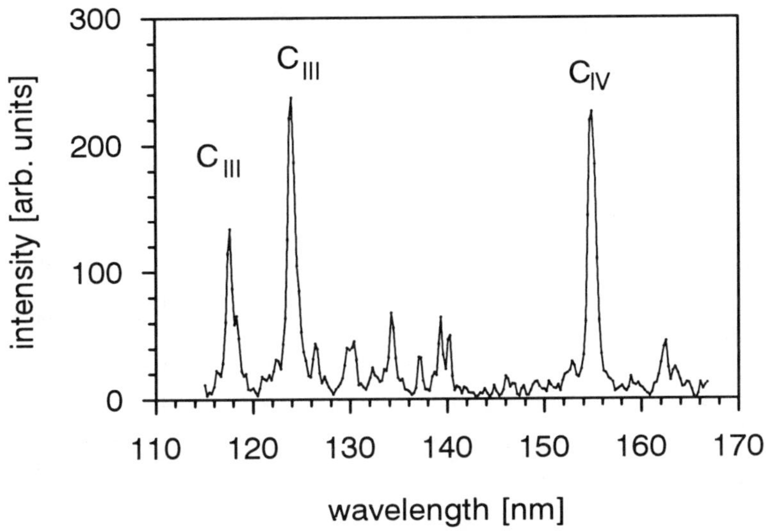

Fig. 5: Time integrated spectroscopic measurements in a schock wave generated carbon plasma.

To get a first impression of the thermalization of the shock wave on axis the line intensity ratio C_{III}(432.6nm)/C_{II}(426.7nm) was used to determine the electron temperature time and space resolved. Due to reabsorption in outer layers of the plasma column the resulting temperatures are a lower limit for T_e [8]. Nevertheless figure 6 hints at the generation of a hot core plasma when the shock thermalizes on the axis. The observed dimensions of the hot central region (500 μm) are in good accordance with previous results in lower-Z gases.

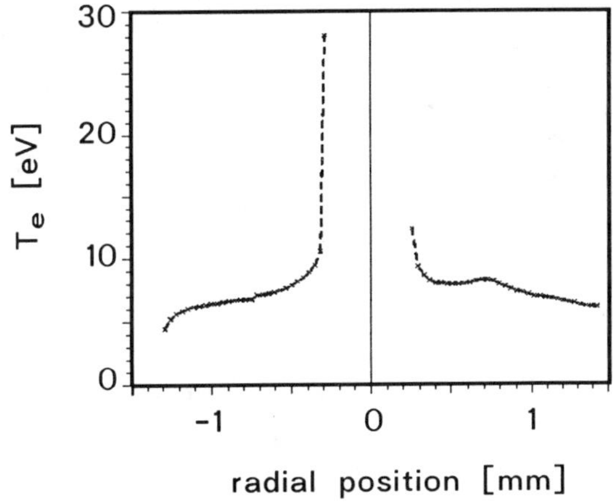

Fig. 6: Electron temperature of a carbon plasma during maximum contraction

Optimization of cooling rate

Once a hot plasma column is generated, rapid cooling is the prerequisite for the building-up of inversion. Three effective processes contribute to cooling of the expanding pinch plasma, as listed in table 2.

maximum cooling rates [K/s]	Helium	Acetylene
radiative cooling	$< 2*10^{12}$	10^{14}
electronic heat conduction	$4*10^{14}$	10^{14}
adiabatic expansion	10^{14}	10^{14}

Tab. 2: maximum cooling rates for: radiative cooling from [9], electronic heat conduction calculated numerically with Helium data, adiabatic expansion from experiment.

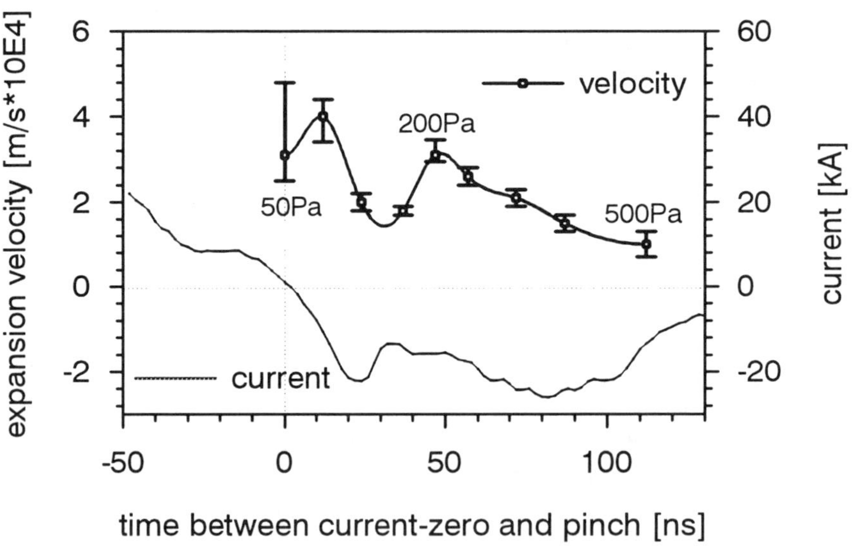

Fig. 7: variation of expansion velocity; possible influence of current during expansion

Unlike with Helium, the contribution of radiative cooling (RC) cannot be neglected in the case of Acetylene. However, the values for RC are an upper limit, since the model assumes a steady-state plasma [10]. Electronic heat conduction is only effective in the first few ns. Adiabatic expansion is less predominant than in helium but still of major importance to cooling the pinch plasma in acetylene.

In order to maximize cooling by adiabatic expansion the possibilities of influencing the expansion velocity were investigated. The expansion velocity changes appreciable with the initial parameters pressure and discharge current, taking values within 1 - $4*10^4$ m/s. First results from measurements in Helium show non-monotonous dependence on initial pressure. Apparently the current in the early expansion phase influences the expansion velocity (fig. 7).

Summary

With the experiment presented, homogeneous pinch plasmas with high temperatures and densities can be achieved for a variety of filling gases (e.g. hydrogen, helium, nitrogen, neon and acetylene).

Proof of principle experiments for generation of dense hot plasmas (N_e~10^{19} cm^{-3}, T_e~100 eV) have been performed in Helium. The achieved plasma parameters are promising for experiments in acetylene. The first spectroscopic investigations in the visible were hampered by opacity and strong continuum radiation. Measurements in the XUV-regime will be carried out soon.

References:

1 C. Steden, H.-J. Kunze, Phys. Lett. A, Vol. 151 p. 534 (1990).
2 J. J. Rocca, IEEE J. Quant. Elec., Vol. 29, 1, p.182 (1993).
3 W. Hartmann et al., Appl. Phys. Lett., Vol. 58, 23, p. 2619 (1991).
4 W. Hartmann, Ph. D. thesis, Erlangen, Germany (1986).
5 T. Wagner, Dipl. thesis, Erlangen, Germany (1991).
6 R. Elton, X-Ray-Lasers, academic press, New York (1989).
7 H. Griem, Plasma Spectroscopy, Mac Graw Hill (1964).
8 H. Zwicker, Zeit. f. Phys. 178, p189 (1964).
9 D. Post, R. Jensen, At. Data Nucl. Data Tables 20, p397 (1977).
10 C. Keane, C. Skinner, Phys. Rev. A 33, p4179 (1986).

X-RAY EMISSION CHARACTERISTICS IN A 1J-50 ns VACUUM DISCHARGE

A. Ikhlef and M. Skowronek
Laboratoire des Plasmas Denses, Tour 12 E5, Univ. P. & M. Curie
F-75252 Paris cedex 05 (France).

ABSTRACT

The mechanism of the X-ray emission from a small vacuum discharge is studied. Two X-ray pulses whose intensities vary versus the anode-cathode distance are identified. They are well reproducible and little dependent on the cathode geometry. Time integrated pinhole images show that the X-ray emission originates from the Teflon insulator, from the metallic anode or from the emitted tunsgten vapor in its vicinity. Three different geometries have been used : hollow cathode, hollow conical cathode and needle or massive cathode. A systematic study of the delay time, the commutation time and the X-ray duration with respect to the cathode geometry allows us to build some hypothesis on the emission mechanism

INTRODUCTION

The fundamental physics involved in vacuum spark discharges has been described in the reference book by Mesyats and Proskurovsky [1]. It has been shown that very dense plasmas with $n_e > 10^{26}$ m^{-3} are produced in anode and cathode flares. Powerful vacuum sparks are X-ray sources with "hot spots" of very small diameter ($\Phi < 10$ µm) [2] due to the radiative collapse mechanism [3].

Recently, the dense electrode plasma has been studied by means of laser absorption [4]. The measured continuum absorption coefficient corresponds to an electron density n_e in the range 10^{26} to 10^{28} m^{-3} valid for both cathode and anode plasmas. In the arc phase, fragments of 5 to 10 µm diameter of cathode plasma have been observed with changes occurring within few nanoseconds.

We have designed a miniature X-ray source feeded by a small Marx generator, where different electrode configurations and separations can be tested. The stored electrical energy is one Joule. A sensitive imaging device gives the spatial distribution of the emitted

X-ray. It has been associated with the recording of the electrical characteristics (V(t) and di/dt) and that of the X-ray intensity.

We have checked systematically the effect of the replacement of a massive cathode by hollow cathodes of different types.

EXPERIMENTAL SET-UP

The X-ray source has been previously described [5,6]. It consists of a cylindrical vacuum chamber ($\Phi = 50$ mm) on the axis of which the anode is a tungsten needle in one end and the cathode is either a massive stainless steel pointed or a hollow cathode, in the other end. In some cases, they have been replaced by iron or copper electrodes. The anode is insulated from the ground cylinder by a Teflon envelope.

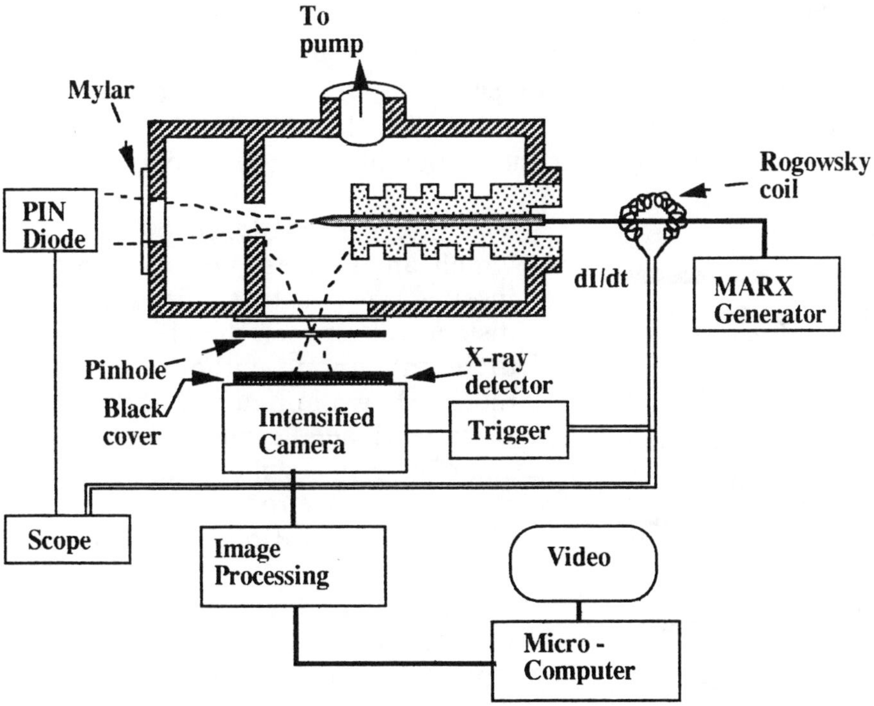

Fig.1. Experimental set-up

The distance between the electrodes can be varied by steps of 0.1 mm until a maximum of 7 mm. In another series of experiments, the cathode is replaced by a metallic ring ($\Phi_{int} = 6$ mm) and then the source presents a hollow cathode geometry. The anode is connected to a small 4-stage Marx generator delivering a 50 ns pulse of maximum voltage $V_0 = 70$ kV in a 50 Ω load. The value of di/dt is obtained by means of a Rogowski coil placed around the anode and recorded on a 4-channel oscilloscope (500 Mhz bandwidth) together with the X-ray intensity.

The X-rays are recorded either end-on, when the hollow cathode configuration is set, or side-on, through symmetric apertures in the wall chamber. The X-rays are emitted through mylar windows of thickness 70 μm and are detected by means of a silicon PIN diode or by means of a scintillator placed in front of a photomultiplier. A pinhole having a diameter of 0.1 mm is used to image the discharge. Different phosphors are used to convert the X-ray in light : a) a very efficient one (Dup. cronex, quanta III) having a decay time of about 1 μs ; b) a plastic scintillator sheet (type NE 102A) of thickness 0.1 mm having a 2 ns decay time. The low-level light camera is composed by a gated light intensifier in front of an opticon tube (intensified vidicon).

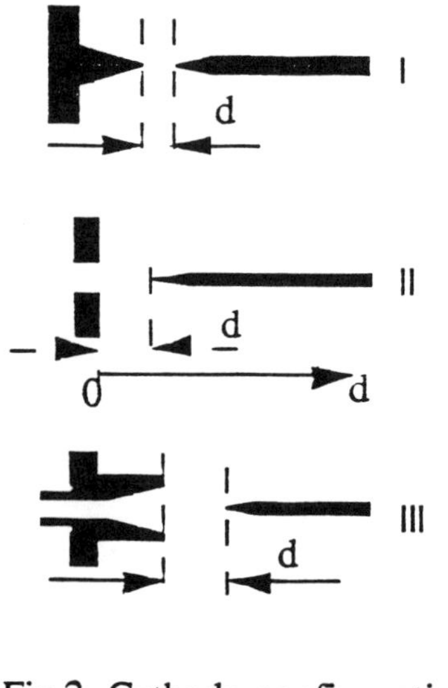

Fig.2. Cathode configurations

The image is analyzed by a PC in which an imaging card (Matrox) is inserted. Systematic studies versus the distance d between the electrodes for different anode-cathode configurations shown in figure 2 are made :
a) case I : needle anode and needle cathode, d gives the actual distance between the electrodes.
b)case II : hollow cylindrical cathode ;
c) case III : hollow conical cathode ;
For these different configurations, the current derivative curves are not very different from each other. The time evolutions of the emissions are similar

MEASUREMENT RESULTS AND DISCUSSION

A typical oscillogram is displayed on figure 3. The upper trace displays the Rogowski coil signal. We have found that the electrical characteristics are nearly independent of the chamber pressure in the range : 10^{-3} Pa to 10 Pa in agreement with [1]. Also di/dt does not depend too much on the distance d. We have marked as A , the prepulse which is seen on each signal and as B the first maximum which corresponds to the breakdown. The time separation between the points A and B is called ΔT. The lower trace displays the X-ray signal which shows a negative part due to a reflection caused by an imperfect impedance matching between the cable and the entrance . Two peaks are observed on the X-ray signal : a) the first one appears before the breakdown and corresponds to the vaporisation and the expansion of the tungsten cloud ; the X-ray is emitted at an instant when the voltage is high, the electron density being low and the current being also very low ; b) the second emission coincides with the breakdown ; it corresponds to the electron interaction with the anode and also to the dense plasma around it. The time separation between the two X-ray peaks is called δt. On figure 4, the variation of the time delay ΔT versus the distance d between the electrode planes is displayed :

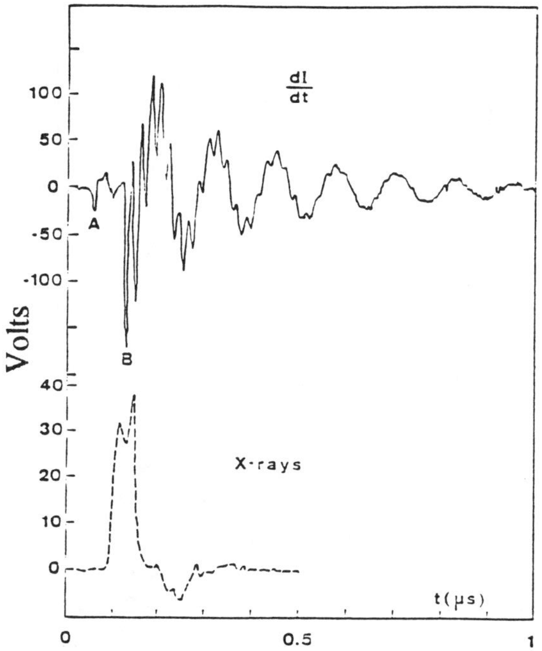

Fig.3.Oscillograms of di/dt and $I_x(t)$

a) in case I, needle electrodes being installed: the time delay ΔT is minimum and increases with d from 30 ns to 70 ns ;

b) in case II, an hollow cylindrical cathode being installed : ΔT varies from 60 ns to 107 ns ;

c) in case III, an hollow conical cathode being installed : ΔT varies from 75 ns to 133 ns.

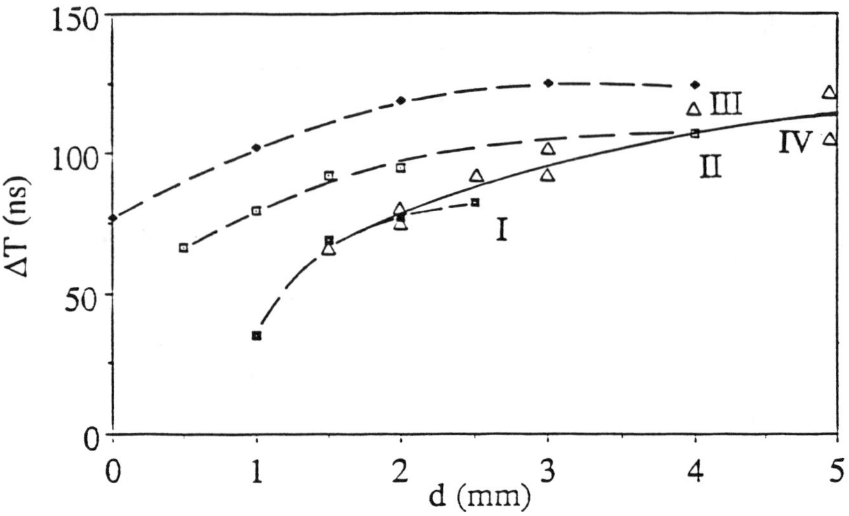

Fig. 4 . Variation of the time delay ΔT versus the distance d

A unique curve IV can be obtained if we consider that, in case II, the actual origin of distances is about 1 mm inside the hollow cathode and that, in case III, the actual origin is about 2 mm inside the hollow conical cathode. This suggests that the starting point of the expansion cloud is situated on the axis of the cathode. The expansion velocity of the cathode plasma deduced from the curve of figure 4 (IV) is $v_{exp} \sim 50$ km/s at the beginning and at the end $v_{exp} \sim 100$ km/s. These values are significantly higher than that given in[1] : $v_{exp} \sim 25$ km/s. The time separation δτ between the two X-ray peaks is displayed on figure 5 . The time duration of the X-ray emission has the same variation and ordering as δt. The mean value of δt is the same for the three cases : 35 ns. This time is related to the commutation time. We obtain a unique curve for the "true"delay time by displaying (ΔT- 35) ns versus d (fig. 6). We obtain an expansion cloud velocity of 50 km/s.

The sensitivity of the PIN diode, given at the X-ray energy of 10 keV is : $S_{10keV} = 0.71$ W/V. It varies with the energy like the absorption coefficient of the Si component of the diode. The diode is situated at a distance of 5 cm from the anode, its surface being 1 cm^2. The intercepted solid angle is about 1/300 sr. The measured signal reach 100 V, corresponding to an emitted power in 2π sr of about 130 kW and an energy of 8 mJ during 60 ns. If the mean photon energy is about 20 to 30 keV, then and the emitted energy is even higher. The formula : $\eta = 1.3\times10^{-6}\ Z\ V$ gives the emission efficacy for a thick target, where $\eta = P_x / P_{el}$ [7] . P_x is the X-ray emitted

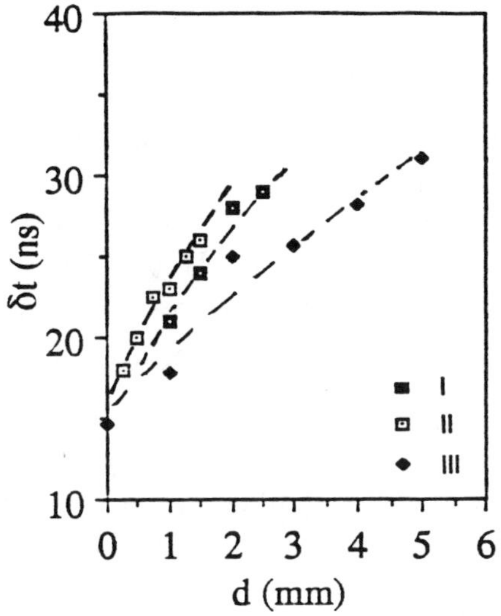

fig.5 : δT versus d

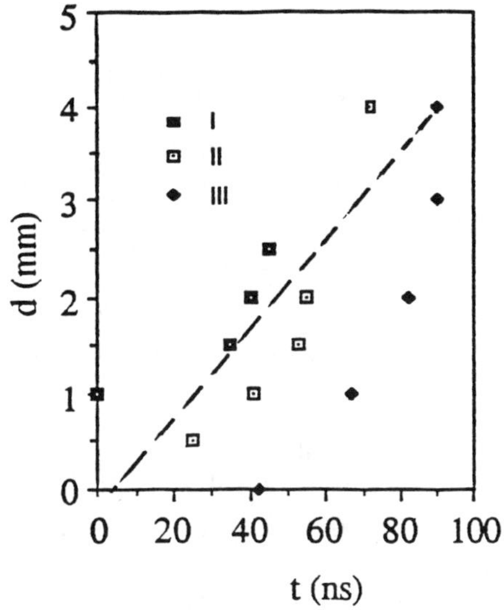

fig.6 : true delay time versus d

power and P_{el} is the electrical power fed in the discharge, V is the applied voltage in kV (here, the delivered voltage is 70 kV), and Z the atomic number of the anode (here Z = 74). For 1 J electrical energy spent in the discharge, the emitted radiation energy in 2π sr would be about 7 mJ, in fair agreement with our measurements. The attenuation coefficient of a Molybdenum foil is only slightly different for the two peaks. The intensity of the X-ray emission is maximum for a small distance d. The figure 7 displays the variation of both X-ray signals (the first and the second peak) with d. The application of approximate relations concerning the

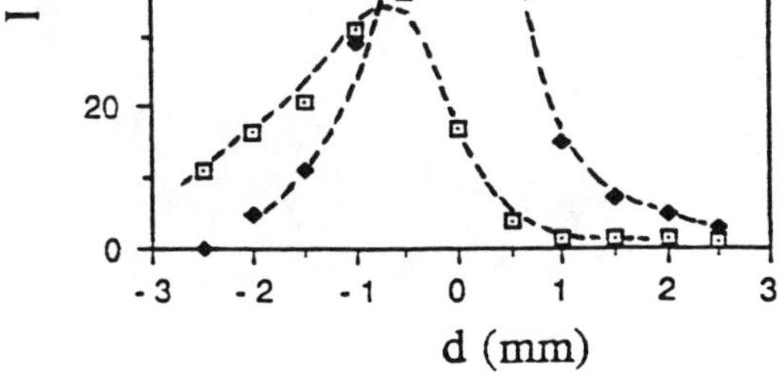

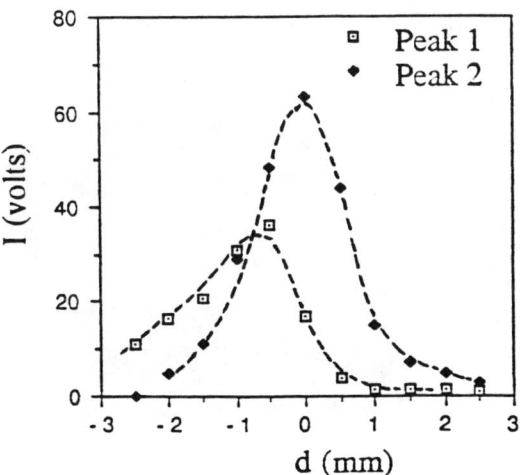

Fig.7.X-ray intensity versus d

electrical characteristics in discharging a line gives for the current intensity, the voltage across the gap and the radiation intensity.

$$i(t) \approx (V_0 / R)\,(v_c t / d)$$

$$V(t) \approx V_0\,(1 - v_c t / d)$$

$$J(t) \approx i(t)\,.\,V^2(t) = (V_0^3 / R)\,(v_c t / d) \times [1 - v_c t/d]^2.$$

Where v_c is the plasma cloud expansion velocity, d the distance anode -cathode. We have observed that J(t) is a decreasing function of the distance d : the maximum values being obtained for d = 0, the same behaviour is found for the three configurations I, II, and III.

The absorption coefficient of a Mo sheet of thickness 30 µm has been measured respectively for the two emission pulses and the transmission coefficient : $T = J_t/I_0 = \exp(-\mu x)$ is calculated for each measurement.

a) the first pulse (mainly from the Teflon insulator region) gives a mean transmission coefficient of 0.11 on axial direction and 0.17 on side-on direction.

b) the second pulse (coming mainly from the anode point region) gives a mean transmission coefficient of 0.096 in axial direction and 0.148 in side-on direction.

These coefficients correspond to the characteristics lines L of tungsten and are the most predominant radiations .

SPATIAL DISTRIBUTION OF THE X-RAY

On the figure 8, the X-ray image of the anode is shown. On this

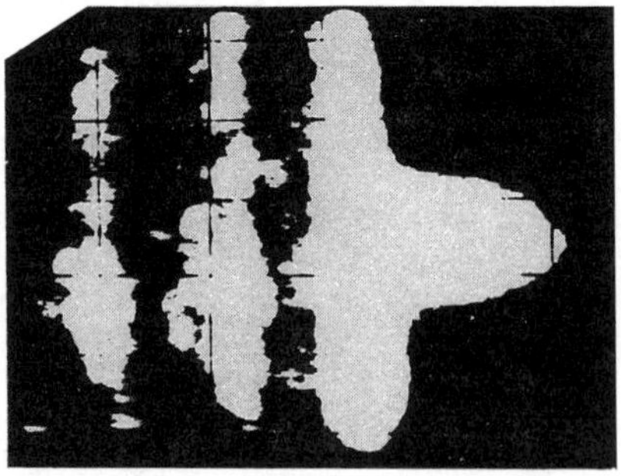

Fig.8.Image of the X-ray emission

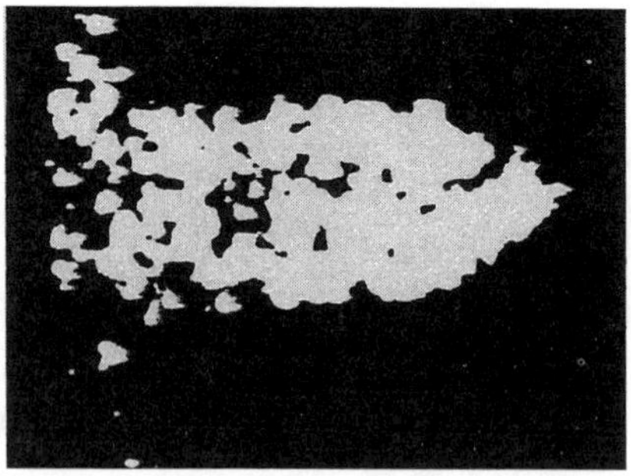

Fig.9.Image through Mo foil

image, the characteristic shape of the Teflon insulator around the anode is observed : three vertical bands, corresponding to the grooves existing on the Teflon insulator, are seen. On figure 9, an image is obtained through a Mo foil. Very intense spots appear from time to time in the discharge gap, near the anode tip.

In this case, the shape of the tungsten anode is only visible. The Teflon emission zone and individual spots have almost disappeared.

CONCLUSION

The spatial distribution may be described as follows : the main X-ray emission originates from the anode and its vicinity ; another important emission comes from the Teflon insulator. The temporal history of the X-ray emission shows that the two predominant X-ray peaks seem to come from the the tungsten anode and its immediate vicinity, whereas a less energetic X-ray comes from Teflon.

No very important change in the X-ray intensity, nor in the X-ray time evolution between the three different anode-cathode configurations. The hollow cathode seems without any strong effect.

REFERENCES

[1]G.A. Mesyats and D.I. Proskurovsky, Pulsed Electrical Discharge in Vacuum, (Springer- Verlag, Berlin, 1989).

[2]J.L. Schwob and B.S. Fraenkel, Phys. Lett., **40A**, 83 (1972).

[3]K.N. Koshelev and N.R. Peirera, J. Appl. Phys., 69, R21 (1991)

[4]A. Anders, S. Anders, B. Jüttner, W. Bötticher, H. Lück and G. Schröder, IEEE Transactions on Plasma Science, **20**, 466 (1992).

[5]M. Skowronek and P. Roméas, IEEE Transactions on Plasma Science, PS-15, 589 (1987)

[6]M. Skowronek, P. Roméas and P. Choi, IEEE Transactions on Plasma Science, PS-17, 744 (1989)

[7]N.A. Dyson, X-rays in Atomic and Nuclear Physics, Cambridge University Press (1990)

AKNOWLEDGEMENTS

We are grateful to Dr Charles (Univ. P. et M. Curie) and Dr Bourdinaud (CEA-Saclay) for helpful discussions.This work is supported by CNRS (URA1096) and by DRET (n°86/038) contracts.

RESULTS OF THE STUDY OF THE NEW TYPE OF COMPACT GAS-PUFF PLASMA SOURCE OF SXR (SOFT X-RAY)

V.L.Kantsyrev, K.I.Kopytok, A.S.Shlyaptseva
Scientific Research Institute of Technical Glass
Moscow, Russia, 117218

ABSTRACT

The results are presented dealing with the study of the SXR compact source of the new type based on the typical " gas-puff " one. The source has the value of the conversion coefficient of initial energy supply into SXR - $\eta =(1-2)\cdot 10^{-2}$ and averaged effective size of emission region- d = 2-3 mm , what is better the traditional ones of " gas-puff " have. The new one has also greater resource (in two-three orders greater) comparing with widely-used low-inductance vacuum spark.

An improvement of the source characteristics was achieved as result of realization of mixed regime of micropinching of heavy - current discharge when plasma points are appeared both in plasma of compressed gas liner and plasma of anode vapor.

INTRODUCTION

Among all compact plasma sources of SXR the most powerful are low-inductance vacuum spark (LIVS) and the source with impulsive gas injection named "gas-puff". However, LIVS has the small resource caused by anode destrúction (100-200 impulses). The source with impulsive gas injection has the smaller value of η , but the better effective size - d than for LIVS. Besides that this source has the greater resource comparing with LIVS.

For many practical applications of compact sources (for microroentgenograpy, X-ray lithography) the source is needed posseeing the maximal value of η, minimal value of d and sufficient resource (not less than 10000 impulses)[1].

The purpose of our investigations was the study of the possibility of compact plasma source of SXR construction which would include the best characteristics of both types of sources : LIVS and "gas-puff".

EXPERIMENTAL SET UP

The experimental set up consists of gas plasma source (GPS) of SXR (Fig.1) and X-ray diagnostic complex[2].

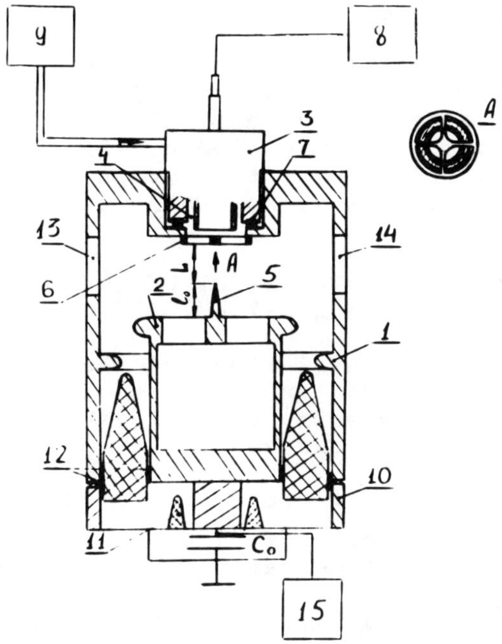

Fig. 1. The scheme of the GPS source
1. Vacuum chamber ; 2. Anode; 3. Batcher of ionized gas liner; 4. Cycle nozzle; 5.Anode edge; 6. Cathode; 7, 12. Vacuum packing; 8. Control block; 9. Gas block; 10. Conductor; 11. Low-inductance capacity; 13. Diagnostic window; 14. Vacuum pumping; 15. Charge block.

GPS presents itself simplified modification of "gas-puff" set up with impulsive gas injection into vacuum chamber of weak ionized gas-like liner. This allowed to get rid of controled intermediate discharger between anode and basic low-inductance capacity. Anode making as radial grazing was plane or had central edge similar with one in LIVS. The following materials were used for anode : Al, Ti, Fe, Cu, Mo and W. Using impulsive batcher supersonic "hollow shape " weak ionized gas jet was created between electrodes which reserved discharge spacing , caused heavy-current discharge and provided formation of plasma rope in GPS. The duration of the injection front of gas-like liner wasn't greater 50...100 mcsec. The volume of injected gas is equal to few cm^3. The vacuum was created in GPS chamber by turbomolecular unit. The initial energy supply reaches 2.9-3 kJ at the capacity voltage equal up to 23 kV, an amplitude of discharge current Im=220 kA, discharge half-period - 2.6 mcsec.

The diagnostic complex includes sensors with absorbtion filters and X-ray film UF-VR or UF-SHC, thermo-luminescence detectors (TLD), X-ray calorimeters , X-ray pin diodes, three- channel vacuum pinhole camera, survey spectrograph with convex mica crystal ($\lambda < 1.8$ nm), Rogowsky coil or magnetic probe.

THE RESULTS OF THE STUDY OF THE PECULARITIES OF X-RAY EMISSION

During experimental study of the dependences of η, d and spectra on the electrode geometry, composition and geometry of gas-like liner and the value of discharge current Im the following was established:

1) The value η increases from 10^{-4} (for GPS with anode without edge) up to (1-2) 10^{-2} (for GPS with anode having edge) in spectral range $\lambda < 1.5$ nm;
2) The value d decreases from 3-6 mm up to 2-3 mm accordingly in spectral range $\lambda < 1.2$ nm.

It was observed , that at the same time of weak shining in SXR diffusion region with size of a few mm, two kinds of the small size shining formations (micropinches) - subplasma points with size 500...1000 mcm and plasma points with size less than 200 mcm (the lower limit - 30 mcm). By that , plasma points were observed only for anode with edge but subplasmas ones - in both cases. The pecularities on current waveforms were observed in 80-90 % discharges.

Spectrum of GPS was dependent on composition of gas liner, the value of pressure in batcher and the presence of anode edge. Without edge spectrum contains intense spectral lines of multiply-charged ions of liner plasma (for example, He-like Ar) and sometimes weak lines of ions of anode (for example, Ne-like Cu). The presence of the edge caused the appearance of intense lines of anode plasma (Fig. 3) which could exceed in intensity the lines emitted by gas-like liner.

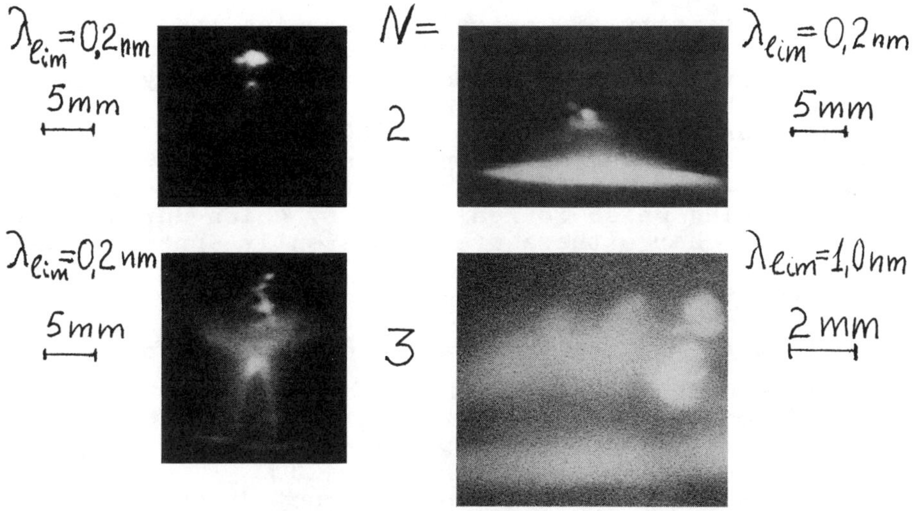

Fig. 2. Pinhole photographs of heavy-current discharge in GPS. Anode - Fe, gas - Ar, Im = 200 kA. On the right - anode without central edge, on the left - with central edge.
Im - maximal discharge current.

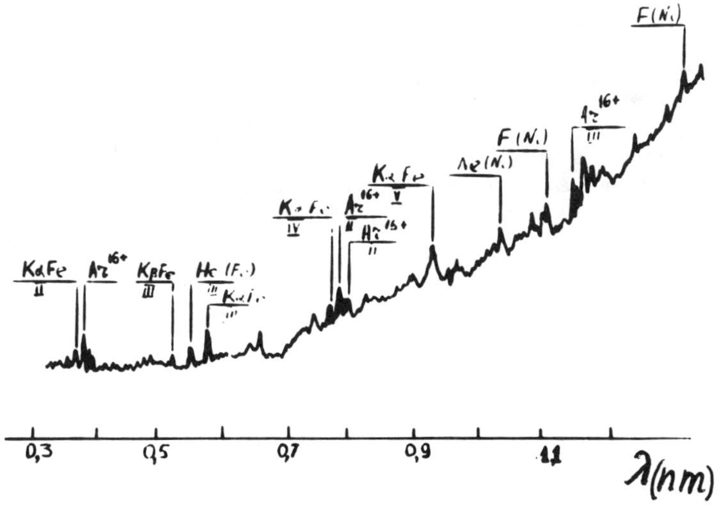

Fig. 3. Spectrum of plasma of GPS.
Anode - Fe, with edge; gas - Ar,
Im=220 kA.

Maximal intensity of emission was in the spectral range λ = 1.0-1.2 nm.

Temporal impulse waveform of SXR was very complex and nonstable from discharge to discharge. Average impulse duration of SXR - 50 - 100 nsec.

An explanation of phenomenons observed may be caused by the fact that there is the mixed regime of micropinching which is realised ; by which during the typical discharge there are consistently appeared both subplasmas points in plasma of injected cycle-like liner and the plasmas ones in plasma of anode material vapor when anode having the edge. This is pointed also on with the fact of weak dependence of η on discharge numbers $N > 10^4$ and therefore on the wear of anode edge , what makes the difference of GPS from LIVS (where the number of plasmas points and also the value η abruptly fall after N=100-200).

The GPS of SXR which is constructed and studied in detail is used in high-resolution impulsive micro-roentgenography and X-ray lithography investigations.

ACKNOWLEDGEMENTS

The authors thanks B.Yu. Sharkov, A.N. Golubev, V.V.Dubenkov for help in experiments and useful discussions.

REFERENCES

1. N.P.Economou, D.C.Flanders , J. Vac Sci.Techn., 19, p. 868 (1981).
2. V.L.Kantsyrev, A.A.Kologrivov, K.I.Kopytok, A.S.Shlyaptseva, Reports of XIY Intern.Conference of coherent and non-linear optics , Leningrad, p. 103 (1991)

ION BEAMS FROM AXIAL DISCHARGES AND THE X-RAY LASER PROBLEM

K. N. Koshelev
Laboratory of Plasma Spectroscopy, Institute of Spectroscopy, Russian Academy of Science, Troitzk, Moscow region, 142092 Russia

H.-J. Kunze
Institut für Experimentalphysik V, Ruhr-Universität, 44780 Bochum, Germany

ABSTRACT

Recent estimates and a few available data for charge exchange between colliding ions reveal that for specific systems cross-sections can become considerable even at relatively low impact energies despite Coulomb repulsion. Lasing in the x-ray region by respective charge transfer pumping thus appears feasible employing a scheme, where a suitably prepared ion beam interacts with a proper target plasma. Axial discharges can produce such beams and various configurations are discussed. Finally the possibility is analyzed that lasing recently reported for a capillary discharge is due to such a mechanism.

INTRODUCTION

X-ray lasers require high-density high-temperature plasmas as the active lasing medium. In general, the necessary pump powers are enormous, and high-power long-wavelength lasers themselves were applied as drivers in those pumping schemes, which were most successful so far. For a survey of the fundamentals in x-ray lasers and of the results until 1990 the monograph of Elton[1] may be consulted, the most recent advances being given in respective conference proceedings.[2] On the other hand, z-pinches driven by pulsed-power electrical generators are reported to emit the most powerful incoherent x-ray radiation[3] but have not resulted yet in lasing. Collisional and recombination pumping as well as photopumping were attempted. These discharges, however, can produce intense beams of multiply charged ions, and their interaction with a neutral or weakly ionized gas target offers the possibility of obtaining population inversion via charge exchange.

CHARGE TRANSFER

Charge transfer was originally proposed by Vinogradov and Sobelman[4] as a promising method for producing population inversion for the x-ray region. Since it is a resonance process, only a few levels are populated selectively, and combined with very high cross-sections this indeed results in high pumping rates for the upper level. The pumping probability P is simply given by

$$P = n_t \sigma v , \tag{1}$$

where n_t is the density of the donor atoms (or ions), σ is the respective cross-section, and v is the relative velocity between the interacting particles. Large cross-sections having typical values of $\sigma = 10^{-14}$ cm^2 are reported for collisions of highly charged ions with neutral atoms,[5] and hence practically all respective pumping experiments focused on the interaction of hot plasmas with a neutral target gas.[1] The hot plasma was usually produced by focusing high-power laser beams onto suitable solid targets and the plasma expanded into a background gas. Although population inversion could be observed,[6] no gain in the x-ray region has been reported so far. Several problems can be advanced, one being for example the creation of shock waves in the gas, which cause heating and perhaps ionization prior to the arrival of the hot plasma.

The possibility of ion-ion charge transfer pumping has been more or less dismissed on grounds that the cross-sections are too small due to the strong Coulomb repulsion. However, a few theoretical calculations reported in the literature (see for example Ref. 7) reveal that such cross-sections are not small at all, and specific calculations[8] for collisions of bare carbon nuclei with CIII and CIV ions indicate cross-sections of up to larger than 10^{-15} cm^2, which is the magnitude of ion-ion Coulomb collision cross-sections. At proper impact velocities pumping probabilities thus can become substantial. If we consider a velocity of the order of $v = 10^7$ cm/s and a target ion density of $n_t = 10^{17}$ cm^{-3}, we have already $P = 10^9$ s^{-1}. New experiments become conceivable, where plasma beam and target plasma are tailored independently to optimize a specific pumping scheme. Implications for the interpretation of recent experiments in capillary discharges will be discussed in the last section.

AXIAL DISCHARGES

High-velocity plasma beams can be formed by axial discharges, and several configurations are relevant. We start with ***annular gas-puff z pinches.***[9,10] A fast-acting valve injects the initial filling gas into the anode-cathode gap through an annular nozzle. The discharge through the hollow gas cylinder drives the plasma towards the axis, where the stagnation of the imploding shell results in a plasma of high temperature and high density. In contrast to standard z pinch discharges no material is initially on the axis, and the implosion velocity and hence the final temperature can in principle be higher.

However, as the injected gas flows across the electrode gap, the initial gas annulus spreads radially, which can cause the plasma near the nozzle to implode earlier than the plasma further away. The effect is clearly seen on framing pinhole pictures and is called descriptively "*zippering*".[11] The current path becomes conical resulting in an axial component of the **J** x **B** force. As a consequence plasma is axially ejected and, if the electrode has a proper hole, the beam can be used for the interaction with a target plasma or gas.

Previous investigations were concerned with a reduction of the zippering time.[11,12] Tailoring of the initial gas-puff density distribution and inclining the nozzle was employed. Now, these methods can be used to optimize the zippering effect with respect to the plasma ejection.

The specific geometry of **plasma focus devices** makes them especially suited for the ejection of plasma during the final collapse phase, where the initially axial motion of the plasma is converted into a radial collapse and "zippering" occurs quite naturally. Substantial amounts of the plasma can be ejected and the velocities of the axial jets reach values of a few times 10^7 cm/s.[13] It is certainly possible to further optimize these plasma jets with the present application in mind. Geometry of the discharge arrangement, electrical discharge parameters and filling pressure can be varied. In this context we do not consider the extremely energetic ions (100 keV and higher), which are observed in most devices; their number usually being too small to obtain gain for an x-ray laser. Several mechanisms are advanced to explain this energetic ion beam formation, one arising from high electric fields connected with a violent m = 0 instability.

However, the **m = 0 instability** itself and the resulting **micropinch** in the neck of the instability produce already a small plasma jet of highly ionized atomic species. The reason is straightforward. Due to the higher magnetic pressure in the neck of the instability the plasma pressure is higher there than in the neighboring region and plasma flows out of the neck while this is being compressed further. The velocity of the plasma jet is approximately given by the ion-sound speed c_s.[14] For a plasma of 150 eV with fully stripped carbon atoms one obtains $c_s \cong 8 \times 10^6$ cm/s, i. e. the velocity is of the desired magnitude.

SOME CONSIDERATIONS ON PUMPING AND LASING

The necessity to transport the hot beam plasma to a cooler target plasma imposes some restraints on the parameters of the beam, but those of the target have to be chosen properly, too. First of all, conditions must be such that recombination remains negligible in order to avoid loss of beam ions. Neither in the beam nor in the target should the electron density be that high that upper and lower laser levels of the beam ions are coupled by collisions; even collisions from the upper level to the next higher one be preferably small, since they represent a loss channel. If we consider, for example, the Balmer-α transition between the n = 3 and n = 2 levels in hydrogenlike ions of nuclear charge Z as lasing transition, this implies an electron density

$$n_e\,(\mathrm{cm}^{-3}) \le 5 \times 10^{12}\, Z^6\, (kT_e)^{1/2} \qquad (2)$$

where kT_e is in eV. For a beam plasma with Z = 6 ions the electron density of the target plasma should therefore be lower than 10^{18} cm^{-3} at a temperature of 20 eV. In such a plasma the collsion limit is about 3.5 for the carbon beam ions and the collisional-recombination pumping probability P_c into the host of all high-lying levels[15] is obtained as $P_c \cong 10^7$ s^{-1}. This is considerably lower than the possible pumping rate by charge exchange discussed above.

The density $n_b(3)$ of the upper level of the beam ions (b) is simply obtained by equating the radiative decay rate A with the pumping rate:

$$n_b(3) = n_b\,(P/A) = n_b\, 10^{-8}\, Z^{-4}\, P, \qquad (3)$$

or $n_b(3) \cong 0.008\ n_b$ for carbon ions and the assumed charge exchange. This is indeed very efficient pumping.

The pumping volume is determined by the penetration depth d of the beam ions. It is given by the macroscopic cross-section σn_t, i.e.

$$d = 1/\sigma n_t\ . \tag{4}$$

For our conditions this yields d ≅ 100 μm, which is probably acceptable. It certainly can be varied by changing the target density, by selecting other reactions or by simply changing the beam velocity which changes the cross-section. The penetration depth by ion-ion Coulomb collisions is of the same magnitude for the above example.

For an estimate of the possible gain we neglect the population of the lower level and take the inversion factor F = 1. The line profile of the lasing transition is assumed to be Gaussian with a width determined by the temperature of the initial plasma. Emission transverse to the plasma beam may display narrower profiles. For kT = 150 eV the peak gain coefficient G for the Balmer-α line of CVI at 182 Å is estimated to [16]

$$G \cong 1.3 \times 10^{-17}\ n_b\ . \tag{5}$$

Beam densities of the order of 10^{17} cm^{-3} thus can result in gain coefficients of 1 cm^{-1}. An active zone of 100 μm is traversed by the beam on time scales of the order of nanoseconds and this may permit even lasing during longer times. Finally, the influence of possibly existing magnetic fields has to be carefully analyzed.

CAPILLARY DISCHARGE

Recent studies of the emission from discharges through narrow capillaries made of polyacetal revealed features of the Balmer-α line of CVI at 182 Å, which are characteristic of lasing.[17,18] Although the discharge energies are very low, usually of the order of 10 J, the Lyman and the Balmer series of CVI can be recorded. Spectroscopic studies in the visible and in the soft x-ray region employing a pinhole transmission grating indicated that the emission from the highly ionized species originates in the center. At current maximum of the second half-cycle of the discharge, the Balmer-α line shows a spike on its emission, which appears always at the same time with a jitter of less than 10 ns. The emission is directed: if the spectrometer is not aligned along the axis of the capillary, no spike is recorded. This collimated emission is also observed on the density distribution along the photographically recorded line, which allows to deduce a divergence angle. This angle decreases with the length of the capillary whereas the magnitude of the spikes increases. This increase is not exponential, which would be the case for a homogeneous plasma with homogeneous inversion. However, a multi-layer x-ray mirror positioned at one end of the capillary increased the spikes and reduced the divergence.

Considering first recombination pumping after rapid cooling for an explanation was obvious, but known recombination and cooling rates were not sufficient. The occurrence of the spikes at the second current maximum gave the clue to the present interpretation, which favours a possible m = 0 instability.

During the first half-cycle of the discharge a dense carbon plasma containing also oxygen is produced from wall material, and it cools and thins at current zero. During the second half-cycle the current is lower and the carbon ions are predominatly in the CIII and CIV ionization stage. At current maximum the m = 0 instabilty develops, the plasma is compressed and heated in the neck region, carbon ions are fully stripped, and the ejected hot plasma jet interacts with the cool plasma outside the neck. This interpretation is supported by the considerations above.

ACKNOWLEDGMENT

This work was supported by the Deutsche Forschungsgemeinschaft.

REFERENCES

1. R. C. Elton, X-ray Lasers (Academic Press, San Diego, 1990).
2. E. E. Fill, ed., X-ray Lasers 1992 (Inst. Phys. Conf. Ser. No. 125, London, 1992).
3. N. R. Pereira, J. Davis, N. Rostoker, eds., Dense Z-pinches (AIP Conf. Proc. 195, New York, 1989).
4. A. V. Vinogradov and I. I. Sobelman, Sov. Phys. JETP 36, 1115 (1973).
5. R. K. Janev, L. P. Presnyakov, V. P. Shevelko, Physics of Highly Charged Ions (Springer, Berlin, 1985).
6. R. H. Dixon and R. C. Elton, Phys. Rev. Lett. 38, 1072 (1977).
7. R. K. Janev and D. S. Belic, J. Phys. B: At. Mol. Phys. 15, 3479 (1982).
8. D. Uskov, private communication.
9. J. Shiloh, A. Fisher, and E. Bar-Avraham, Appl. Phys. Lett. 35, 390 (1979).
10. C. Stallings, K. Childers, I. Roth, and R. Schneider, Appl. Phys. Lett. 35, 524 (1979).
11. T. W. Hussey, M. K. Matzen, and N. F. Roderick, J. Appl. Phys. 59, 2677 (1986).
12. W. W. Hsing and J. L. Porter, Appl. Phys. Lett. 50, 1572 (1987).
13. V. F. D'yachenko and V. S. Imshennik , in Reviews of Plasma Physics, Vol. 8, ed. M. A. Leontovich (Consultants Bureau, New York, 1980), p. 199.
14. V. V. Yankov, Sov. J. Plasma Phys. 17, 305 (1991).
15. H. R. Griem, Plasma Spectroscopy (McGraw-Hill, New York, 1964).
16. Ref. 1, p. 26.
17. C. Steden and H.-J. Kunze, Phys. Lett. 151, 534 (1990).
18. C. Steden, H. T. Wieschebrink, and H.-J. Kunze, in X-ray Lasers 1992, ed. E. E. Fill, (Inst. Physics Conf. Series No. 125, London, 1992), p. 423.

INFLUENCE OF DEVICE PARAMETERS ON THE NUMBER OF N VII LYMAN-α PHOTONS EMITTED DURING ONE SINGLE PINCH EVENT

D. Rothweiler[1], J. Hannawald[1], R. Lebert[1], W. Neff[2], and A. Tusche[1]
[1]Lehrstuhl für Lasertechnik, RWTH Aachen,
[2]Fraunhofer Institut für Lasertechnik,
Steinbachstraße 15, 5100 Aachen, Germany

ABSTRACT

A plasma focus x-ray source is optimized experimentally to produce N VII Lyman-α line radiation for its application to pulsed x-ray microscopy. The brightness in this x-ray line increases proportional to the nitrogen pressure peaking at pinch currents that match the energy per ion input into the plasma with respect to proper ionization and excitation. Its upper limit is due to the onset of filamentary streamer discharges that give rise to loss currents and thus interfere with an appropriate adjustment of pinch current and gas pressure. A stationary gas flow scheme with two different gases for plasma ignition and radiation production is applied to extend the pressure range where ignition occurs undisturbed and an improvement of the brightness as expected in view of its scaling with pinch current and gas pressure is demonstrated.

INTRODUCTION

To supplement research work at electron storage rings [1] at higher flexibility and at low costs an x-ray microscope based on a pinch plasma x-ray source is under development as a joint effort of the Forschungseinrichtung Röntgenphysik, Universität Göttingen, the Carl Zeiss Company, the Fraunhofer Institut für Lasertechnik and the Lehrstuhl für Lasertechnik, Aachen, Germany [2]. The proposal [3] of the microscopes x-ray optical arrangement, i.e. a high resolution micro zone plate forming the image and a mirror condenser providing the illumination of the specimen, and its improvements [4,5] are presented elsewhere.

The spectral features of the pinch plasma source are discussed with respect to the requirements by the properties of the x-ray optical components. Emphasis is on the brightness in the N VII Lyman-α line used for imaging and its dependence on device parameters, namely, pinch current and gas pressure. The experimental results reported support theoretical considerations [6] on maximizing the brightness.

X-RAY SOURCE

The x-ray source (Fig. 1) of the microscope consists of a Mather type [7] plasma focus electrode system that is powered by a 2.5 kJ capacitor bank. If the voltage is applied to the electrode system a plasma layer is formed by a sliding discharge upon the cylindrical insulator. The plasma layer lifts off the insulator and runs through the electrode system driven by magnetical forces due to the rapidly increasing discharge current. At the end of the electrode system the plasma layer is imploded onto the axis of symmetry, gathering kinetic energy that is thermalized and radiated away during stagnation on axis.

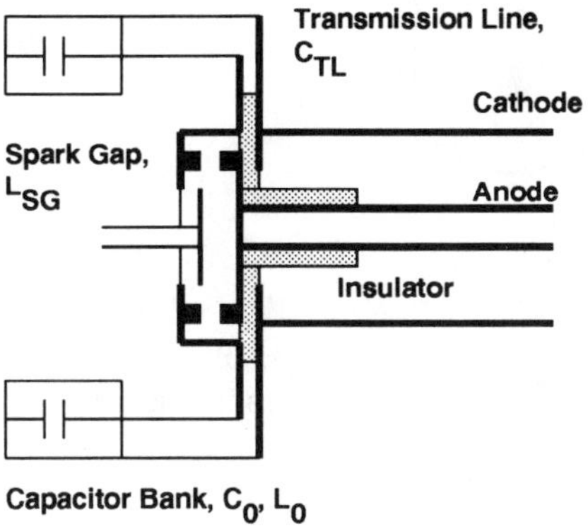

Fig. 1. Scheme of the plasma focus x-ray source

C_0 = 40 μF
C_{TL} = 20 nF
L_0 = 17 nH
L_{SG} = 7 nH

The low inductance spark gap is mounted to the electrode system directly, such that the parallel plate transmission line acts as an additional high impedance $((L_{SG}/C_{TL})^{1/2} = 600\ m\Omega)$ driver powering plasma ignition faster than the capacitor bank does, according to $(L_{SG}C_{TL})^{-1/2} = 100$ MHz >> $((L_0+L_{SG})C_0)^{-1/2} = 1$ MHz (Fig. 2). As the spark gap inductance is small compared to the inductance of the capacitor bank, this design also establishes a higher current rise $U_0/L_{SG} > 10^{12}$ A/s $> U_0/L_0$ to sustain the formation of the plasma layer. After ignition, the discharge is mainly powered by the capacitor bank. Its impedance of $((L_0+L_{SG})/C_0)^{1/2} = 24\ m\Omega$ dominates the (inductive) load impedance $dL_p/dt = \Lambda v = 9\ m\Omega$ (Λ: inductance of the electrode system per unit length, v: velocity of the plasma layer during run down phase, typically $5*10^6$ cm/s), to achieve an apropriate current efficiency in terms of peak current per storage energy.

SPECTRUM

The x-ray source is operated with nitrogen gas discharge loads producing x-ray lines and recombination radiation of both hydrogenic and helium like ions (Fig. 2). For microscopy experiments, the N VII Lyman-a line at 2.48 nm wavelength is chosen. A second transition, i.e. the N VI $1s^2$-1s3p at 2.49 nm, coincides within a reciprocal relative bandwidth of $\lambda/\Delta\lambda = 210$ and can be used simultaneously, in agreement with the requirement on monochromaticity $\lambda/\Delta\lambda = 150$-200 [8] by the microscopes zone plate optics. The spectral location of these lines within the water window wavelength range [9] close to the oxygen K-edge (Fig. 3) is best suited to study thick biological specimen in their wet natural state [10].

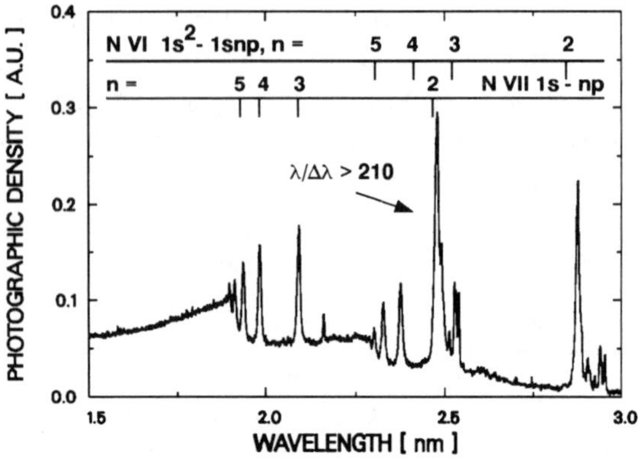

Fig. 2. Emission spectrum recorded with a grazing incidence spectrograph[11] in Rowland geometry

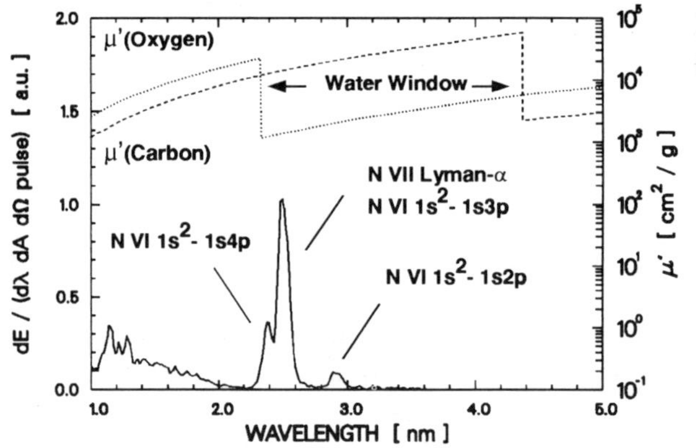

Fig. 3. Laterally resolved spectrum (radial position: r = 0)

Laterally resolved spectra (Fig. 3) are measured with a pinhole transmission grating spectrograph (Fig. 4). Although sensitive to x-rays, a Gd_2O_2S:Tb x-ray to visible converter covers its CCD detector for radiation hardening. Since the pinhole grating is fabricated [12] as the microscopes zone plate is [13] and since the CCD corresponds to the microscopes CCD detector [5], the spectral distribution of radiation is recorded as it effectively contributes to image formation in the dedicated x-ray microscope; including the spectral dependence of the zone plate efficiency, of the beam line transmission [14] and of the detection sensitivity. The reflection efficiency of the mirror condenser (gold coating) can be considered constant within the wavelength range of interest [15].

The analysis of the pinhole grating spectrum shown in Fig. 3 gives a ratio of line intensity over continuum intensity of $I_{\text{N VII Lyman-}\alpha} / I_{\text{continuum}} > 550$ within the reciprocal relative bandwith of $\lambda/\Delta\lambda = 210$ defined by the spectral spacing of the two x-ray lines used for microscopy experiments; sufficient so that a reduction of imaging resolution is not to be expected [8]. Also neglectable are image contrast losses due to recombination radiation below 2.0 nm wavelength [16].

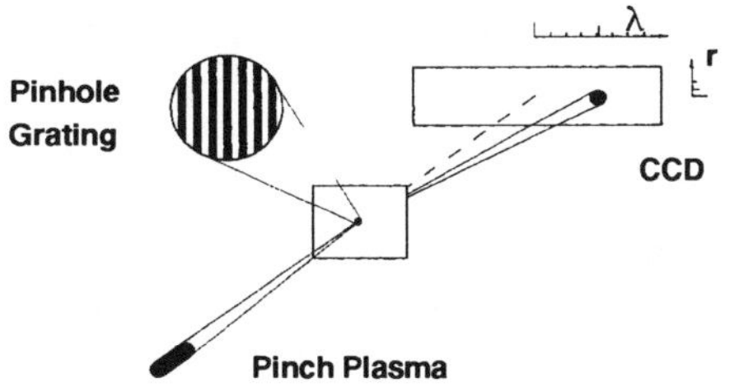

Fig. 4. Pinhole Grating Spectrograph (10000 lines/mm, $\lambda/\Delta\lambda$ =20, Δr = 150 μm)

INFLUENCE OF DEVICE PARAMETERS ON BRIGHTNESS

The main goal is to produce a sufficient number of photons in one single pulse, such that x-ray microscopy images are formed before microscopic motion within the specimen or radiation damage can interfere. However, only photons originating from a finite spot of the source can effectively contribute to image formation. The corresponding spot diameter is determined by the size of the image field in the microscopes object plane and the demagnification of the source profile by the mirror condenser. Therefore, the brightness, i.e. the energy yield in the N VII Lyman-α line per pulse, per unit source surface and per steradian, is the figure of merit (in view of broadband cw sources the brightness is actually given in terms of photons per second per unit source surface, per steradian and per bandwith).

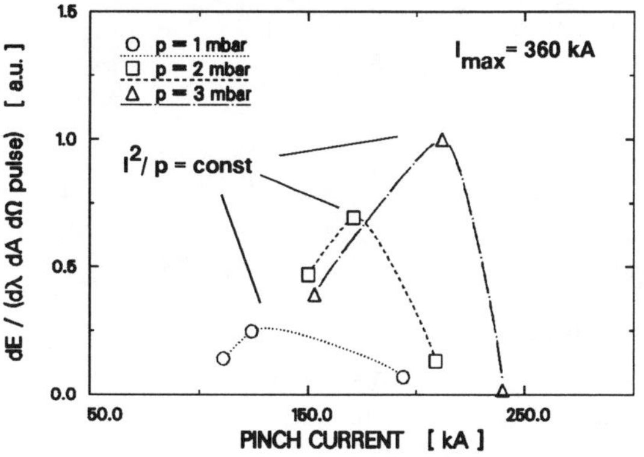

Fig. 5. Dependence of brightness on pinch current and gas pressure

Its dependence (Fig 5) on pinch current and gas pressure is studied with the pinhole grating spectrograph (Fig. 4). The corresponding pinch currents are measured with a calibrated magnetic probe placed at the end of the electrode system within the coaxial electrode gap so that the integration of its dI/dt signal gives the pinch current solely, without contributions of loss currents.

The brightness increases proportional to the pressure p of the nitrogen gas and peaks at pinch currents I according to I^2/p = 14575 kA^2/mbar = const (Fig. 5). As the kinetic energy gained during implosion (to be thermalized during pinch phase) goes with I^2 and as the total number of particles in the pinch is proportional to the gas pressure, this result is interpreted [6] in terms of the ratio of the energy per particle input into the pinch plasma and the thermal energy per particle effecting ionization and excitation. At $I^2/p > 14575$ kA^2/mbar the resultant average plasma temperature is too high and at $I^2/p < 14575$ kA^2/mbar the average temperature is too low as to achieve maximum N VII 2p population. According to theoretical considerations [6] the N VII 1s-2p excitation is maximized at $I^2/p = 16670$ kA^2/mbar, in good agreement with experiments.

CONSTRAINTS BY STREAMER DISCHARGES

An increase of the density of radiating ions with gas pressure results in an corresponding increase of the brightness, in case of optically thin plasmas. Consequently, the results shown in Fig. 5 suggest that a further improvement of the brightness with gas pressure can be expected until either the plasma becomes optically thick [6], or the pinch current neccessary according to I^2/p = const exceeds the peak pinch current that can be delivered by the capacitor bank under investigation.

Although the capacitor bank is capable of producing peak currents up to $I_0 = U_0(C_0/(L_0+L_{SG}+L_P))^{1/2} = 360$ kA (the coupling of the load to the pulse generator is roughly estimated taking $L_P = \Lambda z$, Λ: inductance per unit length of the

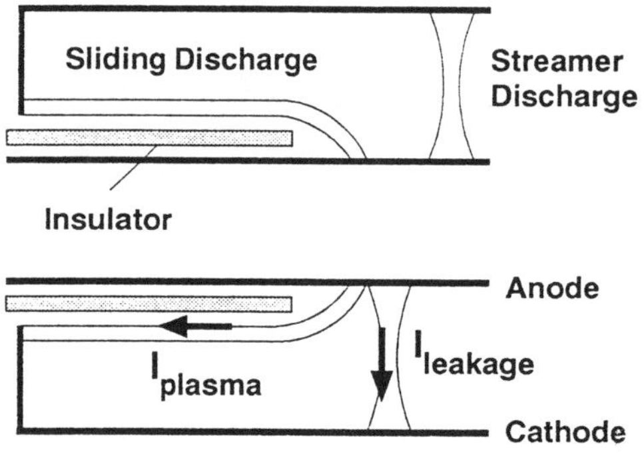

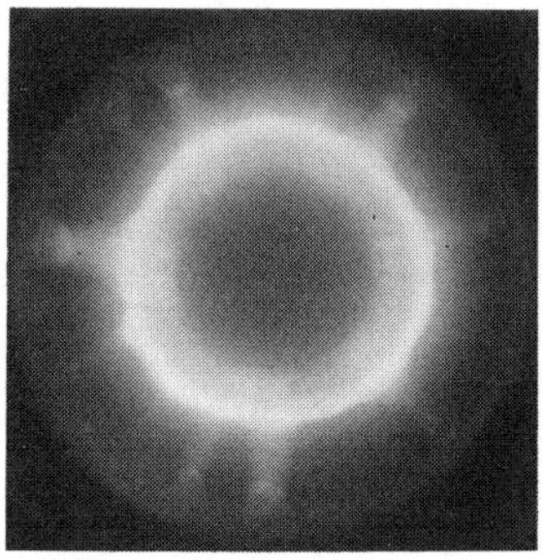

Fig. 6. End on view of the discharge at 300 ns after breakdown (4 mbar)

electrode system, z: length of the electrode system) an increase of the brighness at pressures above 3 mbar is not observed. Instead, the brightness drops as a result of loss currents carried by filamentary streamer discharges during plasma ignition (Fig. 6). Any prerunning filament produces magnetic stray fields in front of the plasma layer which will lower the kinetic energy gained during implosion and thus interfere with adjusting the relation $I^2/p = 14575$ kA^2/mbar.

According to the Raether criterion [17] streamer discharges are avoided if the product of the Townsend primary ionization coefficient α and the flight distance d of electrons along electrical field lines is kept below α*d = 18. Theoretical calculations on α*d by means of Monte Carlo simulation taking into account the differential scattering cross section [18] of the discharge gas as well as the electrical field distribution inside the coaxial electrode system confirm the onset of streamer discharges at the threshold pressure of 3 mbar in nitrogen gas (Fig 7). In contrast, helium gas promises undisturbed plasma ignition up to pressures above 15 mbar.

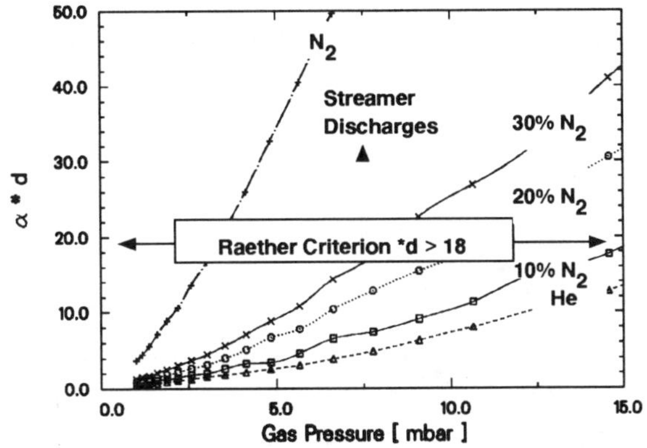

Fig. 7. Comparisson of α*d values for for nitrogen gas and helium gas as computed by means of Monte Carlo simulation according to the electrical field distribution within the coaxial electrode system

DECOUPLING OF PLASMA IGNITION AND RADIATION PRODUCTION

To take advantage of both the emission spectrum of nitrogen and the ignition properties of helium a modified gas flow scheme (Fig. 8) is applied to the elctrode system. Helium gas is let into the electrode gap next to the insulator where ignition and run down take place and the nitrogen gas is fed through the center electrode there where the implosion of the plasma and radiation production occur.

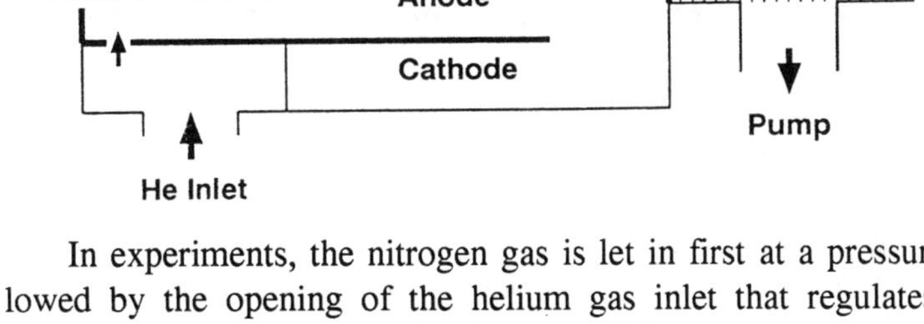

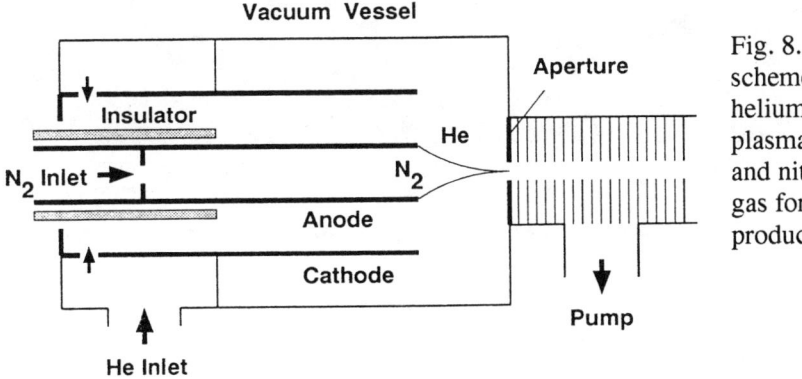

Fig. 8. Gas flow scheme with helium gas for plasma ignition and nitrogen gas for radiation production

In experiments, the nitrogen gas is let in first at a pressure $p_{\mathbf{nitrogen}}$ to be followed by the opening of the helium gas inlet that regulates the total pressure $p > p_{\mathbf{nitrogen}}$. At appropriate throughputs of the vacuum pump, a helium gas flow is established that entirely drives the nitrogen gas out of the coaxial electrode gap. Ideally, both gases are seperated in the resultant stationary flow with no partial pressure gradient and at a total pressure p > 3 mbar.

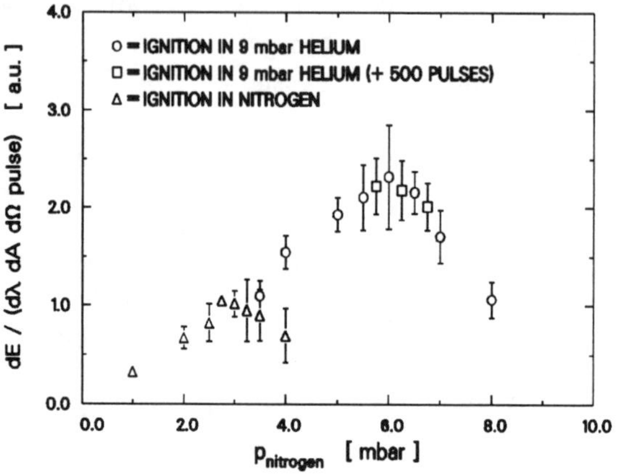

Fig. 9. Increase of brightness with gas pressure. The error bars give the pulse to pulse reproducibility which amounts to 15 %

Best results are achived at a pressure $p_{\mathbf{nitrogen}}$ = 6-7 mbar and a total pressure of 9 mbar (Fig. 9). As expected in view of its scaling discussed above, the brighness in the N VII Lyman-α line increases linearily with gas pressure above 3 mbar; improving appreciable for microscopy experiments. However, nitrogen particle diffusion into the coaxial electrode gap causes streamer discharges if the pressure $p_{\mathbf{nitrogen}}$ is chosen too close to the total pressure p controlled by the helium gas regulation (Fig. 10).

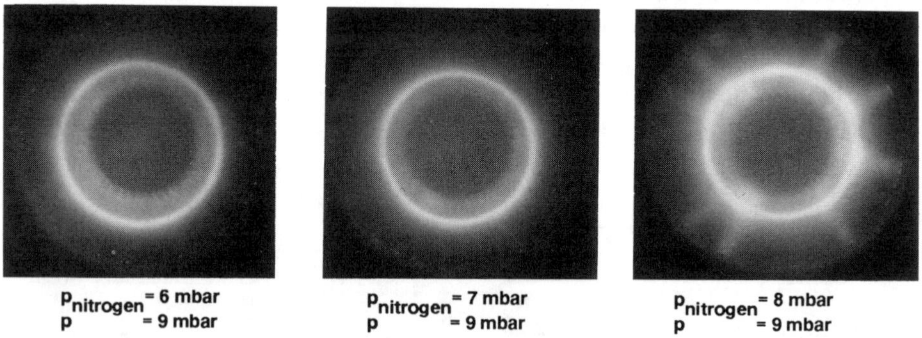

Fig. 10. End on view of the discharge at 300 ns after breakdown

CONCLUSION

The brighness of the pinch plasma x-ray source in the N VII Lyman-α line scales proportional to the nitrogen gas pressure or, with respect to the results on the coupling of kinetic implosion energy to ionization and excitation, proportional to the square of the pinch current. An improvement of the maximum brightness of the device is achieved by a reduction of loss currents carried by filamentary streamer discharges. So far, a modified gas flow scheme is applied to avoid undesired streamer discharges, in view of the differential scattering cross sections of the gases involved. To achieve a further improvement of the maximum brightness it is intended to optimize the electrical field distribution within the coaxial electrode system that also effects the growth of electron avalanches.

ACKNOWLEDGEMENTS

This work was funded by the German Ministry of Research and Technology (BMFT) under contract number 13N5838. The cooperation with the Forschungseinrichtung Röntgenphysik, Universität Göttingen, and with the Carl Zeiss Company is gratefully acknowledged.

REFERENCES

1. A variety of results from microscopy work at electron storage rings is reported in X-Ray Microscopy III (Springer, Berlin, 1990)
2. B. Niemann, D. Rudolph, G. Schmahl, M. Diehl, J.Thieme, W. Meyer-Ilse, W. Neff, R. Holz, R. Lebert, F. Richter, and G. Herziger, Optik 84, No 1(1990), p 35
3. D. Rudolph, B. Niemann and G. Schmahl, and J. Thieme, X-Ray Microscopy II, (Springer, Berlin, 1987), p 216
4. G. Schmahl, B. Niemann, D. Rudolph, M. Diehl, J. Thieme, W. Neff, R. Holz, R. Lebert, F. Richter, and G. Herziger, X-Ray Microscopy III (Springer, Berlin, 1990), p 67
5. J. Thieme, G. Schmahl, D. Rudolph, B. Niemann, and P. Guttmann, Inst. Phys. Conf. Ser. No 130, (IOP Publishing, Bristol, 1993), p 555
6. R. Lebert, D. Rothweiler, and W. Neff, this volume
7. J.W. Mather, Methods of Experimental Physics, (Accademic Press, New York, London, 1971), p 187
8. J. Thieme, X-Ray Microscopy II, (Springer, Berlin, 1987), p 70
9. P.J. Duke, and A.G. Michette, Modern Microscopies, (Plenum Press, New York, London, 1990), p 1
10. D. Rudolph, G. Schneider, P. Guttmann, G. Schmahl, B. Niemann, and J. Thieme, X-Ray Microscopy III (Springer, Berlin, 1990), p 392
11. W. Neff, J. Eberle, R. Holz, F. Richter, and R. Lebert, X-Ray Microscopy II, (Springer, Berlin, 1987), p 22
12. C. David, Forschungseinrichtung Röntgenphysik, Universität Göttingen, Germany
13. J. Thieme, C. David, B. Kaulich, P. Guttmann, D. Rudolph, and G. Schmahl, Inst. Phys. Conf. Ser. No 130, (IOP Publishing, Bristol, 1993), p 527
14. R. Holz, J. Eberle, R. Lebert, W. Neff, and F. Richter, X-Ray Instrumentation (1989), SPIE Vol. 1140, p 133
15. B. L. Henke, P. Lee, T. J. Tanaka, R. L. Shimabukuro, and B.F. Fujikawa, Atomic Data and Nuclear Data Tables, Vol 27, No 1, Jan 1982
16. M. Diehl, X-Ray Microscopy IV, (Springer, Berlin, to be published)
17. E.E Kunhardt, and L.H. Luessen, Electrical Breakdown and Discharges in Gases, (Plenum Press, New York, 1983), p 25
18. L.J. Kiefer, A Compilation of Electron Collision Cross Section Data for Modelling Discharge Lasers, Joint Institute for Laboratory Astrophysics Boulder, Colorado (1973)

TIME-INTEGRATED AND TIME-RESOLVED MEASUREMENTS OF X-RAYS AND CHARGED PARTICLES FROM PF-360 FACILITY

M.Sadowski, J.Zebrowski, and E.M.Al-Mashhadani*

Soltan Institute for Nuclear Studies,
05-400 Świerk n. Warsaw, Poland

ABSTRACT

The paper presents results of recent experimental studies performed with a large 360-kJ plasma-focus facility PF-360 operated at energy levels ranging from 120kJ/30kV up to 176kJ/35kV. Particular attention has been paid to the fine structure of X-ray emitting regions, and to the emission of fast deuterons, impurity ions, as well as fusion-produced protons. Simultaneously, there have been performed measurements of relativistic electron beams emitted in the upstream direction. Time correlations of X-rays and the charged particle pulses have also been investigated.

INTRODUCTION

High-current discharges of the Plasma-Focus (PF) type have been known for many years as powerful sources of X-ray pulses[1-3], accelerated deuterons[4,5], fast impurity ions[6,7], nuclear fusion reaction products[8,9], and relativistic electron beams[10,11]. Experimental studies, which make possible the determination of X-ray and particle fluxes, as well as their angular and energetic distributions, are necessary in order to explain physical phenomena occurring within the PF discharges. Such studies are also needed in order to determine possibility of various technological applications of PF facilities.

The main aim of this paper was to collect and summarize results of experimental research on characteristics of the X-ray and particle emission from PF-type discharges, which were performed in Świerk, Poland.

EXPERIMENTAL SET-UP

The studies have been carried out with the PF-360 facility[12-13] equipped with a 120-mm-dia. inner electrode and 170-mm-dia. outer electrode. The two electrodes were manufactured of 300-mm-long copper tubes. A tubular insulator embracing the beginning of the inner electrode was made of a special ceramics of 80 mm in length. Initial gas filling pressures were 2-8 torr D_2 with various admixtures. The electrodes were supplied by a 288-μF condenser bank charged up to 176kJ/35kV.

The main experimental chamber of the PF-360 device was enlarged by an additional expansion chamber and it was equipped with different diagnostic tools, as shown in Fig.1.

DIAGNOSTIC EQUIPMENT

In additional to the standard measuring equipment, applied for V(t), I(t), dI/dt, Y_n, and Y_n, and Y_n (t) registration, in order to study the X-ray emission use was made of the VAJ-type and two different pinhole cameras. To measure the charged particle emission there were applied various arrangements of nuclear track detectors and miniature light-protected scintillators. Fast electron beams were investigated with a magnetic analyzer and Čerenkov-type detectors[11].

---

* Participation within a framework of a fellowship granted by the Polish Ministry of Education.

EXPERIMENTAL RESULTS

On the basis of previous optimization tests of the PF-360 facility, as performed for various electrode configurations[11-14], it was easy to determine the optimal operating conditions for the chosen electrode set described above. Routine X-ray and neutron yield measurements confirmed that for PF discharges at the energy level of about 150 kJ the optimum value of the initial D_2 filling pressure is equal to about 4.0 torr, as shown in Fig.2. It was a little strange that for a series of shots performed under p_o =3.0 torr D_2 there was observed a strong decrease in the X-ray emission, but it probably can be explained by smaller amount of impurities than for discharges carried out previously under higher initial pressures.

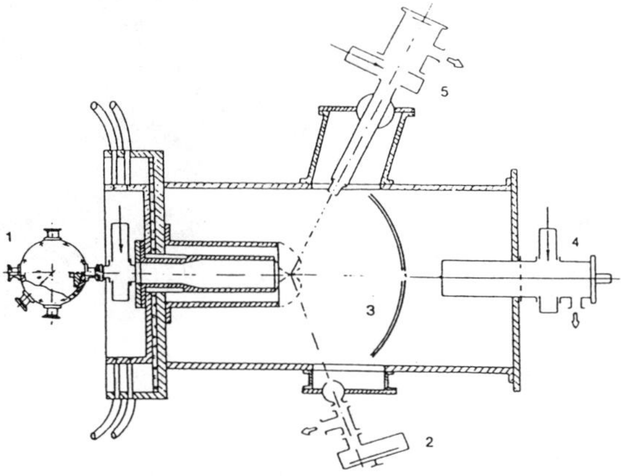

Fig.1. Scheme of the PF-360 facility, with a modified experimental chamber and some diagnostic equipment: 1 - magnetic analyzer of electron beams, 2 - X-ray pinhole camera with a rotated film, 3 - support of nuclear track detectors for measurements of the ion angular distribution, 4 - ion pinhole camera for axial measurements.

X-ray emission measurements, as performed with a vacuum pinhole camera equipped with a 10-μm-thick Be-filter (E_x >800eV) and a rotated X-ray film, have revealed that the current sheath during the radial compresion phase is not uniform since on the inner electrode surface there are visible numerous bright spots, as shown in Fig.3. These regions of the increased X-ray emission, which are relatively weak for a single shot but quite strong and distinct for series of shots, can correspond to places where current filaments interact with the electrode surface. Also within a pinch column there appear numerous "hot-spots", and their characteristics depend evidently on an amount of heavy-gas admixtures.

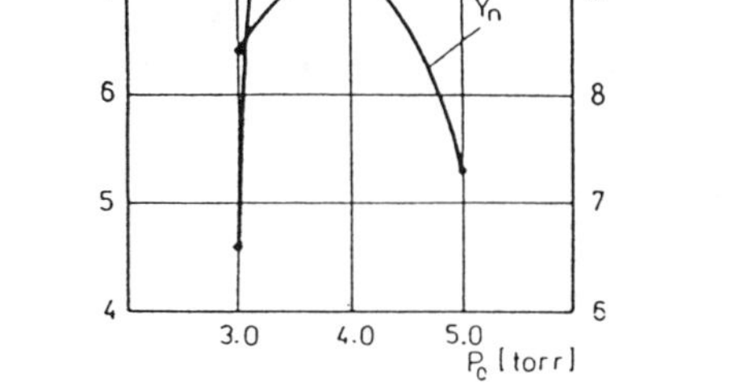

Fig.2. X-ray emission and neutron yield as a function of the initial deuterium filling for the PF-360 facility equipped with a given electrode system and operated at the energy level of 144 kJ/34kV.

In order to investigate the fine structure of the PF pinch column use was made of the other pinhole camera equipped with very thin (0.75-μm or 1.5-μm) Al-filters, an auxiliary pumping stand, and a long tubing enabling to obtain enlarged images.

Soft X-ray pinhole pictures, taken under different operating conditions, have demonstrated the appearance of quasi-axial filaments and tiny "hot-spots" within the pinch column, as shown in Fig.4. Similar to other experiments[10,12] it was observed that "hot-spots" number depends on an amount of heavy-gas admixtures. Microscopic irreproducibility of the "hot-spots" can be explained by their stochastic origin.

Time-resolved hard X-ray pulses were registered together with fusion neutron signals by means of scintillator detectors protected from scattered radiation by paraffine shieldings. The waveforms registered were similar to those obtained for the PF-360 facility run under other operating conditions[13], as shown in Fig.5. The first peak can be identified with hard X-rays only, but the second peak is possibly the superposition of the second X-ray pulse and neutron induced signals. The third (and sometimes the fourth) peak is again neutron signal, and in the PF-360 device it is usually lower than the second peak. For the present electrode system and applied operating conditions the FWHM of the two neutron pulses is equal to 20-80 ns, and the separation of the peaks is 100-150ns.

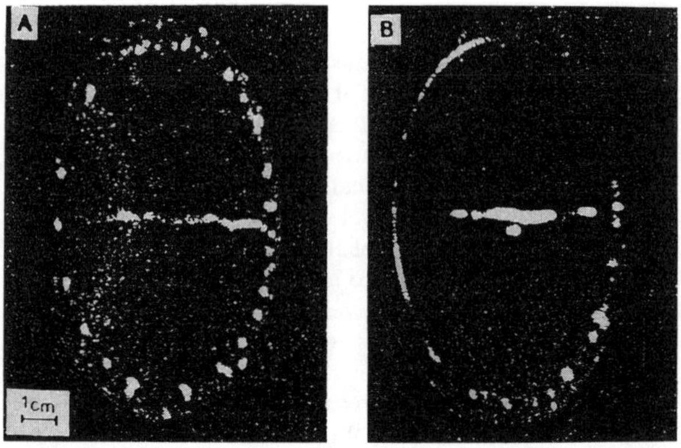

Fig.3. Time-integrated X-ray pinhole pictures, showing distinct regions of an increased emission on the inner electrode rim and some "hot-spots" within the pinch column, as registered in the PF-360 Facility: A - for 10 shots at p_o=5.0 torr D_2, U_o=34 kV, W_o=144 kJ; B - for 1 shot at p_o=3.5 torr D_2 + 0.1 torr Ar, U_o=32 kV, w_o=127 kJ.

Time-integrated measurements of fast ions and fusion produced protons, which were performed by means of the axial pinhole camera and various nuclear track detectors, revealed numerous ion species, as shown in Fig.6. Using different absorption filters made of Al-foils, it was possible to perform a rough energy analysis. Ion flux density maps (similar to that shown in Fig.6), as obtained on the basis of irradiated track detectors after their etching, have proved that among medium-energy ions 30 keV< E <300 keV) there are emitted narrow beams of high-energy(>300 keV) deuterons. Using 80-μm-thick Al-filters and CR-39 plastic detectors it was also possible to observe fusion produced protons (E_D>3.9 MeV, E_p>2.9MeV). For 150-kJ shots there was found about 4.5 x10^3 protons/cm^2 at the distance of 34 cm from the focus region.

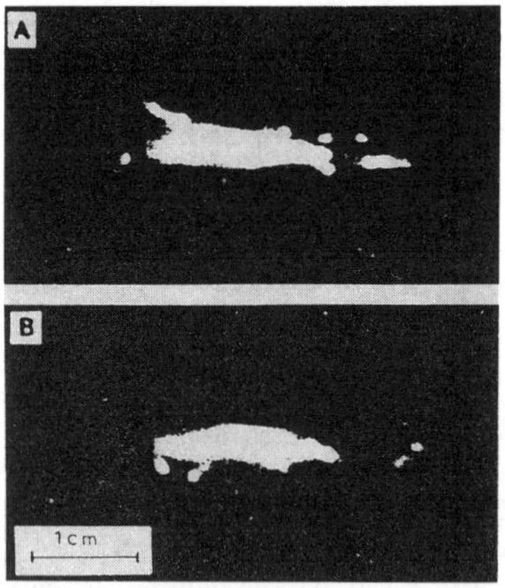

Fig.4. Soft X-ray pinhole pictures of the pinch column, showing the appearance of filamentation and hot spots. The pictures were taken for 10 successive shots performed at U_o=34 kV i W_o=154 kJ under different initial pressures: A - p_o=3.7 torr D_2 + 0.15 torr Ar, B - p_o = 3.9 torr D_2 + 0.2 torr Ar. Black vertical strips correspond to a scaling grid.

Measurements of angular distributions of high-energy deuterons and fusion protons were performed by means of nuclear track detectors shielded with different foil-filters and fastened to a semicircular support in front of the electrode outlet. Tracks registered in the detectors (after irradiation and etching) were classified in several groups corresponding to particles of different energy. In general, the fast deuteron and fusion proton distributions have revealed a distinct anisotropy, but the results averaged over a series of PF shots can be described by relatively smooth curves, as shown in Fig.7. Since calibration data for the CR39 detectors, i.e., characteristics of track diameters vs proton energy and etching conditions, are scarce and hardly available, it was necessary to perform special calibration measurements[15]. Using results of the calibration and integrating the proton angular distribution for 12 shots performed under identical conditions (p_o = 3.5 torr D_2, U_o = 30 kV, W_o = 130 kJ, the fusion proton yield was estimated to be Y_p = (1.7-2.5) x 10^{10}, while the total neutron was Y_n = (2.8-4.7) x 10^{10}. Taking into account stochastic character and anisotropy of the particle emission, the discrepancy in Y_p and Y_n values is within the limits of an experimental error.

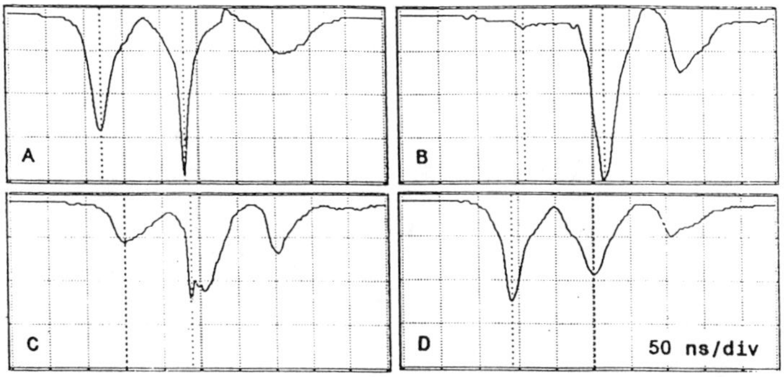

Fig.5. Time-resolved hard X-ray and neutron pulses, as obtained from successive PF discharges performed under identical experimental conditions (p_o=5 torr D_2, U_o=30 kV, W_o=130 kJ), but with different neutron yields: A - Y_n=9.8x10^{10}, B - Y_n=1.2x10^{10}, C - Y_n=5.7x10^9, D - Y_n=5.2x10^9. The signals were registered side-on, at a distance of 2.5 m from the z-axis.

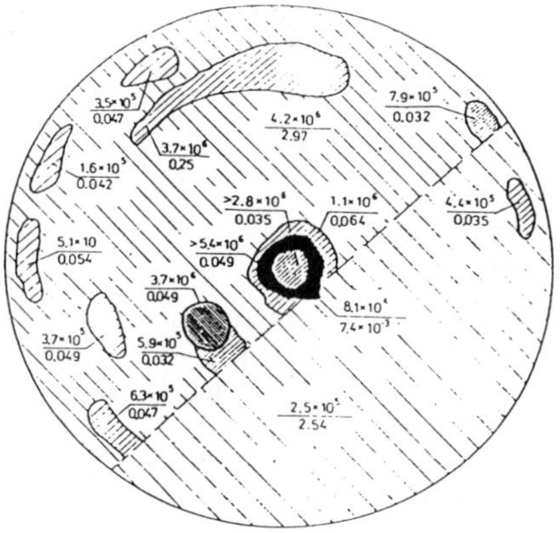

Fig.6. Map of ion beams emitted along the symmetry axis of the PF-360 facility, as obtained from a single shot performed at p_o=8 torr D_2 , U_o=30 kV, W_o=126 kJ, Y_n=4.3x10^{10}. The upper-left part of the CN track detector was covered with a 0.75-μm-thick Al-foil (E_D>80 keV), and the lower-right part shielded with a 1.5-μm-thick Al-foil (E_D>220 keV).

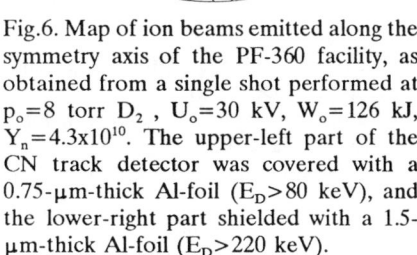

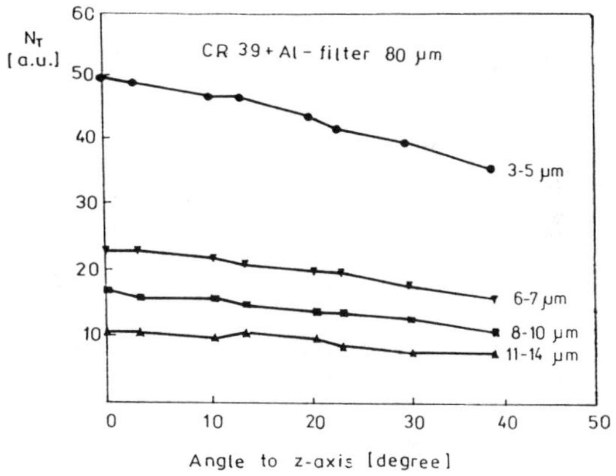

Fig.7. Angular distribution of high-energy ions and fusion produced protons emitted from the PF-360 facility, as measured with CR39 track detectors shielded with the 80-μm-thick Al-foils (permeable for deuterons above 3.9 MeV, and protons above 2.8 MeV). The tracks were classified in 4 groups depending on crater diameters. The experimental points correspond to numbers of tracks registered for 30 shots performed at p_o=5 torr D_2, U_o=34 kV, W_o=144 kJ, Y_n=1.67x10^{11}.

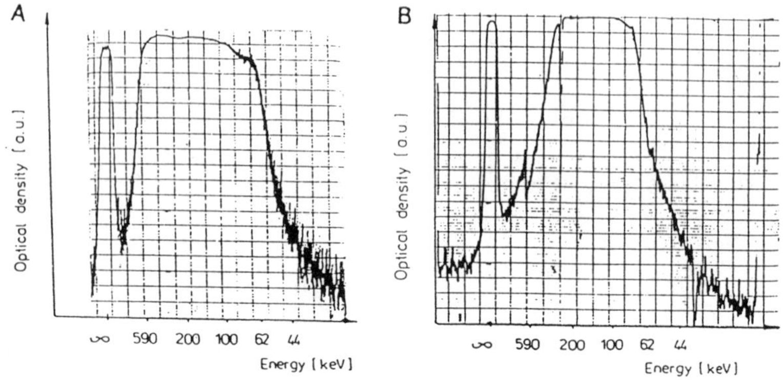

Fig.8. Time-integrated energy spectra fast electron beams emitted from the PF-360 facility, as measured for different shots: A - at p_o=4.0 torr D_2, U_o=32kV, W_o=127kJ, Y_n=1.1x10^{10}, B - at p_o=3.7 torr D_2 +0.1 torr Ar, U_o=34 kV, W_o=154kJ, Y_n=3.1x10$_9$. The detector films were protected with a 10-μm-thick (A) or 16-μm-thick (B) Al-foils.

Time-integrated measurements of fast electron beams emitted through the inner electrode opening in the up-stream direction, as performed with a magnetic analyzer and X-ray films, proved that electron energy spectra range from about 50 keV up to about 800 keV, as shown in Fig.8. Densitometric analysis of the films exposed has appeared to be cumbersome and it has not given accurate results.

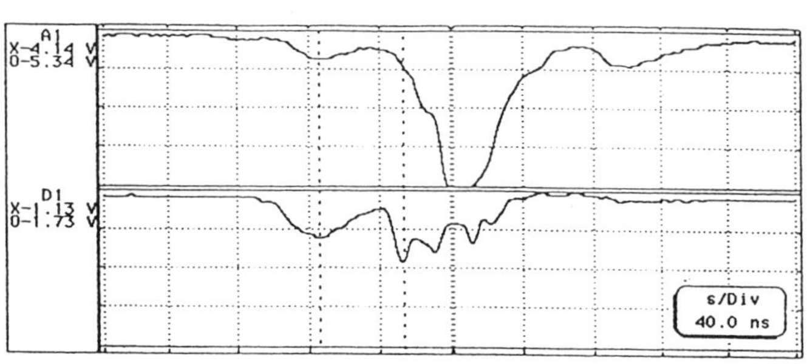

Fig.9. Time-resolved X-ray and fast-deuteron signals from an ion detector (lower trace) in comparison with X-ray and neutron induced signals from the neutron monitor (upper trace), as registered for a single shot at p_o=5 torr D_2, U_o=33kV, W_o=145kJ, Y_n=3.0x10^{10}. The ion detector was placed on the z-axis at a distance of 112 cm from the focus, while the neutron monitor was located side-on at a distance of 250 cm.

Time-resolved studies of fast deuterons were first performed with a miniature scintillation detector placed behind the axial ion pinhole camera. A comparison of deuteron detector signals with hard X-ray and neutron signals from the side-on scintillation monitor, as shown in Fig.9, demonstrated that the deuteron detector was sensitive to the hard X-rays, but it evidently registered also ion pulses. The deuteron signal (with RWHM=60 ns) had a distinct multi-spike structure.

Some efforts were also undertaken to perform time-resolved measurements of fusion protons by means of shielded track detectors, placed at different angles to the z-axis but at the same distance (about 15 cm) from the focus region. A comparison of signals from those detectors with the hard X-ray and neutron signals from the side-on monitor, as presented in Fig.10, has shown waveforms with many spikes which were induced by X-rays, neutrons and fusion-protons. Time-of-flight analysis is practically impossible since proton trajectories are strongly influenced by a local magnetic field. Taking into account the FWHM of the main neutron pulses one can however conclude that proton peaks (with FWHM = 4 - 10 ns) have been registered.

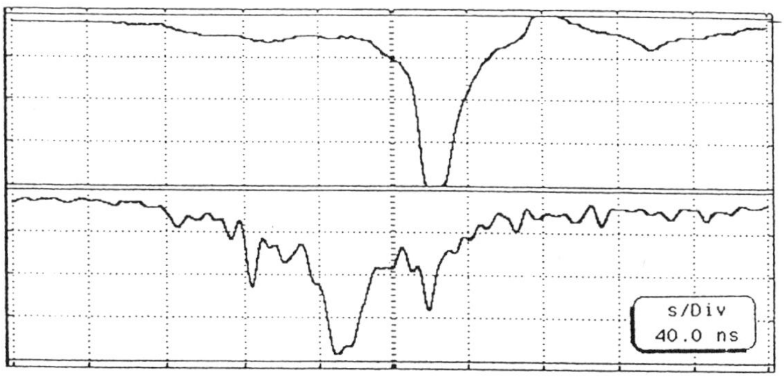

Fig.10. Time-resolved X-ray and fusion proton signals from a proton detector (lower trace) in comparison with X-ray and neutron-induced signals from the neutron monitor (upper trace), as registered for a single shot performed at p_o=5 torr D_2, U_o=33 kV, W_o=157 kJ, Y_n=8.4x10^9. The proton detector was situated at angle of 10° to the z-axis, at a distance of 15 cm from the focus, while the neutron monitor was located side-on at a distance of 250 cm.

Time-resolved measurements of fast electron beams were carried out with various Čerenkov-type detectors, as shown in Fig.11. For the experiment described the FWHM of individual electron pulses was 30-50 ns, and the population of electrons with energy >360keV (detected with plastic radiators) was about 3 times smaller than that of energy > 215 keV (detected with diamond radiators). In the both cases electron energy losses in the 30-μm-thick Cu-filters and real energy detection thresholds were taken into account.

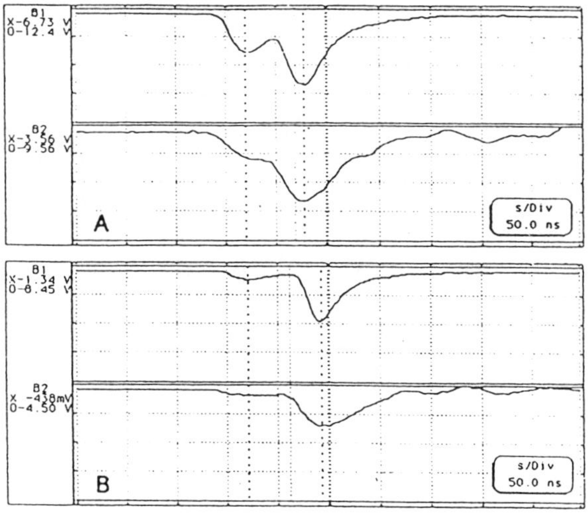

Fig.11. Time-resolved fast-electron signals registered with Čerenkov-type detectors made of diamond (upper traces) and transparent plastic (lower traces), as measured in the PF-360 facility for different shots: A - at p_o=3 torr D_2, U_o=30 kV, W_o=130 kJ, Y_n=3,9x10^9; B - at p_o=5 torr D_2, U_o=33 kV, W_o=157 kJ, Y_n=1.5x10^{10}. The both detectors were placed at z-axis behind the main collector plate.

SUMMARY

The experimental studies described above can be summarized as follows:

1. For the PF-360 facility equipped with a chosen electrode configuration the optimal operating conditions are in agreement with predictions of scaling laws and mhd modelling.
2. Results of X-ray measurements confirm appearance of a fine structure in the X-ray emitting regions observed on the inner electrode rim and within the pinch column.
3. Time-integrated and time-resolved characteristics of the charged particle emission from the PF-360 facility differ slightly from those measured for other PF experiments of a similar scale.
4. Filamentation and hot-spots observed within the pinch column require more detailed studies. In particular, influence of the filamentation on the magnetic field distribution and charged particle trajectories should be investigated, e.q., by computer modelling.

REFERENCES

1. W.H.Bostick, V.Nardi, W.Prior, J.Plasma Phys.8, 7 (1972).
2. K.H.Schönbach, L.Michel, H.Fisher; Appl. Phys. Lett.25, 547 (1974).
3. M.Sadowski, H.Herold, H.Schmidt, M.Shakhatre, Phys.Lett.105A, 117 (1984).
4. R.I.Gullickson, H.L.Sahlin, J.Appl.Phys.49, 1099 (1978).
5. M.Sadowski, H.Schmidt, H.Herold, Phys.Lett.83A, 435 (2981).
6. A.Mozer, M.Sadowski, H.Herold, H.Schmidt, J.Appl.Phys.53, 2959 (1982).
7. M.Sadowski, J.Żebrowski, E.Rydygier, J.Kuciński, Plasma Phys. Contr.Fusion 30, 763 (1988).
8. A.Bernard, Atomkernergie 32, 73 (1978).
9. U.Jager, L.Bertalot, H.Herold, Rev.Sci.Instrum.56,77 (1985).
10. P.Choi, C.Deeney, Phys.Lett.A128, 80 (1988).
11. M.Sadowski, L.Jakubowski, J.Żebrowski, Proc.18th Europ.Con. Contr.Fusion and Plasma Phys. (EPS, Berlin 1991), Part II, p.233.
12. M.Sadowski, A.Szydłowski, J.Żebrowski, E.M.Al-Mashhadani, Proc.1992 Int.Conf.Plasma Phys. (EPS, Insbruck 1992), Part I, p.691.
13. H.Herold, A.Jerzykiewicz, M.Sadowski, H.Schmidt, Nucl.Fusion 29, 1255 (1989).
14. M.Sadowski (Edit.), Dept. Thermonuclear Research Annual Report 1987 (SINS, Otwock-Świerk 1988), Report SINS 2056/P-V/PP/A.
15. M.Sadowski, E.M.Al-Mashhadani, A.Szydłowski, M.Jaskóła, T.Czyżewski, M.Wieluński, to be published in Nucl.Instr.Meth.

Macroscopic Behavior and X-ray Radiation Characteristics of SHOTGUN Z-pinch Plasma

Keiichi Takasugi and Tetsu Miyamoto
Atomic Energy Research Institute, Nihon University, Tokyo 101, Japan

Kinya Moriyama and Hideaki Suzuki
College of Science and Technology, Nihon University, Tokyo 101, Japan

ABSTRACT

Gas-puff z-pinch plasmas were produced using Ne, Ar, Kr and Xe as pinch materials in the SHOTGUN device. The hardening of x-ray radiation with atomic number Z of the operating gas was confirmed. It results from higher energy input into the higher Z plasma due to stronger compression, which is attributed to radiation cooling in the contraction phase.

The contracting plasma showed nonuniformity along the axis, which results from $m = 0$ unstable mode. As the nonuniformity develops, the plasma column is divided into several sections, and each of them collapses by turns from the anode. The x-ray is emitted at several localized spots which corresponds to the nodes of the rippling.

INTRODUCTION

The gas-puff z-pinch[1] is now a facile and simple method of producing high density and high temperature plasmas. It has an advantage as an intense pulsed x-ray source of repetitive operation with various combinations of the operating gas.

The importance of radiative energy loss in contraction of the z-pinch was suggested,[2] and it was shown experimentally on a mixed-gas z-pinch.[3,4] Spectral characteristics of soft x-ray and XUV radiations have been investigated by many authors.[5-7] In this experiment, the global characteristics of soft x-ray radiation and its relation to the radiation dynamics are investigated.

The z-pinch is full of extremely unstable MHD modes. The stabilization of z-pinch and understanding of mechanisms of neutron emission have been important subjects in the nuclear fusion research. Particle acceleration due to the instability was shown,[8] and accelerated electron beam was detected from a z-pinch.[9] The instabilities play important role not only for nuclear fusion but also for x-ray generation. Micro-pinches formed in a z-pinch column should have some relation to the instabilities. The understanding of the z-pinch micro-dynamics is important for both purposes. Macroscopic behavior of the collapsing z-pinch is observed in the experiment, and the relation between the axial nonuniformity and the x-ray spot formation is investigated.

EXPERIMENTAL SETUP

The experiment was carried out using the SHOTGUN z-pinch device at Nihon University.[10] Figure 1 shows the schematic diagram of the device. The bank energy is 4.8 kJ (24 μF - 20 kV). The return current plate is located inside the vacuum chanber to minimize circuit inductance. The plasma currents are measured by Rogowski coils winded around two electrodes. The anode current expresses total current into the discharge section, and the cathode current expresses the current that passes through the z-pinch plasma. The difference in two currents indicates current leakage to the chamber wall.

The operating gas is injected through the hollow nozzle on the anode using a high speed gas valve. The inner diameter of the nozzle is 26 mm and the outer one is 30 mm. The gas used here are Ne, Ar, Kr and Xe whose plenum pressure is fixed to 5 atm. The electrodes are made of carbon, and the spacing is 25 mm.

Scintillation probes with various x-ray filters are used to detect radiations from the pinched plasma. A NE-102A scintillator is installed in each probe. A four-pinhole x-ray camera is used to take x-ray images of the pinched plasma with different filters. The Kodak DEF x-ray film placed in atmospheric pressure helium is used for taking soft x-ray image. These diagnostics are located azimuthally to the plasma axis.

A nitrogen laser shadowgraphy is used for monitoring the plasma density profile. The laser light is once focused on a pinhole before a camera after passed through the plasma in order to exclude plasma light.

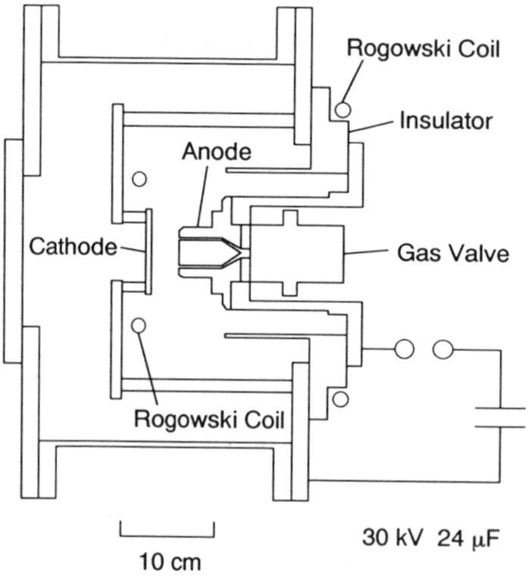

Fig. 1. Experimental setup of the SHOTGUN z-pinch device.

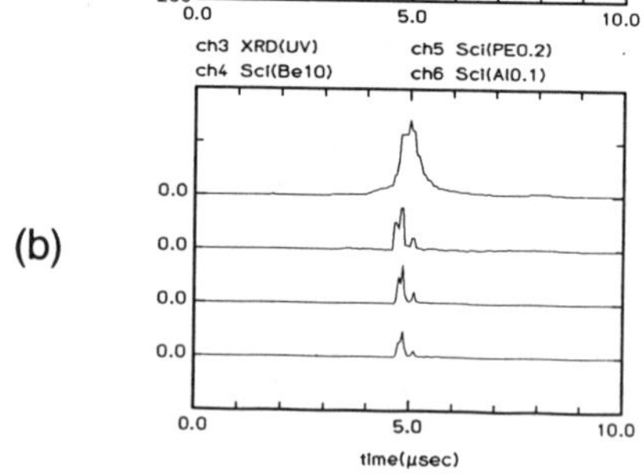

Fig. 2. Typical current waveforms and soft x-ray signals.

X-RAY RADIATION

Typical current waveforms, XUV and x-ray signals are shown for Ar z-pinch in Fig. 2. The maximum plasma current is about 200 kA. The XUV and x-ray radiation occur simultaneously with current dips, which indicate rapid increase of circuit inductance due to radial collapse of the plasma. The XUV signal shows temporally broad profile compared with x-rays.

Figure 3 shows typical x-ray photographs of Ne, Ar, Kr and Xe z-pinches taken with four filters, Be 5 μm, Al 15 μm, Al 0.1 mm and Al 0.5 mm. The photographs of each gas are taken in one shot. In the Ne z-pinch, the x-ray image is observed in left two photographs. In the Ar, x-ray is observed in left three photographs. In the Kr and Xe, it is observed in all photographs, and the latter shows more intense image. Although the film is not sensitive to x-rays with the energy exceeding 5 keV, it is a clear observation that the hard component of x-ray increases with *Z* of the gas. The data indicates that the x-ray radiation has broad spectrum, and is not dominated by characteristic radiation of neutral atoms. The x-ray radiation mainly comes from spots of the pinched plasma. Cloud structure of the x-ray image is again observed between the spots.[10]

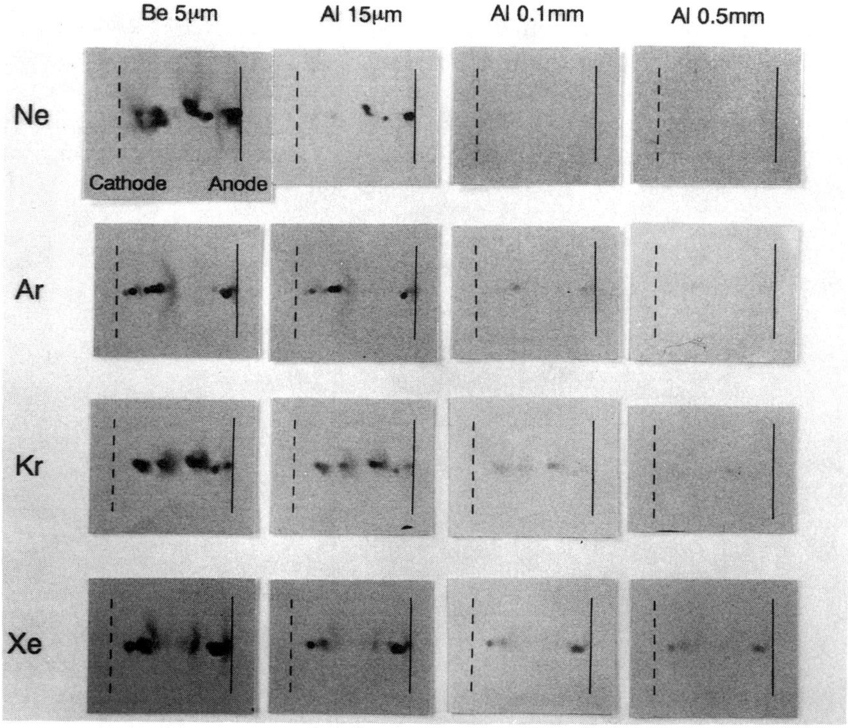

Fig. 3. X-ray pinhole photographs of Ne, Ar, Kr and Xe z-pinches taken with different x-ray filters.

In Fig. 4 the soft x-ray intensities of Ne, Ar and Kr z-pinches are compared using four x-ray filters. The material and the thickness of the filters used here are Be 10 μm, Al 50 μm, Al 0.1 mm and Al 0.5 mm. All the intensities with fixed filter increase with Z, and the gradient of the lines become steeper for harder x-ray. The radiation temperatures of the plasmas are evaluated as 1.5 keV for Ne, 2.9 keV for Ar and 7.8 keV for Kr. The temperature is not necessarily true for optically thin plasmas, but it is a direct expression of radiation characteristics for the plasma in which the detailed radiation process is not clear.

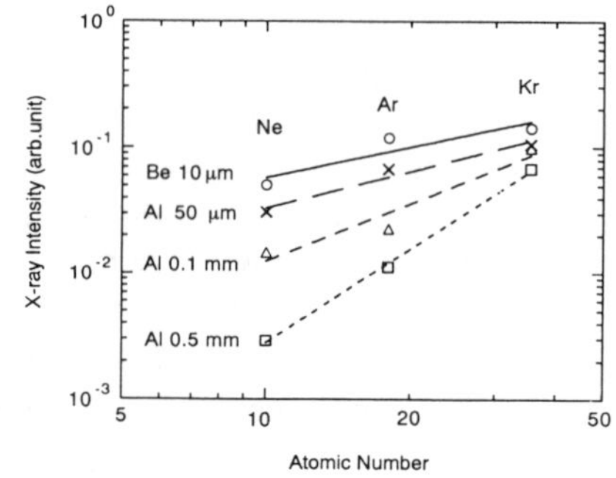

Fig. 4. Scintillation probe intensities of Ne, Ar and Kr z-pinches with four x-ray filters.

ENERGY INPUT

Figure 5 shows the typical current profiles of Ne, Ar and Kr z-pinch discharges. As the operating regions of the three z-pinches are fixed, the peak current levels and the contraction times are almost the same. The current dips, which correspond to maximum pinch, become deeper as Z of the gas increases. In the rough assumption that the magnetic flux is constant during the short interval of the final contraction, the magnetic energy converted into the kinetic energy of the plasma ΔE is[3]

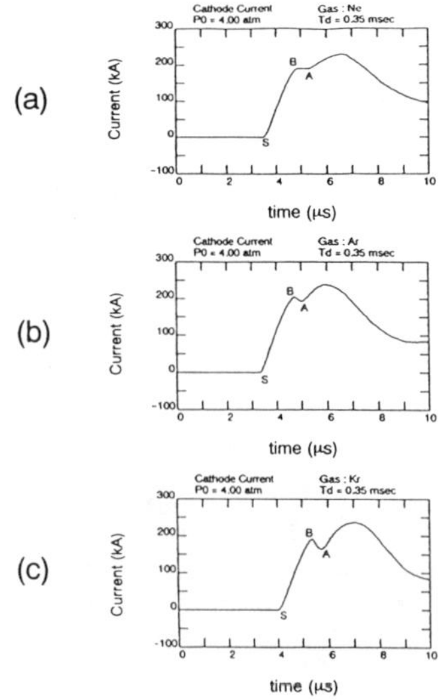

Fig. 5. Current waveforms of (a) Ne, (b) Ar and (c) Kr plasmas. The dip depth changes with the gas.

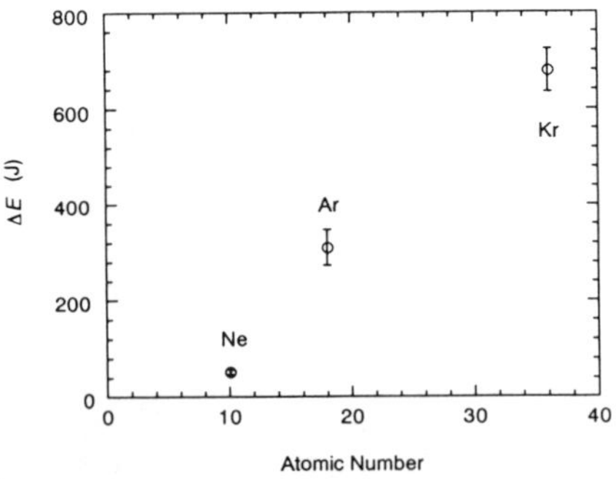

Fig. 6. Magnetic energy converted into kinetic energy of the plasma as a function of atomic number Z of the z-pinch.

$$\Delta E = \frac{1}{2}L_B I_B^{\ 2} - \frac{1}{2}L_A I_A^{\ 2} = \frac{1}{2}L_B I_B(I_B - I_A)$$

Here L is the total circuit inductance and I is the plasma current. The suffixes B and A denote before and after the maximum pinch respectively. In Fig. 6 the energy ΔE is plotted as a function of Z of the gas, and it is shown to increase rapidly with Z. These energies are still small fractions of the total energy of the power supply. The estimation may possibly includes large error for a plasma with a shallow dip. More detailed analysis of the waveforms is necessary for more precise comparison.

HOT SPOT FORMATION

The dynamic behavior of Kr plasma in the contraction phase is shown by a series of shadowgraphs of Fig. 7. The photographs were taken in different shots. The phenomenon was fairly reproducible, and the moment of laser firing is pointed by the arrows in the top trace of the figure for a typical discharge. The time sequence of the plasma behavior was observed as follows. (a) The contracting thin plasma layer is observed. The plasma layer has already exhibited ripples in the z-direction ($m = 0$ mode). The wavelength is smaller than that observed in the following photographs. (b) The contraction continues, and the first node appears at about 5 mm apart from the

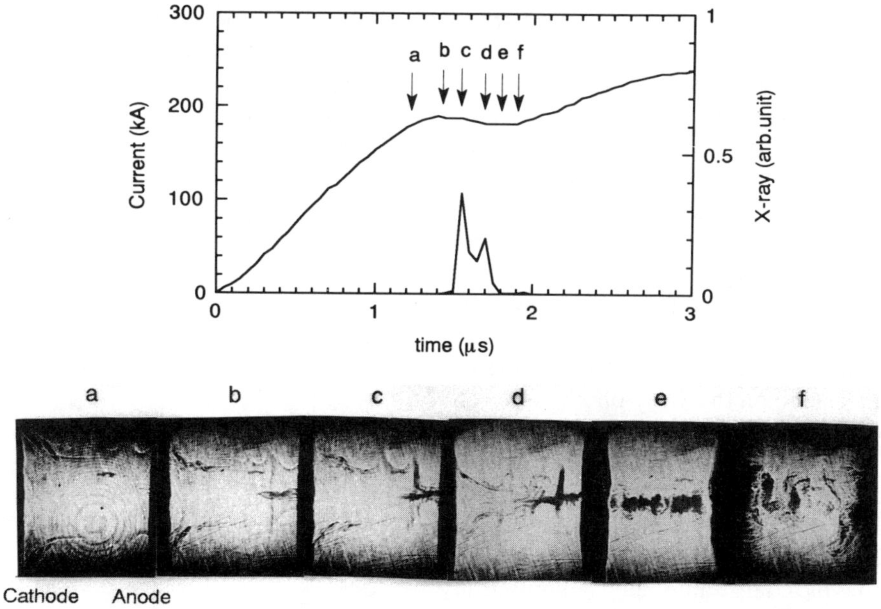

Fig. 7. Time sequence of contraction of the z-pinch plasma. The plasma is divided into sections, and each of them collapses by turns.

anode. (c) The plasma between the node and the anode collapses rapidly and forms a divided z-pinch column. This contraction corresponds to the first x-ray pulse signal. (d) The second divided contraction is observed in coincident with the second x-ray pulse. The divided contraction occurs one after another from the anode (gas nozzle). It may be understood as a kind of zippering effect due to initial gas distribution. (e) After passing through the maximum pinch (current minimum), an averaged z-pinch column is formed between the electrodes. The column diameter is about 3 mm. Expansion of the column has already started. The locally constricted part of the column grows up to form helical structure (m = 1 mode) (f) The instability grows up further, and its amplitude becomes comparable to its length. Finally the plasma shadowgraph disappears from the view, while the plasma current continues.

Figure 8 compares a x-ray photograph (bottom) with a shadowgraph (top) taken in a same shot of Kr discharge. Four x-ray spots are observed with the separation of 4 - 6 mm. Three nodes exist in the shadowgraph, and each of them spatially corresponds to the x-ray spots. X-ray clouds are also seen between the spots. The intervals of the spots are measured on the x-ray photograph. The event number of the spot is shown for its interval in Fig. 9. The data is summed over eight shots of typical Kr z-pinch discharges. The interval is peaked at 4 - 5 mm, which corresponds to typical large amplitude nonuniformity in contracting Kr z-pinches.

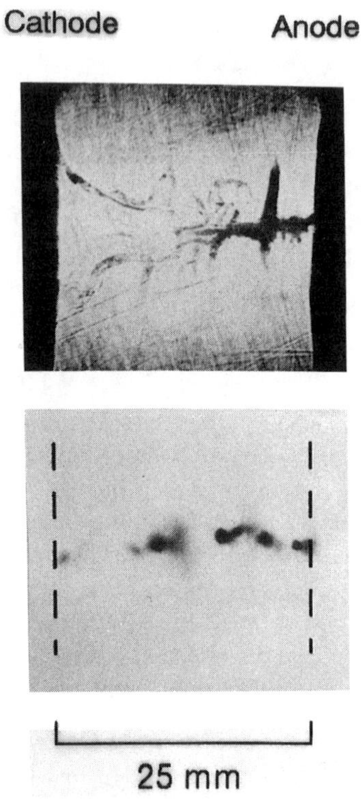

Fig. 8. Comparison of an x-ray image with a shadowgraph taken in a same shot.

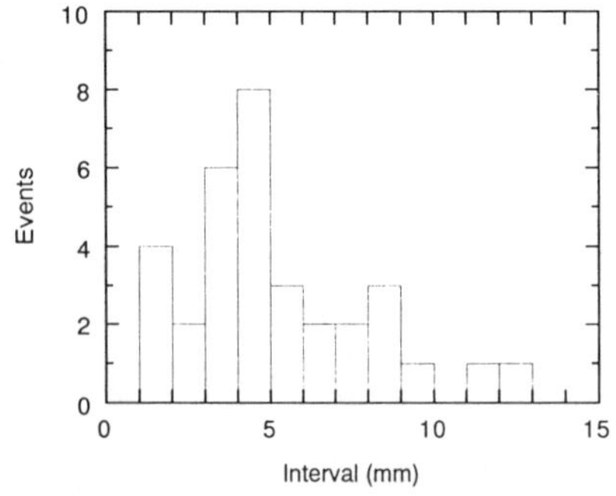

Fig. 9. Spatial intervals of hot spots observed in Kr z-pinch.

Summary and Discussion

In the gas-puff z-pinches the x-ray radiation characteristics were investigated with various operating gas. The hardening of x-ray radiation with Z of the gas was confirmed. The x-ray mainly comes from the hot spots existing in the pinched plasma. The x-ray radiation was shown to increase with Z in all range of observed energy spectrum. The radiation temperatures of the spots were evaluated, and it was shown to increase with Z.

The magnetic energy converted into the kinetic energy of the plasma was calculated from the current waveform assuming flux conservation during the final contraction. The energy was shown to increase with Z of the gas. This was the direct result of formation of smaller average radius at the maximum pinch. The high Z gas plasma is radiatively cooled in the radial contraction phase, and it is compressed to a smaller radius, then accelerated further to become a higher density and higher temperature state.

The dynamic behavior of the z-pinch plasma was investigated using a nitrogen laser shadowgraphy. A large amplitude nonuniformity develops during the contraction phase, which forms nodes between the electrodes. The plasma is divided into several sections, and they contracts by turns from the anode. The spatial position of x-ray spots coincides with that of the nodes. At the node the plasma is three dimensionally compressed and forms high density and high temperature state. This micro-pinch would be the source of the x-ray spot. On the other hand the x-ray cloud is not so intense and is formed between the spots. It may have some relation to $m = 1$ mode instability observed in the expansion phase. The observations of the node and the divided pinch would be owing to relatively slow z-pinch discharge among the contemporary fast z-pinches.

References

1. J. Shiloh, A. Fisher and N. Rostoker, Phys. Rev. Lett. 40, 515 (1978).
2. J.W. Shearer, Phys. Fluids 19, 1426 (1976).
3. J.E. Bailey, Ph.D. thesis, U.C. Irvine (1983).
4. J. Bailey, A. Fisher and N. Rostoker, J. Appl. Phys. 60, 1939 (1986).
5. P.G. Burkhalter, J. Shiloh, A. Fisher and R.D. Cowan, J. Appl. Phys. 50, 4532 (1979).
6. R.E. Marris, D.D. Dietrich, R.J. Fortner, M.A. Levine, D.F. Price and R.E. Stewart, Appl. Phys. Lett. 42, 946 (1983).
7. G. Mehlman, P.G. Burkhalter, S.J. Stephanakis, F.C. Young and D.J. Nagel, J. Appl. Phys. 60, 3427 (1986).
8. M.G. Haines, Nucl. Instrum. Methods 207, 179 (1983).
9. D.R. Kania and L.A. Jones, Phys. Rev. Lett. 53, 166 (1984).
10. K. Takasugi, A. Takeuchi, H. Takada and T. Miyamoto, Jpn. J. Appl. Phys. 31, 1874 (1992).

PLASMA FOCUS

ANISOTROPY OF THE NEUTRON EMISSION IN A MEDIUM ENERGY PLASMA FOCUS DEVICE

R. Aliaga-Rossel[*] and P. Choi
The Blackett Laboratory, Imperial College
London SW7 2BZ U. K.

INTRODUCTION

The results of a series of experiments carried out on the DPF-78[1], a 60 kV, 28 kJ plasma focus device, are presented. The main objectives were to investigate the anisotropy of the neutron emission and to correlate the total neutron yield with other phenomenological plasma parameters, like the appearance of hot spots. To influence the plasma parameters, gas filling of deuterium with admixtures of neon, argon and krypton were used. A set of time resolved neutron detectors were placed at different angular position with respect to the pinch, in order to investigate the polar distribution in the neutron emission. To detect neutron signals at short distance from the focus, so as to diminish the dispersion of the signal due to the time of flight, a novel technique for neutron detection was developed. It was found for the first time that in a medium energy plasma focus device, the neutron emission is composed of two periods, similar to that reported in high energy experiment on the Poseidon plasma focus device[2]. The first period was measured to start immediately before the maximum compression of the focus plasma and lasts less than 50 ns while the follow on second period lasts between 150 and 200 ns. The polar distribution of the neutron emission showed a strong anisotropy during the second period which depends on the Z of the doping gas. No correlation was found between the total neutron yield and the hard x-ray emission or the appearance of hot spots.

THE EXPERIMENT

The experiments were carried out on the medium energy, high voltage plasma focus device DPF-78 at Institute für Plasmaforschung, University of Stuttgart, operated at 60 kV, 28 kJ. With a base gas of deuterium, doping with different combinations of neon, argon and krypton was used to study their influence in the polar distribution of the neutron emission as well as the total neutron yield. In order to keep the dynamics of the plasma approximately constant, a mass density equivalent to a filling gas of deuterium at 5 mbar was used throughout most of the experiments. Correlation between neutron emission, hard x-ray and hot spots was also investigated.

The following diagnostics were used: current and voltage monitors, LSADRRM x-ray streak technique[3], x-ray multipinhole photography, single and double frame schlieren photography, visible plasma radiation, time integrated and time resolved neutron measurements. Plastic scintillator with photomultiplier detector is used to perform time resolved neutron measurements. To resolve the conflicting requirements between good time resolution from the scintillator and low hard X-ray induced signal from the photomultiplier, the two devices should be well separated, with the former as close to the pinch plasma as is practical and the latter sufficiently far away and well shielded. Some form of optical coupling is thus required between the two. A new technique was developed in order to improve the optical coupling between the plastic scintillators and remote photomultipliers. This technique allowed the monitoring of the neutron pulses at distances as small as 40 cm from the focus. The novel technique makes use of a fluorescent fibre which is wrapped around a rod of plastic scintillator NE102A. The blue light emitted by the scintillator when it is excited by recoil protons from interaction with neutrons, is then collected by the fluorescent fibre and converted into green light. The ends of the fluorescent fibre are coupled to a larger diameter normal plastic fibre and thus carried to a shielded photomultiplier placed 6 m away from the focus. Details of this system will be published elsewhere. A set of 9 such detectors arranged in different polar locations with respect to the pinch column were used to monitor the neutron emission. Time of flight information was also obtained at 0° and 90° with respect to the pinch axis. Single and double frame Schlieren photographs were used to investigate the dynamics of the plasma sheath and the effect of the doping gas on the structure of the plasma sheath in order to correlate with the neutron data.

EXPERIMENTAL OBSERVATIONS

a.- Time integrated neutron measurements.

Time integrated neutron measurements were carried out using a silver activated counter, placed at 100 cm from the pinch in the end-on direction. Figure 1 shows the neutron yield for discharges in deuterium with doping of either neon, argon and krypton at different mass fraction.

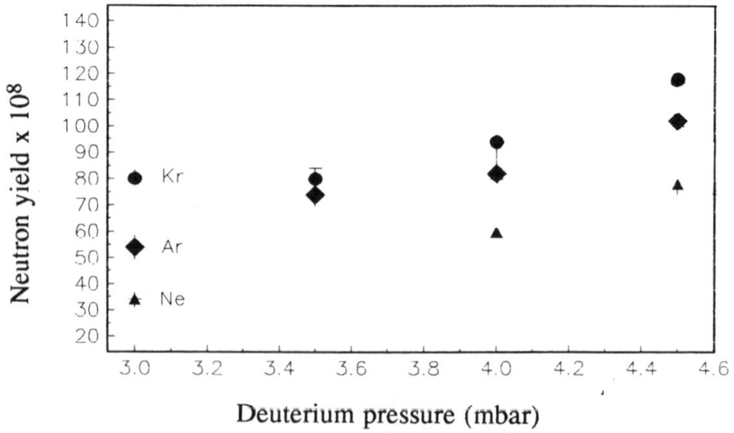

Figure 1. Neutron yield for different doping levels

The total mass of the filling was maintained at an equivalent value to 5 mbar of deuterium in order not to affect the dynamic of the discharge. Each data points represents the average of 25 shots. In all the discharges taken with a base deuterium pressure between 3 and 4.5 mbar, it is observed that the neutron yield depends on the Z of the doping gas. When more than one impurity is admitted to the discharge, neutron production is found to depend on both the percentage of the admixture and the Z of the actual mix. Table I shows the total neutron yield for doubly doped discharges at two different base pressures of 3 and 4 mbar deuterium, representing 40% and 20% by mass fraction of noble gases admixture. In both cases, strong dependence on the Z of the actual constituent of a given mix is seen. This is best illustrated when one considers the conditions with argon as one of the constituent, and contrast the neutron yield with neon as the other component, which is lowest, to the neutron yield with krypton as the other component, which is highest. Similar trend can be found when krypton is taken as the common component and one considers the neutron yield variation when neon and argon are introduced into the mix. The effect is strongest when the impurity content is highest.

Table I. Total neutron yield in doubly doped admixtures with neon, argon and krypton.

Admixture (pressure in mbar)	Total neutron yield (x 10^8)
3 D_2 + .20 Ne + .10 Ar	44 ± 10
3 D_2 + .20 Ne + .05 Kr	44 ± 10
3 D_2 + .10 Ar + .05 Kr	66 ± 12
4 D_2 + .10 Ne + .05 Ar	84 ± 14
4 D_2 + .10 Ne + .025 Kr	88 ± 10
4 D_2 + .05 Ar + .025 Kr	94 ± 12

b.- Time resolved neutron polar distribution measurements.

Time resolved neutron polar distribution measurements were performed with the novel technique previously described. Five of such detectors were located radially with respect to the plasma column at 65 cm from it. In discharges with a gas filling of pure deuterium, without doping, the polar distribution of the neutron emission is shown in figure 2, where all signals are normalized to a peak value of 100 a.u. The neutron emission lasts longer for angles smaller than 45 degrees, FWHM about 170 ns. For larger angles it is about 24 ns shorter.

In discharges with impurities the temporal anisotropy is stronger, as is shown in figure 3, a discharge carried out with an admixture of 20% of neon. In this case the difference in the neutron FWHM pulse length is 65 ns between end-on and side-on directions. For admixtures with a higher doping level, as the one shown in figure 4 the difference is clearly marked. In figure 4 the admixture is doped at 40% of neon and the presence of a double structure in the neutron emission can be identified in the

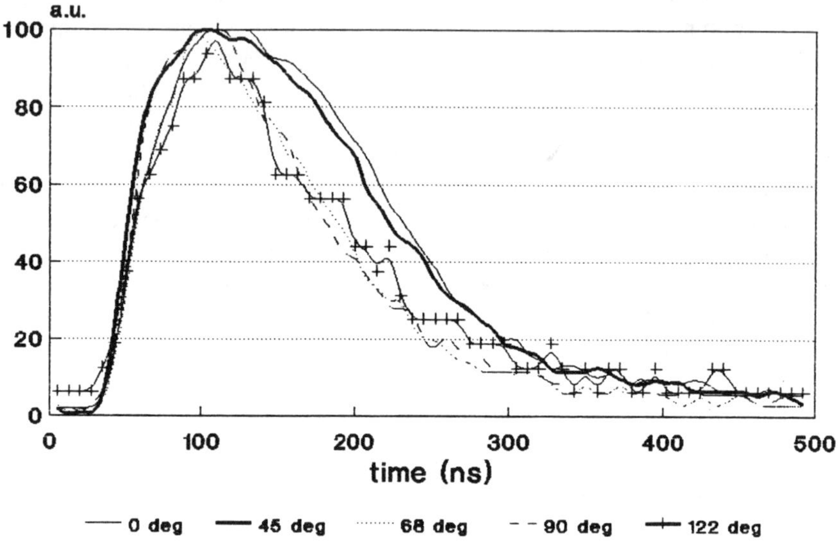

Figure 2. Polar distribution of the neutron emission in pure deuterium discharges.

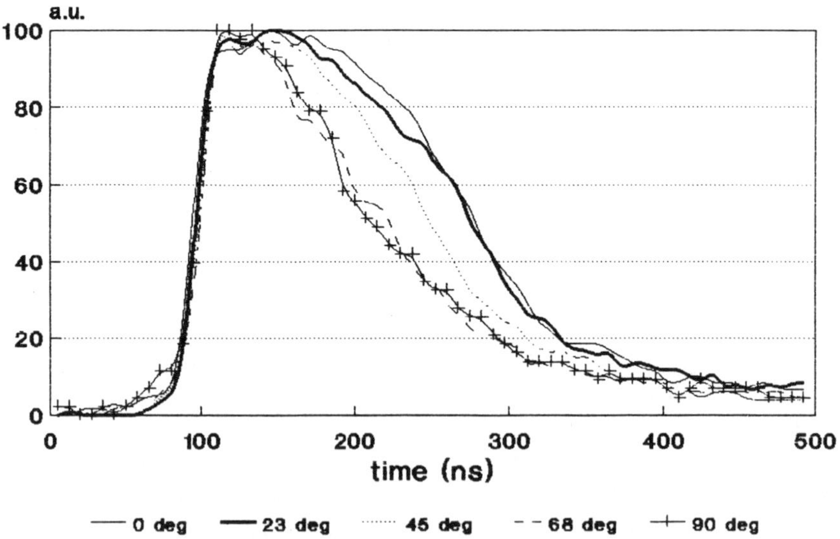

Figure 3. Polar distribution of the neutron emission, doping level of 20% neon.

side-on direction. In the forward direction the neutron emission still appears as a single pulse with a relatively flat top. The neutron diagnostic used here does not have the capability of resolving the starting time of the second pulse, but it has enough time resolution to show the two separated pulses without doubt.

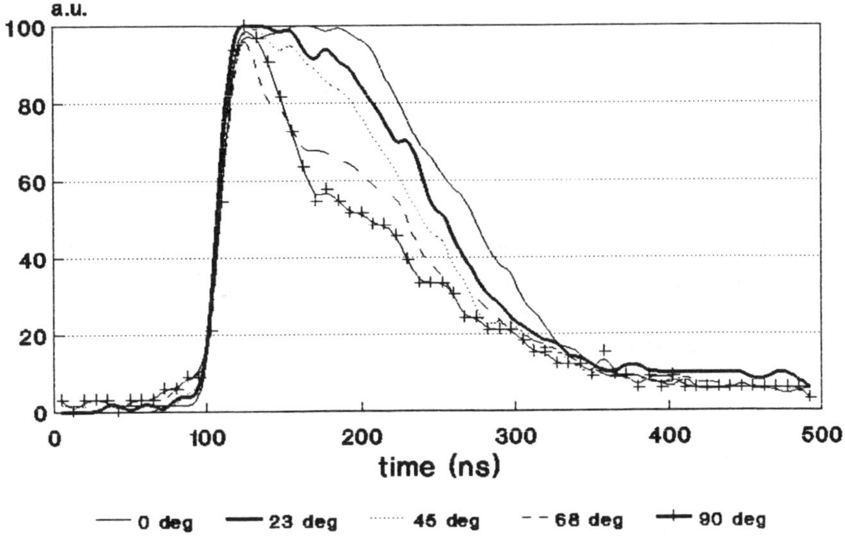

Figure 4. Polar distribution of the neutron emission, doping level of 40% neon.

The increase in impurity level causes the neutron emission in the first period in the side-on direction to fall rapidly once it has reached peak value and allows the second period to be clearly seen.

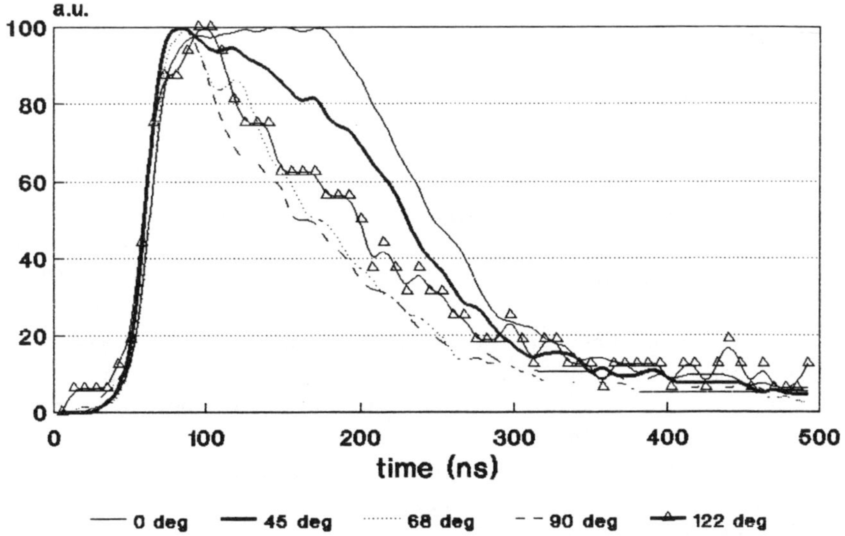

Figure 5. Polar distribution of the neutron emission, doping level of 20% argon.

In discharges with an argon doping, the end-on emission does not change significantly with respect to that in pure deuterium discharges. However, in other directions (45, 68 and 90 degrees) the neutron pulse decreases when the angle increases from 0 to 90 degrees. The emission in the backward direction (at 122 degrees) also shows the difference in time and intensity with respect to the end-on direction, as is shown in figure 5. Similar narrowing of the side-on neutron emission time is also observed at high impurity concentration, as in the case of neon in figure 4.

The important influence of the Z of the doping gas was seen in the time integrated neutron yield measurements. The time evolution of the neutron yield, however, clearly shows even greater variations. With high Z doping gas, the second part of the neutron pulse begins to be diminished earlier for a given impurity concentration compared with low Z doping. At 40% doping with krypton the neutron emission disappears almost completely in the side-on direction, as is shown in figure 6.

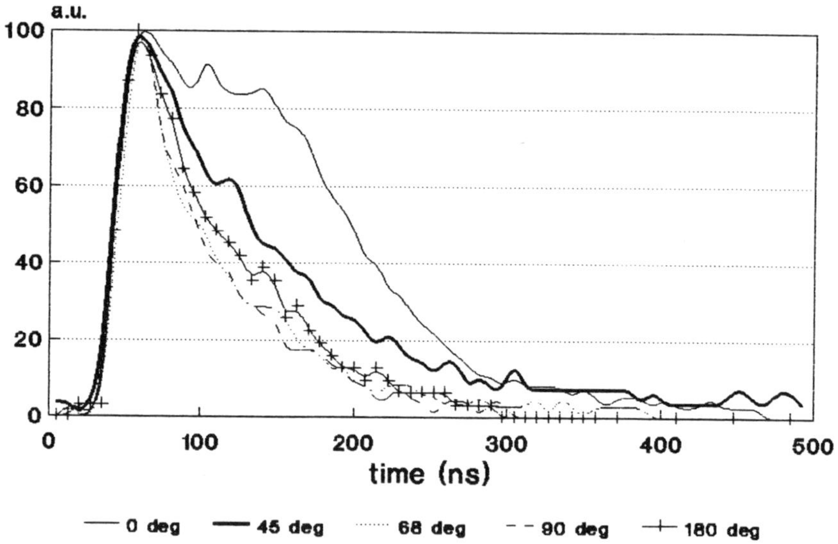

Figure 6. Polar distribution of the neutron emission, doping level of 40% krypton.

The neutron emission in the backward direction is also composed of only one period of emission. Furthermore, it is observed that as the Z of the doping gas is increased, the angle at which the second part of the neutron pulse is strongly reduced begins to be smaller. In other word, the polar pattern of the neutron emission becomes more and more peaked in the forward direction in the second period with increasing Z. The neutron emission in the forward direction lasts roughly the same time in all the different admixtures.

TIME CORRELATION OF NEUTRON EMISSIONS

The exact time of the neutron emission is correlated to a number of other circuit and plasma parameters in the experiment to an accuracy of ± 2.5 ns. A typical set of data is shown in figure 7, showings $\acute{I}$, the rate of change of current measured on the collector plate of the focus device (channel C1); r_{min}, the optical emission from the pinch column imaged onto a PIN photodiode together with the laser pulse for the schlieren diagnostics (C2 and C3, with r_{min} at 5 and 12 mm from the anode respectively), C4, C5 and C6 shows the signal from the neutron detectors placed in the side-on direction at 45, 165 and 260 cm respectively. C7 shows the signal from a conventional close coupled scintillator-photomultiplier (PM5) at 8 m from the pinch. A gamma ray signal is also shown in C5, which is timed with the dip of $\acute{I}$. The discharge is in 4 mbar of deuterium 20% doping of krypton. It is clearly seen that the beginning of the neutron emission, as well as that of the gamma ray signal, takes place during the final moment of, but preceding, the tight pinch formation, as shown by the r_{min} signal. This is further confirmed by the simultaneous schlieren photography of the density structure of the plasma column[4]. The timing is further checked by displaying the hard X-ray induced signal from a suitably shielded scintillator and the optical emission signal, through two fibre optics cables, onto the streak camera for either optical or X-ray streak photography[1].

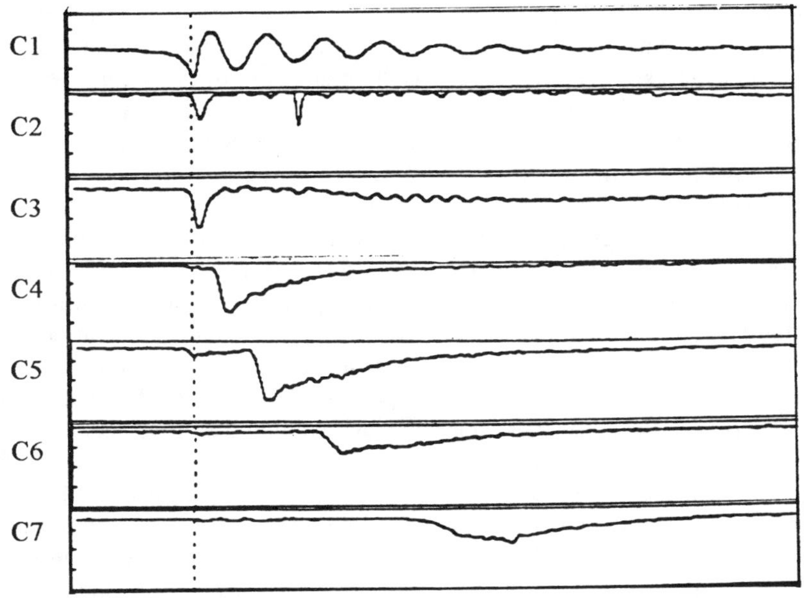

Figure 7. Evolution of circuit and plasma parameters with respect to neutron emission.

The gamma ray induced signal from PM5 was used to compared with the time integrated neutron yield and no direct correlation was found. The occurrence of hot spots was monitored from a 9 pinholes time integrated X-ray camera. Different filters were used with each pinhole to provide information on the energy structure of the hot spots emission. This time integrated picture was supplemented by time resolved measurements of the hot spots evolution using the LSADRRM X-ray streak technique[3] in part of the experiment. The hot spots were found to be produced at positions close to the anode within a macroscopically stable pinch column during the first twenty nanoseconds after the first compression. Within the time resolution of the neutron detectors, no correlation of the hot spots and the neutron yield could be identified.

DISCUSSION

The measurements of the neutron signals obtained in this experiment show that the neutron emission is composed of two identifiable pulses or periods. These two periods of the neutron emission have a spatial anisotropy which depends on the level of doping as well as the Z of the doping gas. The half-width of the neutron emission in the forward direction is longer than that of the emission in the side-on direction and its emission time is approximately independent of the doping gas but its intensity, in term of the total neutron yield, is strongly dependent on the doping level and the Z of the doping gas. On the contrary, in the side-on direction, the second period lasts for a time that depends strongly on the level of impurities and of the Z of the doping gas or gases. For a high Z gas doping (figure 6) the second period of the neutron pulse is practically absent in the side-on direction. This suggests that there are two different mechanisms in the neutron production, which can be affected selectively by the doping level as well as the Z of the doping gas.

The first period starts during the final phase of the radial compression of the plasma column and before the formation of the hot plasma pinch column. The most likely origin of neutron production is that due to the radial acceleration of ions from a Fermi mechanism, gaining energy by bouncing off the incoming sheath[5]. It was calculated that an ion can gain energy from thermal to 100 keV in a matter of four or five bounces from the sheath, given the usual sheath velocity of few 10 cm /μs. The peak neutron production during this first period would occur when these energetic ions interacts with the dense plasma in the pinch column that is subsequently formed. The high magnetic field from a tightly pinch column would lead to an effective increase in the interaction time between a gyrating energetic ion with the hot plasma[6]. This can then explain the fast rising emission time detected. The rapid decay of the side-on signal with high Z and high doping level suggests that the neutron producing interaction during this period is supported by a hot plasma pinch column.

The second period of the neutron emission which last more than 150 ns, starts well after the first compression. At this time, the dense hot part of the plasma column is moving in the forward direction away from the anode[1]. From the schlieren measurements, the original tight pinch column has now expanded to form a diffuse plasma column. The most likely mechanism for the observed neutron production would be due to an axially directed ion beam-hot target interaction. The strong dependence on doping level and on Z of the mix, would suggests that relatively little contribution originates from the interaction of the ion beam with the background gas target.

Both time integrated and time resolved neutron emission results show the strong dependence on the actual Z of the doping gas, with krypton playing an important role in the admixture compared to neon. This is not obvious as the average Z is maintained roughly constant when the total mass of the admixture is kept constant. Schlieren photographs of the different doped discharges, however, show that the thickness of the plasma sheath depends on the actual Z of the doping gas. In the case of krypton doped discharges, it was found that the plasma sheath is thinner than in neon doped discharges and as a result, a better compression and a much tighter pinch is produced. The surprising time integrated results with best neutron yield from krypton doped discharge could then be explained by a much more efficient Fermi acceleration process and a few more bounces from the creation of a tighter pinch. The present of a higher Z component in the admixture would lead to a higher ionization and radiation loss, both at low temperature during the radial compression phase and at high temperature during the pinch phase. The increase in losses would effectively reduce the shock to piston separation during the compression and thus leading to the thinner plasma sheath observed. After the formation of the dense pinch, the much higher level of radiation losses would then cool the pinch column quickly, leading to the rapid fall of the neutron signal in the side-on direction. What has not been investigated adequately at this stage, is the exact mechanism of neutron production in the second period which leads to the highly forward peaked distribution.

With regard to hot spots and hard x-ray emission, the other diagnostics used during this experiment show no correlation between the number or even the existence of hot spots with the total neutron yield. The hard x-ray emission was also found to have no correlation with the neutron emission.

To summarise, it was found for the first time in a medium energy plasma focus, the presence of two period of the neutron emission. The first period occurs when the tight plasma column is being formed, well before the disruption of the plasma column, and lasts less than 50 ns. The second period is much longer and lasts between 150 and 200 ns. The second period shows a strong polar anisotropy (with respect to the pulse length) between the side-on and end-on directions. This suggest the presence of two different mechanism in the neutron production, which can be affected selectively by the level of impurity and the Z of the doping gas.

ACKNOWLEDGEMENTS.

Part of the work performed was under the support of a CEC Stimulation Action Contract ST2J-59. The authors are indebted to Professor Hans Herold, of the

Institute fur Plasmaforschung, University of Stuttgart, for the use of the apparatus and for extensive discussions on the results.

REFERENCES

1 Choi, P., Wong, C.S. and Herold, H., " X-ray Studies of the Spatial and Temporal Evolution of a Dense Plasma Focus.", Laser and Particle Beams, 1989, 7(4), 763-772.

2 Herold, H., "Physics of Alternative Magnetics Confinement Schemes", S. Orlandi and E. Sindoni, (Eds), SIF, Bologna 1991, pag 371.

3 Choi, P. and Aliaga, R. (1990) Rev. Sci. Instrum. **61**(10) 2747.

4 Choi, P. and Aliaga, R. this Proceeding.

5 Deutch, R. and Kies, W. (1988) Plasma Physics and Controlled Fusion **30**(3) 263.

6 Jäger, U. and Herold, H. (1987) Nuclear Fusion, **27**(3) 407.

✠ Permanent address: Facultad de Fisica, Pontificia Universidad Catolica de Chile, Casilla 306, Santiago 22, CHILE

HIGH POWER SELF-COMPRESSING DISCHARGE WITH A SPATIAL CURRENT PEAKING ON THE BASIS OF TSP ENERGETIC

E. A. Azizov, A. P. Lototsky, A. F. Nastoyashchy, T. I. Filippova, and N. V. Filippov
TRINITY, RRC "Kurchatov Institute"
123182 Moscow Russia

In the high power discharge, megaampere range, at their final stages ("pinching") the production of a hot dense plasma with high parameters,[1,2] is observed. In this connection, the similar discharge are successfully used as an effective source of neutron and X-ray radiations. The best results, from the viewpoint of attaining high parameters, were observed at the facilities of plasma focus (PF) type[2].

Essential advantages of the PF-type systems is in that the current pulse peaking occurs due to spatial cumulation of the plasma-current shell (PCS) in the discharge chamber of rather large size. Hence, a comparative simplicity and an opportunity to use a comparatively "slow" energetics with the times of the order of a few tens of microseconds (in megaampere range of current) that allows one, in principle, to use inductive high power-energy storages instead of capacitor banks.

The discharges under consideration occur, as a rule, at three stages: 1) breakdown and the initial phase of plasma shell formation in the vicinity to the insulator surface; 2) mode of electrodynamic plasma shell acceleration; 3) stage of plasma column selfcompression ("pinching") in the near-axis zone.

From the viewpoint of achieving a high compression degree, the first two stages play an extremely important role. Thus at the first stage it is necessary to organize a homogeneous plasma shell possessing of high axial symmetry.

Another important problem is to solve the matching between the power supply source and the "plasma loading". At high energy contributions greater than 10 MJ, magnetic-high-power-storages (the sources of similar type, in principle, can provide the best matching of the system in a wide range of initial discharge parameters) seen to be promising instead of capacitor banks. Development and implementation of coded current pulses[3] are an additional possibility of matching with the loading.

An opportunity to create a facility of PF-type with the maximal current of $\sim$100 MA, with the usage of the T-14 tokamak energetics is discussed below.

At the first stage it is expedient to create a facility with the maximal current $I \geqslant 10$ MA and with the plasma energy content $W \geqslant 20$ MJ (let us call it PF-10), these parameters exceed the opportunities of PF-3 (RRC "Kurchatov Institute") by about the order of magnitude. The similar facility would provide an essential progress in comprehension of physical cumulation mechanism at high currents and an advance on the way of creating the conditions for the fusion reaction ignitition. It would be a unique (in its characteristics) source of X-ray and neutron radiations.

The creation of the facility with the current $I \geqslant 100$ MA-which would allow one to approach the solution to the problem of fusion ignition—would be the next step.

Let us discuss the engineering problems in implementation of the T-14 energetics.

The level of power consumption by PF-10 ($\sim$10 GW) should exceed the power consumption by T-14 ($\sim$1 GW); hence a principally—another inductive storage (IS) structure follows. The necessary approach to the energy removal from the inductive storages has been developed in[5–8]. In case of the current interruption in a large scale IS, the main difficulties emerge because of the necessity in provision of long-lasting charge current ($\sim$5–10 s), e.g., from the electric machine. At the massive and rather heat capacitive contacts of the interrupter, these contacts will be under high voltage at the energy removal into the loading. In order to avoid it the distance between the contacts should be rather long and this, in its turn, would require a considerable energy for the drive[5]. In a number of cases, this energy of the drive becomes comparable with the energy stored in IS and the interrupter design turns out to be hardly feasible. The way out of this situation is in the transition from the multistage cumulation of current to the technique of multistage power transfer and peaking (power amplifier circuit-diagram[6]). The required electric strength of the interrupter in each stage here is conformed with the time of energy removal (transfer) to the next IS-stage. Thus the provision of long-lasting IS-charge procedure and that of high electric strength under operation of the loading is functionally distributed among various groups of devices[7–8].

The similar circuit-diagram of magnetic energy transfer can be realized in the known sectioned IS with the current multiplication at the coded sequence of operations for the N-interrupters (N is the number of sections in IS). In this case, an average magnitude of the current over the time of the process through the inductive loading is considerably higher than an average magnitude of the current through the interrupter. The losses are $\sim 1/N$ of the usefully-transferred energy. These conclusions were confirmed by a number of experiments with inductive storages having the stored energy ~ 100 kJ.

Then, it is necessary to match energy unit with the PF-10 chamber. The preliminary estimating calculations have shown that it is possible to produce quite a versatile PF-10 power supply source, applicable to the power supply of both the neutron source and the X-ray radiation one. At the same time, the efficiency of energy transfer (matching with the plasma loading) is not very high: it is a fee for the high current amplitude.

According to the calculations realized within the framework of the "snow-plough" model, the discharge current and the time of PCS collapse at the axis of discharge are increased proportionally to the anode radius, and the required current switching time is not too small (e.g., at the current ~ 10 MA and at the anode radius $R = 1.25$, corresponding to it, the current switching time, $t \sim 30\ \mu s$, is quite permissible).

If the T-14 coils, 8 spare pieces in number, are taken as a base for the primary IS, the highest power which can be taken from them will not exceed $P = 40\ kV \cdot 150\ kA \cdot 8$ pieces $\simeq 5 \cdot 10^{10}$ W. At the same time, the presence of a single peaking stage is sufficient for raising this power to the necessary magnitude $\simeq 10^{12}$ W which can provide the drive and collapse of the PCS of heavy noble gases.

The expected structure of the facility provides a two staged IS with the first stage of a transformer type: amplifier of a current (from 150 kA to 2 MA). The second IS-stage, with the time of energy introduction and conservation $t = 20$ ms, can be a mechanically stressed and small-sized single turn toroid with the total current ~ 12 MA. The PF-10 chamber is performed as a monounit with the IS-stage.

Thus the realized analysis has shown a real opportunity to create the plasma focus facility, based on the T-14 inductive storages, with the energy contribution up to ~ 20 MJ into the plasma. The creation of similar facility would allow one to make an advance in comprehension of the processes in pinches at the level of the currents ~ 10 MA and to have an experience in matching of the magnetic storage with the pinch loading.

The characteristic feature of the PF-facility is a rise in the required current pulse duration with an increase in the facility size (anode diameter) and alleviation of the discharge organization. The creation of a high power fusion PF-60 facility with the maximal discharge current up to ~ 60 MA, on the basis of T-14 energetics, seems to be quite realistic. At this facility, according to our estimates, one can directly approach the realization of conditions for the fusion reaction ignition.

REFERENCES

1. B. A. Trubnikov, Plasma Phys., v.2, p. 454 (1986).
2. N. V. Filippov, Plasma Phys., v.9, p. 25 (1983).
3. A. F. Nastoyashy, High Temp. Phys., n.1, p. 5 (1989).
4. A. F. Nastoyashy, L. P. Shevchenko, Atomic Energy.
5. I. A. Ivanov, A. P. Lototsky, Preprint IAE 3498/14 (1988).
6. I. A. Ivanov, A. P. Lototsky, 1st USSR Conference of pulse power sources, Urmala (1988).
7. A. P. Lototsky, Preprint IAE 3714/14 (1982).
8. A. P. Lototsky, Electricity, n.6, p. 64 (1985).

OBSERVATION OF PLASMA FOCUS DISCHARGE X-RAY SPECTRA FOR DIAGNOCTICS OF ENERGETIC-PARTICLE FLUXES

Baronova E.O., Rantsev-Kartinov V.A.,
Stepanjenko M.M., Tykshaev V.P., Filippov N.V.

Russian Research Centre "Kurchatov Institute"
123182 Moscow Russia

ABSTRACT

The presence of fast (E ~ 1 MeV) Ta^{+46} ions moved towards the anode has been shown in conical hollow anode geometry. The electron beams with energy of about 60 keV were registered in case of flat anode. T_e, N_e of high temperature dense plasma, produced under non-cylindrical compression of a pulsed pinch discharge were evaluated.

INSTRUMENTATION

The experiments were done in the Plasma Focus Test Stend (PFTS)[1] - facility in the geometry shown in Fig.1.

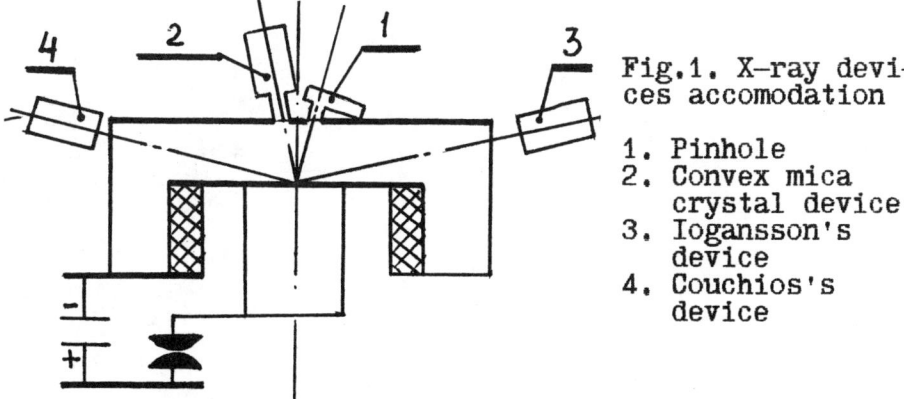

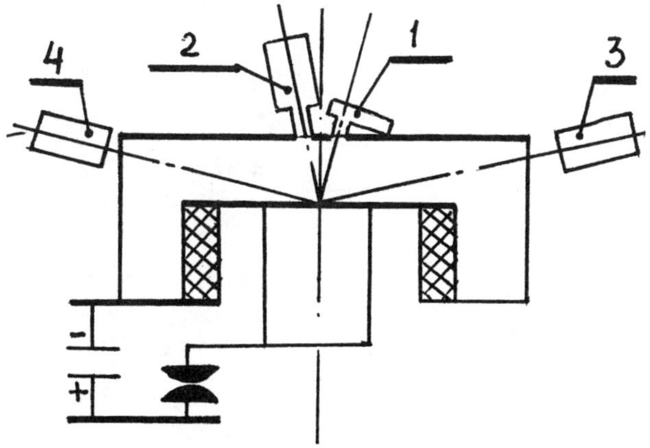

Fig.1. X-ray devices accomodation

1. Pinhole
2. Convex mica crystal device
3. Ioganssons's device
4. Couchios's device

The discharge chamber was filled with argon up to the pressure P = 30 Pa, anode was made of copper with Ta,

Mo, Fe – insertions at the centre. The diagnostic instrumentation shown in Fig.1 had the following main characteristics:

1. Pinhole camera with $d_{orif} \sim 100\ \mu m$, the distance from the plasma to the orifice in the camera is 320 mm, magnification is 1 : 8.

2. The X-ray spectrograph with a convex mica crystal[2]. The plasma-crystal distanse is $L \sim 490$ mm, magnification 1 : 2.

3. The X-ray spectrograph used the Iogansson's circuit- diagram. Roland's radius is R = 250 mm; quartz-crystal is $2d = 6.68$ Å, $\Delta\lambda/\lambda \cong 10^{-4}$.

4. The Couchios spectrograph is for the energy range 10 ÷ 150 keV. The plasma – crystal distance is about 600 mm, $R_{cr} = 600$ mm.

A film with amplifying screen was used as a detector. The measurements were done under two operating conditions of the facility:

a) ion beam mode of operation, the anode has a small, conical hollow;

b) electron beam mode of operation is realized in case of a flat anode.

EXPERIMENTAL RESULTS

I. The ion beam regime. Fig.2,3 show the spectra of He-like Ar and its satellites, which were registered in the third order of reflection on the spectrograph 2 (Fig.1). To treat the spectra and to plot spectrograms a standard manual X-ray film reader was used. The step to measure the film darkening was 10 μm. The absence of H-like Ar was noted. The analysis of the spectra shown in Fig. 2–3 support the well known fact of none-reproducibility of plasma parameters in the PF-discharge (particularly, the dimention of emission region of He-like ions).

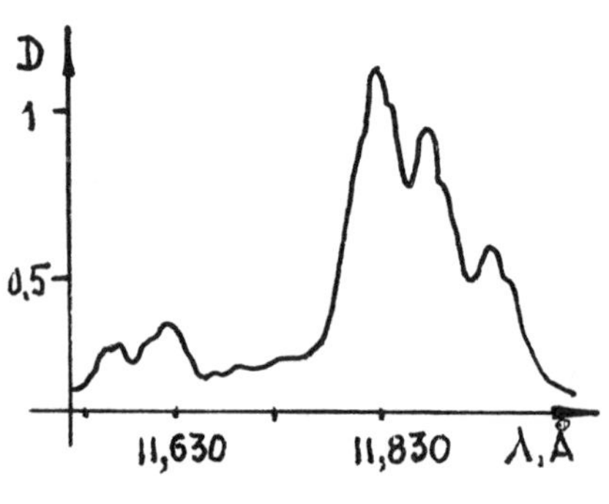

Fig.2. Ar^{+16}, Ta^{+45} spectra

Practically in all the shots in the second order of reflection Ta lines were observed from the short-wave side in the vicinity of Ar-lines (see Fig.2). Using the published data[3,4] the brightest of Ta-lines were identified as transitions ($3d^9 - 3d^8 4f$) in Ta^{+46} ion with the wavelengths of $\lambda_1 = 5.886$ Å and $\lambda_2 = 5.916$ Å. The Ar and Ta spectra in single shot were registered with space resolution of about 300 μm. The plasma images obtained in Ar and Ta wave-lengths allow one to estimate the sizes of the pinch regions, emitting these lines. The usually observed radius and height dimensions of the emission regions of these ions are located in the range of 0.8 - 1,2 mm. The space resolution spectrogram show Ta-lines that are located lower than the He-like Ar-lines. Taking into account the experiment geometry it is possible to make the following conclusion: the emission regions of Ar^{+16} and Ta^{+46} lines are spreaded in the height by ~ 10 mm. At the same time the obscurograms due to an X-rays of $E_\gamma > 0.8$ keV show that the radiating plasma is consist of 7 mm radius and 40 mm height pinch region and 10 mm in diameter semisphere glow region which is located in the vicinity of the anode. The above mentioned experimental data were the basis for the following conclusion: the lines of the He-like Ar glow from the pinch and the lines of Ta glow from the near-to-anode region. The pinch electron density $N_e > 5 \cdot 10^{19}$ cm^{-3} is estimated from the ratio of the intensities of intercombination to resonance lines of He- like Ar (see Fig. 3); the pinch temperature under coronal equiliubrium $T_e > 0.8$ keV; for the near-to-anode region $T_e \sim 1$ keV was obtained.

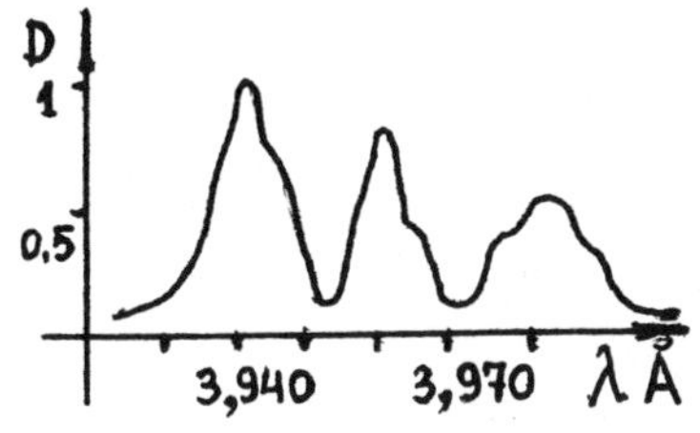

Fig.3. Ar^{+16} spectra

The analysis of the spectra related to different discharges has shown the pair of Ta lines shifted with regard of the He-like Ar group of lines. Taking into account the none-reprodusibility of the plasma parameters and the dependence of the lines position on the position of the source, this effect could have been explained by the change of the distance between the Ar and Ta lines sourses from one discharge to another. An accurate estimation of the distance of the sourses and calculation of the anticipated lines position have shown that a simple increase of the distance between the sources did not explain the observed shift of Ta lines. Another reason of the mentioned shift of lines could be the Doppler's effect caused by Ta ions movement towards the cathode. Understanding of these factors allowed one to estimate $V_i=(0.9 \div 1.8)\cdot 10^8$ cm/s in different discharges.

II. The electron beam regime. In this regime the absence of Ar-lines was registed, the transitions $(3d^n - 3d^{n-1}kf)$, $k_o = 5,6,7,9$ in the Ni-like Ta in the range $3 \div 5$ A were observed. In addition to Ta-ion-lines, most brightist lines of M-series of Ta atom $M_{\alpha,\beta}$, $M_{III} - N_I$ are present. The lines 1, 2 were not interpreted (Fig.4). Using transitions in Ni-like

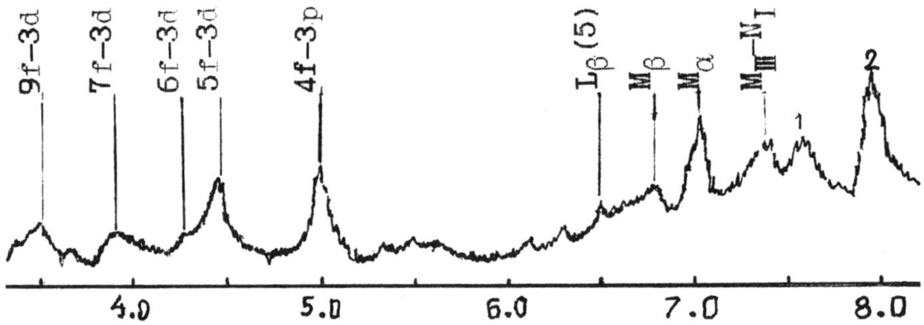

Fig.4. Ta^{+46} and Ta-spectra

Ta as marks it is possible to plot an empirical calibration curve for the spectrograph 2. $M_{\alpha,\beta}$, $M_{III} - N_I$ lines coinside with this curve only under the condition that their emission region is located 1 cm lower than the Ni-like Ta emission region. The spectra obtained with space resolution are the evidence for this statement. The width of Ta^{+46} lines and of Ta atom lines is similar and in different shots no displasement of the groups of these lines realitive to each other were observed. The spectrograph 3 (Fig.1) registered K_αTa and $L_{\alpha,\beta,\gamma,\eta}$Ta and $K_{\alpha,\beta}$Mo. Similarly, $K_{\alpha,\beta}$ for Fe were obtained by Jogansson's spectrograph. These measurements prove the presence of an electron beam of energy not less than 60 keV in the plasma. The sourse of the atomic lines could be the plasma itself as well as the anode surface bombarded by the electron beam. Non-coinsidence of $M_{\alpha,\beta}$Ta and Ta^{+46} emission regions to each other, as well as the absence K_α–lines of weakly ionised Fe ions are the evidence that the atomic spectra is the result of the interaction of a powerfull electron beam with the anode surface. The estimation of the height of the $K_{\alpha,\beta}$Ta–lines allows one to evaluate the size of their emitting region as < 1 mm. Similar conclusions were made in reference[5] for the vacuume spark. The ratio of K_β to K_α lines intensities of Mo were mesured. The value of this ratio, which is known for the X-ray tubes is less than the same one in our experiment. The opposite value of this ratio has been observed in the experiments with the exploding wires[5] for Fe. The explanation of such a fact is a small number of ions with 3p electrones along side with the presence of a noticible quantity of Fe^{+17} ions. The high value of $I(K_\beta)/\ I(K_\alpha)$ in our experiment is a result of different absorbtion of K_α and K_β lines radiated from the anode surface and transported through a dense Mo plasma. Having evaluated volume of plasma from the obscurograms $N_e > 10^{19} cm^{-3}$ were obtained.

PROCESSING OF THE RESULTS.

For plotting the empirical calibration curve for the spectrograph 2 the following correct formula was used to estimate the distance between two lines of λ_1 and λ_2:

$$X = R\pi^{-1}(\alpha/2 - \theta + \arccos(r\cdot\cos\theta/\sqrt{r^2 + L^2}))\Big|_{\theta_1}^{\theta_2}, \quad (1)$$

where

$$\cos\alpha/2 = (r/R)\cos^2\theta + \sin\theta\sqrt{1 - (r/R)^2\cos^2\theta}\ , \quad (2)$$

X – film distance between λ_1 and λ_2,
R = 110 mm, film radius,
r = 35 mm, crystal radius,
θ – Bragg-angle for λ,
L = 490 mm plasma-crystal distance.

The corresponding fomula for the dispersion is:

$$dX/d\theta = R\pi^{-1}(d\alpha/2d\theta - 1 + (\sqrt{r^2+L^2}\cdot\sin\theta)/\sqrt{r^2\sin^2\theta + L^2}). \quad (3)$$

REFERENCES

1. V.D.Ivanov, V.A.Kochetov, M.P.Moiseeva et al. Plasma Physics and controlled nuclear fusion research, v.1, pp.573-600 (1971).
2. N.J.Peacock, R.J.Speer and M.J.Hobby, J. Phys. B., v.2, ser.2, pp.798-809 (1969).
3. P.G.Burkhalter, D.J.Nagel, R.R.Whitlock, Phys. Rev. A 9, p. 2331 (1974).
4. N.Tragin, J.-P.Geindre, P.Monier et al., Phys. Script., v.37, pp.72-82 (1988).
5. P.G.Burkhalter, C.M.Dozier, D.J.Nagel, Phys. Rev. A, v.15, n.2, pp.700-717 (1977).

OPERATING REGIME IN LOW ENERGY PLASMA FOCUS DEVICES

H. Bruzzone*, D. Grondona+, H. Kelly*, A. Márquez+, C. Moreno+, R. Vieytes+

Instituto de Física del Plasma (INFIP), CONICET-
FCEyN, UBA, Ciudad Universitaria, Pabellón 1,
1428, Buenos Aires, Argentina.

ABSTRACT

It is a well know fact that Plasma Focus devices suffer from several drawbacks which affect their efficiency as soft X-ray, electron and ion beams or neutron sources. Among others, current leakages in the vicinity of the insulator and bad shaped current sheaths during the early stages of the discharge have been signaled as causes which can lead to an irreproducible behaviour and also to limit the useful operating pressure range of the device. In order to put into evidence the presence of the quoted phenomena in relatively small energy devices (PFII: 6 KJ, 30 kV; PFI: 0.2 KJ, 20 kV), we will present measurements of the current density distribution during the breakdown stage of the current sheath, of leakage current and current sheath structure during the rundown stage and, as a measure of the quality of the device performance, neutron and ion beam production. Magnetic probes, Rogowski coils, plasma light sensing photodiodes, neutron detectors and ion track detectors were employed as diagnostics. The implications of the presented results on Plasma Focus operating conditions are discussed.

INTRODUCTION

The presence of leakage currents, that is, sizable currents flowing elsewhere than into the current sheath (CS) which implodes to the axis forming a dense pinch in Plasma Focus (PF) devices has been frequently reported[1,2]. It has been suggested that these currents have a deleterious effect on the performance of the device as a neutron source[3]. It has also been claimed that "thin" CS are fundamental for optimizing the device operation, and that the initial phase of the discharge is important in this respect[3,4]. In order to look for a better understanding of these problems, we set an experiment for measuring leakage currents and CS structures, as a function of the Deuterium filling pressure in a small energy PF device. The neutron and the deuteron ion beam productions were used as measures of the device performance.

DESCRIPTION OF THE EXPERIMENT

Two PF devices (PFI and PFII) were employed for the reported measurements. The PFII is a conventional Mather-type device, which currently produces a neutron yield (Y) in the range 10^7-10^9, while the PFI is a very low energy device (~ 0.2 kJ) designed in such a way[5] that only the breakdown phase of the discharge takes place in it. Both devices have their outer electrodes made out of bars, providing good access for the magnetic and optical diagnostics within the inter-electrode region. A list of their parameters is given in Table I, which is also used for defining them.

The diagnostics employed on the PFII device were: a Rogowski coil, for registering the discharge current I(t); two magnetic probes (MP1 and MP2) for measuring the magnetic field in the

	PFI	PFII
Voltage V_0 (kV)	17	30
Capacity C (μF)	1	10.5
Stray Inductance L_0 (nH)	65	33
Inner electrode radius a (cm)	5	1.8
Outer electrode radius b (cm)	10	3.6
Electrodes length l (cm)	25	12
Pyrex insulator radius r_a (cm)	5.5	2.3
Pyrex insulator length l_a (cm)	6	3.5
Pressure range p (mbar)	0.5 - 5	0.5 - 5

Table I: List of the parameters of PFI and PFII devices

vicinities of the insulator and at the middle of the gun length, respectively; an activation neutron detector; an end-on ion pinhole camera (PIC) or, alternatively, a Thomson ion spectrometer (TS)[6]. Both the PIC and the TS were aligned with the electrode axis. The 100 μm entrance pinhole of the PIC (this is also the first pinhole of the TS), was located at 20 cm from the focus region resulting in a magnification of 2.3. The TS deflection region, in which a magnetic (0.17 T) and electric ($\approx 4\ 10^5$ V/m) parallel fields were applied, was 2.5 cm long. Both instruments employed CR-39 film as the ion detector, which was later on etched with a NaOH 6.25N solution at 70°C during 1 hour.

Besides, an optical system which registered the emitted plasma light from a region located at the same axial and radial position as the MP2 probe was assembled. This system was described elsewhere[7], and allows to measure, in a small zone within the electrodes, the angle ϕ between the normal of the CS front and the z axis of the device; the normal velocity of the CS, v_n, and within certain limits, the plasma electron density, n(x), where x is a coordinate perpendicular to the CS front at the measuring position. A scheme of the experimental arrangement employed in the PFII device is shown in Fig. 1.

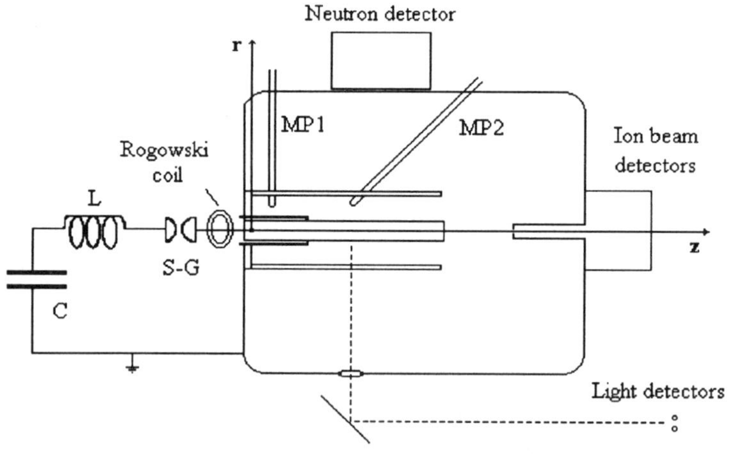

Figure 1: Experimental setup and diagnostics in the PFII device.

To detect leakage currents, a careful "in situ" calibration between the Rogowski coil and the probes was performed. This was accomplished by replacing the PF electrodes by a coaxial short-circuiting piece which has the same geometry as the electrodes system. This piece provided a well-defined path for the discharge current, and the calibration procedure allowed for an experimental determination of the dependence of the magnetic field with the radial coordinate (r), which was particularly necessary for the MP1 probe measurements at radial positions near the barred outer electrode. The relative error of this calibration procedure was better than 5% in all cases.

In the PFI device, a Rogowski coil and a magnetic probe (MP3) axially located 1 cm apart from the bottom of the electrodes and which can be radially moved above the insulator were employed.

RESULTS

a) Neutron and Ion Measurements.

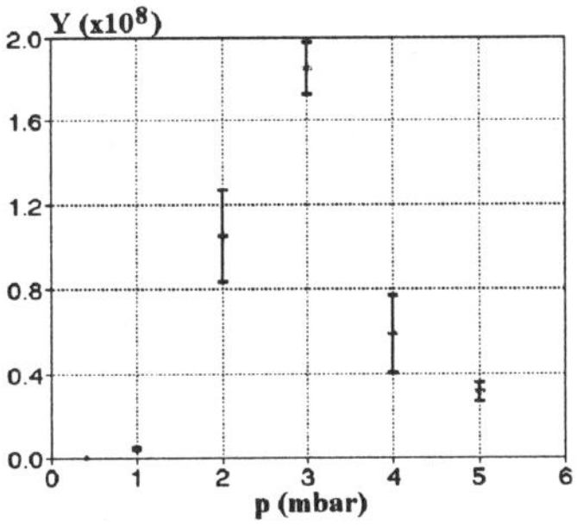

Figure 2: <Y> as a function of p

With the neutron detector located at 90° from the electrode axis, Y was registered for several hundred of shots varying p in the PF II device. In Fig. 2 the average neutron yield <Y> as a function of p is shown. It presents a maximum of (1.8 ± 0.1) 10^8, and the best shots reach 5 10^8. The presence of the probes (particulary the MP2) decreases the yield and shifts the maximum to a lower pressure value (≈ 2 mbar, at which <Y> ≈ 6 10^7).

The images obtained from the PIC showed a circular region impinged by ions with a size of 2.5 cm in diameter, which corresponds to that of the exit pupil of the PIC. Taking into account the magnification of the camera, this fact implies a radial dimension of the ion source > 1 cm, which is larger than the focus size. We will later on show that the ions giving this image have relatively low energies, so that the image enhancement can be explained by changes in the ion trayectories due to the pinch magnetic field.

Shots with larger Y produced denser damaged images, as revealed by a larger opacity of the film. For the larger shots (Y ≥ 8 10^7), an irregular (and irreproducible) pattern composed of several darker zones with typical sizes of some mm appeared within the circular region. For bad shots (Y ≤ 10^6) no ion damage on the CR-39 film was observed.

Several measurements were done covering the CR-39 detector with different film absorbers (Mylar of 2.5, 3.6 and 6.3 μm thickness). In these cases, the circular region disappeared (thus indicating that it is produced by ions with energies lower than 200 keV), but the darker zones remained on the film for the best shots. For instance, using the 6.3 μm absorber (which stops ions with energy lower than 500 keV), ions tracks were found on the detector only when Y > 1.6 10^8.

Under the assumption that the opacity of the PIC images is proportional to the number of ion tracks on the film, the CR-39 plates were analyzed by a light digitizing camera, and thus an average opacity (δ) was obtained for each ion beam image. In Fig. 3, δ is plotted as a function of Y for individual shots. A clear correlation between both quantities is apparent.

The results obtained with the TS essentially confirmed those of the PIC above described: the ion production is correlated with the neutron production in the PF II device. The spectral distribution of the ion energy is formed by two separate components, one with energies between ≈50 keV and ≈ 200 keV, which follows a power law of the type $E^{-\alpha}$ (α ≈ 1.5), similar to that found in a previous work[8], and a higher energy component (>500 keV), whose spectra cannot be resolved with our TS.

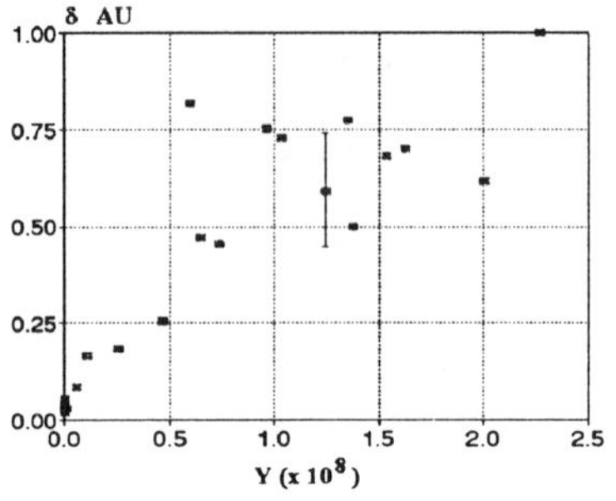

Figure 3: δ as a function of Y

Besides of the deuterons, a parabola of H^+ ions (amounting to ≈ 5 % of the deuterons) was discernible in the TS register, and smaller amounts of impurity ions (charge to mass ratio between 1/4 and 1/8 of that of H^+) were found.

b) Current Measurements.

To determine the behaviour of the currents circulating between the PF II electrodes, the discharge current was compared with those obtained from the MP1 and MP2 probes (I_1, I_2, respectively). This was done for different p values and for several different radial positions of MP1 (r/a = 1.4 , 1.5, 1.6 and 1.7). By assuming an axially symmetric discharge, the time integral of each probe signal is proportional to I - I_p, where I_p is the current circulating in the plasma below and behind the position of the probe. In Figs. 4 a), b), c) and d), the quantity $\xi_{1,2} = 1 - I_{1,2} / I$ is shown as a function of the time, for p = 0.5, 1, 3 and 5 mbar, respectively. We have chosen this non dimensional variable because it depends only on the relative calibration between the Rogowski coil and the magnetic probes. .

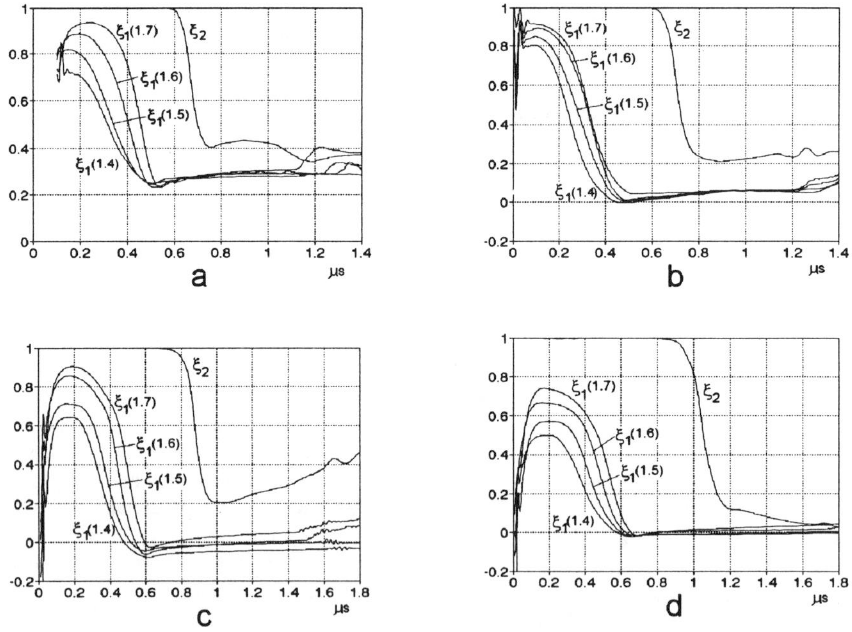

Figure 4: ξ_2 and ξ_1 for several radial positions, as functions of time

Note that ξ represents, once the CS has surpassed the probe, the fraction of the current which remains circulating in the plasma below and behind the probe position, that is, the relative leakage current. Each curve in these Figs. represents an average between several shots performed under identical conditions (although a good shot to shot reproducibility was found).

The initial peak observed in the ξ_1 curves represents the passage of the forming CS. The end of the peak of the signal corresponding to r/a = 1.4 (that is, the position closer to the insulator) defines the CS lift-off time, and results ≈400 ns for the lower pressures, reaching ≈600 ns for p = 5 mbar, which is quite reasonable. The different heights of the peak maxima at the different radial positions, and their relative separations are related to the thickness of the current distribution within the lifting CS. It is not simple to derive current structures from these data, but we can say that cualitatively the thinnest current structure is found at p = 1mbar. After the passing of the CS through the probe, leakage current remains, slowly increasing with time.

The amount of the relative leakage current is pressure dependent, resulting in ≈30% at 0.5 mbar; ≈10% at 1mbar and negligible at 3 and 5 mbar. On the other hand, the MP2 signals show larger relative leakage currents of ≈40% at 0.5 mbar; ≈30% at 1 mbar; ≈20% at 3 mbar and ≈5% at 5 mbar (and decreasing with time in the last case). This behaviour puts into evidence the existence of relevant currents circulating behind the CS, and not only on the surface of the insulator. However, within the explored pressure range the relative value of these "stray" currents tend to decrease as long as p increases.

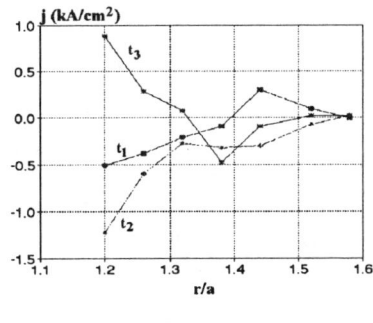

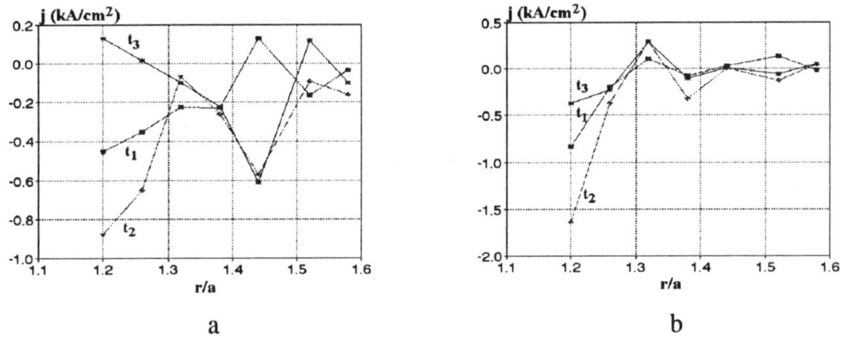

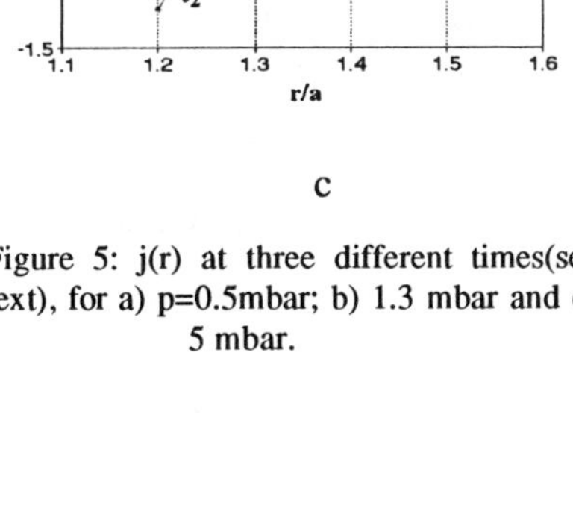

Figure 5: j(r) at three different times(see text), for a) p=0.5mbar; b) 1.3 mbar and c) 5 mbar.

The relatively small size of the PFII electrodes, together with the CS movement make very difficult to obtain current density distributions from our measurements. For this reason, we tried to measure the current density distribution in the larger sized PFI device. The results which will be presented in what follows must be taken, however, only as indicative because, on the one hand the relevant similarity parameter[4], $(dI/dt)/r_a$ is ≈5 times smaller in this device than it is in the PFII; on the other hand, the evaluation of the current density, j(r,t) is still affected by a very large relative error.

A series of measurements using the MP3 probe were done, changing its position in 3 mm radial steps, from r = 6.0 cm (the position closest to the insulator) up to r = 7.9 cm, for several p values. The results obtained at each position and pressure value were normalized using the peak value of the corresponding I(t) measured with the Rogowski coil (which essentially did not change) and averaged. From these data, the currents $I_p(r_i,t)$ circulating in the plasma below the probe position, r_i were evaluated, and used to calculate $j(r_i,t)$ as:

$$j(r_i,t) = 2[r_{i+1}I_p(r_{i+1},t) - r_iI_p(r_i,t)]/(r_{i+1}^2 - r_i^2)$$

In Figs. 5 a), b) and c), we give the obtained results for p = 0.5 mbar, 1.3 mbar and 5.0 mbar, respectively. In each Figure, three instants of time are presented: t_1(=140 ns), the time corresponding to a current value equal to one half of its maximum; t_2(=400 ns), the time of maximum current and t_3(=675 ns), the subsequent time instant at which the current dropped to one half the maximum. Negative j values mean that the current flows in the plasma from the central electrode to the back plate. One can see that the current distributions are broader at 0.5 mbar and 5.0 mbar than the one found at 1.3 mbar. Also, j reverses in the vicinity of the insulator are seen in the upper and lower pressures cases, at t = t_3, (inverse skin), and not in the 1.3 mbar case, probably meaning that the region with reversed j values is too small in this case to be revealed by the probe.

c) Plasma Structure.

Using the MP2 probe and the light detecting system, the current density profile, j(x) and the plasma electron density profile, n(x) in the CS were determined at the positions r/a = 1.5 cm and z = 9.5 cm (according to the coordinate system of Fig 1), for a large number of shots performed at p values between 0.5 and 5 mbar. As an example, in Fig. 6 we give a typical set of n and j profiles obtained at p = 3 mbar. The n(x) profile has been interrupted at x = x_m ≈ 0.3 cm, because for x > x_m several limitations in the procedure used for deriving n(x) (mainly the curvature of the CS and the finite field depth of the optical system) make the results dubious. However, the total thickness of the plasma zone (Δ_n) can be estimated by making the reasonable assumption that its trailing edge must coincide with the magnetic piston position, (that is, with the peak value of j(x)), because of the balance between the kinetic and magnetic pressures. In Table II the maximum of the n(x) profile (n*) and Δ_n as functions of p are shown. The angle ϕ, the normal velocities of the CS and the thickness of the current distribution (Δ_j) are also given. For comparison purposes, the theoretical value of the density (n_{rh}) predicted by the Rankine-Hugoniot jump condition (including ionization) has been also calculated. It can be seen that the values of n* and n_{rh} are in reasonable agreement, excepted in the lowest p values. In these cases, however, the errors in n* are very large due to the large noise to the signal ratio in the photodetector signals, so that the disagreement is not relevant. The structure of the leading edge of the plasma and the values of Δ_n and Δ_j resulting from our measurements are essentially independent on p.

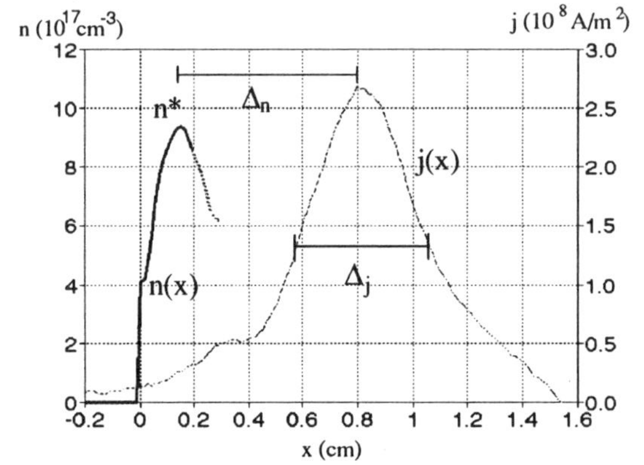

Figure 6: Plasma electron and current densities profiles for the case p = 3 mbar

p (mbar)	n* (cm-3)	n_{th} (cm-3)	Δ_n (cm)	v_n (cm/μs)	φ (deg)	Δ_j (cm)
0.5	5 10^{17} ?	1.3 10^{17}	0.7±0.2	9±1	40±10	0.5±0.1
1	5 10^{17}±5 10^{17}	2.7 10^{17}	0.7±0.2	8±1	40±10	0.6±0.1
2	7 10^{17}±4 10^{17}	5.5 10^{17}	0.7±0.2	7.6±0.9	46±5	0.6±0.1
3	10 10^{17}±5 10^{17}	9.5 10^{17}	0.7±0.2	6.7±0.7	46±6	0.5±0.1
4	14 10^{17}±5 10^{17}	14 10^{17}	0.7±0.2	6.2±0.8	47±7	0.5±0.1
5	19 10^{17}±5 10^{17}	20 10^{17}	0.9±0.2	5.7±0.5	48±4	0.5±0.1

Table II: n*, n_{th}, Δ_n, v_n, φ and Δ_j as functions of p

FINAL REMARKS

In our PFII device, the neutron (and ion beams) yield is negligible at 0.5 mbar, has a maximum around 2 to 3 mbar and drops at 5 mbar. The current structure of the forming CS seems to be at ≈1 mbar thinner than at the other pressures, but the structure of the CS (both electron density and current density) at the middle of the run down phase appears to be independent of the pressure. In turn, the amount of the leakage currents have a diminishing behaviour with p. Therefore, no correlation can be find in our device between leakage currents and/or CS structure on the device performance.

ACKNOWLEDGMENTS

We express our recognition to Dr. L. Bilbao for his valuable suggestions and help with the data analysis. This work was partially supported by grants from the Buenos Aires University and from CONICET.

REFERENCES

1. C.Gourlan, H.Kroegler, Ch.Maissonier, T.Oppenlander, J.P.Rager, Proc. 8th Conf. on Contr. Fus. and Plasma Phys.,Prague 1977, Vol II, p.247.
2. T.Oppenlander, G.Pross, G.Decker and M.Trunk, Plasma Phys. 19, 1075 (1977).
3. J.P.Rager, "The Plasma Focus", Int. Rep. 81.19cc, Assoc. EURATOM-CNEN sulla fusione, Frascati, 1981.
4. G.Decker, W. Kiess and G.Pross, Phys. Fluids, 26, 671 (1983).
5.H.Bruzzone, Proc.II Latin Am. Workshop on Plasma Phys. and Contr. Therm. Fusion, Medellin 1987, Fus. Res., CIF series, Vol 12, Singapore 1989, p.313.
6 M.J.Rhee, Rev. Sci. Instrum. 55,1229 (1984).
7. L.Bilbao, H.Bruzzone, H.Kelly and M.Esper, IEEE Trans. Plasma Sci., PS-13, 202 (1985).
8. H.Bruzzone, H.Kelly, A.Marquez, Proc. V Latin Am. Workshop on Plasma Phys., Mexico 1992, p.43.

DYNAMICS OF HOT SPOTS FORMATION IN A DENSE PLASMA FOCUS OPTICAL AND X-RAY OBSERVATIONS

P. Choi & R. Aliaga-Rossel[*]
The Blackett Laboratory, Imperial College London SW7 2BZ, U.K.

C. Dumitrescu-Zoita
Faculty of Physics, University of Bucharest, Bucharest, Romania

C. Deeney
Physics International Co., San Leandro, California, USA

ABSTRACT

Hot spots are dense plasma formations emitting strongly in the medium to hard X-ray regions. Their presence can be detected in a dense plasma focus when high Z impurities are added to the main gas of hydrogen or deuterium. Experiments have been carried out on DPF 78, a medium size, 28 kJ, 60 kV Mather type Plasma Focus operating with heavy gas admixtures to the hydrogen or deuterium fill gas. Using a range of time and space resolved diagnostics in the optical and X-ray region, the dynamics and properties of hot spots were investigated. The complementary nature of these diagnostics working in conjunction allowed us to construct a physical description on the evolution of these dense plasma formations. The first appearance of hot spots was detected within 5 ns of first compression, at a time when the bulk of the plasma column is grossly stable. However this gross stability coexists with strong local density gradients as observed in Schlieren photographs which are correlated to the hot spots. The existence of a macroscopically stable plasma column at this time suggests that the large scale m=0 MHD instability, which is observed to take place at a later time at high z position and leads to the disruption of the integrity of the column, is not responsible for the hot spots formation in the initial period.

INTRODUCTION

The existence of very small regions (< 0.2 mm) of intense X-ray emission within a larger region of hot, dense plasma in a Plasma Focus was long observed and reported.[1] It was observed that these plasma spots were positioned roughly along the current channel and were considered to be the point of focus of the collapsing plasma sheath. It was mentioned that they resembled the m=0 instability in a pinched discharge. Further studies on the nature of plasma spots observed in the Plasma Focus devices indicate that they emit strongly in the characteristic K shell region of the electrode material and could have originated from highly ionized metallic vapour ejected from the electrode [2], resembling the action of the vacuum spark. Their reproducibility was significantly improved when high Z noble gases were used as the main working gas, or when they were introduced into the main working gas of deuterium or hydrogen as an admixture.[3,4,5,6] However, it was found that low Z impurities, like neon and nitrogen, do not help in the formation of these plasma spots.[7] Other studies reveal that these localized emission region could exist as bright spots or elongated micropinches.[8] Detailed time integrated measurements in the soft X-ray region indicate that they have radial dimensions of 50 μm or less and electron temperatures above 1 keV. The electron densities in these hot spots are often in excess of $10^{21} cm^{-3}$ as evaluated from time integrated spectra. Their life time is extremely short, often less than 1 nsec. The combination of short life time, very small size and the randomness of when and where they appear makes it difficult to set up any measurements with the necessary high spatial and temporal resolutions to study these hot spots in detail.

Over the past few years, an extended series of experiments have been carried out on the DPF-78 Plasma Focus device, [9,10], at the Institut für Plasmaforschung, Universität Stuttgart to study the evolution and to measure the properties of the Plasma Focus in the presence of various controlled amounts of high Z admixtures. One of the aim of these experiments is to develop suitable diagnostics to investigate the properties and to better correlate the appearance of hot spots to other plasma parameters in a Plasma Focus. A number of questions have to be answered. In particular, where does a hot spot originate from and when. Are they hotter or cooler than the surrounding plasma. What is the role of radiation cooling in their creation. Are they simply the late phase manifestation of the m=0 instabilities. Some of the findings have been reported before and indicate that the hot spots are features within a macroscopically stable plasma. In the following, some of the novel diagnostics developed are described and the results obtained in characterizing the hot spots presented. The general limitation due to traditional diagnostics are highlighted. Based upon the findings from the experiment, a model to describe the properties of the hot spots is discussed.

EXPERIMENTAL TECHNIQUES

The present experiments were carried out on the DPF-78 Plasma Focus operating at 60 kV, 28 kJ, providing a bank current of about 880 kA. The discharges were made in hydrogen and deuterium, with different admixtures of noble gases, including neon, argon, krypton and xenon, at different mass fractions of up to 60%. An extensive range of diagnostics have been used throughout the experiments. Apart from the routine measurements of the electrical parameters with voltage and current probes, Silicon PIN diodes were used with collimation to look at the X-ray emission in the 1-20 keV range with various foil filters, axial and radial optical streak photography were used to look at the collapse dynamics of the emitting plasma shell, and silicon photodiodes were used to image different positions along the plasma column on axis to detect the time of maximum compression. For operation in deuterium, a silver activation counter was used to obtain the total neutron yield and a set of 9 scintillator-photomultiplier detectors was used to record the neutron yield rate. Other more specific diagnostics are discussed in the following. They were developed to investigate the hot spots and the evolution of the plasma structure associated with them.

Single and double frame Schlieren photographs were taken with a nitrogen laser pumped pulsed dye laser (Rhodamine 6G), with a nominal 1 ns pulse length. A delay of 8 ns was used in the double frame system. Pinhole aperture was used in the focal stop of the laser probe beam to provide bright field Schlieren pictures. Different aperture diameters were used to provide different Schlieren sensitivities, ranging from .3 mm to 1 mm. The Schlieren photography provides a good method of visualizing the structure of the plasma sheath as well as its dynamics during the different stages of the discharge. The double frame system is invaluable in revealing some of the gross dynamical features of the plasma column in the formation of the hot spots.

The evolution of the hot spots was monitored directly in the soft X-ray region with a number of techniques involving streak photography. A 1-D X-ray streak technique was used to monitor the X-ray emission over two different spectral bands.[5] A new technique, Axially Displaced Radially Resolved Multipinhole (ADRRM) streak , was developed to provide a continuous picture of the plasma structure in 2-D.[11] Further refinement of this technique led to an X-ray technique for continuous time- and space-resolved observation of the hot spots in the Large Slit Axially Displaced Radially Resolved Multipinhole (LSADRRM) streak.[12] This technique provides redundancy information in the recording to absolutely locate the occurrence of the hot spots, with mm spatial and ns temporal resolution.

To isolate the evolution of the hot spots at different energetic stages, a technique of multiple doping was introduced in part of the experiments. Different admixtures of Ne, Ar and Kr were introduced simultaneously into the discharge in controlled amount. The differences in the ionization energy required to bring each impurity to the K and L shell allowed us to study the hot spots at different stages of heating. By monitoring over several different spectral bands, and taking advantage of the fact that the hot spots exist for a very short duration, a 2-D snap shot of the hot spots could be obtained in a particular spectral band using filtered pinhole photographs, even if the recording is integrated in time over the whole duration of the discharge.

A multi-filter K-edge diagnostic technique was used to diagnose the high temperature plasmas with impurities. This technique uses a number of filters to obtain several energy windows with quite acceptable spectral resolution. The essence of the technique is in selecting a K-edge filter with the absorption edge located about the recombination edge of a particular impurity. The spectral selection is achieved by additional low Z filters, having the transmission characteristics matched to the long and the short wavelength sides of the K-edge filter. Such filters can be chosen to cover a large energy band, overcoming most disadvantages that occur when using Ross filters. [13] Since they are used to cover a broad band, the measured signal is large and errors induced by out of band noise signals are thus smaller. The deconvolution process is done by fitting an emission spectra, for different plasma temperatures and densities, into the filters and detectors, and comparing the result with the measured response. A collisional radiative equilibrium code, RATION [14] is used to provide an accurate estimate of the level population of the impurities.

In the multiple doping experiment, time integrated, spatially resolved measurements were made with an 8 channels multipinhole camera, using the multi-filter K-edge technique, to record the X-ray emission from the K-shell radiation of Ne, Ar and Kr. Different pinhole diameters, between 50 μm and 750 μm, were used with the different filters in order to optimize the range of exposure obtained from the different amount of radiation emitted by each of the impurities.

To accurately study the structure of the hot spots, simultaneously information in at least 2 spatial dimensions, 1 temporal dimension and 1 spectral or energy dimension is needed. This is obviously very difficult if not impossible, in view of the short life time and small dimension of the hot spots. Correlation of some of the previous presented diagnostic methods, however, can provide a more complete picture of the behaviour of these dense plasma structures. For example, in order to obtain information when the hot spots occur, we have to consider simultaneously the Schlieren images, the 2-D X-ray streak and the axial X-ray streak. For information about the spatial position where the hot spots are formed, time integrated spatially resolved multipinhole camera, Schlieren pictures, collimated PIN diodes measurements and LSADRRM images have been considered. For informations about a continuous history of hot spots formation and development we looked at the LSADRRM streak and the multipinhole photography results.

THE EXPERIMENTS

General Behaviour

The dynamical behaviour of the DPF-78 Focus plasma has been reported in a number of articles previously.[5,15,16] The evolution of the hot pinch column can best be summarized with the help of a time and axially resolved study in the X-ray region, as shown in the digitized output of a two channel X-ray streak photograph in Fig.1. The discharge is in 4.6 mbar H_2 with 0.4 mbar of Ar. The top image is filtered to look at X-rays emission below 1 nm while the bottom image views mainly radiation between 3.2 nm and 10 nm. The curvature of the collapsing plasma sheet leads to the creation of a dense pinch, initially at a small region along the axis, which then expands both into and out of the hollow anode within 10 ns to a length greater than 10 mm, as the other part of the curved sheath stagnates onto the axis. Thereafter the axial movement of the hot, dense location continues in both axial directions. In the direction away from the anode, an average velocity of 30 $cm.\mu s^{-1}$ is measured. Localized dense, hot structures could be observed when harder filters are introduced in the X-ray streak system.[5] They tend to form among the dense moving plasma, though the resolution of the system is not sufficient to provide spatial details of the hot spots. Different hot spots are observed at different locations at different times.

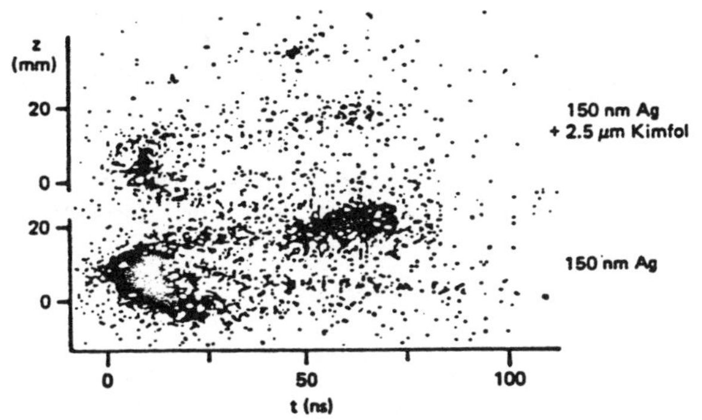

Fig.1 Axial soft X-ray streak using double pinhole camera

The moving dense plasma lasts until disruption associated with the m=0 necking of the bulk plasma takes place, in general at a high z location away from the anode. In some instances, a second hot, dense region is observed to evolve after the disruption, in a region close by but beyond that of the point of necking. Hot spots are again observed in this dense region. Correlation with electron beam measurements indicate that this sudden creation of a hot, dense plasma is associated with the onset of an intense electron beam with an energy below 50 keV.

The evolution of the electron temperature of the first pinch has been studied in some detail from collimated PIN diodes measurement in an argon doped discharge in hydrogen.[17] Using different filters and fitting to an emission spectra including the H- and He-like resonance lines and recombination continua, the plasma is found to be dense at first compression with strong activities of energetic electron beam, before being heated rapidly to above 1 keV in about 10 ns, mirroring the observations from the 2 channel X-ray streak system.

Effect of high Z-impurities

The introduction of high Z impurities into the operating gas has little effect on the evolution of the plasma sheath up to the point of the radial collapse phase, as long as the mass of the discharge gas is kept constant in the admixture and the mass fraction of the impurities is below 40%. In the radial collapse and the pinch phase, a number of features are observed which could be correlated with the Z of the admixtures. These effects on the plasma dynamics are discussed in detail in another paper in this proceeding.[18]

In general, Schlieren measurements show that the radially converging sheath is thinner as the Z of the admixture is increased. This resulted in a significantly tighter pinch for a Kr doped discharge compared with a discharge in pure deuterium. The surface of the sheath is also affected by doping, with more noticeable rippling observed in discharges with high Z doping. These surface perturbations tend to grow at the beginning of the radial collapse phase when the sheath turns round the edge of the anode, and appears to be frozen in the plasma sheath throughout most of the radial collapse . However, at the final stage of the formation of the dense pinch, their amplitude becomes comparable to or larger than the dimension of the tightly pinched column. It is important to note that while the surface disturbances of the collapsing sheath in a krypton doped discharge lead to the formation of large amplitude flares at pinch time, the tight pinch column itself is not significantly disturbed for some 10-20 ns after the first compression.

Experimental results

Figure 2 is an X-ray streak picture taken with the LSADRRM technique, showing a discharge in 4.6 mbar D_2 with 40 μbar Kr. A total of nine channels was used, with each channel being a radial streak picture of the discharge and centred on an axial region which is separated from the next region by 2.5 mm. Channels 2, 5 and 8 were filtered to look at radiation between 3.2 nm and 10 nm from the dense pinch plasma and the rest filtered to look at harder X-rays below 1 nm from the hot spots. The camera was positioned in such a way that the presence of a single hot spot would produce a set of 4 images simultaneously in 4 channels, with the images aligned at 16° to the vertical.

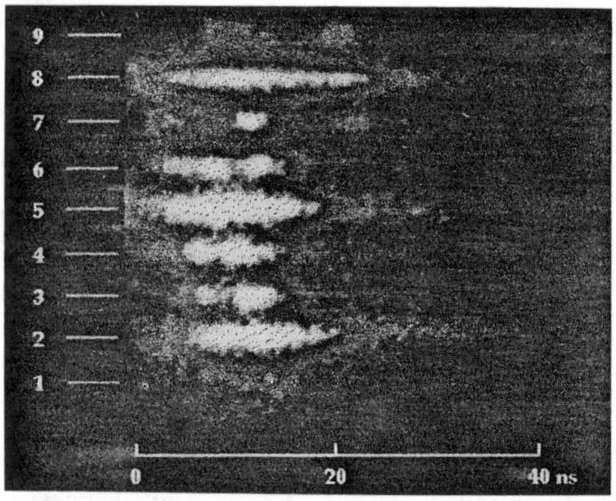

Fig.2 LSADRRM X-ray streak image

The formation of the pinch column appears first at channel 5, centred at 5 mm from the end of the hollow anode. This is followed closely at locations on either side, as channel 2 and 8 show. The 3 mm diameter pinch is grossly stable and lasts for 30 ns, after which the intensity of the radiation fades. The first hot spot is observed within 5 ns of first compression, at 5 mm from the end of the anode. The second appears about 3 ns later, at 1.5 mm, following within the next 2 ns by two other hot spots, one at 12.5 mm and one at the edge of the anode. The fifth one can be seen 19 ns after the pinch first formed, at 16 mm. It is important to note that the pinch column is macroscopically stable throughout the whole period when the hot spots are formed. The positions of the hot spots are determined to within ±1 mm while their durations are all about 2.5 ns as recorded. Given the response and resolution of the detection system, the actual lifetime of all hot spots are below 1 ns.[12] Similar behaviours are observed at various levels of doping, with the first hot spot appearing within a few ns after pinch formation. The limited resolution of the system, however, does not provide information as to whether the hot spots are stationary, nor to their exact sizes. That information is provided by time integrated pinhole photographs.

Table 1
Filter and pinhole characteristics for the multipinhole camera images presented in Fig.3 & 4 (Fig.4 in ( ) where different)

Channel number	1	2	3	4	5	6	7
Filter material	Zn	Mylar	Al	Al	Mylar	Ti	Fe
Filter thickness (μm)	12.5	430	17	31	100	10.5	8
	(20)	(1000)	(5)	(10)			
Pinhole diameter (μm)	200	200	200	200	200	200	200
	(400)	(400)	(50)	(50)			

Figure 3 shows two digitized images obtained from the multipinhole camera, using the multiple K-edge filter technique to look at different spectral band between .1 nm and 1 nm. The discharge is in 4 mbar D_2 with .2 mbar Ne in (a) and with .1 mbar Ar in (b), the mass fraction of the impurities being the same in both cases. The set of 7 filters and pinholes used are listed in table 1. In Fig.3(a), a straight pinch column about 2 mm diameter is seen in channel 3 & 4, where the Ne K-shell radiation is transmitted through. The column is barely visible in channel 5 & 6, where radiation softer than

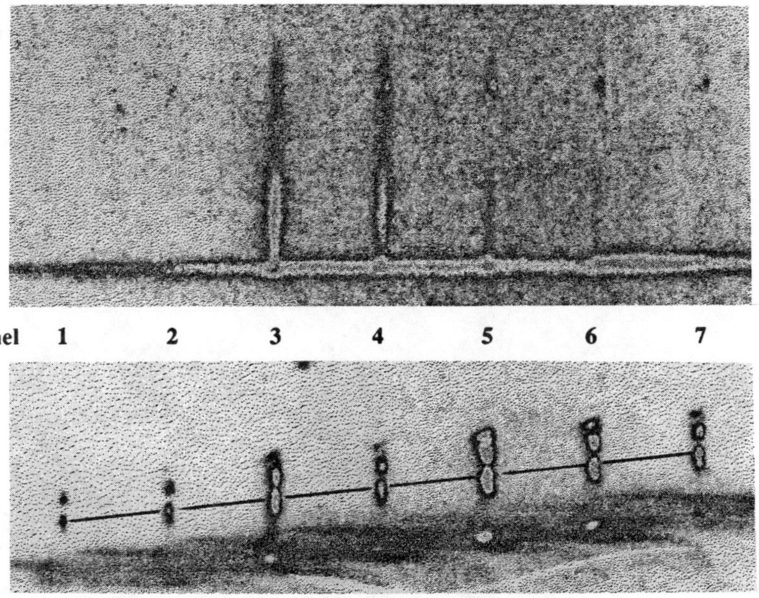

Fig.3b Multipinhole X-ray image for a discharge in 4 mbar D_2 with (a) .2 mbar Ne admixture and (b) .1 mbar Ar admixture

.7 nm is attenuated by more than 2 orders of magnitude. However, the only hot spot observable, located at 20 mm from the end of the anode, is detected on channel 5, 6 & 7. Comparing the signals from all channels suggests that the hot spot emission is composed primarily of the continuum radiation of a fully stripped Ne plasma. The broad emission seen at the base of the pinch originates from the end face of the anode. In general, hot spots are rarely recorded in Ne doped discharges even though Schlieren photographs detect necking and other local instabilities.

This picture of a uniform plasma is in dramatic contrast to that in Fig.3(b), when Ar is used in place of Ne. The pinhole array was rotated slightly with respect to the axis of the Focus so that the 7 channels were displaced vertically from each other and the emission from the anode recorded in one channel extended on one side to the plasma signal in another. A number of hot spots can be seen in the first 10 mm of the pinch. Interesting features about the energetic state of the hot spots can be found by examining the relative strength of the signals from the different channels. Contrary to the case with Ne doping, the strongest emission is now recorded by channel 5 & 6, showing a number of elongated beads with diameter of about 1.5 mm, while channel 3 & 4 shows spot like features with diameter below 1 mm. Even smaller spots are seen in channel 1 & 2, which are filtered to look at radiation below .35 nm. This can be explained by noting that the Ar K-shell emission concentrates between .35 to .4 nm, with little emission in the .4-1.2 nm band. On the other hand channel 3 & 4 has their pass band ending at .78 nm, and the transmission remains low up to .35 nm.

The signals across the channels, together, draw a picture of a plasma which is becoming dense at a certain region of the pinch column, where it is also heated to a temperature to be in the argon K-shell. This dense plasma region is not small, about 1.5 mm in diameter in the radial dimension and extends several mm in the axial direction. The process of compression continues and the dimension of the dense region is further reduced, while the local temperature is increasing at the same time. By the time the dense region is below 1 mm in diameter, it takes on a spot like feature while most of the radiation is now in the argon H-like recombination continuum. The present system could not tell us the point at which the compression and heating processes stop. The lower limit which could in principle be estimated from the size of the image in channel 1 falls into the resolution limit of the system and is below 100 μm.

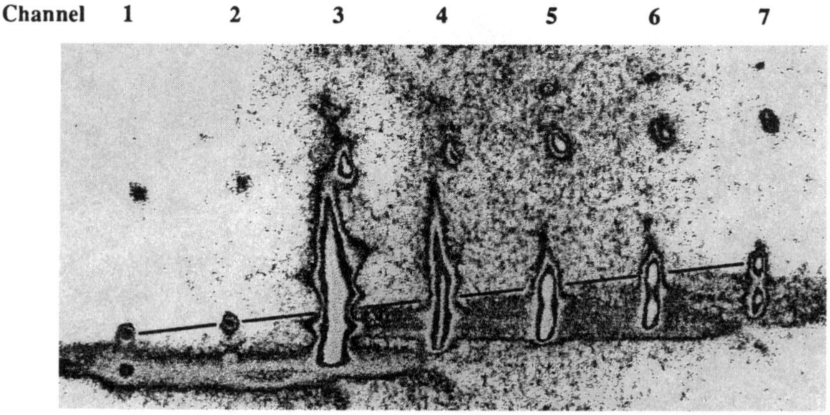

Fig.4 Multipinhole pinhole X-ray image for a discharge in 3 mbar D_2 with 200 μbar of Ne, 50 μbar of Ar and 25 μbar of Kr

To circumvent the limit set by the ionization temperature of the impurities, multiple doping with a mixture of Ne, Ar and Kr was introduced. Figure 4 shows a digitized image from the multipinhole X-ray camera of a discharge in 3 mbar D_2 with 200 μbar of Ne, 50 μbar of Ar and 25 μbar of Kr. Details of the filters and pinholes are in table 1 and the response of the filters are shown in Figure 5. A combination of the features observed for Ne and Ar doped discharge are now seen. A sub-mm size uniform pinch column is seen in channel 3 & 4, surrounded by non-uniform plasma structures of mm size diameters. These channels are filtered to look at the Ne K-shell emission before the ions are fully stripped. The elongated bead structure is clearly seen in channels 5,6 & 7, showing progressive-ly smaller radiation region of Ar K-shell lines and recombination continua as the cut-off wavelength of the filter becomes shorter. Dense structures exist in both low z and high z positions, as far as 25 mm from the anode. The radiation in the harder X-ray region is quite different. In the presence of Kr, strong signals are detected in channel 1 & 2. The larger pinholes of 400 μm allowed thicker filters to be used, providing a 99% cut-off point at below .3 nm. The signals detected would therefore originate from the Kr K-shell emission, as well as the H-like recombination continuum of Ar. Almost circular projection of the pinhole is recorded in both channels. The reference line drawn across the different channels provides a guide to the location of these dense hot spots among other radiation features. The composite signals further support the picture of hot spots evolution described in the last paragraph.

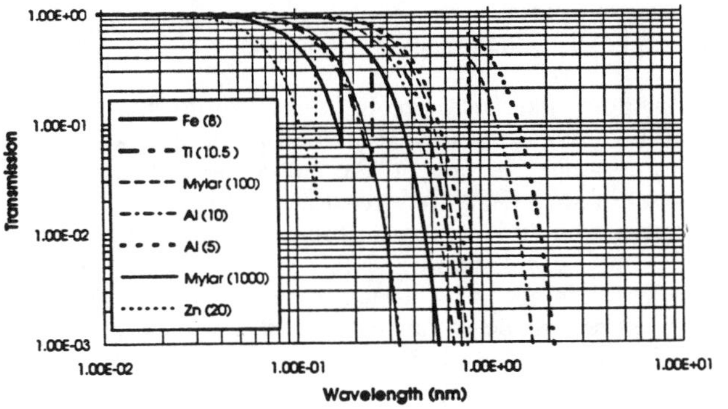

Fig.5 X-ray filter response curves

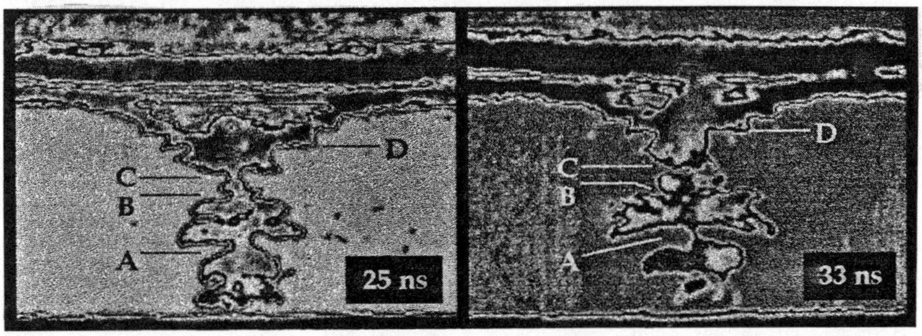

Fig.6 Double frame Schlieren images for a discharge in 4 mbar D_2 with 50 μbar of Kr

The development of the dense plasma structure can be observed directly in the double frame Schlieren diagnostics. The discrete nature of the exposure meant that a large number of shots had to be taken to capture a picture of the compression. The results from the LSADRRM streak helped to narrow down the time window of interests to that of the first 30 ns after first compression. Figure 6 shows a result from the double frame Schlieren system at a moment when the local compression takes place. The two frames are taken at 25 ns and at 33 ns for a discharge in 4 mbar D_2 with 50 μbar of Kr.

Three compression events can be identified, located at positions marked A, B and C on the pictures. The initial size of the plasma column at A and B is about 1.4 mm in the radial dimension and about .7 mm in the axial dimension. The column at C is about 0.7 mm in diameter and in length. After 8 ns, location A and B have both turned into spindle shape objects, with the plasma between them being squeezed out radially from both ends by about 2 mm, extending by 50% of the original size. The compression in both cases are not axisymmetric with respect to the pinch axis. The plasma at A was compressed into a line of at least 1.6 mm long with a diameter below .5 mm. During this time, the end of the plasma column, at D, advanced by 1.3 mm.

It should be noted that no singular point like structure is found, among the localized compression observed, which could act as the point like emission source recorded on time integrated X-ray pinhole photographs. In fact, in the two dozen plus Schlieren photographs, where the development of such tight compressions was recorded, the end point of the compression is always a spindle shape structure. It is possible that the development of the compression is more rapid than the 1 ns exposure time of the dye laser used and the Schlieren pictures represent a summation of the trajectory of the compressing plasma. The average speed of the expansion is below 30 $cm.\mu s^{-1}$. This would lead to an integration of .3 mm of movement in the 1 ns exposure time, and should not be the reason behind the line type rather than point like compression observed, unless the compression due to the instability progresses at a speed much above the average speed of expansion recorded.

DISCUSSIONS

Two unique features can be seen from the data presented. Firstly, the size and shape of a hot spot depends almost entirely on the spectral region through which it is observed. In a discharge with multiple doping with Ne, Ar and Kr, the long duration plasma emission from the pinch column completely covers up any radiation signature from the hot spots in the Ne K-shell channels, while the channels looking at the Ar K-shell region produce details of radiation points when the plasma undergoes further compression. When the observation is shifted to the harder X-ray region beyond the Ar K-shell, bright and very small radiation points are found. The correlation of time resolved signal with limited spatial resolution and time integrated signal with high spatial resolution let us conclude that the events recorded in different spectral channels at one physical point are one and the same event. The variation in shapes between an elongated micropinch and a very small plasma point observed in an experiment could thus simply be related to the energetic level of the plasma impurities and the amount of filtering being used to set the observation region in the detector.

As for the size of the hot spot, the variation recorded would suggest that a spot size measured with a given detector set up is not necessary the true spot size. This is clearly demonstrated in the Ar doped discharge in Fig.3b, if one were to take the data in each of the channels as records over separate but identical discharges respectively. The size of the emission zone is diminishing as the observation is shifted towards shorter wavelengths, indicating that the emission is more efficient in the shorter wavelength region as the size of the emission source decreases, and hence implying a higher temperature plasma at smaller size. For a compression with constant line density, the total radiation would go up as r^{-2} and so, for pinhole limited recording, the exposure would be directly proportional to r^{-2}. This is not observed in the experiment and it indicates that there is a reduction in line density from mass loss, which would also lead to an increase in temperature of the local plasma region. The axial mass flow out of the region of compression is supported by the Schlieren measurements.

Secondly, the non-linear development of a local instability does not lead to a singular point which could be attributed to a hot spot. The Schlieren data captured the growth of the instabilities and show that the further compression of a local neck region results in a tiny pinch column, with a physical feature not unlike a miniature version of the original radial compression of the Focus, with both radial and axial motion involved in the collapsing plasma sheath. Significant increase in size of the flare structure on both sides of the necking, support the idea of the continuous ejection of mass in the compression process. The Schlieren observation, however, is at variant with the measurement from the X-ray emission, which shows clearly the formation of highly localized bright and energetic radiation points. However, the more elongated features observed at longer wavelength indicate that the plasma structure at the end of the compression is in a columnar form but within this column, there is a region of much higher density which forms the very small hot spot observed at short wavelength. Indeed, in some cases, several hot spots can be identified within very small axial separations and these could then be associated with the structure within the column observed in the Schlieren measurement.

Radiation cooling has been suggested to be the driving mechanism which lead to the creation of the hot spots.[3,6,19,20,21] Studies on the non-linear growth of an m=0 instability in a hydrogenic pinch predicts the formation of a spindle shaped pinch column which cannot explain the small localized hot spots observed.[22,23] A recent model for pinch spot formation in radiation dominated, high Z pinches predicts the formation of a V-shape profile pinch when the development of radiation enhanced radial compression exceeds the speed of axial mass loss.[24] The strong density concentration at the base of the V shape is considered to be the origin of the highly localised radiation spots. The Schlieren photographs obtained in the present experiments, do not support the model of V-shape necking proposed, even for Kr doped discharges.

Most of the hot spots observed in the low z position, within 10 mm from the end of the anode, occurs when the Schlieren measurements show that large scale disturbances are absent. A dense core is present and the hot spots are observed to exist within this core. At late time, localized necking begins to appear which lead to the creation of short tight pinch columns with sub-mm diameter. Some of these structures can be identified with bright radiating regions which are also considered to be hot spots. Not all hot spots are the same, even if they appear to be very similar when observed under one spectral region. The emitting temperature of these spots varies from spot to spot, with the smallest size spots emitting hardest radiation. Measurement with large pinholes and hard filtering using the penumbral technique would therefore yield the smallest size hot spots observed.[6]

The size and properties of a hot spot are dependent on how they are measured. Apparently identical hot spots observed in the K-shell emission region of an impurity could have totally different density and temperature characteristics when observed in another spectral region. The life time is below 1 ns though the actual time would again be dependent on the spectral region where the measurements are made. The lack of tight hot spots in X-ray pinhole pictures in a low Z doped discharge is not indicative of the absence of such dense hot regions. It could simply be that the energy content of the plasma is sufficiently high to fully ionize almost all of the dopant during the pinch phase, resulting in a fall in emission in the spectral region under observation. The high Z dopant helps to enhance the radiation but the structure of the final pinch is also modified. In general, the number of observable hot spots increases slightly with Z and with impurity fraction. However, in Ar and Kr doped discharges, there are occasions when an order of magnitude increase in the number of hot spots are observed. These conditions are not yet understood. There are no correlation between the number of hot spots and the neutron yield.

Acknowledgement
The experimental work presented were carried out in the Institut für Plasmaforschung at the University of Stuttgart, part of it under a CEC Stimulation Action Twinning Contract ST2J-59. The authors would like to express their sincere thanks to Prof. Hans Herold at the Institute for his interest throughout and for the use of the facilities.

REFERENCES

[1] J.W.Mather, Phys.Fluids 8, 366 (1965)
[2] V.D.Ivanov, V.A.Kochetov, M.P.Moiseeva, A.A.Palkin, E.B.Svirskii, A.R.Terentiev, N.V.Filippov, T.I.Filippova et al., Proc. 8th IAEA Conf. on Pl. Phys. and Contr. Fus., Brussels, (1980)
[3] N.J.Peacock, R.J.Speer, M.G.Hobby, J.Phys.B: Atom.and Molec.Phys., 2, 798,(1969)
[4] K.H.Schönbach, L.Michel, H.Fischer, Appl.Phys.Lett., 25, 10, 547 (1974)
[5] P.Choi, C.S.Wong, H.Herold, Laser and Particle Beams, 7, 4, 763 (1989)
[6] K.N.Koshelev, V.I.Krauz, N.G.Reshetnyak, R.G.Salukvadze, Yu.V.Sidelnikov, E.Yu.Khautiev, J.Phys.D:Appl.Phys., 21, 1827 (1988)
[7] J.M.Bayley, G.Decker, W.Kies, M,Mälzig, F.Muller, P.Röwekamp, J.Westheide, Yu.V.Sidelnikov, J.Appl.Phys. 69, 613 (1991)
[8] W.H.Bostick, V.Nardi, W.Prior, J.Plasma Phys. 8, 7, (1972)
[9] T.Oppenländer, Institut für Plasmaforschung der Universität Stuttgart, Report IPF-81-2
[10] H.Herold, et al., Int. Conf. on Pl. Phys. and Contr. Nucl. Fus. Res., London, 1984, paper CN-44/D-III-6-3
[11] C.Deeney, P.Choi, Rev.Sci.Instrum., 60, 11, 3558, (1989)
[12] P.Choi, R.Aliaga, Rev.Sci.Instrum., 61, 10, 2747, (1990)
[13] D.J.Johnson, Rev.Sci.Instrum., 45, 191, (1974)
[14] R.W.Lee, Manual for RATION, Lawrence Livermore National Laboratory, 1990
[15] P.Choi, C.Deeney, H.Herold, C.S.Wong, Laser and Particle Beams, 8, 3, 469, (1990)
[16] H.Herold, H.J.Käppeler, H.Schmidt, M.Shakhatre, C.S.Wong, C.Deeney, P.Choi, Proc. 12th Int. Conf. on Pl. Phys. and Contr. Nucl. Fus. Res., Nice, 1988, paper CN-50/C-IV-5-2
[17] P.Choi, C.Deeney, C.S. Wong, H. Herold, A. Wieder, Proc. 5th Int. Workshop on Plasma Focus and Z-pinch Research, Toledo, Spain, 1987, p.76.
[18] P.Choi, R.Aliaga, This Proceedings, London, (1993)
[19] J.W.Shearer., Phys.Fluids, 19, 1426 (1976)
[20] V.V.Vikhrev, JETP Lett., 27, 95 (1978)
[21] K.N.Koshelev, N.R.Pereira, J.Appl.Phys., 69(10), 21 (1991)
[22] D.L.Book, E.Ott, M.Lampe, Phys.Fluids 19, 1982, (1976)
[23] V.V.Vikhrev, V.V.Ivanov, G.A.Rozanova, Sov.J.Plasma Phys. 15, 44 (1989)
[24] D.Mosher, D.Colombant, Phys.Rev.Lett., 68, 17, 2600 (1992)

✠ Permanent address: Facultad de Fisica, Pontificia Universidad Catolica de Chile, Casilla 306, Santiago 22, CHILE

Dynamics of a medium energy plasma focus

P. Choi and R. Aliaga-Rossel*
The Blackett Laboratory, Imperial College
London SW7 2BZ, U. K.

INTRODUCTION

More than 30 years after it was first reported by Mather [1] and Filippov [2], the Plasma Focus remains one of the most intriguing high density and high temperature transient plasma structure. A relatively small spread of plasma parameters, in terms of electron temperature and density, have been observed over some two and half order of magnitude variation of the electrical stored energy up to .5 MJ systems. Favourable neutron scaling were observed at low to moderate bank energies, though these scaling were obtained only under a narrow range of operating conditions. A number of explanations have been put forward to explain the neutron production mechanism though no universal model appears to hold. Early measurements on the anisotropy of the neutron yield lead to the conclusion that most of the neutrons produced were not of thermonuclear origin and that a beam-target mechanism resulting from some gross instability of the plasma column was in favour. With more detailed measurements, particularly on large devices, what has emerged is that there are a number of distinct periods in the course of the focus formation, when significant neutrons are produced, and that a number of different mechanisms might be responsible for the neutron production at various times.[3,4]

In the plasma focus, the dense, hot pinch plasma is formed as a result of the convergence of a two dimensional plasma sheath, moving in both radial and axial directions. The direct connection between the formation of this radially collapsing sheath, the quality of the sheath, the sheath velocity and the resulting focus pinch plasma as well as the stability and break-up of this plasma column, in the framework of the various mechanisms that might be responsible for the neutron production, has never been adequately elucidated. Most of the early studies were centred around optimization of the total neutron yield in terms of the machine parameters and the operating conditions. More recent studies with nanosecond precision have revealed the highly dynamical nature in the formation of the dense pinch plasma from optical refractivity measurements[5] and from X-ray measurements [6]. Efforts to correlate the measured sheath parameters with the neutron yield rate indicate that optimal performance is related to the formation of a stable pinch column from well defined narrow sheath.[7]

In the present experiment, high Z noble gas impurities were introduced into the focus plasma in a controlled fashion to modify the discharge parameters. Different levels and types of impurities, either singly or in combinations, were used in the investigation on the influence of impurities upon the sheath thickness, the sheath velocity, the sheath stability and the resulting pinch formation. The plasma density structure was studied using either single or double frames laser Schlieren photography. A number of other time and space resolved diagnostics were used simultaneously to study the quality of the focus plasma and the polar distribution of the neutron yield rate. The aim is to investigate the neutron production mechanisms through a correlation with the plasma structure. Detailed measurements of the neutron emission properties are presented in a companion article [8] and in the following, various properties of the plasma structure obtained from the Schlieren diagnostics are presented.

THE EXPERIMENT

The experiments were carried out on DPF-78, a medium energy high voltage plasma focus device at the Institut für Plasmaforschung at the University of Stuttgart.[9] The focus was operated at 60 kV, 28 kJ throughout the experiments, operating in either pure deuterium or in admixtures of different combinations of neon, argon and krypton. The fill pressure of the discharge ranges below 1 mbar to over 5 mbar of deuterium. When different admixtures were used, a near constant equivalent fill mass was adopted to minimize the effect on the collapse time so that valid comparisons can be made between the different impurity types and concentrations.

Single and double frame Schlieren photographs were taken with a pulsed dye laser (Rhodamine 6G), 1 ns pulse length. A delay of 8 ns was used in the double frame system. The Schlieren photography provides a good method of visualising the structure of the plasma sheath as well as its dynamics during the different stages of the discharge. Pinhole aperture was used in the focal stop of the laser probe beam to provide bright field Schlieren pictures. Different aperture diameters were used to provide different Schlieren sensitivities, ranging from 1 mm to .3 mm to provide a sensitivity in the electron density gradient of 2.4 x $10^{18}cm^{-4}$ at the latter. This lower limit was set by the quality of the probe beam and the associated optics.

Other diagnostics were also used during the experiments, which include time resolved and time integrated neutron measurements, current derivative and voltage monitors, 2D X-ray streak [10], X-ray multiple pinhole photography and visible plasma radiation.

Deuterium Discharges

A sequence of single frame Schlieren photographs of discharges carried out in deuterium filling gas, are shown in figure 1. Apertures of 1 and .5 mm were used with a sensitivity in the electron density gradient of 8 and 4 x $10^{18}cm^{-4}$ respectively. The Schlieren photographs were taken at different times during the compression and expansion phase of the focus. In the pictures, the shadow cast by the 5 cm diameter hollow anode can be seen on the left, while on the right the envelope of the umbrella shaped plasma sheath can be seen, advancing with time away from the anode. The classical formation of the dense pinch column, some 3 mm in diameter, at the end of the radial collapse, is observed to begin first near the anode and then extending to a length of about 12 mm in front of the anode. After the compression stage, the expanded plasma maintains the cylindrical structure whose border is detected in the Schlieren photographs. At late time, the expanded plasma column appears to be disrupted and no longer confined. However, with the smaller aperture in the Schlieren system, the photographs clearly show that a large plasma column, some 15 to 20 mm in diameter, still remains at this time.

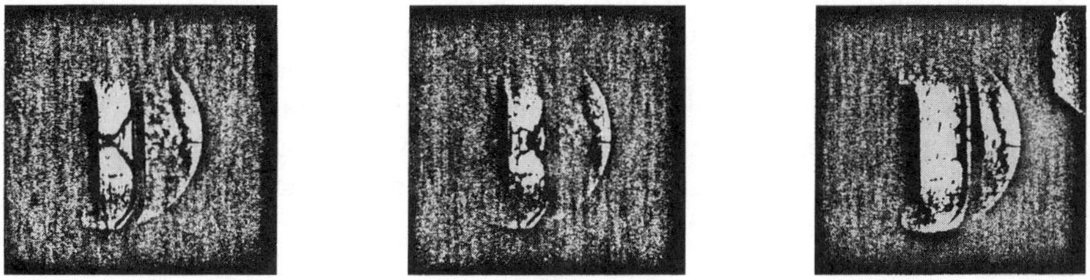

Aperture: 1 mm.

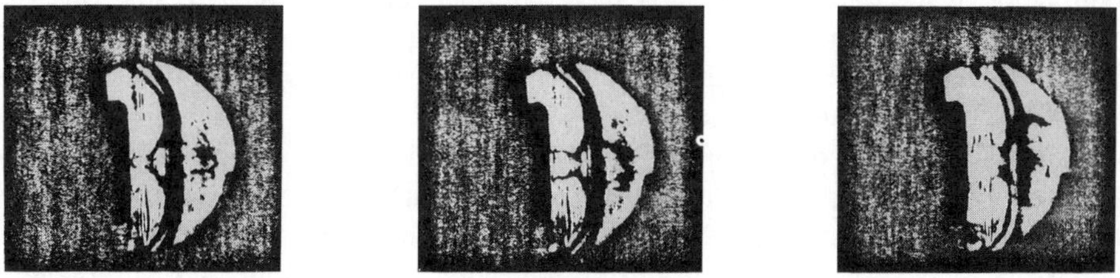

Aperture: 0.5 mm.

Figure 1: Discharges in pure D_2 (Sequential time: left to right)

The evolution of the pinch column is better illustrated by the double frame Schlieren system. A typical set of photographs, taken in a discharge with a filling gas of pure deuterium at 4 mbar, is shown in figure 2. The shadow of the anode is in the lower part of the picture. The plasma sheath is not uniform in thickness but decreases down to about 1 mm at the plane of the anode. The thickness also decreases as the sheath

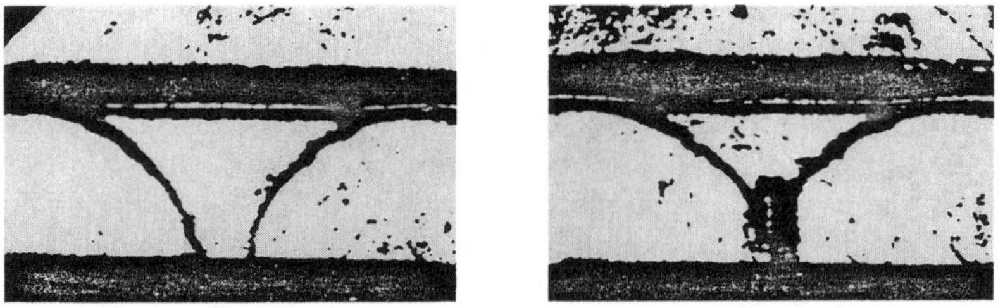

Figure 2: Discharge in pure D_2 with 4 mbar deuterium

collapses towards the axis. In most discharges with pure deuterium, the electron density gradient, and hence the refractive index gradient is much stronger on the outside of the sheath compared with that from the inner surface, closer to the pinch axis. As a result, the inner surface of the sheath is less prominent in the Schlieren photographs. The double frame system allows an average value of the sheath velocity to be measured. The plasma sheath reaches a maximum radial velocity of the order of 30 cm/μs during the collapse phase.

Discharges with High Z Admixtures

In admixtures of deuterium doped with neon up to 60% by mass, it is found that the stable plasma column extended to about 15 mm in length, longer than in discharges with pure deuterium at an equivalent mass. The sheath structure is clearly defined on both sides of the collapsing plasma shell, with a thickness similar to that observed in pure discharges. There is a smaller variation in thickness along the length of the collapsing column. In general, the plasma column remains grossly stable for over 30 ns, as is shown in figure 3 for discharges 24704 and 24705. This is longer than that in discharges with pure deuterium. The rapid expansion observed in pure discharges is absent at high doping level, but strong surface perturbations can be found through out the evolution of the pinch. These surface perturbations could be traced back to that appearing early on in the sheath during the radial compression. Measurements from the double frame system indicate that they are formed as the plasma sheath emerges from the axial run down phase and turns around the edge of the anode. These perturbation remains more or less frozen in until pinch time and tends not to destroy the integrity of the collapsing column. The upper part of the column appears to be less sensitive to disturbances and hence more stable. These perturbations either have a ring like shape around the surface of the plasma column, as is shown in discharge 24704 or are helical around the plasma column as is shown in discharge 24705. The appearance and evolution of disturbances during this phase can be clearly identified.

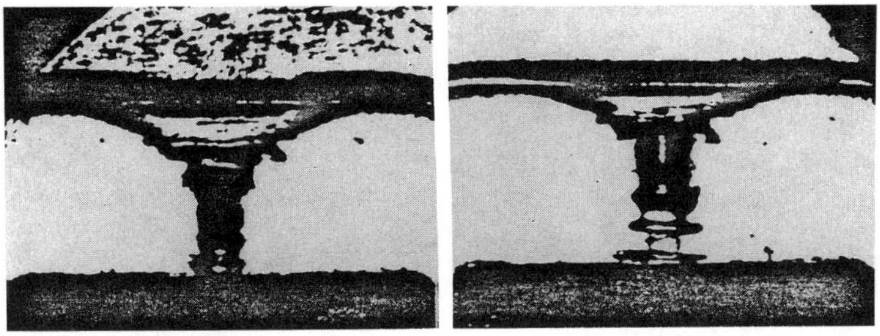

t= 26 ns a) t= 34 ns

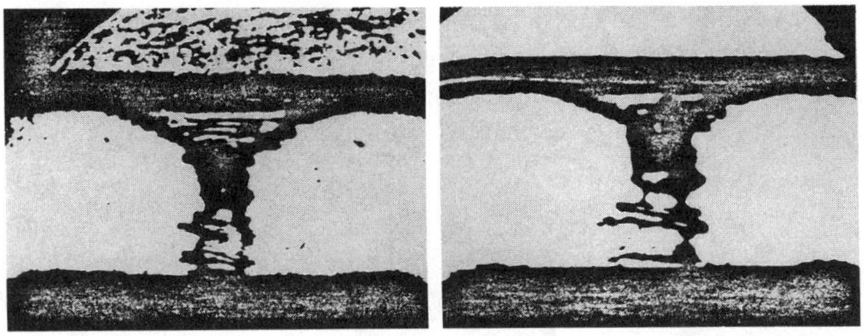

t= 32 ns b) t= 40 ns

Figure 3: a) Discharge 24704, b) Discharge 24705. 4 mbar D_2 and .2 mbar Ne.

These types of disturbance evolve during the collapse phase and can lead to an m=0 type instability which disrupts the plasma column as shown in figure 4, for discharge 24711. The second photograph of this discharge shows an early expansion of the plasma column at the high Z position. The white area that appears near the bottom of the column (absence of density gradients) gives clear evidence that the plasma is flowing along the axis as a result of local compression. This leads to a lower density plasma with a density gradient below the detection threshold.

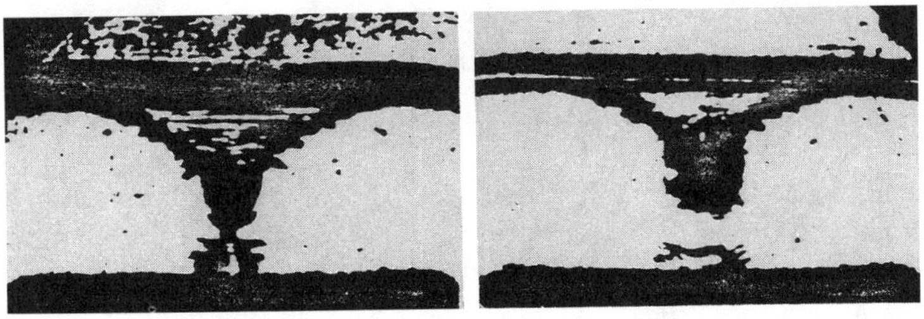

t= 28 ns t= 36 ns

Figure 4: Discharge 24711, 4 mbar D_2 and .2 mbar Ne.

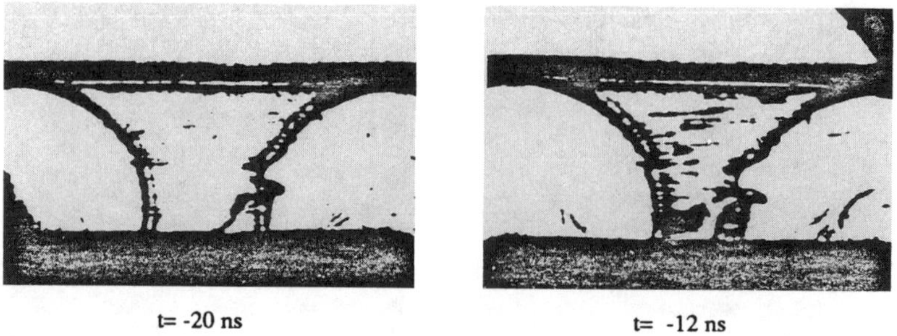

Figure 5: Discharge 24243 with a filling gas of 4 mbar D_2 and .2 mbar Ne.

In some discharges, the Schlieren photographs show a second plasma sheath moving behind, at roughly the same speed as the main plasma sheath, as shown in figure 5 for discharge 24243,. This second incoming plasma sheath can increase the lifetime of the focus because it slows down the expansion of the plasma column.

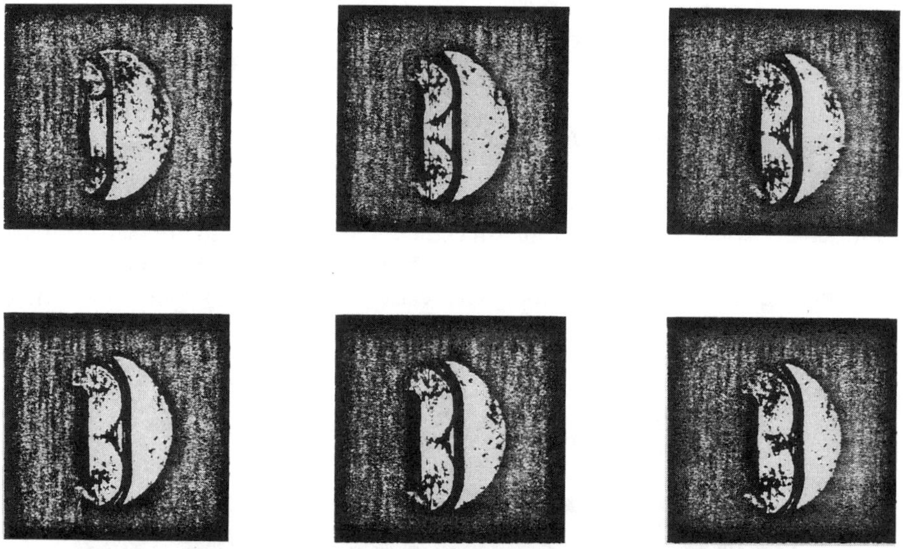

Figure 6: Discharge in D_2 + Kr , Schlieren aperture 1 mm.

Figure 6 shows a sequence of single frame Schlieren photographs taken in an admixture of 4.6 mbar of deuterium and .06 mbar krypton. These photographs show the typical characteristic of the focus development as the Z of the doping gas is increased. A sub-mm thick sheath is observed in the collapsing plasma shell, compressing to a final dense pinch column with a radius about the same as that given by the sheath thickness. The pinch column is thus more compressed than for the other admixtures. The maximum collapse velocity in general is of the same order as that for the other admixtures, 30 cm/μs.

Significant surface instabilities are observed to appear during the first cm or so of radial movement of the sheath, but again, as in the case with neon doping, remains frozen in during the rest of the radial compression. These helical disturbances and the flares are easily detectable when the Schlieren aperture is reduced to 300 μm diameter. As the diameter of the plasma shell decreases, these flare like structures become much more prominent. At pinch time, the radial dimension of these structures are often much larger than that of the pinch column. There are no uniform re-expansion of the pinch column but instead the whole column appears to be in the form of a twisted rope at late time.

Compared to discharges with neon and argon mixtures, the helical disturbances has a smaller pitch and are more pronounced at a given point in the radial compression. This is clearly shown in figure 7, for discharge 25147, an admixtures of 4 mbar deuterium with .05 mbar krypton.

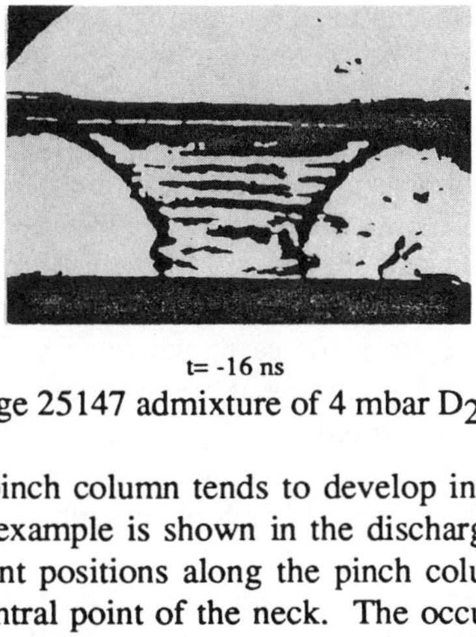

t= -16 ns

Figure 7: Discharge 25147 admixture of 4 mbar D_2 and .05 mbar Kr.

With high Z doping, the pinch column tends to develop into multiple necking as the pinch evolves. A typical example is shown in the discharge 25195 in figure 8 The necking develop at different positions along the pinch column, each pushing plasma material away from the central point of the neck. The occurrence of two such necks would produce a jet like expansion of plasma at the junction between the two. This can be seen referring to the positions marked A, B and C in figure 8. A and B mark the positions of the two necking points while C clearly develops into a strong flare. This observation is in line with the model put forward by Vikhrev.[11] These positions have been correlated with regions of intense X-ray emission from time integrated filtered X-ray pinhole pictures.

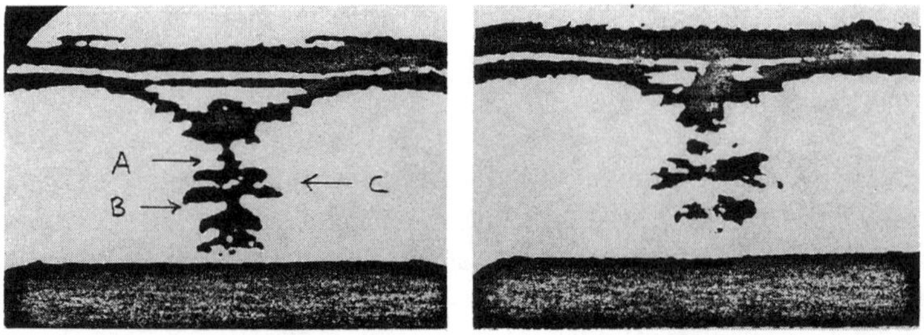

Figure 8: Discharge 25195, 4 mbar D_2 and .05 mbar Kr.

DISCUSSION

The Schlieren photographs clearly show the instabilities that take place in a plasma focus device and give quantitative information about their temporal evolution. It Is found that in general, for low Z doping (neon) the plasma column is grossly stable for more than 30 ns but multiple necking begins to occur some 15-20 ns after first pinch formation in high Z doping (krypton). The plasma column is observed to expand after first compression at low level of low Z doping, with a larger expansion for pure discharges. Rayleigh Taylor type instabilities are observed to grow in the expanding plasma column. With pure discharge, a repinching is often observed at high Z position and is correlated to the onset of a second period of neutron emission. In the case of high Z doping, no global expansion of the plasma column is observed.

The appearance of instabilities or perturbations on the plasma shell is detected during the radial collapse phase. The perturbations that grow near the surface of the plasma sheath are detected early on and have a helical form which has a smaller pitch for a high Z doping gas. These perturbations do not grow significantly during the radial compression stage and there are some cases where they are apparently frozen in the plasma sheath. With krypton doping, initial perturbations on the plasma sheath lead to the growth of strong flare like structures at pinch time. Multiple necking is observed on the much tightly pinched dense plasma column and are correlated to regions of enhanced X-ray emission from filtered X-ray pinhole photographs.[12]

Among discharges with high Z admixtures, it is found that neon doped discharges produce a longer and more stable plasma column and they have a mm thick plasma sheath, similar to that in pure deuterium discharges. The final pinch diameter tends to be a factor of 2 to 3 larger than that given by the thickness of the compressing sheath. This is in line with the Schlieren observations on SPEED 1 [5], which shows that the radial collapse stops before the plasma sheath reaches the axis. If the Z of the doping gas is increased then the plasma sheath becomes thinner and a better compression is achieved, though less stable. In krypton discharges the final pinch radius is about the same as that given by the thickness of the sheath.

Correlation with time integrated and time resolved neutron measurements carried out in doped discharges indicate that the tighter pinch diameter from a krypton doped discharge leads to a better neutron yield, even though it is stable for a shorter period. This improved neutron production rate is entirely associated with the emission which begins at the time of first pinch formation The neutron emission is peaked in the forward direction along the pinch axis and this anisotropy is observed to increase strongly with the Z of the admixture.

Acknowledgement
The experimental work presented were carried out in the Institut für Plasmaforschung at the University of Stuttgart, part of it under a CEC Stimulation Action Twinning Contract ST2J-59. The authors would like to express their thanks to Prof. Hans Herold at the Institut for discussions and for the use of the facilities.

REFERENCES

1. J. Mather, The Physics of Fluids, **8**, 366 (1965).
2. N.V. Filippov et al., Nucl. Fusion Suppl.2, 577 (1962).
3. F. Pecorella, M. Samuelli, A. Messina and C. Strangio; The Physics of Fluids, **20**(4) 675 (1977)
4. H. Herold et al., Plasma Physics and Controlled Nuclear Fusion Research, 1984, IAEA, London, UK. Vol. II, p.579.
5. G. Decker, R. Deutsch, W. Kies and J. Rybach, Plasma Phys. & Cont. Fusion, **27**, 609 (1985).
6. P. Choi, C.S. Wong and H. Herold, Laser and Particle Beams, **7**, 763 (1989).
7. R. Deutsch, W. Kies and G. Decker, Proc. Workshop on Plasma Focus and Z-pinch, 1987, p.119.
8. R. Aliaga-Rossel and P. Choi, this Proceeding.
9. T. Oppenlander, Institut für Plasmaforschung der Universität Stuttgart, Report IPF-81-2, 1981.
10. P. Choi and R. Aliaga, Rev. Sci. Instrum., **61**, 2747 (1990).
11. V.V. Vikhrev, Sov. J. Plasma Phys. **12**(4)265 (1988).
12. P. Choi et al, this proceeding.

✠ Permanent address: Facultad de Fisica, Pontificia Universidad Catolica de Chile, Casilla 306, Santiago 22, CHILE

STUDY OF ACCELERATED IONS IN THE FUEGO NUEVO DENSE PLASMA FOCUS

J.J.E. Herrera and F. Castillo
Instituto de Ciencias Nucleares, UNAM
Apdo. Postal 70-543,4510 México, D.F.,MEXICO

ABSTRACT

One of the most interesting features of the Dense Plasma Focus is the acceleration of ions due to large electric fields produced during the compression. The importance of the issue to determine the origin of neutrons has been long recognized, and its implications for applications other than fusion are of utmost interest. The purpose of this work is to analize accelerated ions in the device currently under operation at UNAM, where some evidence of MeV ions in 54 kV discharges has been obtained. This is a low energy device (~6kJ) with a 3.0 cm. diameter copper inner electrode, 10 cm. long. The outer electrode is 10.0 cm. in diameter. The device is operated with two 1.85μf capacitors in parallel. The accelerated particles are analized with filtered CR-39 plastic track detectors.

1. INTRODUCTION

Certain magnetic confinment devices used in Controlled Nuclear Fusion research share common properties with some particle accelerators. Both store charged particles in varying electric and magnetic fields, and the essential difference is that in fusion research the purpose is to confine a quasineutral, fully ionized, high temperature plasma, in the presence of higher currents. Thus it is not surprising that high neutron yields, up to 20 X 10^6 per shot, produced by accelerated ions were obtained since the early days of fusion research, back in 1958 in pulsed power devices[1-5]. Although these neutrons were first thought to be of thermonuclear origin, they were soon recognized to come from a beam-target effect[5], and it became clear that the instabilities responsible for ion acceleration needed to be supressed for reactor purposes. Furthermore, it was found that a high yield of pulsed neutrons was not enough to achieve energy breakeven. Longer confinement times, enough to compensate for energy losses through radiation and other sources, were also needed. Simultaneous advancements in accelerator technology, where higher and well resolved energies were being achieved, rendered further studies of these phenomena unnecessary, and interest in them faded, although it did not disappear entirely.

The Dense Plasma Focus[6,7], is one of such pulsed powered devices, which up to date provides the best cost/neutron yield ratio, and it is an interesting subject of study because of its relation to astrophysical phenomena, such as solar flares, as well as for certain applications, such as ion implantation[8]. Understanding the acceleration mechanisms and characterizing the energy spectrum of the ions is of fundamental interest. Yet, the pulsed character of the device, with shots of the order of 10-100 *ns*, renders conventional techniques useless. A simple alternative is the use of plastic nuclear track detectors, which in combination with other diagnostics, such as a time-resolving plasma Thomson spectrometer, can provive valuable information about the temporal evolution of the ion beam energy, as Rhee, Weidman and Schneider have shown[9,10].

The purpose of this work is to show some preliminary results, where protons in a hydrogen plasma are accelerated in a small Dense Plasma Focus, driven by a 6 kJ capacitor bank at 54 kV. Evidence suggests some protons may be accelerated up to MeV energies, although, as it will be discussed, further studies are needed in order to have conclusive results.

Extensive theoretical and computational work on the acceleration mechanisms has been pursued. Trajectories of the ions have been studied in the crossed electric and magnetic fields generated by a collapsing current[11,12], considering the collapse of the plasma sheath[13],such as in the case of the compressional Z-pinch, and the neck of the sausage instability, when finite Larmor radii effects become relevant[14]. In a different approach, reflections on the conical collapsing plasma sheath are held responsible for the acceleration[15]. So far, a selfconsistent three dimensional model, incorporating all these effects, is lacking. The problem is further complicated by the fact that there is no relevant theory to describe the nature of the plasma resistivity during the pinch phase, which might be highly turbulent. This problem will need further attention, and should be elucidated by better diagnostics that could allow a clearcut discrimination of the possible mechanisms. It might very well be possible that different mechanisms may be relevant to different devices, depending on their geometry and operational features.

The experimental device will be described in section 2, while sections 3 and 4 will be devoted to the detection method, and the discussion of results. Some conclusions will be drawn in section 5.

2. THE EXPERIMENTAL DEVICE

The device used in this work is of the Mather type, where the outer electrode is a 10 *cm*. diameter brass pipe which, welded to a 30 *cm*. brass header forms the vacuum chamber[16]. The inner electrode is a 3 *cm*. diameter copper bar, 10 *cm*. long. The insulator is a 9 *cm*. Pyrex tube which covers 4 *cm*. of the inner electrode into the chamber. There are portholes for diagnostics 10 *cm*. away from the breech (Figure 1). Eight pipes, spaced 3 *cm*. along the outer electrode, provide access for magnetic probes.

Power is supplied by 12 coaxial cables, described below, whose outer braids are clamped to the header, while their inner braids are soldered to brass bananas which plug into a brass ring tightly attached to inner electrode. The ring is insulated from the header by a sheet of glass, which glued to the Pyrex tube, provides vacuum sealing via an O-ring. Additional insulation is provided by a stack of mylar an polyethylene foils. Further sealing is met by O-rings between the copper inner electrode and the Pyrex glass.

The main feature of the electric circuit is the capacitor bank which consists in two Sangamo 1.85 μf capacitors at 60 kV, which can store an energy of 3.33 kJ each. They are charged by the power supply through a 10 MΩ resistor, and triggered independently by spark gaps pressurized at 28 psi with nitrogen. The sparks are provided by a six stage micromarx trigger, each stage consisting in a 2.5 nf capacitor at 40 kV and a 10 MΩ resistor. These stages are charged in parallel at 20 kV by a power supply, and shifted to series by spark-gaps. The micromarx trigger is pressurized by nitrogen at 19 psi within an acrylic lined chamber. It is coupled to the capacitor spark gaps by a 2.5 nf capacitor, and triggered by an ordinary car coil whose power is supplied by a 12 V car battery. It has been found this assembly provides pulsed power for the spark-gaps with a rise time smaller than 50 ns.

The whole setup is carefully shielded, and connections are made with RG8 and RG58 coaxial cables in order to prevent radio frequency noise into the environment. Further shielding has proved necessary.

The capacitor bank is connected to the gun by 12 low inductance (45 nH per foot) coaxial cables, 6 coming from each capacitor. They have a foamed polyethylene core, with a braid of RG14/U cable as central conductor. The main insulator is extruded over the inner braid, and the outer conductor is braided over the insulator. A final jacket is applied, so that the size of the whole cable is approximately the one of a RG17/U. The outer braid, as mentioned above, is attached to the gun header, which is

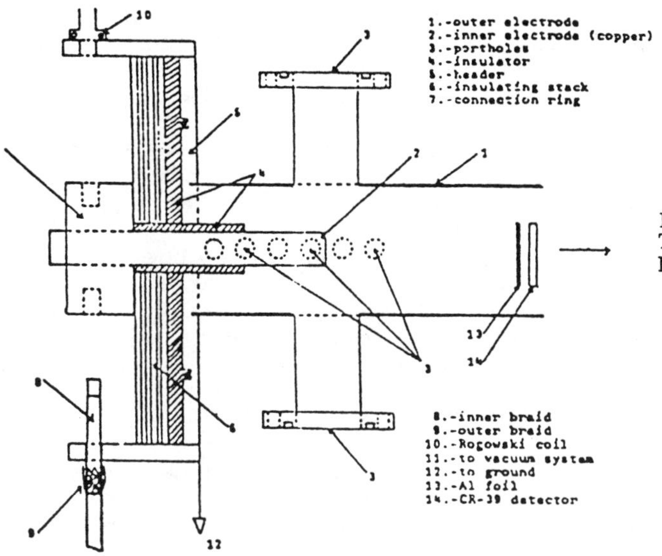

FIGURE 1.
THE FUEGO NUEVO
DENSE PLASMA FOCUS

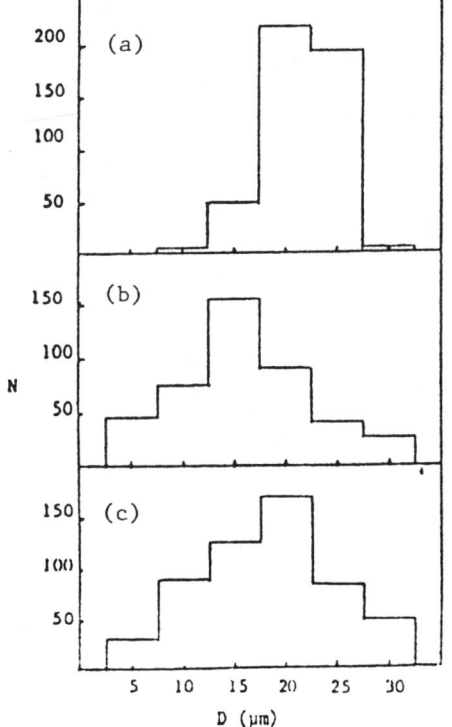

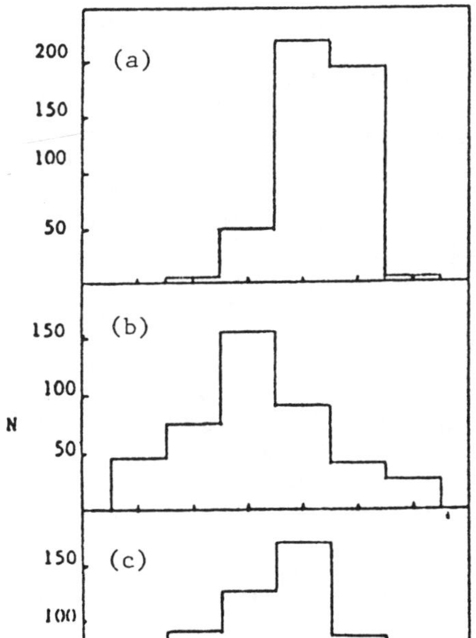

FIGURE 2.
DIAMETER DISTRIBUTION OF ETCH PITS PRODUCED BY ACCELERATED IONS. (a) 1.0 torr of Hydrogen, (b) 1.5 torr of Hydrogen, (c) 2.5 torr of Hydrogen.

grounded, and on the opposite side to the spark gap cans, which are mounted over the capacitors. Since the shield of the trigger system is also attached to the spark gap cans, the whole electric system is grouded this way.

Usual diagnostics of this device are a capacitive voltage divider and Rogowski coils to monitor the voltage an current signals respectively. Other simple diagnostics are magnetic probes along the gun, which trace the advancement of the plasma sheath.

On the axis of the gun, at 80 *cm.* from the tip of the inner electrode, plastic nuclear track detectors are placed, filtered by an aluminum foil, in order to get information about accelerated ions. They will be discussed in the following section.

3. ION DETECTION

The detector used in this work is the polymer carbonate allyl diglycol, known commercially as CR-39. Its composition is $C_{12}H_{18}O_7$, and is manufactured by Pershore Mouldings Ltd., at Pershore, England. It is compatible with high vacuum conditions, insensitive to electromagnetic radiation and electrons, and well suited for track analysis.

The detectors used in this experiment were 40 X 40 mm slabs 500μm thick., and were placed at 80cm. from the tip of the internal electrode, perpendicular to the axis, filtered by a 750μm Aluminum foil.The device was operated at 54kV, with Hidrogen as filling gas, at pressures in the ranging of 1.0, 1.5 and 2.5 torr. After being exposed to 10 shots, the track detectors were etched with KOH in solution at 6.25N, in a constant temperature bath at 70°C, during 16 hours. All the detectors reported in this work were immersed simultaneously, along with an unexposed witness, in order to account for spurious tracks, and then washed and dried for observation. The reason for the choice of concentration for the KOH solution, temperature and time of etching was based on the fact that they prove to be adequate for proton detection, as shown by the reference curves obtained by Khan el al[17].

Measurements were made with an optical Carl Zeiss microscope, using a x100 objective and a x10 eyepiece.The diameters of the etch pits at the surface of the plastic were measured by means of an eyepiece with a diafragm attached to it, and a grating calibrated in μm. The shapes of the etch pits were straight cones. For each slab, used for a different filling pressure, 500 tracks were measured.

4. RESULTS AND DISCUSSION

Figure 2 displays the frequency of occurrence of trace diameters for each filling pressure, the upper one corresponding to 1.0, the middle one to 1.5, and the lower one to 2.5 torr. The energy associated to each diameter can be infered from the reference curves of Khan et al.[22], taking into account the aluminum filter can only allow protons with a threshold energy of 11MeV. Some of the tracks show significantly larger diameters than those expected from protons, so it is clear some of them are produced by impurities. The reference curve, on the other hand shows two branches, which yield two posible energies for a given diameter. It is found by means of a filtered surface barrier detector that some of the tracks can indeed belong to the higher energy branch. Although this is not an appropriate detection method, since the recovery time of the detector is much larger than the time at which the ion beam is produced, after a few shots enough counts can be recorded in order to estimate the range of energies of the particles.

As a result, an energy vs. diameter curve, as that shown in figure 3 can be drawn, assuming that the tracks of size compatible with the reference curve are produced by protons.

5. CONCLUSIONS

This results of this work can only be considered as preliminary and inconclusive, since there is no possibility of determining if the tracks are produced by protons, or impurities. In order to solve this problem, a Thomson parabola has been designed, and is currently being tested in a modified device. However, the very fact of recording a large amount of tracks when a 750μm Aluminum filter is used, is a promising result, since the energies of the ions needed to produce them is far larger than expected for the operation parameters of our device.

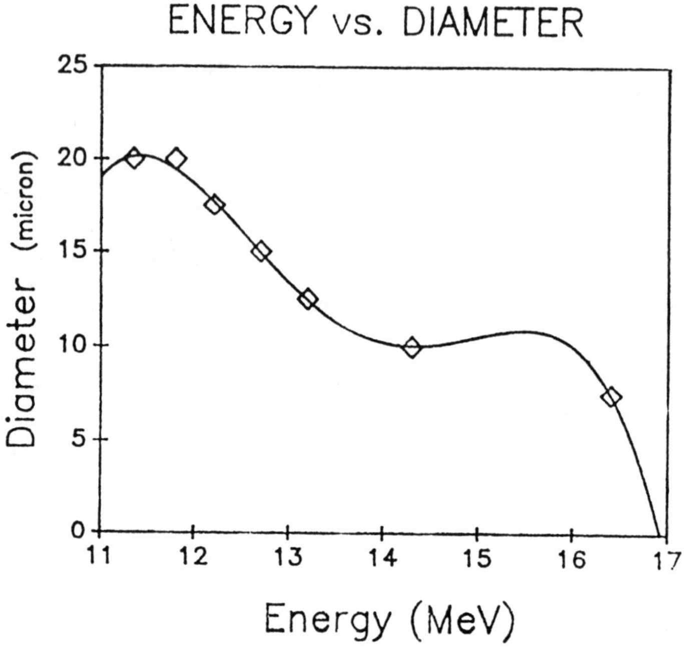

Figure 3. Variation of etch pit diameter as a function of proton enrgy. The threshold of 11 MeV is established by the Aluminum filter.

ACKNOWLEGEMENTS

This work was partially funded by the Programa Universitario de Energía, UNAM and the IAEA research contract 3603/R2/RB.

REFERENCES

1. P.C.Thoneman et al.,Nature 181, 217 (1958).
2. D.C.Hagerman and J.W.Mather, Nature 181, 226 (1958).
3. L.C.Burkhardt and R.H.Lovberg, Nature 181, 228 (1958).
4. J.Honsaker, H.Karr, J.Osher, J.A.Phillips and J.L.Tuck, Nature 181, 231 (1958).
5. B.Rose, A.E.Taylor and E.Wood, Nature 181, 1630 (1958).
6. J.W.Mather, Methods in Experimental Physics (Academic Press, N.Y.,1971)Vol.9B, p.187.
7. N.V.Filippov, Sov.J.Plasma Phys. 9, 14 (1983).
8. J.N.Feugeas, E.C. Llonch, C.O.de González and G.Galambos, J.Appl.Phys. 64, 2648 (1988).
9. M.J.Rhee, R.F.Schneider and D.J.Weidman, Rev.Sci.Instrum. 58, 240 (1987).
10. M.J.Rhee and D.J.Weidman, Phys.Fluids 31, 703 (1988).
11. M. Bernstein, Phys.Fluids 13, 2858 (1970).
12. S.P.Garry and F.Hohl, Phys. Fluids 16, 997 (1973).
13. Y.Kondoh and K.Hirano, Phys.Fluids 21, 1617 (1978).
14. M.G.Haines, Nucl.Instrum.Meth. 207, 179 (1983).
15. R.Deutsch and W.Kies, Plasma Phys. and Controlled Fusion 30, 263 (1988).
16. F.Castillo and J.J.E.Herrera, Small Plasma Physics Experiments (World Scietific,Singapore, 1988) p.3.
17. H.A.Khan et al. Nucl.Tracks 7, 129 (1983).

CRITERIA FOR MAXIMIZING THE SINGLE-LINE EMISSION OF THE PINCH PLASMA IN PLASMA FOCUS DEVICES

R. Lebert[1], K. Bergmann, D. Rothweiler[1], W. Neff[2]

[1]RWTH Aachen, Lehrstuhl für Lasertechnik
[2]FHG-Institut für Lasertechnik
Steinbachstraße 15, D-5100 Aachen, Germany

ABSTRACT

The number of photons emitted into one emission line (i.e. NVII 1s-2p at 2.48 nm) per unit area and per unit solid angle per pulse (TISB) is the relevant figure of merit for pulsed x-ray sources used in combination with x-ray optics for x-ray microscopy. Applying a pseudo-two-dimensional plasma bag simulation allows for microscopic understanding of the conditions responsible for x-ray emission and show that maxima of the TISB are achieved at constant kinetic energy per ion with the corresponding values nearly independent of the pinch current. Experiments show that the TISB is scaling proportional to I^2 under the condition, that the kinetic energy per particle is constant. These two results are discussed in terms of the published I^4 to I^2 transition of the scaling of total x-ray yield.

INTRODUCTION

An x-ray microscope with an pulsed plasma x-ray source is developed in a cooperation for imaging of living biological material providing sub optical resolution[1-7]. According to the optical system and the application, several physical and technical restraints for the x-ray source have to be considered:

- Short pulses ($\tau < 10$ ns) pulses to avoid blurring.
- Emission wavelength range in the "water window" (2.3 nm-4.38 nm) to use natural contrast mechanism.
- Narrowband ($\lambda/\Delta\lambda < 200$) line radiation with low continuum and other lines in its vicinity for high zone-plate resolution and low noise level.
- The total device has to be small enough for biological laboratory.

1s-2p lines of hydrogen-like ions are suited as they have few neighbors and, if helium-like recombination radiation is avoided, low background level. For the spectral range of interest the 1s-2p line of Nitrogen VII and of Carbon VI are suited. The nitrogen line at $\lambda = 2.48$ nm is chosen as it promises higher transmission through water. Temperatures below 1 keV and electron densities below 10^{21} cm^{-3} are favorable as this avoids too high Doppler and Stark broadening.

The source is generated in a Plasma Focus device as these plasmas come close to point like sources and reach high enough energy density. For high resolution single pulse exposure of a wet sample the number of photons emitted per pulse into one line per unit source area per unit solid angle (in axial direction) has to be maximized. This number is equivalent to the spectral brightness integrated over the line profile and the duration of the emission (t_{plasma}) (TISB) .

$$TISB = \int\int L_{\lambda}(\tau) \cdot d\lambda \cdot d\tau \approx L_{\lambda} \cdot \Delta\lambda \cdot \tau_{plasma} \quad (1)$$

For dynamic pinches the total x-ray yield Y emitted into K-shell radiation is published to scale roughly according to I^4 with a transition to I^2 with current (I) [8-16]. Scaling of the TISB is not yet published and thus subjects of this paper.

THE PINCH PLASMA

The Plasma focus is a kind of a dynamic z-pinch. In a first step energy from a driver is stored in the magnetic field inside a coaxial electrode system. These electrodes - usually called plasma accelerator - allow for matching the current rise to plasma dynamics in a manner to time the start of the radial collapse phase to the peak of the current pulse.

With respect to x-ray generation all dynamic pinches can be treated in a similar manner. During collapse predominately kinetic energy is stored by particle collection and acceleration due to jxB forces. Pinches in gaseous background pointing at K-shell emission are usually operated in a manner that during the collapse movement in low-Z gases the He-like ions dominate (but most of them are in ground state). The kinetic energy is thermalized when the collapsing plasma collides with itself on the axis (pinch). The plasma is rapidly ionized and radiates.

As the time constants of relaxation processes in the plasma are larger by up to one order of magnitude than the time scale on which the plasma changes, the plasma is in an essentially non equilibrium state. This is especially true for the ion species distribution. Stationary estimates of what might be the proper set of plasma parameters are often misleading. Plasma parameter change within one order of magnitude during the short (<10 ns) pinch duration. Even more complicate: the rates of the most important relaxation processes are at least proportional to density.

LIMITS OF THE TISB

The upper limit of the TISB is reached, if a plasma having high abundance of the ions of interest radiates optical thick in the lien of interest during all of the existence of the pinch.

The duration of radiation emission is assumed to be equal to the inertial confinement time of the plasma with equal electron and ion temperature. The maximum brightness is given by Planck's law with the tendency that higher temperatures gives higher brightness.

Using a plasma, the abundance of the ions of interest has to be high. For a stationary plasma the hydrogen-like ion (NVII) abundance is > 30% for the temperature range of 100 eV $< kT_e <$ 180 eV.

For a dynamic plasma higher abundance of H-like ion species at higher temperature (and thus higher brightness of a Planck's emitter) is achieved, if during the life time of the pinch ionization from N VI to NVII ions can occur whilst ionization from NVII to NVIII is avoided. Combining this argument with the lifetime, gives a range from $3*10^{17}$ to $2*10^{18}$ cm^{-3} for the ion density where the hydrogen-like ion abundance is still high for 250 eV. Thus the upper limit of the TISB for an ideal situation is at about 300 mJ/μm^2/(4 π sr)

Having a Planck' s emitter in a line means that the optical thickness in the center of the line is high ($\tau_c \gg 1$). Calculating of the optical thickness according to [18,19,20] and assuming τ_c=5 gives an upper limit of the number of emitting ions per unit area ($N_f * l_{plasma}$: density of ions in lower level, l_{Plasma} :length of plasma)

$$N_f \cdot l_{plasma} = 4.7 * 10^{15} \cdot cm^{-2} \cdot \left(kT_i / eV\right)^{1/2} \tag{2}$$

for the 1s-2p line of interest. With the range of ion densities the best suited aspect ratio of the plasma is 3 to 8. (For disk-like plasmas like laser produced plasmas short pulses, low temperature and high density are better suited).

THEORETICAL ANALYSIS AND SIMULATION OF COLLAPSE

The collapse phase is described in snow plow model with the equation of motion for the radial collapse (r: radius of sheath, a: starting radius) and the circuit equation for the current I

$$-\rho_0 V_r \ddot{r} = \frac{\mu_0 I^2}{2\pi r}(a-r) + \rho_0 \dot{V}_r \dot{r} \tag{3}$$

$$-L\ddot{I} = 2\dot{L}\dot{I} + \left[\frac{1}{C_0} + \ddot{L}\right] I \tag{4}$$

Vr is the volume passed by the radially collapsing sheet ($V_r \propto$ particles collected in Z-Pinch like geometry in neutral gas density ρ_0; C_0 : capacitance of the bank and L the inductance of driver and moving plasma sheet. Numerical analysis shows, that a typical time scale t_k^2 is the dominating similarity parameter.

As the kinetic energy per ion (ε_{kin}) should exceed the energy necessary for ionization from helium to hydrogen-like ions (ε_i(He)), to excite to upper radiation level (ε_{ex}) and to heat the plasma to a temperature (kT_e) high enough to achieve high abundance of the ions of interest. For nitrogen VII 1s-2p emission, e_{min} is about $3.35*Z^2*Ry \approx 2$ keV corresponding to a final collapse velocity $v \approx 2.5*10^7$ cm/s (equal to measured values [17]).

The snowplow calculations show, that the ratio between kinetic energy and ε_{min} depend on the product of the current at the beginning of radial collapse (I_0), on the current at pinch time (I_P), on the number of collected particles ($\propto n_0 a^2$) and on a function of the final radius $\alpha(R_e)=\varepsilon_{kin}/\varepsilon_{min}$. If $I_0 \propto I_p$ for constant velocity, the neutral gas density n_0 is proportional to the neutral gas pressure p and R_e is nearly constant, ß only depends on device parameters:

$$\beta = \frac{\epsilon_{kin}}{\epsilon_{min}} \cdot \frac{1}{\cdot \alpha(R_e)} = 1.3 \cdot 10^6 \cdot \frac{I_0 \cdot I_P}{n_0 \cdot a^2} \cdot \frac{1}{A^2 cm} \approx 6 \cdot 10^{-8} \cdot \frac{I_0^2}{p \cdot a^2} \cdot \frac{mbar*cm^2}{A^2} \tag{5}$$

How the kinetic energy is thermalized and how the radiation is generated must be determined from a detailed analysis of the dynamic processes. The processes during collapse and pinch phase are simulated by a pseudo-two-dimensional plasma-bag model cylindrical geometry and the electrical circuit[1,22-24]. The current is assumed to flow on the outer boundary of the sheet and drives the supersonic movement of the plasma sheet bounded by the shockfront. The plasma sheet is described by a bag of plasma with homogeneous ion and electron densities and temperature and linear decrease of the velocity between shock front and magnetic piston. The shock dynamics leads to collection of neutral atoms by the plasma layer and shock heating of the plasma[25,26,1,17].

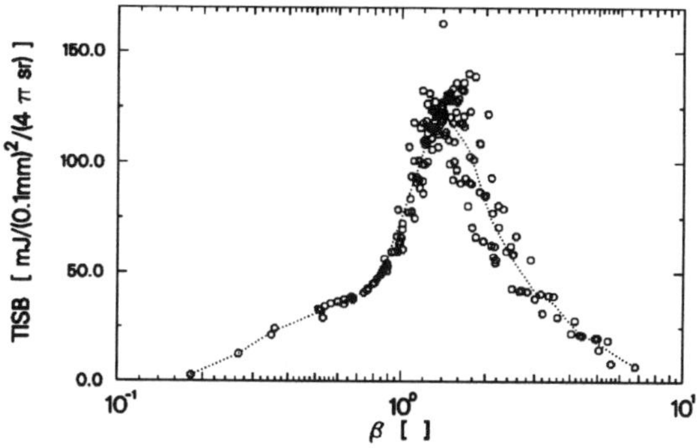

Fig. 1. Simulated TISB as a function of the parameter β from equation 5. The TISB reaches it maximum close to β = 1.

The internal properties of the plasma are calculated by accounting for the thermal, potential and kinetic energy. Thermalization in the pinch is due to ion-ion collisions. The ionization of K-shell ions is calculated time dependent, for the L- and M shell ions in equilibrium with electron temperature. Resonance lines and continuum emission is separately calculated taking radiation transport into account (bulk optical thickness[27,28]). Population of radiative levels is assumed to be in equilibrium with the respective ion ground state distribution [29]. Ohmic heating and losses due to thermal conductivity and particle losses are accounted for. Even using this simulation finding optima of the TISB as a function of all the accessible device parameters is quite complex. The complexity is significantly reduced if the result are plotted as a function of a parameter ß. As shown in figure 1 the maxima of the TISB peak at ß= 1 for a large variety of device parameters.

In figure 2 the maximum values of ion and electron temperature are plotted as a function of ß. At ß = 1 the maxima of the electron and ion temperatures are nearly equal. Additionally, the optical thickness at the very moment of maximum emission is close to one at ß=1[1]. These results allow to summarizing the picture obtained for the influence of ß on the TISB:

- ß = 1: electron and ion temperature are nearly in equilibrium at the instance when H-like ions dominate. Time constants of relaxation are adopted to plasma dynamics. At the instance of maximum H-like abundance and high excitation the optical thickness is close to 1. A large amount of the kinetic energy is radiated away, thus reexpansion is slow and duration of emission is larger.

- ß < 1: the plasma has low electron temperature and thus the optical thickness is high.
- ß > 1 the ions become very hot (up to 10 keV) whereas the electrons nearly stay at kT_e = 500 - 700 eV. In this case the pinch dynamics is too fast for thermalization.

Fig. 2. Calculated maxima of the ion and electron temperatures and of the mean temperature as a function of β The temperatures group according to β. At β = 1 equilibrium-like conditions for electron and ion temperature maxima are observed. For β >> 1 maximum ion temperature exceeds 10 keV.

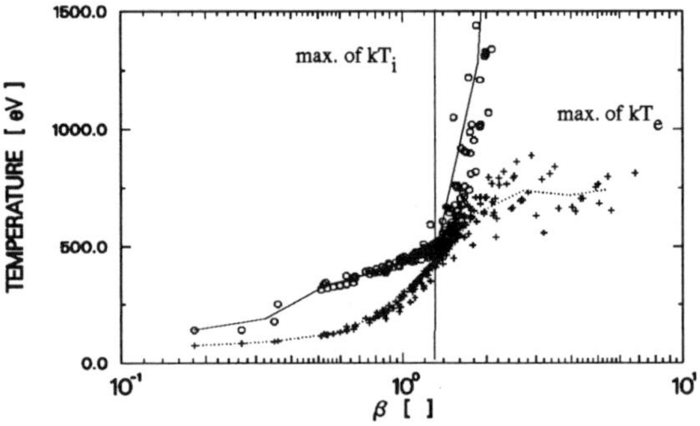

COMPARISON WITH EXPERIMENTS

To compare the theoretical predictions with experiments the current through the pinch was measured with a calibrated inductive probe located close to the end of the anode. Details about this experiments are discussed elsewhere in this volume[7]. The current measured outside the electrode system reaches 320 kA at maximum and 290 kA at pinch time. The pinch current measured inside the electrode system is 220 kA The 70 kA current losses (discussed in [7] in more detail) reveal that measuring of the total device current is quantitatively misleading and conceals the knowledge of scaling with current. The TISB is measured with a pinhole grating. The respective maxima are found at 0.9 < ß < 1.2 for 1, 2 and 3 mbar gas pressure, which is in fairly good agreement with prediction (Fig. 3).

The TISB increases nearly linear with the gas pressure (∝ line density) in contrast to the simulations where $TISB_{max}$ is nearly independent from line density. This discrepancy can be discussed in terms of I scaling found at larger pinch devices. According to[15,16] there are two cases for the Yield of z-pinch array implosions.

Although the boundary conditions are different, the arguments used can be applied for our case, as the main processes (i.e. kinetic energy transformed to thermal energy; radiation transport etc.) are identical. A transition between two regimes for the total K-shell yield is predicted:

$$Y \propto I_{Pinch}^4 \propto m^2 \propto N^2 \quad (6) \qquad\qquad Y \propto I_{Pinch}^2 \propto m \propto N \quad (7)$$

Y: is the total x-ray yield ; m: mass per unit length ;

N: number of particles per unit length.

Fig. 3. Measured TISB as a function of beta shows maxima close to β =1.

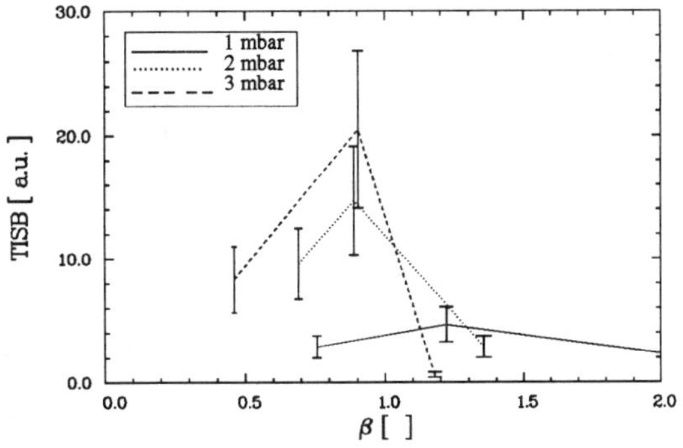

To apply those predictions to the studied case some assumptions are necessary:

1. Line radiation is up to 80 % of the total yield for low mass density[16].
2. As nearly constant length of the radiating pinch is observed for fixed geometry, the mass per length of the pinch is proportional to the number of particles collected and thus proportional to the working gas pressure.

It may be assumed that the proposed increase of the yield Y with current cannot be attributed to higher TISB but rather to the enlargement of the pinch size

The total x-ray yield and the brightness are - if the emission into one line is proportional to the total emission - proportional to one another with the square of the pinch radius as the proportionality factor.

$$Y \propto \int_0^{r_{pinch}} \int_0^{\tau_{em.}} \int B_\lambda \cdot 2 \cdot \pi \cdot r_{pinch} \cdot d\lambda \cdot d\tau \cdot 2 \cdot \pi \cdot r_{pinch} \cdot dr \approx TISB \cdot \pi \cdot r_{pinch}^2 \tag{8}$$

$$TISB \propto \frac{I_{pinch}^x}{r_{pinch}^2} \quad mit \; x \in \{2,4\} \tag{9}$$

There are two reasons to assume, that rise of the yield (especially into one line) is often due to a rise in pinch size rather than due to a rise of the TISB:

- For maximizing the yield of a optically thick emitting pinch the number of emitting ions per unit area ($N_f * l_{plasma}$) is at best kept constant as otherwise the emitting diameter and thus the total yield would decrease.
- To conserve proper matching of all relaxational, collisional and radiative rates (partly $\propto n_i$ partly $\propto n_i^2$) n_i is at best kept constant.

Both reasons have the consequence that $r^2 \propto N_i \propto I_{pinch}^2$. Using this argument leads to two scaling regimes, that can be proposed for the TISB for the two scaling regimes of the TISB respectively:

$$TISB \propto I_{pinch}^2 \quad and \quad TISB = const\,(I_{pinch}) \tag{10}$$

Consequently in order to optimize TISB the aim is reached, if the breakpoint between the two regimes is reached. This regime can be calculated using the breakpoint parameters given in[16] (however, as we are only interested in line emission for smaller mass load.) Comparing of the cited formulas with our experimental data for nitrogen [7], we find, that they match well (for the parameter η used in reference[16] equal to $\alpha(R_e) = 3.2$, corresponding to a final radius of 0.08*anode radius). The breakpoint parameters are a mass density of about 1.7 µg/cm and a pinch current of 180 kA. The mass density is identical to that we reach (corresponding to gas pressure of 4 mbar). The difference to the 220 kA current we measured is negligible having in mind the two absolute different device concepts.

Thus, the discrepancy between theory and experiment can be interpreted, that the experiment is working in the I^2 regime, whereas theoretical situation has already surpassed the breakpoint. One reason for this gap might be, that a cylindrical sheet is simulated whereas it is curved in experiment. The curved sheet leads to a radiation front traveling in z-direction as published elsewhere[30].

CONCLUSION

Experimentally as well as theoretically maximum TISB of the N VII 1s-2p emission is found to occur at an optimized kinetic energy per ion (ß = 1). Experimentally the maxima are proportional to the pinch current I^2, modeling predicts a TISB independent on the pinch current I. We assume, that this discrepancy is due to a physical limit in increasing the TISB which is not yet reached in experiment, but in modeling. Probably due to the curved sheet not been taken into account in the simulation. The postulate, that the TISB reaches a region where it is independent on pinch current is supported by Whitney et all where a breakpoint is suggested where the total K-shell Yield changes from $I_{pinch}{}^4$-scaling to $I_{pinch}{}^2$-scaling.

ACKNOWLEDGEMENTS

Parts of this work has been supported by the Bundesministerium für Forschung and Technologie (BMFT) and by the Deutsche Forschungsgemeinschaft (DFG) under the contract numbers 13N53290, 13N5680 and He 979/17-1 respectively. Cooperations with the Forschungseinrichtung Röntgenphysik at Göttingen are acknowledged.

REFERENCES

1. R. Lebert, D. Rothweiler, W. Neff, submitted to Journal of X-Ray Science and Technology
2. G. Schmahl,.B. Niemann, D. Rudolph,M. Diehl, J. Thieme, W. Neff, R. Holz, R. Lebert, F. Richter, G. Herziger, in [32] , 66 (1992)
3. B. Niemann, D. Rudolph, G. Schmahl, M. Diehl, J. Thieme, W. Neff, R. Holz, R. Lebert, F. Richter, G. Herziger, Optik, **1**, 35 (1989)
4. R. Lebert, R. Holz, D. Rothweiler, F. Richter, W. Neff, in [32] , 62 (1992)
5. R. Lebert, R. Holz, D. Rothweiler, W. Neff, ICPIG XX 3 (1991)
6. M. Diehl, in [32] , 290 (1992)
7. D. Rothweiler, R. Lebert, A. Tusche, W. Neff, (in this volume)
8. N. R. Pereira, J. Davis, J. Appl. Phys. 64 (3), R1-R27 (1988)
9. P. Apruzese, D. Duston and J. Davis, NRL Memorandum Report No. 5406 (1984)
10.J. P. Apruzese, G. Mehlman, J. Davis, J. E. Rogerson, V. E. Scherrer, S. J.

Stephanakis, P. F. Ottinger and F. C. Young, Phys. Rev. A **35 (11)**, 4896 (1987)
11.S. Wong, C. Gilman, P. Sincerny, T. Young, in Proc. of the 1982 IEEE Plasma Science Conference
12. J. P. Apruzese, P.C. Kepple, K. G. Whitney, J. Davis, D. Duston, Phys. Rev. A **24(2)**, 1001 (1980)
13.J. P. Apruzese, J. Davis, D. Duston, K. G. Whitney, J. Quant. Spectrosc. Radiat. Transfer **23**, 479 (1980)
14.B. E. Meierovich, Sov. Phys. Plasma Phys., **11,** 831 (1985)
15.J. W. Thornhill, K: G: Whitney, J. Davis, J. Quant. Spectros. Radiat. Transfer
16.K. G. Whitney, Davis , J. Appl. Phys. 67(4) 1735 (1990)
17.R. Lebert: "*Pinchplasmen als gepulste Röntgenquellen hoher spektrale Strahldichte*, PhD thesis, RWTH Aachen (1990)
18.R. C. Elton, *X-Ray Lasers*, Accademic Press, San Dieago, London 1990
19.H. R. Griem, *Plasma Spectrocopy*,Mc Graw Hil, New York 1964
20.R. C. Elton, in *Methods of Experimental Physics*, Vol. 1, Chapt. 4, H. R. Griem and R. H. Loveberg, (Eds.), Accademic Press, New York, 1970
21.I. H. Hutchinson, *Principles of Plasma Diagnostics*, Cambridge University Press, 1987
22.J. Davis, R. Clark, AIP Conference Proc., Dense Z-Pinches, 2nd Int. Conference, Laguna Beach, California, Plenum Press **195**, 81 (1989)
23.R. W. Clark, J. Davis, F. L. Cochran, Phys. Fluids 29(6), 1971 (1986)
24. J. Bailey, A. Fisher, N. Rostoker, J. Appl. Phys. 60(6), 1939 (1986)
25.Y. B. Zeldovich, Y. P. Raizer, *Physics of Shock Waves and High Temperature Hydrodynamic Phenomena*, Vol. 1 and 2, Academic Press, New York (1966,1967)
26.M.A. Libermann, A. L. Velikovich, *Physics of Shock Waves in Gases and Plasmas*, Springer Series in Electrophysics **19**, Springer Verlag Berlin 1986
27.C. J. Cannon, *The Transfer of Spectral Line Radiation*, Cambridge University Press, Cambridge, MA, USA 1985
28.I. I. Sobelman, "Atomic Spectra and Radiative Transitions", Chemical Physics 1, Springer Verlag Berlin, Heidelberg, New York, 253 1979
29.I. I. Sobelman, L. A. Vainsthein, E. A. Yukov, *Excitation of Atoms and Broadening of Spectral Lines*, (Springer Series in Chemical Physics; 7), Springer-Verlag Berlin 1981
30.R. Lebert, T. Rohe, D. Rothweiler, R. Lebert, in *X-Ray Lasers*, IOP Conference Series **125,** E.E. Fill (ed.), IOP Publishing Bristol, p. 411-414 (1992)
31.D. Sayre, M. Howells, J. Kirz, H. Rarback, G. R. Morrison, C. J. Buckley (eds.) *X-Ray Microscopy II*, Springer Series on optical Sciences, Vol. 56, Springer, Berlin 1988
32.A. G. Michette, G. R. Morrison, C. J. Buckley (eds.) *X-Ray Microscopy III*, Springer Series on optical Sciences Vol. 67, Springer, Berlin, 1992

INVESTIGATIONS ON THE TRANSITION BETWEEN COLUMN AND MICROPINCH MODE OF PLASMA FOCUS OPERATION

R. Lebert[1],A. Engel[1],K. Gäbel[3],D. Rothweiler[1],E. Förster[3],W. Neff[2]

[1]RWTH Aachen, Lehrstuhl für Lasertechnik, [2]FHG-Institut für Lasertechnik Steinbachstraße 15, D-5100 Aachen, Germany; [3] MPQ-AG Röntgenoptik, Friedrich-Schiller-Universität Jena, Max-Wien-Platz 1, D-O-6900 Jena, Germany

ABSTRACT

In Plasma Focus devices operated with pure nitrogen, neon or argon ($Z \leq 18$) with pinch current from 200 to 400 kA the emitting region is a column with several 100 μm in diameter and several mm in length (column mode). For $Z \geq$ 18 micropinches are observed. Usually micropinches and column emission is examined independently. Small Plasma Foci feature both modes simultaneously. In this paper the transition regime between column and micropinches is investigated with particular regard to Z=18 (argon). Peculiarities in the emitted spectra are interpreted by the interaction of simultaneous existing column and micropinch plasma.

INTRODUCTION

The pinch effect - first investigated by Bennett[1] - generates hot plasmas. These plasmas are strong radiation sources in the x-ray region, if the discharges are operated with heavy gases. Such sources find diverse applications in areas of x-ray lithography, x-ray microscopy, the pumping of x-ray lasers, materials studies and fundamental spectroscopy. These x-ray sources are known as Z-Pinch, vacuum sparks, and Plasma Focus, all having similar x-ray characteristics.

With concern to applications bulk x-ray production is emphasized where more than 10% of the plasma energy is radiated in soft x-rays[2] so that x-ray bursts of up to 100 kJ per pulse are emitted. Typical data of these plasmas are diameter 100 mm to 0.1 mm, density from 10^{18} to 10^{21} cm^{-3} and temperature from 30 eV to 2 keV . In Z-pinch like geometry these sources are column-like. When the pinch remains cylindrical, the characteristics of the plasma and the emitted spectra are well described by radiation-hydrodynamic models[3].

The phenomena of micropinches - also called "hot spots" - are now subject of intense investigations[4]. Typical values of micropinches are diameter 300 - 1 μm, density higher than 10^{20} - 10^{23} cm^{-3} and temperature up to 500 eV - 10 keV .Many aspects of these plasmas are well interpreted by the radiative collapse model[5-9].

Usually micropinches and column emission is examined independently. In this paper we describe investigations on the transition regime between the two phenomena with particular regard to Z=18. Hereby we use the fact, that Plasma Focus[10,11] devices operated with pure gases feature both modes of operation simultaneously

DEPENDENCE ON GAS PRESSURE

At our institute small Plasma Focus devices operated with pure nitrogen and neon and pinch current of 200 to 400 kA are developed as sources for x-ray microscopy and lithography. Table 1 shows typical results obtained at the same device operated with different working gases. For light elements ($Z \leq 18$) the volume that emits K-shell radiation is a column with several 100 µm in diameter and several mm in length (column mode). Operating with heavier elements ($Z \geq 18$) micropinches are observed (micropinch mode). Argon features both modes and thus is of special interest for investigations[12].

Tabelle 1 : Typical values of the length, the diameter, and of wavelengths of the most prominent soft-x-ray lines for different gases obtained at the same device. Micropinch-mode is observed for Z ≥ 18

Element	Shell	Wavelength [nm]	Length [µm]	Radius [µm]	Aspect Ratio
Nitrogen	K	2.5	5000-10000	600	20-50
Neon	K	1.2	5000-10000	200	20-50
Argon	K	0.4	5000-10000	150	20-50
Argon	K	0.4	60-100	20-30	3-5
Krypton	L,K	0.7	50-80	10-20	3-5
Xenon	M,L	0.9	50-80	10-20	3-5

Prerequisites for column-mode is:

- the collapsing plasma has nearly the temperature, which is necessary to ionize the next completed shell.
- kinetic energy per ion is high enough to achieve complete ionization
- relaxation-rates between electrons and ions are high enough to ionize the pinch-plasma during the compression phase of the pinch.

Conditions for micropinch-mode are:

- kinetic energy per ion is not high enough to get complete ionization
- further ionization is only reached after further magnetic compression
- or two shells can be ionized during pinch duration. One by means of transfer of kinetic energy, the other by magnetic compression in the micropinch.

COMPARISON OF MICROPINCH AND COLUMN

Figure 1 shows streaks of < 1 nm emission from argon pinches taken at the same device operated at two different gas pressures. An imaging slit perpendicular to the entrance slit of the streak camera[13] allows for spatial resolved measurements of the axial or the radial extension of the radiating volume integrated over the other extension. For a working gas pressure of 1 mbar the plasma emits for up to ten ns. Emission starts at one z0,r0 position and extends in radial direction to a maximum of about 500 μm and to a length of about 5 mm (half of the pinchlength is obscured by the anode).

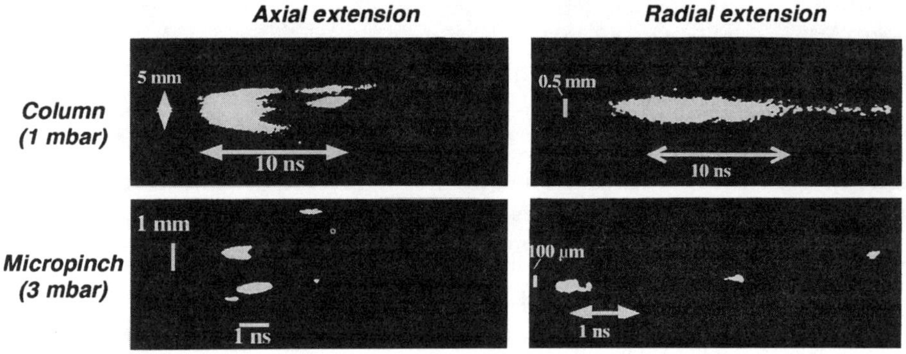

Fig. 1 : Streak images of radial and axial extension of argon emission. At p = 1 mbar working gas gas pressure column mode is observed, at p =3 mbar micropinches. The differences in temporal and spatial properties are obvious.

The situation looks different at 3 mbar. The plasma only emits in pulses of about 1 ns duration. Subsequent pulses occur at different z-positions. Individual small spots emit for up to one ns from a region that is less than 500 μm long and less than 100 μm in diameter [= resolution limit].

DECISION PARAMETER

The occurrence of micropinches strongly depend on the working gas, the gas pressure and the pinch current. Plasma Focus devices with pinch currents of 200 - 400 kA operated with heavy gases feature both modes of operation. To decide, whether micropinch or column mode can be expected three "currents" play a significant role:

- Pinch current I_{Pinch}:
  Current at the moment of the development of the bulk plasma
- Shell-current I_{shell} :
  Current estimated from Bennett-equilibrium[1] at a temperature necessary to ionize a noble gas shell. I_{shell} depends on line density and ionization potential. For $I_{Pinch} > I_{Shell}$ the shell is ionized.
- Critical current I_{Cr} :

Generalized to Pease-Braginskii Current[14] for not totally ionized Plasma
For $I_{pinch} > I_{Cr}$ radiative collapse is possible.

Experimentally it is found, that the occurrence of micropinches or column in the various gases is determined by a parameter ε

$$\epsilon = \frac{I_{cr}^2}{I_{pinch} \cdot I_{shell}}$$

with *the property that for* $\varepsilon < 1$ *micropinch-mode and for* $\varepsilon > 1$ *column-mode is observed*

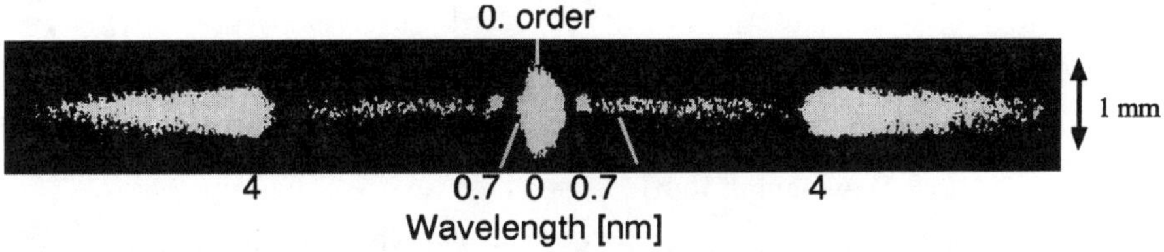

Fig. 2 : Spatial resolved spectrum of a krypton-plasma taken from axial direction with a pinhole-grating spectrograph. Two different types of emission are observed: M-shell radiation λ>4 nm is emitted from a region with about 500 μm in diameter in column-mode, L-emission λ ≈ 0.7 nm from small spots with diameter <100 μm in micropinch-mode. This behaviour is predicted by the decision parameter.

Figure 2 shows that column and micropinch mode occur simultaneously in two different modes for two different shells. M-shell radiation λ>4 nm is emitted from a region with about 500 μm in diameter pointing to column-mode, L-emission λ≈0.7 nm from small spots (diameter < 100 μm) in micropinch-mode. This proves, that micropinches develop from a higher shell bulk plasma. Streaks of this spectra show, that that the M-shell lasts for more than 5 ns, whereas the L-shell "hot-spots" emit in pulses of about 250 ps.

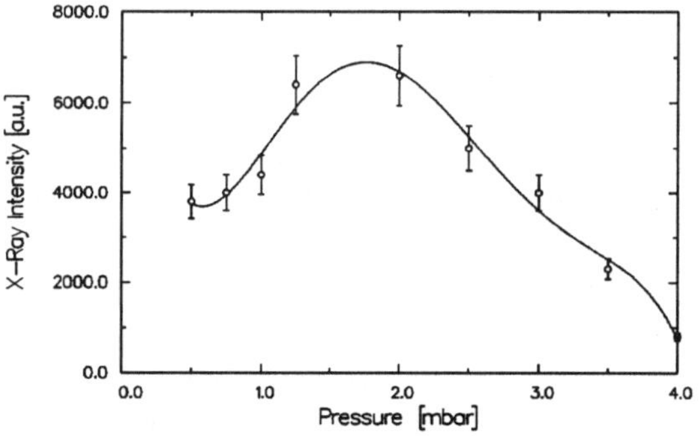

Fig. 3 : The argon x-ray yield has a peaks in the transition regime from column-mode to micropinch-mode

INVESTIGATION OF THE TRANSITION REGIME FOR ARGON

For argon both modes can be selected by the device parameters i.e. gas pressure. The transition regime between column-mode and micropinch-mode (ε≈1) has been studied. For $p < 1$ mbar nearly no micropinches occur, for 1 mbar $< p <$ 1.5 mbar there is a transition regime and for $p > 1.5$ mbar only micropinch is found. It showed, that the total emission of the pinch event peaks at the onset of micropinches figure 3.

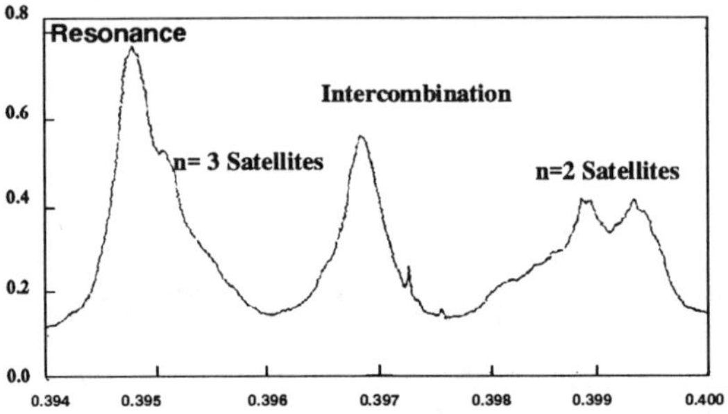

Fig. 4 : Emission Spectrum of the argon He-like line emission in the transition regime taken from radial direction

Figures 4 and 5 show two spectra of the He-like Ar XVII emission for gas pressures in the transition regime taken from radial and axial direction respectively. Analysis[15,16,17] of spectrum from radial direction gives T_e = 850-900 eV and $n_e = 3*10^{21}$ cm^{-3}. They are similar except to a peculiarity on the long wavelength wing of the resonance line.

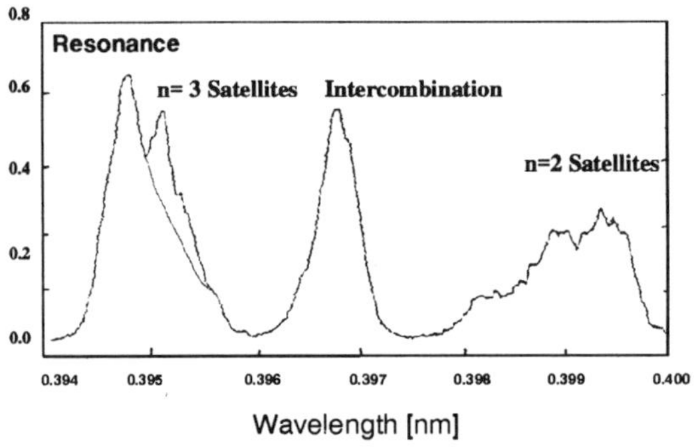

Fig. 5 : Emission spectrum of the argon He-like emission in the transition regime taken from axial direction. The spectrum shows a peculiar peak on the long wavelength wing of the resonance line compared to the spectrum from radial direction.

This peculiarity is studied in more detail. It occurs at the same wavelength as the n=3 satellites (transition in double excited lithium-like Argon XVI). Whilst the small peak observed into radial direction at this location can be explained by thermal excitation of these n=3 satellites (amplitude ratios between n=3 satellite and He-like resonance line = 0.15 for kT_e=1000 eV and $n_e=10^{23}$ cm^{-3}) the high amplitude of the ratio in axial is not explainable.

Additionally a fit of this lines with Lorentz profiles requires two different widths of the resonance line for the short wavelength wing (= 6 mA) and the long wavelength wing (= 3 mA).

This peak is also observed in simultaneously time ($\lambda/\Delta\lambda \geq 1000$) and spectral ($\Delta\tau \leq 100$ ps) resolved measurements from axial direction. As result is shown in figure 6 the position of the center of the resonance line is nearly constant

(Δλ<2 mA). The n=3 emission occurs at the same moment when high intensity of resonance line emission is observed and partly exceeds resonance line emission.

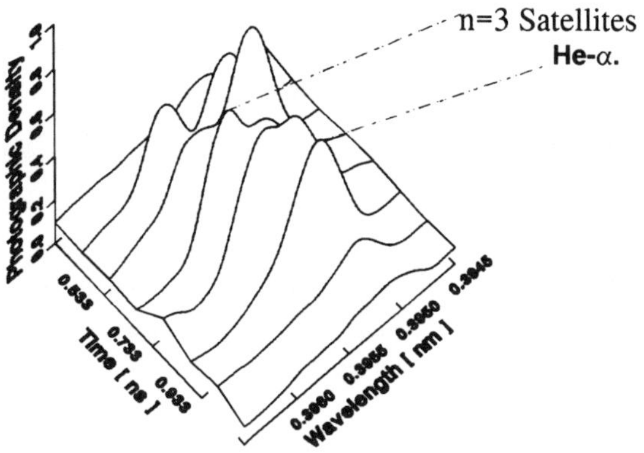

Fig. 6 : Temporal and spectral resolved resonance line emission. The position of the center of the resonance line is nearly constant. n=3 emission occurs at the same moment when high intensity of resonance line emission is observed. n=3 emission partly exceeds resonance line emission.

The following approaches are suggested to explain this additional peak in the spectrum:

- 10 mA Doppler broadening due to high temperature or high velocities. In the vicinity of the emitter plasma the radiation is absorbed by cool Ar XVII ions in the line center of the line is reabsorbed
- Differential movement of an optical thick, fast collapsing or expanding micropinch shows similar features[18,19,20].
- Optical thick emission of both resonance and n=3 satellite with the same intensity.
- Photo pumping of the n=3 transitions[21]

The first two approaches do not explain the anisotropy and the temporal behaviour of the emission (no line shift and no increase of the width of the resonance line is observed) The third approach is in contradiction with the fact, that the n=3 emission exceeds the resonance line.

In order to check the fourth approach, it is assumed, that the emission of a micropinch in axial direction has to pass a cooler bulk plasma, whilst it is observed from radial direction without disturbances. Both plasma regions are in close contact (Δx < 100 μm).

In the bulk plasma, the resonance line is absorbed from helium-like ground state and from exited lithium like ions in the $1s^2 3x$ levels. At temperatures below 200 eV lithium-like ions dominate. The $1s^2 3x$ levels are populated to a 10^{-5} fraction. The 6 mA[22,23] difference in wavelength is surpassed due to the width of the pumping resonance line. Absorption leads to a 1s2p3x-state of the ion (the x=p sublevel has negligible autoionization probability). Transition back to $1s^2 3x$ leads to the n=3 satellite photons. As the linewidth of this emission is smaller than 6 mA, these photons escape the plasma. Calculations (using[22,23,24]) on that scheme gives a result as shown in figure 7, which show, that the pumping is a possible explanation of the observed spectra.

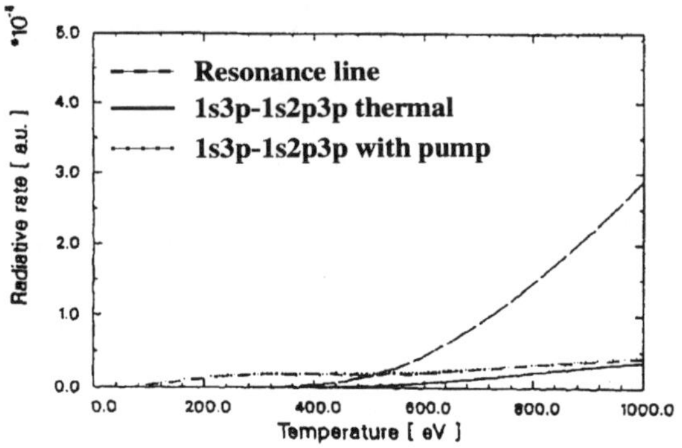

Fig. 7 : Resonance line and n=3 satellite intensity as a function of electron temperature. Whilst the resonance line domitates the emission in a thermal plasma, the $1s^2 3p$-$1s2p3p$ line exceeds the resonance line when pumped.

CONCLUSIONS

Depending on pinch current, line density, and atomic number of a pinch plasma the x-rays emission is observed from cylindrical bulk plasma (column) or from micropinches. A parameter ε is proposed which decides the transition between both modes. At our devices argon features both modes only depending on line density, with a peak of the yield at the transition regime. Peculiarities in the argon emission spectra could be due to the simultaneous occurrence of column and micropinch. A pumping process is a possible description for this phenomena. Problems as absolute wavelength and detailed calculations on reabsorption of n=3 satellites in the cool plasma is planed as future work.

ACKNOWLEDGMENT

The discussions and cooperations with Prof.. Kunze from the Bochum university, Prof. Koshelev from Troitzk university are gratefully acknowledged.

REFERENCES

1. W. H. Bennet, Phys. Rev., **45**, 890 (1934)
2. N. R. Pereira, J. Davis, J. Appl. Phys. 64 (3), R1-R27 (1988)
3. R.W. Clark, J. Davis, F.L. Cochran, Phys. Fluids 29(6), 1971 (1986)
4. K. N. Koshelev, N. R. Pereira, J. Appl. Phys., **69**, (1991) R21
5. K.N. Koshelev, Y.U. Sidelnikov et al., Sov. J. Plasma Phys. 15 (9), (1989)
6. V.V. Vikhrev, V.V. Ivanov and K.N. Koshelev, Sov.J.Plasma Phys. 8(6), (1982)
7. G. Decker, W. Kies, M. Mälzig, P. Röwekamp, F. Müller, J. Westheide, Y.U. Sidelnikov, J.Appl. Phys. 69, (1991)
8. K. N. Koshelev, V.I. Krautz, N.G. Reshetniak, R.G. Salukvadze, Y.V. Sidelni-

kov, E.Y. Khatiev, J. Phys. D.: Appl. Phys. **21** (1988) 1827-1829
9. V. V. Vikhrev, V. V. Ivanov, K. N. Koshelev, Sov. J. Plasma Phys. **8** (1982) 688 J.W. Mather, Vol. 9B, Academic Press w York p. 187 , (1971)
10. J.W. Mather in "Methods of experimental Physics, Plasma Physics", Vol. 9B, H. R. Griem and R. H. Loveberg, (Eds.), p. 187, Accademic Press, New York, 1971
11. N.V. Filippov, T.I. Filipova, V.P. Vinogradov, Nucl. Fusion suppl., part 2 577 , (1962)
12. R. Lebert, Pinchplasmen als gepulste Röntgenquellen hoher spektraler Strahldichte, PhD Thesis TH Aachen (1990)
13. G. D. Tsakiris, R. Sigel, Phys. Rev. A **38**, 5769 (1988)G.D. Tsakiris, SPIE Vol. 1358 (1988)
14. S.I. Braginskii, Plasma Physics and the problem of uncontrolled Thermonuclear Fusion, Vol. 1, edited by M.A. Leontovich, Pergamon Press, 1961, p. 135 (1990)
15. A. V. Rodé, A. Maksmchuuk, G. V. Sklizkov, A. Ridgeley, C. Danson, N. Rizvi, R. Bann, E. Forster, I. Uschmann, J. X-ray Sci. and Techn. **2**, 149-159
16. A. Schulz, R. Burhenn, F.B. Rosmej and H.J. Kunze, J. Phys. D: Appl. Phys. **22** (1989) 659-662
17. E.Ya. Kononov, K.N. Koshelev, U.I. Safranova, Yu.V. Sidelnikov and S.S. Churilov, JEPT Lett., Vol. 31, No. 12, (1980)
18. A.Schulz, Der Mikropinch im Vakuumfunken als Quelle harter Röntgenstrahlung, ThD Thesis, RU Bochum, (1992)
19. M. Hebach, A. Engel, A. Schulz, R. Lebert, H.-J. Kunze, Europhys. Lett, **21**(3), 311-316 (1993)
20. K.N. Koshelev, Y.U. Sidelnikov, G. Decker, W. Kies, M.Mälzig, P.Röwekamp, F.B.Rosmij, A.Schulz, H.-J. Kunze, J. Phys. D , (1991)
21. P. Monier, C. Chenais-Popovics, J. P. Geindre and J.C. Gauthier, Phys. Rev. A 30(5), 2508 (1988)
22. L.A. Vainshtein, U.I. Safranova : Atomic Data nd Nuclear Data Tables **25**, 311-385 (1980)
23. L. A. Vainstain, Atomic data and nuclear data tables 21, 49-68 (1978)
24. M.H.Chen : Dieletronic satellite spectra for He-like ions, Atomic Data and Nuclear Data Tables **34**, 301-356 (1986)

SPECTRAL INVESTIGATIONS OF MICROPINCHES IN THE SPEED 2 PLASMA FOCUS

P. Röwekamp, G. Decker, W. Kies, F. Schmitz, G. Ziethen
Inst. f. Experimentalphysik, Heinrich-Heine-Universität, 4000 Düsseldorf 1, FRG

J.M. Bayley
Imperial College of Science, Technology and Medicine, SW7 2AZ London, UK

K.N. Koshelev, Yu.V. Sidelnikov
Inst. f. Spectroscopy, Russian Academy of Sciences, 142092 Troitzk, Russia

F.B. Rosmej, A. Schulz[a]
Inst. f. Experimentalphysik V, Ruhr-Universität, 4630 Bochum, FRG

D.M. Simanovskii
A.-F.-Ioffe-Physico-Technical-Institute, 194021 St. Petersburg, Russia

ABSTRACT

Soft X-radiation mainly emitted by 'micropinches' is generated in the driver SPEED 2 by injection of heavy rare gases (e.g. argon) into a discharge of pure deuterium. It has been proved by a fast diagnostic in the VUV range based on a microchannelplate, that the short-living micropinches with sub-mm-size are developing where neckings in the plasma column can lead to local radiative collapses as predicted by the collapse model.

Spectra of the whole pinch column in the wavelength range close to the ArXVII-resonance line (λ = 0.3948 nm) were used to improve temperature and density determination. The results of computer simulated fits using a Monte-Carlo-method for the photon transport including effects of optical density for evaluation of plasma parameters (electron density 10^{28} m^{-3} ... 10^{29} m^{-3}, temperature 1 keV ... 1.5 keV) agree well with estimates based on the corresponding model. Spatially resolved spectra reveal that about 90 % of the line radiation stems from micropinches.

The energy output in small channels of the water window is up to 45 Joules per discharge assuming an isotropic emission.

INTRODUCTION

Due to the rising interest in radiation sources in the vacuum ultraviolet (VUV, 100 nm ... 1 nm) and soft x-ray (SXR, 1 nm ... 0.05 nm) range, micropinches (hot

---

[a] present address: Auburn-University, Auburn, Alabama, United States

spots, plasma points) are subjects of intensive studies. At the high performance driver SPEED 2, used as a plasma focus with a relatively short accelerator[1], the experiments were carried out in a slightly different parameter range than mostly reported. The driver layout characteristics are summarized in table I. Note the comparatively fast current rise time of $\tau_0/4$ = 400 ns and the high pinch current in the megampere range.

Table I: SPEED 2 driver characteristics

bank energy	187 kJ	current rise rate	2×10^{13} A/s
voltage	300 kV	current rise time	400 ns
capacitance	4.16 μF	pinch current	2 MA
inductance	15 nH	short circuit current	5 MA
impedance	60 mΩ		

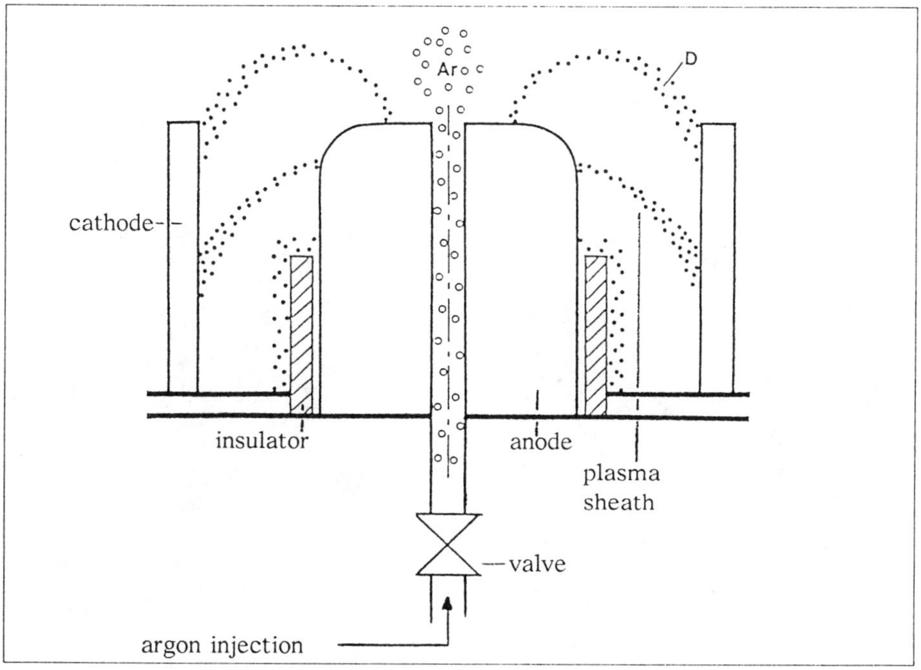

Fig. 1: Experimental set-up of the SPEED 2 driver

The experimental set-up of the plasma focus is schematically shown in figure 1. The driver layout requires an extraordinary operation mode working with heavy gases: the high power input leads to sheath break-up and filamentation, even if only a small percentage of heavy gas is added to the deuterium filling of several hectopascal. Therefore a cloud is generated on the top of the anode by injecting the heavy rare gas (e.g. argon) through a hole in the anode by opening a fast valve a few milliseconds (typically about 6 ms) before ignition of the discharge. The first phase, sheath formation by a sliding discharge along the insulator, and the second phase, rundown, are taking place in pure deuterium. In the third phase, beginning approximately 400 ns after the ignition, a mixture of deuterium and argon is compressed to form the pinch. This technique provides reproducible and appropiate plasma conditions during operation[2].

MICROPINCH CHARACTERISTICS AND STRUCTURE

Investigation of the generated plasma in the SXR range starts with time integrated pinhole pictures, which are showing the pinch structure: figure 2a ($\lambda < 2$ nm) shows a strongly emitting cylinder-like structure embedded in a diffuse, cone-like, weakly radiating cloud. The stronger filtered picture (fig. 2b, $\lambda < 0.5$ nm) reveals that the cylinder-like structure consists of a few micropinches, which differ from the cloud by both intensity and emitted wavelengths. From the dot-like structure of these spots it can be concluded that these dots are images of the pinhole (diameter 300 μm) and therefore smaller than its diameter. The size can be evaluated by the penumbra method to values of 250 μm ... 40 μm depending on the observation wavelength, which is in accordance with the collapse model[3,4]. Looking from the top of the discharge chamber in axial direction along the pinch the diameter of the plasma column is about 1 mm.

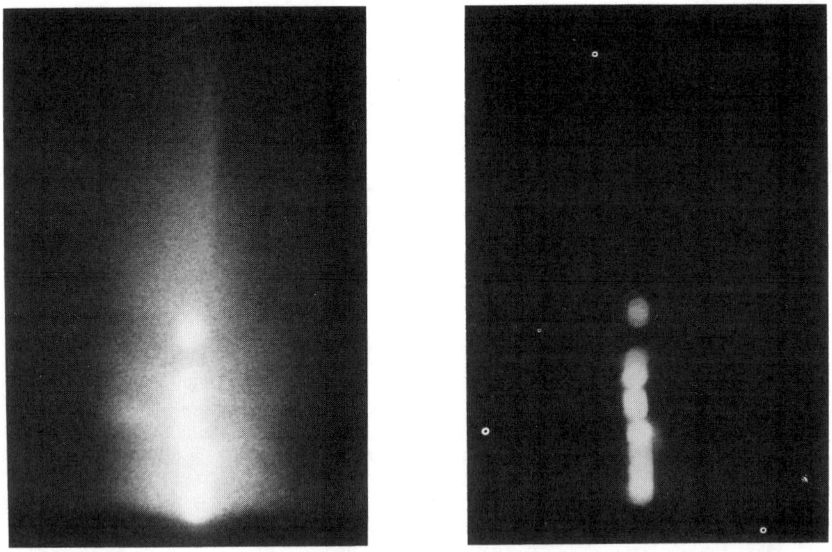

a) $\lambda < 2$ nm b) $\lambda < 0.5$ nm

Fig. 2: X-ray pinhole pictures

Applying a fast x-ray streak camera in combination with the pinhole enables the resolution of the temporal behaviour of the micropinches. First the overall time interval of micropinch generation lasts for 20 ns ... 30 ns. Secondly it can be evaluated that the generation of micropinches starts - as guessed from the shape of the plasma sheath - at the bottom of the pinch, close to the anode, and later at higher z-positions.

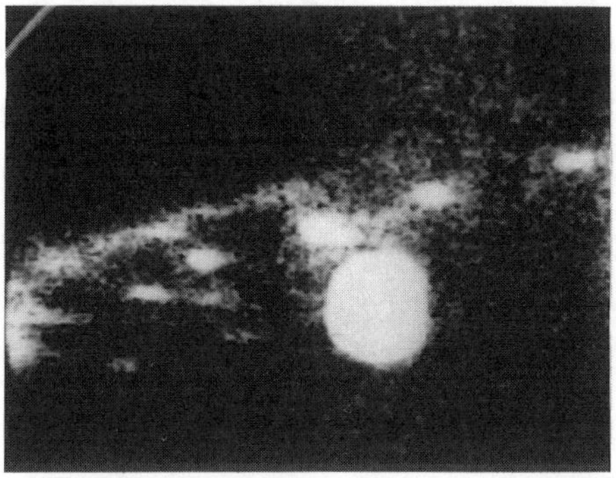

——— 5 ns ———

Also the life time of micropinches can be measured from the pictures with the fastest available streak velocity (fig. 3) to 0.3 ns ... 0.7 ns. (The large dot in the center is an artefact caused by hard x-rays.)

Fig. 3: X-ray streak

Since these measurements give no information about the triggering mechanism(s) of micropinch generation and their early stages a new diagnostic tool was used to broaden the detectable wavelength range. A one-stage microchannelplate system (MCP), which is divided into four independently triggerable sectors, was applied. The pinch is imaged by four pinholes onto the sectors of the MCP. Each sector is separately gated by a high voltage pulse of about 5 ns halfwidth resulting in approximately 4 ns ... 5 ns exposure. The time of exposure can be chosen and changed by the cable length between the pulse generator and the MCP. By comparison of pinhole pictures recorded in the soft x-ray range (like fig. 2) and visible range CCD images with MCP pictures and the use of filters within the pinholes it was found that this diagnostic preferentially provides information of wavelengths following the SXR range (i.e. $\lambda \geq 1$ nm).

From a sequence of four pictures of a single discharge (fig. 4a - d) and the corresponding time integrated pinhole picture (fig. 4e) the positions of developing micropinches two of which are marked by arrows can be identified. It can be seen that the micropinches are initiated where constrictions of the plasma column (due to m = 0 – instabilities) occur. The fact that the micropinches (strongly radiating in the SXR range, especially helium-like argon lines) are hardly visible on the microchannelplate phosphor screen is due to the lower sensitivity of the MCP system at these SXR-wavelengths[5]. These final state structures can be detected by using quite strong filters (10 μm beryllium) and higher gain of the MCP system.

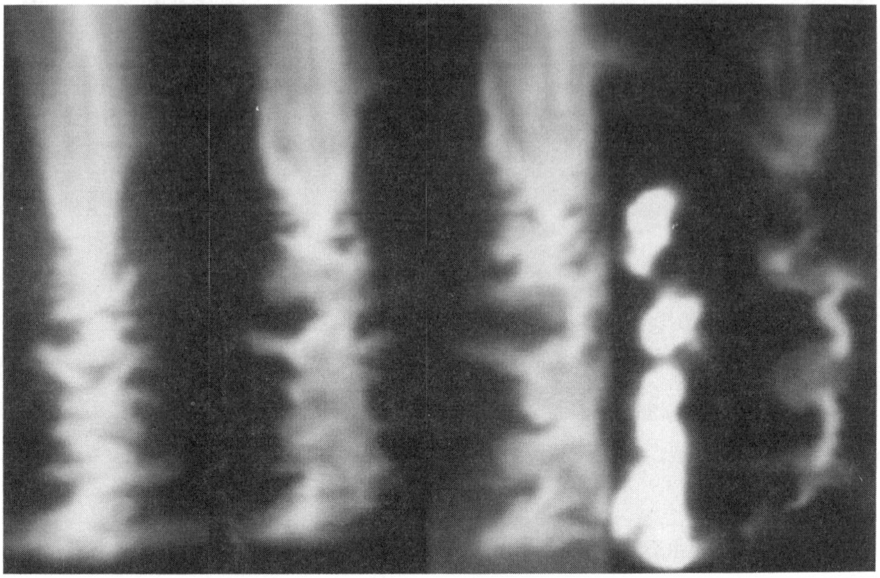

a) b) c) e) d)
Fig. 4: MCP pictures (a - d) and corresponding SXR pinhole picture (e)

The analysis of several series of these MCP and pinhole pictures leads to the conclusion that such a constriction is a necessary but not sufficient condition for micropinch generation.

SPECTRA

To obtain more informations about the micropinch radiation the spectral resolution was improved by using a crystal spectrometer in semi-focussing Johann set-up. The achievable resolution (quartz crystal {10$\bar{1}$0}-layer) of $\lambda/\Delta\lambda \approx 10^4$ enables different measurements in the wavelength range around the He-and H-like lines of argon (ArXVII and ArXVIII, λ = 0.37 nm ... 0.40 nm). While the helium-like lines were observed in all discharges, the hydrogen-like spectrum (Lyman-α line and satellites) was only detected in the most efficient discharges[6] (concerning sheath formation and pinch voltage).

The intensity scan of a typical side-on-spectrum of helium-like argon (fig. 5b) consists of resonance line (W)[b] and intercombination line (Y) as well as multiple (dielectronic and innershell excited) satellites. Several plasma parameters were evaluated from these spectra by computer simulated fits (program codes:

---

[b] Notation according A.H. Gabriel, Mon. Not. R. Astr. Soc., 160, 99, 1972

SATELLITE and TRACE)[7,8] using a Monte-Carlo-method for the photon transport including effects of optical density. The main results (compressed plasma regions with a maximum electron density $n_e = 10^{29}$ m^{-3} and an electron temperature T_e = 1 keV ... 1.5 keV) agree well with estimates based on the radiative collapse model.

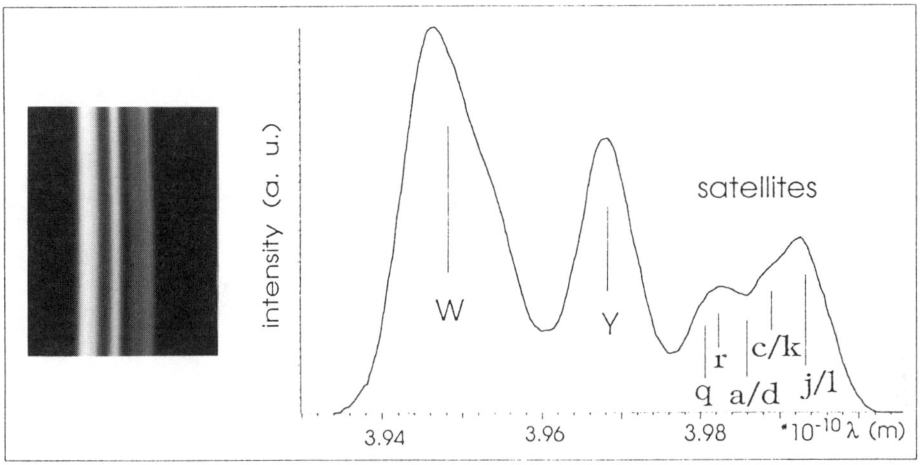

a) spectrum b) scan

Fig. 5: side-on spectrum and scan of helium-like argon

Adding a slit perpendicular to the pinch axis (between plasma column and crystal) provides spatial resolution within the spectrum in z-direction. The comparison of the obtained spectrum (fig. 6a) with the corresponding x-ray pinhole picture (fig. 6b) shows that nearly 90 % of the radiation in this wavelength range is emitted by micropinches.

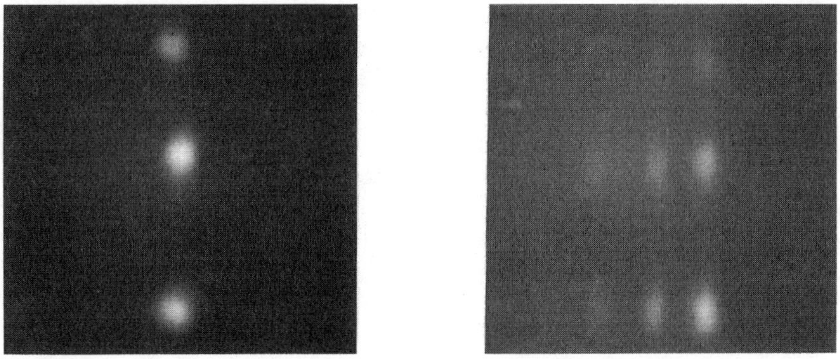

a) pinhole picture b) spectrum (<- λ --)

Fig. 6: comparison of pinhole picture and spectrum with spatial resolution

Time integrated end-on-spectra (taken along the pinch axis) are showing intensive self absorption phenomena on the blue wing of the strongly broadened helium-like resonance line (W) close to the line center. Also the intensity ratio between resonance (W) and intercombination line (Y) has changed from about 2 : 1 (see fig. 5b) to less than 1 : 1. This "distortion" of the line profile is caused by the (moving) plasma column above the region where the micropinches are generated (fig. 7).

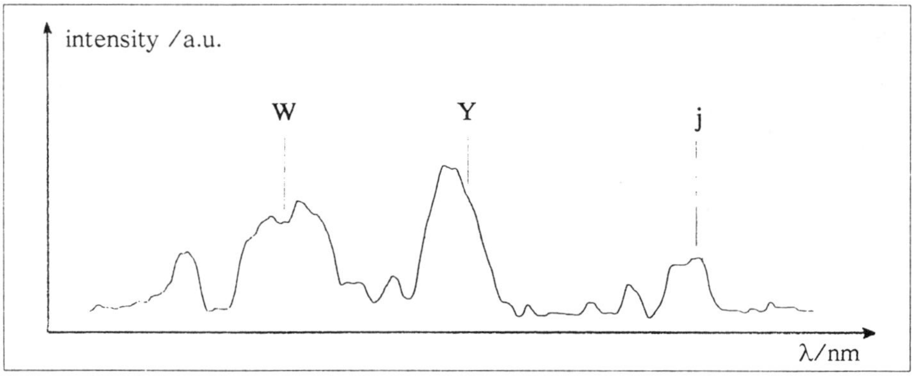

Fig. 7: end-on spectrum of helium-like argon lines

EMITTED ENERGY

Radiation sources in the described wavelength range of SXR and VUV (here: investigations up to about 8 nm) are under discussion for application in microscopy and lithography. The emitted radiation energy is a figure of merit and should therefore be determined.

Measurements with a pinhole grating (diameter 50 μm and 1000 l/mm) performed with poor spectral resolution in the region of He-like lines gave several ten joules between 0.3 nm ... 1.3 nm. But for x-ray microscopy line radiation from the water window (2.33 nm ... 4.36 nm) is of primary interest for obvious reasons (studies of living objects in aqueous solution). Thus the following diagnostic set-up was chosen: a plane tungsten-antimony-multilayer mirror (W-Sb-MLM) was mounted in a distance of approximately one meter in radial direction from the pinch. By changing the angle of incidence a wavelength selection with a resolution of $\lambda/\Delta\lambda \approx 50 \ldots 150$ was achieved. The radiation was detected by a calibrated PIN-diode (which was protected by a filter of 0.3 μm aluminium on an 1.5 μm kimfoil substrate against visible light) and a fast scope. Taking into account diode sensitivity, filter transmission and the small registerable solid angle together with the assumption of isotropic emission the energy emitted in selected wavelength windows was calculated. Up to 45 Joules per discharge in a single channel lasting for about 30 ns were obtained although optimum discharge conditions could not be established (fig. 8).

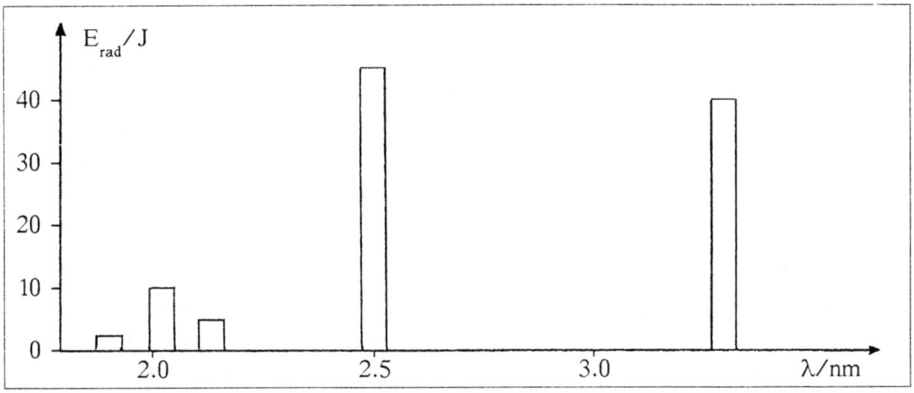

Fig. 8: calculated radiation energy

REFERENCES

1 G. Decker, W. Kies, M. Mälzig, C. van Calker, G. Ziethen, Nucl. Instr. and Meth., A249, p.477, 1986

2 J.M. Bayley, G. Decker, W. Kies, M. Mälzig, P. Röwekamp, G. Ziethen, Yu.V. Sidelnikov, J. Appl. Phys., 69 (2), p.613, 1991

3 V.V. Vikhrev, V.V. Ivanov, K.N. Koshelev, Sov. J. Plasma Phys., 8 (6), p.688, 1982

4 K.N. Koshelev, V.I. Krauz, N.G. Reshetnyak, R.G. Salukvadze, Yu.V. Sidelnikov, E.Y. Khautiev, Sov. J. Plasma Phys., 15 (9), p.619, 1989

5 C. Martin, S. Bowyer, Appl. Opt., 21 (23), p.4206, 1982

6 W. Kies, Plasma Phys. and Contr. Fus., 28 (11), p.1645, 1986

7 F.B. Rosmej, dissertation, Ruhr-Universität Bochum, FRG, 1991

8 A. Schulz, dissertation, Ruhr-Universität Bochum, FRG, 1992

ANNULAR GAS PUFF TARGET EXPERIMENTS WITH POSEIDON PLASMA FOCUS

H. Schmidt
Institut für Plasmaforschung, Universität Stuttgart, Germany

L. Jakubowski, M. Sadowski, E. Składnik-Sadowska, J. Stanisławski, A. Szydłowski
Soltan Institute for Nuclear Studies, Swierk, Poland

ABSTRACT

Gas puff target experiments as performed with the POSEIDON plasma focus (PF) device enable partial decoupling of the initial PF phases from the final compression phase. The influence of deuterium and argon gas puffing on pinch dynamics, neutron and X-ray emission in 135 kJ discharges was investigated. Various diagnostic techniques including optical streak and framing as well as Schlieren and X-ray pinhole photography, neutron, X-ray and electron beam detection were applied. It was found that a large plasma focus can be operated in a controlled way with the injection of an additional gas target. Gas puffing causes a rise in the level of turbulence and results in an increased tendency of hot spot formation. With deuterium gas puffing the neutron yield could be increased by at least a factor of 1.3.

I. INTRODUCTION

First experiments with injecting an additional gas target in front of the inner electrode of the POSEIDON 500 kJ plasma focus [1,2] have shown that the formation of the pinch and the characteristics of the pinch phase can be altered considerably [3].

If an additional amount of working gas (D_2) or gaseous admixtures (e.g. Ar) are injected through a fast acting electromagnetic gas valve into the pinch region, the initial phase of the current sheath formation and the final compression and pinch phases can be decoupled. Thus only the final phase of the plasma focus discharge is affected by the gas puffing. The interaction of a well-formed current carrying plasma sheath with a gas-puffed target, which exhibits density gradients in radial and axial as well as in azimuthal direction, involves complex physical phenomena. But it has been shown [3] that gas puffing through radially arranged axially directed nozzles can be used for modifying plasma focus discharges.

In order to investigate emission characteristics of such discharges various series of plasma focus experiments with gas puffing have been performed.

Various electrical, optical and radiation diagnostics have been applied in order to allow conclusions on the influence the gas-puffed target exerts on the plasma focus formation as well as on its neutron, X-ray and electron beam emission characteristics.

II. EXPERIMENTAL SET-UP

The POSEIDON plasma focus facility (500 kJ, 80 kV, $\tau/4 = 3\ \mu s$) [2] was operated mainly at a 135 kJ (41.6 kV) level with a maximum discharge current of 1.8 MA. The inner 131 mm dia. copper electrode had a central circular hole of 21 mm in diameter and contained a fast-acting high pressure gas valve (see Fig. 1). The outer 280 mm diameter electrode (squirrel cage) was made of 24 copper rods of 12 mm in diameter. Both electrodes were 420 mm long. The main 210 mm long tubular insulator was made of a special ceramic material.

The fast electrodynamic gas valve [3] had an annular form with a movable tubular piston opening a gas plenum of 5 cm^3, which may be filled with the working gas under pressure up to 5 MPa. When the piston is shifted, the working gas can flow out through two types of annularly arranged nozzles (A: 16 conical 5 mm diameter holes with parallel symmetry axes arranged symmetrically at r = 28 mm in the front plate of the inner electrode, see Fig. 1; B: 24 conical holes of 2 mm diameter, arranged at r= 24 mm, with their axes directed convergently towards the pinch axis with an angle of 15 degrees).

Two important parameters for the experiments were the initial pressure (p_v) of the working gas and the time delay (t_d) of the PF discharge in relation to the triggering of the valve. The injected gas forms a tubular gaseous curtain or a quasi-spherical gas cloud in front of the inner electrode plate, which depends on the kind of the injected gas, the initial pressure p_v and the delay time t_d as well as to some degree on the basic filling pressure p_0.

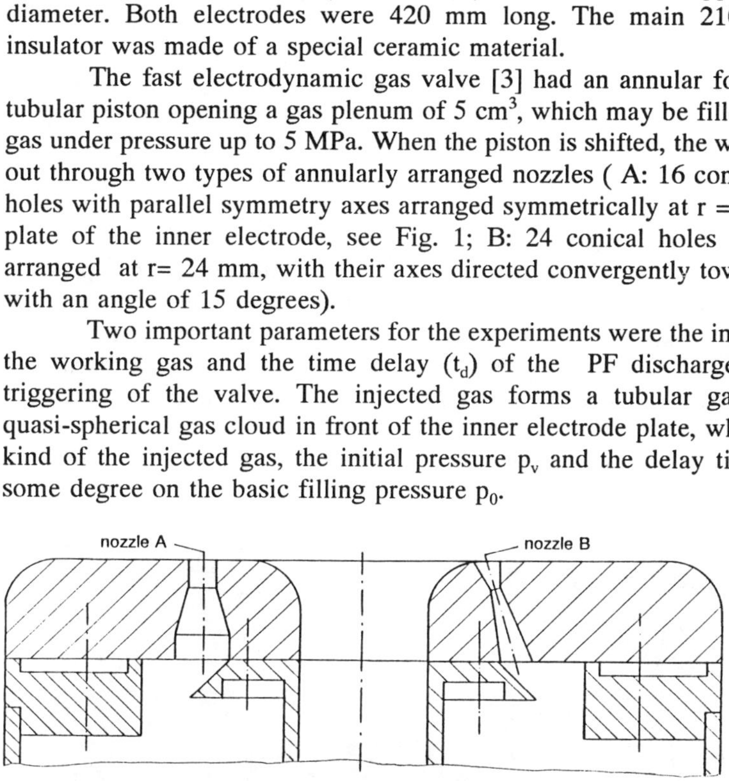

Fig. 1: POSEIDON inner electrode (anode, 131 mm dia.) showing the two types of nozzles (A: 5 mm dia., B: 2 mm dia.) and the central e-beam diagn. hole (21 mm dia.).

III. DIAGNOSTIC EQUIPMENT

Various diagnostics were applied for the investigations. The pinch formation was observed with a high-speed image converter camera operated in the streak mode. Radial streaks were taken through a slit located perpendicular to the electrode axis 20 mm above the anode top end. Additional frame pictures were taken side on with a pulsed planar diode (exposure ≈ 3 ns). The pinch phase was also studied by N_2-laser Schlieren photography (exposure 1 ns). Soft X-ray emission was observed side-on with an evacuated pinhole camera through a 0.2 mm dia. ,10 μm thick Be window. Hard X-rays and neutrons were registered by two scintillator/photomultiplier detectors positioned side-on and end-on at a distance of 4.5 m from the pinch. The time of reaching the minimum radius r_{min} of the pinch was detected by means of a photodiode: an area of 8 mm in diameter, located at (r,z) = (0,25) mm was imaged to the entrance of an optical fibre which was connected to the photodiode. All the voltage- and current- wave forms, timing-,X-ray and neutron signals were registered with two HP16500A multi-channel digital oscilloscopes connected to a computer system.

IV. EXPERIMENTAL RESULTS

Experiments were performed with a static D_2 -filling and additional Ar- or D_2 - puffing through two different nozzle type arrangements (A or B). In the 135 kJ/41.6 kV discharges mainly the dynamics and micro-structures of the pinch phase as well as X-ray and neutron emissions were investigated.

1. Compression phase studies

A comparison of streak pictures (Fig. 2) shows that radial compression velocities for gas-puffed discharges do not appear very different. The compression in the case of the dense gas puffing is slightly slower and weaker. The level of turbulence and MHD instabilities are increased (Fig. 3). In contrast to discharges without gas-puffing one can observe distinct density filaments in the compression phase ($t < 0$) as a consequence of the gas jets flowing out through the nozzles. At later times ($t > 0$) these differences between discharges with and without puffing are small (see Fig. 4). For higher pressure p_v the level of turbulence is strongly increased

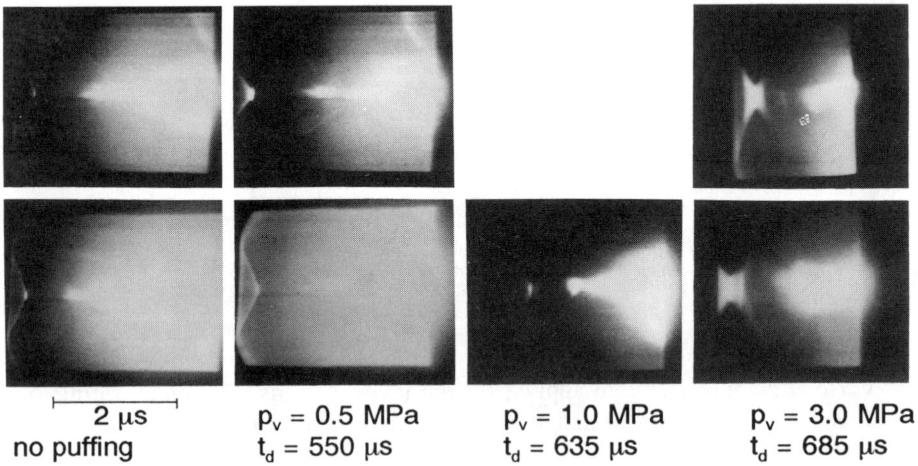

Fig. 2:
Streak camera pictures taken from 135 kJ discharges with different basic D_2 filling with and without additional D_2 puffing. The visible radiation was observed through a slit perpendicular to the electrode axis at z = 20 mm above the anode end plate (upper row: p_0 =2 hPa D_2, lower row: 1 hPa).

and higher density gradients occur (Fig. 5). With argon puffing a further strong increase of turbulence and instabilities is observed (Fig. 6).

Important information on the dynamics of the compression phase can be gained from an analysis of dI/dt wave forms (Fig. 7). One can define a time interval t_c (compression time) between the moment when the current (plasma) sheath hits the dense gas-puffed target and the moment of the first strong peak (dip) in the dI/dt signal which corresponds to the maximum compression or minimum radius r_{min} of the pinch. This compression time t_c depends on p_v and t_d and is plotted in Fig.8.

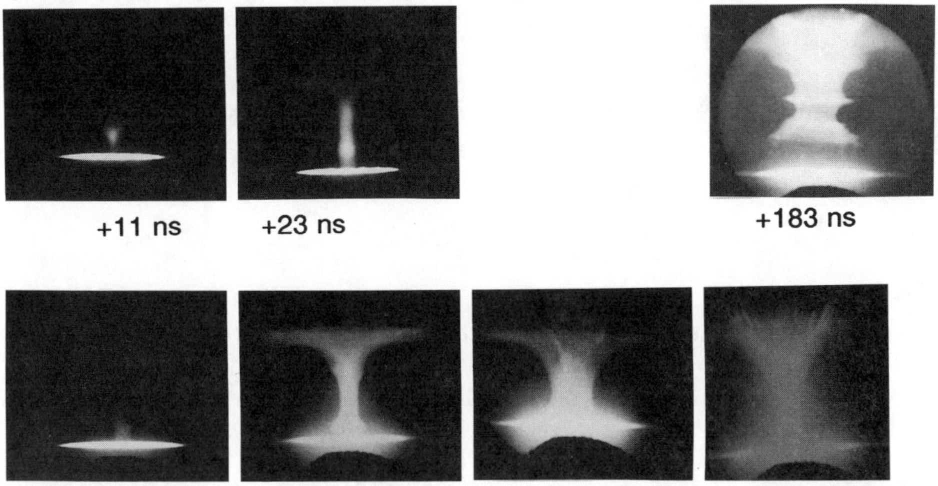

+5 ns	+11 ns	-43 ns	+9 ns
no puffing	p_v = 0.5 MPa	p_v = 1.0 MPa	p_v = 3.0 MPa
	t_d = 550 μs	t_d = 635 μs	t_d = 685 μs

Fig. 3:
Visible radiation pictures taken from 135 kJ discharges with an image converter camera (3 ns exposure). The radiation was observed through different apertures in order to get legible photos. Upper row: 2 hPa D_2, lower row: 1 hPa. Scale (also Fig.4-6 and 11):width of frame ≡ 106 mm.

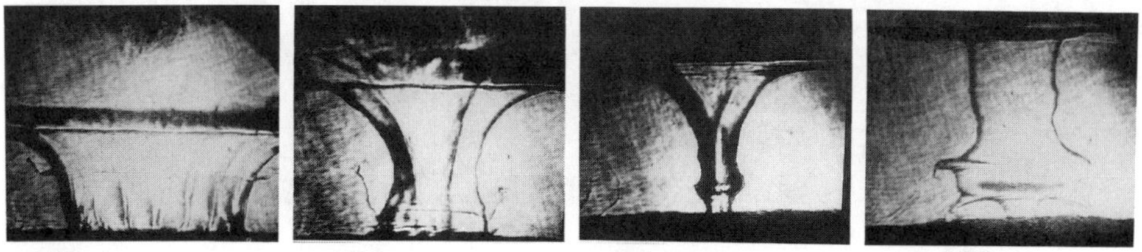

-169 ns	-43 ns	+1 ns	+123 ns

Fig. 4:
Schlieren pictures (exposure 1 ns) taken from 135 kJ discharges with additional D_2 puffing (p_v= 0.5 MPa, t_d = 500 to 550 μs) showing various phases of the radial compression (t=0: maximum compression).

2. Neutron emission

In the past the optimization of neutron yields from PF discharges in the POSEIDON facility has been performed by an appropriate choice of the electrode/insulator configuration, the inital filling pressure p_0, as well as the material for the insulator [1,2]. In the experiments carried out recently particular attention has been paid to the neutron emission as a function of the additional gas puffing. As can be seen from Fig. 9 the neutron yield increases up to a factor of about 1.3 when an

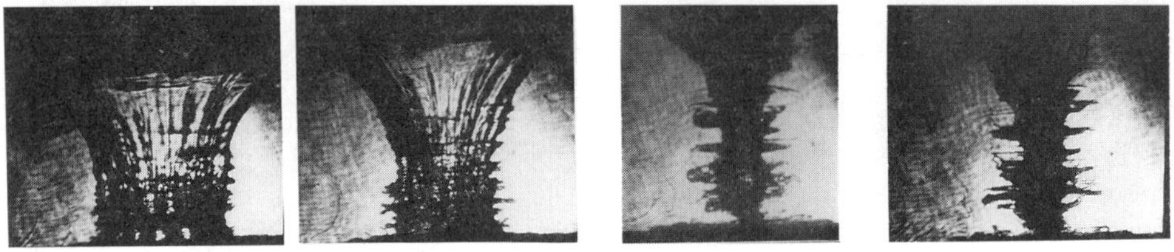

-103 ns | -81 ns | +30 ns | +42 ns

Fig. 5:
Schlieren pictures taken from 135 kJ discharges showing various phases of the radial compression at similar gas conditions (2 hPa D_2-filling, p_v = 3 MPa D_2, t_d = 500 μs). The high density gas target (with large density gradients) results in increased turbulence and instabilities.

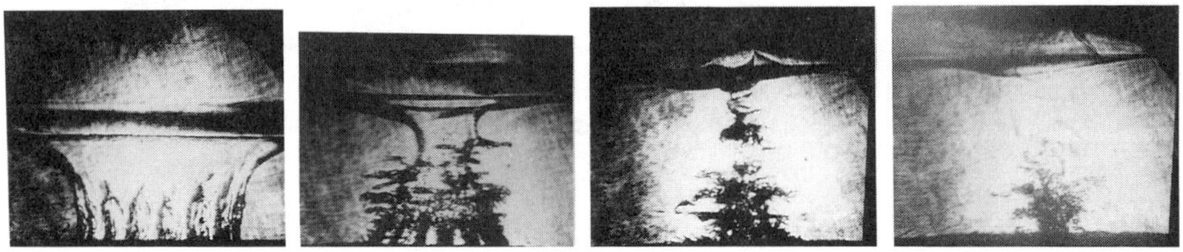

-131 ns | -13 ns | +61 ns | +95 ns

⇑(p_v = 0.5 MPa, t_d = 470 μs)

(p_v = 3.0 MPa, t_d = 485 μs)⇒

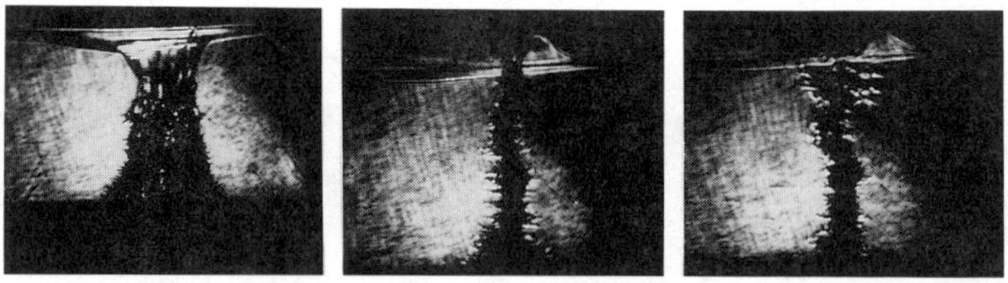

-51 ns | +18 ns | +49 ns

Fig. 6:
High turbulence and instabilities as observed in Schlieren pictures taken from 135 kJ discharges (2 hPa D_2-filling) with heavy gas (Argon) puffing (uppper row: p_v = 0.5 MPa Ar, t_d = 470 μs, nozzle A; lower row: p_v = 3 MPa Ar, t_d = 485 μs, nozzle B).

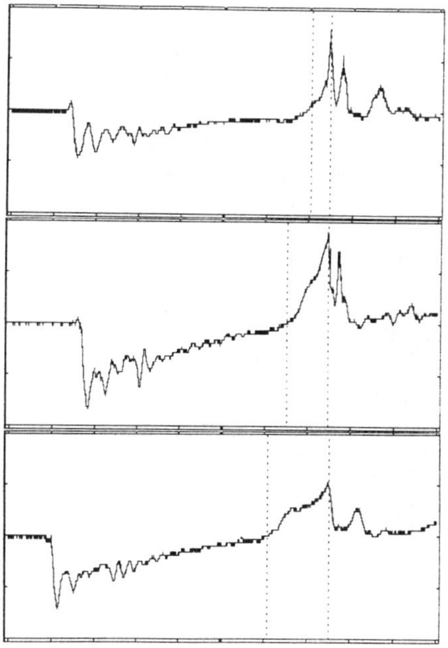

Fig. 7:
Comparison of various dI/dt wave forms (polarity reversed!) from 130 kJ discharges (p_0 = 2 hPa D_2) with additional gas puffing (a: no puffing; b: p_v = 0.6 MPa, t_d = 550 μs; c: p_v = 3 MPa, t_d = 500 μs).

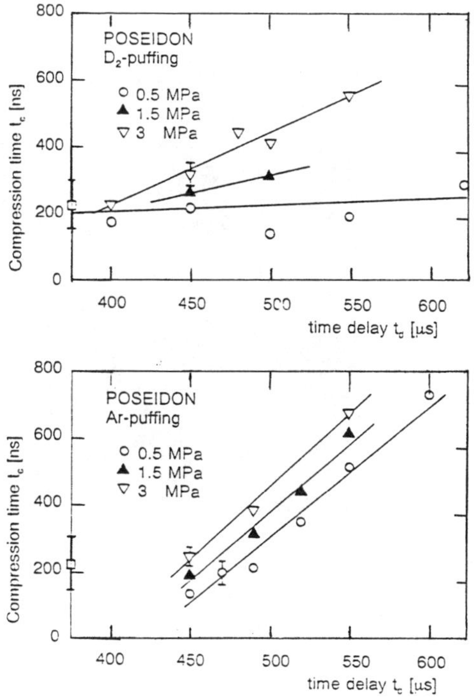

Fig. 8:
Compression time t_c (from dI/dt signals) as a function of the valve delay time t_d for 135 kJ discharges (p_0= 2 hPa D_2) for D_2 and Ar puffing (p_v = 0.5, 1.5 and 3 MPa).

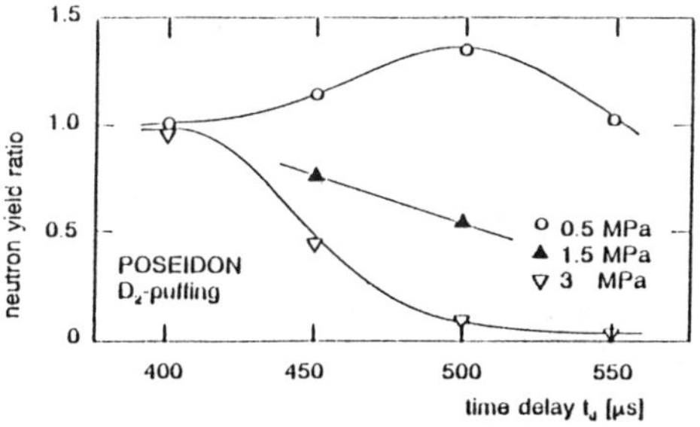

Fig. 9:
Average neutron yield ratio from 135 kJ discharges (p_0 = 2 hPa D_2) with D_2 gas puffing (p_v = 0.5, 1.5 and 3 MPa, nozzle A) as a function of the valve delay time t_d. The reference unit (Y_n = 1) corresponds to the neutron yield without gas puffing.

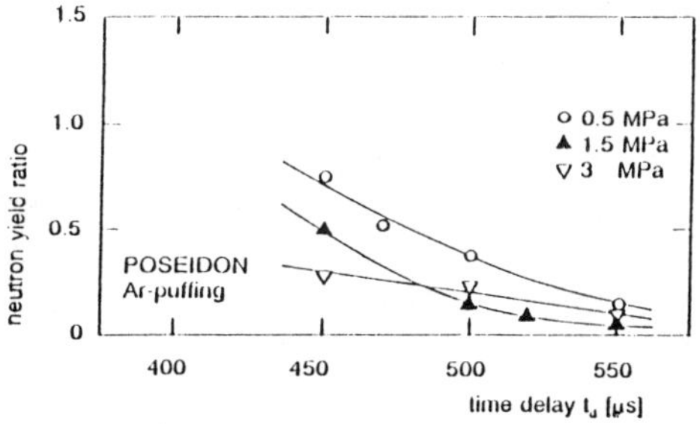

Fig. 10:
Average neutron yield ratio from 135 kJ discharges (p_0 = 2 hPa D_2) with additional Ar puffing (p_v = 0.5, 1.5 and 3 MPa, nozzle A) as a function of the valve delay time t_d.

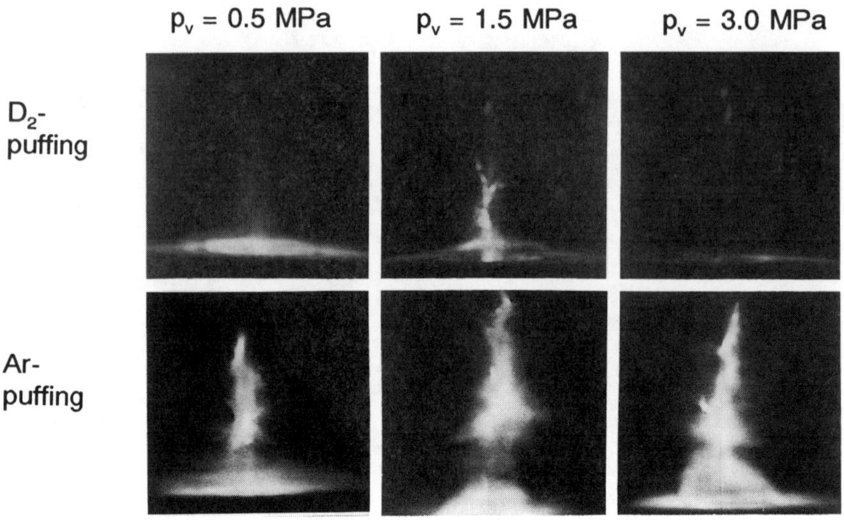

Fig. 11:
Soft X-ray pinhole pictures (0.2 mm dia., 10 μm Be window) from 135 kJ discharges (p_0 = 2 hPa D_2) with additional D_2 and Ar puffing (t_d = 500 μs).

appropriate D_2 gas target (p_v = 0.5 MPa, t_d = 500 μs) is injected. As expected, the neutron yield decreases when argon is injected (Fig. 10).

3. X-ray emission

Routine measurements of time-integrated hard (> 100 keV) X-ray yield have been made. Comparing the yields registered for several series of 135 kJ shots, one finds that gas puffing causes a distinct decrease in the average X-ray yields, up to one order of magnitude, depending on p_v and t_d. Measurements of the time integrated soft X-ray emission were performed with a vacuum pinhole-camera (10 μm Be-foil). Pinhole pictures taken under different experimental conditions are presented in Fig. 11. Small regions of increased X-ray emission (hot spots) are often observed with D_2 as well as with Ar puffing. These hot spots predominantly occur in shots with large dips in the dI/dt signal where also high neutron yields are registered.

V. CONCLUSIONS

1. In general, the injection of an additional gas target in front of the anode of the POSEIDON plasma focus - operated at the 135 kJ level - increases the level of turbulence and MHD instabilities.

2. Neutron and hard X-ray yields are rather decreased than increased. The neutron yield could, however, be increased by a factor of 1.3 with D_2 gas puffing when appropriate values for the valve pressure p_v and the valve delay time t_d were chosen.
3. In high neutron yield shots one finds the tendency that X-ray hot spots are observed. The observation that no direct relation between neutron emission and hot spot formation exists is due to the fact that effective pinch formation results in various pinch phase characteristics (e.g. beams, instabilities) which are not directly related to each other.
4. Experiments at higher energy levels (above 135 kJ) should be performed for investigating the energy scaling (as to neutron and X-ray emission) of plasma focus experiments with additional gas puffing. This is of importance as conclusions from PF operation can not be transferred from low energy to high energy devices [3].

ACKNOWLEDGEMENTS

The investigations have been performed within a German-Polish scientific co-operation agreement. The authors acknowledge the support of the Karlsruhe office for International Co-operation.

REFERENCES

[1] H. Schmidt and the POSEIDON Team: in Plasma Focus Research (Proc. 3rd Int. Workshop, Stuttgart, 1983), Institut für Plasmaforschung, Report IPF 83-6, p.63.

[2] H. Herold, A. Jerzykiewicz, M. Sadowski, H. Schmidt: Nuclear Fusion 29, 1255 (1989).

[3] H. Schmidt, M. Sadowski, et al.: submitted for publication in Plasma Phys. and Controlled Fusion.

X-RAY EMISSION FROM MICROPINCHES IN THE DPF78 PLASMA FOCUS

H. Schmidt, D. Schulz
Institut für Plasmaforschung, Universität Stuttgart, Germany

P. Antsiferov
Institute of Spectroscopy, Troitzk near Moscow, Russia

ABSTRACT

Heavy gas admixtures of up to several percent to the basic deuterium filling of the DPF78 plasma focus (28 kJ, 60 kV, $\tau/4 = 1.5$ μs) results in the formation of micropinches inside the 25 mm long pinch column. Dimensions of the micropinches, which range from 20 to 500 μm, were determined from X-ray pinhole pictures and space resolved X-ray spectra by means of the penumbra method. PIN-diode array measurements have revealed that micropinches occur during a period of about 50 ns near the time of current maximum of 0.9 MA. Space resolved line emission of helium- and hydrogen-like argon from single micropinches was measured side-on and end-on using two quartz crystal spectrometers. Observed Doppler shifts up to 8 mÅ between spectra of adjacent micropinches indicate velocities of up to $6 \cdot 10^5$ m/s. Temperatures in the keV range and densities up to $3 \cdot 10^{21}$ cm^{-3} were determined from the X-ray spectra.

INTRODUCTION

The appearance of micropinches (hot spots, plasma points) with dimensions down to 20 μm within the pinch plasma column of a plasma focus (PF) discharge has been established on various PF devices. The proper formation of such micropinches in a PF depends primarily on appropriate heavy gas admixtures to a basic hydrogen/deuterium filling which support rapid radiative collapse of localized regions to high density/temperature plasmas.

As the X radiation of multi-charged ions is an important diagnostic tool for hot dense plasmas, heavy gas admixtures in a PF discharge have the twin function of enabling temperature and density diagnostics as well as strongly supporting the micropinch formation.

For the investigations of micropinches in a medium size plasma focus spatially resolved argon X-ray spectra were registered on film using transmission grating and quartz crystal spectrometers. Model calculations for the spectra allow the evaluation of plasma parameters. The fact that the spectra could be taken with spatial resolution allows also the determination of axial and radial micropinch plasma velocities which were deduced from measured Doppler shifts of adjacent micropinches during one PF discharge.

EXPERIMENT SET-UP

DPF78 is a Mather-type plasma focus device (28kJ, 60 kV, $\tau/4 = 1.5$ μs). The copper anode diameter is 50 mm, the copper cathode (squirrel cage) diameter 100 mm. Both electrodes are 150 mm, the pyrex insulator is 50 mm long. At the current maximum of 0.9 MA a 25 mm long pinch column is formed. Experiments were performed with basic D_2-filling (3 to 5 hPa) mostly with Ar admixtures (0.1 to 0.2 hPa).

Two quartz curved-crystal spectrometers of the Johann type (2d = 6.687 Å and 8.512 Å, resp.) positioned side-on and end-on with slits for spatial resolution were used [1,2]. The dispersion plane of the side-on positioned spectrometer was perpendicular to the pinch axis.

The time-integrated structure of the pinch column in the wavelength region $\lambda < 10$ Å was studied with a pinhole camera (0.4 x 0.4 mm^2 , 50 μm thick Be-window, 4-fold magnification). Radial and axial sizes of the pinch could be determined by means of the penumbra technique [3]. This technique was also applied to the spatially resolved spectra.

Time- and space-resolved information on the micropinch was obtained from a 10-channel PIN-diode array in the detection plane of a second pinhole camera.

EXPERIMENTAL RESULTS

1. X-ray emission with spatial and temporal resolution

The amount of heavy gas admixtures to the deuterium strongly influences the final pinch [1], especially the probability of micro-pinches to occur. The number of micropinches increases with the atomic number as well as with the percentage of the heavy gas admixture [4] up to an upper limit. This limit is probably due to a strong influence of the heavy gas admixtures on the initial (breakdown) phase of the discharge which affects the final (pinch) phase.

Measurements of the time of appearance of various micropinches along the 25 mm long pinch column were

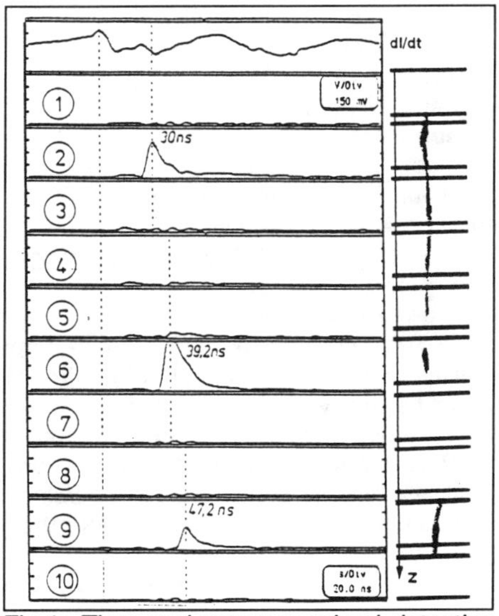

Fig.1: *Time and space resolved detection of micropinches.*

performed by means of a linear PIN-diode array which consists of 10 single X-ray sensitive PIN-diodes [5] with dimensions of 2 x 2 mm^2. The pinch column was imaged to this array by means of a 0.3 mm diameter pinhole with a 3 μm thick polycarbonate window covered with a 0.6 μm thick silver layer. It was found that micropinches occur earlier near the anode than at more distant positions. Usually two to four micropinches appear in a discharge during a time interval of up to 50 ns. In Fig. 1 an example of a pinhole picture together with the signals of the PIN-diode array and the dI/dt trace is shown. On the right three micro-pinches are visible on the X-ray pinhole picture (electrode on top, the horizontal bars indicate the corresponding positions of the ten PIN-diodes within the one-dimensional array). On the left the dI/dt signal and the ten PIN-diode signals are shown. The indicated time values (30; 39.2 and 47.2 ns) are given relative to the first dip in the dI/dt signal. As the resolution of the PIN-diodes is about 10 ns (signal rise and fall times 3.5 to 4.5 ns) the registered signals do not give the lifetime of single micro-pinches. By means of an X-ray streak camera, however, lifetimes of micropinches (heliumlike argon line emission, 4 Å) in the DPF78 between 200 ps and 400 ps were measured [1]. The PIN-diode array which we used is suitable for determining the time of appearance of micropinches with sufficient accuracy although the true time behaviour of soft X-ray emission cannot be resolved. The time of occurence of micropinches (relative to the dip in the dI/dt signal) registered during many discharges is plotted in Fig. 2 as a function of the distance z from the anode end. The zippering effect is evident. As will be shown below, plasma parameters of micro-pinches depend on the location where they appear in the pinch.

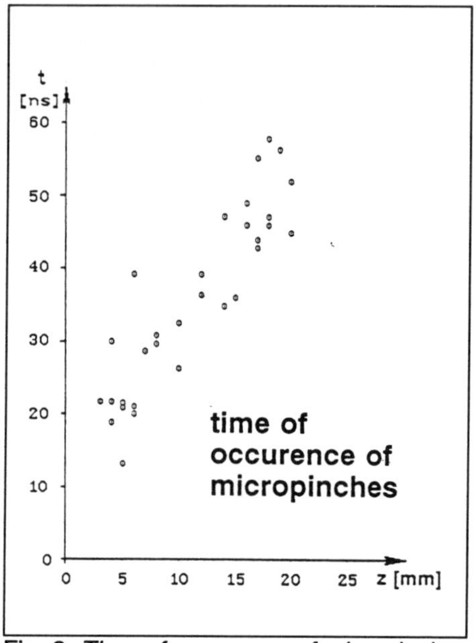

Fig. 2: *Time of occurence of micropinches*

2. Spatially resolved X-ray spectra

X-ray spectra of hydrogen- and heliumlike argon were studied by means of two curved crystal spectrometers positioned side-on and end-on. The side-on positioned spectrometer was equipped with a quartz crystal (50 x 8 x 0.5 mm, 2d = 8.51 Å, radius of curvature R = 378 mm) oriented with its dispersion plane perpendicular to the pinch axis. The distance between the crystal and the plasma column was 575 mm. A slit (1.4 mm wide) between the plasma and the crystal allowed a spatial resolution along the pinch axis. The position of this slit (50 mm from the crystal) resulted in a 2.35 fold demagnification of the pinch in the plane of the spectrum registration. A 10 μm thick Mylar foil with a thin Al deposit blocked the visible light emitted by the pinch from the film. The end-on positioned spectrometer was also equipped with a slit for spatial resolution in radial direction. In Fig. 3 an example of the spectrum of a single micropinch is shown. The calibration of the spectrometers by means of an X-ray tube with different metals (Cd,Ag,Ho,Tb) allows an absolute accuracy of about ±3.5 Å. The limited accuracy is due to the positioning of the film tangentially to the Rowland circle. The relative accuracy for one spectrum is in the range of ±0.5 Å and is also limited by the reproducibility in positioning of the film. In Fig.3 one can clearly discern the resonance and intercombination lines (W,Y) of heliumlike argon as well as their satellites. The two Lyman α lines of hydrogenlike argon are clearly visible, but the intensity of their satellites (in Fig.3 near the reference mark) cannot be determined with high accuracy because of the noise level.

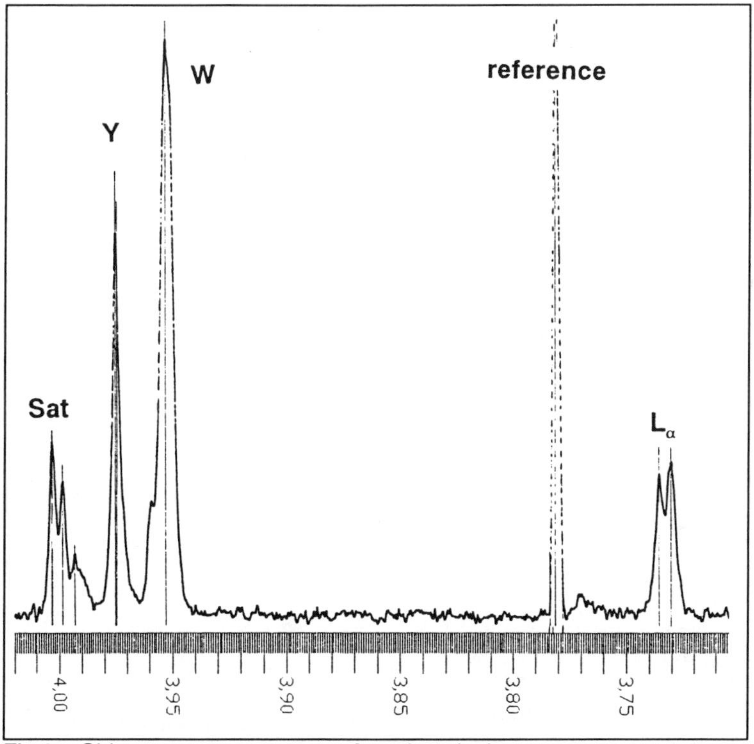

Fig.3 : *Side-on argon spectrum of a micropinch*

Due to the possibility of spatial resolution of the X-ray spectrometers differences in temperature and density of different micropinches along the pinch column could be found. Generally micropinches which occur near the anode are characterized by higher temperature and density than those which occur at larger z-positions. In Fig. 4 an example of argon spectra from two micropinches (A and B) at different z-positions (z_A = 8 mm, z_B = 25 mm) is shown. Main differences are found in the widths of the resonance (W) and inter-combination (Y) line. From the intensity ratios of satellite to resonance lines one finds electron temperatures $T_{e,A}$ = 1500 eV and $T_{e,B}$ = 900 eV. The electron densities which are determined from a comparison of simulated (program SATLI [6]) and measured spectra, range around $n_{e,A} = 2\cdot10^{22}$ cm^{-3} and $n_{e,B}$ $9\cdot10^{21}$ cm^{-3}. An example of the comparison between a measured and a simulated spectrum emitted from a single micropinch far from the electrode in the region of He-like Ar-lines with their satellites (DS, IS) is shown in Fig. 5. The dashed line represents the experimental spectrum. The discrepancy between the two spectra on the right wing of the resonance line (W) is due to $1s2l3l'-1s^23l'$ transitions which are not included in the modelling program. Best fit para-

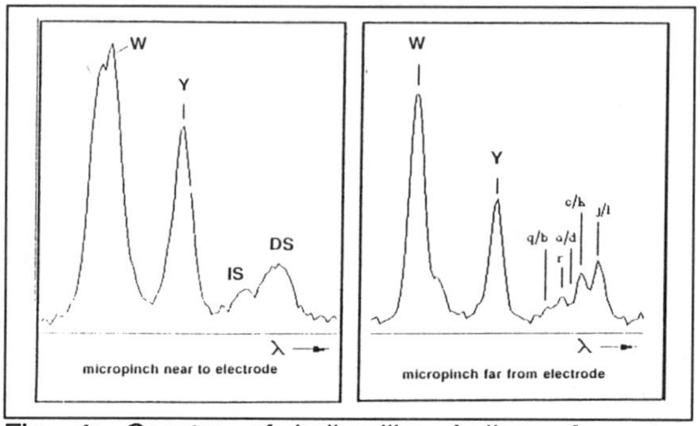

Fig. 4: *Spectra of heliumlike Ar-lines from two micropinches.*

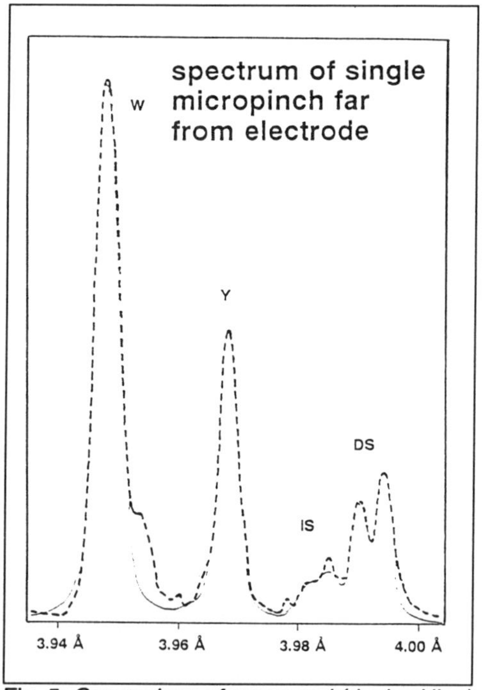

Fig. 5: *Comparison of measured (dashed line) and simulated spectrum.*

meters for the simulation were T_e=900 eV and n_e=$9\cdot10^{21}$ cm^{-3} .

As the micropinch plasma is highly transient (lifetime < 400 ps) one should expect different plasma parameters in its development. This dynamic behaviour of the micropinch plasma was checked in a case where hydrogen- as well as heliumlike argon lines were analyzed. By means of the penumbra method, which was applied to the spectra, dimensions between 370 μm and 150 μm were deduced. Our interpretation assumes that the micropinch decreases its dimensions and simultaneously increases its plasma density and temperature, although some effects according to density and temperature profiles (e.g. a hot core) may also contribute to the measured results. In an example for which intensity scans (for λ = const) are shown in Fig.6, the following data were found:

line	dimension [μm]	T_e [eV]	n_e[cm^{-3}]
[He]	370	600	$2\cdot10^{21}$
[H]	150	2000	?

Dimensions which were deduced from pinhole pictures are slightly larger than those deduced from the heliumlike spectral lines.

3. Doppler shifts

In spectra registered during the same discharge Doppler shifts between heliumlike lines of up to 8 mÅ were found from adjacent micropinches. Radial and axial velocities of $3\cdot10^5$ m/s and $6\cdot10^5$ m/s, respectively, were found. The Doppler shifts observed from side-on spectra are always in a direction which corresponds to radially outward moving micropinches.

SUMMARY AND CONCLUSIONS

1. During one plasma focus discharge performed with heavy gas admixtures to the basic deuterium filling usually several micropinches occur during a time interval of about 50 ns following the compression of the pinch column to a minimum radius. The lifetime of a single micropinch in the DPF78 plasma focus is 0.2 to 0.3 ns as was determined earlier by X-ray streaks.

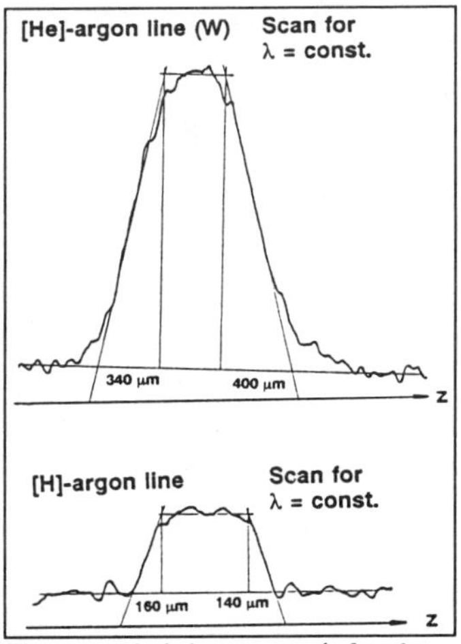

Fig. 6: *Axial scans (for λ = const) of [He]- and [H]- argon lines.*

2. Spatially resolved spectra of hydrogen- and heliumlike argon could be registered from single micropinches.
3. Analysis of these spectra results in differences in temperature and electron density between micropinches which are located near to or far from the electrode.
4. Electron temperature rises in the development of a micropinch from 600 eV, as determined from heliumlike lines, to 2 keV, as determined from hydrogenlike lines.
5. Simultaneously to the rise in temperature the linear dimensions (in z-direction) decrease by a factor of ~ 2.
6. Electron densities up to $3 \cdot 10^{21}$ cm^{-3} were evaluated from heliumlike spectra.
7. Time-integrated line emission ([H]- and [He]-like Ar, ~ 3 keV) from one single micropinch (volume < $5 \cdot 10^{-3}$ mm^3 , lifetime ~ 0.2 ns !) was determined from X-ray pinhole pictures to be 0.1 to 0.3 J into 4π sr.
8. Measured Doppler shifts between lines of adjacent micropinches of up to 8 mÅ correspond to axial velocities of $6 \cdot 10^5$ m/s.
9. Doppler shifts (up to 4 mÅ) from radially observed spectra are always in a direction which corresponds to radially outward mowing micropinches.

REFERENCES

[1] P. Antsiferov, D. Franz, R. Pfau, A. Rupasov, H. Schmidt, D. Schulz: Hot Spot X-ray and Microwave Emission of the DPF-78 Plasma Focus. Proc. 19th EPS Conference on Controlled Fusion and Plasma Physics, Innsbruck 1992, Europhysics conference abstracts, Vol. 16 C, Part I, 683 (1992).

[2] D. Schulz, Thesis, Institut für Plasmaforschung, Universität Stuttgart, 1992.

[3] E.V. Aglitskii, P.S. Antsiferov, A.M. Panin, Sov. J.Plasma Phys. 10, 726 (1985)

[4] P. Antsiferov, D. Franz, H. Herold, L. Jakubowski, A. Jonas, M. Sadowski, H. Schmidt: Hot Spot Formation and Emission Characteristics of the Plasma Focus. Proc. 18th EPS Conf. on Contr. Fusion and Plasma Physics, Berlin 1991, Europhysics conference abstracts, Vol. 15 C, Part II, 221 (1991).

[5] L. Pina, Cs. Casopis pro fiziku A35, 363 (1985)

[6] F.B. Rosmej, PH.D. Thesis, University of Bochum, 1991

MICROPINCH FORMATION IN THE SPEED 2 PLASMA FOCUS

F. Schmitz, P. Röwekamp, G. Decker, W. Kies, G. Ziethen
Inst. f. Experimentalphysik, Heinrich-Heine-Universität, 4000 Düsseldorf 1, FRG

J.M. Bayley
Imperial College of Science, Technology and Medicine, SW7 2AZ London, UK

K.N. Koshelev, Yu.V. Sidelnikov
Inst. f. Spectroscopy, Russian Academy of Sciences, 142092 Troitzk, Russia

D.M. Simanovskii
A.-F.-Ioffe-Physico-Technical Institute, 194021 St. Petersburg, Russia

ABSTRACT

Micropinch formation is described as local collapses initiated in necked regions of a pinch plasma column by enhanced plasma radiation and particle outflow [1]. While the final stage of this formation is well known from soft X-ray measurements ($\lambda < 2$ nm) [2] the local and temporal conditions of the initial phase has not yet been clarified.

Therefore a fast (ns) four frame pinhole camera system has been applied to the SPEED 2 plasma focus in order to investigate the micropinch formation process in a wavelength range from VUV to the soft X-ray (0.4 nm).

The experimental setup consists of a configuration of four pinholes each imaging the pinch plasma. These pictures are taken with a special microchannel plate which is divided into four independent sectors. Each sector can be independently gated by a short triggerpulse (FWHM 5 ns) leading to an exposure time of a few nanoseconds. Therefore four differently time delayed pictures are taken during a single discharge.

Different filters allow to determine the spectral range of the detected radiation (4 nm - 8 nm) and to observe single micropinches (0.4 nm) in the plasma column. Comparing the images with time integrated soft X-ray pinhole pictures it is realized that local neckings ($m=0$ type) are necessary but not sufficient conditions for micropinch formation. It seems that only those neckings lead to micropinches where the plasma density and temperature are high enough to intensively emit line radiation.

For details compare: P. Röwekamp, F. Schmitz et al. "Spectral Investigations of Micropinches in the SPEED 2 Plasma Focus".

REFERENCES

[1] V.V. Vikhrev, V.V. Ivanov, K.N. Koshelev, Sov. J. Plasma Phys. 8 (6), p.688, 1982
[2] J.M. Bayley, G. Decker, W. Kies, M. Mälzig, F. Müller, P. Röwekamp, J. Westheide, Yu.V. Sidelnikov, J. Appl. Phys. 69 (2), p.613, 1991

DEPENDENCE OF THE NEUTRON FLUENCE ANISOTROPY ON THE SOURCE AXIAL EXTENSION IN A MEDIUM ENERGY PLASMA FOCUS

I.Tiseanu, N.Mandache
Institute of Atomic Physics, Bucharest-Magurele, Romania

ABSTRACT

Angular and energetic characteristics of the reacting deuterons were evaluated using a fitting procedure applied to the neutron fluence anisotropy data measured on different axial emissive zones of a plasma focus device. A relative large angular distribution and medium energy values (50÷100 keV) characterize the forward escaping deuterons from pinch zone, which produce ≈60% from total neutron yield. In the pinch region the dominance of the reacting deuterons with large radial velocity components and a backward moving deuteron component were put into evidence.

INTRODUCTION

Investigations of the angular characteristics of the neutron emission from IPF-2/20 plasma focus device were carried out. In this paper a particular attention was paid to the dependence of the neutron fluence anisotropy on the source axial extension. The experimental data were analyzed in order to obtain the angular and energetic characteristics of the reacting deuterons (the deuterons producing neutrons).

As compared with previous measurements[1] the experimental data are corrected taking into account the axial extension of the neutron source and using more accurate data of the differential fusion cross sections .

EXPERIMENTAL SET-UP AND DATA PROCESSING

The experiments described in this paper have been performed on the IPF-2/20 plasma focus device at 16 kV bank voltage, 300 kA peak current and at a deuterium filling pressure of 3 torr.

Absolute neutron fluence values have been obtained using seven silver neutron activation battery detectors ("in situ" calibrated with radioisotope neutron sources located in the pinch formation region) placed at seven angles with respect to the electrode axis: 0° to 180° in steps of ≈30°, at distances of 1÷1.5 m from the pinch. Each battery detectors consists of three to five Geiger-Mùller tubes coupled in parallel. A sketch of the experimental arrangement is presented in Fig.1.

The relative statistical errors of the anisotropy measurements were bellow 3-4%. This relative high accuracy was obtained by equilibrating the counting rates of all battery detectors at values sufficiently high for a good counting statistics but under the level from which the "death time" effects become important[3].

A fast photomultiplier AVP 56 - scintillator NE102A neutron detector placed ≈8 m from the pinch in the forward direction has been used for monitoring neutron pulses.

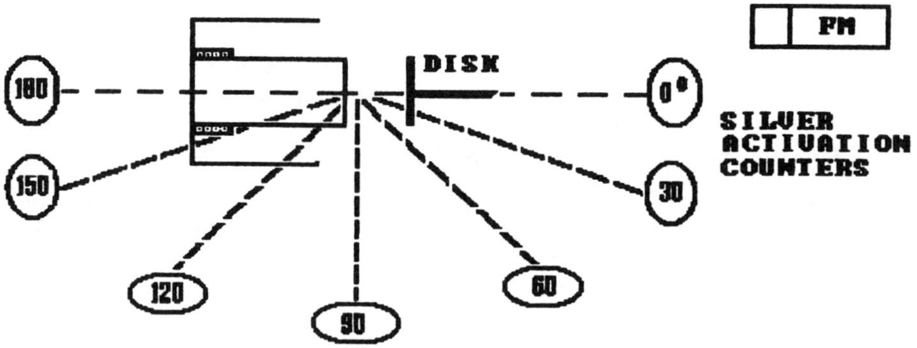

Fig. 1. Experimental Arrangement

In order to determine the dependence of the neutron fluence anisotropy (the ratio of the neutron fluence at different angles to that at 90°) on the source axial extension a metal disk obstacle, with the same diameter as the anode (45 mm), was placed on the axis in front of the anode at five distances: 3, 5, 7, 9 and 13.5 cm.
The minimum distance was chosen, based on shadowgraphic images of the pinch, in such a way to avoid the perturbation of the final stage evolution of the plasma focus by the obstacle.

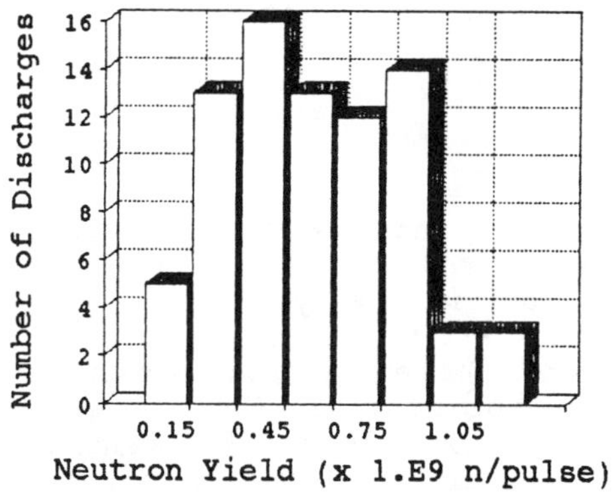

Fig. 2. Neutron yield distribution

Series of discharges with improved neutron yield reproducibility have been run for each of these disk obstacle positions. This improvement reproducibility was achieved using a 1 mA D.C. glow discharge for preionization and handling the disk obstacle from exterior without affecting the vacuum in the discharge chamber.

In Fig. 2 is exemplified the neutron yield distribution measured over six series of ≈15 discharges every, the disk obstacle being placed at 13.5 cm. The discharge chamber was refilled after every series. This histogram show that the neutron yield distributions are roughly gaussian shaped with standard deviation values less than 50% of the average neutron yield. It is remarkable that the average neutron yield values of every series of discharges do not differ with more than 4-5% from the neutron yield average value over all series. In addition it was noticed that for the discharge series with the disk obstacle placed closer to the anode the standard deviations have decreased.

Based on these results the neutron angular distribution of a certain emissive axial zone in front of the anode is obtained by subtracting the counting rates measured for the two obstacle positions limiting the respective zone.

For a quantitative study, within a fitting procedure on the experimental anisotropy data, an effort was done for deriving analytical expressions of the neutron fluence anisotropy[3]. In our model one consider that the largest part of the neutron yield is due to the interaction of a fast deuteron population and a stationary target. The angular distribution of the reacting deuterons was described by a Gaussian shaped profile, centered on the Z-axis (the full width at half-maximum Θ_0) and the energetic characteristics by a mean energy E_0, based on the direct ion energy analysis results[3,4].

RESULTS AND DISCUSSION

The experiments performed by placing the disk obstacle in different axial positions provided the following informations:

- the axial distribution of the neutron yield;
- the angular neutron fluence distribution on seven directions for different emissive zones;
- the correlation between the neutron fluence anisotropy and the neutron yield.

For the purpose of data analysis the discharges were divided in three domains for each of the obstacle positions:

- medium neutron yield discharges $\overline{Y}_n-\sigma < Y_n < \overline{Y}_n+\sigma$;
- high neutron yield discharges $Y_n > \overline{Y}_n+\sigma$;
- low neutron yield discharges $Y_n < \overline{Y}_n-\sigma$

where $\overline{Y}_n$ and σ are the statistical mean and standard deviation, respectively. The discharges featuring neutron pulses (as recorded by the photomultiplier-scintillator detector) with many peaks or large temporal widths are not included in the statistical analysis.

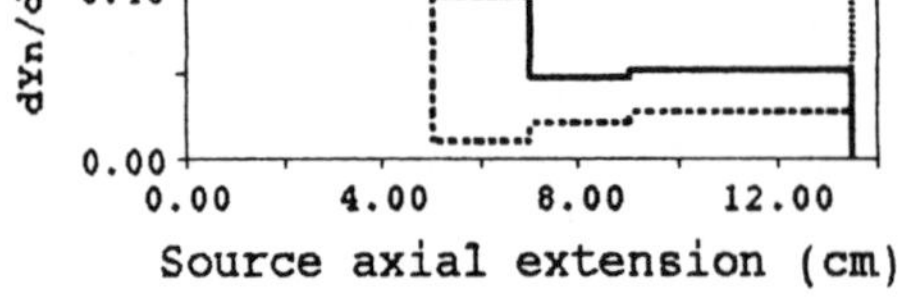

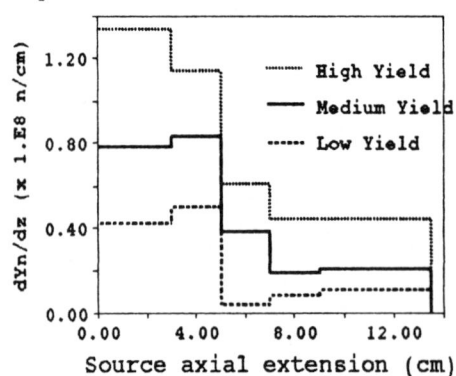

Fig. 3. Neutron yield axial distribution.

Their number is less than 20% of the total number of discharges.

For these three domains the axial distributions, measured by the detector placed at 90°, are presented in Fig. 3.

It results that a major part of the neutron yield (≈60%) is originated outside the 0-3 cm axial zone. Moreover, the axial neutron source length is larger when the neutron yield is higher, suggesting that the higher neutron yields are due to relative higher mean energy of the reacting deuterons.

The quasi-uniform neutron emissivity displayed by the first

two axial zones, (0-3 and 3-5 cm) could be explained by the presence of a relative dense axial target which extends well outside the pinch region[5]. Similar results were obtained, both by neutrons and protons measurements with spatial resolution, by many groups[6,7,8]. Interferometric measurements performed on our device have shown that a high density shock front ($\approx 10^{18}$ cm^{-3}) is forward propagating through these zones during the neutron emission period[9].

The relative large extension of the neutron source impose to correct our experimental anisotropy data taking into account the differences between pinch to battery detectors distances, and axial emissive zone centers of gravity to detectors distances.

The dependence of the forward neutron anisotropy on the source axial extension is presented in Fig. 4 for low, medium and high neutron yield domains.

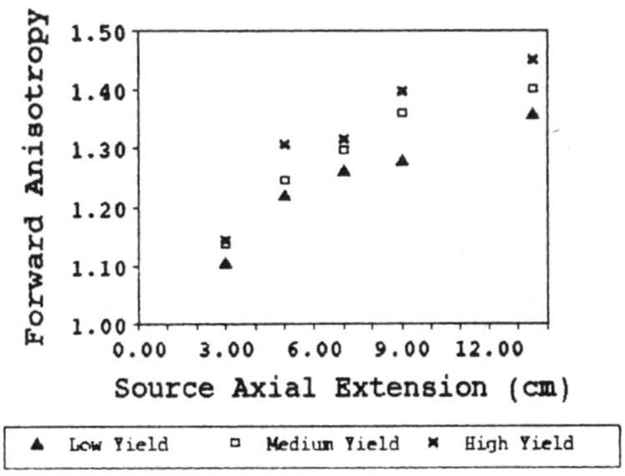

Fig 4. Dependence of the forward anisotropy on the source axial extension.

The increase of the anisotropy with the neutron yield can be explained by an increase of the reacting deuteron mean energy, in agreement with the above results.

The relative low anisotropy value of 0-3 cm zone reflects that the majority of the neutrons are produced by fast deuterons with dominant radial velocity components, a characteristic feature of the pinch zone[6,7,10]. On the other hand the increase of the forward anisotropy with the source axial extension is explained by an increased contribution of the forward moving deuterons to the neutron yield due to longer paths through the neutral gas.

A fitting procedure is used to evaluate the mean energy E_0 and the full width at half-maximum θ_0 of the angular distribution of the reacting ion population for different axial zones.

The experimental neutron fluence anisotropies (points) of the whole emissive zone (the disk obstacle removed) is presented in Fig. 5. On the same figure is plotted the fitting curve (continuous line).

For a medium neutron yield of $Y_n \approx 6.1 \cdot 10^8$ n/pulse the fitting procedure provided the following results: E_0=70±12 keV and θ_0=61°±2°.

For the high neutron yields (Y_n=9.3$\cdot 10^8$ n/pulse) it resulted E_0=92±11 keV, θ_0=63°±2° and for low neutron yields (Y_n=3.0$\cdot 10^8$ n/pulse) E_0=40±5 keV, θ_0=52°±2°. The increase of the neutron yield with the mean energy of the reacting deuterons confirms the above mentioned results and is roughly explained by the variation of the fusion cross section with deuteron energy.

Applying the fitting procedure on the averaged neutron fluence anisotropies measured on the 3-13.5 cm axial zone (Fig. 6) which does not contain the pinch region, one gets: E_0=73±10 keV and θ_0=55°±2°. As expected the exclusion of the pinch zone, characterized by an important contribution of the gyrating deuterons to the neutron yield, is accompanied by a slight reduction of the FWHM of the reacting deuteron angular distribution.

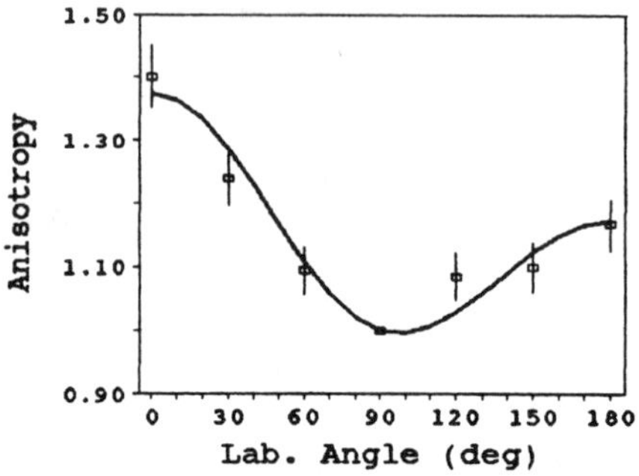

Fig. 5. Neutron angular distribution; disk removed.

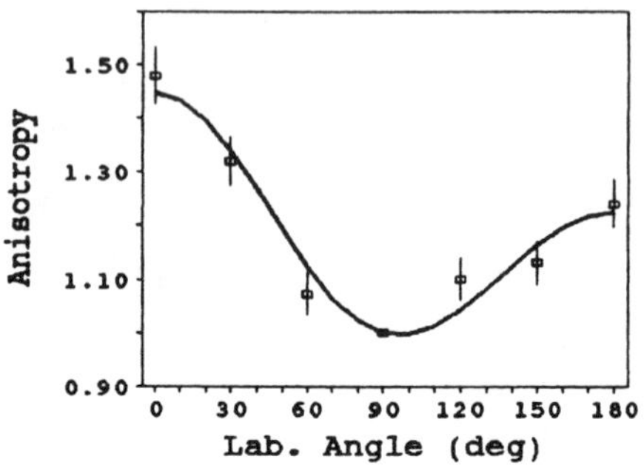

Fig. 6. Neutron angular distribution; extra pinch zone.

As it was mentioned before an interesting emissive zone is the 3-5 cm one. The fitting parameters obtained for this zone (Fig. 7) are: E_0=78±13 keV and θ_0=62°±3°. From Fig. 3 one can see that the 3-5 cm axial zone accounts for about 25% of the total neutron yield. Supposing an average density number of the target in this zone 2-3 times greater than the neutral gas density, as an effect of the presence of the shock front, we can estimate the total number of fast deuterons responsible for this neutron yield.

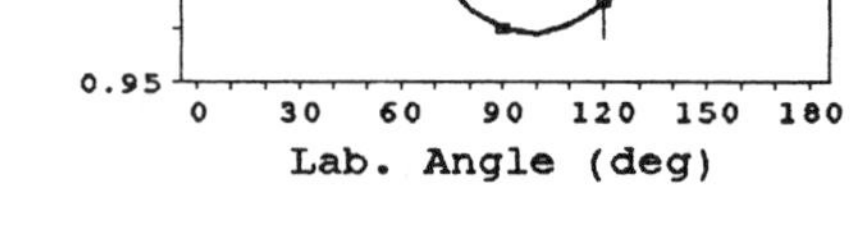

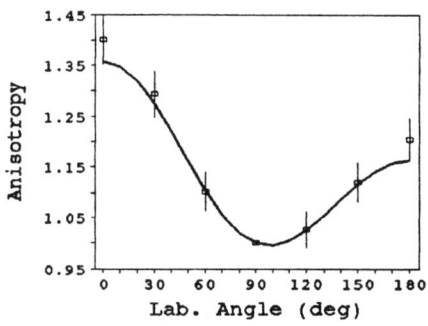

Fig. 7. Neutron angular distribution; 3-5 cm zone.

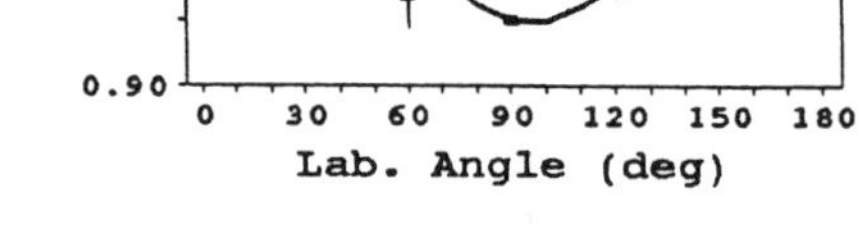

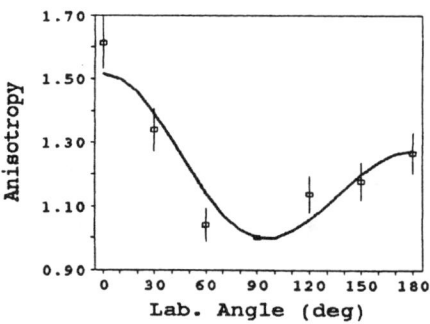

Fig. 8. Neutron angular distribution; 5-9 cm zone.

Thus, one obtains that $\approx 1.3 \cdot 10^{16}$ fast deuterons with a mean energy of 78 keV interacting with a target of $\approx 5 \cdot 10^{17}$ deuterons/cm^3 and 2 cm length, produce $\approx 1.5 \cdot 10^{8}$ neutrons. This number of fast deuterons represents only a few percents of the total particles content in the pinch.

For the adjacent emissive zone (5-9 cm) the obtained fitting parameters are (Fig. 8): E_0=72±8 keV and θ_0=49°±2°. The slight reduction in the mean energy can be explained by the deuteron energy loss in the target. On the other hand the narrowing of the angular distribution is mainly a consequence of the finite diameter of the disk obstacle.

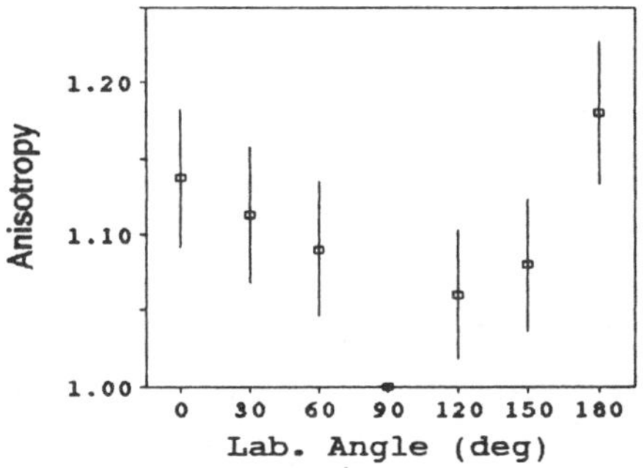

Fig. 9. Neutron angular distribution; 0-3 cm zone.

These energetic characteristics of the reacting ion population are in good agreement with the axial direct ion spectra measurements performed earlier on the same device[4].

In discharges using the metal disk obstacle placed at 3 cm from the anode some interesting features were remarked: a) a significant decrease of the forward anisotropy and an increase of the backward one, their values getting practically equal (Fig. 9), b) while the forward anisotropy features the usual dependence on the neutron yield (Fig. 10) the backward anisotropy - neutron yield correlation shows a non-typical dependence (Fig. 11), the backward anisotropy getting higher at lower neutron yields (negative slope of the regression line).

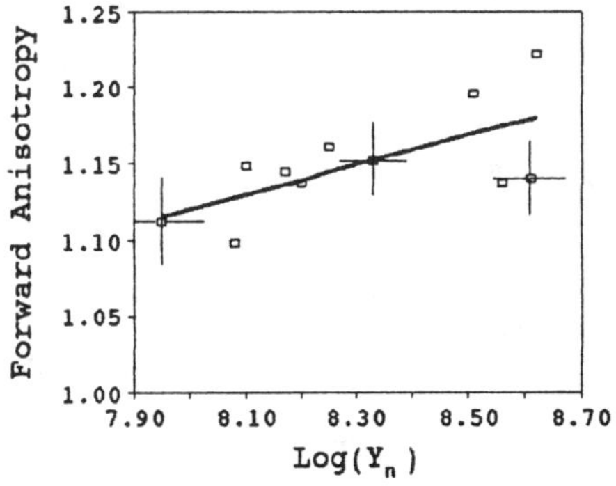

Fig. 10. Forward anisotropy neutron yield correlation; 0-3 cm zone .

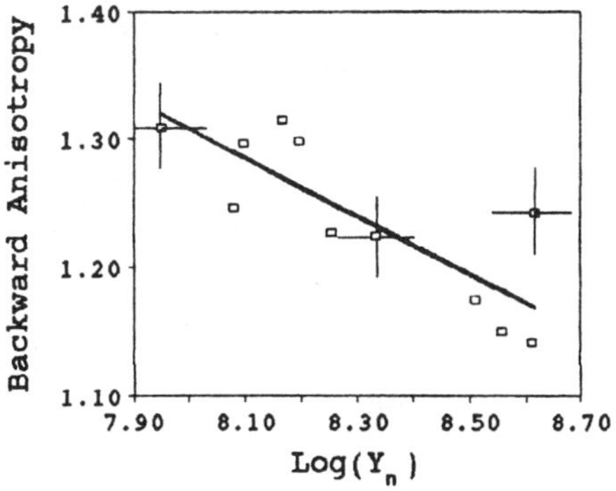

Fig. 11. Backward anisotropy neutron yield correlation; 0-3 cm zone .

The following arguments can be taken into account to understand these results:

- in the 0-3 cm zone the majority of the neutrons are produced by fast deuterons with dominant radial velocity components;
- the forward moving ions contribution to the fusion processes is reduced in the presence of the disk at 3 cm;
- if one accept that the fast deuterons are produced in the

region of m=0 instability, when the m=0 instability develops far from the anode, any backward moving deuterons[11] have larger paths, through plasma target, to massive-end anode. Some experimental evidences which support this behaviour of the m=0 instability are reported in References 3 and 12.

CONCLUSIONS

The axial distribution of the neutron fluence shows that a major part of the yield (≈60%) originates outside of the pinch region.

The angular and energetic characteristics of the deuterons escaping forward from the pinch were determined. These deuterons represent a sample of the fast deuteron population of the pinch. By analyzing the anisotropies measured on the 0-3 cm axial zone it resulted that the gyrating deuterons are responsible for the largest part of the neutron yield in the pinch.

By corroborating these results with the direct ion energy spectra measurements, performed on the same device, we proved that medium energy deuterons (50-100 keV) with relative broad angular distribution produce the major part of the total neutron yield.

REFERENCES

1. N.Mandache, I.Tiseanu, V.Zambreanu, V.Zoita, A.Serban, C.Doloc, Proc. of the 18th European Conf. on Controlled Fusion and Pl. Phys., Berlin II-333(1991)
2. M.Drosg and O.Schwerer, Handbook on Nuclear Activation Data, T.R.S. 273, (ed. IAEA, Vienna, 1987)
3. I.Tiseanu, N.Mandache, V.Zambreanu, Energetic and angular characteristics of the reacting deuterons in a plasma focus, (sent to Pl. Phys. and Controlled Fusion)
4. A.Bocancea, N.Mandache, V.Zambreanu, Study of the plasma focus ionic component, Internal Rep. LOP-27-1982, Central Institute of Physics, Bucharest (1982)
5. A.Bernard, P.Cloth, H.Conrads, A.Coudeville, G.Gourlan, A.Jolas, Ch. Maisonnier, J.P.Rager, NIM **145**, 191 (1977)
6. K.Hubner, J.P.Rager, K.Steinmetz, Proc. of the 10th European Conf. on Controlled Fusion and Plasma Phys., Moscow (1981) Vol.1, D2
7. U.Jager, H.Herold, Nuclear Fusion, **27**, No. 3, 407(1987)
8. H.Schmidt, Atomkernenergie 36, 161 (1980)
9. V.Zambreanu, Ph.D. Thesis, Bucharest University (1988)
10. G. Deker, W. Kies, and G. Pross, Phys. Fluids, 26, 571(1983)
11. M.G.Haines, Nucl. Instr. and Methods, **207**, 179(1973)
12. B.Ruckle, Uber den Zusammenhang zwischen Neutronenproduktion und Instabilitaten am Plasmafokus, IPF-81-1, Stuttgart, (1981)

IMPLOSIONS

IMPLOSION OF MULTILAYER LINERS

R.B. Baksht, I.M. Datsko, A.V. Luchinsky, V.I. Oreshkin,
A.V. Fedyunin, Yu.D. Korolev, I.A. Shemyakin,
and V.G. Rabotkin
Institute of High Current Electronics
Russian Academy of Sciences Siberian Division
Tomsk, Russia

ABSTRACT

Experiments on the implosion of multilayer cascade light liners were carried out on the IMRI-3 (0.3 MA, $T/4$ = 700 ns) and IMRI-4 (0.24 MA, 900 ns) installations. It has been demonstrated that when using three sequantial cascades, owing to the suppression of the Rayleigh-Taylor instability, the ratio $r_{10}/r_f \sim 70$ (where r_{10} is the initial radius of the outer cascade) can be achieved. The input energy density and the X-ray output therein are observed to increase (respectively, up to 1 kJ and 200 J for Kr). Measurements performed with the use of a grating spectrograph for different gases have shown that the temperature in the plasma column formed in liner implosion increases with atomic number.

Introduction

Earlier[1,2,3] it has been shown that the use of a multilayer (cascaded) liner for the production of radiation in the region 0.1-1 keV is highly efficient and given a two- or three-fold increase in radiation energy as compared to a single-layer liner. The idea of cascading consists in using several (two or more) coaxial cylindrical shells. In a cascaded liner, spatial instabilities appearing in the outer liner during acceleration are quenched on striking the inner liner. Estimates performed by Velicovich[4] have shown that in order to efficiently suppress a spatial Rayleigh-Taylor instability it is necessary that the mass of the outer shell be somewhat lower than the mass of the inner one. In this case the shock wave having appeared in the outer shell in its collision with the inner one will quench the R-T instability and, in further implosion, the sum mass of both shells will start from a smaller radius.

The very first experiments with a multilayer liner have shown that a certain contribution to the heating of the liner plasma is from the resistive heating of the plasma column formed as a result of implosion. Actually, for a linearly increasing current the kinetic energy per unit length is given by

$$W = a I_{imp}^2 \Delta L,$$

where I_{imp} is the current at the instant of implosion and ΔL is the change in inductance.

For I_{imp} = 250 kA, $\Delta L = 2 \ln r_{01}/r_f$, where r_{01} is the initial radius of the outer shell (r_{01} = 3.5 cm) and r_f the final radius of the liner (r_f = 0.05 to 0.1 cm) we have W_k = 130 to 160 J.

In the experiments described in[3] the total energy radiated by the three-layer liner was 300 J which exceeds W_k. In order to elucidate the features of the energy balance and plasma heating in a multilayer liner we have investigated how the atomic weight of the liner substance affects its parameters. In the context of these investigations both measurements and 1-D MHD calculations taking into account radiation processes in the plasma liner have been carried out.

The Experimental Setup

The experiments were performed on the IMRI-3 and IMRI-4 installations. The IMRI-3 installation based on two capacitors of totalcapacitance 5.6 μF used three-electrode gas switchs operating at atmospheric pressure[5]. With the discharge chamber inductance equal to 13 nH (Fig. 1) the total circuit inductance was 40 nH and the current was 310kA at 30 kV. The IMRI-4 installation assembled on the same capacitors used low-pressure sealed-off pseudospark switchs[6]. Using

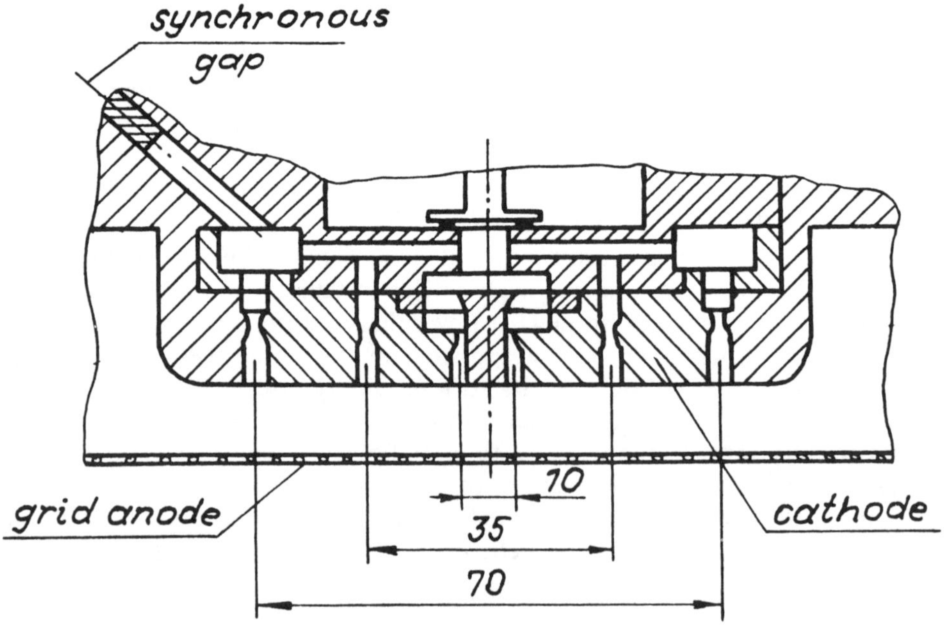

Fig. 1. Three-layer liner

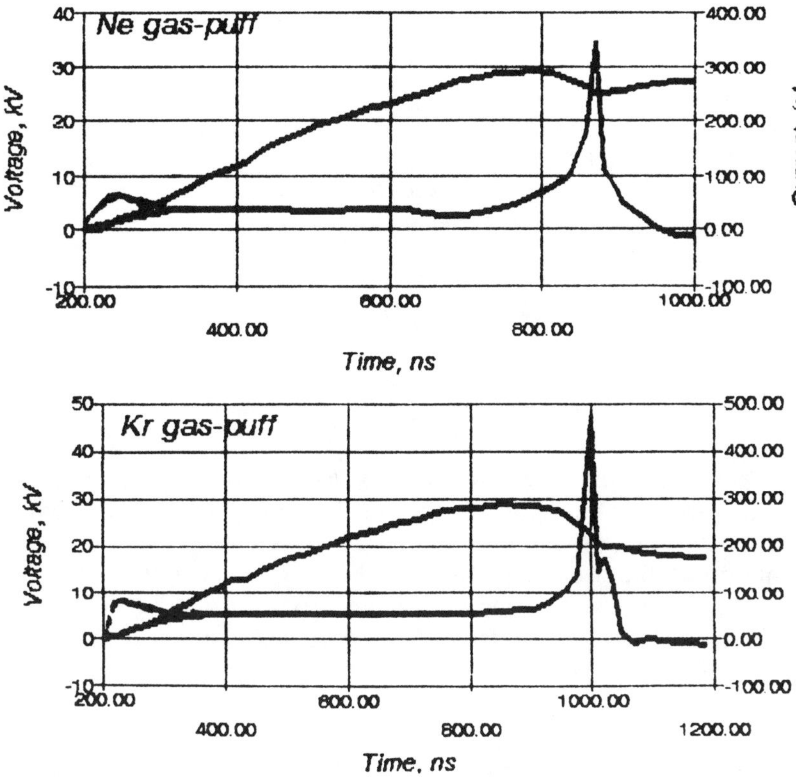

Fig. 2. Current and voltage waveforms for Ne and Kr liner implosion

pseudospark gaps substantially lengthens the installation lifetime (up to 10^4 to 10^5 shots), makes its operation noisless and offers a conceptual possibility for the transition to the repetitive operating mode. The experiments have shown that with the same discharge chamber inductance the sum circuit inductance increased not essentially (up to ~ 45 nH) and with the charge voltage equal to 23 kV the short-circuit current was 240 kA.

The investigation subject was a three-layer liner with the following shell radius: r_{01} = 3.5 cm, r_{02} = 1.7 cm, and r_{03} = 0.5 cm; the liner length was 1 cm, except in specially mentioned cases. Used for the operating material were nitrogen, neon, argon, and krypton.

In addition to conventional electrophysical measurements (Rogowsky coil and a pinhole camera), used in the experiment was a spectrograph of the Microtex company (Russia) which allows radiation measurements in the region 3 to 70 nm. For all gases the liner mass was chosen such that the liner implosion occured near the current maximum, which permitted us to keep the mass constant to a certain degree, since $m \sim a I_0^2 t^2 r_0^{-2}$.

Energy consumption and radiation output

The experiment have been shown that the energy consumed by the liner sharply increased with the atomic number of the liner substance. To illustrate this Fig. 2 gives current waveforms obtained during the implosion of Ne and Kr liners. It can be seen that for Ne the current dip is not significant, 8-10 % of the peak current, while for Kr it reaches 25-30 %.

The total energy deposited in the liner plasma increases from 200-300 J for nitrogen to 1 kJ for krypton. We have calculated the maximum impedance, Z_m, for a 1-cm long liner. The Z_m being the sum of the liner ohmic resistance and $\dot{L}$ is defined as

$$Z_m = R + \dot{L} = V_{max}/I_{max}, \text{ Ohm}$$

where V_{max} and I_{max} are, respectively, the voltage and the current at the point where the absolute value of the current derivative reaches its maximum, which corresponds to the maximum degree of compression for the liner. The Z_m as a function of the atomic number of the liner gas is given in Fig. 3. It should be noted that increasing the liner length to 2 cm for Kr decreases Z_m. This seems to be related to the fact that the energy stored in the chamber inductance was not sufficient for efficient implosion of the Kr liner.

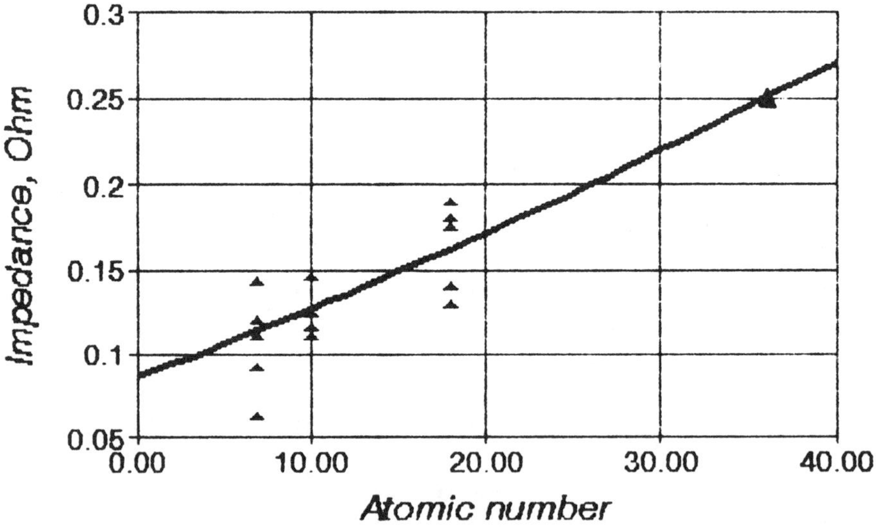

Fig. 3. Impedance Zm versus the atomic number.

The Z_m (A) dependence is shown on the Fig. 3. This curve coincides with the radiation yield dependence from atomic number (Fig. 4). The radiation was measured by X-ray diode with My filter, the approximate range

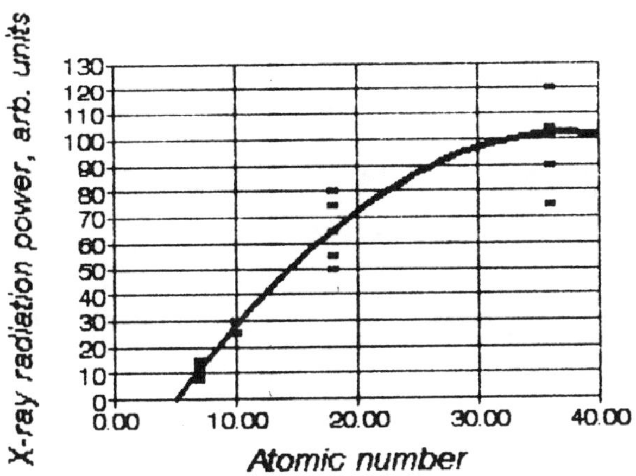

Fig. 4. X-ray radiation power versus atomic number, the range of the energy recording radiation is 0.19-0.288 keV.

of wavelength 4.4-6 nm. Note that moderately small increasing radiation between Ar and Kr may be connect with the change of radiation spectrum distribution.

Plasma Temperature Measurements

Let us dwell on the features of the liner radiation spectra we measured in the region 35 to 25nm.

The spectra of the Ne and N_2 (air) contain lithium- and berilliumlike ion lines, though rather inrease lines of nitrogen and neon heliumlike ions ($^3P_0$ - ^3D transitions) fall within the investigated spectral region. This suggests that the high-temperature radiation from "hot spots" contributes not essentially to the spectrum under investigation. This makes it possible to use the measured spectra to evaluate the temperature of the bulk plasma. The Ar spectrum is featured by a large number of nitrogenlike and carbonlike ion lines. No lines of ions with higher ionicity were indicated.

In determining the temperature for light elements, such as N_2(air), Ne, and Ar, we calculated the temperature dependencies of relative line intensities for ions of different ionicity in terms of a collisional-radiative model[6]. The line pairs were chosen such that the distance between the lines would be no more than a few angstroms. So the absence of a calibration curve for the UFS-5 film used to record spectra introduced only a little error to the evaluation Te.

In the experiment on temperature determination, the air was used instead of nitrogen, which was dictated by the fact that convenient line pair was not available in the nitrogen spectrum. Table 1 lists the experimentally found line pairs used for temperature determination, the relative line intensities, and the temperature estimates for the air, Ne and Ar.

Table 1. Plasma column temperature

Line pairs	T_{exp}, eV	T_{calc}, eV
$\frac{\text{OV } ^3P\text{-}^3D}{\text{NV } ^2P\text{-}^2D}$	20	140-165
$\frac{\text{Ne VIII } ^2P\text{-}^2S}{\text{Ne VII } ^3P\text{-}^3D}$	50	170-190
$\frac{\text{ArXI } ^1D_2\text{-}^1P_1}{\text{ArXII } ^2D_{5/2}\text{-}^2P_{3/2}}$	100	160-180
Kr	100-150	240-260

As the calculation of relative intensities for Kr involves considerable difficulties, the Kr temperature was found from the transparent gratice spectrograph data.

Result and Discussion

From the experiments performed it follows that in going from nitrogen to krypton the energy input per one heavy particle increases more than 20 times. Thus, the observed increase in temperature with atomic number well agrees with the data showing an increase in input energy. At the same time, a 1-D calculation with allowance made for radiation does not predict this abrupt change in temperature as compared to that observed in experiment (see Table 1). It would make sense to propose that increasing the atomic number a more profound compression of the liner takes place. However, from the time-investigated pinhole records taken in the region of quantum energies below 0.4 keV a substantial decrease in final liner radius does not follow.

The final radius varies for various shots between 0.3 and 0.5 mm and it in fact does not decrease in going from nitrogen to krypton. It is possible that with increasing the atomic number there occurs more intense micropinch formation and the related ohmic heating. At least, a number of experimental observations reported in the literature[7,8] suggest that micropinches are formed more intesively with increasing *Z*. Also it is not improvable that pinhole camera records taken at a higher degree of time resolution would show higher-compression degrees for argon and krypton than for neon and nitrogen. Thus, the problem concerning the reason for the increase in liner impedance with the atomic number of the gas requires further studies.

Conclusion

Implosion of a three-cascade liner resulted in a sharp, 3- to 5-fold increase in input energy and radiation intensity in the region 1 to 10 nm. In the experiment it was found that the effective impedance of the liner and the temperature of the plasma column increased with the atomic number of the liner material.

References

1. R.B. Baksht, A.V. Luchinsky, A.V. Fedyunin, Preprint Tomsk Science Center, N 30 (1990)
2. A. Fisher, Chang and A van Prie, J. Appl. Phys., **69**, 3447 (1991)
3. R.B. Baksht, A.V. Luchnsky, A.V. Fedyunin, Zh. Tekh. Fiz., **62**, (1992)
4. A.L. Velicovich, Private communication
5. K.A. Klimenko et al. Pribory i Tekhnika Eksper., N 6, 135 (1992)
6. K.N. Koshelev and N.R. Pereira, J. Appl. Phys., **69**, R21-R42 (1991)
7. K.N. Koshelev et al., J. Phys **D21**, 1827 (1988)

PLASMA DRIVEN IMPLOSIONS OF A BUBBLE-LINER

A Bortolotti and J.G. Linhart
Department of Physics, University of Ferrara,
Via Paradiso 12, 44100 Ferrara, Italy

J. Kravárik and P. Kubeš
Czech Technical University, Prague, Czechoslovakia

ABSTRACT

Thin bubbles ($\delta < 1\,\mu m$) are imploded by a coaxial snowplough. The current sheet remains separated from the dense part of the bubble-liner, the latter compressing a deuterium plasma to ion densities $> 10^{20}$ ions/cm ^{3}. Cylindrical and quasi-spherical implosions are obtained. Results of Schlieren and soft X-ray photography are reported. The application of the spherical implosions to the generation of an intense hohlraum radiation and to fibre super-pinches are discussed.

INTRODUCTION

The concept and theory of acceleration of liners by plasma drive has been described in several papers[1,2,3] and some preliminary experimental results were published[4,5].

In this paper we would like to report some recent experimental results on the implosion of bubbles in a coaxial discharge apparatus (Fig. 1).

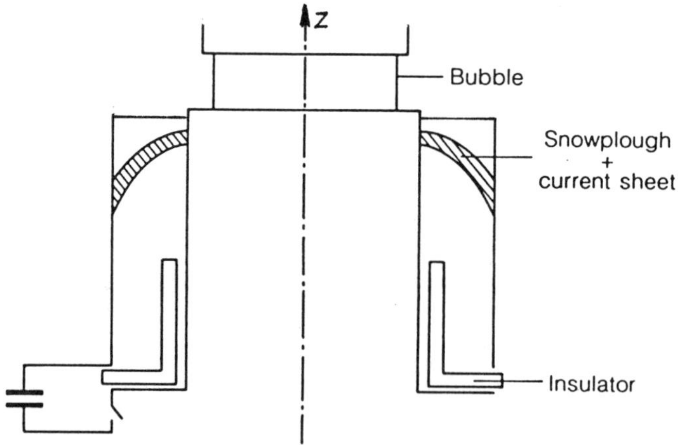

Fig. 1 Sketch of the concept of a plasma drive

In order to know the properties of these bubble-liners, interferometric studies of their thickness have been carried out and we shall report first on the results of these measurements.

The coaxial gun was operated in hydrogen and helium at pressures of $3 \div 10$ Torr, bank voltages of $20 \div 35$ kV. The bank energy at 35 kV was 34 kJ, the current half-period about 5.2 μ sec. We have taken streak photographs of the discharge with and without bubbles.

The process of the plasma drive was then investigated through Schlieren photography, both in a cylindrical and in a quasi-spherical geometry.

Pin hole time integrated pictures were taken showing soft X-ray emission at the time and place of maximum collapse.

The results of these measurements are encouraging and we shall discuss how we intend to apply the spherical liner implosion to the production of intense X-ray field in a spherical hohlraum and to the generation of super-pinches on short fibres.

1. THIN BUBBLES AT LOW PRESSURES

We have decided to use liquid film bubbles as liners for two reasons:

a) the mass of the liner must be small and therefore the liner thickness $\delta < 1\,\mu m$ when condenser-bank energies below 100 kJ are envisaged;

b) the nature of a liquid film guarantees a high degree of uniformity for $\delta < 1\,\mu m$, a uniformity which is difficult to obtain by other methods of liner-production.

In the past both the life-time and the stability of such bubbles have been plagued by two phenomena.

The first was a formation of holes which results sometimes in the complete fold-up of the bubble. In some cases the hole expanded only to a certain diameter, the bubble form remaining almost unaltered in testimony to the membrane-like nature of the bubble film at that stage.

The second reason for the malformation or even the destruction of the bubble was the growth of secondary bubbles - usually at the metal lips where the main bubble was attached.

We came to the conclusion that the hole formation was favoured by the high density of micelles in the liquid used and consequently it was decided to work with liquids with a low concentration of surfactants (less than 1% by volume) and a relatively high percentage of stabilisers such as Agar-Agar.

In order to minimise the growth of secondaries, it was decided to draw the bubbles at pressures very much smaller than atmospheric, e.g. at $p \leq 20$ Torr and using liquids that have been degassed down to a pressure of ~ 10 Torr.

The adoption of these measures improved noticeably the reproducibility of formation and quality of shape of the bubbles.

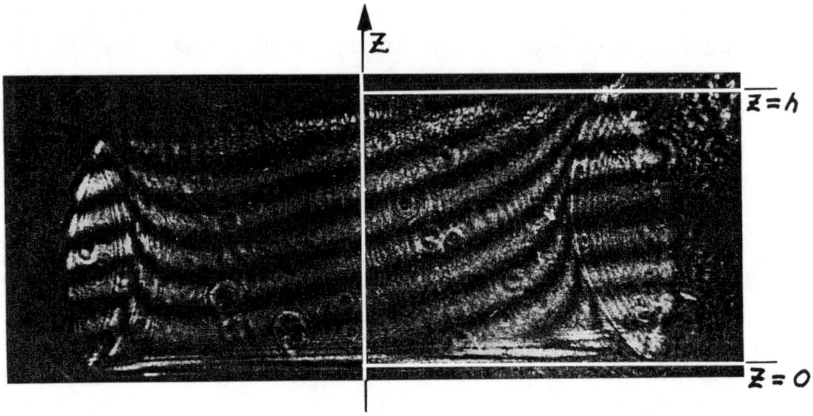

Fig. 2

In order to understand the processes of evolution of the bubble thickness and its uniformity, it was decided to use diagnostics based on interferometry. The fringe distortion and shift have provided us with the means of determining the history of the film-thinning.

A systematic study of the uniformity of the thickness d in cylindrical bubbles drawn and maintained at $p = 20$ Torr was made. A typical interferogram is shown in Fig. 2. During the first 1/2 minute after the bubble has been drawn, nonuniformities in it are observed both axially (z) and azimuthally (ϕ). The initial thickness (even at the top of the bubble) often exceeds 1 μm. Within 1 min the film becomes azimuthally uniform and the only variation in d is in the axial direction. The thickness was evaluated (assuming that $\partial d/\partial\phi = 0$) observing at what radius b a fringe is shifted by 1/2 fringe-width with respect to its position on axis. The thickness d at a given height z can then be calculated from

$$d = \frac{\lambda}{8N}\left[\left(1 - \frac{b^2}{a^2N^2}\right)^{-1/2} - 1\right]^{-1} \qquad (1)$$

where N is the refraction index of the liquid (~ 1.34), λ the wavelength of the laser light and a the radius of the bubble .

A typical results is shown in Fig. 3. The difference between the thickness at the top and bottom of the bubble tends to disappear and the average $< d >$ converges towards 0.2 μm after about 3 minutes.

At pressures < 20 T and after about 10 minutes the liquid film changes (due to evaporation of unbound H_2O) into an elastic membrane - conserving the homogeneity of the liquid film and preserving a small degree of elasticity (up to ~ 10% of extension).

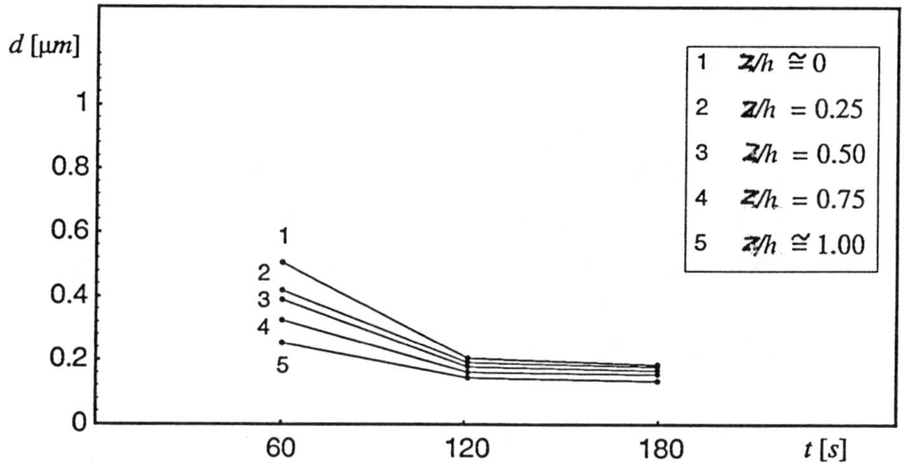

Fig. 3

2. IMPLOSION OF THE BUBBLE LINER

From a one-dimensional numerical analysis[5] and from simple heat-diffusion calculations[4] it follows that the heat energy input into the bubble is mainly by heat conduction from the snowplough zone in which the Joule's heat is dissipated - meaning that only a small portion of the latter reaches the liner during the implosion. However, the situation near the contact between the current sheet and the electrodes is far from one-dimensional and some diffusion of the B_{ϕ} field may occur here.

We have taken streak photographs in the mid-plane of the bubble (Fig. 4). The gas inside and outside the bubble was H_2, pressure $p \sim 7$ Torr, $V = 28$ kV. The implosion is symmetric but rather slow ($V_{max} \sim 2.5$ cm/μ sec). The light intensity on the photographs was processed by computer, showing a fine structure in which the compression of the hydrogen plasma inside the bubble is clearly visible. The corresponding soft X-ray pinhole picture (6 μm Mylar + 800 Å Al filter, Fig. 5) indicates that the $T_{max} > 300$ (eV) and $r_{min} \sim 1$ (mm). This agrees reasonably well with our 1-D numerical simulation, if anything it suggests that the thickness δ of the liner is smaller and the temperature higher than that given by the computer[5]. The density of the compressed hydrogen plasma must be higher than 10^{20} ions/cm^3 (from $\rho v^2 \sim 2nkT$).

The frame photos were not satisfactory (Fig. 6) owing mainly to some technical difficulties which will be mentioned in the conclusion.

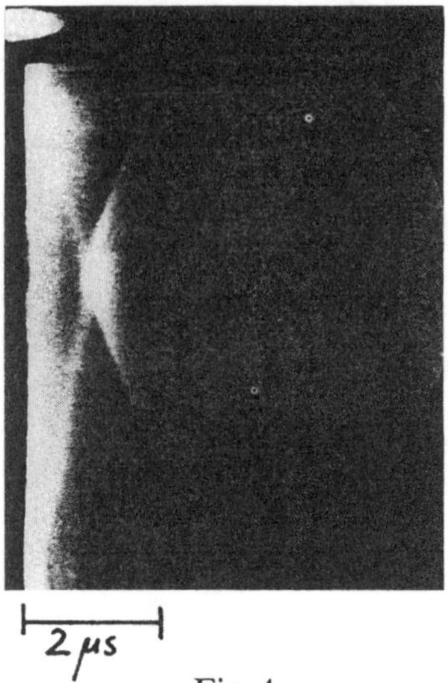

Fig. 4

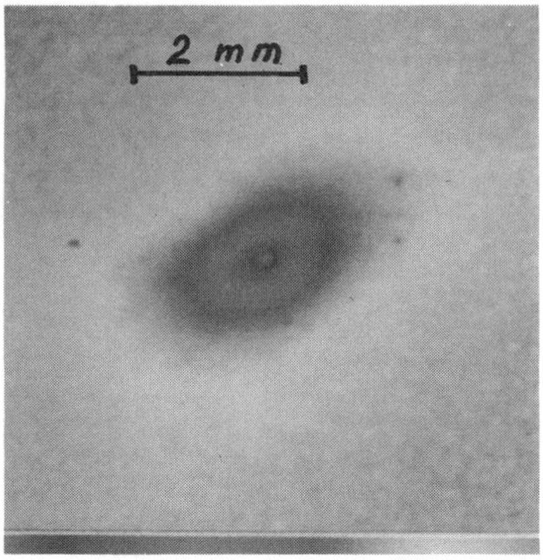

Fig. 5

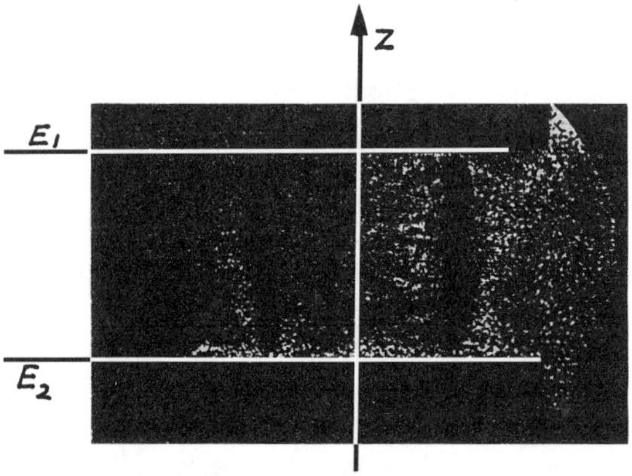

Fig. 6 Frame ~ 1 μsec after bubble-SP contact.

From the I and I traces it was possible to obtain the time τ of the arrival of the current sheet at the position of the bubble (Fig. 7).

The Schlieren pictures show very clearly the plasma layer which pushes the bubble (Fig. 8). The snowplough and the current sheet in hydrogen is never symmetric and homogeneous, yet when it piles up on top of the bubble it becomes a rather uniform plasma piston (Fig. 9).

The implosion appears symmetric and again, as in the case of the streak pictures, slower than it should be according to the numerical analysis.

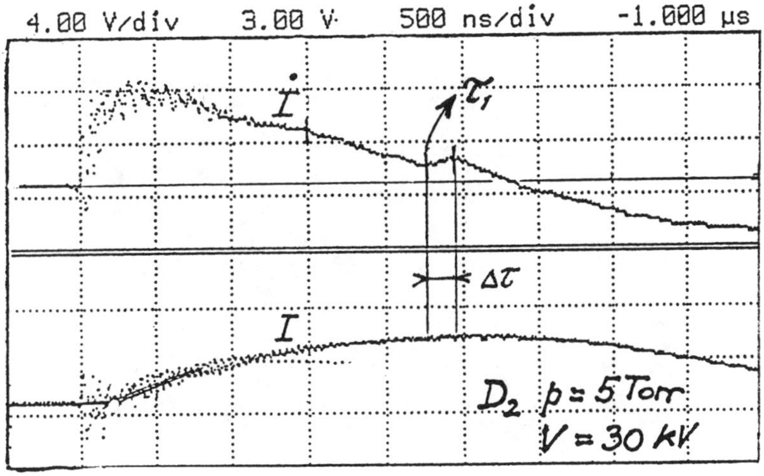

Fig. 7 I and I traces; $\tau_1 \cong$ time of B-SP contact;
Δt the period of SP bounce off on B[5].

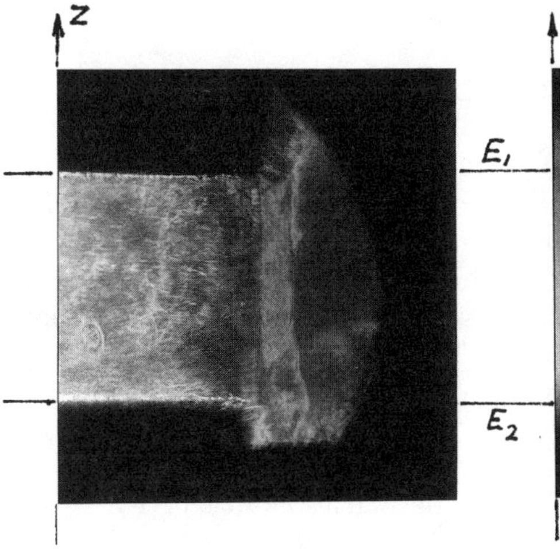

Fig. 8

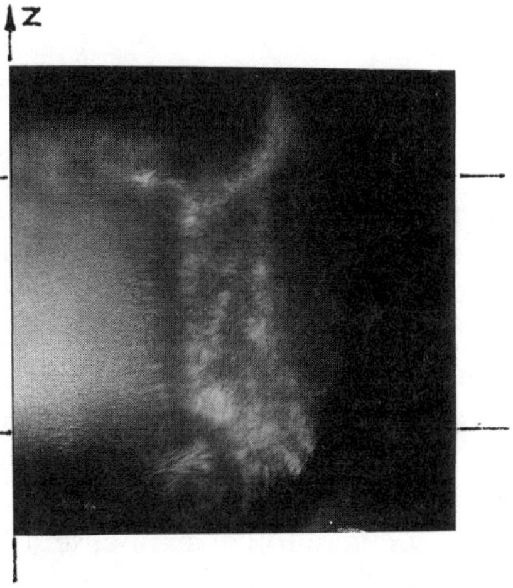

Fig. 9

In the last series of experiments we used conical electrodes and convex bubbles. The Schlieren shows that a quasi-spherical implosion has been achieved (Fig. 10).

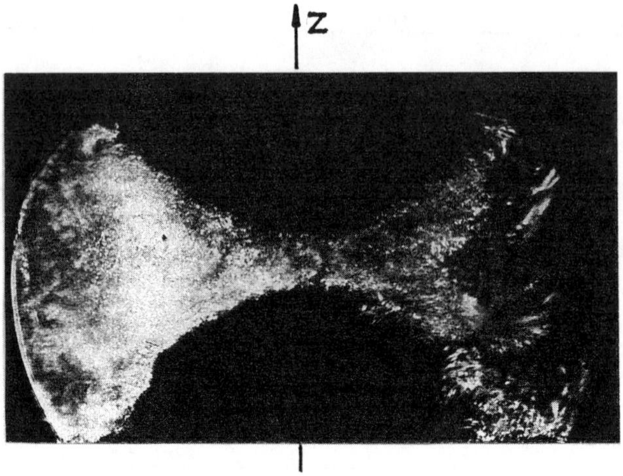

Fig. 10 Left side overexposed. R.h.s. shows the bubble just before a point collapse.

When (by mistake) a concave bubble has been imploded, the last stage of the collapse resembles a plasma focus. The estimate of the bubble thickness in these cases is $\delta < 1$ (mm). Considering the mass-concentration due to a quasi-spherical convergence we estimate the $\rho_L \approx 0.05$ g/cm ^{3}.

CONCLUSION

The liner speeds deduced from the streak and Schlieren pictures are low compared to the computed speeds. This is almost certainly due to some of the current remaining within the coaxial gun region (impurity caused breakdown on the insulator) and consequently the current in the current sheet visible in the Schlieren is only about 70% of the total current. In spite of this defect the plasma propulsion of the bubble is as it was hoped to be, i.e. the liner remains dense and collapses in a quasi-spherical fashion when convex bubbles are employed.

There are at least two interesting applications of the quasi-spherical liner implosion:

1) Generation of an intense soft X-ray flash in a spherical cavity.
2) Induction of a Mega-ampere Z-pinch in a short fibre.

The concept of the 1st mechanism is shown in Fig. 11. A portion of the kinetic energy of the liner L is converted during impact on the hollow sphere (radiator) into X-ray energy[6]. About 1/2 of this energy will appear in the X-ray field inside the sphere S. If S is heated to a temperature T, the radiation energy inside S will be

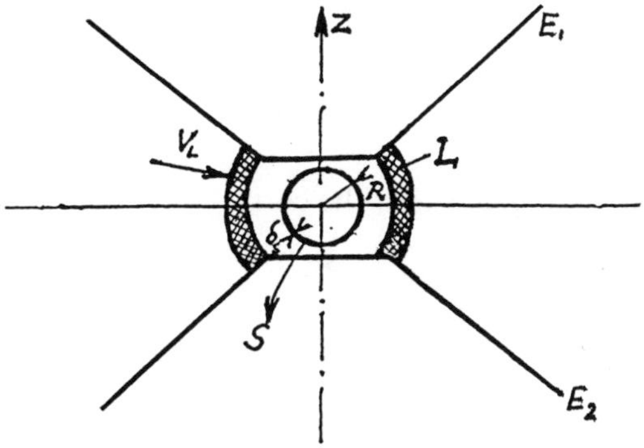

Fig. 11 Liner before impact on radiator S.

$$W_r = \frac{4}{3}\pi R^3 \cdot c^{-1} \cdot \sigma T^4 \tag{2}$$

whereas the thermal energy of the S (heated by the impacting L) will be

$$W_s = 4\pi R^2 \cdot \delta_s \cdot \frac{3}{2}(1+Z)kT \cdot n_s \tag{3}$$

In order that the kinetic energy W_L of the liner L is transformed into W_r efficiently it is necessary that $W_r \gg W_s$ or

$$T > 10^7\left[(1+Z)\cdot\frac{\delta_s}{R}\right]^{1/3} \tag{4}$$

Ex: $R = 0.2$, $\delta_s = 10^{-3}$, $(1+Z) = 4$ we get $T > 2.7 \cdot 10^6$ (°K).

We will further require that $V_L > 10^7$ cm/sec and the mass of δ at least as large as L. It seems that all these requirements could be met in future experiments. If so, the radiation inside δ should be large enough to consider an indirect drive of pellets containing a DT target.

The second application (Fig. 12), i.e. the compression of a B_ϕ flux of a Z-pinch by a quasi-spherical collapse of a liner, has been mentioned in Ref. 8.

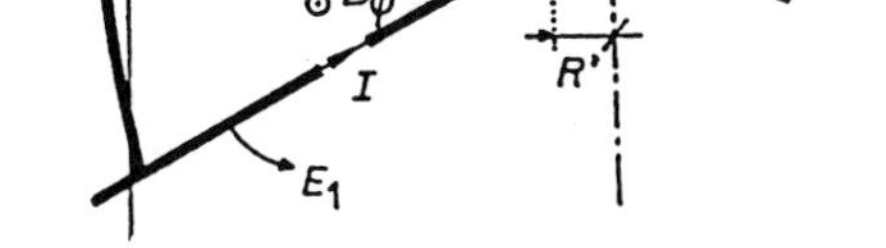

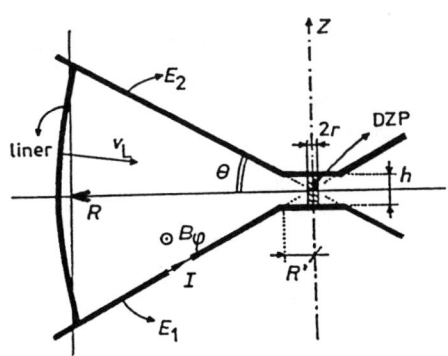

Fig. 12 B_ϕ flux compression in a conical geometry.

The inductance of the conical system is

$$L_c = 2 \cdot 10^{-9}(R - R')f(\theta) + 10^{-9} h \ln \frac{R'}{r}, \; f(\theta) = \ln \frac{1 + \sin\theta}{1 - \sin\theta} \tag{5}$$

Assuming flux conservation we get for the current I' at the moment when the liner reaches R'

$$I' = \frac{\phi}{10^{-9} h \ln R'/r} \approx I_0 \left[\frac{2R f(\theta)}{h \ln R'/r} + 1 \right] \tag{6}$$

The current amplification is $2(R/h) \cdot (f(\theta)/\ln R'/r) + 1$ and could be at least equal to 10. Consequently if $I_0 = 100$(kA) the pinch currents corresponding to the Z-pinch compression in the small cylindrical space $r < R'$ will be > 1 (MAmp), the final compression will be effected within a time of R'/V_L which could be shorter than 100 (nsec).

ACKNOWLEDGEMENTS

The authors wish to thank ENEA and EURATOM for supporting this research. Recognition is due to Dr. R. Appartaim, P. Lee and F. Beg of Imperial College for their help with the streak and pinhole photograph and to Ing. J. Hakr for technical assistance with the interferometric measurements.

REFERENCES

1. J.G. Linhart, Nucl. Fusion 19, 264 (1979).
2. J.G. Linhart, IEEE Trans. Plasma Sci. 16, 438 (1988).
3. J.G. Linhart, Nucl. Instrum. Methods in Phys. Res. A278, 114 (1989).
4. H. Kilic et al., 19th EPS Conf. on Controlled Fus. and Plasma Phys., part 2, p.330 (Innsbruck 1992).
5. L. Bilbao et al., Kerntechnik, 57, 330 (1992).
6. J.G. Linhart, Laser and Particle Beams 2, 87 (1984).
7. J.G. Linhart, Czech. J. Phys. 42 (1992).

DIRECT DRIVE FOIL IMPLOSION EXPERIMENTS ON PEGASUS II

J. C. Cochrane, R. R. Bartsch, J. F. Benage, P. R. Forman,
R. F. Gribble, M. Y. P. Hockaday, R. G. Hockaday,
L. S. Ladish, H. Oona, J. V. Parker, J. S. Shlachter, F. J. Wysocki
Los Alamos National Laboratory, Los Alamos, NM 87544

ABSTRACT

Pegasus II is the upgraded version of Pegasus, a pulsed power machine used in the Los Alamos Above Ground Experiments (AGEX) program. The goal of the program is to produce an intense (>100 TW) source of soft x-rays from the thermalization of the KE of a 1 to 10 MJ collapsing plasma source. The radiation pulse should have a maximum duration of several tens of nanoseconds and will be used in the study of fusion conditions and material properties. This paper addresses z-pinch experiments done on a capacitor bank where the radiating plasma source is formed by an imploding annular aluminum foil driven by the JxB forces generated by the current flowing through the foil.

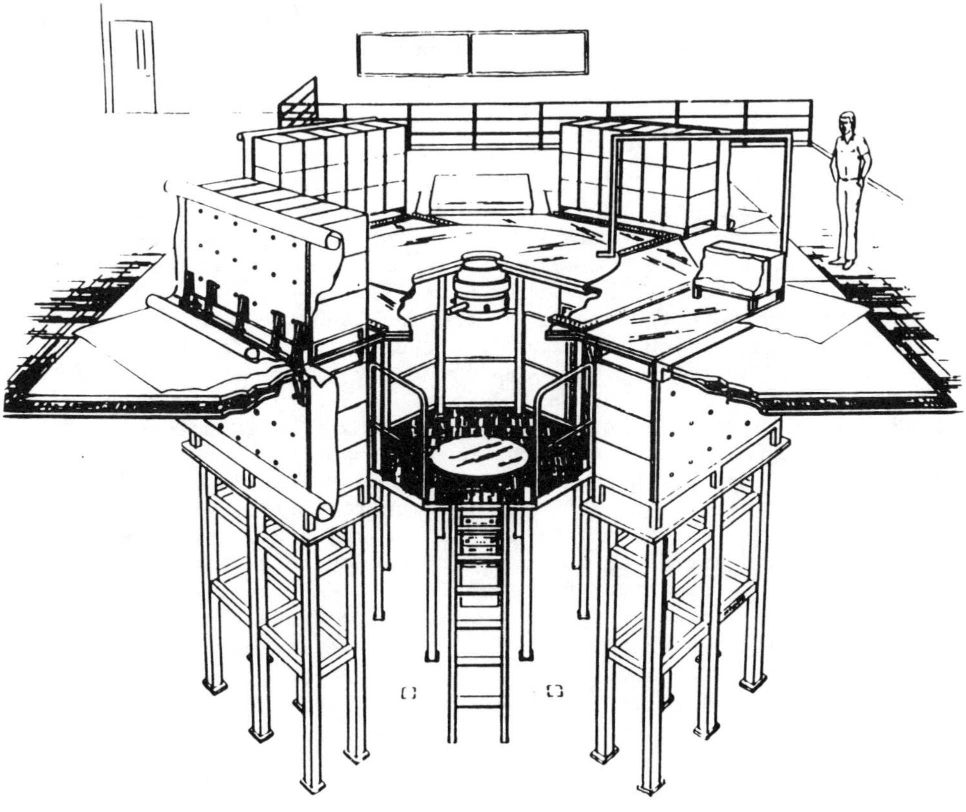

Fig. 1. Pegasus II capacitor bank facility.

FACILITY DESCRIPTION

Pegasus II machine parameters include a stored energy of 4.3 MJ at 100 kV, a system inductance of 30 nH and current capability of 15 MA. This quadruples the energy of Pegasus I at this voltage. The upgrade was accomplished by replacing the capacitors rated at 10 kJ stored energy at 60 kV with capacitors rated at 30 kJ at 50 kV. The new capacitors have a current capability of 250 kA/capacitor and can stand up to 20% reversal at full charge for a rated lifetime of over 3000 shots. To stay within this voltage reversal specification, series fuses are employed to shut off the current after peak current. The bank itself is composed of two halves charged to opposite polarity. Each half has four modules with eighteen capacitors each. The modules are placed around a radial transmission line with the load in the center of the line (see Fig. 1). Detonator switches that form an annular aluminum jet that penetrates the polyethylene switch insulation are used to switch the bank. The facility has been used in "direct drive" z-pinch implosions of thin aluminum foils, high magnetic field diffusion experiments, pulse sharpening and switching experiments using a plasma flow switch, and most recently, in liner experiments where the load is an aluminum cylinder with a 0.4 mm thick wall.

EXPERIMENTAL RESULTS

The purpose of the "direct drive" z-pinch implosions is to experimentally test the theoretical predictions of instabilities and their growth rates and to try various means to suppress these instabilities.[1] Direct drive shots also allow us to separately study foil implosion dynamics that would otherwise be possibly hidden in interactions between the plasma of the foil and the plasma of the pulse sharpening switch. In the direct drive mode, the bank is fired directly into the foil without employing any switching to sharpen the driving current pulse. The foil dimensions are typically 2 cm in height, 5 cm in radius, and 2.5×10^{-5} to 7.5×10^{-5} cm in thickness. The foils are manufactured by evaporating aluminum onto a polyvinyl alcohol (pva) form. The pva form is then dissolved in a bath of water leaving the foil attached to the mounting rings which are themselves on an insertion fixture. Discharging Pegasus directly into such a load yields an implosion time much less than the quarter cycle time of the capacitor bank meaning that a relatively small fraction of the stored energy is available to drive the implosion. Foil implosion times from 1.8 to 2.5 μs are observed depending on the foil thickness. The quarter cycle time for Pegasus I was 4 μs and for Pegasus II is 8 μs with a static system inductance of 30 nH. The current waveform is shown in Fig. 2 for a Pegasus II "direct drive" shot.

Earlier work on Pegasus I used a pure aluminum foil 2.5×10^{-5} cm in thickness that was secured both top and bottom by steel fingers attached to the foil mounting rings.[2] This configuration gave a radiation pulse of about 200 ns in width with some interesting structure. The difficulty of mounting a foil in this configuration has led us to using a foil that is suspended by its top mounting ring only. A 1-mm gap around the bottom ring is bridged by plasma when the bank fires. More recent foils have also had a thin parylene coating on their inside surface. This was originally added for strength in the early experiments. Calculations have shown that using a 3x thicker foil than the 2.5×10^{-5} cm thickness would result in a more stable implosion and of course couple more of the bank's energy into the implosion. The enhanced stability should actually give a hotter plasma even with the extra mass. These experiments were done both on Pegasus I and Pegasus II and are discussed below.

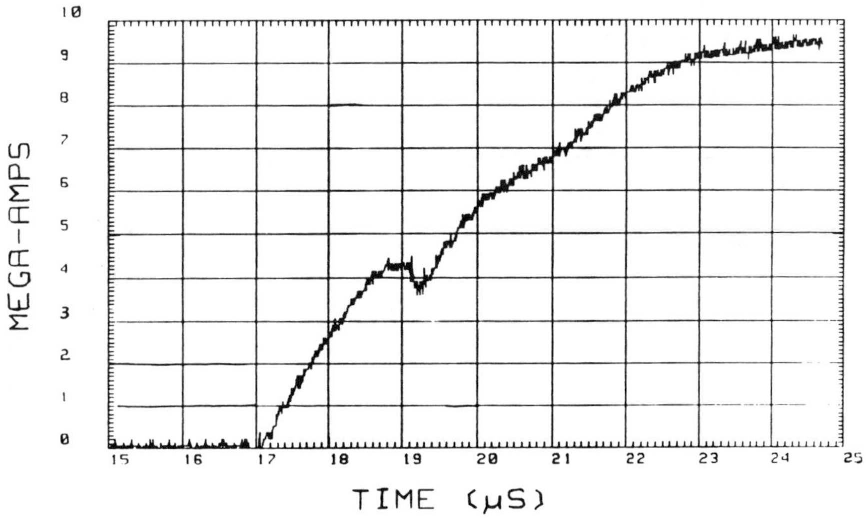

Fig. 2. Current waveform for a Pegasus II "direct drive" shot.

In Fig. 3, a typical array of electrical diagnostics is shown for a plasma flow switch shot. The array is the same for a direct drive shot except that the number of B-dot probes in the coaxial region of the power flow channel is reduced. Note that B-dot loops are located in the load slot at radii both outside and inside of the foil radius. Diagnostics fielded on the foil implosion shots consist of current and voltage diagnostics mentioned above in Fig. 3 and also include a four channel filtered bolometer, a four channel filtered XRD array, a grazing incidence spectrometer, a time resolved multichannel transmission grating spectrometer, x-ray pinhole cameras, and fast framing and streak cameras. All diagnostics that are recorded electrically have their signals transmitted over fiber optic links to the shielded screen room. All diagnostics are on battery power when the machine is fired to avoid ground loop problems. This method of having no electrical connections to the machine when it is charged and fired has given essentially no noise fed into the screen room or building via the AC mains.

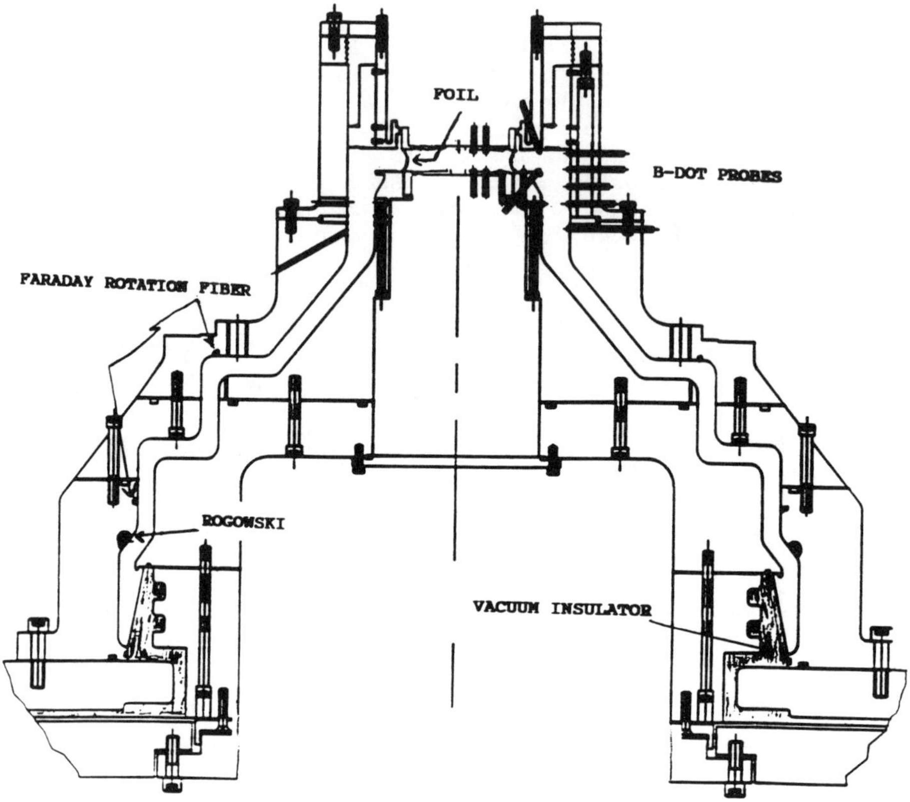

Fig. 3. Pegasus II power flow channel.

The implosion on Pegasus II produced a total radiated energy of 125 kJ as measured by a thin film bolometer. The peak current before the I-dot reversal, due to the inductive voltage produced by the implosion, was 4.3 MA. The charge voltage on Pegasus II was chosen to match the I-dot on the Pegasus I experiment (i.e., 80 kV for Pegasus II vs. 84 kV for Pegasus I). Fast framing camera photographs of the actual implosion (Fig. 4) show very similar structure for the two shots and are in good agreement with theoretical predictions. This structure is expected to broaden the radiation pulse when the foil stagnates on axis.

The framing camera photographs in Fig. 4 have an interframe time of 167 ns. The two vertical lines are vanes. Time runs alternately bottom to top, left to right. The instabilities are clearly visible by the distortion of the luminous edge of the foil. The unfiltered XRD channel gave a pulse width of 500 ns. The signal begins as the current begins to "roll over" (see Fig. 2) and peaks near the current minimum. The filtered XRD channels had a FWHM of about 200 ns with a single peak on the Pegasus II shot. On the identical Pegasus I shot, the XRD signals had two peaks, which indicate two separate implosions. This was predicted in simulations as a hot tenuous plasma implodes ahead of the main mass of the plasma. The difference between these supposedly two identical shots is still being investigated theoretically. The filtered XRD signals for a Pegasus II shot are shown in Fig. 5 and their filters are identified in Table I. Major time divisions are 500 ns for Fig 5.

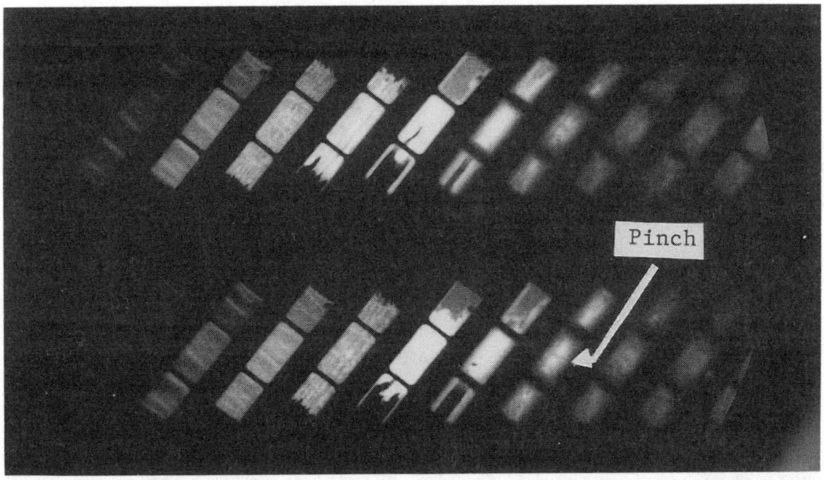

Fig. 4. Framing camera photographs of a "direct drive" foil implosion.

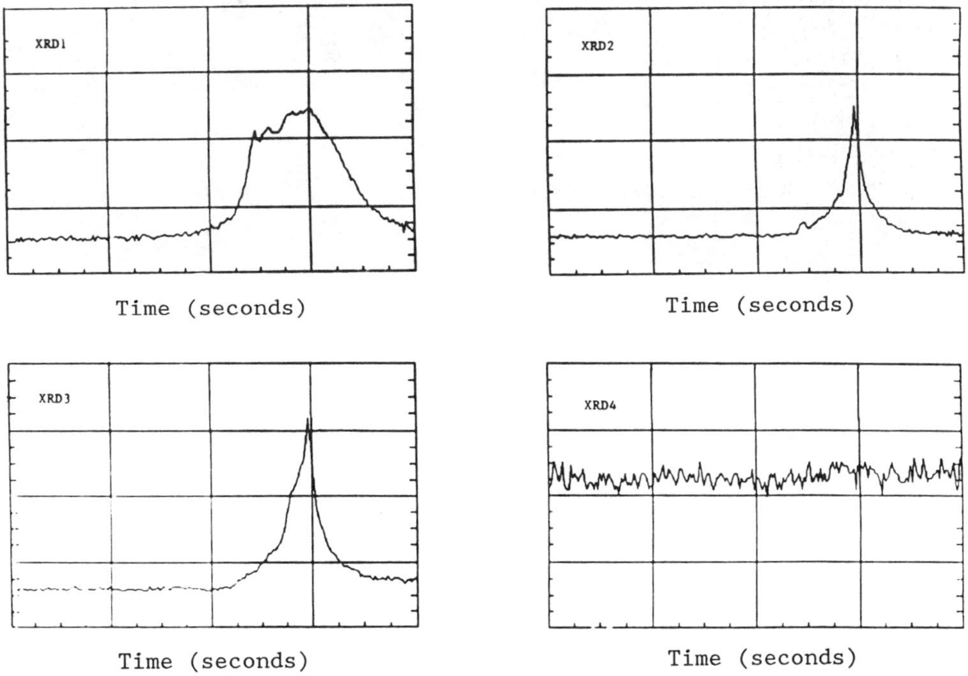

Fig. 5. The filtered XRD signals for a Pegasus II shot.

Table I. XRD filters

XRD1	100 lpi Au mesh (1.79% transmission)
XRD2	158 $\mu g/cm^2$ Saran + 20 $\mu g/cm^2$ Al + 150 lpi SS mesh
XRD3	600 $\mu g/cm^2$ Al + 150 lpi SS mesh
XRD4	400 $\mu g/cm^2$ Mylar + 24 $\mu g/cm^2$ Al + 150 lpi SS mesh

Spectra were obtained on both a grazing incidence and a transmission grating spectrometer. A calibration problem with the transmission grating spectrometer at low photon energies prevents agreement on a blackbody temperature or on whether the spectrum has a blackbody shape. The grazing incidence spectrometer data yields a temperature between 37.5 and 50 eV. A different cutoff filter should provide a closer estimate of the fitted blackbody temperature.

Figure 6 is the time resolved spectrum from the transmission grating spectrometer. Intensity is proportional to the energy in a wavelength interval and the photon energy increases from right to left (65 eV to 1896 eV). Time is streaked from top to bottom. The total streak duration is 2 microseconds.

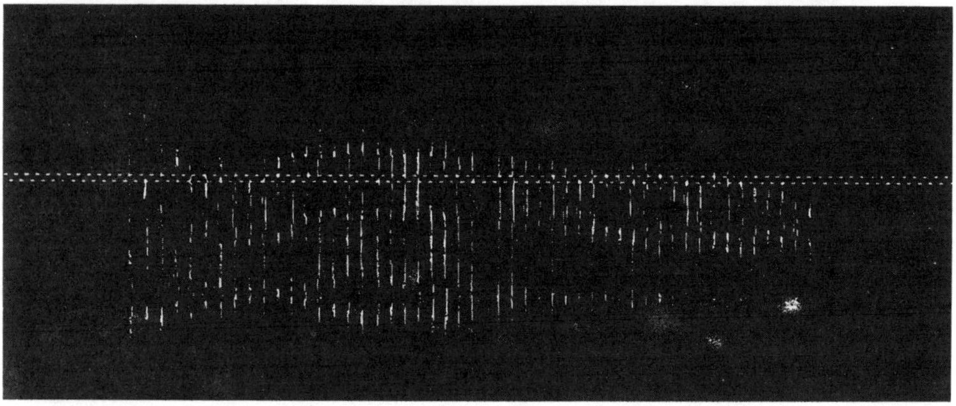

Fig. 6. The time resolved spectrum from the transmission grating spectrometer.

SUMMARY

Direct drive foil implosion experiments have been performed on the Pegasus facility to investigate the validity of theoretical predictions. To date, the predictions have accurately predicted the radiation output and instability growth in the imploding foils. Future shots are planned to investigate the effect of the ratio of the parylene thickness to the aluminum thickness on the stability of the implosion and to compare these results to pur aluminum foil implosions. The results of these tests will determine the best configuration for an imploding load to be used with a plasma flow switch. The plasma flow switch should increase the energy coupled into the load and produce a more narrow radiation pulse.

ACKNOWLEDGMENTS

We acknowledge the valuable and expert work of the Pegasus mechanical crew of F. Garcia, M. Garcia, D. Rodriguez, P. Roybal, G. Sandoval, and B. Stone. The B-dot probes, Rogowski belts, data acquisition links, and machine electrical maintenance are the responsibility of B. Anderson and F. Garcia.

This work is performed under the auspices of the U. S. Department of Energy under contract W-7405-ENG-36.

REFERENCES

1. D. L. Peterson et al., these proceedings.
2. J. C. Cochrane et al., "Foil Implosion Studies on Pegasus," Proceedings of the 7th IEEE Pulsed Power Conference, Monterey, California (1989).

TWO-DIMENSIONAL SIMULATIONS OF FOIL IMPLOSION EXPERIMENTS ON THE LOS ALAMOS PEGASUS CAPACITOR BANK

D. L. Peterson, R. L. Bowers, J. H. Brownell,
A. E. Greene, H. Lee, and W. Matuska
Los Alamos National Laboratory, Los Alamos, NM 87545

ABSTRACT

A number of z-pinch experiments have been conducted at Los Alamos on the Pegasus capacitor bank in which 2-cm high, 5-cm radius, thin foil loads were imploded with currents in excess of 3 MA. Two-dimensional (2-D) radiation magnetohydrodynamic (RMHD) simulations of these implosions have been performed to model the implosion dynamics and subsequent generation of an x-ray pulse. Comparisons of the simulation instability development with visible light framing camera photographs show good agreement in wavelength and amplitude and illustrate the instability evolution from short to long wavelengths and a final disruption of the imploding plasma shell. The calculations also show good agreement with experimental timing and measured current and voltage waveforms, and also reproduce features characteristic of the x-ray output. These include a broad pulsewidth, and the presence of multiple peaks and small time scale structures, features which cannot be reproduced by one-dimensional (1-D) models. X-ray spectra obtained from the calculated pinch also reproduce qualitative features in the measured spectra.

INTRODUCTION

The magnetic implosion of cylindrical, conducting loads has been used in pulsed power experiments for decades as a source of soft x-rays.[1] The Los Alamos Foil Implosion program is developing imploding plasmas as an x-ray source, with the ultimate goal being a 1 to 10 MJ radiation source, delivered in a pulse of greater than 100 TW power. The prime power source could be a large capacitor bank or an explosive flux-compression generator.[2] Experiments have been designed and fielded with thin (several thousand angstrom thick) cylindrical aluminum foil loads.[3, 4, 5] In this paper we discuss computational modeling and the results of experiments in which 4.9 mg and 13.0 mg aluminum loads were imploded on the Pegasus capacitor bank facility.[6]

One-dimensional models of these implosions generally predict efficient conversion of magnetic energy into x-ray output and radiation pulsewidths that are very short (full width at half maximum under 5 ns). The imploding plasma is, however, subject to the development of magnetically driven Rayleigh-Taylor instabilities. As a result of this instability growth, the radiation pulse is broadened and the maximum power is reduced. Understanding the development of the instabilities and how they effect the radiation pulse, as well as being able to properly simulate experiments which show the effects of the instabilities, is the first step in designing implosion loads which will produce higher quality radiation pulses.

We have studied the physics of the imploding plasmas using two codes: a 1-D RMHD Lagrangian code, which can simulate the development of the solid foil through melt, vaporization, and the formation of the plasma; and a 2-D RMHD Eulerian code which is well suited to examining instability growth. Both codes use SESAME equation-of-state tables and are self-consistently coupled to ladder circuit networks, which simulate the capacitor bank and associated power-flow hardware. The 2-D simulations begin with profiles of density, temperature, velocity, and magnetic field provided from the 1-D simulation at a point after the plasma has expanded from the original thin solid foil (typically, the point of maximum expansion of the plasma is used). Perturbations in the density of the plasma are then imposed to mock-up variations which arise in the experiment. These perturbations then seed the growth of magnetically driven Rayleigh-Taylor instabilities. The simulations can then be compared with experiments where the effects of the instability development are seen not only in the observed radiation pulse, but in current and voltage waveforms, in visible light framing camera photos of the imploding plasma, in time-integrated x-ray pinhole photos of the stagnation region, and in time-dependent spectroscopy measurements.

2-D INSTABILITY DEVELOPMENT

In a typical 2-D simulation, the instability development is seeded by the imposition of random density variations on the 1-D density profile in a range between $-\delta$ and $+\delta$ (typical values of δ are between 5% and 20%). Random perturbations allow the development of a range of instability wavelengths, with shorter wavelengths growing fastest. In general, the results are insensitive to the exact pattern of the initial random perturbations.

Figure 1 shows the instability growth in a simulation using a 20% random density perturbation. Shown in Fig. 2 are the current and radiation pulse profiles for a 1-D simulation and for the 2-D perturbed simulation. The peaks of the two radiation pulses in Fig. 2 have been normalized to a value of 2.0 in each case for ease of comparison (the 2-D peak is much lower than the 1-D peak, representing a lower peak temperature). Figure 2 also shows structure in the 2-D radiation pulse and current waveforms which is absent in 1-D.

As may be seen in the density contours in Fig. 1, the initial short wavelength modes saturate and give way to longer wavelengths, with a dramatic "bursting" of a Rayleigh-Taylor bubble through the plasma shell shown at the last time. From the viewpoint of the external circuit drive, the inductance change in the implosion remains essentially like that in 1-D until about 1.5 μs (as evidenced in the rB_θ contours in Fig. 1). Note that the linear instability growth regime occurs during the initial part of this time, and thus it is the nonlinear development that is paramount in the dynamics of the imploding shell. Significant effects begin to arise at the point of bubble burst when magnetic field and a small amount of

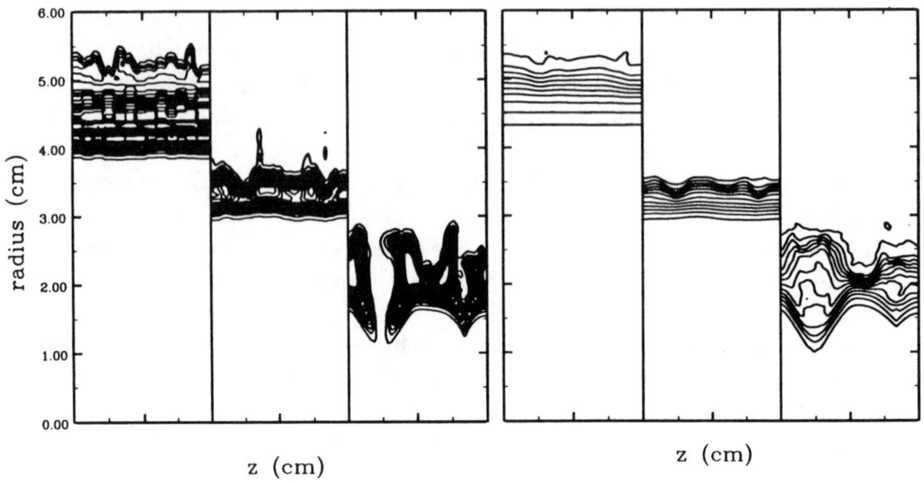

Fig. 1. (Left) 2-D simulation isodensity contours in the $r-z$ plane for a perturbed implosion of a 4.9 mg load at times $t = 1.1, 1.5$ and 1.7 μs. (Right) Current streamlines (contours of rB_θ) at the same times.

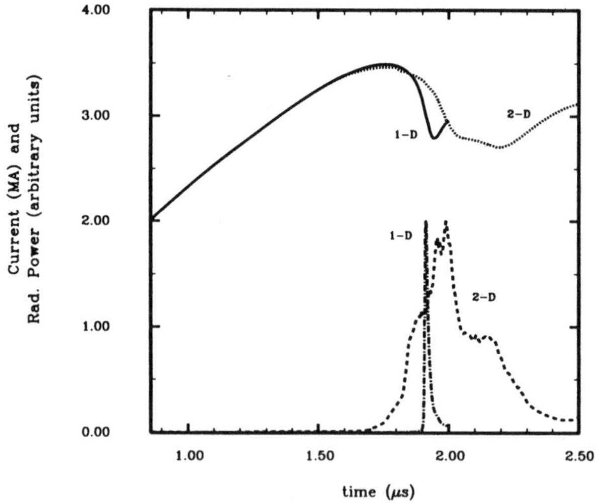

Fig. 2. Currents (solid and dotted curves, in MA) and radiation power (dot-dash and dash curves, arbitrary units, scaled to a peak value of 2.0) for 1-D and 2-D simulations of a 4.9 mg load.

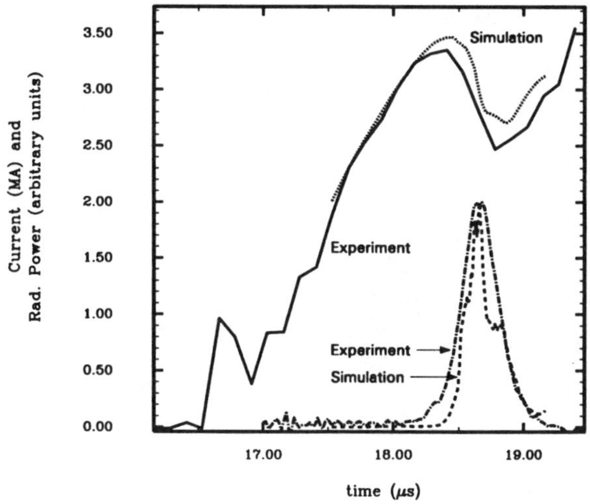

Fig. 3. Experimental current (solid) and x-ray diode (XRD) radiation measurement (dot-dash) compared with a 2-D simulation current (dot) and radiation power (dash) for the 4.9 mg implosion experiment. Current is shown in MA, and the XRD signal and radiation power have both been scaled to a peak value of 2.0 in arbitrary units.

material are driven to the axis. The change of inductance in this process causes the current to drop, or rise at a reduced rate, and at the same time the radiation pulse begins (as seen at $t = 1.7\ \mu s$ in Fig. 2). The current path then changes to allow flow between the instability spikes, which effectively saturates further development of the instability. When sufficient material drifts into the region between the spikes to support current flow, a re-acceleration occurs of the main plasma shell, resulting in a final pinch and radiation pulse, and a local current minimum. This may be seen at $t = 2.2\ \mu s$ in Fig. 2. The relative timing, radiation pulse shape and width, and current waveform thus reveal information about the instability growth.

EXPERIMENT WITH A 2500 Å Al FOIL

The first experiment to be discussed here was conducted with a 5-cm radius, 2-cm high, 2500 Å thick aluminum foil with a 1000 Å backing of parylene (for a total mass of 4.9 mg). Simulation comparisons with this experiment (Fig. 3) show good agreement in timing and peak current and in the radiation pulsewidth. Note in particular the beginning of the x-ray pulse coincides with the point at which the current begins to turn over, in agreement with the mechanism of the bubble burst through the plasma shell described earlier.

Figure 4 shows a comparison between the computational model and the optical framing camera data at several times during the implosion. The comparison

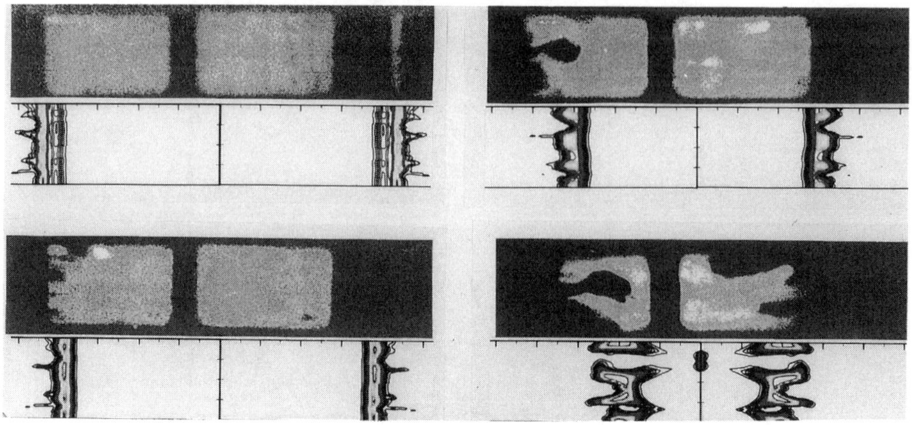

Fig. 4. Visible light framing camera pictures (above) paired with density contours from the corresponding 2-D simulation (below) at four times during the 4.9 mg implosion (time proceeds from upper left, lower left, upper right, to lower right). The pictures and plots are to scale and the pinch is viewed side-on. Three return conductors block part of the view in the photos (at extreme left, near the center, and at the right).

shows the development of short to long wavelength modes in both the experiment and the simulation with good agreement in the instability wavelength and amplitude. In the last frame shown, bubble burst has occurred, coincident with the start of radiation and the turn-over in the current. Also at this time, a tilt in the pinch can be perceived, exhibiting 3-D effects. We find excellent agreement between amplitude and lengthscale of the features in the instability growth prior to the onset of 3-D effects.

The data from the 4.9 mg implosion demonstrates that magnetic instabilities are present, that the x-ray pulsewidth is broader than predicted by 1-D models, and that a simple randomly perturbed initial model can reproduce many of the observed features of the implosion. In this model the adjustable parameter is the amplitude of the density perturbations, and the best fit to the data is obtained with an amplitude of $\delta = 20\%$.

EXPERIMENTS WITH 7500 Å Al FOILS

Based upon the experience of the first experiment, simulations were performed for foils of increased mass. These simulations indicated that for a similar initial perturbation level, there should be substantially less instability development and a narrowing of the radiation pulse. Two experiments using an Al thickness of 7500 Å with a backing of 1000 Å of parylene (for a total mass of 13.0 mg) are discussed in this section.

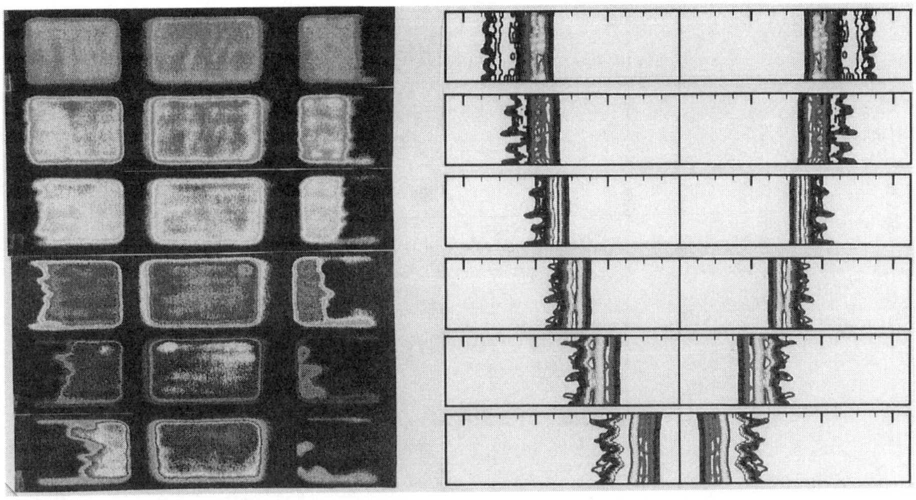

Fig. 5. Visible light framing camera pictures (left) paired with density contours from the corresponding 2-D simulation (right) for the first 13.0 mg implosion. The pictures and plots are to scale and the pinch is viewed side-on. Three return current conductors block part of the view in the photos (at extreme left, center left, and center right).

The first of these experiments suffered a short (probably caused by a vacuum leak, which may also have introduced additional material into the system) in the power flow system. This resulted in a reduced current delivered to the load, especially at late time when the instability growth is most dramatic. The 2-D simulation of the experiment therefore imposed the measured current to drive the load, rather than using the self-consistent circuit model, since the nature of the short is unknown. The instability development, as captured by the framing camera photography, provided an excellent comparison with that of the simulation as can be seen in Fig. 5. The photos show clear evidence of banding, which may be interpreted as variations associated with 2-D axisymmetric instabilities. Again, the evolution from short to long wavelength growth can be seen, as well as good correlation between the observed and calculated instability amplitudes. We do not expect the exact distribution in the $r-z$ plane to be the same in the experiment and simulation as this distribution would depend on the exact random density pattern imposed initially. This implosion seemed to have better stability properties, though the reason for this may have been due to the reduction of the drive current.

The second experiment, also using a 7500 Å Al foil, failed to produce framing camera data, but was otherwise successful diagnosed. The radiation pulsewidth was not reduced, but rather increased, suggesting a higher level of initial perturba-

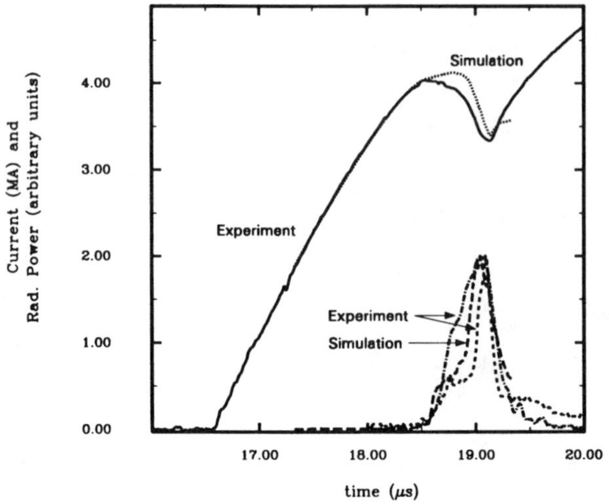

Fig. 6. Experimental current (solid) and two XRD radiation measurements (dot-dash and short-dash, representing two energy filterings) compared with a 2-D simulation current (dot) and radiation power (long-dash) for the the second 13.0 mg implosion experiment. Current is shown in MA, and the XRD signals and radiation power have been scaled to a peak value of 2.0 in arbitrary units.

tion than had been seen previously. In this case, the experiment was best matched computationally by assuming an initial perturbation consisting of four "notches" imposed early in the implosion. These may correspond to wrinkles which were observed in the foil. Comparison between this simulation and measured current and two XRD traces (representing differing filters) is shown in Fig. 6. Timing, peak current, and features in the radiation pulse are all in very good agreement. Comparisons of $\frac{dI}{dt}$ and the inferred voltage at the load (Fig. 7) also show agreement with the features of bubble burst and re-acceleration of the shell evident in their structure. Time-integrated x-ray pinhole photos also indicate two bright regions near each electrode, consistent with an examination of time-integrated radiation energy on-axis in the simulation.

Another diagnostic, the time-dependent x-ray spectrometer, is currently undergoing a period of development as applied to Pegasus experiments. Side-on spectra from the cylinderically symmetric 2-D simulation were calculated using ray-tracing and SESAME multi-group opacities. This comparison showed an encourging correlation of major features (main peak, valley and high $h\nu$ peak) in the experimental spectrum which also appear in the calculation. A more detailed discussion of this comparison is given in a companion paper.[7]

CONCLUSIONS

Comparisons between the experimental results of Pegasus implosion data with 2-D simulations reveal good agreement in timing (especially relative timing

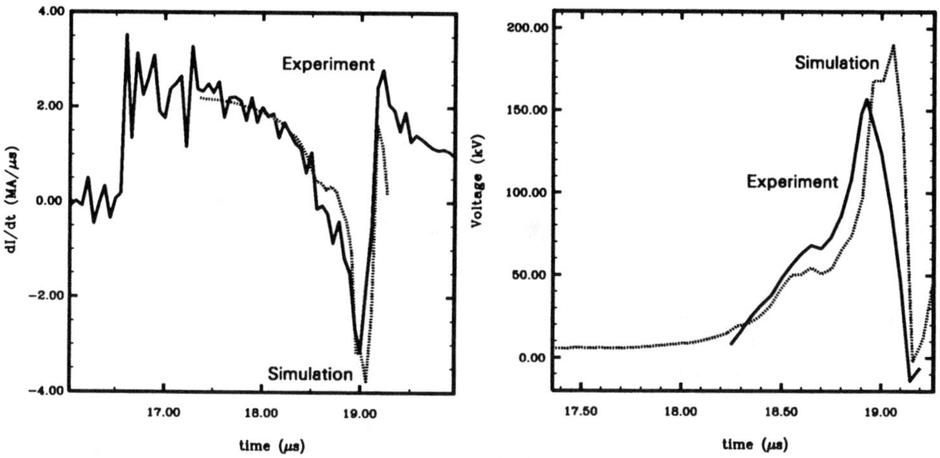

Fig. 7. Experimental (solid) and 2-D simulation (dot) curves of $\frac{dI}{dt}$ (left) and voltage (right) for the second 13.0 mg implosion experiment.

between features in the drive current and the radiation pulse), waveform shape, and radiation pulsewidth. There is substantial difference between 2-D and 1-D results, due to the development of the Rayleigh-Taylor instabilities which show a pattern of growth from short to long wavelength, a "bursting" of the bubble regions through the plasma shell, and finally electrical shorting across the spike regions allowing a re-acceleration of the plasma shell. These features are evidenced in the current and radiation pulse shapes. In addition, framing camera pictures have verified the presence of instabilities as indicated in the 2-D simulations. Time-dependent spectroscopy measurements have also been compared with the simulation with encouraging results. The experiments have thus provided an important benchmark in verifying our capability to simulate the implosion physics, and the understanding provided by the simulations has helped to explain the complex development seen in these experiments.

REFERENCES

1. P. J. Turchi and W. L. Baker, J. Appl. Phys. 44, 4963, (1973).
2. J. Brownell *et al.*, Proceedings of the 9th International Conference on High-Power Particle Beams (1992, to be published).
3. D. L. Peterson *et al.*, Proceedings of the 6th International Conference on Megagauss Field Generation and Related Topics (1992, to be published).
4. R. L. Bowers *et al.*, ibid.
5. H. Oona *et al.*, ibid.
6. J. C. Cochrane *et al.*, these proceedings.
7. W. Matuska *et al.*, ibid.

EFFECT OF INITIAL CONDITIONS ON GAS-PUFF Z-PINCH DYNAMICS

G. G. Peterson, F. J. Wessel, and N. Rostoker
Physics Department, University of California
Irvine, CA 92717

A. Fisher
Naval Research Laboratory
Washington, D.C. 20375

ABSTRACT

The implosion dynamics for a neon gas-puff z-pinch is investigated to study the correlation between the initial and final conditions. Several nozzle material combinations are tested along with other variable parameters such as valve plenum pressure and capacitor bank voltage. The 5-cm diameter annular nozzle consists of an inner-core surrounded by an outer-mantle. The pinch is driven by a 5 kJ, 30 kV, 11 μFd capacitor bank which delivers 0.4 MA in 1.4 μs. The effects of different initial conditions on the final compression and x-ray yield are described, based upon the following diagnostics: dI_{load}/dt, filtered XRDs, x-ray filtered pinhole photos, framing camera photos, and N_2 laser interferometer.

INTRODUCTION

Previous gas-puff z-pinch experiments at U.C. Irvine investigated methods to reduce Rayleigh-Taylor instabilities using a trapped axial magnetic field. The radius of the nozzle and anode were increased from 2.0 cm to 2.5 cm in order to magnify the effects of this instability. This modification had the desirable effect of increasing the run-in time of the implosion, which allowed for a better coupling of the machine's rather slow current rise-time to the time of maximum compression. X-ray yield was improved by a factor of > 2.5 over previous experiments.[1,2] At large values of injected field the pressure of the axial magnetic field that was trapped in the interior of the imploding plasma annulus inhibited peak compression and x-ray yield. This motivated the design of a system that mostly excludes the axial magnetic field from the interior while allowing the field to be embedded within the plasma shell. The system consisted of a fast rise-time magnetic coil in combination with a nozzle that consisted of a highly-conducting (copper) inner-core and a teflon outer-mantle. The results using the teflon outer-mantel with and without applied axial magnetic fields were substantially different from those using a graphite outer-mantle. X-ray yield further increased with the teflon mantle compared to the smaller diameter nozzle without a Bz and the dynamics of the imploding shell changed as seen with our N2-laser interferometer. These results have led us to investigate the effect of initial conditions upon pinch dynamics and final compression. This paper reports on our initial studies of the correlation between initial and final states, without a Bz field, using a framing camera to take axial photographs of the initiation phase of the pinch and an x-ray pinhole camera to investigate the final stages. Neon gas is used because its distinct spectral lines simplify x-ray yield analysis.[3]

NOZZLE AND VALVE

The electrodes consist of a 2.7-cm radius hollow stainless-steel anode located 1.3-cm from the nozzle (cathode); the design of the hollow electrode is similar to those used on plasma focus machines.[4] The nozzle is 5-cm in diameter with a 0.5-mm throat width and consists of three parts; the base, the inner-core, and the outer-mantle. Different combinations of copper, graphite and teflon are used for the inner-core and outer-mantle as illustrated in Fig. 1.

The gas is puffed into the center of the nozzle base, via a simple diamagnetic-impact puff valve,[5] with rise time ~50 μs (Fig. 2). Valve plenum pressures are varied between 20 and 60 PSIG. At pressures above 40 PSIG the shot-to-shot jitter in the pinch's time to maximum compression began to increase from 25 ns to 60 ns. The jitter at higher pressures was reduced significantly (to ≈ 30 ns) after decreasing the diameter of the valve's aluminum hammer to give 2.5-mm clearance with the valve wall.

DIAGNOSTICS

The diagnostics include: 1) dI_{load}/dt which consists of a wire loop placed inside the vacuum chamber ~ 5 cm away from the pinch axis. It is useful for timing and qualitative assessment of implosion dynamics. 2) Framing Camera which is used to study the initial gas breakdown. It is located outside the vacuum chamber, looking from behind the hollow anode (Fig. 1) and takes 3 photographs per shot at variable time intervals. The camera takes axial photographs through the center of the anode and is attenuated by a 400 nm, $\Delta\lambda = 80$ nm interference filter. 3) X-ray Diodes which are negatively biased polished nickel photo cathode cylinders filtered with combinations of 2μ mylar, 12μ saran and 25μ beryllium. 4) Filtered X-ray Pinhole Camera which has two lead foil pinholes (50-100μ ϕ), filtered with graphite powder coated 2μ mylar, and 25μ beryllium +12μ saran filters. 5) Nitrogen Laser ($\Delta t = 8$ ns) Interferometer which is used to take radial photographs of the plasma shell near maximum compression.

RESULTS

The framing camera photographs (Fig. 3) are enhanced to show light intensity in the third dimension. It is assumed that the photographs are mainly of line radiation produced during electrical breakdown of the z-pinch gas and that the intensity of the emitted light is a relatively good indication of current flow. If an increase in current uniformity is associated with decreased spacing between the light-intensity-spikes, then the photographs show more uniform gas breakdown for the following conditions: the conducting graphite outer-mantle as opposed to the teflon mantle, higher plenum pressure (50 psig) as opposed to lower pressure (20 psig), H-Kr gas mixture as opposed to Neon, and to higher main-capacitor voltage (30kV) as opposed to 20 kV.

The nitrogen laser interferograms and shadowgrams (Fig.5) are taken by firing the laser radially through the pinching plasma into a 0.337 μm filtered polaroid camera. The interference pattern is created by splitting the laser beam into a main beam and a reference beam and then recombining the beams after the pinch region. The nitrogen laser interferograms (Fig. 5) show a tendency for zippering from the anode to the nozzle for shots with pinch times < 1120 ns, for both the teflon and the graphite outer-mantles; with more extreme zippering for graphite.

The x-ray pinhole photos (Fig.4) show that when zippering occurs from the anode, x-ray production near the nozzle (cathode) is suppressed. A comparison between x-ray pinhole photo shots 4147 and 4235, which have comparable neon k-shell yield, suggests that use of the teflon nozzle results in higher compression and more bright spots than the graphite nozzle.

It is worth noting that in order for the framing camera to take photos of the initial gas-current discharge process it was necessary to increase the anode diameter from 5.0 cm to 5.4 cm to increase the viewing solid angle. This anode modification resulted in a decrease in neon k-shell yield by more than a factor of three.

DISCUSSION

The results show that nozzle materials significantly affect the initial gas breakdown of small-scale gas-puff z-pinches. The difference in initial gas breakdown between the teflon and graphite outer-mantle nozzle parts may explain why the zippering tends to be more extreme for graphite. Zippering is mainly caused by an axially non-uniform mass / length gas distribution. The better initial gas breakdown due to graphite over teflon probably results in a greater amount of gas capture in the plasma shell at the nozzle than at the anode, resulting in a longer run-in time at the nozzle. Furthermore, the apparent correlation between shorter run-in times and more extreme zippering for both teflon and graphite is again probably due to a non-uniform mass / length gas distribution, which can result from firing the machine too early when the valve-to-nozzle gas pressure is rising most rapidly.

The framing camera was also used to photograph the plasma light from a puff-on-puff z-pinch, with the hope of seeing the central puff light up during the implosion. The results so far are negative; application of an axial field may show otherwise.

Another use for the framing camera diagnostic is to measure the plasma shell radius as a function of time (Fig. 6). This along with the B-dot trace (Fig. 6) and an appropriate equivalent electronic circuit model for the z-pinch can be used to obtain a quantitative estimate of the ratio between the initial and final current shell radii. Assumptions must be made though about plasma uniformity and current flow. Also, the exponential dependence of the current radius on the circuit inductance introduces a large component of error.

Future experiments will continue the survey of using different nozzle materials to include teflon inner-core and copper outer-mantle combinations.

REFERENCES

1. T. F. Chang, A. Fisher, and A. Van Drie, J. Appl. Phys. **69**, 3447(1991).
2. G. Peterson, F. Wessel, N. Rostoker, and A. Fisher, Bull. Am. Phys. Soc., **37**(6), 1579(1992).
3. R. W. Clark, J. Davis, and F. L. Cochran, Phys. Fluids **29**, 1971(1986); Phys. Fluids **B2**, 1698(1990).
4. P. Choi, C. S. Wong, and H. Herold, Proceeding of the 7th Intl. Conf on High Power Beams, p. 1193(1988), W. Bauer and W. Schmidt editors, Kernforschungszentrum Karlsruhe, GmbH.
5. J. Kriesel, R. Prohaska, and A. Fisher, Rev. Sci. Instrum., vol. 62, 2372 (1991).

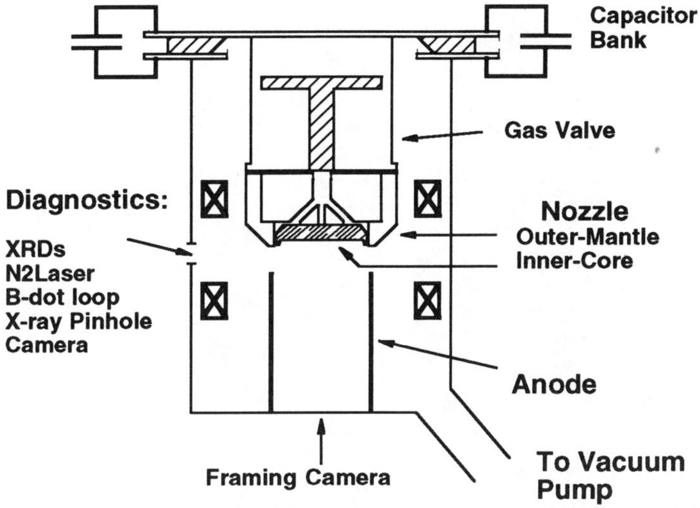

Figure 1. Schematic of the UC Irvine Z-pinch Machine

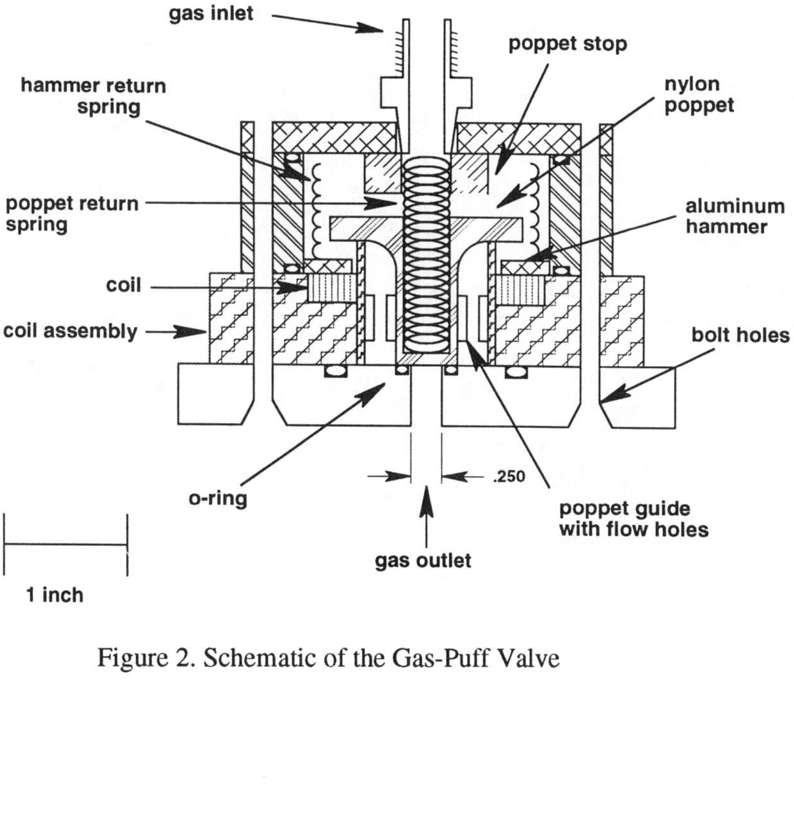

Figure 2. Schematic of the Gas-Puff Valve

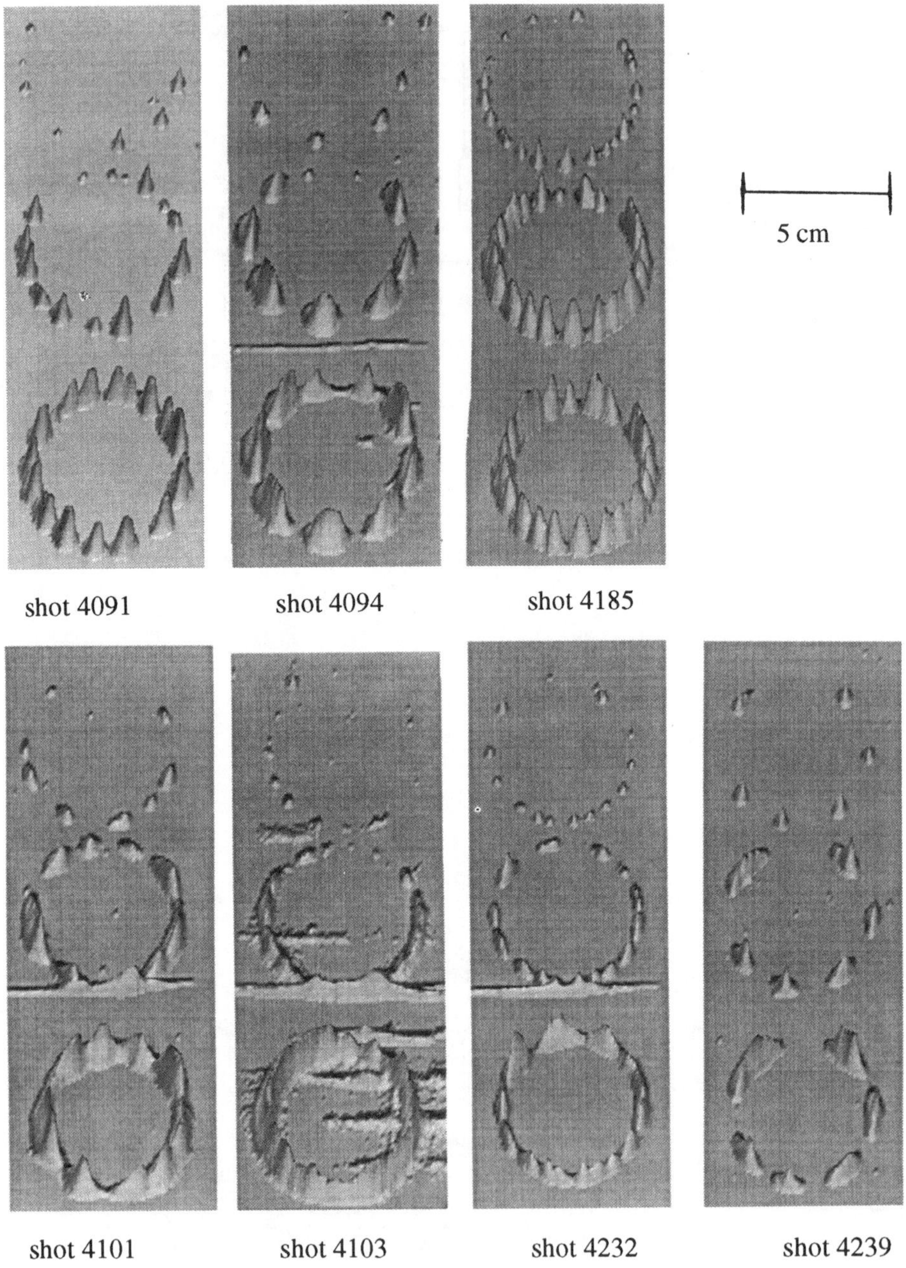

Fig 3. Framing Camera Photos: Axial view of the filtered plasma light
interference filter: 400 nm, 80 nm width
frame time = 200 ns, 300 ns, 400 ns after initiation, 20 ns duration
computer enhanced: 3rd dimension = luminous intensity

Shot	Nozzle		Valve Plenum Pressure	Gas	Capacitor Bank Voltage
	Core	Mantle			
4091	Cu	teflon	50	Neon	30 kV
4094	Cu	teflon	14	Neon	30 kV
4185	Cu	teflon	20	95% Kr+5% H	30 kV
4101	Cu	graphite	14	Neon	30 kV
4103	Cu	graphite	50	Neon	30 kV
4232	graphite	graphite	24	Neon	30 kV
4239	graphite	graphite	20	Neon	20 kV

Table I. Z-pinch Parameters for Framing Camera Photographs

Filters: 25μ Be+12μ saran, 2μ graphite+2μ mylar

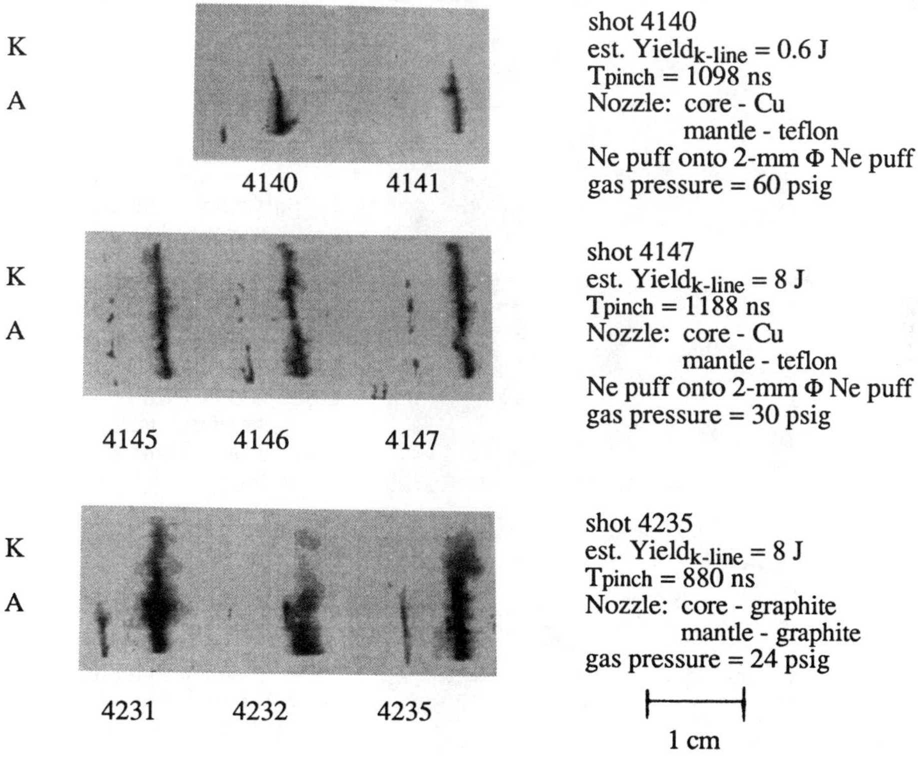

Figure 4. Double Pinhole X-ray Images of a Neon gas pinch, left image: Ecutoff > 0.8 keV, 100μm φ pinhole, right image: Ecutoff > 0.2 keV, 50μm φ pinhole

T_{laser} is relative to the maximum in hard x-ray pulse

K
A

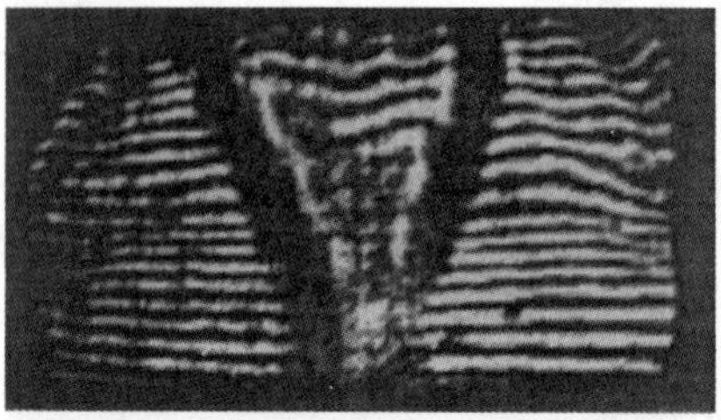

shot 4139
T_{laser} = -20 ns
Trigger delay = 220 μs
Tpinch = 1113 ns
Nozzle: core - Cu
mantle - teflon
Ne puff onto 2-mm Φ Ne puff
gas pressure = 60 psig

K
A

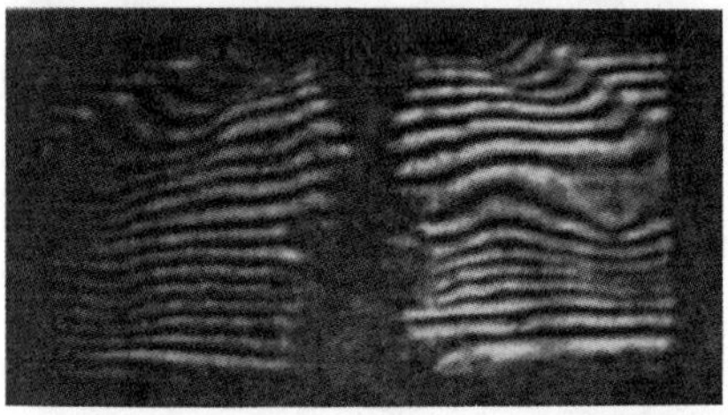

shot 4142
T_{laser} = +20 ns
Trigger delay = 250 μs
Tpinch = 1040 ns
Nozzle: core - Cu
mantle - teflon
Ne puff onto 2-mm Φ Ne puff
gas pressure = 30 psig

K
A

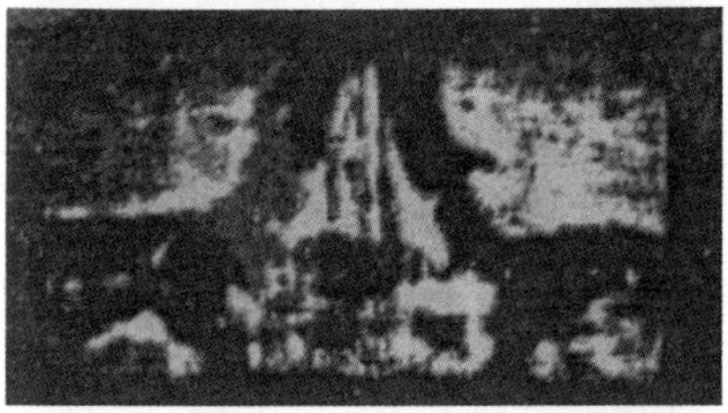

shot 4147
T_{laser} = -42 ns
Trigger delay = 270 μs
Tpinch = 1188 ns
Nozzle: core - Cu
mantle - teflon
Ne puff onto 2-mm Φ Ne puff
gas pressure = 30 psig

K
A

shot 3748
T_{laser} = -90 ns
Trigger delay = 235 μs
Tpinch = 1138 ns
Nozzle: core - Cu
mantle - teflon
Ne puff onto 3-mm Φ Ne puff
gas pressure = 38 psig

K
A

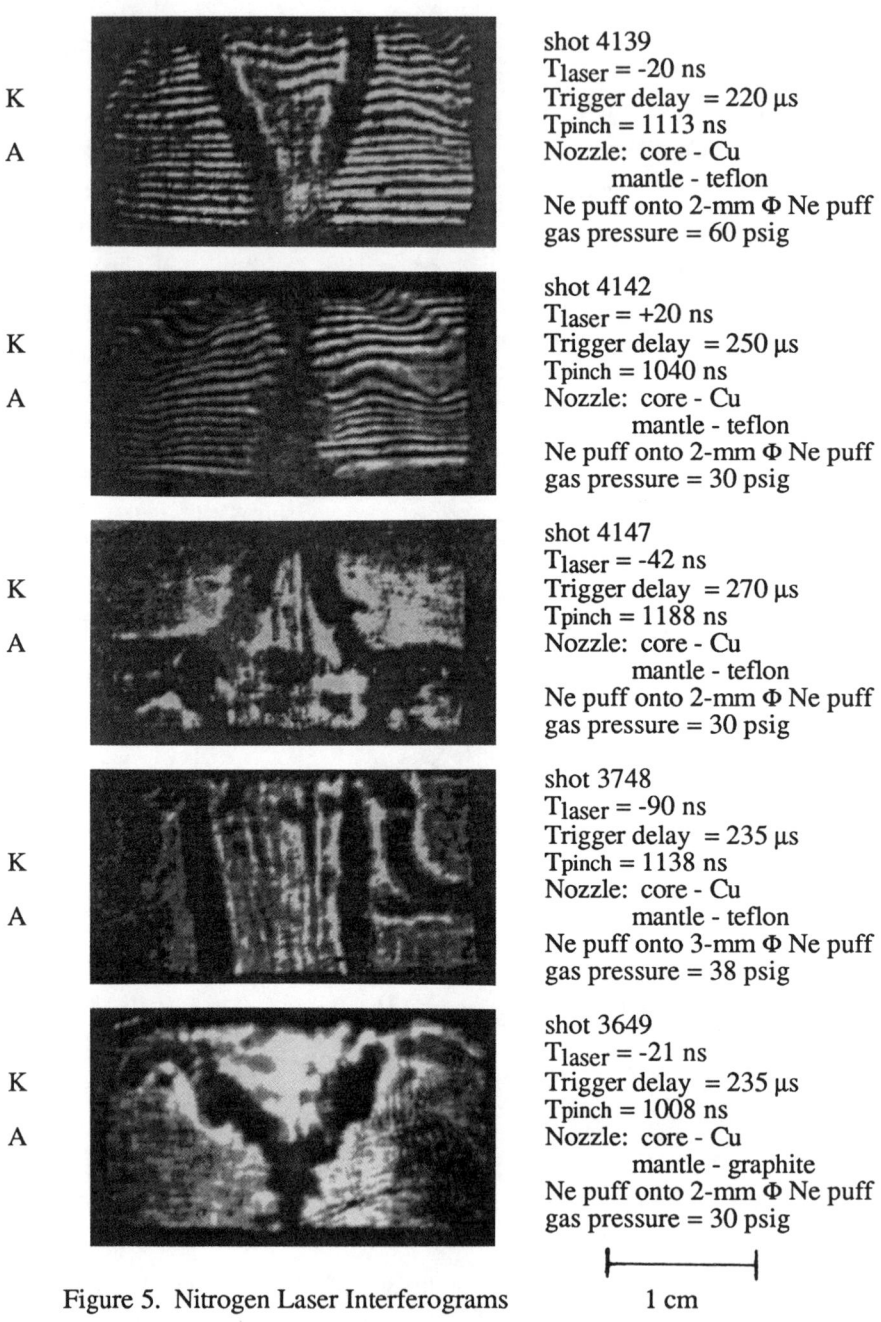

shot 3649
T_{laser} = -21 ns
Trigger delay = 235 μs
Tpinch = 1008 ns
Nozzle: core - Cu
mantle - graphite
Ne puff onto 2-mm Φ Ne puff
gas pressure = 30 psig

1 cm

Figure 5. Nitrogen Laser Interferograms

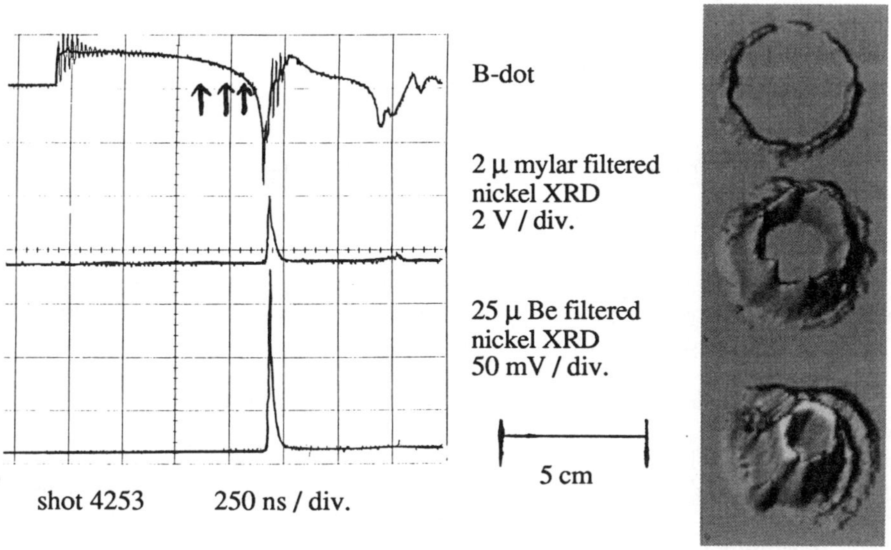

Figure 6. B-dot (upper trace) with framing camera exposure times (denoted by arrows) 670 ns, 790 ns, 880 ns after initiation, soft x-ray pulse (middle trace), and hard x-ray pulse (bottom trace)

STAGNATION DYNAMICS AND HEATING MECHANISMS FOR WIRE ARRAY Z-PINCH IMPLOSIONS*

R. B. Spielman, J. S. De Groot,† T. J. Nash, J. McGurn, L. Ruggles, and M. Vargas
Sandia National Laboratories, Albuquerque, NM 87185

K. G. Estabrook
Lawrence Livermore National Laboratory, Livermore, CA 94550

ABSTRACT

We have conducted experiments with aluminum, copper, and gold wires arrays to study x-ray production, z-pinch stability, and heating mechanisms. Time-resolved x-ray pinhole camera data and PCD data clearly show that the maximum in the keV x-ray production occurs after peak compression during an expansion phase.

The data are consistent with a sheath whose width is determined by the number of wires of the initial array (azimuthal uniformity) and the array diameter. The final stagnated size of the pinch is set by that sheath thickness. The stagnation event is characterized by a relatively uniform initial pinch reaching, in some cases, a diameter < 1 mm, followed by a disruption and increased x-ray production. The source of this energy is apparently not kinetic. The kinetic portion of the energy should have been converted to thermal energy near the time of maximum compression. For aluminum and copper wire arrays, heating from classical Spitzer resistivity is insufficient to explain the observed x-ray yields. While our present data does not show the actual location of current sheath nor does it give the position of non-radiating plasmas and, hence, is incomplete, the data when taken as a whole is strongly suggestive of enhanced resistive dissipation and an MHD instability with a helical shape.

INTRODUCTION

Magnetic forces rapidly implode the plasma in a high current z pinch. The plasma is predicted[1] to collapse to very small radii ($r_f \sim 10\ \mu m$) because the radiated power exceeds Joule heating. One-dimensional MHD calculations confirm this prediction. However, the imploding plasma is unstable to MHD interchange instabilities.[2] The evolution of the collapsing, unstable plasma is not understood because the required computational tools (3-D MHD and PIC codes) are not available. One strategy is to deal with this problem is to perform 1-D MHD calculations and terminate the calculations at the time the plasma radius corresponds to the minimum radius (or maximum compression) observed in time resolved pin-hole photographs. We have previously seen[3] that the x-ray energy radiated can significantly exceed the implosion energy calculated in this way. There are only two possible explanations for the behavior. The magnetic field must do more pdV work on the plasma or the resistance must be anomalously large. The magnetic field

* This work supported by the U.S. Department of Energy under Contracts DE-AC04-76-DP00789 and W-7405-ENG-48.
†Present address: University of California, Davis, CA 95616

can do pdV work on the plasma by continuing to compress the plasma to very small radii as predicted in 1-D MHD simulation calculations. This result does not agree with our data. The other way for the magnetic field to perform work on the plasma is for the plasma to develop a helical or other complex structure. Load inductance must increase as the effective plasma length increases. Theory indicates that ohmic heating is anomalously large and the data suggests that the plasma develops a helical structure.

A more complete model of the z-pinch event must include the physics of three temporal phases: implosion, stagnation, and post-stagnation. Initially the z-pinch plasma consists of an annular column. Current is applied to this plasma over a time scale of several tens of nanoseconds. The current is initially carried on the surface of the plasma, but the current rapidly diffuses inward due to the intrinsic resistivity. The z-pinch plasma is accelerated radially inward by the magnetic field during this implosion phase. The plasma sheath snowplows, accreting mass. Magneto Rayleigh-Taylor instabilities grow. The electrons are compressed by the magnetic field, heated due to Joule heating, and radiatively cooled. The plasma assembles on axis and the implosion kinetic energy is converted to ion and electron thermal energy during the second, stagnation phase of the event. Finally, the plasma expands and develops large amplitude MHD instabilities after peak compression.

We have gathered a considerable amount of data that impacts our basic understanding of z pinches and constrains our models of the z-pinch implosion. We have found that the total x-ray radiation emitted by the pinch exceeds the calculated kinetic energy. In all wire array implosions there is significant keV x-ray radiation emitted following the peak compression of the pinch that is not consistent with a continued conversion of implosion kinetic energy to thermal energy and to x-rays. Spatially-resolved data also show that the plasma develops a complex structure following the peak compression of the pinch.

This paper presents experimental data and models that include some of the physics necessary to interpret the data. Data will be compared with simple models and computer calculations in an attempt to identify the important physics issues of wire array z-pinch implosions.

EXPERIMENTS

The experiments described herein were conducted entirely on the Sandia National Laboratories Saturn facility.[4] Saturn delivers up to 10 MA to a typical z-pinch load with a rise time of 40-50 ns. The inductance of the entire load is approximately 10 nH. An electrical energy of ~ 750 kJ is available in the vacuum section of the machine. Saturn has been used for some time to carry out z-pinch experiments and to use the radiation output from z-pinch implosions for materials effects testing.[5] We have conducted a series of z-pinch experiments on Saturn using a variety of configurations and materials. We describe experiments using aluminum, copper, and gold wire arrays as the z-pinch loads. In our experiments 2-cm long arrays of 8-24 wires with array diameters of 1-2.5 cm were imploded in times of 50-75 ns. Typical wire diameters were 15-35 μm. The imploding plasmas reached calculated velocities greater than 50 cm/μs. A photograph of the anode insert used for these experiments is shown in Fig. 1. The anode insert is fabricated entirely from stainless steel and the current return cage is welded to the main plate for ruggedness.

We used a variety of x-ray diagnostics for these experiments. Electrical diagnostics such as x-ray diodes (XRDs), diamond photoconducting detectors (PCDs),[6] resistive bolometers[7] are routinely fielded for all shots. Fig. 2 shows the relative response of the two diamond PCDs that we used for this paper. PCD215 is 0.5-mm thick diamond with a filter of 6-μm thick beryllium and 1-μm thick Parylene. This is nearly the thinnest filter possible and have good rejection of the x rays below 1 keV. PCD212 uses a robust 8.4-μm thick Kapton filter. This filter is

Fig. 1 A photograph of the anode insert used to hold the wire arrays.

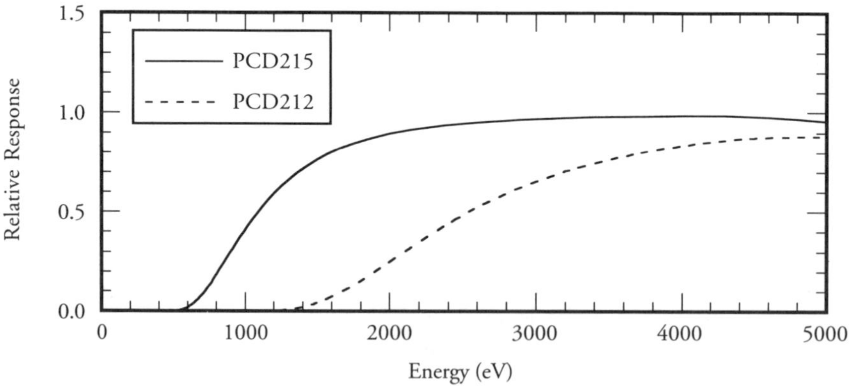

Fig. 2 The relative spectral responses of diamond PCDs, PCD215 (solid line) with a 6-μm thick beryllium and 1-μm thick Parylene filter and PCD212 (dashed line) with a 8.4 μm-thick Kapton filter.

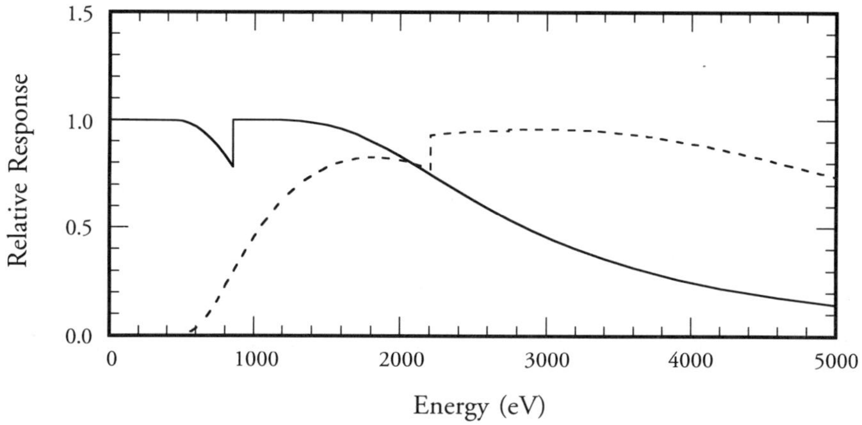

Fig. 3 The relative spectral responses of the bolometers, BT2 (solid line) is unfiltered and BT1 (dashed line) has a 6-μm thick beryllium and 1-μm thick Parylene filter.

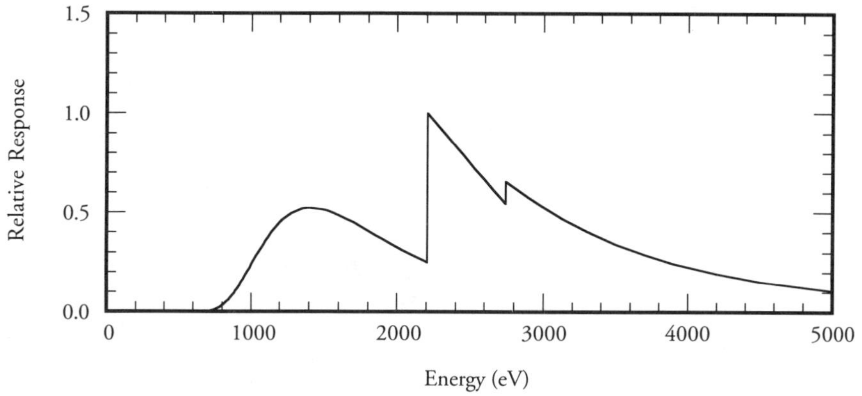

Fig. 4 The spectral response of the time-resolved pinhole camera.

useful as it eliminates the possibility of photons just under 1 keV affecting the signal. Fig. 3 shows the relative spectral responses of the resistive bolometers. BT2 is unfiltered with a 1-μm thick nickel element while BT1 is a 1-μm thick gold element using the same filter as PCD215, 6-μm thick beryllium and 1-μm thick Parylene. Time-integrated and time-resolved VUV and x-ray spectrographs provide ionization information on the pinch plasma. Spatially-resolved information was obtained with both time-resolved, gated microchannel plate x-ray pinhole cameras. Fig. 4 shows the spectral response of our time-resolved x-ray pinhole camera. The camera is filtered with 25.4 μm of beryllium and has gold-coated microchannel plates.

Fig. 5 shows the time-resolved x-ray pinhole photograph of an aluminum implosion, Saturn shot 1606. The initial diameter of the wire array was 1.82 cm and the wire diameter was 25.4 μm. The upper two frames are time integrated and the remaining frames are time resolved. The interframe time was 3 ns and the gating time was 1 ns. The image size is 1:1. We are imaging only the aluminum K-shell radiation. We can see the details of the stagnation. The sheath is visible at 0 ns and the stagnation starts and peak compression is apparently reached by 9 ns. An apparent sheath thickness of 1.5 mm is implied by the pinch radius in the first time resolved frame. The pinch reaches a minimum diameter of 1 mm at 9 ns. The pinch then expands and develops obvious spatial structure. Interestingly, the integrated and apparent brightness of the images increases during the expansion phase. This suggests the continued addition of energy during the pinch expansion. Please note that the microchannel plate is saturated in frames 12 and 15 ns and that the actual x-ray intensity is higher than it appears. Fig. 6 shows a 5X view of the 6 ns frame and the 15 ns frame. The difference in the structure is readily evident. Fig. 7 shows the time history of the keV emission as seen by PCD212. The pulse width is 15 ns FWHM. The location of the 12-ns frame of the pinhole picture is shown. Fig. 8 shows the total x-ray power from Saturn shot 1606. The double pulsed behavior of the PCD is also present in the total power. Bolometers measured a keV yield for shot 1606 of 75 kJ and a total x-ray yield of 250 kJ.

Data from a copper wire array implosion also gives a similar picture of the stagnation dynamics. Fig. 9 is a time-resolved x-ray pinhole photograph of Saturn

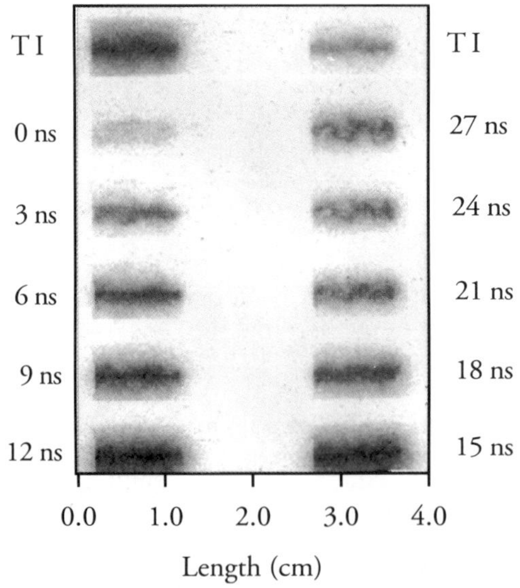

Fig. 5 A time-resolved x-ray pinhole photograph of an aluminum wire array implosion. The upper two frames are time integrated.

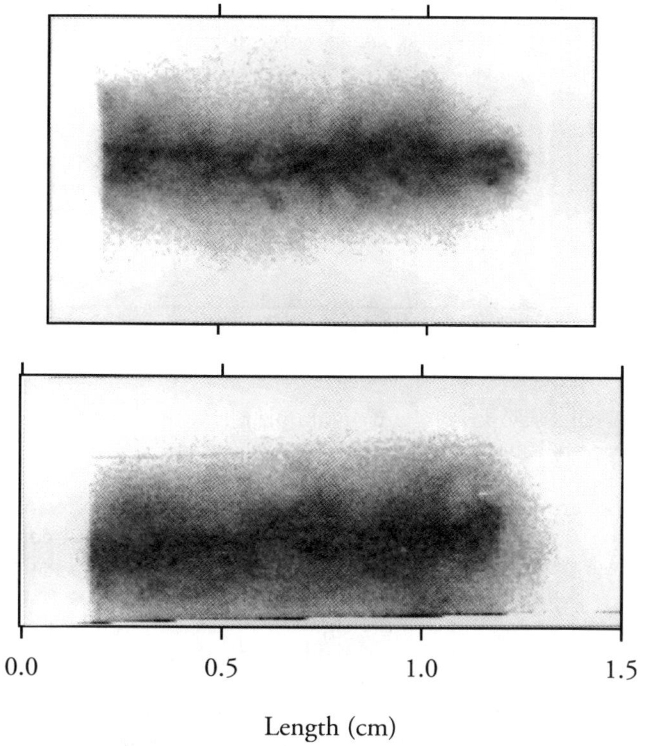

Fig. 6. An enlargement of two frames of Fig. 1, 6 ns (top) and 15 ns (bottom).

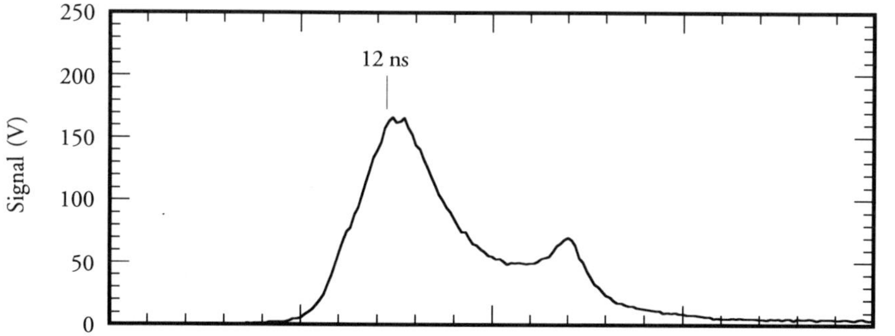

Fig. 7. The keV x-ray pulse from an aluminum wire-array implosion, Saturn shot 1606 recorded with PCD212.

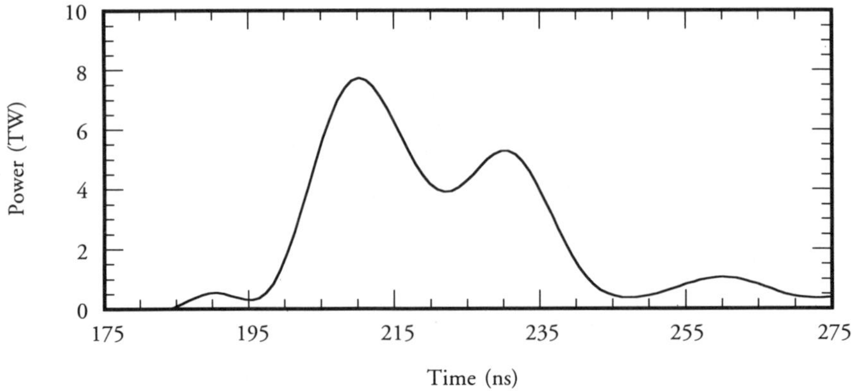

Fig. 8. The total x-ray power from an aluminum wire-array implosion, Saturn shot 1606.

shot 876, a copper wire array implosion with 24, 12.5-μm diameter wires on a 1.5-cm diameter array. The camera was filtered to measure copper K-shell and copper L-shell radiation. The behavior of the copper z-pinch implosion is qualitatively similar to the aluminum data. The peak in the L-shell radiation comes 10 ns after the minimum pinch diamter is attained. Fig. 11 shows the signal from PCD215. The first frame of the pinhole picture is shown. The total x-ray yield on this shot was 440 kJ and the L-shell, keV x-ray yield was 74 kJ. The copper z pinch reaches a minimum diameter of < 1 mm. This pinch is relatively uniform. X-ray emission actually drops for 5 ns while the plasma expands. Large scale length MHD instabilities are evident at 15 ns and the x-ray emission greatly increases. Shots with fewer copper wires replicate this

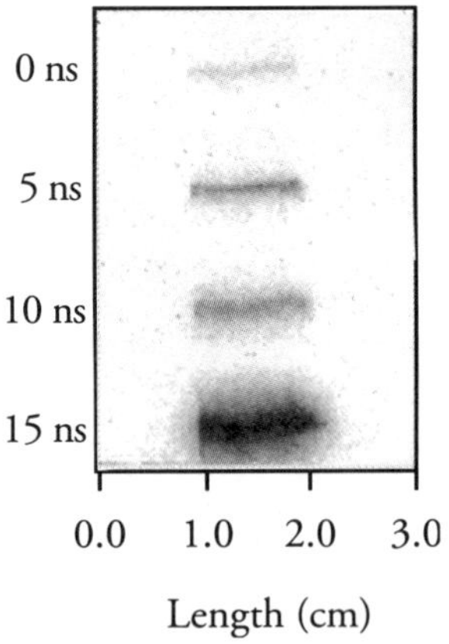

Fig. 9. A time-resolved x-ray pinhole photograph of a copper wire array z-pinch looking at only the L-shell x rays.

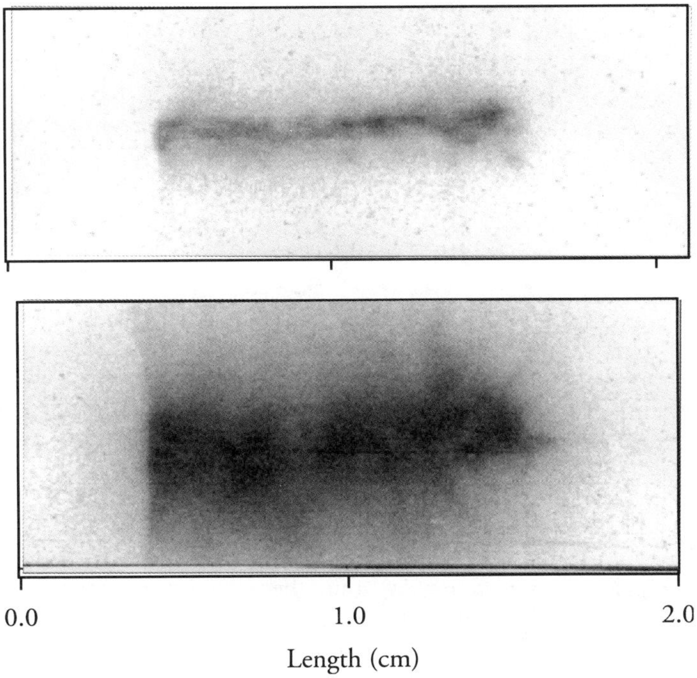

Fig. 10. A 5X view of the peak compression at 5 ns (top) and the increased x-ray output during the MHD disruption phase at 15 ns (bottom).

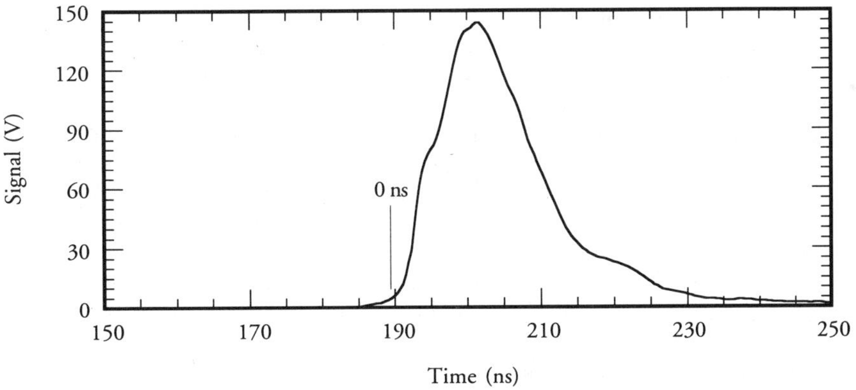

Fig. 11. The keV x-ray pulse from a copper wire-array implosion, Saturn shot 876.

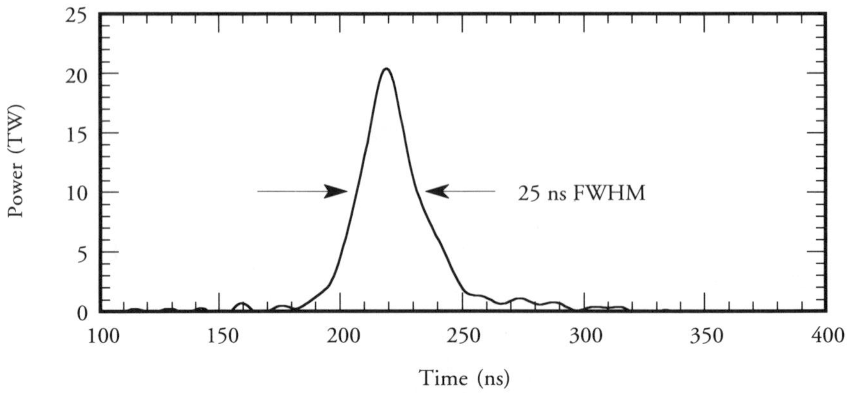

Fig. 12. The total x-ray power from a gold wire-array implosion, Saturn shot 937.

general bahavior but have larger final diameters at peak compression.

Some of the most interesting data comes from gold wire array experiments. We took 5 shots with gold wire arrays covering a number of different wire array parameters. The highest yield shot was Saturn shot 937 in which an array of 24, 12.5-µm diameter gold wires was placed on an initial diameter of 1.25 cm. A total yield of 630 kJ was measured for this shot. Fig. 12 shows the power pulse from this shot. A time-resolved x-ray pinhole photograph of a gold wire array z-pinch implosion is shown in Fig. 13. The photograph is shown at 1:1 magnification. The data show two distinct times of gold M-shell x-ray emission. The first pinch, at 12 ns, occurs at a time expected for the mass of the implosion and is similar to the stagnation time of aluminum and copper arrays of the same mass. There is a second peak in the x-ray emission at 27 ns. The physical structure of these two pinches is quite different. Fig. 14 shows an expanded view of those two frames. The pinch at 12 ns is relatively uniform with a diamter of 200 µm, close to the resolution limit of the camera. The pinch at 27 ns is characteristic of the "hot spot" instabilities seen on many z-pinch devices. The diameter of the emitting spots is ~ 200 µm and, we believe, limited by the resolution of the camera. But more interesting is the apparent helical shape of the pinch. The hot spots are on one side of the turn. The double emission is reflected in the diamond PCD signal for this shot. Fig. 15 shows the signal from PCD215 has two peaks. The location of the 12-ns pinhole camera frame is indicated. The double peaked structure seen in the keV photons is not reflected in the total x-ray pulse. We believe that in the case of gold the rapid radiative cooling quenches the M-shell emission, giving the noticeable double pulse but acts only to broaden the total x-ray pulse.

We believe that the data for all three elements is consistent. Gold is obviously different only because the high radiative cooling rates allow us to more easily discern the two separate heating events. All three implosions follow the same path with energy being released at stagnation and then later during an instability phase.

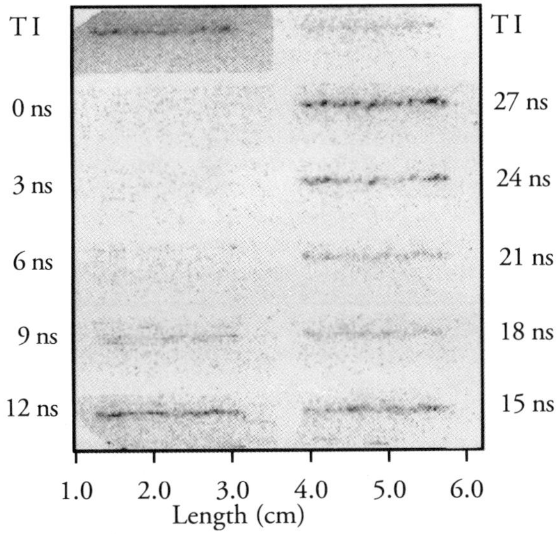

Fig. 13. The total x-ray power from a gold wire-array implosion, Saturn shot 937.

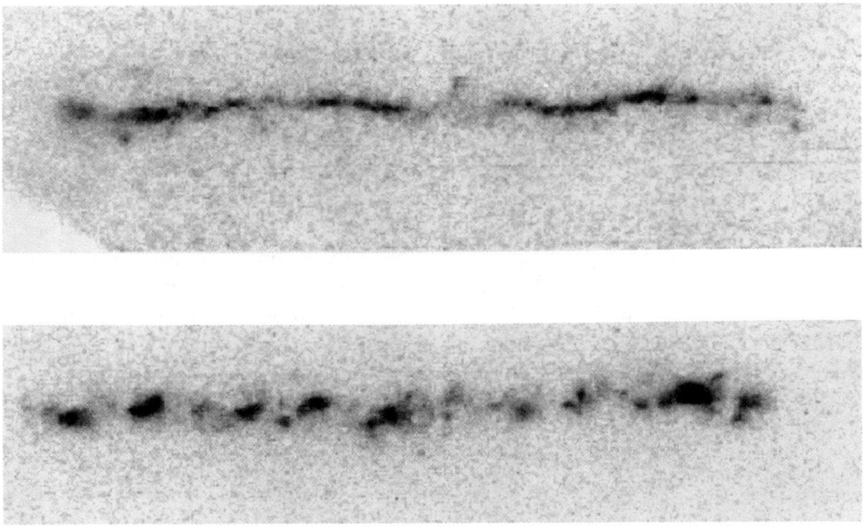

Fig. 14. Frames at 12 ns (upper) and 27 ns (lower) from Saturn shot 1625. The magnification is 5X.

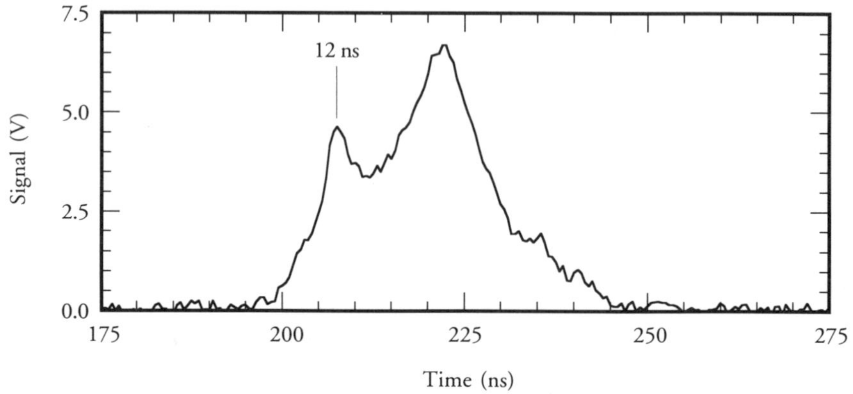

Fig. 15. X-ray pulse from PCD215 looking at gold M- and N-shell radiation.

THEORY AND MODELING

We have used a 0-D model (ZORK), a 1-D MHD code (LASNEX), and theoretical models to predict the plasma energy developed during the implosion phase. These models predict that the implosion energy is ~ 250 kJ for optimum array masses.

The simplest way to understand the dynamics of a z pinch is to use a "0-dimensional" circuit model,[8] described herein, in which the circuit parameters of the driver are included in the model and where the load dynamics are reflected in the inductance. Work is done on the z pinch by the electrical driver with the primary energy sink being radial kinetic energy. Such a simple model does not include loss processes such as radiation or other dissipative mechanisms such as resistive heating. This model is therefore not a function of ionic species. Indeed, these circuit codes contain only the circuit information and straightforward pinch dynamics leaving out all detailed plasma and atomic physics. While severely limited in scope such a simple model can provide significant insight into implosion dynamics and coupling of z pinches to different drivers. We have used "ZORK", a "0-dimensional" circuit code, developed at Sandia National Laboratories,[9] to model z-pinch experiments on the Saturn accelerator. ZORK uses the measured current and voltage waveforms together with the load parameters to simulate z-pinch implosions on Saturn.

This Zork simulation used an initial wire array diameter of 1.25 cm. The total mass of the 2-cm long array was 0.66 mg. The ZORK calculations are terminated when the plasma radius is 10% of the initial radius ($r_f = 0.06$ cm). This radius is consistent with the measured radii of the radiating plasmas. These calculations gave an implosion kinetic energy of ~ 250 kJ. This kinetic energy provides us with a minimum estimate of the x-ray radiation expected from the z-pinch stagnation. Figs. 16 and 17 show the temporal behavior of the radius and the magnetic and kinetic energy for this ZORK calculation. The measured x-ray yield is ~ 20% higher for aluminum, ~ 50% higher for copper, and ~ 100% higher for gold than the calculated ~ 250 kJ implosion kinetic energy. More magnetic field energy than predicted by ZORK must be converted to electron energy and thence to radiation. There are only two ways to obtain the additional energy. The plasma must be reisistively heated and/or the pdV work done by the field on the plasma must be increased.

We have constructed an approximate theory to evaluate the importance of resistive heating. Our theory is based on the theory of Hussey and Roderick.[10] The radial part of Ampere's law and Faraday's law's are combined to obtain Equation (1) for the azimuthal magnetic field,

$$\frac{\partial B_\phi}{\partial t} = \frac{\eta c^2}{4\pi} \frac{1}{r} \frac{\partial}{\partial r}\left[r \frac{\partial B_\phi}{\partial r} \right] \tag{1}$$

where η is the conductivity perpendicular to the magnetic field. We can find a self-similar solution if the magnetic field can be written in the form,

$$B_\phi = B_M(t)\, f(\zeta),$$

where: $B_M(t)$ is the magnetic field evaluated at the radius of the plasma surface, r_0, $B_M = I(t)/5r_0$, $I(t)$ is the current, and ζ is the self similar variable,

$$\zeta = \frac{r_0 - r}{c}\left[\frac{4\pi}{\eta t}\right]^{1/2}.$$

The collisional Spitzer resistivity perpendicular to the magnetic field is,

$$\eta = 1.15\text{X}10^{-14}\,\frac{Z \ln \Lambda}{T^{3/2}}.$$

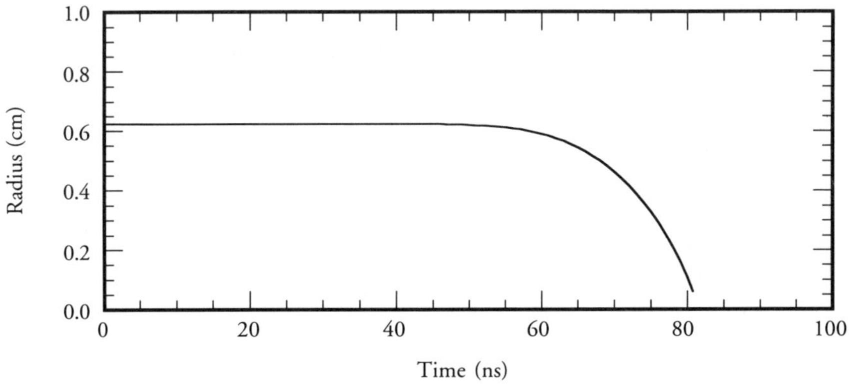

Fig. 16 The pinch radius is shown as a function of time. Most of the radial motion occurs in the final 20 ns.

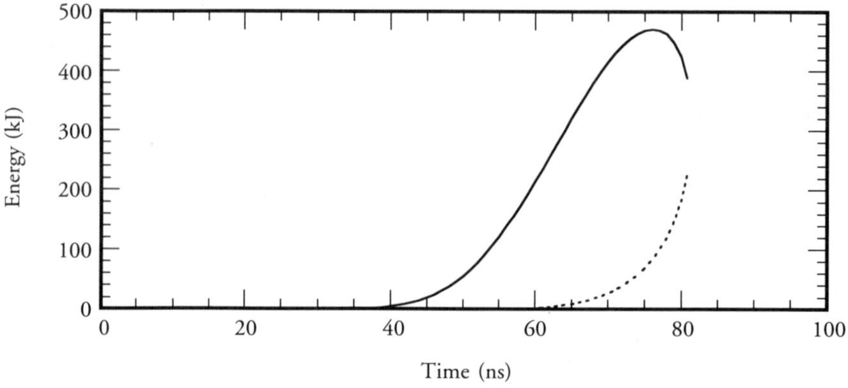

Fig. 17 The total magnetic energy stored in the simulation, $1/2\ LI^2$, (solid line) and the kinetic energy (dotted line) is shown as a function of time.

LASNEX simulations show that the plasma is approximately isothermal during the implosion phase. The predicted time history of the electron temperature from the LASNEX calculations is shown in Fig. 18. The simulations also show that the electron temperature is low, Te < 150 eV and is well approximated by equating x-ray emission to resistive heating. Since the electrons are approximately isothermal, the conductivity is independent of radius so that Eq. (1) becomes,

$$\frac{d^2f}{d\zeta^2} + \frac{df}{d\zeta}\left[\frac{\zeta}{2} - \frac{r_0-r}{r\zeta} - \frac{r_0}{r_0-r}\frac{t}{r_0}\frac{dr_0}{dt}\right] - \frac{t}{B_M}\frac{dB_M}{dt} f = 0 . \tag{2}$$

This equation has an approximate self-similar solution for radii close to the plasma surface, $r \sim r_0$, if the radius of the plasma surface is changing slowly enough in time,

$$\frac{t}{r_0}\frac{dr_0}{dt} << 1,$$

and B_M varies as a power of time, $B_M \sim t^m$. For a linear current rise, $I \propto t$, the solution is well approximated by the exponential form,

$$B = B_M \exp\left[-\frac{r_0-r}{\delta_{ss}}\right], \tag{3}$$

where $\delta_{ss} = 0.9\, c\left[\frac{\eta t}{4\pi}\right]$. Comparing the magnitudes of the terms in Eq. (2), we find that the first and last terms dominate late in the implosion. Also,

$$\frac{t}{B_M}\frac{dB_M}{dt} \cong \left|\frac{t}{r_0}\frac{dr_0}{dt}\right| = g(t),$$

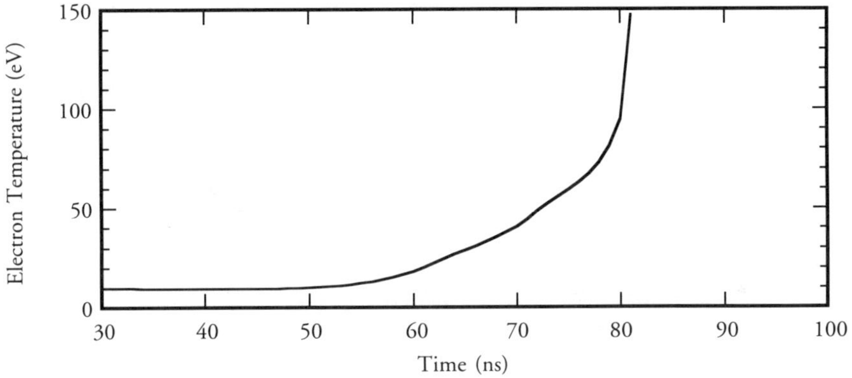

Fig. 18 LASNEX calculations of the time-dependent electron temperature for an aluminum wire array with optimum mass

thus an approximate solution late in the implosion is again Eq. (3) with,

$$\delta \sim \frac{\delta_{ss}}{\sqrt{g(t)}} \; . \qquad (4)$$

Late in the implosion the velocity of the plasma surface increases so that $g(t) >> 1$. The skin depth is then significantly smaller than the self similar value. The radial profile of the LASNEX-calculated magnetic field late in the implosion (Fig. 19) is closely exponential in agreement with Eq. (3). The scale length from LASNEX is $\delta_{LASNEX} \approx 0.011$ cm. The self-similar scale length is significantly larger, $\delta_{ss} \approx 0.027$ cm (the LASNEX-calculated temperature was used). The scale length from Eq. (4) more closely agrees with the LASNEX calculated value $\delta \approx 0.008$ cm. LASNEX simulations of the energy developed in resistive heating compares well with the normalized theory shown in Fig. 20.

We see that LASNEX simulations and the theoretical model predict that for the optimum wire array mass and Spitzer resistivity, the resistive heating energy is about, $E_J \sim 10$ kJ, when the plasma radius has reached the minimum radius observed in the experiments. Thus, an additional energy of ~ 30 kJ is required to obtain the x-ray yield from aluminum. The results do not depend on the plasma species for the wire materials tested (aluminum, copper, and gold). Thus, the 1-D model of a z pinch with Spitzer resistivity does not agree with our data. Please note that our calculations for resistive heating stop at $0.1\ r_i$. Thus, we are neglecting heating that might occur after peak compression. Such heating could be significant for gold.

The only parameter in the theory is the resistivity, $E_J \sim \eta^{1/2}$. Thus, the resistivity must be increased by a factor of about sixteen to obtain the additional heating required in aluminum pinches. However increasing the resistivity can decrease the resistive heating. This counter intuitive result is due to the fact that the electron temperature increases with resistivity. If the electron temperature is less than about

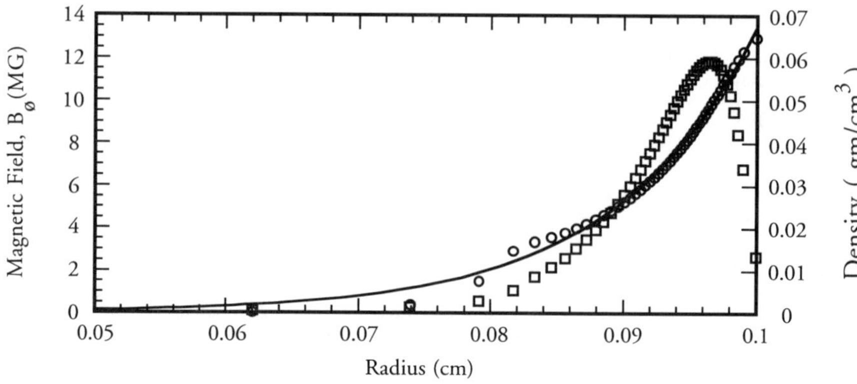

Fig. 19 LASNEX results for the radial profile of the plasma density (squares) and the azimuthal magnetic field (circles) for an aluminum wire array. The line is an exponential fit to the field.

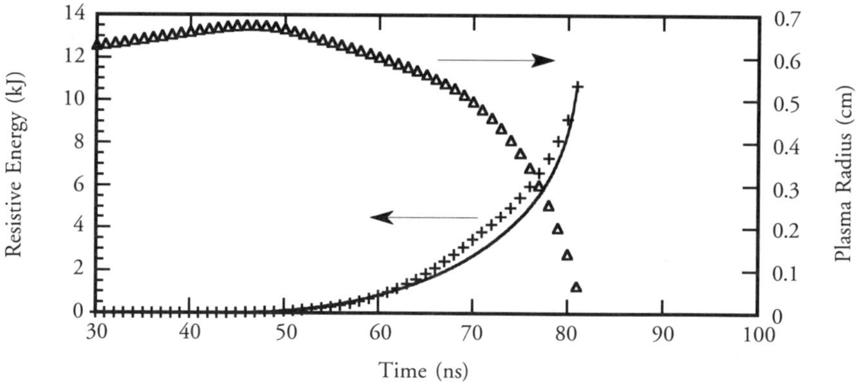

Fig. 20 LASNEX calculations of the plasma radius (triangles) and the energy developed in resistive heating (crosses) are shown for an aluminum wire array of optimum mass. The theoretical prediction (line) has been normalized to the 81 ns LASNEX point.

100 eV, then the x-ray emission rate increases rapidly with temperature, $P_{x\text{-ray}} \sim T_e^{3\text{-}4}$ so increasing resistivity only slightly increases the temperature. But the x-ray emission increases slowly or decreases with temperature for higher temperatures. Thus, the increase in resistivity can be compensated by an increase in the temperature. The net effect is that resistive heating can decrease as the resistivity increases.

We have used our model to evaluate the importance of anomalous DC resistivity due to ion acoustic turbulence.[11] The calculations show that the threshold for the ion acoustic instability is not exceeded in aluminum. Although the electron drift velocity, the ratio ZT_e/T_i is low (~ 10) so ion Landau damping dominates, resulting in a threshold drift velocity of several times the sound speed. The drift velocity is typically several times higher than the sound speed in copper and as much as six time higher in gold and the ratio ZT_e/T_i is much higher (~ 20). Thus, anomalous resistivity plays an important role in copper and gold z pinches.

The other way that energy can be added to the plasma is through PdV work that the magnetic field does on the plasma. We can construct a simple theory that captures the essential physics. The plasma inductance is approximately given by,

$$L_p = \frac{\mu_0}{2\pi} l \ln\left[\frac{r_m}{r_0}\right],$$

where $\mu_0 = 4\pi X10^{-7}$ H/m, r_m is the radius of the current return and l is the length of the plasma. The work the magnetic field does on the plasma, E_m, is proportional to the change in the inductance, $E_m \sim I^2 \Delta L_p$,

$$\Delta L_p = L_p\left\{\frac{\Delta l}{l} + \ln\left[\frac{r_{0max}}{r_{0min}}\right]\right\},$$

where r_{0max} is the initial radius of the plasma surface and r_{0min} is the radius of the plasma surface at the end of the implosion. The inductance increases due to the decreasing plasma radius during the implosion phase. the inductance also increases if the effective plasma length increases. The inductance must be increased by about a factor of two so that the pdV work done on the plasma is consistent with the x-ray yield for gold. Thus, either the plasma diameter decreases to ~ 10 µm or the effective length of the plasma is doubled. The x-ray pin-hole photographs indicate that minimum diameter of the radiating region is ~ 0.05 cm, so collapse to smaller diameters does not appear likely. X-ray pinhole photographs give a hint of a complex structure after peak compression. This is weak for the aluminum wires where an additional energy of only ~ 10% is required. This 3-D structure is clearer for gold and plasma length increases by about a factor of two as required to increase the plasma energy to the x-ray yield of ≈ 500 kJ. Time resolved, stereoscopic x-ray images are required to resolve this issue.

SUMMARY AND CONCLUSIONS

This paper presents experimental data and theoretical and computational models that show that the simple 1-D MHD model of high current z pinches with Spitzer resistivity is fundamentally incomplete. The z-pinch radiates more energy than is developed in the simple 1-D model. We have shown that resistive heating with Spitzer resistivity is far to small to account for the discrepancy. Thus, either anomalous resistivity and/or the pdV work must be increased beyond the simple 1-D model. We showed that anomalous heating is unlikely in aluminum but strong in copper and gold. We also showed that the time resolved x-ray pin-hole photos give a weak indication of a helical structure-weak in aluminum and stronger in gold. This helical structure increases the effective length of the plasma and therefore increases the pdV work that the magnetic field does on the plasma. Time resolved, stereoscopic x-ray images are required to resolve this issue.

It is likely that both anomalous resistivity and enhanced inductance operate to some degree, but confirmation will require both time- and space-resolved Thomson scattering measurements that are presently underway at our laboratory.

ACKNOWLEDGEMENTS

We wish to thank the Saturn operations crew for their help with all aspects of the experiments. The help of Richard Klingler, KTECH Corp. and of Pat Reilly, Mark Hedemann, and Richard Pepping of Sandia was invaluable. I want to thank Ken Struve, Mission Reseach Corp., for his assistance with the piezoelectric current diagnostics.

REFERENCES

[1]J. P. Chittenden, Proc. of the 2nd. Int. Conf. on Dense Z-Pinches, AIP Conf. Proc. #**195**, 118 (1989); L. Turner, Proc. of the 2nd. Int. Conf. on Dense Z-Pinches, AIP Conf. Proc. #**195**, 134 (1989); T. W. Hussey, M. K. Matzen, and N. F. Roderick, Proc. of the 2nd. Int. Conf. on Dense Z-Pinches, AIP Conf. Proc. #**195**, 147 (1989).
[2]N. F. Roderick, T. W. Hussey, R. J. Faehl, and R. W. Boyd, Appl. Phys. Lett. **32**, 273 (1978); N. F. Roderick and D. A. Kloc,J. Appl. Phys. **51**, 1452 (1980).
[3]R. B. Spielman, Gold Z-Pinches on Saturn, Proc. of the 9th International Conference on Particle Beams, Washington DC (May 1992), to be published.
[4]D. D. Bloomquist, R. W. Stinnett, D. H. McDaniel, J. R. Lee, A. W. Sharpe, J. A. Halbleib, L. G. Schlitt. P. W. Spence, and P. Corcoran, Proc. of the Sixth IEEE Pulsed Power Conf., Arlington, VA edited by P. J. Turchi and B. H. Bernstein (IEEE, New York, 1987), p. 310.
[5]R. B. Spielman, R. J. Dukart, D. L. Hanson, B. A. Hammel, W. W. Hsing, M. K. Matzen, and J. L. Porter, Proc. of the 2nd. Int. Conf. on Dense Z-Pinches, AIP Conf. Proc. #**195**, 3 (1989).
[6]R. B. Spielman, Rev. Sci. Instrum. **63**, 5056 (1992).
[7]L. P. Mix, E. J. T. Burns, D. L. Faehl, D. L. Hanson, and D. J. Johnson, edited by D. T. Attwood and B. L. Henke, AIP Conference Proceedings #75, 25 (1981).
[8]P. J. Turchi and W. L. Baker, J. Appl. Phys. **44**, 4936(1973).
[9]M. J. Clauser, L. Baker, D. H. McDaniel, R. W. Stinnett, and A. J. Toepfer, Magnetic Implosion of Plasmas with Short Pulse, High Power Generators, Sandia Report, SAND78-1387C.
[10]T. W. Hussey and N. F. Roderick, Phys. Fluids **24**, 1384 (1981).
[11]A. A. Galev and R. Z. Sagdeev, Basic Plasma Physics, A. A. Galeev and R. N. Sudan (eds.), North-Holland, Amsterdam 1984.

PREDICTION OF Z-PINCH IMPLOSION SHAPE FROM GAS JET NOZZLE GEOMETRY

Waisman, E., Ingermanson, R., Murphy, H.
Maxwell Laboratories, Inc.
S-Cubed Division
3398 Carmel Mountain Road
San Diego, CA 92121

Loter, N. Rix, W.
Maxwell Laboratories, Inc.
Balboa Division
9244 Balboa Avenue
San Diego, CA 92123

ABSTRACT

A technique for predicting the shape of Z-pinch implosions based on gas jet nozzle geometry has been developed and tested against data from argon gas experiments on BLACKJACK 5. The prediction technique involves use of a code which calculates the gas distribution, as a function of nozzle geometry and initial gas input pressure, at the start of the implosion. The calculated gas distribution is then snowplowed to about 0.1 of its initial radius using the measured current waveform. When the snowplow calculation is stopped, the shape of the implosion is compared to experimental data from the associated pinch.

The code 2-D RZ DELTA calculates the 2-D axisymmetric time dependent solution for the viscous compressible Navier-Stokes equations on a triangular gridded mesh. At the time corresponding to the arrival of pulsed power at the gas jet, the density profile of the expanding argon gas is frozen. This 2-D density profile is then used as the starting point for the 2-D snowplow code which computes the pinch dynamics driven by J x B forces in the snowplow approximation.

This technique could be used to predict implosion shapes from nozzle geometry for any driver.

INTRODUCTION

Supersonic jet nozzles have been widely used as sources of gas to be imploded, in the Z-pinch configuration, by the very high currents produced by pulsed power generators. Argon Z-pinch experiments on the BLACKJACK 5 (BJ5) generator and modeling efforts have been undertaken with the ultimate goal of optimizing K-shell radiation output from the argon plasma by modifying the gas nozzle profile.

The K-shell X-ray yield depends on the current driving the pinch and on the gas density distribution in the inter-electrode space, as well as on the geometric details of the pinch region. The geometry of the load region determines its inductance and ideally it should be designed so that all of the current is delivered to the pinch.

For the experiments under consideration, the implosion time, of the order of 100 ns, is much shorter than the typical time that it takes for the injected gas density to vary significantly, which is of the order of tens of microseconds. Given the gas density profile at the time of current injection into the gas, the geometry of the load region, and the characteristic driving voltage and impedance of the generator, the whole pinch process, including current history, is determined.

Thus, in principle, if one had total control of the experimental variables, and a self-consistent circuit code driving the radiation MHD equations for the Z-pinch, detailed meaningful comparison could be made between experiment and theory. Some efforts in this direction have been reported by other researchers[1].

In this paper a new approach to approximate the above procedure is described: (1) The gas dynamic profiles are obtained by a 2-D computer code. (2) Based on the experimentally observed time, a calculated gas profile is selected as the initial condition for the implosion. The experimental current is used to drive the implosion. The pinch dynamics are approximated by 2-D snowplow equations solved by a second computer code. The snowplow approximation produces the shape of the outer envelope of the pinch, r(Z) at all times, and it should be accurate until significant amounts of plasma stagnate on axis. (3) The onset of K-shell radiation observed experimentally by a temporally and spatially resolved X-ray diode (XRD) imaging array is compared with the arrival of the snowplow front at a prescribed radius. Arrival time differences are interpreted to be what is usually termed zippering.

The paper describes the 2-D gas dynamic code, the snowplow code, the experimental test stand gas profile measurements, the actual K-shell radiation measurements performed on BJ5, and the comparison between experimental data and calculations.

GAS DYNAMICS

The gas dynamic flow field from the 3.5 cm diameter annular nozzle on BJ5 was modeled using the code 2-D RZ DELTA. This code calculates the 2-D axisymmetric time dependent solution for the viscous compressible Navier-Stokes equations on a triangular gridded mesh. It is particularly useful in modeling axisymmetric compressible flow through complicated geometries.

The geometry of the 3.5 cm diameter annular nozzle is outlined in Figure 1 which shows part of the grid used in the calculation. This grid was generated in a few hours of operator's time using S3MESH, an automatic triangular mesh generator which directly interfaces to 2-D RZ DELTA. The bottom line in the figure represents the centerline (r=0) of the apparatus. The large region on the right represents the vacuum chamber into which the nozzle exhausts and the large region on the left represents a high pressure plenum. At time equals zero, a "diaphragm" between the high pressure plenum and the secondary plenum of the nozzle is broken and argon, at 70 psi and 298 K, begins flowing into the nozzle assembly. The gas flows into the nozzle secondary plenum region, through the nozzle throat, down the diverging nozzle section and out into the vacuum region where the Z-pinch can be initiated after a chosen delay time. This delay time is selected by using

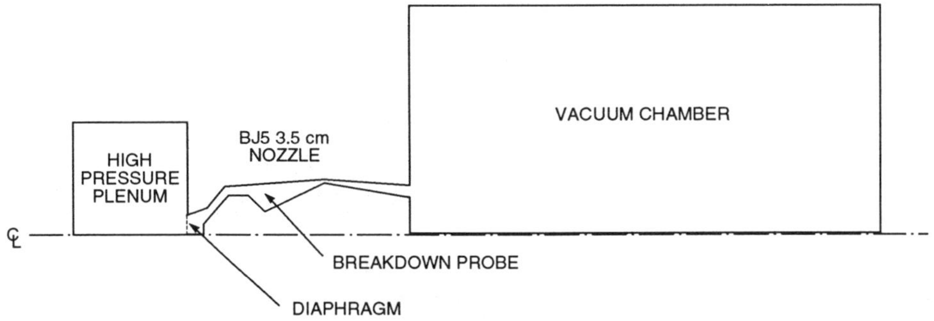

Figure 1. The geometry of the gas delivery system studied.

the experimentally measured time difference between gas detection by a breakdown probe in the secondary plenum and BJ5 firing. This time difference is correlated to a calculational time. The calculation can then generate a density profile which is taken as the initial condition for the implosion. The initial density profile is handed to the snowplow code for use in predicting the implosion dynamics.

SNOWPLOW MODEL

The 2-D snowplow model was first developed by Hussey, Matzen and Roderick[2] to study the effects of nonuniform plasma distributions on Z-pinches. The model had limitations, for two reasons; (1) No boundary conditions were supplied, so the ends of the snowplow could and did drift away from the physical boundaries during calculations; (2) Segments of the snowplow front could expand or contract without limit, degrading the resolution of the computation late in time.

We have recently solved both of these problems, and have developed a fast, robust code implementing the 2-D snowplow model with plausible boundary conditions and adaptive mesh refinement along the snowplow front. The code has been used to study both the plasma opening switch and the Z-pinch[3].

The equations of motion are the same as those of Ref. 2. One partitions the snowplow front initially into N segments, as shown in Figure 2. For each segment, five ordinary differential equations must be solved, involving mass, momentum, and position. No energy equation is solved, since the snowplow model does not conserve energy. The equations taking into account azimuthal symmetry are:

$$\dot{\vec{X}}_i = \frac{P_i}{M_i}$$

$$\dot{M}_i = 2\pi r_i \Delta_i \rho(\vec{X}_i) \left[\dot{\vec{X}}_i - \vec{U}(\vec{X}_i)\right] \bullet \hat{n}_i$$

$$\dot{\vec{P}}_i = 2\pi r_i \Delta_i \left[\frac{B^2(\vec{X}_i)}{8\pi} \right] \hat{n}_i + \dot{M}_i \vec{U}(\vec{X}_i)$$

where: $\vec{X}_i = (r_i, Z_i)$ are the coordinates of the segment midpoints

M_i and $\vec{P}_i$ are the segment masses and momenta
B is the magnetic field which is in the ø direction
ρ is the plasma mass density distribution
$\vec{U}$ is the background plasma velocity distribution
Δ_i is the length of each segment
$\hat{n}_i$ is the unit normal to each segment at its midpoint
and a dot indicates differentiation with respect to time.

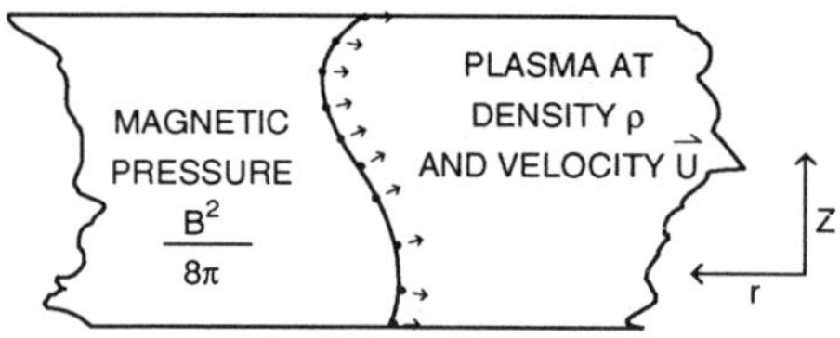

Fig. 2. Segmented snowplow front used in modeling implosion.

Three issues must be resolved in order to solve these equations: initial conditions, boundary conditions, and maintaining resolution.

Initially, the snowplow front has zero mass and momentum, which leads to a removable singularity in the equations above. We solve this problem by using the analytic approximate solution to the 1-D snowplow model to estimate the mass and momentum of each segment of the snowplow front at some very small initial time.

Boundary conditions must be specified for all five quantities of the snowplow front endpoints. The coordinates are restricted to lie on the boundary, and the velocities of the endpoints are adjusted to account for a continuous shortening or lengthening in the end segments, maintaining them at constant angle to the boundaries. Mass and momentum may be lost or reflected at the boundaries if the endpoint segments have a velocity component into the boundaries.

To maintain the initial resolution, the snowplow front is reparametrized after each time step. New snowplow segments can be added or lost, and the lengths may change. Segments are made shorter in regions where the density gradient is higher. Mass and momenta are then reassigned to each segment to regain the distribution which held before reparametrization.

GAS JET TEST STAND MEASUREMENTS

Gas jet pressure profiles were measured on a test stand using a calibrated piezoelectric probe. The probe was covered with a shroud which had a pinhole whose area was less than four percent of the probe's active area; the probe signal was thus proportional to stagnation pressure. Stagnation pressure versus time was measured at various points in space with several shots taken at each location to check reproducibility. Timing reference was provided by the breakdown probe located just past the puff valve. Timing of the breakdown probe relative to the puff valve coil waveform was also measured to verify reproducibility of valve operation.

Stagnation pressure measurements were performed for the 3.5 cm nozzle at a plenum pressure of 70 psi (4.83 x 10^6 dynes/cm^2). Figure 3 shows typical measured stagnation pressure versus time at a radius of 1.6 cm at distances 1 and 3 cm from the nozzle exit. Gas dynamic pressure calculation results from the model described in Section 2 are overlayed for comparison. The measured and calculated pressures are in good agreement which justifies using the gas dynamics model to specify the initial mass distribution for snowplow calculations.

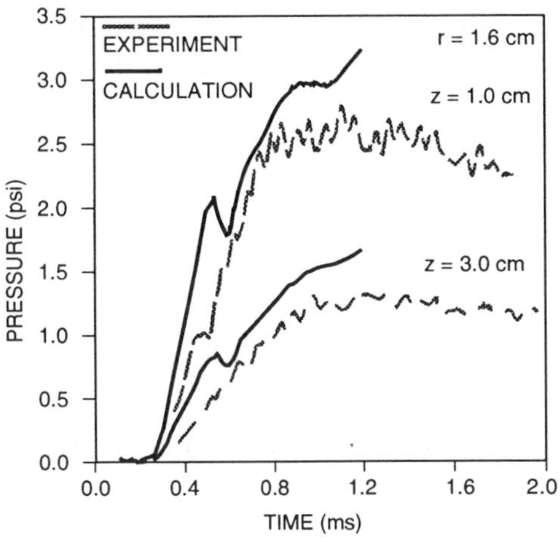

Fig. 3. Measured and calculated stagnation pressures for the BJ5 argon nozzle.

EXPERIMENTAL MEASUREMENTS OF Z-PINCH IMPLOSION SHAPE

Experimental measurements of Z-pinch implosion shape have been performed on argon discharges from a gas nozzle mounted on the BLACKJACK 5 pulsed power driver. The typical current waveform used to drive the implosion reaches a peak of about 4 MA as shown in Figure 4. The driving current exhibits some prepulse with the rapidly rising part occurring in about 80 ns. No preionization was used. The discharge implosion length was 4 cm. Azimuthal variations in the nozzle throat opening were checked with feeler gauges and found to be less than 15 percent of the opening dimension.

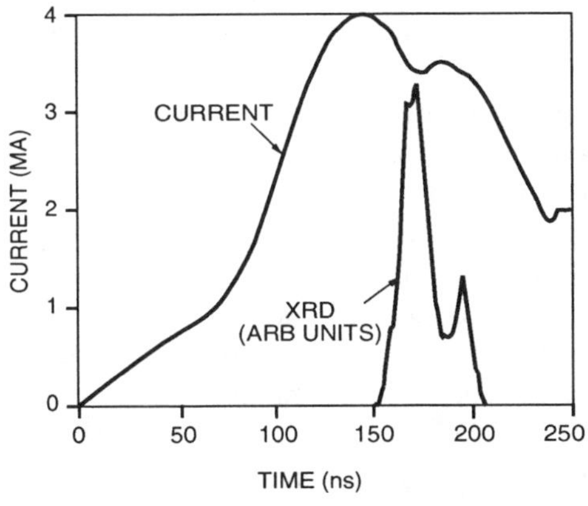

Fig. 4. Typical BJ5 implosion current with overlayed XRD signal.

Plasma implosion shape data were obtained with a one-dimensional X-ray diode (XRD) imaging array. Individual XRDs were masked to record K-shell X-rays from a 3 mm wide slice of the discharge. Center-to-center spacing of the XRD viewing areas was 6.8 mm. Five useful channels of data were obtained with the first channel recording X-rays from within a few mm of the cathode. Data from the XRD imaging array for a typical shot is shown in Figure 5. The relative timing of the XRDs and the current waveform is accurate to within ±2 ns. The shape of the implosion can be characterized by plotting the time of K-shell X-ray onset as a function of the position of the XRD.

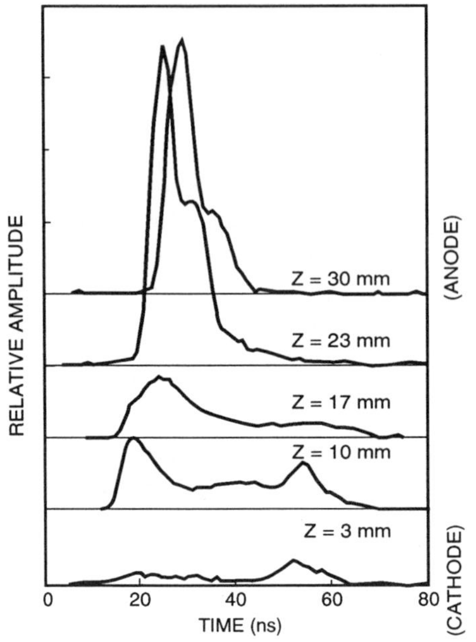

Fig. 5. XRD traces from the five different Z positions observed along the discharge indicating K-shell X-ray onset time and relative amplitude.

COMPARISON OF MODELING AND EXPERIMENTAL IMPLOSION SHAPES FROM K-SHELL X-RAY ONSET

Since the onset of K-shell X-rays was the most straightforward measurement available of implosion shape, a methodology for calculating K-shell X-ray onset from the snowplow code was developed. The technique chosen is similar to that used in Ref. 2. The results of the 2-D gas dynamic calculation provide the initial mass distribution of the plasma being snowplowed toward the Z-axis. The measured current for BLACKJACK 5 is used to drive the implosion. For comparison purposes, the onset of K-shell X-rays in the model is defined to occur when the plasma at any axial position reaches a radius of 2 mm. This radius was chosen because it resulted in good agreement with the measured implosion time and X-ray pinhole photographs of K-shell radiation indicate that 2 mm is about the outer radius of significant K-shell emission. The snowplow calculation is stopped when the leading edge of the snowplow region reaches a radius of 2 mm. The positions and velocities of points along the remainder of the snowplow front are then used to compute the onset of K-shell X-rays (arrival at 2 mm radius) as a function of position along the Z-axis.

Results of this approach for a specific BLACKJACK 5 argon shot are shown in Figures 6 and 7. The leading edge reaches the 2 mm radius at 156 ns after the current pulse begins. This is reasonably close to the measured value of 149 ns for the onset of K-shell X-rays. Figure 7 overlays the onset time of K-shell X-rays as a function of Z as measured by the XRD array and as calculated with our model for the BJ5 shot. The calculated spread in implosion time along the length of the discharge is about 15 ns while the spread measured by the XRD array is about 13 ns. For this shot, the experimental data indicated that the implosion begins near the cathode and progressed to the anode. The calculation indicated a different implosion shape. These results are typical of the small number of comparisons made to date in that while the gross features of implosion time and spread in implosion time are in good agreement between experiment and modeling, implosion shape is not.

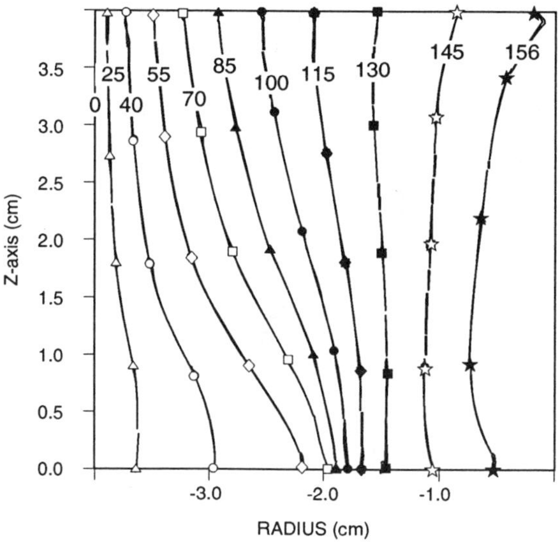

Fig. 6. Successive snowplow fronts at 15 ns intervals.

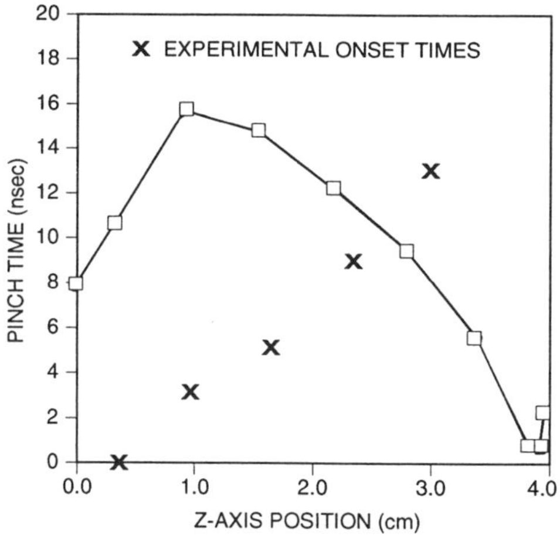

Fig. 7. K-shell X-ray onset time predicted by snowplow model (solid line) with experimental data overlayed.

CONCLUSIONS

To validate the technique described in this paper for predicting implosion shape from nozzle geometry, two comparisons between experiments and calculations were performed:

1. Test stand measurements of the gas pressure were compared with the 2D gas dynamic calculations with adequate agreement.
2. The onset of K-shell radiation from argon was compared with the prediction of the arrival of the 2-D snowplow front at a prescribed radius. The comparison is only good in the gross features: implosion time, and spread of onset time.

The work presented here is preliminary in the sense that additional efforts in several areas are required to refine these conclusions. The stagnation pressure must be measured over the entire volume of interest. Additional independent measurements of gas parameters are also needed to permit an experimental determination of the initial gas mass distribution. Reproducibility of the experimental results for K-shell onset time as a function of Z has not been sufficiently tested. Also, criteria for the required degree of similarity of gas jet performance between BLACKJACK 5 and test stand have to be developed. In addition, sensitivity studies are needed in understanding and implementing boundary conditions in the 2D snowplow code. For instance, no account was taken of the fact that the actual anode in the experiment is a wire mesh.

The comparisons made to date, although not exhaustive, indicate that the procedure utilized will result in a useful methodology to optimize nozzle design for X-ray production in pulsed power generators. Future work will include the addition of a circuit model to obtain the current at the pinch for any generator. This model will self-consistently incorporate the time dependent inductance of the pinch.

REFERENCES

1. C. Deeney, et al., IEEE Conference Record, 91CH3037-9, 205, (1991).
2. T. Hussey, M. Matzen, N. Roderick, *J. Appl. Phys.,* **59**, 2677, (1986).
3. E. Salberta, D. Parks, E. Waisman, *Phys. Fluids B*, 4, 3440, (1992).

ANALYSIS OF RECENT SATURN ALUMINUM PRS EXPERIMENTS

K. G. Whitney and J. W. Thornhill
Plasma Physics Division
Naval Research Laboratory, Washington D.C. 20375
R. B. Spielman, T. J. Nash, J. S. McGurn, and L. E. Ruggles
Sandia National Laboratories, Albuquerque, NM 87185
M. C. Coulter
Berkeley Research Associates, Springfield, VA. 22151

ABSTRACT

A set of experiments was recently completed at the SATURN plasma radiation source (PRS) facility at Sandia National Laboratories. The purpose of the experiments was to compare the results of K-shell yield measurements made on SATURN from aluminum wire array z-pinch implosions to a similar set of measurements made at the Double EAGLE facility at Physics International, Inc.. These experiments were designed to study the behavior of the kilovolt x-ray yield from aluminum z-pinches when a tradeoff is made between the amount of mass imploded and the maximum velocity that is achieved in the implosion. A comparison of the experimental results, however, suggests that a tradeoff between the mass imploded and the implosion initial conditions may be involved in determining the K-shell emission capabilities of these aluminum array implosions; namely, as the mass of the wires is increased, the coupling between the explosion of each single wire and the implosion of the array itself is increased. A simple wire explosion model is employed and applied to each of the SATURN and Double EAGLE experiments to make this case.

INTRODUCTION

A series of aluminum PRS experiments was designed for the SATURN machine at Sandia National Laboratories[1] to test the K-shell yield scaling that had been predicted for aluminum PRS loads[2] and that had been partially confirmed in experiments at Physics International Inc..[3] The basic concept being tested by the PI experiments was that, for efficient production of K-shell emission, a z-pinch array load had to be imploded to a final kinetic energy per ion, K_i, in excess of (and usually a small multiple of) a minimum energy, E_{min}, which scales with the atomic number, Z, of the load:

$$E_{min} = 1.012 Z^{3.662} \quad eV/ion.$$

For aluminum, this minimum energy is approximately 12 keV/ion. The dimensionless parameter, η, defined by $\eta \equiv K_i/E_{min}$, thus should satisfy the criterion, $\eta > 1$, i.e., $K_i > 12$ keV/ion for aluminum.

In Ref. (2), MHD calculations had been carried out that showed how the K-shell yield from aluminum would scale with m, the imploded array mass per unit length, when η was held constant or with η when m was held constant. These calculations, which were carried out using classical plasma conductivities, produced hard implosions, i.e., implosions in which the plasma assembled on axis into a tight pinch of radius on the order of 0.1 mm. The experiments that were conducted at PI to test the validity of the calculated yields were not designed to hold either m or η constant. In them, mr_0^2 was held roughly constant, where r_0 is the initial radius of the imploded array. This condition corresponds roughly to holding either K_T, the total kinetic energy of the implosion, or the product $m\eta$ constant. Scaling relations had been derived from the MHD calculations which predicted that the yields in the PI experiments would be approximately constant as well, corresponding to a fixed fractional ($\sim$ 30%) conversion of kinetic energy into K-shell x-rays for all implosions of different radii, r_0, for which $mr_0^2 =$ constant.

The PI experiments confirmed that, when the condition, $\eta > 1$, was satisfied, the pinch behaved as an efficient bulk K-shell emitter with, at minimum, the kinetic energy conversion efficiencies predicted by the calculations. However, two important differences between theory and experiment were observed. First, the K-shell yields were not flat, but had a peak for small η values ($\eta \sim 2$ or 3), corresponding to a kinetic energy conversion efficiency of $\geq$100%, which indicates the presence of energy inputs other than kinetic. Second, the experimental implosions were much softer than the calculated implosions, i.e., they assembled on axis with outer radii on the order of 1 mm. The first difference suggests the conjecture that an anomalously high amount of Ohmic heating occurs under certain plasma conditions on axis as the current continues to heat and confine the pinch. The second points to the conjecture that the implosions have anomalously high (MHD turbulence induced) viscosities and heat conductivities as well. Calculations utilizing enhanced electrical resistivities, viscosities, and heat conductivities appear to confirm these conjectures,[4] but more experiments are needed for further confirmation especially in larger current pulsed power machines.

Thus, a major motivation for carrying out experiments on SATURN was to determine how the experimental results of Double EAGLE would scale to higher currents and larger imploded aluminum masses. Because SATURN is a softer machine than Double EAGLE, (i.e., the PRS load implosion dynamics reacts back on the current flowing through the load more strongly in SATURN than in Double EAGLE), it was important to design the SATURN experiments using the lumped circuit model for SATURN. The same procedure is also needed for Double EAGLE if greater control of the theoretical parameters is required than was achieved in the Ref. (3) experiments.

The circuit model that was used to design the SATURN experiments is given in Figure 1 of Ref. (5). The pulsed power generator is described as a time dependent voltage source, $V(t)$, driving a line resistance, Z_0, a line inductance, L_p, and a dynamic imploding z-pinch load with a time dependent inductance, $L(r_a(t))$. By using a slug model description of the dynamic load to determine $r_a(t)$, the location of the array (shell) of imploding plasma, the PRS load dynamics can be initially simplified to the problem of solving the following two nonlinear equations of motion:[5]

$$L_T \frac{dI}{dt} + \left(Z_0 - L_0 \frac{1}{r_a}\frac{dr_a}{dt}\right) I = V(t), \tag{1}$$

$$m\frac{d^2 r_a}{dt^2} = -\frac{L_0}{2\ell} I^2 \frac{1}{r_a}, \tag{2}$$

where I(t) is the current flowing through the load, $L_T \equiv L_p - L_0 ln(r_a/r_0)$, $L_0 \equiv (1 - 1/n_w)(\mu_0 \ell)/(2\pi)$, n_w is the number of wires in the array, ℓ is the length of the array, and $\mu_0 = 4\pi \times 10^{-7}$ henry/m. The dynamic inductance of the load in this model is given by $L = -L_0 ln(r_a/r_0)$.

The SATURN experiments were designed using commercially available aluminum wire sizes, and, in an attempt to achieve maximum implosion symmetry, the arrays were constructed of 24 wires. One set of six experiments was designed to closely parallel the Double EAGLE experiments so that direct inferences as to how the K-shell yields were scaling to higher mass loads could be made. Therefore, as in the Double EAGLE experiments, the initial array diameters were taken to be < 2 cm. In these experiments, the slug model product $m\eta$ was held constant, and the total kinetic energy generated per unit length in each slug model implosion was fixed at 85 kJ/cm. The calculated η's ranged in value from less than 1 to 6. If 30% of the kinetic energy could be converted to K-shell emission, the minimum expected output of K-shell x-rays in these experiments would be 25.5 kJ/cm. Two other experiments using the same mass wires as in two of the 85 kJ/cm experiments were also planned. These wires would be placed at larger initial radii and would implode to higher η values. In these experiments, the slug model kinetic energy generated by the implosion was calculated to be 115 kJ/cm.

Eqs. (1) and (2) describe the dynamics of hard MHD implosions only to the point where the plasma is close enough to the axis to generate a back pressure comparable to the $\mathbf{j} \times \mathbf{B}$

forces, which are described by the I^2 term on the right side of Eq. (2). Thus, the solutions to Eqs. (1) and (2) had to be terminated before this condition ensued. Thus, the parameters of the above eight SATURN experiments, which are given in Table I, were calculated from Eqs. (1) and (2) by terminating the solution in each case at $r_a(t_{imp}) = 1.5$ mm, where t_{imp} is by definition the implosion time.

Table I: SATURN PRS Experiments

Wire Diam (mil)	Array Mass (μg/cm)	Array Diam (cm)	η	K.E. (kJ/cm)	# Wires
2.5	2050	1.28	0.95	85	24
2.0	1310	1.34	1.6	85	24
1.7	950	1.40	2.2	85	24
1.5	740	1.46	2.7	85	24
1.2	470	1.58	4.2	85	24
1.0	330	1.72	6.0	85	24
1.5	740	2.20	3.5	115	24
1.2	470	2.29	5.6	115	24

For purposes of comparison, we present the same parameters in Table II for the Double EAGLE experiments of Ref. (3):

Table II: Double EAGLE PRS Experiments

Wire Diam (mil)	Array Mass (μg/cm)	Array Diam (cm)	η	K.E. (kJ/cm)	# Wires
1.7	474	0.90	0.5	8.5	12
1.3	277	1.25	1.6	12.0	12
1.0	164	1.50	1.7	13.3	12
0.8	105	2.00	3.0	16.5	12
0.7	80	2.50	5.0	20.5	12

EXPERIMENTAL RESULTS

One measure of the accuracy of the implosion model that was used to design the experiments is provided by comparing the observed to the calculated implosion times. This comparison is shown in Fig. (1). The experimental times were inferred from the locations of the peaks of the K-shell emission pulses, and they may be slightly larger than the actual implosion times. Nevertheless, the data show that, for implosions that occur late in the current pulse (i.e., after the peak of the pulse), there is a tendency for the pinches to radiate after the calculated implosion time; whereas, for early implosions, which occur before the current peaks, the pinches radiate before the calculated implosion time. Overall, the agreement between observed and calculated implosion times seems to indicate that the lumped circuit model description of the wire array implosions has some validity.

The K-shell x-ray yields that were measured in both the SATURN and Double EAGLE

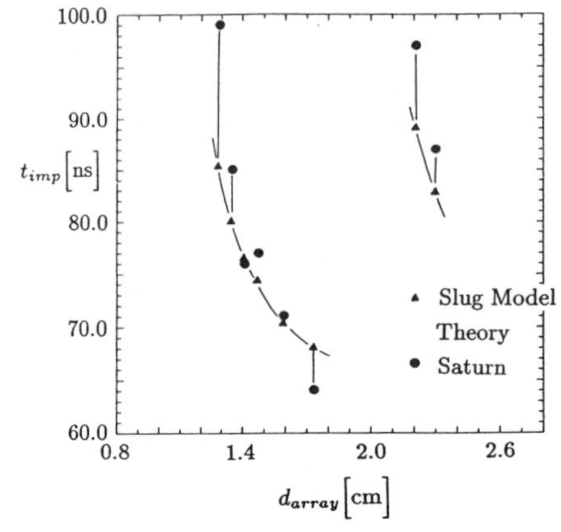

Figure 1. Calculated and observed implosion times for the eight wire array implosions in Table I are shown as a function of array diameter. All the plots in this paper are displayed as a function of the array diameters of the 13 experiments discussed in this paper.

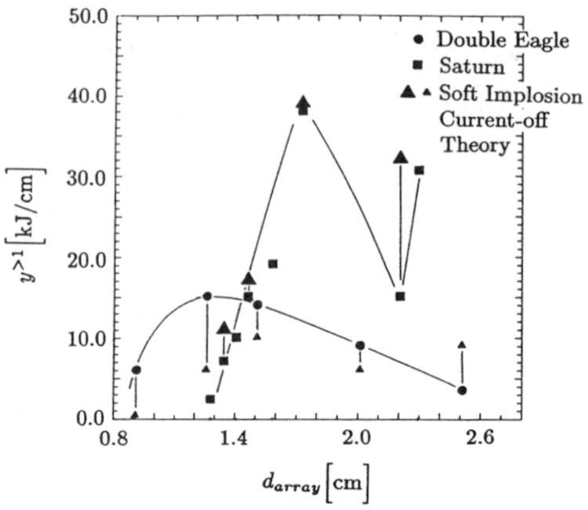

Figure 2. Comparison of the observed and calculated K-shell yields in both the Double EAGLE and SATURN experiments defined in Tables I and II.

experiments are compared in Fig. (2). The Double EAGLE yields peak for η values near 2 and then fall as the array diameter ($\geq$ 2 cm) and η are increased. The comparable SATURN yields, however, continue to rise from $\eta < 1$ to their largest value at $\eta = 6$. Beyond this data from the 85 kJ/cm implosions are the two yield points that represent an extension of the η/mass tradeoff to larger array diameters, but now at the increased kinetic energy input of 115 kJ/cm. However, the large drop in K-shell output to $\sim$ 15 kJ/cm was not expected in going from a mass of 330μg/cm and $\eta = 6$ to a mass of 740μg/cm and $\eta = 3.5$. In other words, a larger η produced a larger K-shell yield for the 470μg/cm implosions as anticipated, but the same behavior was not observed for the 740μg/cm implosions.

Figure 2 also contains the predicted yields from 1-D MHD calculations, which were performed, as in Ref. (2), by terminating the current near the end of the implosion. For these predictions, circuit models for Double EAGLE and SATURN were used, and the current was terminated when the voltage across the pinch exceeded 3 MV. Phenomenological enhancements of the artificial viscosity, heat conductivity, and electrical resistivity by factors of 40, 30, and 20 respectively were also used in these calculations to soften the implosions.[4] Use of these factors reduced the calculated densities on axis in the Double EAGLE experiments roughly to those that had been inferred from the x-ray data.[3] Moreover, when the implosions are softened in this way, the calculated yields follow, roughly, the observed Double EAGLE yield behavior and show a peak at $\eta \sim 2$. When similar calculations were carried out, in advance of the experiments, for four of the eight SATURN experiments using the same enhancement factors as in the Double EAGLE calculations, they also accurately predicted the location of the yield peak, seen in Fig. (2), at $\eta \sim 6$.

Another important difference between Double EAGLE and SATURN, aside from the shift in peak yield, is suggested by the the fact that the calculated yields were less than the observed yields for Double EAGLE, while the calculated yields for SATURN were larger than observed. The significance of this result is shown in Fig. (3), where the measured efficiencies for converting the (calculated) slug model kinetic energy to K-shell emission are plotted. The (hard implosion) conversion efficiencies that were calculated in Ref. (2) are also shown in the shaded region of this figure. They are indicative of implosions in the so-called I^2 scaling regime. By extrapolation from the Double EAGLE data, one would have expected the SATURN points to lie in or above this theoretical region. Figure 3 shows that only the 330μg/cm, $\eta = 6$ implosion fulfilled this expectation. On Double EAGLE by contrast, only the large array diameter, low mass, $\eta = 5$ implosion failed, somewhat, to perform up to expectation.

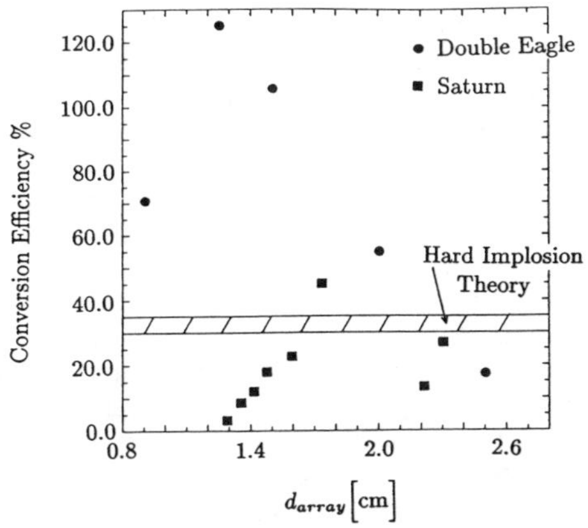

Figure 3. The measured yields in Fig. (2) are plotted relative to the kinetic energies of tables I and II and expressed as percentages.

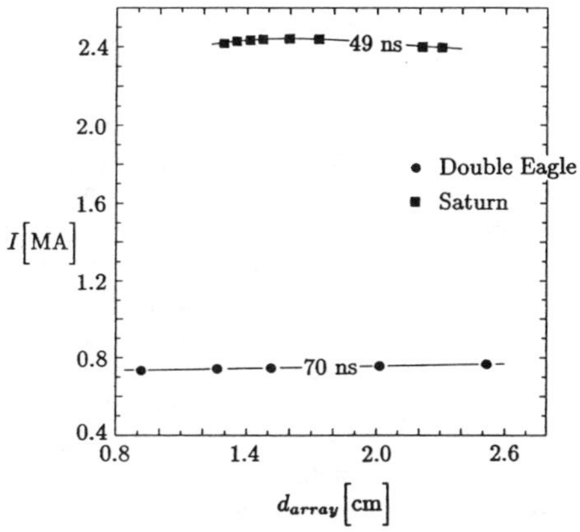

Figure 4. Each wire explosion calculation was carried out over the time intervals listed in the figure: 70 ns for the Double EAGLE and 49 ns for the SATURN experiments. The figure shows the final currents flowing through the arrays in each case.

ANALYSIS

One of the first places to look in order to explain the differences between the SATURN and Double EAGLE x-ray conversion efficiencies is at the initial conditions of the MHD calculations vis-a-vis the experiments. All of the calculations begin by driving current through a cold ($\sim$ 5 eV) preexisting plasma. The experiments, on the other hand, all begin with solid wires. Therefore, in order to unambiguously compare the calculated to the measured yield behavior, one must verify that the wire explosion and the array implosion dynamics are separable and that the explosion approximately establishes the initial conditions that are used in the 1-D MHD calculations. We will show, in fact, by applying an earlier model of wire explosions to the Tables I and II arrays that this assumption apparently is much less correct, in general, for the SATURN experiments than it is for the Double EAGLE experiments.

The explosion model to be used is described in Ref. (6). The model contains momentum and energy equations for a single wire that are coupled to a circuit equation for the pulsed power generator. The equations have the following structure:

$$\frac{1}{2}M\frac{d^2r}{dt^2} = -(\mathbf{j}\times\mathbf{B}\text{ force}) + (\text{pressure expansion term}), \tag{3}$$

$$\frac{d\varepsilon}{dt} = (\text{Ohmic heating}) - (\text{expansion cooling}), \tag{4}$$

$$V(t) = Z_0 I + L_g\frac{dI}{dt} + \frac{R_w}{n_w}I + \frac{d}{dt}(L_a I), \tag{5}$$

where M is the total mass of the exploding wire, r is its radius, L_g is the inductance of the generator, and ε is the internal energy of a wire (thermal plus ionization). The load inductance,

L_a, is that of a circular array of n_w wires of length ℓ:

$$L_a = \frac{\mu_0 \ell}{2\pi n_w}\left[\ln\left(\frac{R_{rc}}{n_w r}\right) + (n_w - 1)\ln\frac{R_{rc}}{r_a}\right], \tag{6}$$

where R_{rc} is the radius of the return current path. Note that, in this model, r is considered the dynamical variable and r_a is fixed. In the slug implosion model, on the other hand, r_a is the dynamical variable and r is fixed. These equations were derived by averaging the hydrodynamic equations for the energy and momentum for a wire over its volume. Self-similar assumptions were made about the hydrodynamics; namely, that the wire density is uniform and that the velocity rises linearly within the wire. The current is also assumed to uniformly penetrate the wire up to a skin depth. Radiation losses are ignored and an ad hoc dependence of the ionization state of the wire on its average temperature is assumed. Futhermore, an ideal plasma equation of state is used, and the ionization energy is assumed to be equal to the thermal energy. The worst assumption of the model appears to be in its use of the Spitzer formula for plasma resistivity, which is roughly four orders of magnitude larger than the measured values for high density, sub-eV aluminum plasmas. However, the expansion dynamics is remarkably invariant to such resistivity changes. The main effect of an artificially large resistivity is to slow the risetime of the current, but it does not otherwise basically alter the wire expansion dynamics. In our calculations, we replaced the resistivity of the Ref. (6) model with resistivities obtained from Ref. (7), which essentially have a linearly rising dependence on the electron temperature rather than the $T^{-3/2}$ dependence of Spitzer.

We used the above model to calculate wire expansion for each of the arrays listed in Tables I and II. Each calculation was begun at solid wire densities and with an initial temperature of 0.1 eV. The circuit models for Double EAGLE and SATURN were used to drive the calculations, and each one was carried to a fixed time, which preceded significant inward motion and coelescence of the arrays. The Double EAGLE calculations were carried to 70 ns and the SATURN calculations to 49 ns. In Figs. (4)-(8), we present some of the final values that were reached by the variables in each calculation.

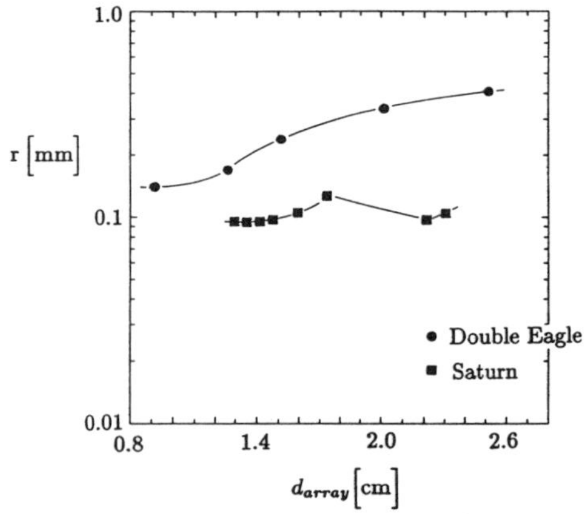

Figure 5. Final wire radii as a function of array diameter. For Double EAGLE, they are the calculated radii at 70 ns; for SATURN, they are the radii at 49 ns.

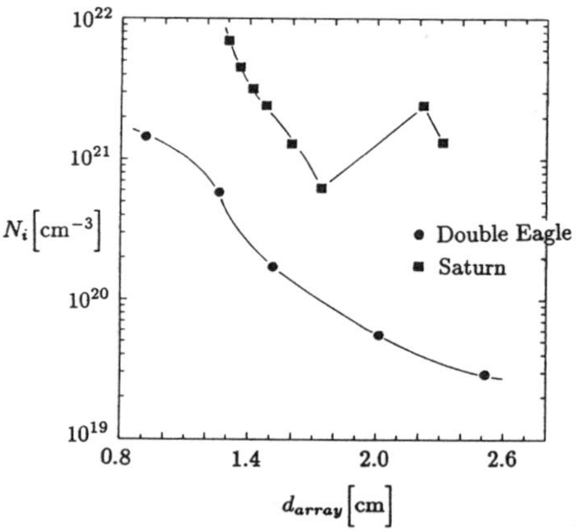

Figure 6. Final wire ion densities as a function of array diameter.

Figs. (4) and (5) give, respectively, the current that flowed through each array and the radii reached by each expanding wire at the end of each calculation. These figures show that the strength of the current, which produces the counteracting effects of the j × B forces that retard the expansion and Ohmic heating that drive it, is not as important as time in determining of the extent of the wire expansion. Note that the behavior of the SATURN kilovolt yields in the different experiments is mimicked by the behavior of the final wire radii. This effect is even more pronounced for the final wire densities and temperatures. When the differences in the initial wire masses are factored in with the differences in final wire radii, one obtains the final wire densities shown in Fig. (6). The final temperatures that were reached by the wires during these expansions are shown in Fig. (7). In this figure, the SATURN wire expansions show a temperature behavior that directly mimicks the yields. It would appear from Figs. (6) and (7) that the SATURN kilovolt yields vary in direct relationship to the temperatures achieved in the wire explosion phase of the pinch dynamics and in inverse relationship to the ion densities achieved.

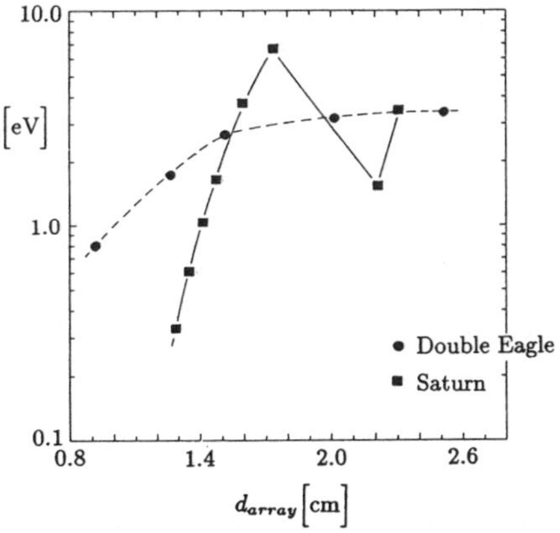

Figure 7. Final wire temperatures as a function of array diameter.

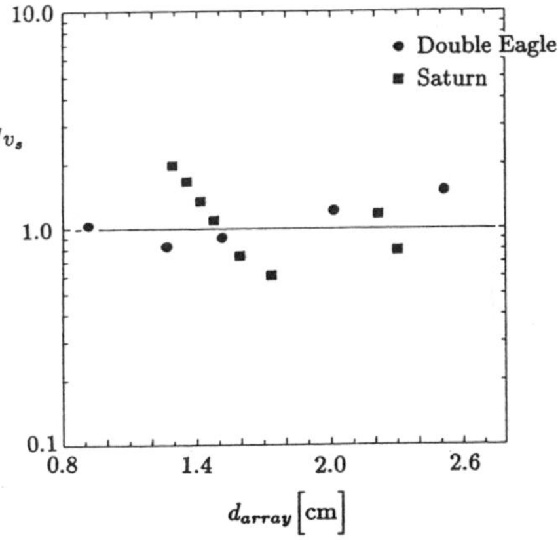

Figure 8. Final values of the ratio of electron drift speed to the plasma sound speed, $\sqrt{ZkT/m_i}$, as a function of array diameter.

Figs. (5)-(7) suggest a stronger coupling of the explosion and implosion phases of array implosion dynamics in SATURN than is present in Double EAGLE. The SATURN data, in particular, show a strong correlation between the plasma conditions reached early on by the wires during explosion and the efficiency of producing kilovolt x-rays later on following assembly of the imploded plasma on axis. The dynamical assumptions of the explosion model have led to the conclusion that SATURN implodes much denser plasmas with more variable temperatures than Double EAGLE. Consequently, wires would coelesce together later in the implosion on SATURN and perhaps be prone to less implosion smoothing and less symmetric assembly on axis than on Double EAGLE.

These wire expansion calculations contain one other piece of information, shown in Fig. (8), of interest in the modeling of z-pinch implosions. In all cases, we found that the electron drift speed and the ion sound speed were comparable at the end of the above calculations. In a majority of cases, moreover, it exceeded the sound speed. Thus, in the early explosive expansion of the wires, conditions appear to be established that are conducive to generating a highly microturbulent (and perhaps MHD turbulent) plasma. This finding apparently provides some justification for the soft implosion MHD modeling that was used in Ref. (4) and in the work presented in Fig. (2).

SUMMARY AND CONCLUSIONS

A set of z-pinch experiments was recently conducted on SATURN in which aluminum arrays were to be imploded to roughly the same final velocities (i.e., to the same η values) as had been achieved in an earlier set of Double EAGLE experiments. However, because SATURN generates a larger current than Double EAGLE, larger massed arrays could be used on SATURN in the expectation of generating correspondingly higher x-ray outputs than had been measured on Double EAGLE. In other words, it was hoped that the x-ray outputs from the two machines would scale in proportion to their input kinetic energies as computed from a slug model description of the implosion dynamics. Aluminum was chosen because it is potentially an efficient converter of input energy, which scales as I^2, into kilovolt x-rays. However, the expected scaling was not generally observed. The efficient conversion of (calculated) kinetic energy into kilovolt x-rays was observed in only one important case: that of an array of 1 mil wires whose array diameter was less than 2 cm.

A self-similar model was then employed to analyze the early time expansion of the wires used in the above experiments in an attempt to understand the SATURN scaling. A close relationship was found between the yield behavior observed in the SATURN experiments and the temperatures and ion densities reached by the wires prior to implosion and assembly. This analysis, which was based on a number of simplifying, somewhat crude, and yet suggestive assumptions about the wire dynamics, suggests a much stronger coupling in SATURN than in Double EAGLE of the explosive and implosive phases of the pinch dynamics.

The model we used to analyze the SATURN and Double EAGLE experiments assumes that the wires are heated and expanded uniformly. Thus, it does not address questions such as: does all of the wire mass participate in the implosion or is some of the wire mass drawn in and imploded early? These effects would vitiate the assumptions made in the slug model design of the SATURN experiments and would also lead to the emission behavior seen in the wires whose diameters were larger than 1.2 mil.

ACKNOWLEDGEMENT

The authors would like to thank J. L. Giuliani, Jr. for providing us with the aluminum resistivity model that was used in this work.

REFERENCES

1. R. B. Spielman, R. J. Dukart, D. L. Hanson, B. A. Hammel, W. W. Hsing, M. K. Matzen, and J. L. Porter, "Dense Z-Pinches, Second International Conference" ed. by N. R. Pereira, J. Davis, and N. Rostoker, AIP Conference Proc. **195**, 3 (1989).

2. K. G. Whitney, J. W. Thornhill, J. P. Apruzese, and J. Davis, J. Appl. Phys. **67**, 1725 (1990); J. W. Thornhill, K. G. Whitney, and J. Davis, J. Quant. Spectrosc. Radiat. Transfer **44**, 251 (1990).

3. C. Deeney, T. Nash, R. R. Prasad, L. Warren, K. G. Whitney, J. W. Thornhill, and M. C. Coulter, Phys. Rev. A **44**, 6762 (1991).

4. K. G. Whitney, J. W. Thornhill, C. Deeney, P. D. LePell, and M. C. Coulter, "Phenomenological Modeling of Argon Z-pinch Implosions", Proceedings of the 9-th International Conference on High-Power Particle Beams, Washington DC, 1992, to be published.

5. J. Katzenstein, J. Appl. Phys. **52**, 676 (1981).

6. H. W. Bloomberg, M. Lampe, and D. G. Colombant, J. Appl. Phys. **51**, 5277 (1980).

7. L. F. Mondolfo, "Aluminum Alloys: Structure and Properties", Butterworths, London, (1976), p. 98.

CHARACTERIZATION OF NEON Z-PINCH PLASMAS FOR SODIUM-NEON PHOTOPUMPING

F.C. Young, B.L. Welch,* and H.R. Griem**
Plasma Physics Division, Naval Research Laboratory
Washington, D.C. 20375-5346

ABSTRACT

Implosions of a small neon Z-pinch with 120- to 230-kA current pulses are studied to establish neon plasma conditions appropriate for sodium-neon photopumping experiments. Plasma compression and heating are determined from measurements of soft x-ray, XUV, and NUV radiations from the ionization states: Ne III, Ne VII, Ne VIII, and Ne IX. Changes in conditions of the imploded plasma with variations of the initial neon gas density and the magnitude of the driving current are described.

INTRODUCTION

The near coincidence of the n=2 to n=1 Na X and the n=4 to n=1 Ne IX characteristic transitions at 11 Å provides an excellent opportunity to create an inversion in the n=4 level of Ne IX and to produce lasing on the 4 to 3 and 4 to 2 transitions in a properly prepared neon plasma.[1] Large sodium pump powers, which are required for this scheme, have been produced with Z-pinch implosions driven by megampere level currents. For a sodium pump power exceeding 200 GW, inversion has been reported in a neon gas bag.[2] For a sodium pump power of 30 GW, inversion has been observed in a neon gas puff.[3] In the latter experiment, the neon plasma was produced by imploding a gas puff with a fraction (150 to 200 kA) of the current (1.2 MA) used to produce the sodium source. To determine the conditions of the neon in this photopumping experiment, a compact inductive-store generator, which delivers up to 250 kA, was developed and used to study similar neon implosions.[4] A plasma opening switch was used on this generator to reduce the current risetime from 1 μs to 0.1 μs. This experiment is described in Refs. 4, 5, and 6. Soft x-ray, XUV, and NUV radiations from these implosions have been measured to determine electron densities, temperatures, and the Ne IX ground-state fraction.[5,7]

For sodium-neon photopumping, the neon plasma should be primarily in the Ne IX ground state. Calculations[1,8] indicate that the ion density should be 10^{18} to 10^{19} cm^{-3} and that the electron temperature should be 50 to 100 eV. Measurements[7] of imploded neon plasmas confirm that a large Ne IX ground-state population exists in a plasma of 10^{18}-cm^{-3} ion density and 75-eV temperature. The density and temperature of the imploded neon are expected to depend on the initial neon gas

* Also at: University of Maryland, College Park, MD 20742.
** University of Maryland, College Park, MD 20742.

density and the magnitude of the driving current. Attempts to control the density and temperature by changing these quantities are discussed in this article.

EXPERIMENT

The neon gas puff is mounted on an inductive-storage generator as shown in Fig. 1. Neon is fed through a fast-opening valve[9] to a 1-cm-exit-diam nozzle to form a confined gas puff across the 4-cm gap between the nozzle and the wire-mesh cathode on the center conductor. The valve is operated with plenum pressures of 50, 100, or 150 psi to vary the initial gas density. The mass load scales linearly with pressure for this puff valve. The current is fed from a capacitor bank through a low-inductance feed to a coaxial vacuum inductor and a plasma opening switch (POS). Without the POS, this system delivers a slow risetime (1 μs) current pulse to the neon as shown in Fig. 2(a) for two different charging voltages. The change in slope near 0.6 μs is attributed to inductive loading of the generator due to the neon being imploded to small diameter.[6] Emission from Ne VIII, shown in Fig. 2(a), peaks at this implosion. With a POS plasma between the center conductor and the current-return rods, a fast risetime (100 ns) current pulse can be produced, as shown in Fig. 2(b). The upstream current (I_u in Fig. 1) increases to 210 kA in 0.5 μs before current is switched to the load. The downstream current (I_d in Fig. 1) increases to 150 kA in 50 ns. Most of I_u is switched to the load when the switch opens. Inductive loading, corresponding to an implosion near 0.8 μs, is apparent in the shape of I_d. This operation of the POS is described in Ref. 6. Downstream currents and Ne VIII emissions are shown in Fig. 3(a) for charging voltages of 35 and 45 kV. Charging voltages of 25 to 45 kV are used to provide peak currents of 120 to 210 kA without the POS and 120 to 200 kA with the POS.

The size and uniformity of these implosions are indicated by the visible-light images in Fig. 3(b). In both cases, emission is observed along the entire 4-cm length

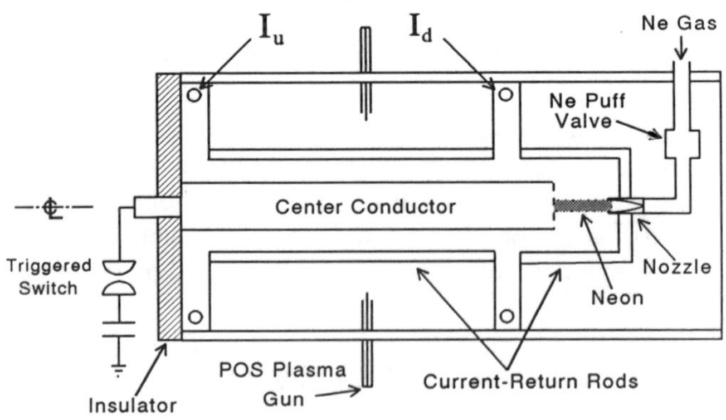

Fig. 1. Inductive-storage generator with neon gas-puff apparatus and POS plasma sources. Monitors I_u and I_d record currents upstream and downstream of the POS.

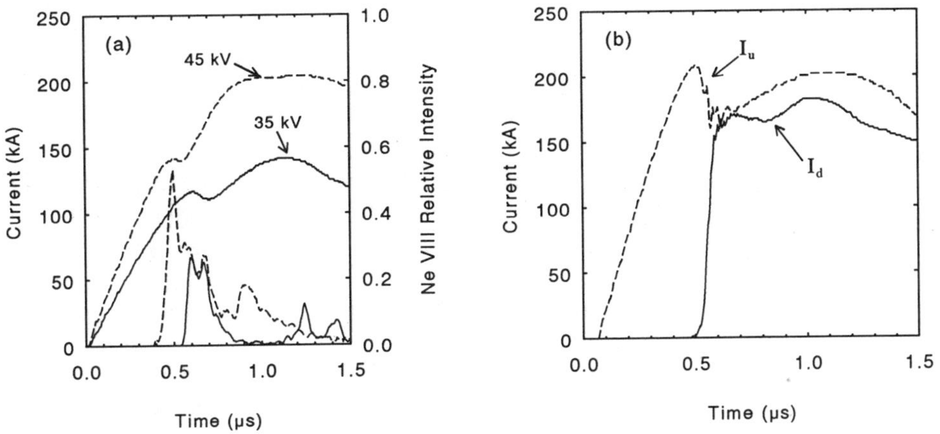

Fig. 2. Driving currents for implosions (a) without the POS and (b) with the POS. Ne VIII 88-Å emissions are also presented in (a).

of plasma with a diameter of about 1 mm. The faster risetime current produces a more uniform image. More uniform images for faster risetime currents have also been observed in soft x-rays from megampere-level neon implosions.[10] The nonuniformities in the image without the POS in Fig. 3(b) are correlated with localized regions of soft x-ray and XUV emission.[7] On the other hand, XUV images show no improvement in uniformity for faster risetime current.[6]

Time-resolved measurements of Ne IX, Ne VIII, Ne VII, and Ne III emissions were used to diagnose these neon plasmas.[4-7] Kilovolt x-rays from Ne IX were

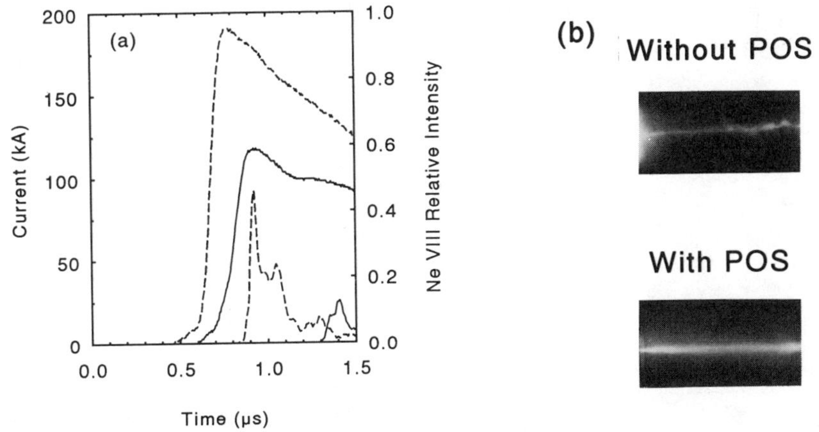

Fig. 3. (a) Downstream current and Ne VIII 88-Å emission for 35-kV (solid) and 45-kV (dashed) charging voltages. (b) Visible-light images for implosions without and with the POS. The nozzle is on the left, and the cathode is on the right.

recorded with a curved-crystal spectrograph and an x-ray vacuum diode (XRD). Ne VII and Ne VIII lines were measured with an XUV grazing-incidence spectrometer used as a spectrograph or as a monochromator. Time-integrated spectra were recorded on film, and time-histories of a few intense Ne VII and Ne VIII lines were measured with a scintillator-photomultiplier detector. Time-resolved Ne VIII lines in the NUV, as well as NUV continuum, were measured with two monochromators. A gated optical-multichannel-analyzer, mounted at the exit of one of the NUV spectrometers, provided spatially resolved NUV spectra.

IMPLODED PLASMA DENSITIES AND TEMPERATURES

Peak powers of soft x-ray and XUV emissions are presented in Fig. 4 as a function of driving current for implosions without and with the POS. XRD detectors filtered to optimize responses for > 1-keV (soft x-rays) and 150- to 280-eV x-rays (XUV) were used.[4] These measurements are for 100-psi plenum pressure. The radiated power increases with current for both emissions. For similar peak currents, less power is radiated by implosions with the faster risetime current.

For larger driving currents (> 150 kA), two successive implosions are identified in time-resolved soft x-ray, XUV, and NUV measurements.[6] For example, a second implosion near 1.3 μs is evident in the Ne VIII emission in Fig. 2(a) for 35-kV charge. For 45-kV charge, the second implosion occurs near 0.9 μs. Electron densities and temperatures at both implosions have been determined from extensive spectroscopic measurements[5,7] for a slow-risetime peak current of 170 kA. The first implosion produces a 1-mm-diam plasma of high density and temperature. This

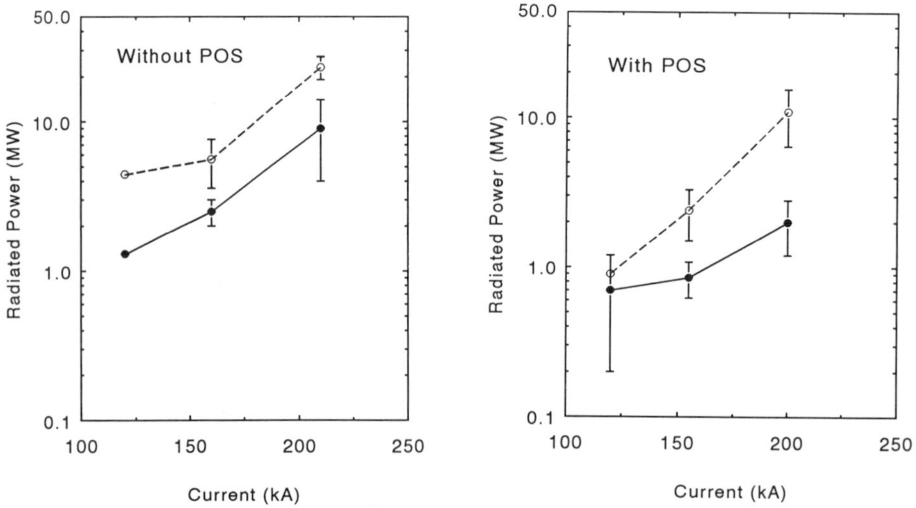

Fig. 4. Scaling of soft x-ray (solid) and XUV (dashed) peak powers with peak current without and with the POS. Vertical bars represent shot-to-shot variations.

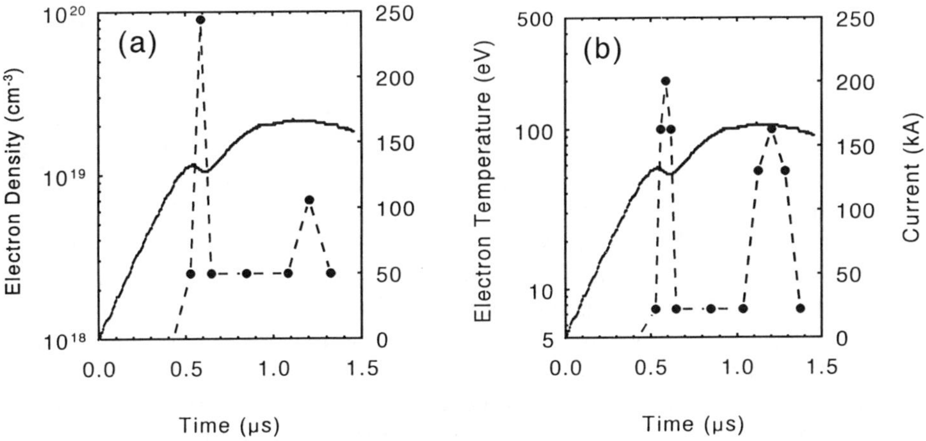

Fig. 5. (a) Electron densities and (b) temperatures for implosions driven with the slow risetime current for 100-psi plenum pressure. The driving current is also shown.

implosion is characterized by weak Ne IX soft x-ray emission, strong Ne VIII and Ne VII XUV lines, and strong NUV continuum. The second implosion produces a 3.5-mm-diam plasma of lower density and temperature. For this implosion, the Ne IX emission is extremely weak, Ne VIII and Ne VII XUV lines are observed, and Ne VIII NUV lines are observed above a less intense continuum. Properties of these two implosions are summarized in Fig. 5. Conditions before the first implosion and between the implosions are based on Ne III measurements as described in Ref. 7.

For the fast-risetime current, radiation intensities are reduced (see Fig. 4), and spectroscopic measurements are more difficult. In this case, electron densities and

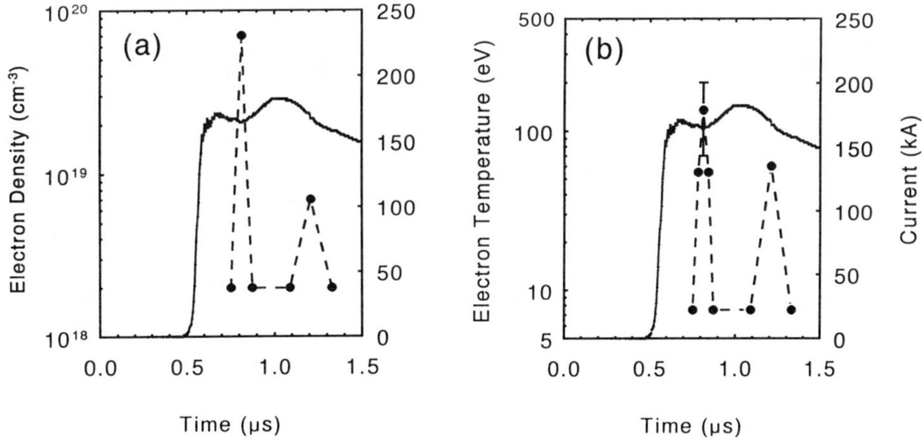

Fig. 6. (a) Electron densities and (b) temperatures for implosions with the fast risetime current for 100-psi plenum pressure. The driving current is also shown.

temperatures are determined by comparing measurements with similar slow-risetime-current measurements. Again, two successive implosions are observed for larger currents,[6] as indicated by the Ne VIII emission in Fig. 3(a) for 45-kV charge. Properties of these implosions for a driving current of 180 kA are summarized in Fig. 6. At the first implosion, the density is determined from an intense NUV continuum by comparison with the same emission measured without the POS.[5] At the second implosion, the width of the Ne VIII 3s-3p 2820.7-Å line, observed in spatially integrated NUV spectra, is used to extract the density as described in Ref. 5. The temperature at the first implosion is not well determined. Soft x-ray emission is too weak to obtain a Ne IX spectrum, and this emission is detected with an XRD only at the first implosion, implying a 200-eV temperature corresponding to localized regions of emission observed in soft x-ray images (see Ref. 6). Both Ne VII and Ne VIII spectral lines are recorded in time-integrated XUV spectra without and with the POS.[4] Comparisons of these spectra indicate that the temperature is lower with the POS, and a value of 75 eV is estimated. A temperature range of 75 to 200 eV is indicated in Fig. 6. At the second implosion, no Ne IX emission is detected, and Ne VII and Ne VIII XUV spectral lines have similar time histories.[4] A temperature of 60 eV is deduced from these measurements.[7]

DRIVING-CURRENT VARIATIONS

As the driving current is increased, the implosions occur earlier in time relative to the start of the current. This trend is apparent for both implosions without the POS in Fig. 2(a) and for the first implosion with the POS in Fig. 3(a). This behavior is consistent with the expected inverse scaling of implosion time with driving current.[11]

The intensity of the NUV continuum near 2810 Å is a sensitive monitor of the electron density and is relatively insensitive to the electron temperature.[5] Time-

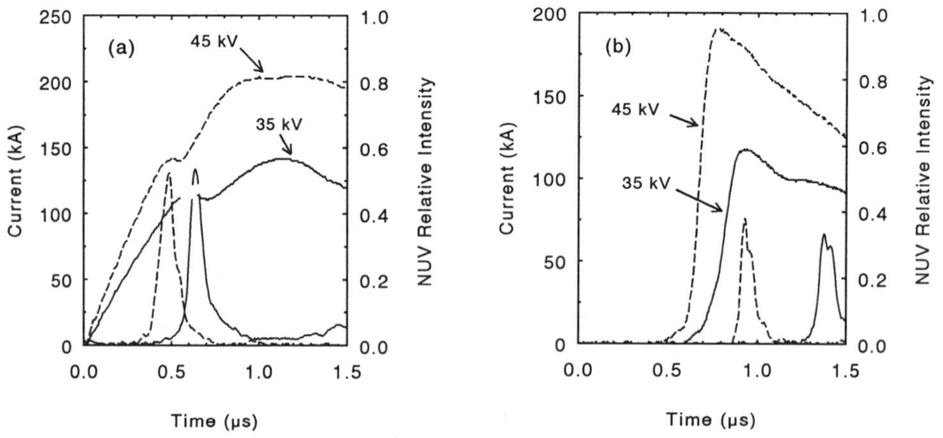

Fig. 7. Driving current and NUV continuum emission near 2810 Å for implosions (a) without the POS and (b) with the POS.

resolved measurements of this continuum for charging voltages of 35 and 45 kV are presented in Fig. 7 for implosions with 100-psi gas-puff pressure. The intensity of these emissions, which correspond to the first implosion, does not vary much with driving current. Therefore, the electron density at the first implosion remains nearly constant as the driving current is varied for either the slow or fast risetime current.

Changes in the temperature of the imploded plasma for different driving currents are evaluated by comparing the relative intensities of Ne VII, Ne VIII, and Ne IX emissions. Spectral lines from Ne VII and Ne VIII have been identified in time-integrated XUV spectra.[4,7] For the slow risetime current, the intensities of these lines are relatively constant as the current is increased from 150 kA to 210 kA.[4] Also, no change in the intensity of a Ne IX line at 78.3 Å (1s2p-1s3d transition) is apparent. These spectra only include emission from a central 6-mm-long region of the plasma. Relative intensities of Ne VII, Ne VIII, and soft x-ray emissions, observed in time-resolved measurements at both implosions, do not change significantly as the current is increased from 140 to 230 kA. Therefore, the electron temperature does not change significantly as the current is varied. For fast risetime currents, the Ne IX line is not observed in XUV spectra. Only Ne VIII and Ne VII lines are observed, which is consistent with somewhat smaller temperatures for these plasmas.

The most dramatic change with driving current is an increase in the absolute intensity of these plasma emissions. This behavior was already noted in Fig. 4 for soft x-ray and XUV emissions and in Figs. 2(a) and 3(a) for Ne VIII emission. These increases are attributed to an increase in the amount of plasma heated as the current is increased and not to an increase in plasma temperature or density. This result is supported by an increase in the axial length of these implosions, observed in visible-light and XUV images, as the current is increased.[4]

MASS-LOAD VARIATIONS

Implosions for mass loads corresponding to gas-puff pressures of 50 and 150 psi are presented in Fig. 8 for 40-kV charging voltage. Without the POS [Fig. 8(a)], the first implosion occurs slightly later (600 versus 550 ns) and the second implosion occurs considerably later (1.25 versus 1.10 μs) for the larger mass load. With the POS [Fig. 8(b)], these implosions occur somewhat later for larger mass, as expected.

Changes in the intensity of the plasma emissions with mass load are not as large as the changes with current. Peak powers of soft x-ray and XUV emissions as a function of gas-puff pressure are presented in Fig. 9 for 45-kV charging voltage. These powers correspond to the first implosion. Without the POS, both emissions are weakly dependent on the mass load. With the POS, these emissions tend to decrease with increasing mass load.

For implosions without the POS, as the gas-puff pressure is raised from 50 to 100 psi, the intensity of Ne VII, Ne VIII, and XUV (see Fig. 9) emissions at the first implosion increase by a factor of two. These increases are consistent with more neon being heated as indicated by XUV images of these implosions.[4] As the pressure is raised from 100 to 150 psi, only the Ne VII emission continues to increase which

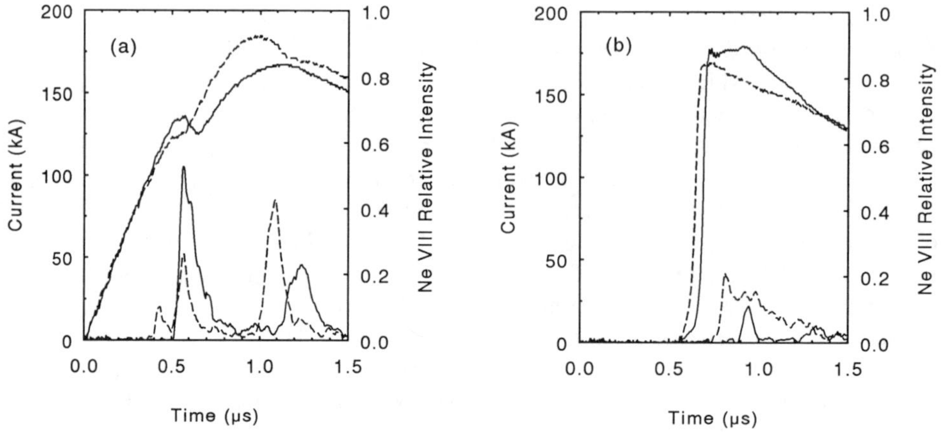

Fig. 8. Driving current and Ne VIII 88-Å emission for pressures of 50 psi (dashed) and 150 psi (solid) for implosions (a) without the POS and (b) with the POS.

indicates a decrease in temperature. As the mass load is increased with the POS, the soft x-ray and XUV intensities decrease [see Fig. 9(b)], the ratio of time-integrated Ne VII to Ne VIII spectral line intensities increases, and the spatial extent of the XUV images decreases.[4] These observations suggest that as the mass load is increased, the temperature decreases and less plasma is heated sufficiently to radiate in the XUV and soft x-ray regions.

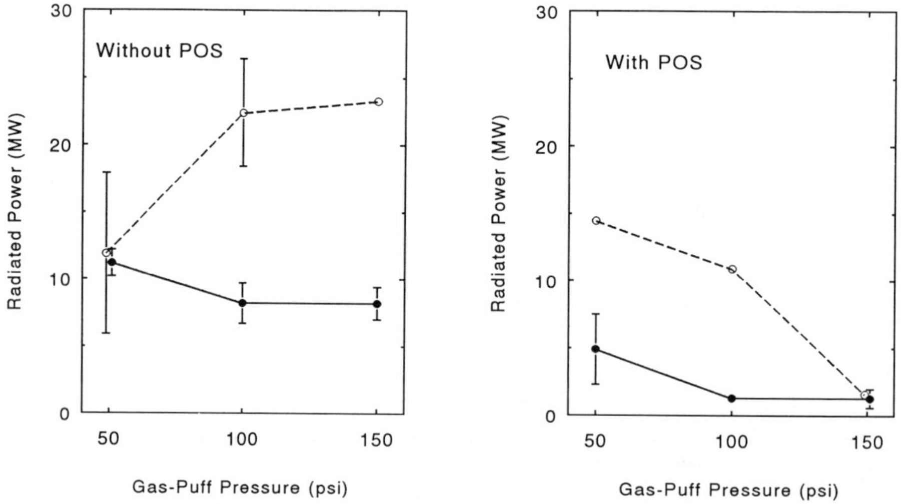

Fig. 9. Scaling of soft x-ray (solid) and XUV (dashed) peak powers with gas-puff pressure (a) without the POS and (b) with the POS. The vertical bars represent shot-to-shot variations.

Relative intensities and time-histories of Ne IX, Ne VIII, and Ne VII emissions are compared with a time-dependent ionization model to determine peak electron temperatures for implosions without the POS.[7] For the first implosion, the peak temperature is 200 eV for all three pressures. For the second implosion, peak temperatures are 200 eV for 50 psi, 100 eV for 100 psi, and 60 eV for 150 psi.

With the POS, electron temperatures can be determined for the second implosion provided lower gas-puff pressures and larger driving currents are used to produce measurable Ne VII, Ne VIII, and Ne IX emissions. For 100-psi pressure and 40-kV charge (180 kA), only Ne VII and Ne VIII emissions are observed, and the electron temperature is 60 eV. For 50 psi and 48 kV (220 kA), all three ionization states are present and the temperature is 100 eV.

CONCLUSIONS

Implosions of uniformly filled, 1-cm-exit-diam neon gas puffs, driven with slow (1 μs) and fast (0.1 μs) risetime currents, are studied for application to the sodium-neon photopumping scheme. A neon plasma with an electron density of 10^{19} cm^{-3} and a temperature of 50 to 100 eV is required for this scheme. For both the slow and fast risetime currents, two successive implosions are observed. At the first implosion, the plasma is too dense (7 to 9 × 10^{19} cm^{-3}) and too hot (200 eV) for either current risetime. With the slow risetime current, the plasma at the second implosion does have an appropriate density (7 × 10^{18} cm^{-3}) and temperature (100 eV). However, the diameter of this plasma (3.5 mm) is too large for lasing.[1] A larger diameter plasma is useful in experiments designed to demonstrate photopumping, such as in Ref. 1. For the second implosion with the fast risetime current, the plasma diameter is not known because emissions from this plasma are too weak to make spatially resolved measurements.

Control of the electron density and temperature by adjusting the magnitude of the driving current is not particularly effective. Increasing the current increases the amount of neon that is heated, but does not change the plasma conditions significantly, particularly at the first implosion. At the second implosion, the electron temperature can be controlled by adjusting the gas-puff pressure. For the slow risetime current, the temperature at this implosion decreases from 200 to 60 eV as the pressure is increased from 50 to 150 psi. For pressures of 100 to 150 psi, this temperature is appropriate for sodium-neon photopumping. To produce similar plasma conditions at the second implosion with the fast risetime current requires a larger driving current and a smaller gas-puff pressure.

The insensitivity of the plasma density to variations in the driving current may be associated with the use of a uniformly filled, rather than annular, gas puff. With a hollow annular puff, most of the plasma can be accelerated and participate in the implosion. For a uniformly filled puff, plasma at small radius undergoes little acceleration and may not contribute much energy to the implosion, but may primarily be an energy sink. As the current is increased, a larger fraction of this plasma may be heated as observed in the experiment.

The uniformity of these plasmas is a concern for x-ray laser applications. Using a faster risetime current reduces the nonuniformities in soft x-ray images, but no improvement in uniformity is apparent in XUV images.[6] However, the nonuniformity may be exaggerated in these time-integrated images which include both implosions. Measurements with a time-resolved pinhole camera[12] should be made to address this issue. Also, the plasma uniformity may be improved by embedding the implosion in an axial magnetic field,[13] if necessary.

ACKNOWLEDGEMENTS

This work was carried out at the Naval Research Laboratory as part of the PhD thesis requirements of BLW at the University of Maryland. Partial support was provided by the Strategic Defense Initiative Organization.

REFERENCES

1. S.J. Stephanakis, J.P. Apruzese, P.G. Burkhalter, G. Cooperstein, J. Davis, D.D. Hinshelwood, G. Melhman, D. Mosher, P.F. Ottinger, V.E. Sherrer, J.W. Thornhill, B.L. Welch, and F.C. Young, IEEE Trans. Plasma Sci. **PS-16**, 472 (1988).
2. J.L. Porter, R.B. Spielman, M.K. Matzen, E.J. McGuire, L.E. Ruggles, M.F. Vargas, J.P. Apruzese, R.W. Clark, and J. Davis, Phys. Rev. Lett. **68**, 796 (1992).
3. F.C. Young, V.E. Scherrer, S.J. Stephanakis, D.D. Hinshelwood, P.J. Goodrich, G. Mehlman, D.A. Newman, and B.L. Welch, in *Proceedings International Conference on Lasers '88*, edited by R.C. Sze and F.J. Duarte (STS Press, McLean, VA, 1989), pp. 98-105.
4. B.L. Welch, Ph.D. thesis, University of Maryland, 1991.
5. B.L. Welch, H.R. Griem, and F.C. Young, J. Appl. Phys. **73**, 3163 (1993).
6. B.L. Welch, F.C. Young, and H.R. Griem, in *Proceedings International Conference on Lasers '92*, edited by C.P. Wang, F.J. Duarte, and D.G. Harris (STS Press, McLean, VA, 1993), to be published.
7. B.L. Welch, F.C. Young, and H.R. Griem, J. Appl. Phys. (to be published).
8. J.P. Apruzese, J.Davis, and K.G. Whitney, J. Appl. Phys. **53**, 4020 (1982).
9. S. Wong, P. Smiley, T. Sheridan, J. Levine, and V. Buck, Rev. Sci. Instrum. **57**, 1684 (1968).
10. S.J. Stephanakis, J.P. Apruzese, P.G. Burkhalter, J. Davis, R.A. Meger, S.W. McDonald, G. Mehlman, P.F. Ottinger, and F.C. Young, Appl. Phys. Lett. **48**, 829 (1986).
11. N.R. Pereira and J. Davis, J. Appl. Phys. **64**, R1 (1988).
12. T. Nash, C. Deeney, P.D. LePell, and M. Krishnan, Rev. Sci. Instrum. **61**, 2807 (1990).
13. F. Venneri, Ph.D. thesis,.University of Illinois at Urbana-Champaign, 1989.

FIBER PINCHES

INVESTIGATIONS OF EXPLODING WIRES AND DIELECTRIC FIBRES DYNAMICS

A.Bartnik*, G.V.Ivanenkov**, L.Karpinski*, A.R.Mingaleev**, S.A.Pikuz**, V.M.Romanova**, W.Stepnewski*, T.A.Shelkovenko**, K.Jach*
*Institute of Plasma Physics and Laser Microfusion, Warsaw, Poland
**P.N.Lebedev Physical Institute, Moscow, Russia

ABSTRACT

Here are discussed the experiments with high-current nanosecond discharges through metal wires and dielectric fibres on NIKE-3 device (100 kA, 50 ns) and BIN one (250 kA, 100 ns). Discharge regime is investigated as a function of load linear density μ by way of example of glass fibres. While mass is increasing discharge through diode changes from regime mismatched regime with electron beam generation ($\mu < 4\ \mu g/cm$) through hot spots forming regime (μ = 8–20 $\mu g/cm$) and then to intense radiating skin-shell regime ($\mu > 30\ \mu g/cm$). Strong influence of surface condition on metal wires explosion character was found. The explosion of the metal wires heated up to 300–500°C for cleaning the surface was the most stable. The experimental results of the wire explosion were explained by the metal-plasma transition model. The results of optical measurements of the plasma movement and column instability development during the discharge process are presented. In several experiments simultaneous development of kink and sausage instability was observed. 2D simulation of the sausage instability according to ideal MHD model corresponds well with experiments.

INTRODUCTION

Nanosecond electrical explosion of metal wires and dielectric fibres in high-current diodes is accompanied by a sequence of complex changes of matter state from room parameters ($T_e \approx 10^{-2}$ eV, $n_e \approx 10^{21}$ cm^{-3}) to plasma with extremal parameters ($T_e \approx 10^2$ eV, $n_e \approx 10^{23}$ cm^{-3}). In this paper an attempt to investigate processes that take place and initial parameters of load in diode influence is made using theoretical and experimental methods.

THEORETICAL PREDICTIONS

They expect that appearance of conductivity during dielectric fibre explosion is caused by breakdown along surface while applying of high voltage to high power diode [1]. Considering of the process in the context of conception [2], we can outline the following three stages: i) charging of dielectric surface during secondary emission of electrons that was initiated by field emission from cathode; ii) electron desorption of atoms and iii) expansion of desorbed gas. Breakdown takes place in case of electron free path being less than gas layer width. According to estimations [1], current about 1 A corresponds to this case. Behind of the breakdown front, the development of avalanche takes place. Joule heat transport to the fibre surface converts the dielectric material to plasma. Further development of the process is propagating of plasma front along the fibre as well as inside it. Depending on ration of transmission times along r and z axis, we can discuss thin and thick fibres. In case of thin fibre the time of shorting out of the diode by plasma is the longest one. During this period of time plasma is able to conduct a current of about several hundred amperes, that is sufficiently higher than pre-breakdown one and in vacuum along the rest part of dielectric current is caused by emitted from plasma electron beam. Estimations for fibre with diameter of 10 μm show that shorting out of 1 cm gap takes place during about 10 ns and while it takes place plasma column radius varies from 200 μm near the cathode to 25 μm near the anode. The explosion of thick fibres is sufficiently more homogeneous but from some diameter fibre skin-effect inhibits its compression.

In case of conductor, an existence of free electrons causes some different development of process [3]. In absence of pre-pulse, there is a quick (several nanoseconds) overheating of metal up to $T \approx 0.1\epsilon_F$ (ϵ_F is Fermi energy of electrons), these temperatures correspond to minimum of conduction. Such transition is not a usual evaporation: phase (n^{-1}, P)-trajectory is on the left from critical point and doesn't cross binodal, bypassing vapor phase. While being in impedance maximum point (2–3 kOhm/cm) wire material stage is indistinguishable from dielectric fibre (electron free path is located in the limits of atomic ranges). In conditions of contaminated surface, it leads to the same process that takes place during surface dielectric breakdown. Plasma forming is accompanied with intensive Joule heating that leads to conductivity rising. It causes skin-effect in which multiplied ionization and radiation have great significance [4]. As a result, high density plasma column with radius > 200 μm is formed. On the "soft" pin-hole frames can be seen the intensive skin-layer luminosity (photorecombination reaches 10 GW/cm). XUV output during 10 ns is about 100 J/cm and more. While current reaches enough magnitude the conditions for plasma column compression occur.

Simple ideal MHD model of such compression [5] uses the large particles method for solving of continuum mechanics axis-symmetric problems. Initial current is taken as homogeneous in section and linearly rising with time up to 300 kA during 100 ns (initial values are taken for the moment of plasma expansion maximum about 10 ns when the current is 10 kA, column radius $a_o = 250$ μm). Wire material is tungsten, linear density $2 \cdot 10^{17}\text{cm}^{-1}$; equation for the state is adiabatic, pressure turns to zero on the plasma surface, instabilities develop from small ($< 5\%$) random initial density non-homogeneity. The computations results for inherent time moments are shown on Fig.1a. These moments are: 21 ns — maximum velocity of first column compression, 27 ns — first compression, 31 ns — maximum of velocity under sequent plasma expansion, 35 ns — maximum of column dimension estimated by minimum of pressure, 39 ns — maximum of velocity of second column compression. As can be seen on Fig.1a, first column compression is rather homogeneous in z-axis. Rather moderate rate of instability development is kept up to acceleration sign change under expansion of been pressed plasma. Deceleration of expansion and sequent accelerated new collapse lead to fast development of MHD instability with reversal increment 1–2 ns (Fig.1b). Up to beginning of

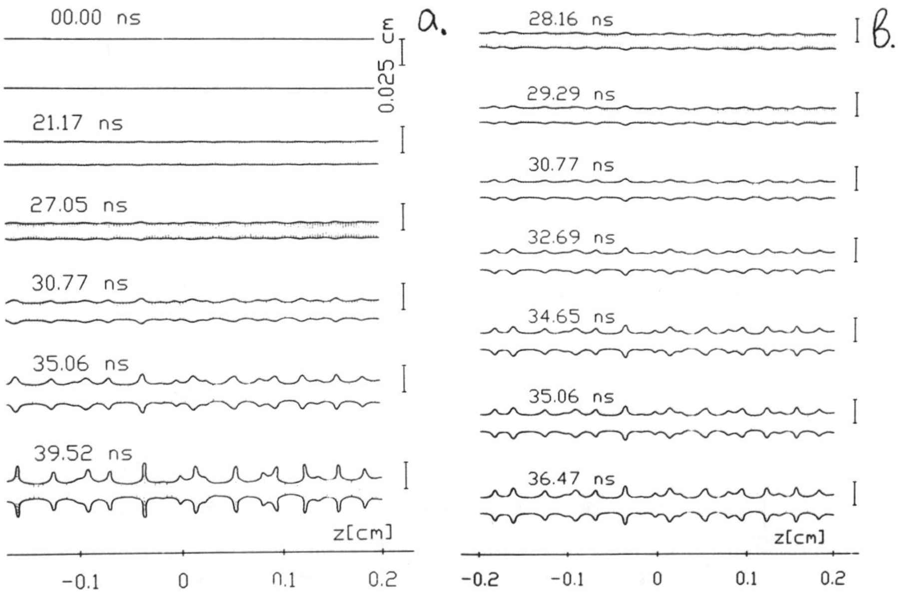

Fig.1: Plasma column MHD instability development simulation: *a*) general picture of necks development from small random non-homogeneity of initial plasma column density; *b*) fast development of MHD instability during plasma expansion after first compression.

deceleration of the second compression, Raleigh-Taylor instability leads to highly non-homogeneous shape of plasma column. It is well known empirically that this moment is characterized by appearance of essential effects of transference and radiation and it makes further calculations unwarranted. In our case it is expressed in small quantity of large particles in maximum compression region, so significance of fluctuations and conditioned by them instabilities rises. This can be observed somewhere after 41 ns when pressure is 4.8 Mbar as calculated. At the same time deceleration rapidly allows to compare calculated picture with experimental one and to ensure in their good correspondence.

EXPERIMENTAL RESULTS

Process dynamics in diode was investigated with the help of 5-frame system of registration of visible and nearest IR plasma radiation and Schlieren system with laser illumination. Frame camera system was created on the basis of five electron-optical converters used as electron shifters controlled by cable generator. Images on photocathodes were formed with the help of prism space divider. Duration of exposition time (8 ns) depended on cable length and intervals between frames (10 ns) depended on brought out cable length. Synchronous accuracy with processes in diode was not more than 5 ns. In Schlieren system second harmonic of NdYaG laser with modulated Q-factor was used for lighting. Pulse duration was 8 ns, interval between frames was 10 ns.

Moreover there were used various X-ray diagnostics (pin-holes, spectrographs, XRD and so on) that allow to observe plasma in final stage, to measure its dimensions, density and temperature.

Fig.2: *Pin*-diode signal for glass fibre explosion.

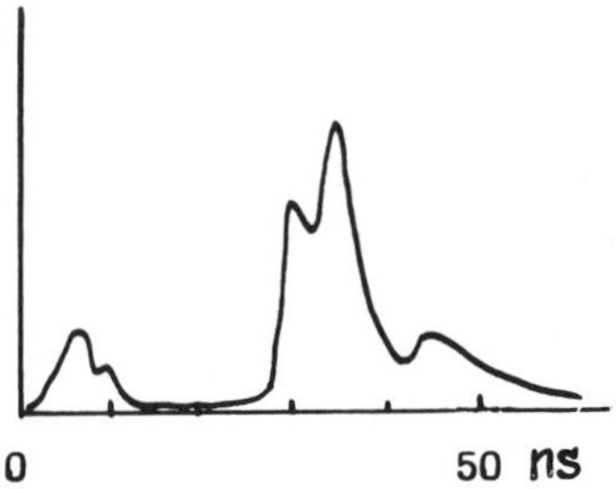

Experiments carried out on NIKE-3 device (100 kA, 50 ns) confirmed proposed scheme of process. There is a *pin*-diode signal peak (Fig.2) caused by bremsstrahlung of electron beam that arrives in the discharge initial stage. After this peak, there is a signal decay corresponding with a stage of current heated skin-layer luminous when plasma radiation energy is less than registration threshold (1 keV). The main radiation peak corresponds with plasma compression. Its shake can be explained by double compression character (the beginning of signal is due to the maximum of hydrodynamics compression and a maximal radiation intensity is due to hot spot).

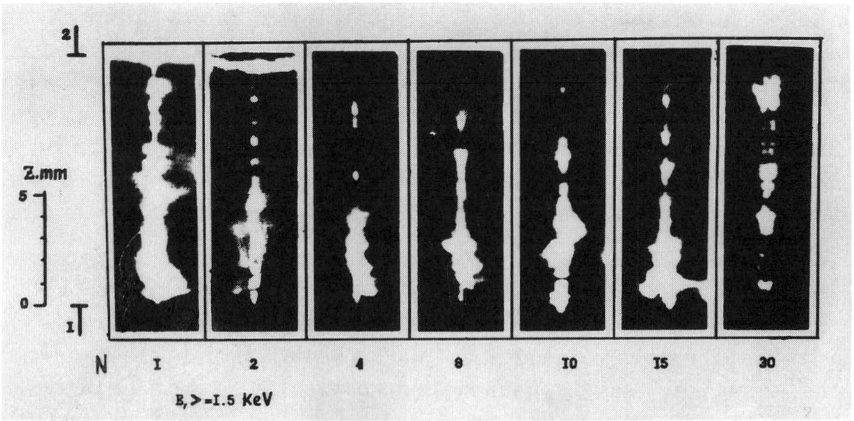

Fig.3: Pin-hole images dependence on glass fibres quantity being exploded in one bundle.

Explosion character dependence on per unit length load mass μ was investigated on BIN device. Per unit length mass was varied by modification of fibres quantity N in bundle of thin dielectric fibres in diode gap (Fig.3). Single glass fibre diameter was 8 μm which corresponds to $\mu = 1.3\ \mu g/cm$. Chemical composition of fibres wasn't especially investigated but spectral data shows a significant quantity of Al in them.

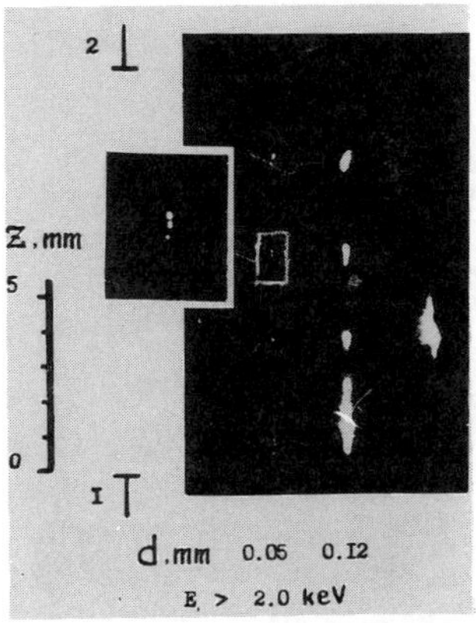

Fig.4: Fine structure of hot spots in radiating plasma region.

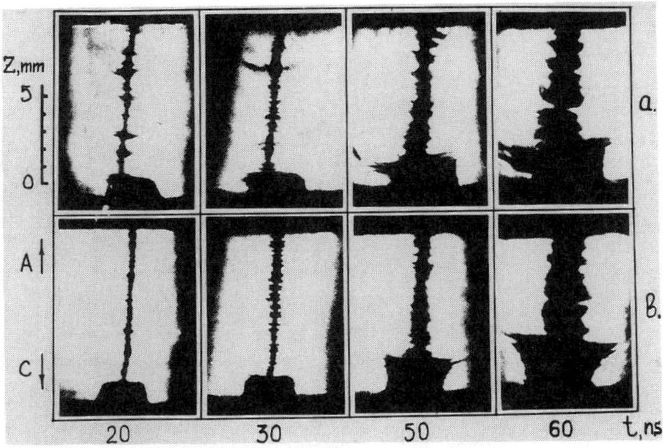

Fig.5: Schlieren images of non-cleaned tungsten wire (a) and a wire initially heated up to 500°C (b).

Under modification of N there were observed the following general rules on the pin-hole frames (Fig.3):

1. For single fibre explosion ($N = 1$), bright hot regions are observed. They have complex structure (Fig.4) and are placed uniformly in anode-cathode gap. Each bright region structure includes 2–4 hot spots with dimension less than 5–10 μm placed with distance of 250–100 μm. H- and He-like spectra of Al and Si ions are absent from each other. Intensive anode fluorescence is observed. Moreover there are observed bright side ejections being rather far placed (up to 2–3 mm) from diode axis (Fig.3). Optical measurements show that this fluorescence can appear only on the last discharge stage after pinch destruction that take place under thin fibres explosion before current achieves its maximum. In this situation, significant fraction of current is due to electron beam.
2. If $N = 2, 3$, discharge character principally remains the same but soft X-ray plasma fluorescence in non-axis regions decreases. Near bright regions cone-like areas are observed.
3. In the range $N = 4$–16 ($\mu = 5$–$20\ \mu g/cm$), bright plasma regions radiate most intensively near the cathode. Complex structure of such regions includes usually several hot spots and emitted quantum energy reaches 5 keV. H- and He-like ions of Si and Al radiation intensity maximum corresponds to $N = 10$–15. In the middle of anode-cathode gap, there can be noticed less bright regions with energy quantum about 2 keV. Finally, the region near the anode emits only soft

diffuse radiation. All this picture is observed on the background of comparably soft tube-like radiation from skin-layer.
4. If $N > 25$ ($\mu > 30$ μg/cm), tube-like fluorescence of skin-layer alternates with separate bright regions ("vertebral column" structure). This case corresponds to explosion of thick 40 μm fiber. Tube-like rectilinear skin-layer fluorescence shows good correspondence with thick metal wires explosion.

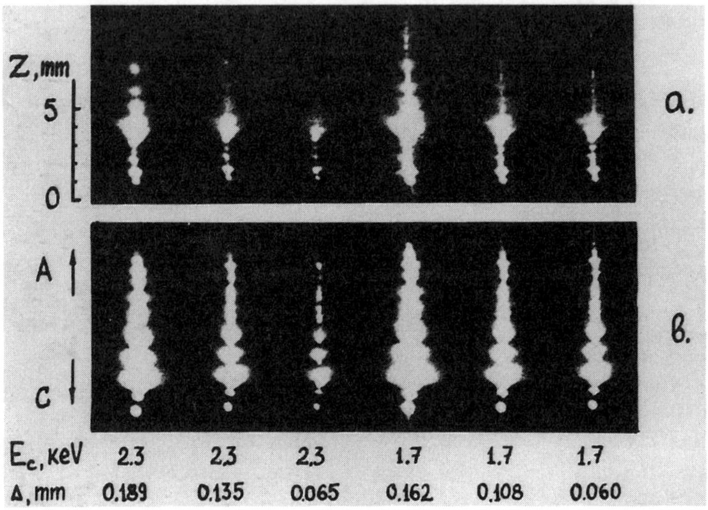

Fig.6: Pin-hole images of exploding non-cleaned tungsten wire (a) and initially heated up to 500°C (b).

For checking of surface condition influence on initial wire explosion process stage character, there were carried out experiments with wires initially heated up to 500°C by direct electric current. Under this temperature, all the adsorbed on the wire surface substances were evaporated and take no part in the process. On Fig.5 there are shown sequent Schlieren images for non-cleaned tungsten wire (a) and initially heated one (b). In the first case, just after process starting, there is an unstable cloud around the wire with just side ejections with velocities up to 10^7 cm/s. In case of heated wire explosion there can be observed the appearance of enough stable plasma column on which instabilities appear under its compression leading finally to appearance of hot spots (Fig.6a) non-regularly placed along pinch axis. Pin-hole frames of non-cleaned wire explosion (Fig.6b) with hot spots localization near the cathode look like ones of dielectric fibres with their optimal mass. All this shows that processes are of one type.

It might be well to point out that there were observed no obvious quality differences between explosions of cleaned and non-cleaned dielectric fibres, nevertheless it can be due to the fact that it was impossible to heat the fibres up to a temperature higher than 200°C (the fibres were heated by IR radiation of heated tungsten wire with the help of an elliptic cylindrical reflector).

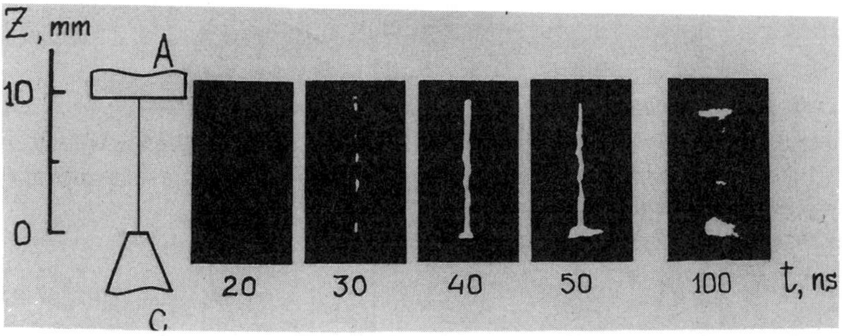

Fig.7: Frame images of tungsten wire with 8 μm in diameter explosion.

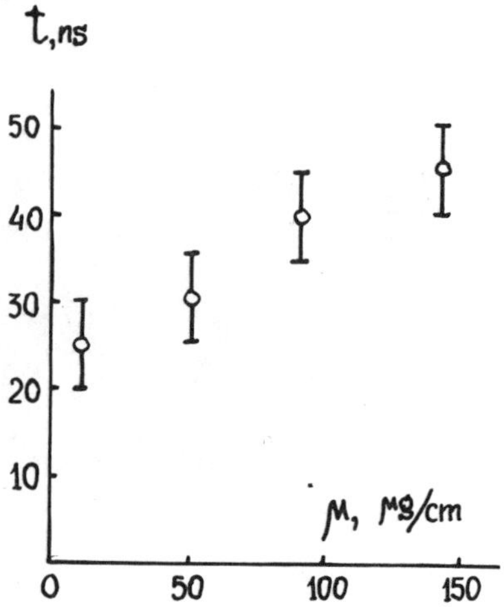

Fig.8: Experimental dependencies of time of plasma instability appearance on per unit length load mass.

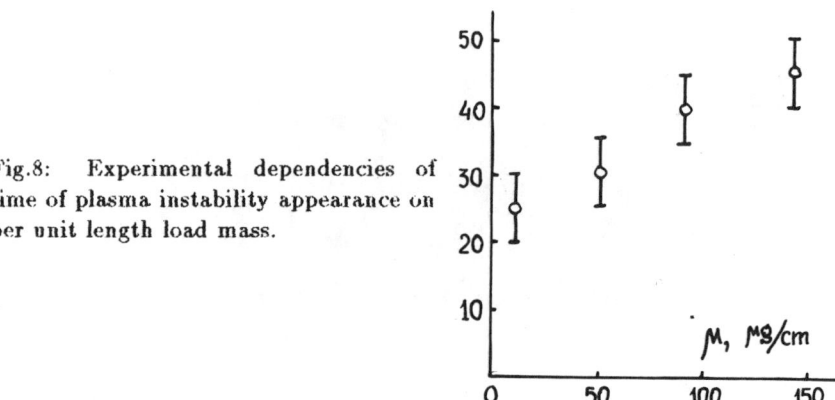

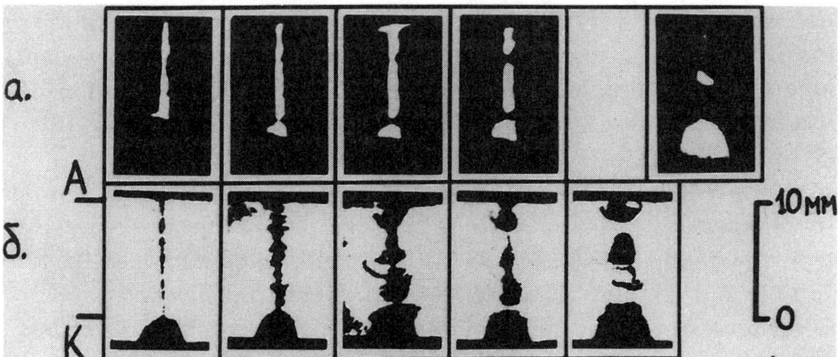

Fig.9: Comparison of frame and Schlieren images of copper wire of 30 μm diameter explosion.

Special features of the plasma generation were observed in the experiments with wires of various metals. On Fig.7 there is presented single tungsten wire explosion. During starting stage of the process on the place of the wire a radiating column appears and, up to 30 ns, it becomes unstable. Then some instabilities appear on it and they destruct up to 40 ns. Instabilities development time depends on initial wire mass and increases from 20 ns for $\mu = 10\ \mu g/cm$ up to 45 ns for $\mu = 150\ \mu g/cm$ (Fig.8).

Comparison of frame photographs and Schlieren images gives approximately equal results. Nevertheless small column instabilities due to surface effects can be seen earlier on Schlieren images (Fig.9).

Experiments show also simultaneous appearance of sausage and kink instabilities from the anode side (Fig.10).

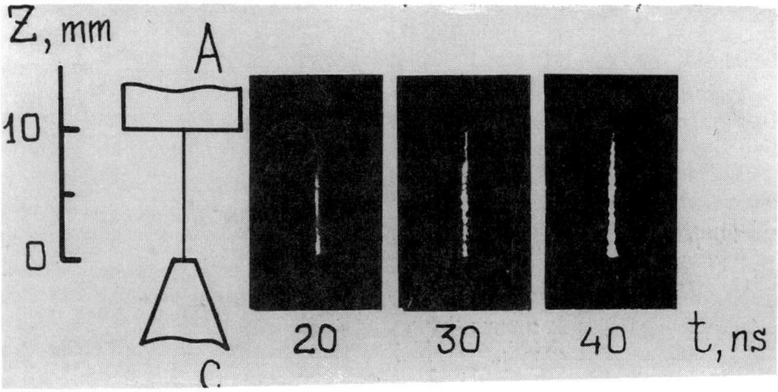

Fig.10: Sausage and kink instability during palladium wire of 30 μm diameter explosion.

On Fig.11 there are shown the results of the comparison of experimental data with computer simulation. Calculating parameters correspond with experimental conditions on BIN device. Tungsten wires with diameter of 8 μm and length of 1cm were used in the experiment. Usual final plasma parameters were obtained such as electron temperature about 1 kV and electron density about 10^{22}–10^{23} cm^{-3}. W^{+46} ions radiation was registered in 4–6 Å range.

Sufficiently non-homogeneous shape of column is formed both in simulation and in the experiment. Deep cavities and disk-like plasma ejects are developed. They show on the tendency to sharpening of the boundary shape. It is very important that in our simulation as well as in the experiment there is observed spontaneous forming of regular-like necks structure. It is worth notice that calculated space period of the plasma column instability and observed one are very close. Very important is also competition separating two zones of compression with middle neck between them. This leads to non-symmetry of plasma flowing from necks. Competition is due to common energy source for necks. Naturally, it leads to less plasma compression. On the whole, correspondence of ideal MHD theory with experiment seems to be very good.

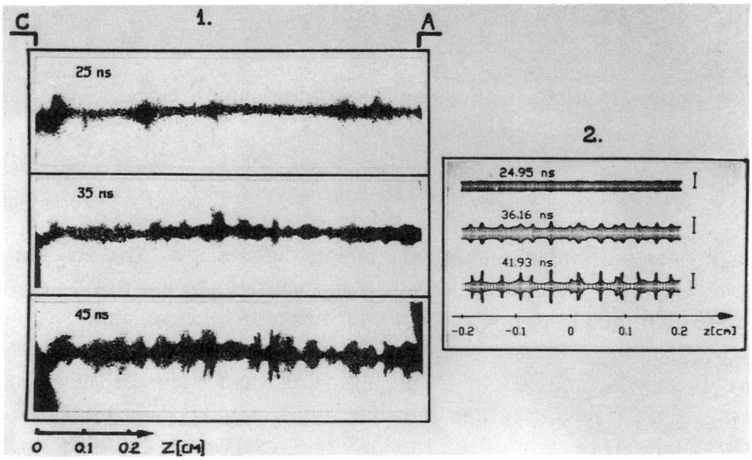

Fig.11: Comparison of experimental Schlieren plasma images of exploded tungsten wire (1) and calculated configurations (2) of plasma column.

CONCLUSION

The results presented in this paper do not give, of course, an absolutely complete picture of processes in high-current diode under nanosecond fibre and wire explosion but they allow to explain many of observed in experiment features.

REFERENCES

1. A.Bartnik, G.V.Ivanenkov, L.Karpinki, S.A.Pikuz, T.A.Shelkovenko. Preprint No 30. M. FIAN (1990); Fiz. plazmy **16**, 1482 (1990).

2. S.P.Bugaev, A.M.Iskol'dskii, G.A.Mesjac. ZhTF **37**, 1855 (1967); E.A.Litvinov, G.A.Mesajc, D.I.Proskurovskii. UFN **139**, 265 (1983).

3. G.V.Ivanenkov, T.A.Novikova. Preprint No 51. M.FIAN (1992).

4. G.V.Ivanenkov, S.B.Taranenko. Preprint No 145. M.FIAN (1987); ZhTF **57**, No 7, 34 (1989); S.M.Zakharov, G.V.Ivanenkov, A.A.Kolomensky, S.A.Pikuz, A.I.Samokhin. Pis'ma v ZhTF **10**, 1145 (1987).

5. G.V.Ivanenkov, A.R.Mingaleev, S.A.Pikuz, V.M.Romanova, W.Stepnewski, T.A.Shelkovenko, K.Jach. Kratkie soobschenija po fizike, to be published.

THE PRODUCTION OF SOLID HYDROGEN AND DEUTERIUM FIBRES FOR DENSE Z-PINCH EXPERIMENTS

J. M. Bayley
Imperial College of Science, Technology and Medicine
London, SW7 2BZ

ABSTRACT

The physics of fibre making and existing devices are reviewed and a new design suggested. The yield stress of the extruded material, the extrusion pressure and the die shape are identified as important factors for successful extrusion.

We propose that the fibre be extruded through a laser drilled aperture held at a temperature T_1 at which the H_2 or D_2 is solid. The yield stress of the solid can be varied by changing T_1. The aperture is connected by a capillary along which there is a temperature gradient. At the top of the capillary is a reservoir of liquid H_2 or D_2 held at a temperature T_2. By varying T_2 and hence the saturated vapour pressure above the liquid, the extrusion pressure can be accurately controlled. Initial results are reported.

REVIEW OF EXISTING FIBRE MAKERS

Controlled thermonuclear fusion programs have parented research into forming solid hydrogen and deuterium. In the 1960s' papers were published describing the production of solid deuterium for laser targets. Latterly research has been driven by the need to refuel tokamaks (eg. Amenda and Lang[1]) with 'pellets' which are typically cylinders or strips 1 mm diameter by 3 mm long. Little work on the production of solid hydrogen has been done because of its worse physical properties.

Recently deuterium fibres have been used in Z-pinch experiments where very small diameters are necessary to achieve the required line density in the plasma. Two groups have successfully fibre under 100 μm. Sethian et al[2] at NRL and Grilly et al[3] at LANL have produced fibres ranging from 40 to 300 μm.

One early paper is by Jarboe and Baker[4] who describe the formation of a fibre from liquid deuterium extruded by its own vapour pressure. The die used is a tapered aperture in a 3000 μm thick plate. The aperture is 100 μm on the side facing the liquid and 70 μm thick. The fibre produced was 50 μm diameter. The temperature of the liquid deuterium was between 19 and 22 K and the vapour pressure 0.5 to 1.6 bar. Normal operation was 21 K and 0.65 bar and the fibre was extruded at 4 mm/s. The die was kept near the triple point of D_2 (18.69 K) to allow the complicated extrusion of three phases to take place. The extruded fibre is deflected by hot gas molecules in the vacuum. Asymmetries in the chamber will result in movement of the fibre. The authors calculate that if these forces are to be small compared with gravity $p \ll \rho g l$ and for a 50 μm fibre the chamber pressure must be less than 7×10^{-4} torr. We note that Sethian et al found an extended nozzle was required as a shield for the fibre to prevent deflection but attributed it to different evaporation on two sides of the fibre rather than momentum transfer. It was found that at too high a pressure or temperature, large amounts of liquid flowed through the hole forming globules.

All bar this one paper describe a similar method of fibre extrusion. Deuterium gas is slowly bled into a cold chamber where it desublimates onto the walls as snow. A ram is then driven into the chamber to extrude the solid through an aperture or capillary of the required diameter. Some experimenters raise the temperature of the snow to make to more plastic before extrusion. It is not necessary to plug the aperture while the gas is being fed in because it is rapidly stopped by frozen gas.

Sethian et al warm the snow to 11 K and then apply pressure with a cranked piston to extrude it. At this temperature the yield stress of deuterium is approximately 1.9 bar. Globules were also observed by Sethian et al but at the much lower temperature of 12.5 K compared with Jarboes' 22K. They state that an a pressure of 6.0×10^{-6} torr is required before extrusion because the presence of nitrogen may make the deuterium brittle and clog the nozzle. To achieve a low extrusion pressure and a straight fibre the thickness of the die was 6 times its diameter. The fibre extruded from the nozzle was protected by a shield formed of 3.3 mm diameter copper tube. Extruded fibres can be up to 10 cm long and exist for 7 minutes before falling off the nozzle. No change in diameter is observed but the vacuum pressure rises to 2.0×10^{-4} and 3.0×10^{-5} for the 125 and 50 μm fibres respectively. Note that the vacuum quality is determined to a large degree by the cryopumping effect of the fibre extruder.

Grilly et al at LANL have produced fibres under slightly different conditions; the deuterium gas is condensed at 8 K rising to 13 K during the process. It is bled into the chamber at a rate of 40 cm^3/min. The snow is then compressed by the plunger at approximately 10 K. The nozzle is then heated (200 mW until the temperature in the condensing chamber is 14 K and extrusion begins. The authors believe that liquid is extruded and that this solidifies due to evaporative cooling. 10 cm long fibres are made 40 μm in diameter. These are stable for 15 minutes before narrowing at the nozzle and falling off. In this experiment there is no shield around the fibre but excessive deflection is not reported. This is possibly due to the cylindrical symmetry of the extruder. The nozzle is 50 μm diameter and 3.2 mm long.

The only paper dealing with the extrusion of long (> 5mm) hydrogen fibres is Markov et al[5]. They describe a helium bath cryostat extruding fibres 300 μm to 4 mm diameter. The hydrogen is solidifies at low temperature but rose to 9-10 K for extrusion. Deuterium was extruded at 15 - 16 K. The lifetime of 800 μm diameter fibre was 30 minutes and it fell off due to excessive thinning at the nozzle.

THE PHYSICS OF EXTRUSION PROCESSES

Some insight into the extrusion process of solid hydrogens may be attained by considering the literature on the extrusion of metals[6] and plastics[7].

In extrusion a metal billet of crosssection A is forced through a die to produce a fibre of crossection A'. The pressure required to do this is $P_{ext} = P_{def} + P_{friction}$.. P_{def} provides the energy for deformation of the material and $P_{friction}$ is required to overcome the friction of the billet with the container walls.

Experiment has shown that P_{def} is of order twice the yield stress (Y) for small aspect ratios. The maximum frictional pressure is given by Tresca's yield criterion which says that the billet will fail in shear when the friction is greater than 0.5Y. The working temperature is bounded by a loss of cohesion at high temperature and excessive stiffness at low temperature. It is found that the minimum extrusion pressure is attained at $T_{melt}/3$. A small change in pressure produces a large change in extrusion rate. For soft metals at much less than $T_{melt}/3$ extrusion velocity goes as P^6.

The extrusion velocity is important in determining the presence of melt fractures in plastics. This is an irregular cracking of the fibre surface when the shear stress in the exudate exceeds the shear strength of the material. The critical parameter is the shear rate during extrusion. This can be reduced by a lower extrusion velocity or a conical inlet to the die.

The die shape is important in the extrusion of metals a square (0°) inlet produces a dead space which may cause instabilities and defects in product. A 90° cone is optimum smaller angles producing excessive friction and thus a higher extrusion pressure.

PHYSICAL PROPERTIES OF HYDROGEN AND DEUTERIUM

An understanding of the physical processes taking place when hydrogen gas is cooled is essential to cryostat design. All data is calculated from information collected by Clark Souers[8].

Hydrogen has two distinct forms at low temperature j=0 and j=1. The difference in the physical properties are negligible for our purposes.

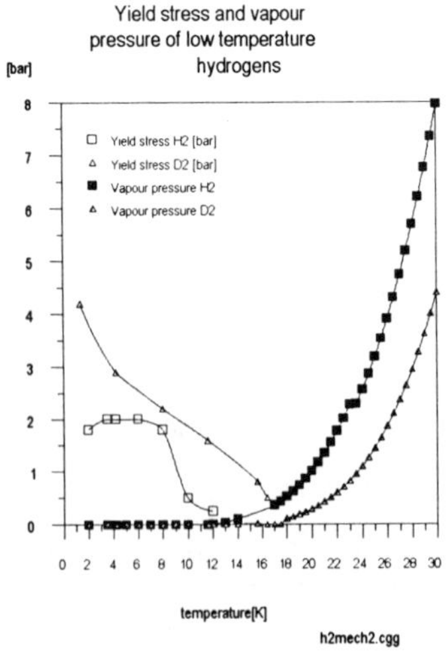

Figure 1

The pressure above the liquid and solid is given by the curve. This is to be compared to the yield stress of the polycrystalline material. Notice that in the temperature range that deuterium is normally extruded the vapour pressure is not sufficient to deform the material. For hydrogen above 12 K the pressures are approximately equal and stable extrusion is not likely to be possible.

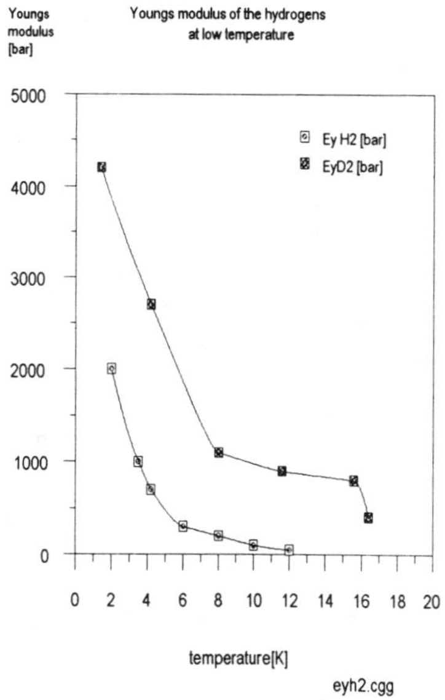

Figure 2

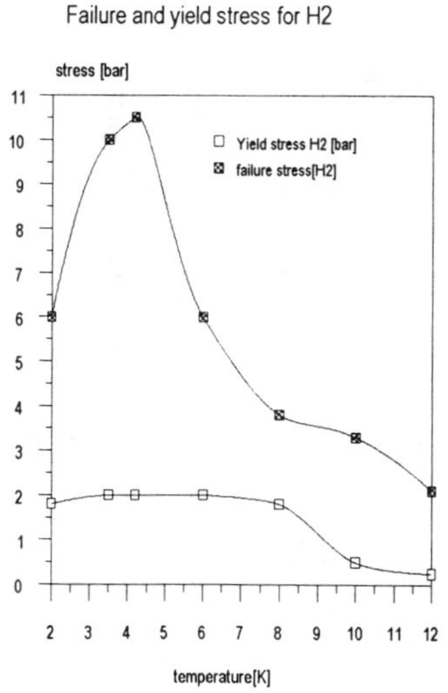

Figure 3

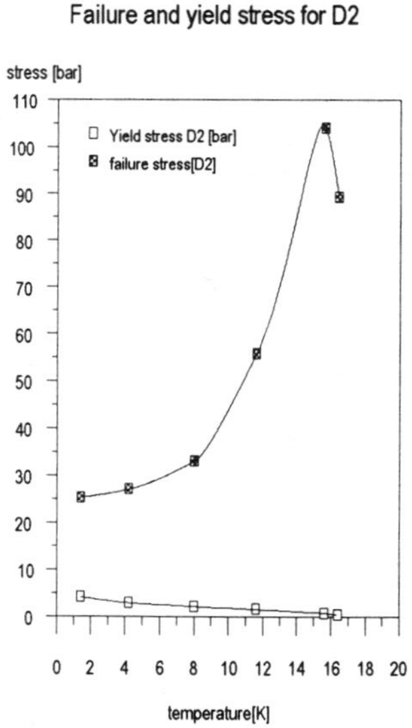

Figure 4

The Young's modulus measures the elastic deformation produced by an applied stress. For deuterium it varies little over the typical extrusion temperature range of 11-14 K and is about 1000 bar.

The failure stress is calculated from the failure strain given in Clark Souers and the Young's modulus.

DESIGN OF A NEW FIBRE MAKER

For our proposed experiments it is desirable to make fibres with 10 μm diameters. This is smaller than that which has been achieved to date and requires a new design of fibre maker. Existing designs have a degree of 'black art'. This is indicative of poor control and instrumentation. Problems which have been identified are: extrusion of globules if the pressure or temperature is to great, catastrophic temperature rise if gas is extruded instead of solid and excessive heat loss up the compaction ram. The extrusion pressure has not been measured however the mechanisms used would be capable of providing pressures of 1-20 bar.

The major question is what is the best temperature for extrusion. Deuterium has been extruded throughout the range 11 - 22 K with static fibres made in the range 11-16 K. Hydrogen fibres have only been made by one group who extruded at 10 K.

From consideration of previous experiments I have concluded that for optimum operation the temperature and extrusion pressure of the fibre must have:

- accurate control
- accurate measurement
- thermal stability

We decided that the best method of extrusion would be to use the vapour pressure of hydrogen liquid to provide an hydrostatic pressure to extrude the solid hydrogen. This requires a two temperature system. An upper chamber contains liquid hydrogen and vapour. This is connected by a capillary to a cooled aperture. The temperature of the chamber and the aperture are independent. The temperature of the upper chamber defines the hydrostatic pressure. The temperature of the aperture defines the mechanical properties of the solid material. From the graphs above we see that in the temperature range reported for extrusion 10-16 K the yield stress of deuterium is between 0.5 and 2 bar. To attain twice these pressures the upper reservoir must be at 25.5 - 29.5 K. For hydrogen at 10 K the yield stress is 0.25 to 0.75 bar and the upper reservoir temperature 16-19 K. The extrusion velocity is a strong function of temperature and pressure. To large a pressure will result in a brittle fibre from melt fracture or exceed the failure stress of hydrogen. Thus accurate control of both reservoirs is essential.

I think that the best method to make a fibre is to extrude it slowly and continuously from a warm nozzle at, approx. 10 mm/min. This will produce fibres with low internal stresses. The constant replenishment of the fibre at the nozzle will mitigate against breakage. The high temperature of the nozzle will guard against brittle fracture at the die. This is a common failure mode.

For our initial experiments it is desirable to produce fibres by extrusion through commercially available pinholes. These are not available with a conical lead-in. Thus we shall be forced to use a square lead-in. The extrusion pressure will be reduced by warming the material to reduce its cohesion. The comparison with metallic extrusion suggests that the minimum extrusion pressure will be approximately 2 times the yield stress. The frictional drag being overcome by additional shear.

These considerations suggest the following conditions:

Table 1: Possible extrusion conditions

	hydrogen	deuterium
Extrusion temperature	10-11 K	10-15 K
Extrusion pressure	0.8 -1.2 bar	1.8 -3.6 bar
Upper reservoir T	19-20.75 K	25-29.5 K
Pressure sensitivity	25 mbar/0.1 K	40 mbar/0.1 K
Vacuum	1-5 x 10^{-6} mbar before cryopumping	

Deuterium presents the more stringent control requirements. A 0.1 K control of the upper reservoir provides 40 mbar resolution in extrusion pressure. This should prove adequate for our needs.

FIBRE MAKER DESIGN

The required temperature and pressure ranges are shown in table 1. The lowest temperature required is 10 K. The cryostat must have sufficient cooling power that it capable of sustaining a temperature differential of 20 K between the two reservoirs.

Three possible cooling mechanisms are possible, continuous flow cryostat, helium pool cryostat and a closed cycle helium refrigerator. The helium refrigerator is the preferred option. Although the purchase cost is large (approx. $14 K) there is no liquid helium usage. A 2W liquid He cryostat typically uses 2 l/hr He @ $5/l + 5 l to cool down. Thus over a six hour day without a vacuum break the cost is $85. This cost although sustainable for a short time on a large experiment is not acceptable for prototyping and small experiments. Refrigerators are mechanically favourable; continuous flow cryostats require cumbersome He connections to a large dewar, and helium pool cryostats are not strong because the pool must be supported by thin wires.

A schematic diagram of a design is shown below:

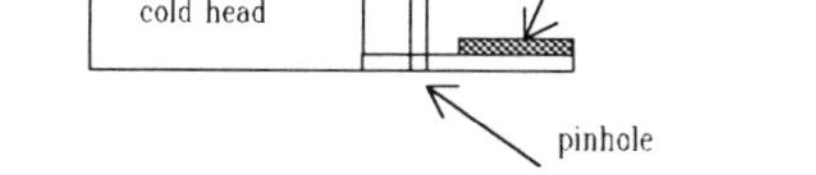

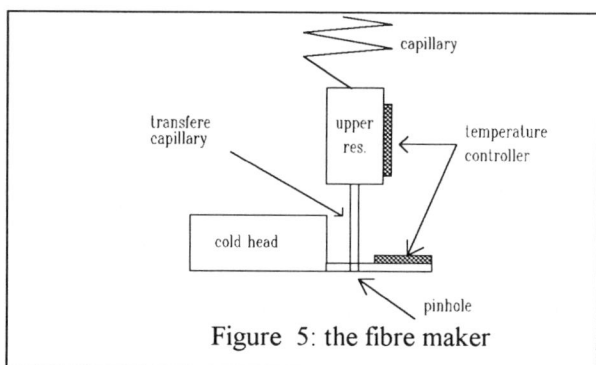

Figure 5: the fibre maker

The upper reservoir and transfer capillary must contain sufficient material to sustain a continuous fibre extrusion of 10 mm/min for half an hour. The transfer capillary must be long enough and have sufficiently thin walls that the heat loss down it does not exceed 2 W for a 10-30 K gradient. The upper reservoir must be firmly supported so that vibrations due to the discharge cannot damage the transfer capillary. The pinhole must be easily replaceable to allow the diameter of the fibre to be changed.

PRELIMINARY RESULTS

A fibre maker has been constructed and initial tests performed. It was found that a 300 μm pinhole at 10 K could be blocked with deuterium and that the upper reservoir could be filled with liquid. As the temperature of the liquid was varied the pressure in the system followed the saturated vapour pressure for deuterium. The extrusion of a fibre has not yet been observed. This may be because the fibre is boiling off due to the proximity of the radiation shield around the nozzle which is at a higher temperature than in previously reported experiments.

ACKNOWLEDGEMENTS

The author would like to thank A. Siddiqui for his assistance in testing the fibre maker and P. Choi for his comments on the initial design.

Table 2: Properties of hydrogens at low temperatures

property	hydrogen H2	deuterium D2	unit
Avogadro's constant	6.022×10^{23}		mol
molar mass	$2.02\text{x}10^{-3}$	$4.03\text{x}10^{-3}$	kg mol^{-1}
Ttriple	13.85	18.67	K
ptriple	70	171	mbar
molar volume of solid at the triple point	23.25	20.37	μ(m3) mol^{-1}
molar volume of liquid at the triple point	26.11	23.16	μ(m3) mol^{-1}
density of liquid at the triple point	38.3	43.18	mol m^{-3}
volume decrease from Ttriple-10K	0.5%	2%	
volume decrease from 10K-0K	0.2%	0.1%	
volume decrease on freezing at Ttriple	12%	14%	

REFERENCES

1. Amenda, W. and Lang, R.S., 'Cryostat for production and fast ejection of deuterium filaments',Cryogenics, July 1982 pp 364-366
2. Sethian J.D., and Gerber K. A., 'Solid deuterium fiber extruder', Rev. Sci. Instrum. 58(4), April 1987 pp 536-538
3. Grilly E.R., Hammel J.E., Rodriguez D.J., Scudder D.W. and Shlachter J.S. 'Production of solid D2 threads for dense Z-pinch plasmas', Rev. Sci. Instrum. 56(10),pp 1885-1887 (1985)
4. Jarboe, Thomas R. and Baker, William R., 'Apparatus for producing laser targets of 50 μm deuterium pellets', Rev. Sci. Instrum., 45(3),pp 431-433 (1974)
5. Markov, A.N, Fradkov, A.B., and Chernetskii, V.D.,' Prepartion of solid hydrogen targets for laser fusion research', Sov. J. Quantum Electron. 7(5), p641-642 (1977)
6. Pearson, C.E., and Parkins, R.N.,*'The extrusion of metals'*, Chapman and Hall, London, 1961
7. Fisher, E. G., *'Extrusion of plastics'*, Newnes-Butterworths, London,(1976), ISBN 0 408 00194 1
8. Clark Souers, P., 'Hydrogen properties for fusion energy', University of California Press (1986), ISBN 0-520-05500-4

THE EFFECT OF VARYING THE FIBER DIAMETER IN PLASMA-ON-WIRE (POW) Z-PINCH CONFIGURATIONS

N.S. Edison, B. Etlicher, P. Zehnter, S. Attelan, C. Rouillé
Laboratoire PMI, Ecole Polytechnique, 91128 Palaiseau, France

A.S. Chuvatin
I.V.Kurchatov Institute of Atomic Energy, 123182 Moscow, Russia

ABSTRACT

We are investigating the dependence of the fiber diameter in POW experiments on the dynamics of the implosion. Recent data from the JEX experiment at Troitsk suggest that the diameter of the fiber plays an important role in the dynamics of the implosion. In general, the smaller fiber diameter permits a more stable implosion possibly due to a higher impedance. High impedance in the fiber forces the current during the initial stages of the implosion to flow preferentially in the outer plasma shell and, thus, prevent the fiber from prematurely exploding. This suggests that there is a maximum diameter fiber that can be used to give a stable core during the compression phase of the implosion. In our experiment, an aluminum plasma jet is created from an exploding foil and then imploded onto a micron sized diameter copper wire (7-50 μm). In addition, an axial DC magnetic field ($B_{z0} \leq 300$ G) is applied externally to stabilize the imploding aluminum plasma and to study the interaction of the magnetic field with different diameter wires. We have found in previous experiments that the load configuration can significantly affect the magnetic field required to optimize the implosion. For example, peak x-ray production for a load consisting of a 25 μm copper wire occurs at fields of 150 G while the aluminum jet alone is optimized at 50 G. The pinch is driven by a 2 Ω, 0.1 TW generator (250 kA in 80 ns). Diagnostics include filtered PIN XRDs, time-resolved schlieren photography, and time-integrated multiple filtered pinholes.

INTRODUCTION

Much research has been done to stabilize z-pinches and improve the uniformity of the compression. Previous plasma-on-wire (POW) configuration results[1] have shown that a fiber immersed in a plasma shell can remain stable and intact during the compression phase of the implosion. However, during recent JEX experiments at Troitsk using gas puffs, several fibers were observed exploding. This suggests that there is a fundamental difference between using imploding gas and plasma shells. The current work seeks to explain the phenomenon described above and explore the regime where the fiber begins to exhibit instabilities.

One possible explanation is that the resistance of the fiber compared to that of the surrounding shell determines the initial current distribution through the load. If the fiber has a low resistance compared to the shell, enough current may flow through the fiber to cause it to explode prematurely. The main difference between our experiments and those of JEX is that the plasma surrounding our fiber has a much higher conductivity than that of a gas puff. In figure 1 we plot the ratios of the fiber resistance and the Spitzer resistance of the surrounding shell as a function of fiber radius. As the fiber diameter increases and its resistance decreases, we would expect to see instabilities associated with the fiber appear.

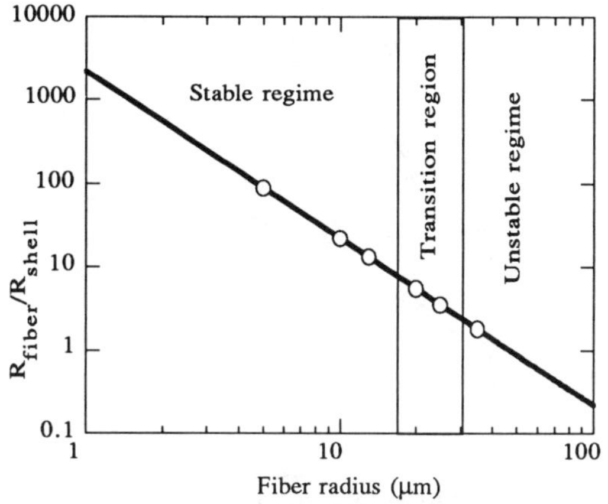

Fig. 1. Ratio of the resistances of a copper fiber to an aluminum plasma shell as a function of fiber diameter. Open circles indicate fibers used.

In addition to the research into the size of the fiber diameter, work is being done to improve the stability of the imploding shell. One method of reducing the growth of instabilities is to freeze a magnetic field into the plasma before the initiation of compression. An axial magnetic field would stiffen the plasma along the pinch axis. Instabilities would be required to expend energy in order to distort the magnetic field, thus, reducing its growth rate. Here we have purposely chosen the applied axial field, B_{z0}, to be small compared to the azimuthal field generated by the pinch, B_θ. Since $B_{z0} << B_\theta$, the work done in compressing the applied magnetic field is negligible compared to the discharge energy of the compression. The average kinetic energy per ion is independent of the applied field. Thus, the final compression diameter should be unchanged but with a more uniform distribution of density and temperature along the pinch axis.

EXPERIMENT

We have conducted this series of experiments on GAEL, a 2 Ω, 0.1 TW generator (250 kA in 80 ns).[2] The load consists of a copper wire immersed in an aluminum plasma. Wire diameters ranging from 10 μm to 70 μm were employed. One should note that the 70 μm diameter fiber is actually two 50 μm fibers twisted together in order to have the same mass as that of a 70 μm fiber. The aluminum plasma is created by exploding a 5 μm foil and injecting the plasma through a hole in the anode. Figure 2 shows a cross section of the load region and the aluminum plasma injection nozzle. The mass in the aluminum plasma is controlled by adjusting the time delay between the firing of the generator and exploding the foil.[3] In addition, a pair of Helmholtz coils have been installed to provide an axial DC magnetic field up to 300 G.

Figure 3 shows the arrangement of diagnostics around the load region. The view in figure 3 is along the load with the generator located behind the load region. The XRD array consists of ten filtered p-i-n diodes. Several of the filters have been chosen as Ross matched pairs in order to determine temperature from the bound-free and bound-bound spectrum. All XRD signals as well as the other electrical signals are recorded on digitizers driven by a single timebase at

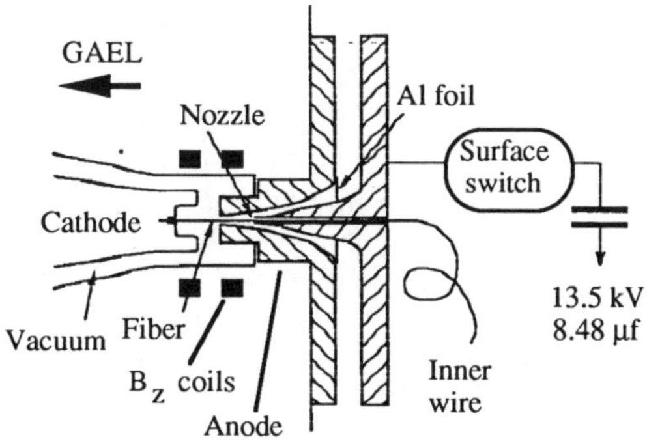

Fig. 2. Load region on GAEL.

500 MS/s or 1 GS/s. A fiber optic picks off part of the laser probe beam in order to determine timing of the schlieren photography. A ∂B/∂t probe monitors the current in the load. Figure 4 shows the generator current, load current, and an XRD channel from a typical shot. The pinhole array consists of nine pinholes utilizing various pinhole diameters and filters to resolve a time integrated image of the pinch.

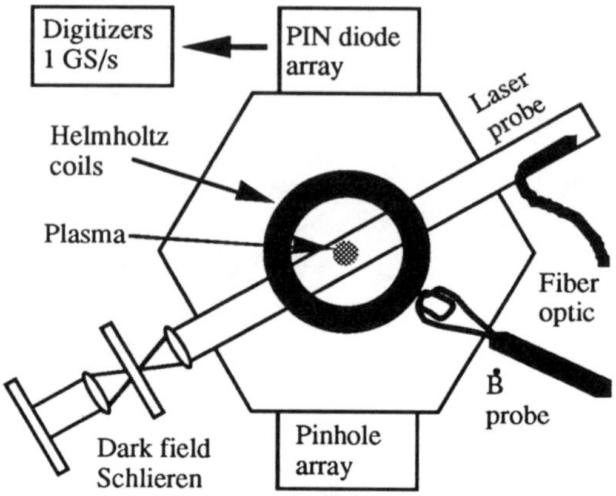

Fig. 3. Arrangement of diagnostics used on GAEL.

RESULTS

The best measure of the stability of the core fiber comes from the pinhole images of the compression. Figure 5 shows the results for various fiber diameters. Visibly the small diameter fibers ($d \leq 25$ µm) appear very uniform. At diameters of 40-50 µm hot spots and inhomogeneities may or may not be present. For example, the 40 µm fiber in figure 5 shows hot spots while the 50 µm fiber appears to be stable. For the 70 µm fiber hot spots are numerous and consistently appear on all data. However, the intensity and distribution of the hot spots is vastly different from those associated with the classical exploding wire. The classical exploding wire tends to have uniformly spaced hot spots with very bright emission. In the present study, the hot spots are irregularly distributed and emit radiation of intensity brighter but comparable to emission from uniform regions of the fiber. The fshell plasma apparently tamps any fiber instabilities and prevents recompression of the exploded fiber from a large diameter.

The data in figure 5 suggests that diameters near 50 µm are in a transition region where non-uniformities in the imploding plasma shell may appear as instability in the central fiber. Additional evidence for the need to create a uniform implosion can be seen in figure 6. In figure 6 we see half of a 25 µm fiber exhibiting hot spot emission. The reason for this emission is due to improper timing of the aluminum jet resulting in a non-uniform distribution of plasma along the pinch axis. Thus, the resistance of the aluminum plasma varies along the axis to the point that a single compression exhibits behavior of both regimes.

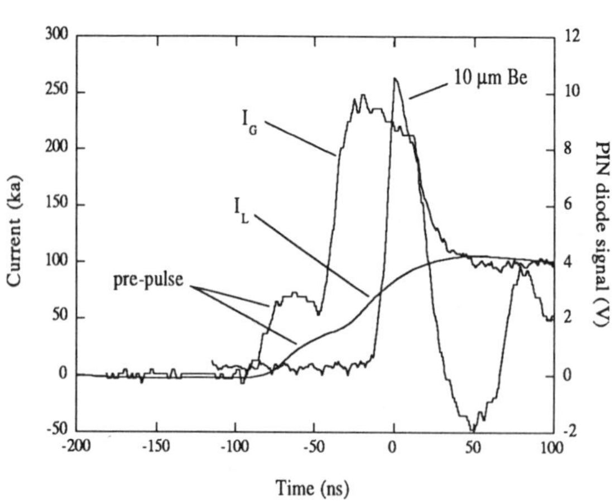

Fig. 4. Timing for generator current, I_G, load current, I_L, and a diode signal.

Figures 7 and 8 show the data from the x-ray diodes for typical hard (Ni) and soft (Be) filters. In figure 7 the peak diode signals have been plotted for various fiber diameters. Considering that the shot-to-shot variation in our experiments is about 20%, the peak x-ray emission is flat for both hard and

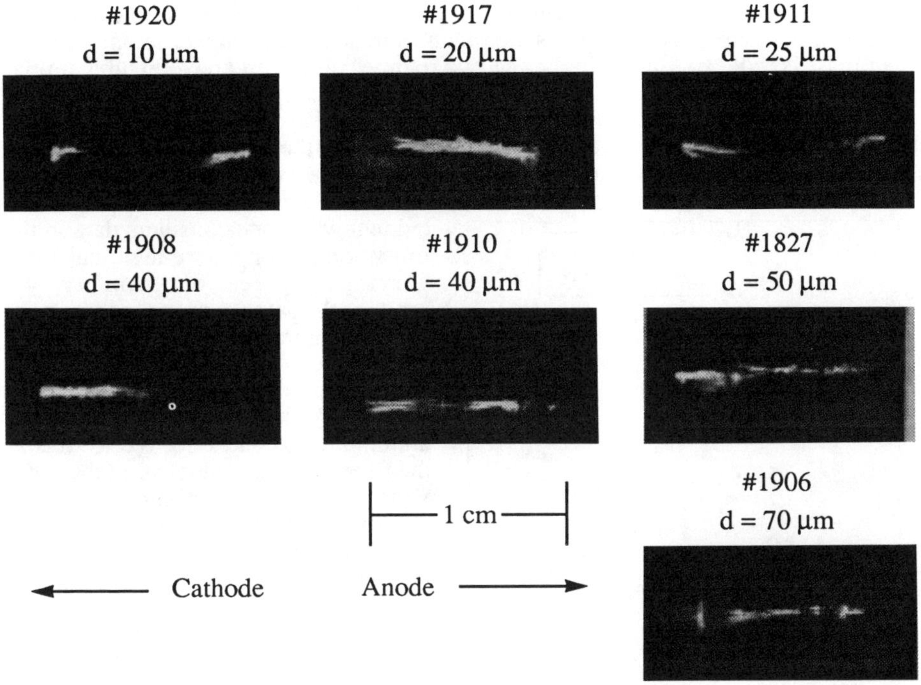

Fig. 5. Pinhole images for various fiber diameters. The filter is 11 μm aluminum and the pinhole diameter is 50 μm.

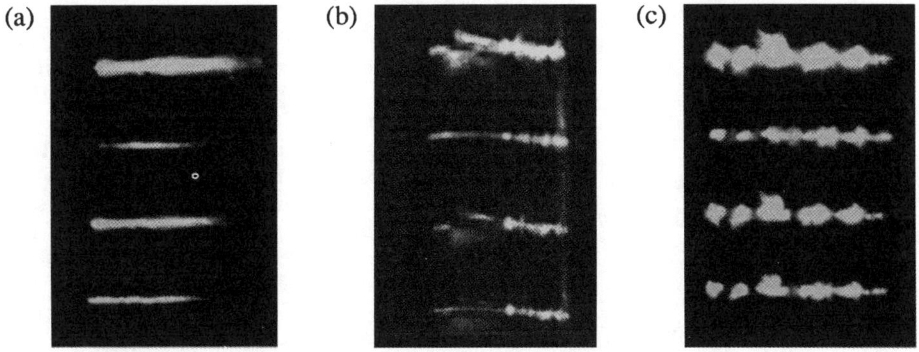

Fig. 6. Pinhole images showing various stages of stability; (a) entire fiber is stable stable, (b) half of the fiber is stable, (c) entire fiber is unstable.

soft x-rays. This is probably a consequence of our loads being over-massed even at the smallest fiber diameters. This prevents the pinch from attaining a higher temperature as the fiber mass is reduced. Surprisingly, the two regimes described above exhibit similar behavior with respect to the peak x-ray emission. Normally, one would expect hot spots to dominate the emission and produce a much harder spectrum. However, because the intensity of the hot spots (see fig. 5) is of the same magnitude as the emission from stable regions, the diodes see relatively the same spectrum whether or not the fiber behaves stably. This is evident when one considers data in the transition region where identical shots varying degrees of stability yet produce the same average spectrum.

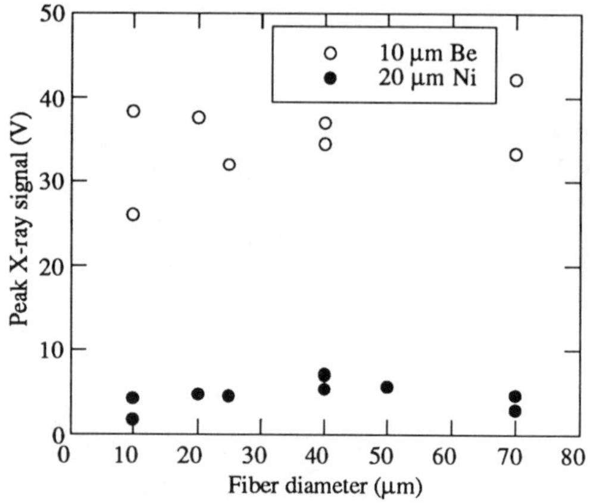

Fig. 7. Peak x-ray signals for fibers of various diameters.

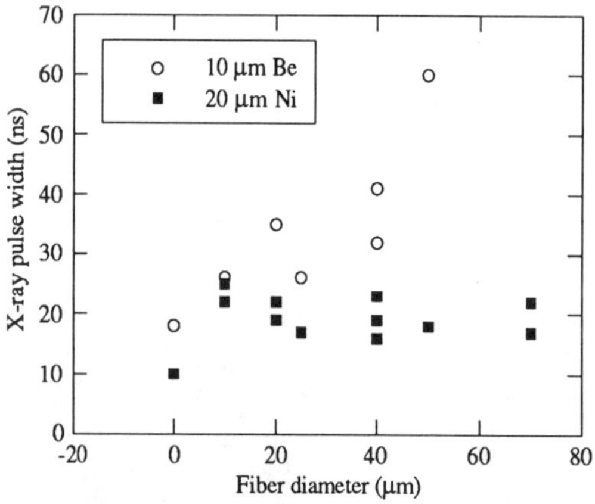

Fig. 8. X-ray diode signal widths for fibers of various diameters. The zero diameter is for the aluminum jet only.

In figure 8 the pulse width of the x-ray diodes has been plotted. Note that the zero diameter fiber is the pulse width of the radiation from the aluminum plasma alone. We see that as the fiber diameter is increased the pulse width of the soft radiation increases and that of the hard radiation remains constant within the limits of reproducibility. This implies that the integrated x-ray yield increases for soft radiation only with the most likely source being UV emission. Again, differences between the two regimes cannot be derived from diode data. Our experience with fiber only loads indicates that the pulse width of the radiation should be much shorter than those of figure 8. This may be due in part to the imploding plasma shell acting to contain and reduce the growth of instabilities associated with the fiber.

When a magnetic field is externally applied parallel to the pinch axis, we observe a more stable compression. Instabilities are reduced with m=1 instabilities negligible compared to m=0 above fields 150 G. The added stability enhances both soft and hard x-ray emission from the pinch. Figure 9 shows the results for various applied magnetic fields for the POW and aluminum jet only configurations. Each configuration is optimized for a different applied field with the POW configuration requiring the higher field. This suggests that the magnetic field necessary to optimize a compression scales with the pinch mass. One can imagine increasing the pinch mass or density but keeping the kinetic energy per unit mass constant. As an instability begins to develop, the higher density configuration has more momentum interacting with the instability and, therefore, requires a larger force to

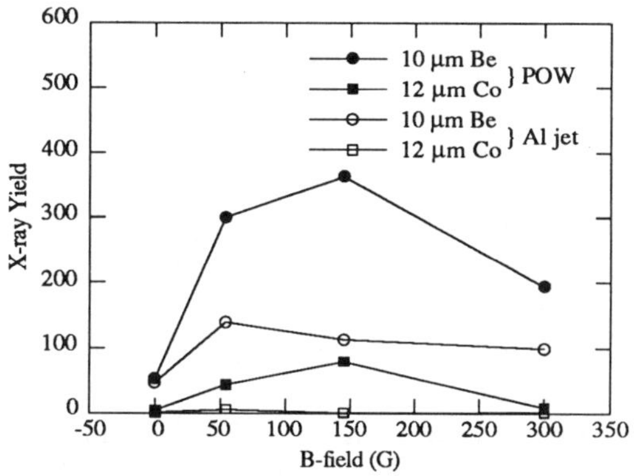

Fig. 9. Integrated x-ray yields as a function of applied magnetic field.

maintain stability. An axial magnetic field provides this restoring force through the magnetic pressure. Our theoretical treatment of the situation, in fact, indicates that the magnetic field scales as the square root of the mass.

CONCLUSIONS

We have observed the onset of the appearance of hot spots as we increase the diameter of copper fibers in the POW configuration. The appearance of the hot spots indicates that there are two regimes of pinches using the POW configuration: one regime where nearly all of the initial current flows through the imploding shell, and the other where a significant amount of current flows through the fiber causing it to explode prematurely.

These results have important implications for target design of staged loads. In order to optimize a pinch and prevent current from penetrating into the core region, the outer stage must have a much lower impedance than the inner stages combined. This suggests that for gas puff loads a means of pre-ionizing the gas is desirable.

In addition, we are continuing to investigate theoretically and experimentally the stabilization of pinches and the optimization of x-rays using externally applied, axial magnetic fields. Our results show that the field required to optimize the compression scales with the pinch mass.

ACKNOWLEDGEMENTS

The authors thank M. Fleury and R. Aliaga for their help and R. Spielman for providing code used to analyze the x-ray data. This work is supported by ETCA/CEG under contract #420/115/01.

REFERENCES

[1]F. J. Wessel, B. Etlicher, and P. Choi, Phys. Rev. Lett. 69, 3181 (1992).

[2]J. Delvaux, H. Lamain, C. Rouille, H. J. Doucet, J. M. Buzzi, M. Gazaix, and B. Etlicher, in *Proceedings of the Fourth International Topical Conference on High Power Electron and Ion Beam Research and Technology* (Ecole Polytechnique, Palaiseau, France, 1981), Vol. 2, p. 775.

[3]M. Gazaix, H. J. Doucet, B. Etlicher, J. P. Furtlehner, H. Lamain, and C. Rouille, J. Appl. Phys. 56, 3209 (1984).

The Dense Z-pinch Programme at Imperial College

M. G. Haines
Blackett Laboratory, Imperial College London SW7 2BZ, UK

ABSTRACT

An extensive programme of research, both experimental and theoretical, into the stability and dynamics of Z-pinches has led to the funding of the DZP Project to study both radiative collapse of Z-pinch plasmas and pinches close to thermonuclear fusion conditions. The MAGPIE (Mega-Ampere Generator for Plasma Implosion Experiments) generator (2.4MV, 336kJ, 200ns) is now being commissioned ready for Z-pinch experiments commencing this summer. The design of the generator has been determined by the perceived requirements demanded by consideration of (a) fusion conditions with end losses to electrodes, (b) radiative collapse at currents well above the Pease-Braginskii limit, and (c) stability studies particularly under large ion Larmor radius conditions. As a result, and in contrast to other generators in the >1TW class this has a long pulse length (200ns) and a final line impedance of 1.25 ohm. The stability regimes together with theoretical and experimental results are reviewed in the framework of the I^4a-N diagram. Our understanding (albeit incomplete) of other phenomena characteristic of Z-pinches, namely the formation of electron beams, dense spots of intense X-ray emission, ion beams and filaments will be summarised.

1. INTRODUCTION

The philosophy behind the Imperial College dense Z-pinch programme, now ready for operation, can be traced back to an early calculation on heat loss to electrodes[1]. When applied to a pinch under fusion conditions[2] it is found that a generator is required in the several megavolt and megampere range; hence the decision to employ pulsed power techniques. But it was noted that fusion conditions require currents close to the Pease-Braginskii critical current[3,4] above which radiative collapse of the pinch is possible[5]. Indeed our one-dimensional calculations[6] indicate that it is possible, in absence of instabilities, to collapse a high pinch to 10^5 times solid density, limited only by electron degeneracy pressure and by opacity effects and cut-off of the characteristic X-ray bremsstrahlung in the high density plasma. Then it is in a physical state similar to that of a white dwarf star. Because of the basic scientific interest in achieving such an extreme condition this became the prime objective of the research programme. The study of a Z-pinch under fusion conditions is a second objective. Encouraged by a scoping study of the technical feasibility of a Z-pinch reactor undertaken by Imperial College Centre for Fusion Studies[7], such a reactor would differ significantly from the proposed tokamak reactor in being pulsed and having a much higher power density. Both research objectives will be very dependent on the achievable stability of the Z-pinch. Accordingly much effort has been put into studying the stability of a Z-pinch in all the varying regimes[8], both theoretically and experimentally. Particularly relevant to our project is the paper to be given at the Conference by Drs. Scheffel, Arber and Coppins on large ion Larmor radius stability[9].

We first review the end loss calculation satisfying Lawson's conditions, and then discuss the particle orbits particularly the singular orbits close to the magnetic nul. The universal diagram for regimes in Z-pinch stability is presented and our latest theoretical results and results on small scale experiments[10] are discussed in terms of this. Simulation of radiative collapse and the inclusion of Nernst and Ettinghausen effects[11] is summarised, the latter confirming the assumptions of some self-similar solutions[12]. The inclusion of anomalous resistivity particularly in the cold start is the subject of another presentation at the Conference. The theory of linear transport has been extended to a degenerate, magnetised electron plasma[13], relevant at radiative collapse. Lastly our contribution to the understanding of thermal instabilities, ion acceleration processes and hot spots will be given. Z-pinches are narrowly confined to laboratory pulsed power experiments, and Dr. A. R. Bell[14] will show how they can occur on the 10μm scale in laser-plasma interactions and on the megaparsec scale in intergalactic jets.

Our DZP project involves the building of the MAGPIE generator (Meg-Ampere Generator for Plasma Implosion Experiments), with characteristic parameters 2.4MV, 336kJ in 200ns. It is funded by the Science and Engineering Research Council with a 50% contribution from the Ministry of Defence. We are also very grateful for much help from Sandia National Laboratory and AWE, Aldermaston.

2. HEAT FLOW TO ELECTRODES: FUSION CONDITIONS

We consider steady heat flow to the electrodes of a Z-pinch. A simple power balance balances ohmic heating against axial heat flux and bremsstrahlung,

$$\underline{J}.\underline{E} = \frac{\partial q_z}{\partial z} + \beta_b n^2 T^{1/2} + \frac{1}{r}\frac{\partial}{\partial r}(r q_r) \qquad (1)$$

This equation is averaged over the cross-section to eliminate the radial heat flux q_r, assumed zero at r = a the pinch radius. The transport equations are of the form

$$\underline{J} = \sigma \underline{E} + \alpha \nabla T \qquad (2)$$

$$\underline{q} = -\kappa \nabla T + \beta \underline{E} \qquad (3)$$

where only the Z components are relevant. In addition radial and axial pressure balance are assumed, together with uniform plasma radius, yielding the Bennett relation for the total current I, line density N and temperature T,

$$16\pi N(z)kT(z) = \mu_0 I^2 \qquad (4)$$

Two extreme cases have been solved[1]: the unmagnetised ($\omega_e\tau_{ei}$=0) plasma and the infinitely magnetised ($\omega_e\tau_{ei} = \infty$) plasma ($\omega_e$ is the electron cyclotron frequency and τ_{ei} is the electron ion collision time). They yield **exactly** the same scaling relations with coefficients that differ by only 30%. This can be seen from the following order of magnitude arguments (neglecting cross phenomena). For $\omega_e\tau_{ei} = 0$, eq (3) is to an order of magnitude

$$q_z = -\lambda T^{5/2}/z_0 \qquad (5)$$

where $\lambda T^{5/2}$ represents the thermal conductivity κ and z_0 is the length of the pinch. Thus, energy balance (eq 1) gives (neglecting bremsstrahlung)

$$IV = -\pi a^2 \lambda T^{5/2}/z_0 \qquad (6)$$

where V is the applied voltage.

Taking the electrical conductivity σ to be $\alpha T^{3/2}$ the resistance of the pinch is

$$\frac{V}{I} = \frac{z_0}{\pi a^2 \alpha T^{3/2}} \qquad (7)$$

Eliminating I and then V between eqs (6) and (7) yields

$$V = \sqrt{\frac{\lambda}{\alpha}}T \qquad (8)$$

$$I = \pi a^2 \sqrt{\alpha\lambda}\, T^{5/2}/z_0 \qquad (9)$$

Taking the other extreme case ($\omega_e\tau_{ei} = \infty$) eqs (2) and (3) combine to give the heat flux as the electron enthalpy flux associated with the current,

$$q_z = \frac{5}{2}\frac{kT}{e}\frac{I}{\pi a^2} \tag{10}$$

so that power balance becomes

$$IV = \frac{5}{2}\frac{kT}{e}I \tag{11}$$

or

$$V = \frac{5}{2}kT/e \tag{12}$$

which gives the same form as eq (8). Employing eq (7) with (12) gives

$$I = \pi a^2 \frac{5}{2} k\alpha T^{5/2}/z_0 e \tag{13}$$

which is exactly the same form as eq.(9). A complete mathematical solution of the two cases reveals only a 30% difference in the coefficients, showing that the results are robust and insensitive to the model. The case of magnetised electrons is more pertinent to the situation, and with the inclusion of bremsstrahlung the results are

$$T_e = 4.64 \times 10^3 V\left(1 - \frac{I^2}{I_{PB}^2}\right) \tag{14}$$

and

$$T_e^{5/2} = 2.04 \times 10^7 \frac{z_0}{\pi a^2} I\left(1 - \frac{I^2}{I_{PB}^2}\right) \tag{15}$$

where I_{PB} is the Pease-Braginskii current given by

$$I_{PB} = 0.433(\ln \Lambda)^{1/2} \times 10^6\ \mathrm{A} \tag{16}$$

The energy confinement time τ_E is the stored plasma energy divided by the power input, ie.

$$\tau_E = \frac{3NkTz_0}{IV} \tag{17}$$

Using eqs (4), (14) and (15) and writing the mean number density $\bar{n}$ as $N/\pi a^2$ we obtain[2]

$$\bar{n}\tau_E = 1.71 \times 10^{-11} NT^{3/2} \tag{18}$$

If we choose $\bar{n}\tau_E$ and T to satisfy Lawson's conditions at $5 \times 10^{20}\mathrm{m^{-3}s}$ and 3×10^8K respectively, eq (18) yields $N = 5.6 \times 10^{18}\mathrm{m^{-1}}$; eq (4) yields $I = 9.7 \times 10^5$A; and eq (14) gives $V = 6.5 \times 10^4$ volt for $I_{PB} = \infty$ and

$V = 9.8 \times 10^5$ for $I_{PB} = 1MA$ ($\ln\Lambda = 5.316$). Whilst N, I, V and $Z_0/\pi a^2$ are determined from the choice of $\bar{n}\tau_E$ and T, one other parameter out of τ_E, n, z_0 or a has to be chosen then to fix the other three. Because we are clearly involved with parameters requiring high voltage pulsed power (I ~ 1MA, V~1MV) as developed by J. C. Martin of AWE, Aldermaston, we choose the energy confinement time to be, say, 100ns; then the other parameters are $\bar{n} = 5 \times 10^{27} m^{-3}$, $a = 19\mu m$ and $z_0 = 8.9cm$ ($I_{PB} = \infty$). We note that for these parameters:

i) the transit time of a thermal deuteron is approximately the same as τ_E; thus the plasma should be free of heavier impurities from the electrodes,

ii) the current I required for fusion conditions is close to the Pease-Braginskii limit; it is therefore possible to study radiative collapse to high density with similar equipment,

iii) the ratio of mean ion Larmor radius $\bar{a_i}$ to pinch radius a, given by

$$\frac{a_i}{a} = \left(\frac{2kT_i}{m_i}\right)^{1/2} \frac{m_i}{ea\bar{B}} = \frac{8.08 \times 10^8}{N^{1/2}} = 0.34 \tag{19}$$

Therefore we would expect that ideal MHD stability theory would not apply, but large ion Larmor radius effects will dominate.

That under fusion conditions the ion Larmor radius is comparable with the pinch radius, and the current is comparable with the Pease-Braginskii current are two remarkable and fortuitous coincidences of nature.

3. PARTICLE ORBITS

The equation describing radial ion pressure balance can be written as [2]

$$v_{zi} = -\frac{1}{Zn_i er}\frac{\partial}{\partial r}\left(\frac{rP_{\perp i}}{B_\theta}\right) - \frac{P_{\perp i}}{Zn_i eB_\theta^2}\frac{\partial B_\theta}{\partial r} + \frac{P_{\| i}}{Zn_i eB_\theta r} + \frac{E_r}{B_\theta} \tag{20}$$

where v_{zi} is the ion centre-of-mass velocity, Z the ion charge number, $P_{\perp i}$ and $P_{\| i}$ the ion pressure perpendicular and parallel to the azimuthal magnetic field B_θ. The first term represents the diamagnetic velocity while the three other terms are the contributions to the mean guiding centre velocity from ∇B, curvature and $\underline{E} \times \underline{B}/B^2$ drifts. On integrating to obtain the mean guiding centre flow in the z direction we find that the diamagnetic term describes the singular flow of ions with snake-like orbits in the z direction within one Larmor radius of the axis. In fact the ion singular current I_{si} associated with this is

$$I_{si} = \frac{4\pi}{\mu_0}\left(\frac{P_{\perp i}}{J_z}\right)_{r=0} \tag{21}$$

and is approximately half the total current. There is a similar formula for the singular electron current. The heat flow associated with singular and guiding centre flow of ions and electrons can therefore be represented as an enthalpy flow associated with the total current flow, leading to identical results as for the magnetised electron case ($\omega_e\tau_{ei} = \infty$) discussed in §2.

4. STABILITY

Early experiments[15-18] and theory[19,20] on stability showed in the ideal MHD regime that the Z-pinch can be violently unstable. However ideal MHD theory neglects much physics such as resistivity, viscosity, finite ion Larmor radius (FLR) effects, pressure anisotropy, the Hall effect and electron pressure gradient in Ohm's law. When we consider each of these effects in turn the relevant dimensionless parameter is found to be a function only of the line density N and I^4a. We can therefore construct a universal diagram[8], fig 1, in which the critical values of these dimensionless parameters are straight lines.

The importance of resistivity on stability is measured by the magnetic Lundquist number

$$S = \mu_0 \sigma \overline{v_A} a \tag{22}$$

where $\overline{v_A}$ is the mean Alfvén speed given for a Z-pinch of mean density $\bar{\rho}$ by $\overline{v_A} = B_\theta(r = a)/2(\mu_0\bar{\rho})^{1/2}$. If the Bennett relation, eq (4), holds this becomes

$$S = 3.86 \times 10^{23}\, I^4 a/N^2 \tag{23}$$

A critical value of order 10^2 was found for stability for the m = 0 mode [21,22], and also the m = 1 mode[23], under a time varying equilibrium of joule heating under pressure balance[24]. In contrast to the resistive tearing mode in a tokamak, in a Z-pinch there are no singular surfaces and the S value has to be so low to affect stability that joule heating is important.

Fig 1. Universal diagram of stability regimes in the Z-pinch

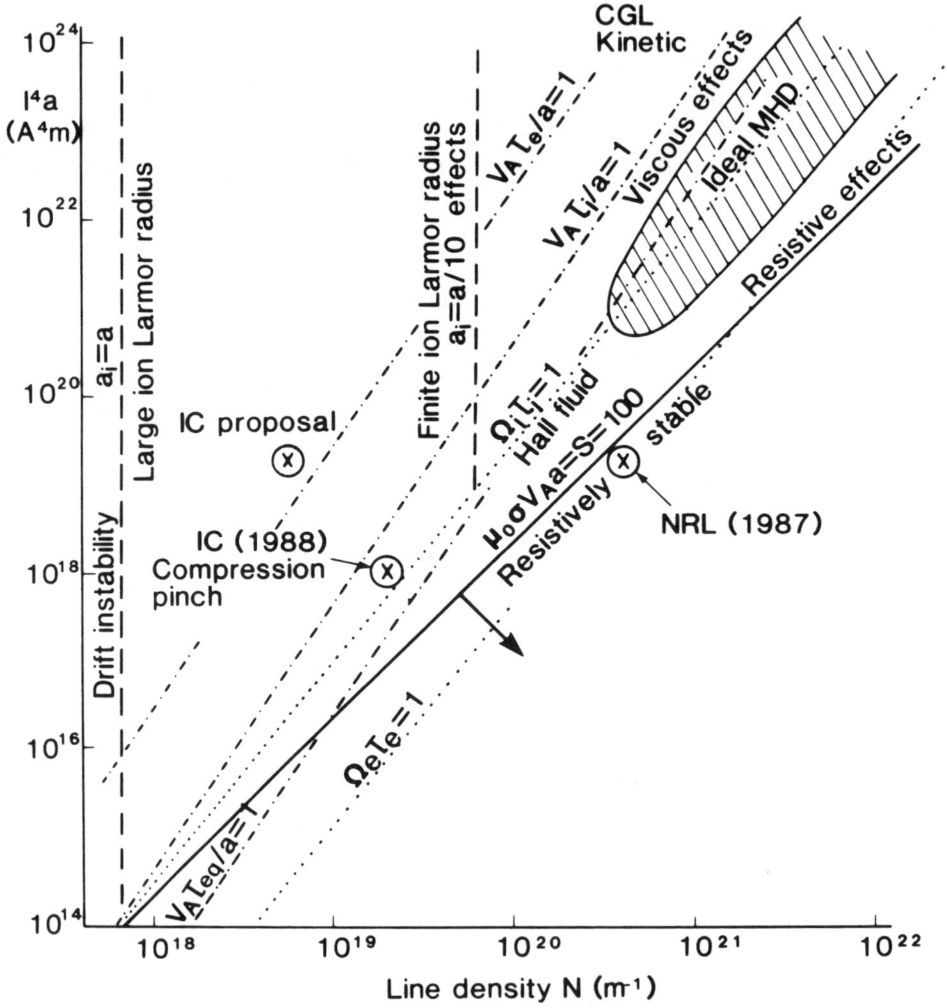

For viscosity a viscous Lundquist number R can be defined

$$R = \bar{\rho}\bar{V}_A a/\mu \tag{24}$$

where μ is the viscosity. Viscous effects are important for $R<1$. This is the same condition as that for the perturbed pressure to become anisotropic, ie. the mean-free-path should be greater than the pinch radius, ie. $R^{-1} = v_A\tau_i/a > 1$. But this can be written as

$$\frac{1}{R} = 2.07 \times 10^{39} \frac{I^4 a}{N^3} \tag{25}$$

Under conditions of $R<1$ it is more appropriate to use the Chew, Goldberger and Low[25] ordering for stability calculations[26].

Eq (19) has already shown that FLR and large ion Larmor radius (LLR) effects depend only on N; $a_i/a = 0.1$ and 1 are plotted in fig 1. Important new results showing in two models how LLR stabilisation is very dependent on equilibrium profiles are presented at this conference[9]. Such effects are only valid if the ions are magnetised, ie. $\overline{\Omega_i}\tau_i>1$ where $\overline{\Omega_i}$ is the mean ion cyclotron frequency and τ_i is the ion-ion collision time. This parameter also depends on I^4a and N

$$\overline{\Omega_i}\tau_i = 3.64 \times 10^{30} \frac{I^4 a}{N^{5/2}} \tag{26}$$

and is plotted in fig 1 for $\overline{\Omega_i}\tau_i = 1$.

5. SMALL SCALE EXPERIMENTS

Three small scale experiments relevant to stability have been carried out at Imperial College in recent years under the direction of A. E. Dangor. In the first of these, a laser-initiated gas-embedded pinch, the results are consistent with the generation of a centrally peaked self-similar current density profile[27] such that the m = 0 instability is not triggered[28], but instead the m = 1 mode is. Holographic interferometry showed that the density perturbation is peaked on the axis[29], consistent with ideal MHD stability theory[30] for this current profile.

In a compression Z-pinch[31] a uniform cylindrical pinch column is formed which at later times is observed by side-on holographic double exposure interferometry to be unstable to a m = 0 mode. However it grows to an amplitude of about 30% and then saturates. Figs 2a and 2b show interferograms taken sequentially in time on different discharges.

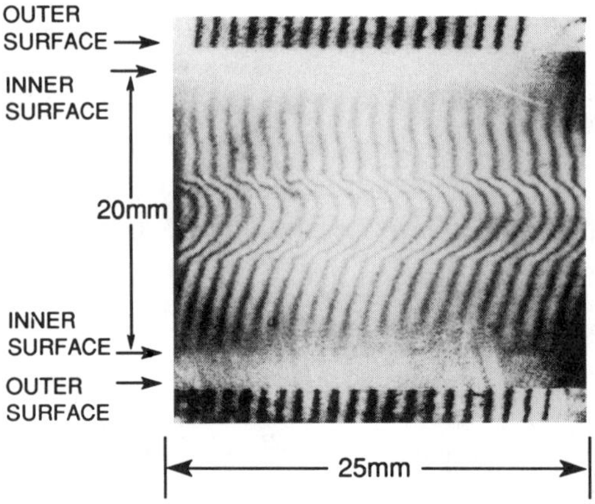

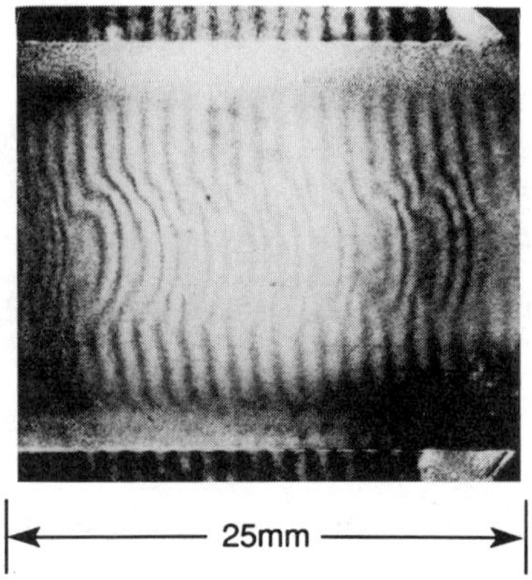

Figs. 2a and 2b

Fig 2a is an interferogram of a compressional pinch (2 torr H2, 150kA) 30ns after it had formed on the axis of a quartz tube. The radius a is approximately 3mm and the line density 4 x $10^{19}m^{-1}$. This corresponds to a Bennett temperature of 85eV and an ideal MHD growth time of 20ns. The fringes show that there is a 5% density perturbation at the surface of the pinch. This is to be expected as the fastest growing modes of an m = 0 instability are localised at the surface. **Fig 2b** is an interferogram of a similar pinch taken 40ns later when the instability has grown to approximately 30%. The periodic variation in the fringes is due to an m = 0 instability with $\lambda = a$. Interferograms showing a pinch similar to this one were obtained for a further 130ns. This shows that the instability has saturated without disupting the pinch.

Large ion Larmor radius effects could be important here since a_i/a is calculated from the measured line density as being 0.18. However the ions are only just magnetised ($\Omega_i\tau_i = 0.8$). A theoretical explanation of the saturation must await a non-linear kinetic simulation of large ion Larmor radius effects.

Lastly carbon fibre Z-pinches with current up to 100kA and a current rise time of 45ns have been simultaneously diagnosed with an X-ray pinhole camera, laser interferometry, and optical framing photography[32]. A clear m = 0 instability arises and is fully developed at 40ns with axial wave number k given by ka = 2.5. The simultaneous observations of the middle 7mm of the pinch show that there is an exact spatial correspondence of the regions of most intense X-ray emission, the radially expanded regions of plasma density, and the optical bright spots. Even though a_i/a was 0.3, the ions were unmagnetised with $\Omega_i\tau_i = 2x10^{-4}$ and the Lundquist number was 8. Resistive MHD should apply, but the instability, we believe, is the faster growing Rayleigh-Taylor instability associated with the radially inward acceleration following the initial expansion of the ionising fibre. It is hoped that in our future experiments the expansion will be reduced because of the higher rate of rise of current.

6. RADIATIVE COLLAPSE

Above a current of $0.433(\ln\lambda)^{1/2}$ MA the hydrogen Z-pinch loses more energy by bremsstrahlung than it gains from Joule heating. Independently Pease[3] and Braginskii[4] derived this critical current for a steady state Z-pinch with uniform current density and temperature and a parabolic density profile. (It was extended to include end losses by Haines[1]) The reason for a unique critical current is the coincidence that the electrical conductivity σ is proportional to the $T^{3/2}$ and bremsstrahlung is $\beta_b n^2 T^{1/2}$ (β_b is a constant). An order of magnitude balance of Joule heating and bremsstrahlung gives

$$\frac{J_z^2}{\sigma} = \frac{J_z^2}{\alpha T^{3/2}} \cong \beta_b n^2 T^{1/2} \tag{27}$$

or,

$$J_z \cong \frac{\sqrt{(\alpha\beta_b)}}{k} p \tag{28}$$

where $\sigma = \alpha T^{3/2}$ and k is Boltzmann's constant. But pressure balance is

$$J_z B_\theta = -\frac{\partial p}{\partial r} \cong \frac{p}{a} \tag{29}$$

and therefore with eq (28), eq (29) states that $B_\theta a$ is a constant. This is just $\mu_0 I/2\pi$, and so the current is a fixed value, I_{PB}.

An analytic model of radiative collapse[5] of a Z-pinch coupled to an external charged pulse forming line gave the combination of parameters that are necessary to obtain a substantial collapse to high density. Besides having a current that can rise to $\sim\sqrt{2}\, I_{PB}$, the product of voltage and time must exceed a certain value. On the basis of this calculation the case for the DZP project was made. There are some interesting features to the calculation; for example, when the pinch collapses, almost catastrophically to a radius of 0.13μm the voltage required to drive ~1MA through this is 100MV, and this is provided by a large negative value of L(dI/dt), where L is the pinch inductance.

A more complete one dimensional simulation was carried out by Chittenden and Haines[6] which included the effects on pressure and transport of electron degeneracy. Radiation transport of the bremsstrahlung was included together with the plasma dielectric response and cut-off at the plasma frequency. With $\ln\Lambda$ varying the radiative collapse is now terminated by the effects of degenerate electron pressure and the effects of opacity when it reaches a density of order 10^5 x solid. In particular we now find that on the axis

$$\bar{h}\omega_p > kT_e \tag{30}$$

where ω_p is the plasma frequency ie. the characteristic X-rays of bremsstrahlung cannot even propagate. Thus the physics conditions are similar to that of a white dwarf star. Fig 3 shows the pinch radius and current as a function of time during the radiative collapse.

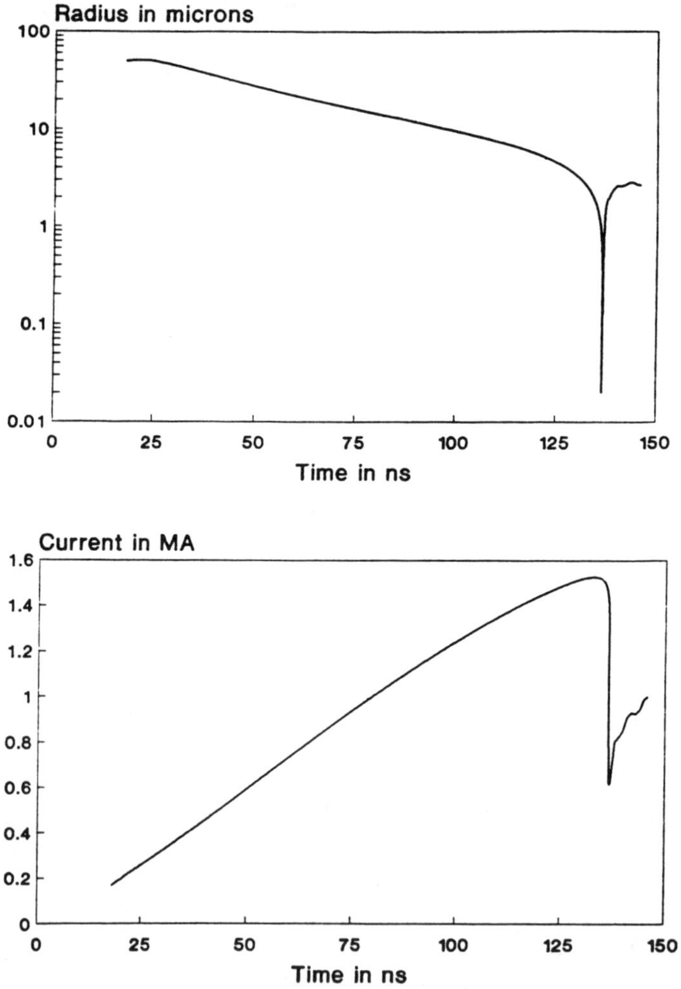

Fig 3. Pinch dynamics with radiation transport

Recently we have extended the model by the inclusion of the Nernst and Ettinghausen effects[33]. Without these effects there is a tendency for the low density outer surface region of the pinch to be thermally unstable with the Joule heating leading to a high, decoupled electron temperature and current density. The Ettinghausen effect which is essentially a E/B inward drift of hot electrons causes an inward heat flux, while the Nernst effect is the advection of the magnetic field (and associated current density) by the heat flux. Both effects have the same coefficient and arise through the velocity dependence of the collision frequency. Inclusion of these effects causes the temperature and current density to flatten.

7. RUNAWAY ELECTRONS AND HOT SPOTS

This preferential inward drift of the hot, collisionless electrons to the axis where the magnetic field is zero is the mechanism which causes runaway electrons to occur. If the pinch is hollow eg. as in a gas puff or wire-array pinch, runaway electrons will be an important feature because of the low collisionality near the axis[33,34]. Low collisionality can also occur in the necking off regions of a m = 0 instability.

Hot spots, or rather dense spots of intense X-ray emission, are associated with the occurrence of electron beams, which incidentally cause hard X-ray emission at the anode[35]. We postulate that a beam can trigger locally an anomalous resistivity through microturbulence; this causes a large local electric field and the partial diversion of the current around the region; then the unbalanced $\underline{J} \times \underline{B}$ of the diverted current causes a transient micropinch to occur; alternatively the converging plasma motion can be described by the large local E/B drift. The dense spot is transient as it is largely inertial and will last for only a fraction of a nanosecond.

8. ION ACCELERATION

The ion acceleration process during a m = 0 instability has had many but doubtful explanations. Bearing in mind that the plasma is quasi-neutral, so that any electrical field with give momentum to ions and electrons equally and in opposite directions we reject a diode mechanism, and instead use our knowledge of singular orbits to show that in a m = 0 necking region there is a resonant transfer of energy through a synchrotron effect to those ions whose total drift velocity is small and which therefore experience a large $-\partial A/\partial t$ electric field for a significant time[36]. Such ions transform into an energetic singular ion beam along the axis away from the anode, while there is an equal and opposite axial momentum in a larger number of off-axis guiding centre ions.

9. THERMAL INSTABILITIES

There is a tendency in both theta pinches and in plasma focus experiments for the current at early times to flow in many separate filaments. We consider that this could be caused by a thermal instability[37] which can occur at early times when the electron mean-free-path is smaller than the collisionless skin depth. A parametric comparison of wavelength of this mode with variation in gas filling pressure gave good agreement.

10. CONCLUSION

In this paper I have reviewed the physics of a Z-pinch and the philosophy behind the DZP project at Imperial College. The MAGPIE generator is now completed and there will be an opening ceremony at this conference at which Drs. Pease and Braginskii will be present, together with the inventor of pulsed power, Mr. J. C. Martin. Besides our principal objectives of studying radiative collapse, fusion conditions and stability, there are many other phenomena, as exemplified above, which we hope to explore on our new experiment.

REFERENCES

1. M. G. Haines, Proc. Phys. Soc. **77**, 643 (1961)
2. M. G. Haines, J. Phys. D; Appl. Phys. **11**, 1709 (1978)
3. R. S. Pease, Proc. Phys. Soc. **70,** 11 (1957)
4. S. I. Braginskii, Zh. Eksp. Teor. Fiz. **33**, 645 (1957); Sov. Phys. JE7P **6**, 494 (1958)
5. M. G. Haines, Plasma Physics & Controlled Fusion **31**, 759 (1989)
6. J. P. Chittenden and M. G. Haines, Phys. Fluids B **2**, 1889 (1990)
7. H. R. Bolton et al., in Plasma Physics & Controlled Nuclear Fusion Research 1986 Vol 3, 367 IAEA Vienna (1987)
8. M. G. Haines and M. Coppins, Phys. Rev. Lett. **66**, 1462 (1991)
9. J. Scheffel, T. D. Arber, M. Coppins, these proceedings

10. M. G. Haines et al., in Plasma Physics & Controlled Nuclear Fusion Research 1990, Vol 2, 769 IAEA Vienna (1991)
11. J. P. Chittenden and M. G. Haines, J. Phys. D: Appl. Phys. **26,** 1048 (1993)
12. M. Coppins. J. P. Chittenden, I. D. Culverwell, J. Phys. D: **25**, 178 (1992)
13. S. R. Brown, Ph.D. thesis, University of London (1993)
14. A. R. Bell, these proceedings
15. S. W. Cousins and A. A. Ware, Proc. Phys. Soc. B: **64**, 159 (1951)
16. R. Carruthers and P. A. Davenport, Proc. Phys. Soc. B **70**, 49 (1957)
17. O. A. Anderson, W. R. Baker, S. A. Colgate. H. P. Furth, J. Ise, R. V. Pyle and R. E. Wright, Phys. Rev. **109**, 612 (1958)
18. I. V. Kurchatov, J. Nucl. Energy **4**, 193 (1957)
19. M. D. Kruskal and M. Schwarzschild, Proc. Roy. Soc. London A **223**, 348 (1954)
20. R. J. Taylor, Proc. Phys. Soc. B **70**, 31 (1957)
21. I. D. Culverwell and M. Coppins, Phys. Fluids B **2**, 129 (1990)
22. F. L. Cochran and A. E. Robson, Phys. Fluids B **2**, 123 (1990)
23. I. D. Culverwell (private communication)
24. M. G. Haines, Proc. Phys. Soc. **76**, 249 (1960)
25. G. F. Chew, M. L. Goldberger and F. E. Low, Proc. Roy. Soc. London A **236**, 112 (1956)
26. M. Coppins, Phys. Fluids B **1**, 591 (1989)
27. M. Coppins, J. P. Chittenden and I. D. Culverwell, J. Phys. D **25**, 178 (1992)
28. B. B. Kadomtsev, in Reviews of Plasma Physics, ed. by M. A. Leontovich (Consultants Bureau, New York) Vol. 2, p153 (1966)
29. P. Choi, M. Coppins, A. E. Dangor, M. B. Favre, Nucl. Fusion **28**, 1771 (1988)
30. M. Coppins, Plasma Phys. & Controlled Fusion, **30**, 201 (1988)
31. J. M. Bayley, PhD. thesis, University of London 1991 See also ref 10.
32. S. N. Niffikeer, Ph.D. Thesis, University of London 1991 See also ref 10 and proceedings of this conference
33. M. G. Haines and J. P. Chittenden, Int. Conf. Plasma Physics, Innsbruck 1992; see also J. Phys. D. (to be published)
34. M. G. Haines and P. Choi, Bull. Am. Phys. Soc. **34**, 1945 (1989)
35. P. Choi, C. Deeney and C. S. Wong, Phys. Lett. A **128**, 80 (1988)
36. M. G. Haines, Nucl. Instru. Methods **207**, 179 (1983)
37. M. G. Haines, J. Plasma Phys. **12**, 1 (1974)

THE MAGPIE GENERATOR

I.H. Mitchell, J.M. Bayley, J.P. Chittenden, P. Choi, J.F. Worley
A.E Dangor and M.G.Haines.
The Blackett Laboratory, Imperial College of Science, Technology and Medicine,
London, UK.

ABSTRACT

The construction of MAGPIE, a terawatt pulsed power facility, has been completed at Imperial College, London. The generator consists of four 2.4MV, 86kJ Marx banks each feeding a 5Ω coaxial pulse forming line. The pulse forming lines are connected, via four trigatron switches, to a vertical coaxial transfer line and hence to the load. The generator is specifically designed to drive high impedance loads, enabling radiative collapse experiments to be carried out in cryogenic Hydrogen fibres. This requires the capability to deliver 1.5MA into a 100nH load in 150ns. A review of the project to date is given. This includes the design philosophy behind the generator, emphasising its unique aspects, and an outline of the tests which have been carried out to optimise its performance. Results from the first stage in the commissioning of the generator are also presented. This involved the firing of the complete generator, charged to 60% of maximum voltage, into a 150nH load. Currents of approximately 700kA have been achieved with an average of 7ns first to last for the four trigatron switches.

GENERATOR DESIGN

The primary objective of the MAGPIE project is to study the physics of radiative collapse[1] in a dense plasma produced from a cryogenic hydrogen fibre load. To reach the conditions leading to radiative collapse, a current exceeding the Pease-Braginskii limit of about 1.5 MA is required to be delivered to a fibre with an initial diameter between 10-100 μm. To deliver this current to such a highly inductive load in a reasonable time means that a high voltage in the Megavolt region is required. A figure of merit for the design of the pulsed power generator is that it should be capable of driving 1.5 MA into a 100 nH load in 150 ns. The decision to adopt a conservative design resulted in the adaptation of an existing 2.4 MV Marx bank design, used on the HERMES III Bremsstrahlung Generator at Sandia National Laboratories[2], as the high voltage generating unit. The high current is then produced by a series of pulse compression units, using water transmission line technology with an intermediate pulsed forming line and a transfer line. Technological consideration dictates that either a multichannel water switch is used to switch the large current required or that multiple gas switches in parallel would be required. Based upon consideration on reliability and good long term performance, a four gas switch system was decided to limit the maximum current through each high voltage switch to 500 kA. This resulted in the final design proposed using four pulse forming lines of 5 ohms impedance, each independently switched by a gas switch to a single transfer line with a matching impedance of 1.25 ohms.

A 200 ns pulse length was adopted for the pulse forming line, striking a compromise between the rate of energy delivery to the plasma load and pulsed power constraints. As it is necessary that the cryogenic load is produced vertically, a vertical transfer line was adopted with a nominal 65 ns transit time. The transfer line section was designed with sufficient flexibility so that future upgrade in the power level of the Generator could be implemented. Coaxial transmission lines are used throughout. These offer the important benefit of self screening against EMP, a very important consideration in a university environment. A schematic of the whole generator is shown in figure 1.

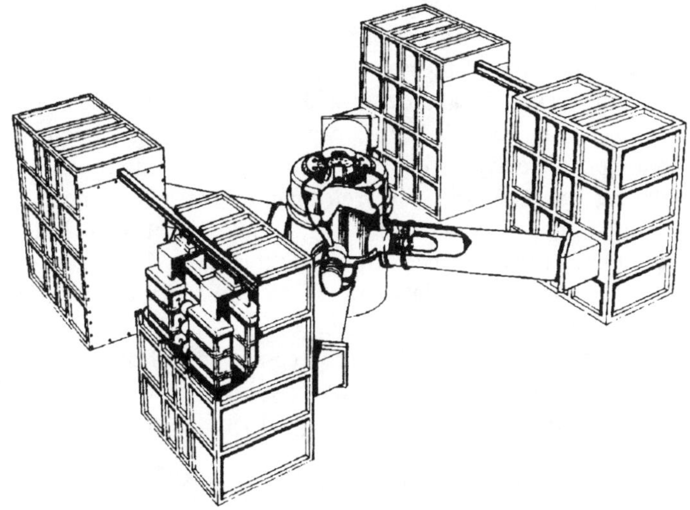

Fig 1 Schematic of the MAGPIE generator

An important and novel feature of MAGPIE is the junction of the four horizontal coaxial pulse forming lines (PFL) to the single vertical coaxial transfer line. Numerical simulation of this junction is a complex 3D problem which would have been very costly in terms of both time and money. Consequently, in order to investigate the junction, a 1/6th size scale model was constructed[3]. Low voltage pulses from a common source were launched along each PFL and voltage probes used to monitor the junction performance. It was observed that the waves launched into the transfer lines from the PFLs did not travel only in the desired upward direction, but instead were launched as spherical waves. Several designs of transfer line inners were tried in an attempt to minimise this and some were found to slightly increase the peak voltage at the load. This beneficial effect was not thought sufficient enough to warrant the extra risks and costs associated with the more complicated design.

At the top of the transfer line the power must be brought in from some 2 m diameter to only a few centimetres to couple into the fibre. It must also pass though a dielectric interface between the water and the vacuum in the load chamber. The electrode separation in the feed must be small in order to limit the inductance. A conventional 45° diode stack is used as the water-vacuum interface. The power flow problem is compounded by the detrimental effect of UV radiation on the voltage hold-off strength of this interface. The solution adopted was to use the principle of magnetic insulation to achieve the low inductance necessary for high rate of high currents and fast rises and a special geometry MITL (magnetically insulated transmission line) section was designed to address these issues.

MARX TESTS

The first construction phase of the project, involved the assembly of a single arm of the generator, incorporating a single Marx bank coupled to a 5Ω, 100ns water filled coaxial line and then to a trigatron switch. Two separate lots of tests were carried out during this phase. The first of these was a series of tests[4] on the Marx generator itself to establish erection, self break, delay and jitter characteristics. The Marx banks consist of twenty four 100kV, 0.7μF capacitors hanging in four columns of six, connected in a bipolar configuration. Two columns of six mid plane spark gaps carry out the switching, gap1 in this text referring to the gap connected to the 'earth' capacitor through to gap12 connected to the output capacitor. A trigger Marx capable of generating 500kV, 1.9kJ was used to trigger the main Marx bank. Resistive voltage monitors measured the trigger Marx output and the Marx output. The Marx output was connected to a liquid $CuSO_4$ resistor of 100Ω. Two fibre optic cables were connected to each spark gap, each fibre looking at one half of the gap, ie. between the trigger and the main electrode. The outputs from these fibres were incident upon photo diodes which were then connected to a digital oscilloscope. In this way the light emitted from each gap could be monitored.

The effect of polarity and magnitude of the trigger pulse on the Marx erection was investigated. Increasing the magnitude of the trigger Marx above 300kV had little beneficial effect on the Marx erection and did not warrant the resulting extra stress on the trigger Marx itself. Hence the majority of tests were carried out with a 300kV trigger. The triggering arrangement was such that the trigger electrodes of the first column of gaps, gaps 1 to 6, were triggered. The trigger electrodes of the second column, gaps 7 to 12, were left floating. (Two amendments were made to this set-up as a result of these tests and shall be detailed later.) Unlike that of the magnitude of the trigger pulse, a change of its polarity had a marked effect on the Marx performance. This is illustrated in figure 2 where the delay between the arrival of the trigger pulse at gap1 and the peak of the output voltage is plotted as a function of percentage of self break for both positive and negative trigger.

It can be seen that, especially at lower percentages of self break, the erection time is significantly reduced with a positive trigger. An insight into the reason for this difference was obtained from the fibre optic outputs. In the case of -ve trigger, the erection sequence begins with gap2 and a rather haphazard, but reproducible, sequence of erection of the first column leads to a sequential erection of the second column. For a +ve trigger, however, gap1 is the first to close, followed by a neat staircase erection through the rest of the gaps. The overall delay is, not surprisingly, reduced by up to 30%.

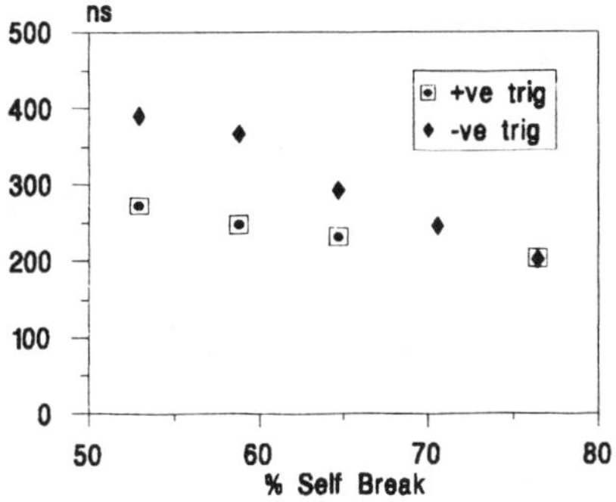

Fig. 2. Marx delay for +ve and -ve trigger.

One detrimental effect of positive polarity triggering is a significant increase in the field stress across the trigger resistors when the Marx erects, which in the case of gap6 is severe enough to lead to failure. Consequently this resistor was removed and replaced by a (longer) resistor tying the trigger electrode to earth, resulting in safe operation. A similar resistor was connected between the trigger electrode of gap7 and earth when it was observed from the fibre optic output, that an increase in delay of some 200ns in a few anomalous shots was due to a 'hang up' of the erection sequence between gap6 and gap7. The first of these two alterations had no detrimental effect on the Marx performance and the second removed the associated problem completely. As a result of these tests, a reliable Marx bank with optimised performance has been obtained.

SWITCH TESTS

The second of the series of tests[5,6] was carried out on the trigatron switches with the object of determining the optimum operating conditions and working lifetime of the switch. These involved the installation of the pulse forming line and the switches. Again a $CuSO_4$ resistor was used as the load, this time with a resistance of 5Ω and therefore matched to the PFL. A schematic diagram of a trigatron switch is shown in figure 3. The switch consists of three electrodes, the earth electrode, the live electrode and the trigger pin

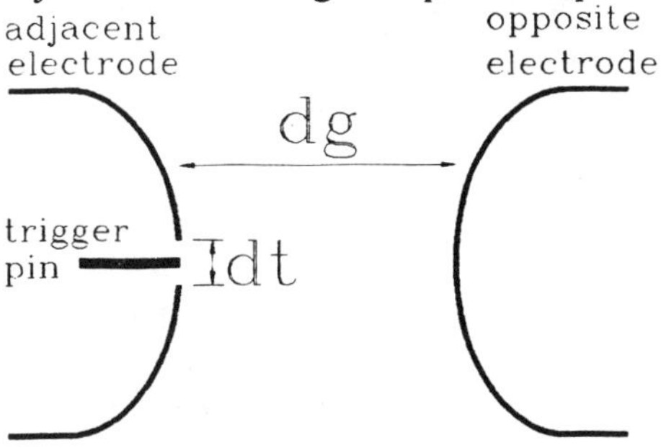

Fig. 3. Schematic diagram of a trigatron.

and two gaps, the main electrode gap, d_g, between the live and earth electrode and the trigger gap, d_t, between the trigger pin and the adjacent, normally earth, electrode. Operation of the trigatron is obtained by applying a trigger pulse to the trigger pin. Closure of the switch can take place via one of two routes depending upon the magnitude of the trigger voltage, V_t. When the electric field across the trigger gap, E_t, is larger than that across the main gap, E_g, the voltage on the trigger pin can cause a breakdown to the adjacent electrode (BAE) with the resulting UV and charged particles initiating the discharge across the main gap. In contrast, when the electric field across the trigger gap is much less than that across the main gap, the voltage on the trigger pin can create a distorted and enhanced field in conjunction with the applied voltage across the main gap. This leads to the formation of breakdown streamers directly across to the opposite gap (BOE).
A series of experiments was carried out to investigate the operation of the trigatrons. A block diagram of the apparatus is shown in figure 4.

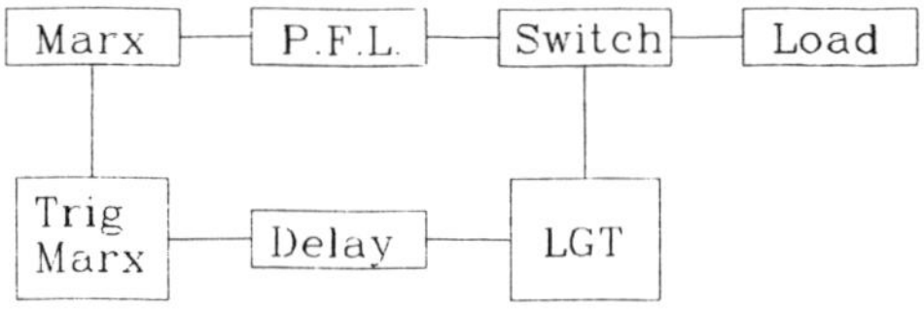

Fig. 4. Block diagram of experimental apparatus.

The output from the trigger Marx is monitored and this signal is used, via a delay box to trigger the line gap trigger (LGT). This consists of a single 0.7μF capacitor and spark gap. The output of this unit can be varied from 20 to 100kV and is fed, via a length of RG218, 50Ω cable to the trigatron trigger pin. The LGT output voltage is monitored by a resistive divider before the feed cable. The signal from this monitor indicated in which of the two modes the trigatron was operating.

a) BOE b) BAE

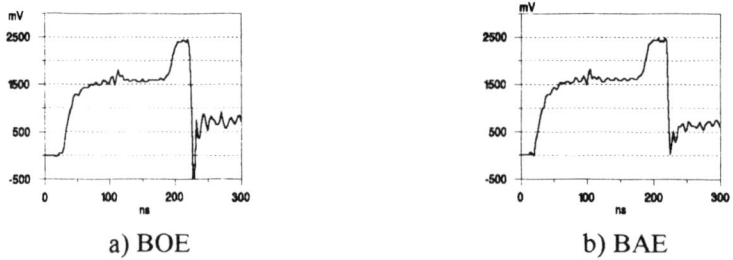

Fig. 5. LGT output traces.

Figure 5 shows two LGT monitor traces. They show the rise in voltage when the unit is triggered and, after a double transit time of the cable, a second rise which is due to the doubling of the pulse upon reaching the trigger pin. From here on the two traces differ. Trace a) shows that the voltage on the trigger pin drops rapidly, well below zero. This is indicative of the BOE mode of operation where the initial breakdown is to the opposite, negatively charged electrode. In trace b), however, the voltage only drops to zero which, after the same argument, is indicative of the BAE mode. The voltage and current in the PFL were monitored using a capacitive probe and a Bdot coil and were used to indicate closure of the switch.

It has been suggested[7] that the optimum operating conditions for the trigatron would occur when both these modes operated simultaneously which would occur if the average electric fields across the trigger and main gaps are equal, i.e. $E_t=E_g$. This requires an optimum trigger voltage, V_t^*, with a magnitude which depends only on the operating voltage and the geometry of the switch which is given by equation 1.

$$V_t^* = -V_g d_t/(d_g-d_t) \tag{1}$$

A series of sixty consecutive shots was obtained with the trigger voltage scanning the optimum voltage V_t^* of 50kV, and keeping the line voltage constant at -1.5MV. Twenty shots were made at each trigger voltage, the results being summarised in Table I below.

Table I Switch performance for varying trigger voltages.

V_t (kV)	E_g (kV/cm)	E_t (kV/cm)	Δt (ns)	σ (ns)
36	161	116	83	16
50	163	161	60	7.5
66	165	213	64	12.6

It can be seen that the minimum delay, Δt, and jitter, σ, was obtained for the case of the trigger voltage equal to V_t^*, i.e. when $E_t=E_g$. In the cases of V_t set at 36 and 66kV, the mode of operation was BOE and BAE respectively. For the case of $V_t=V_t^*$, the switch was observed to operate predominantly in the BAE mode although for some 12% of the shots BOE operation was observed.

An investigation of the switch ageing characteristics under the optimum triggering conditions was made. A switch was operated at 2MV, for 200 shots at approximately 80% of self break. The statistics over all of these shots are shown in Table II below. The running average and standard deviation of the switch jitter are shown in figure 6. The degradation of the switch performance is believed to be due mainly to the erosion of the trigger pin and the surrounding rim of the hole in the electrode.

Table II Statistics for optimally triggered switch.

V_t (kV)	E_g (kV/cm)	E_t (kV/cm)	Δt (ns)	σ (ns)
68	218	219	43	4.4

From these data it was concluded that, under these operating conditions, the switch could be used in excess of one hundred shots before refurbishment should be required.

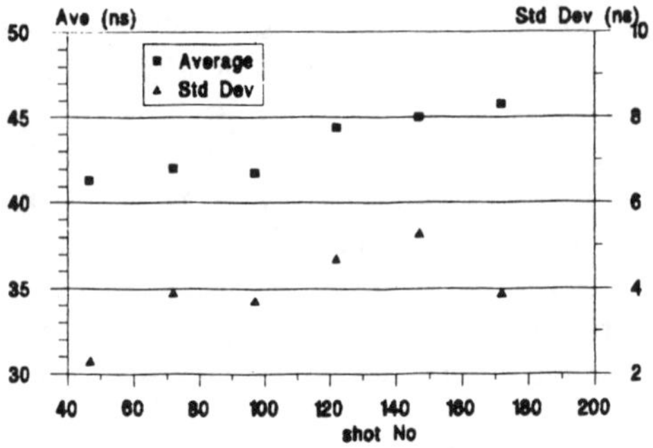

Fig. 6 Performance of optimised trigatron switch.

FULL GENERATOR TESTS

After the switch tests the second, and main, construction phase of the project was undertaken. This involved the erection of the three remaining arms of the generator and of the central vertical transfer line along with the ancillary charging, oil, water and vacuum systems. The initial commissioning tests involved testing the erection of each individual Marx at up to 60kV charge. The Marxes were discharged either into a discrete 90 Ω $CuSO_4$ resistor in place of the line switch or into a distributed ~75 Ω resistance obtained by using slightly conducting water in the pulse forming lines.

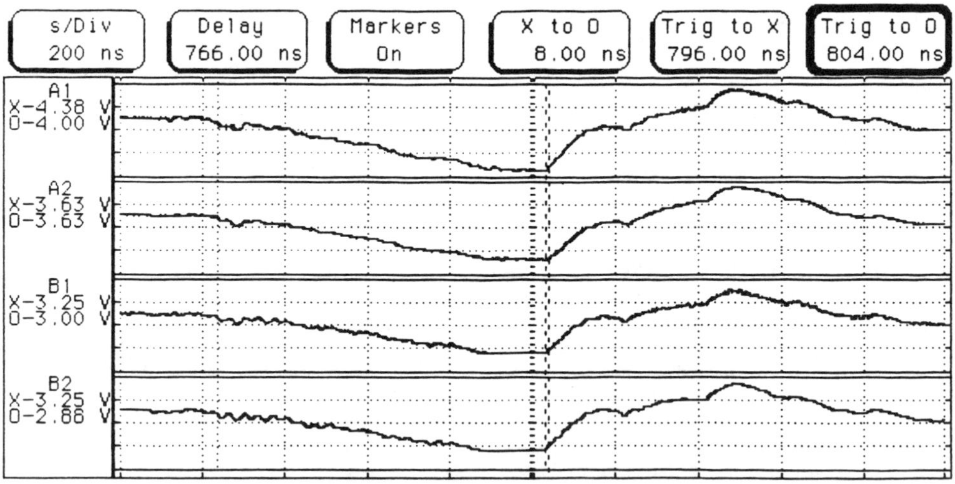

Fig. 7. Voltage monitors at each Marx showing switch closure within 8ns first to last.

Having established the operating conditions for the four Marxes we proceeded to test all four Marxes with all four trigatron line switches. A series of shots at 50kV and

60kV charge were fired with contaminated water in the transfer line providing a 3 Ω resistance down to earth (no vacuum section being present).

By operating with the optimal trigger voltage, V_t^*, a mean first to last for four line switches of 10.8 ns over 9 shots was obtained. A signal on a Rogowski coil located at the far end of each switch is taken as an indication of switch closure. Four Rogowski coil traces from a single shot are shown in figure 7 demonstrating a first to last separation of 8ns. It was found that differences in the preparation of the four switches, particularly in locating the trigger pin, could cause systematic differences in switch delay. By controlling the pressure in each individual switch, the systematic part of the jitter could be eliminated giving a mean first to last of 7.0 ns over 8 shots.

By placing a 150nH inductance (a 4 inch diameter copper tube) coaxially at the top of the transfer line, we were able to produce current levels (~700kA) approaching those expected with an MITL and plasma load. The 3 Ω distributed resistance in parallel with this inductor reduces the peak current by 16% from the loss-free value but provides a convenient dump for the energy over long timescales (~4 μs).

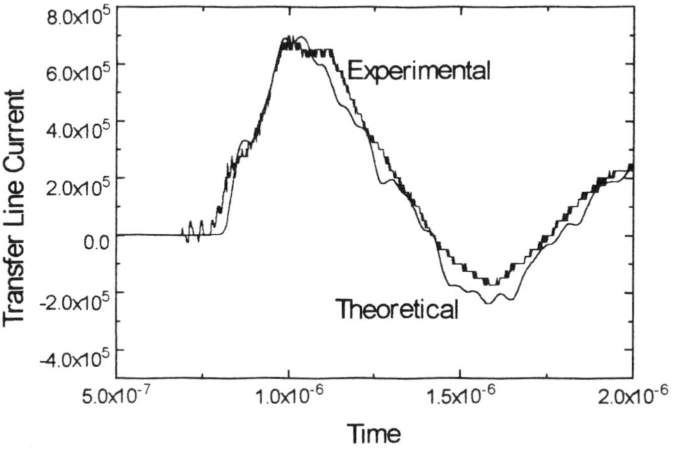

Fig. 8. Comparison of experimental and theoretical transfer line current.

Currently, the machine diagnostics include; resistive divider voltage measurements at each Marx trigger input, each Marx output, each line switch trigger cable and the trigger marx plus Rogowski dI/dt and capacitive divider voltage probes at the PFL-switch interface, the side of the transfer line and the top of the transfer line. By comparison of these probes with a detailed simulation of the generator using the Screamer model, we find that the PFL / transfer line junction is behaving approximately as a simple four into one junction and that full power is being transferred around the right angle bend at the top of the transfer line and on to the load on axis. Figure 8 shows the good agreement obtained in comparing the measured and the theoretical current half way up the transfer line. Replacing the 150nH inductor with an MITL and plasma (~80nH) in a Screamer simulation, we anticipate I ≈ 1 MA for a 60kV charge.

CONCLUSION AND FUTURE WORK

The first stages of the commissioning of the MAGPIE generator have been completed. Extensive testing of a single arm of the generator resulted in a reliable optimised Marx and trigatron switch performance. Positive polarity triggering of the Marx was shown to lead to a neat staircase erection sequence and an associated decrease in Marx delay. Tests on the trigatron switches showed that an optimal performance at the required working conditions was obtained when the trigger voltage was arranged such that the electric fields across the trigger pin to live electrode and the trigger pin to earth electrode gaps were equal. With such a trigger voltage the rms jitter on a single switch over 140 shots was 4.4ns. Using the results from the single arm tests, the completed generator has been tested at 60% of charge voltage. The generator is performing as per simulation and has delivered 700kA into a 150nH load.

Future work will involve the testing of the vacuum load section. The performance of the vacuum diode stack and the MITL into a short circuit non-plasma load will be investigated and compared with SCREAMER simulations. Upon satisfactory results the first plasma physics experiments will be carried out on the generator. These experiments will be with carbon fibres, extending previous experiments at Imperial college[8] from 100kA to the Mega-Ampere regime.

REFERENCES

1. J.P. Chittenden and M.G. Haines, Phys. Fluids B **2** (8) 1889 (1990)

2. J.J. Ramirez et. al., Digest of Technical Papers, 6th IEEE Pulsed Power Conf, Arlington, Virginia, (IEEE, New York. 1987)

3. I.H. Mitchell, P. Choi, J. P. Chittenden and J. F. Worley, (to be published in Proc. 9th Int. Conf. on High Power Particle Beams, Washington DC, USA, 1992)

4. P. Choi, J. Chittenden, I.H.Mitchell, J. Worley, J. M. Bayley, R. Bialecki, A.E. Dangor and M. G. Haines, Digest of Technical Papers, 8th IEEE Pulsed Power Conf, San Diego, California, (IEEE, New York. 1991), p.173.

5. I.H. Mitchell, P. Choi, J. P. Chittenden and J. F. Worley, Proc. 10th Int Conf. on Gas Discharges and their applications, Swansea U.K., 82, (1992)

6. I.H. Mitchell, J. P. Chittenden, P. Choi, J. F. Worley, J. M. Bayley, A.E. Dangor and M. G. Haines,(to be published in Proc. 9th Int. Conf. on High Power Particle Beams, Washington DC, USA, 1992)

7. T.H. Martin, Digest of Technical Papers, 7th IEEE Pulsed Power Conf, San Monteray, California, (IEEE, New York. 1989),p.173.

8. M.G. Haines et al, Plasma Physics and Controlled Nuclear Fusion Research, 2 (IAEA Vienna, 1990), p.769

CARBON FIBRE Z-PINCH

S.L. Niffikeer, F.N. Beg, A.E. Dangor and M.G. Haines
The Blackett Laboratory, Imperial College
London SW7 2BZ, UK

G.H. McCall
Los Alamos National Laboratory
Los Alamos NM 87545, USA

ABSTRACT

We report observations of a 7 μm diameter, 20 mm long carbon fibre Z-pinch driven by a pulsed power generator delivering 85-100 kA with a 50 ns (10 -90 %) rise time. Time integrated soft x-ray pinhole camera images show that the emission is from a series of bright spots distributed non-uniformly along the pinch. The spots are about 100 μm in size and at a temperature of about 100 eV. The tenuous plasma between the spots is at higher temperatures of about 150 eV. x-ray streak photography shows that bright spots do not emit simultaneously or in any particular spatial sequence. The first bright spot emits usually at about 30 ns. Each spot emits for a time less than 1 ns. Harder x-rays are emitted from the anode. In all cases the bright spot emission is seen to occur only during the harder x-ray pulse. Two pulses are seen. The first of 20 ns duration is observed at current initiation and the second 30-100 ns long occurs later at about 20-25 ns. Analysis of the x-ray emission from the anode shows that the first pulses is consistent with a 150 keV, 3 kA electron beam and the second pulse with a beam of about 20 keV, 20 kA. The bright spots are also seen in optical streak photographs. The emission is also transient - the duration of the brightest emission is of the order of 1 ns - and the bright spots move axially. They sometimes bifurcate, occasionally combine. Observations with a 7 ns ruby laser show that the fibre pinch expands radially with a velocity >2.5 10^4 ms^{-1}. Islands of high density form at about 8 ns. These islands are spatially coincident with the x-ray bright spots and persist for a longer time (30 ns to 50 ns). The radial acceleration of the pinch becomes inward at 8 ns, suggesting that Rayleigh-Taylor instability may be responsible for the instability.

INTRODUCTION

The introduction of high voltage pulsed power drivers with fast current rise has lead to a renewed interest in Z-pinch research. Also there is now a better theoretical understanding of the stability properties of the pinch[1]. Recently a number of experiments on pinches generated from fibres have been carried out.

Deuterium fibre experiments have been reported by Hammel & Scudder[2] and Sethian *et al.* [3,4]. Hammel and Scudder[2] used fibres with diameters from 20 to 40 μm. The pinch was driven by a current with a rise time (10-90%) of about 150 ns to a current maximum of 250 kA. At 80 ns $m = 0$ instability was observed which then subsequently disrupted the pinch. The measured temperature was 150 eV which is consistent with that calculated from Bennett's relation. In the experiments by Sethian *et al.* [3,4] linearly rising currents were used (maximum 640 kA in 130 ns, 5 10^{12} A s^{-1}, and 920 kA in 840 ns, 2 10^{12} A s^{-1}). In both experiments the pinch was observed to be stable during the current rise becoming unstable only when dI/dt changed sign. This was independent of the magnitude of the current. The stable period was calculated to correspond to about 100 MHD instability growth times in both experiments.

High Z fibre experiments have been reported by a number of workers. Young *et al.* [5] investigated the neutron production in deuterated carbon fibres with a current of 0.7 MA. Yields greater than 10^{10} were observed. For large diameter fibres greater than 50 μm the neutron emission was thermal whereas for fibres less than 25 μm diameter the emission increased greatly and was due to energetic ion collisions. Ions of energy of the order of 8 MeV were inferred.

Zakharov *et al.* [6] studied discharges in tungsten with currents up to 150 kA. Schlieren observations showed that the pinch column was non uniform consisting of a series of constrictions which coincided with X-ray bright spots. The bright spot emission decreased with larger diameter fibres and was maximum for a fibre pinch length of 5 mm. The electron temperature and density was estimated to be 0.2 - 1 keV and 10^{22}-10^{23} cm^{-3} respectively.

Figura *et al.* [7] used a quartz fibre ranging in diameter from 20 to 125 μm driven by a 150 kA current with a rise time of 50 ns. An $m = 0$ instability was seen to develop during the current rise and at a time when the whole fibre had not yet been fully ionised. Optical bright spots associated with the instability were observed to expand radially and move axially. A coronal plasma was found to exist with density of the order of 10^{16} cm^{-3} at a radius of 500 μm. The bright spots emitted low energy x-rays compared with the rest of the plasma column. The x-ray emission from the anode indicated the existence of an electron beam.

More recently, a Z-pinch experiment with enhanced stability has been reported by Wessel *et al.* [8]. In this experiment a fibre was enclosed within an aluminium metal puff. The observations indicated that a high density, submillimeter diameter pinch was formed on axis which remained stable for 20-40 ns. This time is longer than the instability growth observed in the fibre only discharges or aluminium puff discharges only.

In the experiment reported here, carbon fibre of 7 μm diameter was used. Of interest is the possibility of ionising the fibre completely before any instability occurs. Haines and Coppins[9] showed that different regimes of stability theory can be represented in a universal plot of $I^4 a$ vs. N, where I is the current through pinch, a is the pinch radius and N is the ion line density. In this plot ideal MHD theory occupies only a small wedge shaped parameter space. The other regimes delineated in the universal plot are where resistivity, viscosity and finite ion Larmor radius effects are important. All these effects are likely to lead to enhanced stability. In the present experiment $I = 100\ kA$, $N = 4.2\ 10^{18}$ m^{-1}. Thus we expect resistive effects to be important provided the pinch diameter is less than 10 μm assuming $Z = 3$ in the universal plot.

EXPERIMENTAL DETAILS

The generator, shown schematically in Fig. 1, consisted of a high voltage Marx bank and a water dielectric 3 Ω, 75 ns single transit, coaxial pulse forming line. The line was switched by a sulphur hexaflouride gas insulated self-breakdown spark gap into a matched transfer line and then to the Z-pinch load. The anode and cathode, between which the fibre was suspended, was separated by 20 mm.

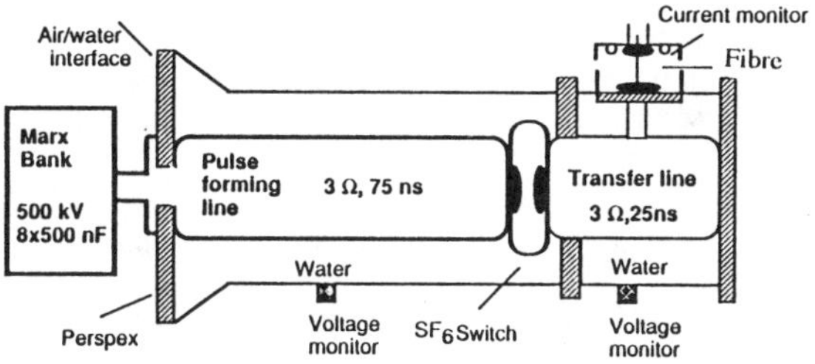

Fig. 1. Schematic of the current generator

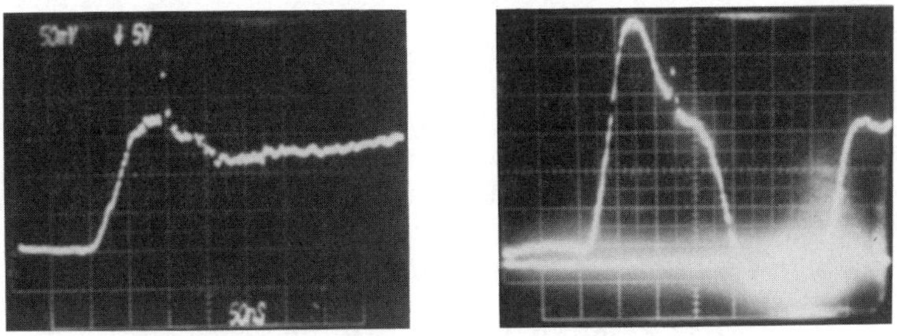

a b

Fig. 2. Current through (a) carbon fibre and (b) short circuit. (Time = 50 ns/div., current = 28.6 kA/div. and 17 kA/div. respectively)

DIAGNOSTICS AND EXPERIMENTAL RESULTS

The x-ray emission from the pinch was observed with pinhole cameras. For time integrated observations the x-ray image was recorded on film; for time resolved observations an x-ray streak camera was used, the pinhole image of the pinch being streaked perpendicular to the axis, so that the evolution of the pinch along the axis was observed. The corresponding image in the optical region was studied with an optical streak camera. Hard x-ray emission was investigated with a scintillator/photomultiplier detector and an array of quartz fibre dosimeters. In addition, the time evolution of the spatial distribution of the electron density in the pinch column was obtained by a holographic interferometry using a technique based on a 7 ns ruby laser.

The typical current waveform in the 7 μm carbon fibre pinch is compared with the short-circuit current in a 2 mm diameter steel tube in Fig. 2. The measured rise time in the pinch is 50 ns which is slightly longer than that in the short-circuit (45 ns). At 150 ns, the pinch current crowbared while the short circuit current reversed direction. We conclude that the crowbar is due to a breakdown across the perspex insulator interface which separates the pulse forming line from the pinch chamber.

Time Integrated X-ray Emission

Double and triple pinhole cameras were used. Each pinhole was separately filtered and the field of view was sufficiently large to include the anode. The images were recorded on Kodak DEF x-ray film.

A typical double pinhole image is shown in Fig. 3. The pinhole images showed that superimposed on a background pinch column is a series of bright spots distributed non-uniformly along the pinch. The typical dimension of the bright spots is 100 μm. The filters used to obtain the images were chosen to transmit only the continuum and recombination bremsstralung from carbon plasma. From analysis of these images, the temperature distribution along the axis shown in Fig.3 was obtained. Clearly, the temperature of the tenuous background plasma column (~150 eV) is greater than that in the bright spots (~100 eV).

Figure 3 shows that more intense and harder x-ray emission was seen from the anode. Assuming that the emission is due to mono-energetic electrons colliding with the stainless steel electrode, an energy of 20 keV and current of 20 kA was inferred for the electron beam. Details of the calculation for the electron beam characteristics are given[10].

Time Resolved X-ray Streak Photography

A Kentech X-ray streak camera fitted with low density cesium iodide photocathode was used. A large pinhole (300 μm diameter) was used with a 2 μm mylar coated with 80 nm aluminium filter, which transmits H-like and He-like line radiation. The magnification was 0.7 allowing observation of the anode emission. A typical streak photograph is shown in Fig. 4. This clearly shows that the bright spots did not emit simultaneously, the first occurring at about 30 ns. The bright spot emission was short lived, typically of duration less than 1 ns. The emission from the anode occurred about 5 ns before the emission from the bright spots. Evidently the bright spots emit only during emission from the anode.

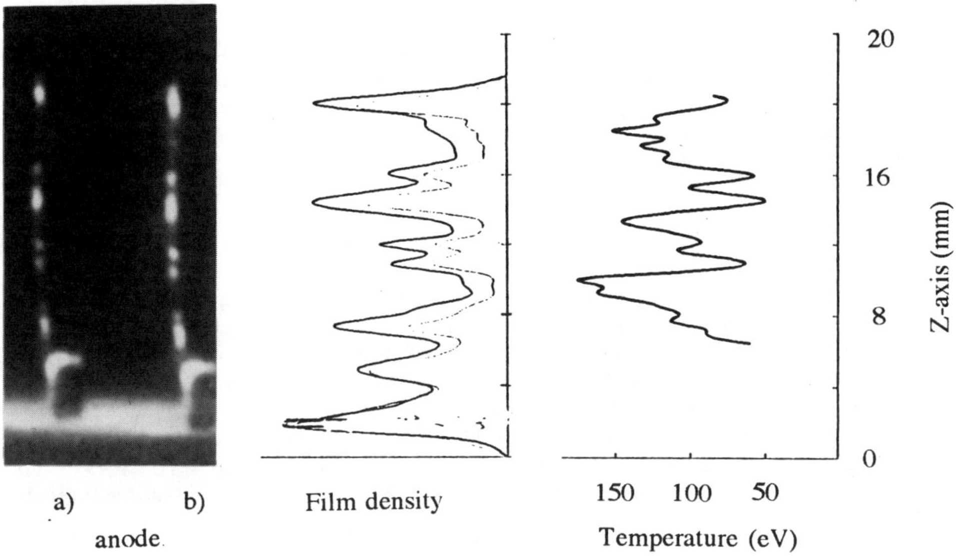

Fig. 3. X-ray double pinhole camera image of the carbon fibre pinch and the distribution of the electron temperature along the axis.[Filters a) 6 μm mylar + 1400 nm aluminium and b) 6 μm mylar +250 nm copper + 800 nm aluminium]

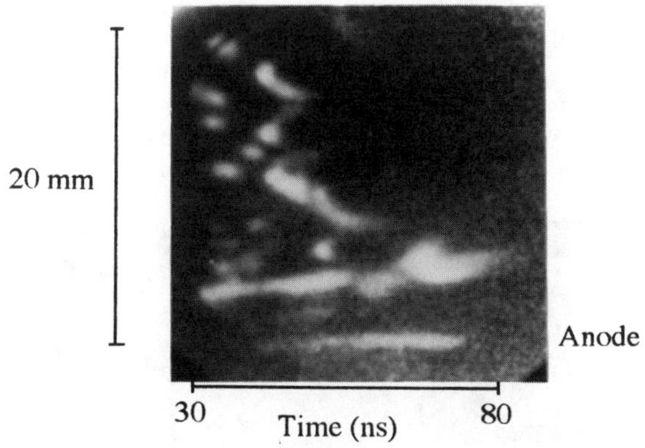

Fig. 4. A x-ray streak.

Hard X-ray Emission

A NE102A plastic scintillator coupled to a photomultiplier was used to monitor hard x-ray emission. The detector was placed outside the 15 mm aluminium wall of the chamber and a 10 mm thick iron filter was used and thus only hard x-rays greater than 100 keV were detected. A large signal was observed only at early times during current initiation and is presumably due to energetic electrons generated by field emission before breakdown of the fibre. The quartz fibre dosimeters measured a radiation dose of 790 μR. Assuming that emission is due to a monoenergetic beam and the beam pulse duration was 30 ns, an electron beam energy of 150 keV and current of 3 kA can be inferred.[10]

Optical Streak Photography

An Imacon 675 streak camera equipped with a S20 photocathode was used. As in the x-ray streak photographs, an extended length of the pinch column was focused onto the photocathode of the camera and streaked perpendicular to the axis. A typical photograph is shown in Fig. 5. This shows that an axial structure in the form of bright emitting spots distributed along the axis developed after 8 ns. The emission from the pinch column is brighter nearer the cathode. There is movement of the bright spots along the axis in either direction the average speed being about 4 10^4 m s^{-1}. The emission from the bright spots varied in time, the brightest emission lasting typically 1-2 ns. The streak photograph shows that some bright spots bifurcated.

Interferometry

A double exposure holographic interferometric technique was used. A 7 ns (FWHM) ruby laser (694.3 nm) was available for these measurements. The laser beam was transverse to the pinch and thus the interferogram obtained after reconstruction had to be Abel inverted to obtain the density distribution as a function of radial and axial position. Details of the interferometric measurements can be found in Niffikeer[10]. Figure 6 shows a sequence of axial interferograms at different times. The first at 4 ns shows a narrow channel of about 50 μm radius. Assuming uniform expansion from the initial 7 μm diameter, this corresponds to an expansion velocity of 1.3 10^4 m s^{-1}. The interferogram at 12 ns shows that the pinch has expanded more nearer the cathode. At 30 ns large scale perturbations are evident in the plasma column. At about 45 ns the pinch column has broken up into series of density islands.

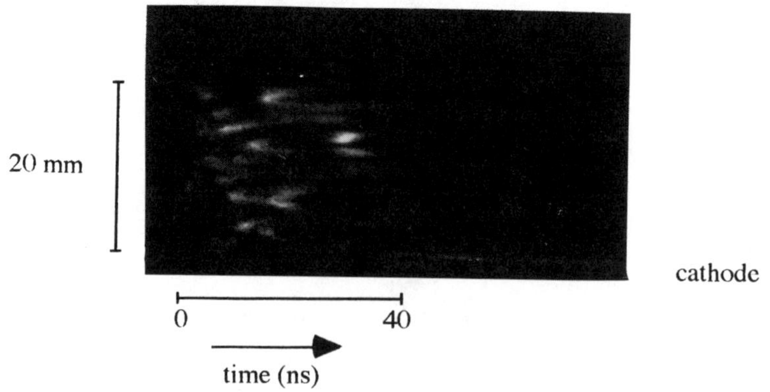

Fig. 5. An optical axial streak.

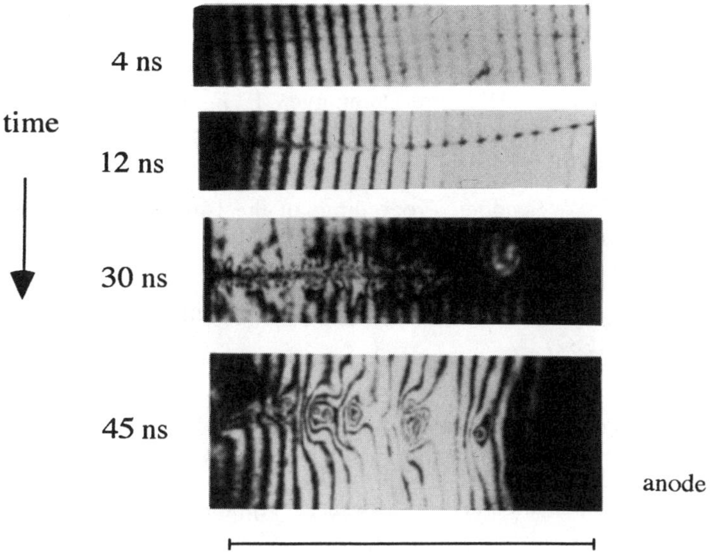

Fig. 6. A sequence of interferograms .

DISCUSSION

For the 7 μm diameter fibre, the line density $N = 4.2\ 10^{18}\ m^{-1}$. Assuming $Z = 6$, for a current $I = 100\ kA$, the Bennett temperaure is 110 eV. This is in good agreement with the 100 eV and 150 eV measured respectively in the bright spots and in the background plasma column. It is thus likely that the fibre is fully ionised.

The pinch is observed to be unstable at an early time when the current is still increasing. This is in disagreement with the observations reported by Sethian *et al.* [2,3]. At the time of first appearence of the instability the current $I = 20\ kA$ and the measured radius is $a = 300\ \mu m$. According to the universal stability plot modified for $Z = 3$, the pinch should be resistively stable at this time. That this is not the case, indicates the existence of other instability mechanisms. But the plasma radial profile is non-uniform in temperature; the coronal plasma can be cosidered as a hotter plasma at a lower line density and the instability could be MHD. Optical streak photography shows that the radial acceleration of the pinch becomes negative at about 10 ns. Instability is first seen at about 8 ns, this strongly suggests that the instability is due to Rayleigh-Taylor instability.

The instability is observed only during the existence of an electron beam. Electron beams were observed in a number of Z-pinch experiments such as quartz fibre pinch[8] gas puff pinch[12,13] and the plasma focus[14]. Indeed, our observations show that the electron beam is of longer duration than the bright spot emision. This suggests that the mechanism for the electron beam production is not associated with the generation of an electric field during the formation of an $m = 0$ instability.

REFERENCES

1. M.G. Haines, A.E. Dangor, A. Folkierski, P. Baldock, C.D. Challis, P. Choi, M. Coppins, C. Deeney, M.B. Favre-Dominguez, E. Figura and J.D. Sethian , in Plasma Physics and Controlled Nuclear Fusion Research, 1986 (IAEA, Vienna, 1987), Vol.2, p.257.

2. J.E. Hammel and D.W. Scudder, Proceedings of the 14th European Conference on Controlled Fusion and Plasma Physics (EPS, Petit- Lancy), Switzerland, pt.2 , p. 450 (1987).

3. J.D. Sethian, A.E. Robson, K.A.Gerber, and A.W. DeSilva, Phys. Rev. Lett. **59**, No. 8, p. 892 (1987).

4. J.D. Sethian, A.E. Robson, K.A. Gerber and A.W. DeSilva, Proceedings of Alternative Magnetic Confinement Schemes edited by Orlanlani S. and Sindoni E., Varena, Italy, October 1 1990.

5. F.C. Young, S.J. Stephanakis and D. Mosher, J. Applied Phys. **48**, No.9 (1977).

6. S.M.Zakharov, G.V. Ivanenkov, A.A. Kolomenskii, S.A. Pikuz and A.I. Samokhin, Sov. J. Plasma Phys. **9**, No.3 (1983).

7. F.J. Wessel, B. Etlicher and P. Choi, Phys Rev. Lett. **69**, No.22, p. 3181 (1993).

8. E.S.Figura, G.H. McCall and A.E. Dangor, Phys. Fluids B**3** (10), p. 2835 (1991).

9. M.G. Haines and M. Coppins, Phys. Rev. Lett. **66**, No. 11, p. 1462 (1991).

10. S.L. Niffikeer, Ph.D. thesis, University of London (1991).

11. F.C. Jahoda and R.E. Siemon , Los Alamos Report LA- 5058- MS (UC-37) 1987.

12. D.R. Kania and L.A. Jones, Phys. Rev. Lett. **53**, 166 (1984).

13. C.D. Challis, P.Choi, A.E. Dangor and M.G. Haines, in Proceedings of the 12th European Conference on Controlled Fusion and Plasma Physics(EPS, Petit- Lancy, Switzerland, 1985), pt. 2, p. 450.

14. P.Choi, C. Deeney and C.S. Wong, Phys. Lett. A **128**, p. 80 (1988).

Fiber Z Pinch Instabilities

D. W. Scudder, J. S. Shlachter, P. R. Forman, R. A. Riley
Los Alamos National Laboratory, Los Alamos, NM 87544

R. H. Lovberg
University of California at San Diego, La Jolla, CA 92093

ABSTRACT

We have studied the development of instabilities, primarily $m = 0$ modes, in CD_2 fiber z pinches. Two simultaneous optical diagnostics have been used in order to obtain an unambiguous picture of the dynamics. The first is two-dimensional interferometer using a short-pulse laser. The second is multi-frame two-dimensional schlieren/shadowgraphy camera. The interferometer serves as a check on the schlieren/shadowgraphy camera helping to overcome uncertainties in interpretation of the schlieren/shadowgraphy results. An x-ray polychromator is used to measure electron temperature. Growth of the unstable modes is compared with predicted growth rates. There are indications that some forms of current prepulse can have a significant effect on the long-term development of instabilities.

INTRODUCTION

Linear z pinches have been studied for several decades for applications ranging from controlled fusion to intense radiation sources. Most z pinches have exhibited robustly unstable modes, which have severely limited their usefulness. Although there have been a number of experiments in which z pinches have been reported to exhibit more stability than expected,[1,2,3] the stabilizing mechanisms have remained obscure or at least controversial and attempts to scale the results have met with limited success. The goal of these experiments has been to perform careful diagnostics of instability behavior in fiber z pinches to better understand the linear and non-linear behavior of these plasmas. Ultimately, we would like to identify stabilizing effects, although we can report only limited success at this time.

Earlier experiments by this group have used the multi-frame schlieren/shadowgraphy technique reported here. Those measurements suffered from ambiguity because of the threshold nature of the technique.[4] In order to improve the value of those measurements, we have added a spatially resolved interferometer observing the same region of the plasma at the time as one of the schlieren/shadowgraphy images. This allows us to calibrate the sensitivity of the schlieren/shadowgraphy technique eliminating much of its ambiquity.

EXPERIMENT

The plasmas for this study are produced by a high-voltage pulsed-power machine. A marx bank storing up to 210 kJ in 41 nF at 3.2 MV is discarged through an

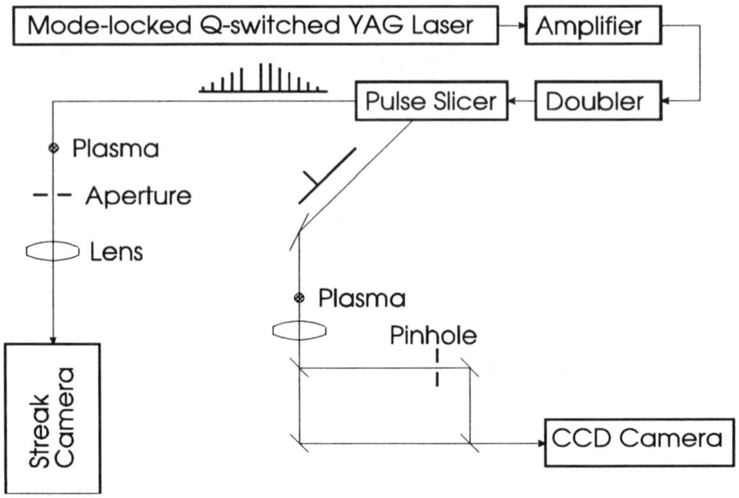

Figure 1. Schematic of the optical diagnostics.

intermediate storage capacitor into a 100 ns, 1.9-Ω pulse line. The pulse line switches into the load through a self-breaking water switch. The load consists of a 5-cm long fiber of CD_2 whose diameter ranges from 15 to 100 μm.

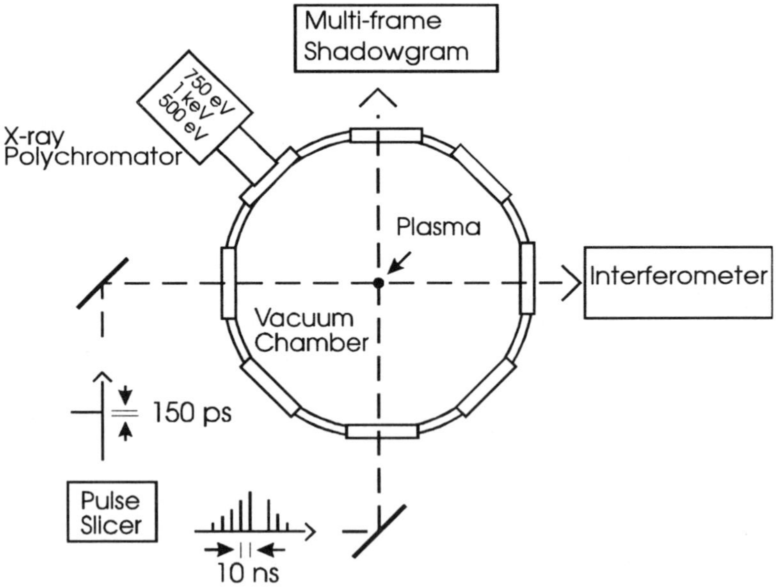

Figure 2. Arrangement of the diagnostics on the experiment.

The optical diagnostics share a mode-locked, Q-switched, frequency-doubled YAG laser which produces a burst of pulses each about 150 ps long separated by 10 ns. Figure 1 is a schematic of the optical diagnostics. The schlieren/shadowgraphy camera obtains multiple images by recording them on a streak camera, making it, in effect, a fast framing camera. The interferometer takes a single pulse from the pulse train, sends it through the plasma and passes it through a point-diffraction interferometer. Its image is recorded on a CCD camera with a frame grabber in a personal computer.

Figure 2 shows the arrangement of the optical and x-ray diagnostics on the pinch chamber. The interferometry and schlieren/shadowgraphy observe the plasma at 90° to one another. An x-ray polychromater observes soft x rays emitted by the plasma in three energy slices. Each channel uses an x-ray crystal set at an angle appropriate to the desired energy.

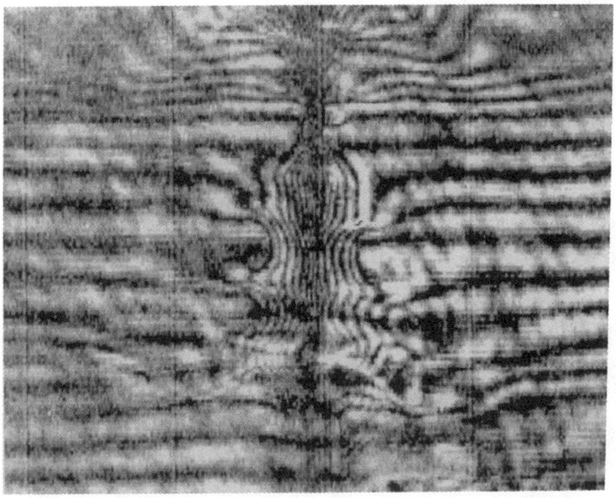

Figure 3. Interferogram of shot 367 taken just at the end of the prepulse.

IMAGING RESULTS

Figures 3 and 4 are an interferogram and a multi-frame schlieren/shadowgram from a shot with vigorous m=0 mode growth. Figure 5 is the current trace for this shot. The dim frame in the schlieren/shadowgram sequence is due to the pulse switched out to the interferometer. Comparison of

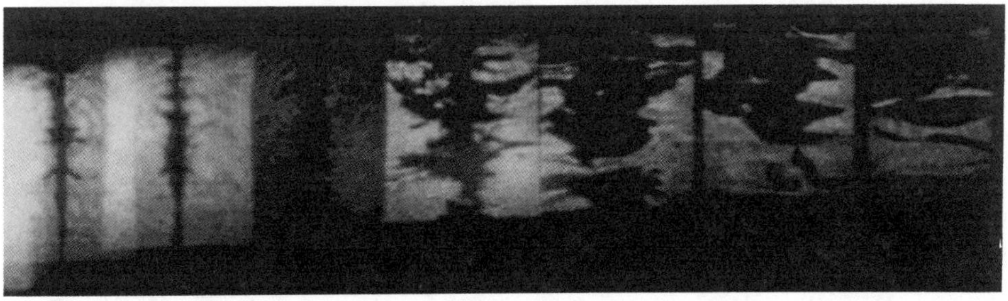

Figure 4. Schlieren/shadowgram for shot 367. The dim frame is 3 ns before the interferogram in Fig. 3.

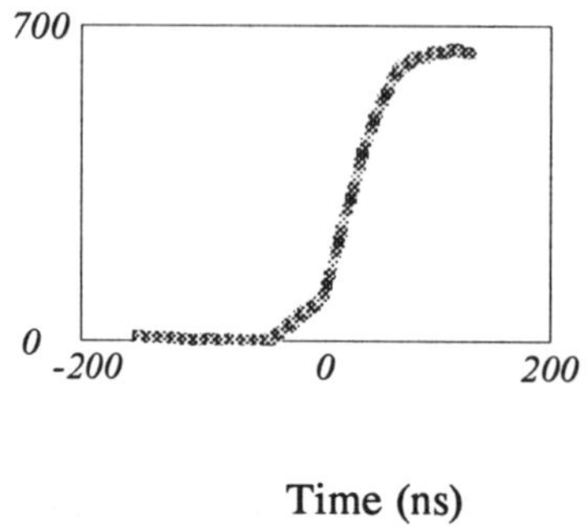

Figure 5. Current for shot 367 in kA.

these images gives a good idea of the value of the schlieren/shadowgraphy for identifying unstable behavior. The schlieren/shadowgram shows nearly all of the features visible in the interferogram. One exception is a broad, diffuse feature near the top of the interferogram.

Most fiber pinch shots on this machine show the type of instabilities shown in figure 3. There are, however, exceptions. The shots with less instability seem to be associated with a prepulse on the current. The current prepulse is caused by capacitor coupling across the self-breaking water switch, and varies substantially from shot to shot. We postulate that this variation is due, not to generator irreproducibility, but to randomness in the breakdown of the fiber.

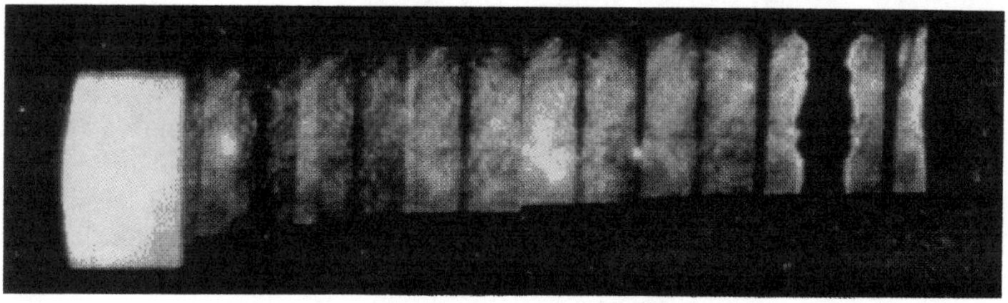

Figure 6. Schlieren/shadowgram for shot 407.

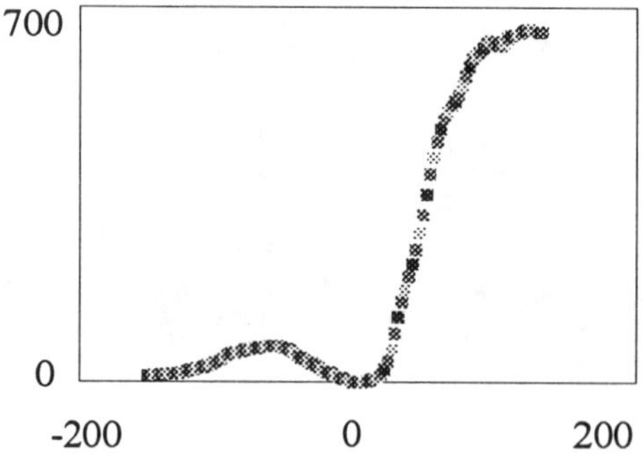

Figure 7. Current for shot 407 in kA.

Figure 6 is a schieren/shadowgram from a shot with a current prepulse (figure 7) which dropped nearly to zero before the main current pulse began. Unfortunately, an interferogram was not obtained. The prepulse appears to have blown out a halo which has almost completely suppressed axial structure. Despite the lack of visible instability, the column grows rapidly later in the current rise. Zero-dimensional modelling without instabilities suggests that the opposite should happen, i.e. the

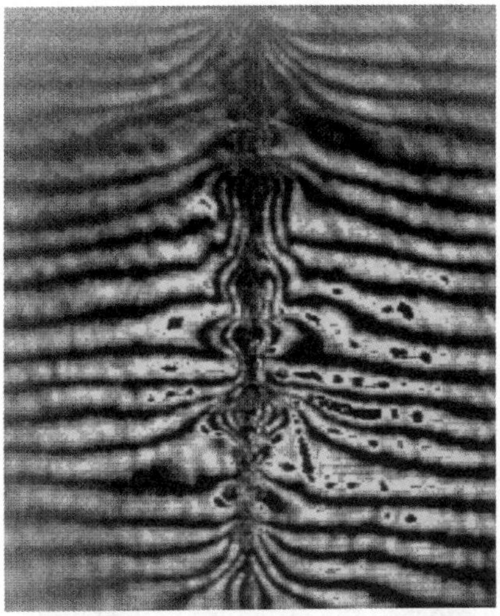

Figure 8. Interferogram for shot 409.

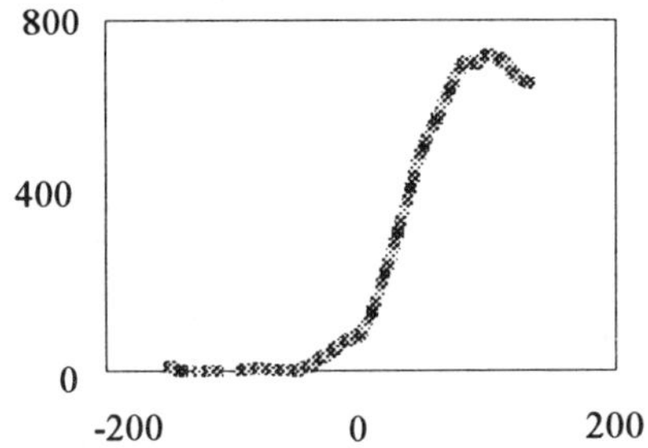

Figure 9. Current for shot 409 in kA.

column should start to compress at this time. Thus the rapid expansion may be due to further ionization of a partially ionized halo.

Figure 10. Schlieren/shadowgram for shot 409 taken at the end of the prepulse.

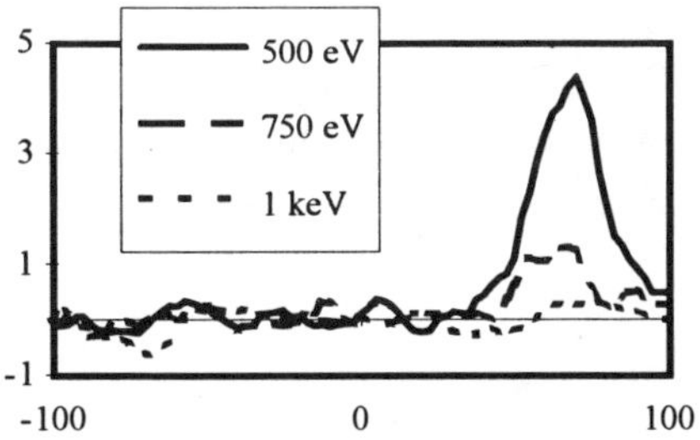

Figure 11. PIN diode signals from x-ray polychromatoror.

Figures 8, 9 and 10 are from a shot in which the prepulse did not drop before the main pulse. The images show the beginning of unstable modes, but instabilities show up noticeably later compared to shots without a prepulse.

X-RAY DATA

Figure 11 shows the signals on the three x-ray polychromator

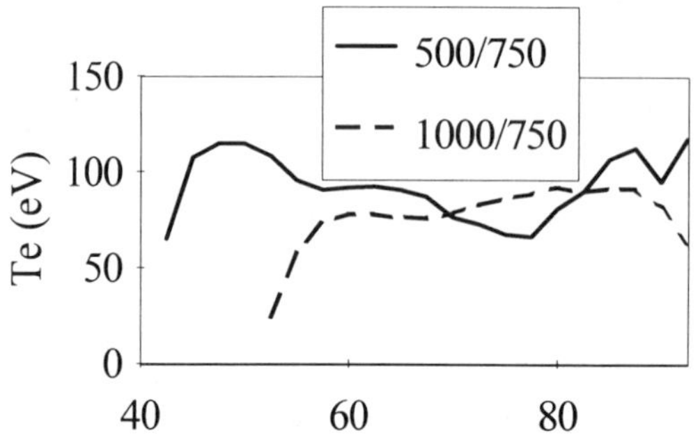

Figure 12. Electron temperature from ratios of polychromator signals.

channels. These traces are typical of the data acquired from CD_2 fibers. Figure 12 shows temperatures derived from ratios of these signals assuming that the spectrum is from bremsstrahlung. The temperature clamps at about 100 eV.

CONCLUSIONS

Simultaneous interferograms and schlieren/shadowgrams of fiber pinches provide a powerful technique for studying m=0 instabilities in these plasmas. We have shown that schlieren/shadowgrams are capable of seeing most of the structure in unstable modes, although diffuse structures tend to be missed.

We have observed a connection between a current prepulse and a suppression of strong axial structure in CD_2 pinches. The optical diagnostics support the thesis that this process is associated with a halo blown off from the fiber by the prepulse. The suppression of m=0 structure does not prevent the columns from expanding at a rapid rate.

The plasma electron temperature clamps at about 100 eV in these experiments. This temperature is consistent with Bennett equilibrium in a plasma with the carbon highly stripped.

REFERENCES

[1] J. Hammel **Dense Z Pinches**, Pereira, Davis and Rostoker, eds. AIP (1989) pp303.

[2] J. D. Sethian, A. E. Robson, K. A. Gerber and A. W. DeSilva, ibid.., pp 308.

[3] J. M. Bayley, P. Baldock and A. E. Dangor, ibid., pp 481.

[4] P. Sheehey, J. E. Hammel, I. R. Lindemuth, D. W. Scudder, J. S. Shlachter, R. H. Lovberg and R.A. Riley **Phys. Fluids B, 4**, (Nov. 1992) pp 3698.

Z-PINCH EXPERIMENTS WITH STYROFOAM FIBRES AND PLASMAJETS

S. Stein, G. Decker, W. Kies, P. Röwekamp, G. Ziethen
Heinrich-Heine-Universität Düsseldorf, Germany

K.Baumung, H.Bluhm, W. Ratajczak, D. Rusch
Kernforschungszentrum Karlsruhe, Germany

J.M.Bayley
Imperial College of Science, Technology and Medicine, London, UK

ABSTRACT

Z-pinch plasmas created from fibres are less prone to macroscopic instabilities than predicted by ideal MHD theory. However, solid fibre experiments at the pulseline KALIF (2 MV, 900 kA) gave disappointing results with respect to driver-load coupling and pinch plasma confinement. High power discharges led to current leaks and plasma expansion presumably due to lacking initial conductivity and compressibility. Therefore two alternative schemes have been investigated: Solid fibres were replaced by styrofoam fibres with about 1 % solid density and plasmajets of a deuterium-argon mixture. Analyses of the experimental results showed no major differences between styrofoam and solid fibres. However, the plasmajet experiments resulted in significant improvements with respect to the initial discharge behaviour. Unfortunately, severe pinch disruptions about 40 ns after discharge initiation prevented pinch formation and confinement. The high electric field of the pulseline KALIF and the low density of the jet plasma ($E/n > 10^{-15}$ Vm^2) resulted in runaway electrons and plasma erosion rather than in magnetic confinement to the pinch.

INTRODUCTION

The generation of a quasi-static thermal pinch plasma and its confinement at sub-mm radial dimensions are the main goal of z-pinch experiments applying the terawatt pulseline KALIF. Z-pinches show a better stability than predicted by ideal MHD theory.[1] This encourages the investigation whether a global radiative collapse could be achieved. Radiative collapse of a quasi-static z-pinch requires a pinch current higher than the Pease-Braginskii current (~ 1.6 MA $\times$ Z_{eff}^{-2}, where Z_{eff} is the effective charge number).[2]

In solid fibre experiments at KALIF pinch disruptions could be prevented if the fibre radius was larger than 10 μm. Since the current should be carried by thermal electrons it is necessary that the electron thermal velocity is much higher than their drift velocity. This condition can be critical in case of current concentration in a skin layer or a plasma halo where the line density is much lower than in the core.[2]

Solid fibres showed a tendency of increased intensity, homogeneity and lifetime with increased power input during the hot-pinch plasma phase. This is a contradiction to the requirements for the initial phase, i.e. during fibre vaporization, fibre ionization, and plasma heating. High power input in this phase leads to current leaks and a fast global expansion of the plasma presumably due to lacking initial compressibility and conductivity. Obviously, solid fibres do not provide for a good driver-pinch coupling. The term "good" is characterized by (i) no or only small breakdown delay, (ii) no or only small current leaks, and (iii) no or only small hard x-ray pulse that means no or only a few runaway electrons. Therefore experiments with styrofoam fibres and plasmajets have been carried out, providing both initial compressibility and conductivity.

EXPERIMENTAL SET-UP

Figure 1 shows the plasmajet experiment connected to the vacuum feed of the water pulseline KALIF. The vacuum feed consists of a water-vacuum interface, a fixed anode, and a variable cathode, which renders the adjustment of the gap width possible. Total voltage and current are measured at the water-vacuum interface. Possible current leaks between the interface and the pinch can be partially detected by a second current probe located at a radial position about 12 cm from the symmetry axis. Possible runaway electrons are stopped at the anode surface and cause intensive hard x-ray emission. The signal from a scintillator-photomultiplier combination therefore provides time-resolved information about the runaway electrons.

Metallized (thickness of the metal layer ~ 10 nm) and unmetallized styrofoam fibres with about 1 % solid density were used. Typical dimensions are a diameter of 1 mm and a length of 30 mm. The fibres are mounted on the symmetry axis between anode and cathode.

The deuterium (90 vol.-%) - argon (10 vol.-%) plasmajets are ejected through a nozzle (i.e. a hole of 8 mm in the center of the cathode) into the anode-cathode gap of the pulseline. These jets are generated with a plasma focus discharge in a gas-puff mode. The jet diameter is 8 mm at the nozzle and expands up to about 30 mm towards the anode. The adjusted gap widths were 19 mm and 35 mm.

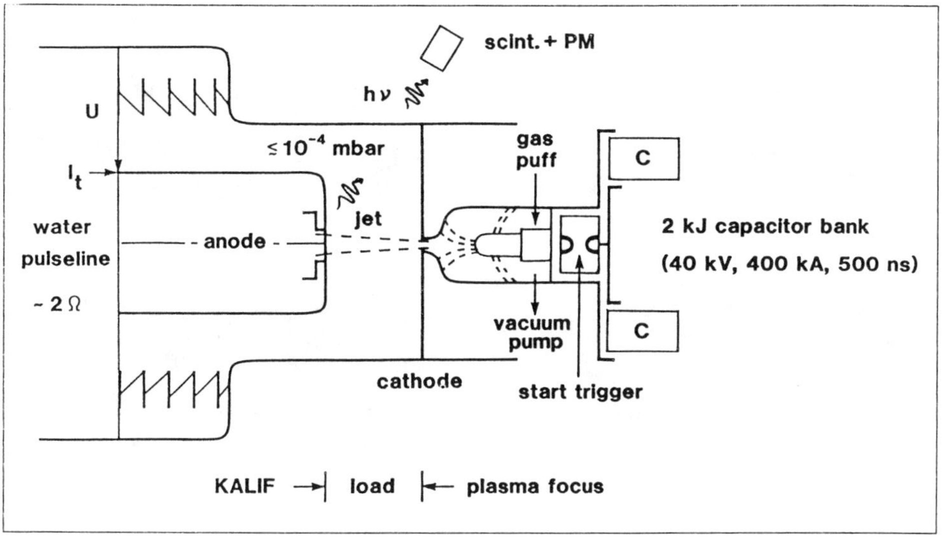

Figure 1: Plasmajet experiment connected to the KALIF vacuum feed.

EXPERIMENTAL RESULTS

A. STYROFOAM FIBRES

The experiments with styrofoam fibres did not provide any significant differences compared with solid fibre experiments. Similarly, no major improvements have been achieved using metallized fibres. Presumably neither the metallizing provides enough conductivity nor the foam is enough compressible to improve the driver-load coupling and to prevent the discharge expansion.

Figure 2 shows optical framing pictures (a) and optical streaks (b) of styrofoam fibre and solid fibre experiments, respectively. Obviously there are no major differences to be seen. The plasma is well concentrated to the axis during the initial phase of the discharges. But at later times the discharges show a fast global expansion. It starts about 50 ns after breakdown with expansion velocities up to 200 μm/ns.

Figure 3 shows two differently filtered time integrated soft x-ray twin pinhole pictures. The hot plasma is radially extended in both cases (a). The stronger filtered pictures (b) show locally compressed regions, so-called micropinches that are even more distinct with styrofoam fibres.

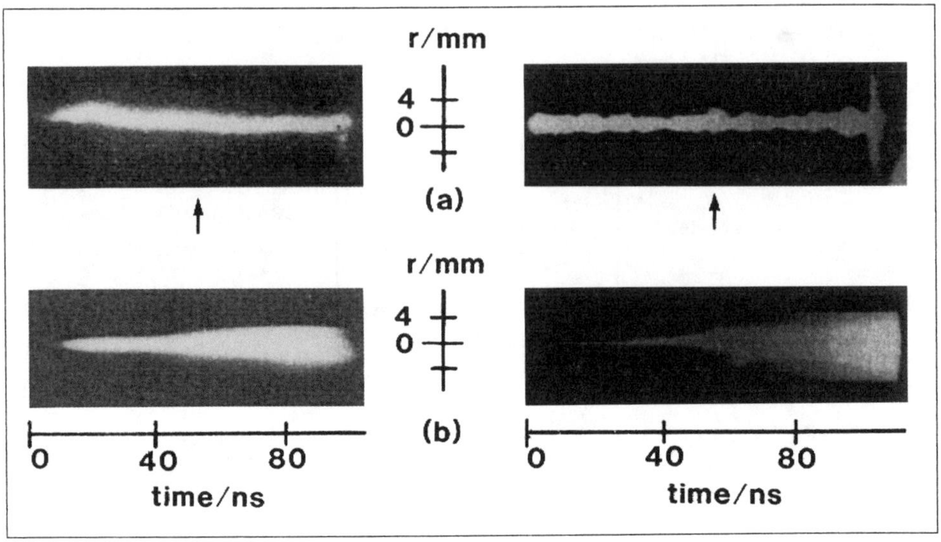

Figure 2: Optical framing pictures (a) and optical streak pictures (b) of discharges on a styrofoam fibre (left) and a solid fibre (right). The framing pictures are taken at about 30 ns after breakdown. The arrows mark the slit position of the streaks.

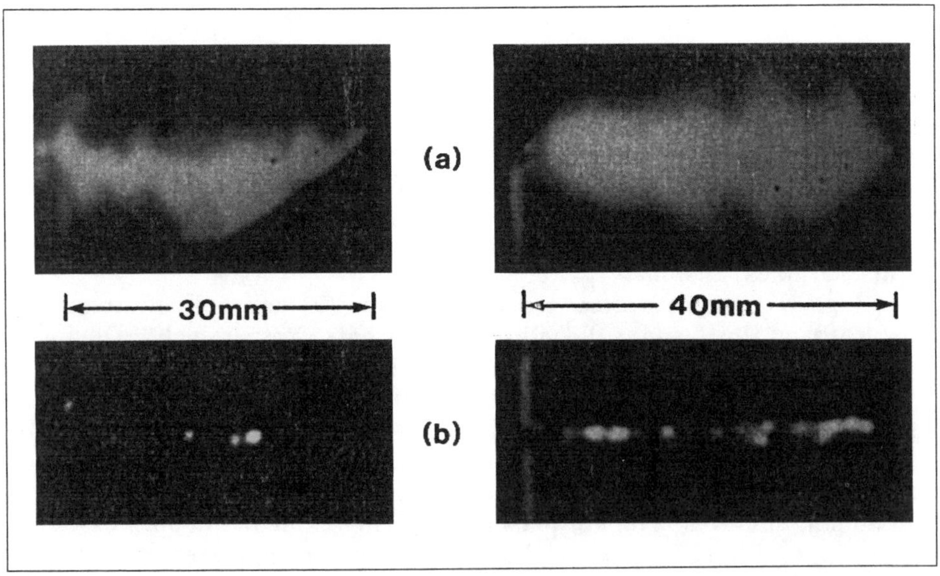

Figure 3: Time integrated soft x-ray twin hole pictures of a styrofoam fibre (left) and a solid fibre (right). The pictures are filtered by 10 μm beryllium (a) and 10 μm beryllium plus 100 μm polyester (b).

B. PLASMAJETS

Figure 4 shows the jet plasma between anode and cathode of the pulseline KALIF. The plasma data have been estimated by spectroscopic methods. The density estimate ($n_e \cong 10^{22}\,m^{-3}$) results from Stark broadening of the first three Balmer lines and the temperature estimate (T_e = 0.5 eV...1 eV) from line-to-continuum intensity ratios. These values are about one order of magnitude below typical values for plasma sheaths of plasma foci.

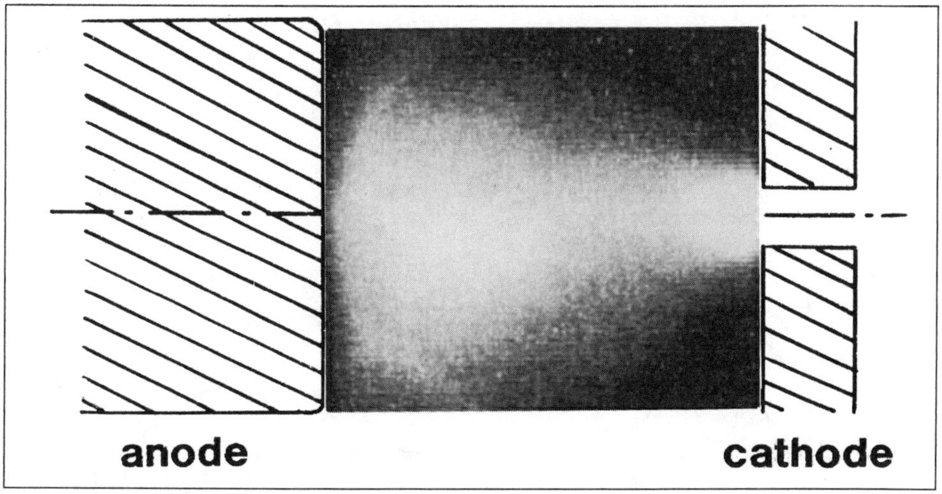

Figure 4: Jet plasma between anode and cathode of the vacuum feed.

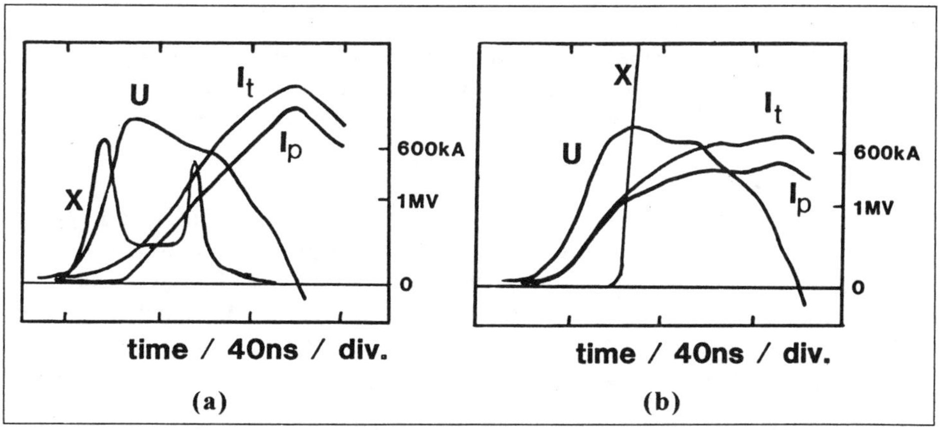

Figure 5: Time correlated signals from a solid fibre discharge (a) and a jet plasma discharge (b) with a gap width of 19 mm.

Plasmajets provide a perfect driver-load coupling in the initial phase. Unfortunately, a strong decoupling takes place after several tens of nanoseconds which prevents the plasma from pinching.

Figure 5(a) shows typical signals of a high power solid fibre discharge. The long breakdown delay (> 20 ns) results in (i) strong current leaks ($I_p < I_t$) persisting also in later phases of the discharge and (ii) an intensive hard x-ray pulse that already occurs before the current rises to high values. The runaway electrons in the initial phase of the discharge are due to a lack of carriers. For comparison figure 5(b) shows typical signals from a jet plasma discharge with a gap width of 19 mm. The breakdown delay is negligible (< 10 ns) and the pinch current is identical to the total current up to about 40 ns after breakdown. Only then hard x-rays are measured with such a high intensity that the detectors are saturated. Simultaneously current leaks appear ($I_p < I_t$) and a strong decoupling takes place via disruption.

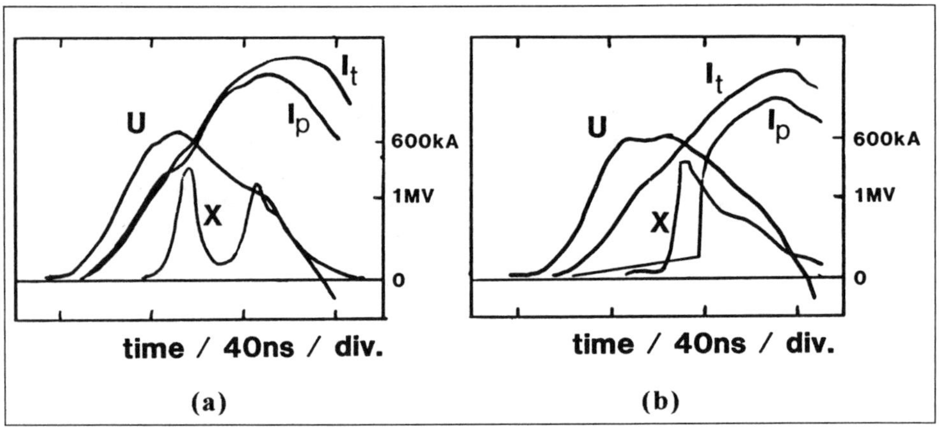

Figure 6: Time correlated signals from plasmajet discharges with a gap length of 35 mm. The time delays between the ignition of the plasma focus discharge and the ignition of the KALIF discharge are 3,5 ms (a) and 4,5 ms (b) respectively.

Extending the gap to 35 mm a decrease of the current was expected due to the higher load impedance. Surprisingly figure 6(a) shows the opposite: The extended gap leads to higher current values. Further, the pinch disruption is delayed (≅ 80 ns after breakdown) and the hard x-rays are strongly reduced. These changes are not due to the enhanced line density. This has been proved by a larger time delay between the jet inlet and the ignition of the KALIF discharge that means a further enhanced line density. Figure 6(b) shows that the larger delay results in a strongly delayed pinch current rise compared to the total current. This delay is due to a hollow current profile: The current flow starts in outer regions with respect to the position of the pinch current probe. The current carrying region is then moving towards the symmetry axis. Hard x-rays occur even before this region reaches the position of the pinch current probe. High values of the ratio E/n (> 10^{-15} Vm^2, where E is the electric field and n the particle density) cause plasma erosion via electron runaway.[3] This fast erosion prevents sheath formation and plasma pinching.

The collision mean free path in the jet plasma is so large that electrons gain enough energy from the electric field to run away the more so the smaller the gap width is. This explains why the disruption is delayed and occurs at higher values of voltage and current in case of the wider gap. Table I gives typical parameter values at the time of the starting plasma erosion for the two gap widths. Although $I_{dot} \propto E$ is expected the opposite holds in this case. This indicates that the discharge is much stronger hindered using the smaller gap.

gap (mm)	U (MV)	I_p (kA)	E (10^7 Vm^{-1})	$I_{p,dot}$ (kA ns^{-1})
19	1.56	276	8.2	14.5
35	1.75	430	5	18

Table I: Parameter values at time of the starting disruption.

CONCLUSIONS

The main goal of our z-pinch experiments, the generation of a homogeneous quasi-static thermal plasma and its confinement to sub-mm radial dimensions has not yet been achieved. In case of fibres the unfavourable start conditions, i.e. the lack of compressibility and conductivity, seem to set a limit. In case of plasmajets the high ratio of electric field and plasma particle density results in a fast plasma erosion preventing sheath formation and compression. However, the jets provide remarkable improvements with respect to driver-load coupling. Therefore, in further experiments it will be investigated whether runaway electrons can be prevented in favour of magnetic confinement. Discharges applying the driver SPEED 1 (200 kV, 900 kA) will show whether the disruptions can be avoided with reduced electric field. Furthermore, efforts will be made to improve the jet plasma data, i.e. to increase the density by at least one order of magnitude.

REFERENCES

1. M. G. Haines et al., The inapplicability of ideal MHD stability theory to the dense z-pinch, AIP Conference Proceedings, Laguna Beach, Ca, 1989.

2. W. Kies et al., Terawatt fiber pinch experiments, J. Appl. Phys. 70 (12), p.7621, 1991.

3. H. Dreicer, Electron and ion runaway in a fully ionized gas.I, Phys. Rev. 15 (2), p.238, 1959.

SPECTROSCOPY

GENERALIZED ESCAPE-PROBABILITY METHOD IN THE THEORY OF HIGH-INTENSITY RADIATIVE TRANSFER IN CONTINUOUS SPECTRA.

A.B.Kukushkin
Russian Research Center "Kurchatov Institute"
123182 Moscow Russia

ABSTRACT

Universal formula for total power losses by a bounded medium due to emission of electromagnetic waves, both transverse and longitudinal, by free/bounded electrons in the regimes of nonlocal (non-diffusion) heat transport in continuous/discrete spectra is obtained. The derivation is based on a method which generalizes the "escape probability method" in the theory of radiative transfer in resonance atomic lines. An analytic description for the deviations from Maxwellian distribution, caused by the transfer of intense radiation, is obtained within the framework of self-consistent kinetic description of plasma electrons and wave intensity.

1. INTRODUCTION

Heat transport by electromagnetic (EM) waves, both transverse (i.e. photons) and longitudinal waves (i.e. plasmons), in the wide range of parameters (which includes, in particular, hot dense plasmas and magnetically confined thermonuclear plasmas) is characterized by its nonlocal (non-diffusion) nature which manifests itself in nonlocal correlation of plasma temperatures and, correspondingly, non-diffusion law of heat propagation.

The phenomenon of nonlocal, non-diffusion transport has been revealed and thoroughly investigated in the theory of radiative transfer in resonance atomic lines (RTRAL) in the late forties and early fifties in the pioneer works by Biberman, Holstein and Sobolev (BHS). Further advances of their analytic approaches are known in literature as "escape probability method" [1]

An extension of the BHS-approach to emission/absorption by plasma electrons gives universal formulae for total power losses by a bounded medium due to emission of electromagnetic (both transverse and longitudinal) waves in the regimes of nonlocal (non-diffusion) heat transport.

The present paper analyzes also the effect of deviations from maxwellian distribution, produced by the propagation of intense waves spontaneously emitted by plasma, within the framework of self-consistent treatment of radiative transfer and electron quasi-linear (Q1) diffusion . A simple description is obtained for combined action of two effects, (a) "enlightenment" of a media due to the QL-reduction of absorption and (b) depletion of electron velocity distribution "tail" due to radiation losses.

2. UNIVERSAL FORMULA FOR TOTAL ENERGY LOSSES

Let us consider the transfer of EM energy which is describes by equation of the well-known form for the intensity $J(\phi,\vec{r},t)$,

$$\left(\frac{1}{v_g}\frac{\partial}{\partial t}+\vec{n}_g\frac{\partial}{\partial \vec{r}}\right)\left\{\frac{J(\phi,\vec{r},t)}{N_r^2}\right\} = -\kappa\left\{\frac{J(\phi,\vec{r},t)}{N_r^2}\right\} + \frac{Q}{N_r^2}, \qquad (1)$$

where $\phi \equiv \{\omega,\vec{n},\zeta\}$, ω and $\vec{k}$ are the frequency and wave vector respectively, $\vec{n}\equiv\vec{k}/k$, parameter ζ describes polarization state of the wave. Here $\kappa(\phi,\vec{r},t)$ is the absorption coefficient, $Q(\phi,\vec{r},t)$ is the source function, N_r is ray refractive index, v_g and n_g are the group velocity and its direction. In a medium with dispersion the substitution of the dependence $k=k(\omega,\vec{n},\zeta)$ from the dispersion relation is implied.

For the radiative transfer by emission/absorption of EM waves by plasma electrons, the quantities Q and κ contain averaging over electron distribution function (EDF) $f(\vec{r},\vec{p},t)$:

$$Q(\phi,\vec{r},t) = \int q_1(\vec{r},\vec{p},\phi,t)\, f(\vec{r},\vec{p},t)\, d\vec{p} \qquad (2)$$

$$\kappa(\phi,\vec{r},t) = \int \kappa_1(\vec{r},\vec{p},\phi,t)\, f(\vec{r},\vec{p},t)\, d\vec{p} \qquad (3)$$

where q and κ_1 are the corresponding rates for an individual electron. Here absorption coefficient allows for "true" absorption and stimulated emission and hence can be reduced in the long-wave limit ($\hbar\omega \ll \varepsilon$, ε is the electron energy) to a differential, in momentum p, operator.

Source function $Q(\phi,\vec{r},t)$ may implicitly depend on the intensity $J(\phi,\vec{r},t)$ via corresponding distortions of the EDF, caused by radiative transfer. If these distortions are small then Q doesn't depend on J and eq.(1) appears to be a closed equation for energy carriers. The multiple reflection at the medium boundary prevents from straightforward use of analytic solution of eq.(1) in the form of the integral over ray path.

Another example of the reduction of transport problem to a closed equation is the Biberman-Holstein equation for accumulators of energy, namely excited atoms, in the theory of radiative transfer in resonance atomic lines (RTRAL) in gases and plasmas for the case of complete redistribution in photon frequencies in individual act of its scattering by a medium's atom.

In both above-mentioned cases the energy transport is characterized by the following general features.

(*) The dominant contribution to energy loss stems from long-free-path energy carriers whose long flights are of medium size Lo in length (or greater, Lo/(1-R), in case of reflection at boundaries, R is reflection coefficient).

(**) Each of the variables of the total phase space $\Gamma \equiv \{\phi, \vec{r}\}$ manifests one of the two limiting forms of its evolution ("redistribution") along the trajectory of energy carrier from its birth to its death (by convertion into medium's heat): either no redistribution ("independent" variables) in which case for each value of this variable the energy transport takes place independently (e.g., variable ω for absorption/emission by free electrons), or complete redistribution (complete loss of memory in elementary act of absorption/emission) in which case for each of those variables the transport equation may be properly averaged.

The whole set of first-type variables we shall denote as $\{\Gamma_{ind}\}$ and the second-type, $\{\Gamma_{crd}\}$.

The properties (*) and (**) enable us to obtain finally the following general result for total power losses:

$$\frac{dE}{dt} = \int d\Gamma_{ind} \left(1 + \nu_{que}/\nu_{esc}\right)^{-1} \int d\Gamma_{crd}\, Q(\Gamma) \,, \qquad (4)$$

Here ν_{que} is the rate of such an absorption of EM energy by the medium which converts transported EM energy into medium's heat (temperature), (this is the quenching of atom excitation by, e.g., medium particle's impact or absorption of the photon(plasmon) by plasma electrons), and ν_{esc} is the rate of free escape of EM energy out of the medium. Both quantities ν_{que} and ν_{esc} are averaged over Γ_{crd}.

For the RTRAL we have $\{\Gamma_{ind}\} = \{\vec{r}\}$ and $\nu_{esc} = \langle T(\vec{r})\rangle/t_{rad}$, where $\langle T(\vec{r})\rangle$ is Holstein function averaged over angles of photon escape, t_{rad} is radiative lifetime of excited atom. Thus we arrive at well-known result in the RTRAL theory (see, e.g., [1,2]). Therefore formula (4) may be interpreted as a generalization of this methods to the case of heat transport via emission/absorption of (transverse and longitudinal) EM waves by free electrons with fixed velocity distribution.

Note that formula (4) covers both limiting regimes of energy loss, namely purely volumetric loss due to free escape from the whole phase space ($\nu_{que} \ll \nu_{esc}$) and purely surface loss ($\nu_{que} \gg \nu_{esc}$). In the latter case the intensity of escaped EM field is close to the equilibrium Planck distribution with some effective, spatially averaged temperature, the heat transport inside the medium being characterized in corresponding domain of phase space by diffusion-type regime of radiative thermoconduction Nevertheless, the total losses (4) are determined dominantly by those part of the total phase space in which the process of energy transport has essentially nonlocal character, namely the non-diffusion regime of free escape. The latter statement just constitutes the essence of a "generalized escape--probability" (GEP) method which enables us to obtain eq.(4) within the framework of principles (*) and (**).

Let us apply formula (4) to radiative transfer in continuous spectra for those geometries where complete redistribution takes place for photon angles and (under condition of multiple reflection from plasma boundary) for space variable (this includes, in particular, tokamak with highly reflecting walls, hot plasma surrounded by dense matter etc.). Neglecting the mixing of different modes in reflections from the boundaries, we have $\{ \Gamma_{ind} \} = \{ \omega, \zeta \}$ and eq.(4) reduces to the form:

$$\frac{dE}{dt} = \sum_{\zeta} \int d\omega \frac{\int dV \int d\Omega_n \; Q(\phi,\vec{r})}{1 + \tau_{ef}} \tag{5}$$

$$\tau_{ef} = \frac{\int dV \int d\Omega_n \; \kappa(\phi,\vec{r})}{\int d\Omega_n \int (\vec{n}, dS_s)(1-R(\phi,S_s))} , \tag{6}$$

where $\tau_{ef}(\omega,\zeta) = \nu_{que}/\nu_{esc}$ is the effective (dimensionless) optical length which describes the imprisonment of photons/plasmons . Here V and S_s are plasma volume and surface, respectively, $R(\phi, S_s)$ describes the dependence of reflection coefficient on wave parameters (first of all, frequency). The comparison of formula (5) with the results of numerical calculations [3,4] and formalae[4,5] which have the character of the fit of numerical results and pertain to (a) specific mechanism of emission, namely cyclotron radiation, and (b) specific profiles of temperature and density, shows good agreement in the regions of applicability of these results. The GEP method allows also to obtain a universal analytic description for spatial profile of wave energy balance and for arbitrary degree of mixing of different polarization modes at plasma boundaries, both for stationary and non-stationary cases.

The formula (5) generalizes the well-known Trubnikov's formula[4] to the case of (a) arbitrary elementary process of radiation emission by high-temperature plasma with highly reflecting walls and fixed (e.g., maxwellian) velocity distribution and (b) inhomogeneous and non-stationary plasma parameters (density, temperature etc.).

3. EFFECTS OF QUASI-LINEAR DIFFUSION AND TAIL-DEPLETION IN RADIATIVE TRANSFER BY INTENSE WAVES

Radiative transfer in continuous spectra in dense plasma may distort equilibrium electron distribution at those velocities which may be responsible for dominant contribution to radiative transfer and total radiation losses.

The evolution of the EDF, $f(\vec{r},\vec{p},t)$, can be described by the Fokker-Planck equation which allows for distant Coulomb collisions (C), spontaneous emission of waves (E_j) and their absorption (A_j which allows for quasi-linear (QL) diffusion). In the long-wave limit, $\hbar\omega \ll \varepsilon$, ε is the electron energy, emission and absorption terms may be expressed in a divergence form :

$$\left(\frac{\partial}{\partial t} + \vec{v}\frac{\partial}{\partial \vec{r}} + \vec{F}\frac{\partial}{\partial \vec{p}} \right) f(\vec{p},\vec{r},t) = \frac{\partial}{\partial p_j}\left[E_j\ f - \hat{A}_j f \right] + C[f] \quad (7)$$

where $\vec{F}$ is (macroscopic) external force, and

$$\hat{C}[f] = - \frac{\partial}{\partial p_j}\left[\hat{A}^c_j(f) - E^c_j f \right] \quad (8)$$

where operators $\hat{A}^c_j$ and E^c_j describe "absorption" and "emission" of plasma electric microfield by an individual electron at distant Coulomb collisions (with proper account of the screening, if any).

According to general laws of conservation the following relations hold:

$$\int d\phi\ J(\phi,\vec{r},t)\ \hat{\kappa}_1(\vec{r},\vec{p},\phi,t) = v_j \hat{A}_j \equiv \hat{Q}_{abs} \quad (9)$$

$$\int d\phi q_1(\vec{r},\vec{p},\phi,t) = v_j E_j \equiv Q_{em}, \qquad d\phi \equiv d\omega\ d\Omega_n \sum_{\zeta} \quad (10)$$

where operators Qabs/Qem gives the value of energy absorbed/emitted by individual electron of momentum $\vec{p}$.

We also assume that the Coulomb (electron-ion) collisions assure isotropic form of the EDF, $f = f(\varepsilon)$. Isotropic form may be additionally maintained by isotropic form of radiation intensity J. In this case

$$\hat{Q}_{abs} f \propto - \int d\phi \left\{ \frac{J(\phi,\vec{r},t)}{N_r^2} \right\} \sigma_{abs}\ \hbar\omega \frac{\partial f}{\partial \varepsilon} \equiv Q^{(1)}_{abs}\ T_e \frac{\partial f}{\partial \varepsilon}, \quad (11)$$

where $\sigma_{abs}(\vec{r},\vec{p},\phi)$ is the cross-section of absorption of wave quantum ϕ of energy $\hbar\omega$, and the quantity $Q^{(1)}_{abs}$ characterizes the (averaged in ϕ) energy absorbed by individual electron and is precisely equal to this mean energy in case of Maxwellian plasma.

For magnetized plasmas the action of the term with Lorentz force in left side of eq.(7) is incorporated into the quantity σ_{abs} which appears to be the cross-section of absorption by an elementary Larmor circle, so that operator $\partial/\partial p$ in eq.(7) and, consequently, eqs.(9) and

(10) should be properly modified. However, for the case of isotropic EDF the implications of this modification can be sometimes unnoticeable, e.g., the quantities

$$\delta(\omega-\vec{k}\vec{v})\ \vec{k}\ \partial f/\partial\vec{p}, \qquad k_{||}\partial f/\partial p_{||}+\left[(\omega-k_{||}v_{||})/v_{\perp}\right]\partial f/\partial p_{\perp}$$

are both reduced to $\omega\ \partial f/\partial\varepsilon$.

Neglecting the effects of inhomogeneity described by operator $v\ \partial/\partial r$ and solving eq.(7) in stationary case, we arrive at the result

$$f(\varepsilon) = f_0(\varepsilon)\ A\ \mathrm{EXP}\left\{ \int_0^{\varepsilon} \left[1- \frac{Q^c_{em} + Q_{em}}{Q^c_{abs} + Q^{(1)}_{abs}}\right] \frac{d\varepsilon}{T_e}\right\}, \qquad (12)$$

where all the quantities Q_{em} and Q_{abs} are averaged in velocity-angles, superscript "c" distinguishes contributions of distant Coulomb collisions, the Q^c_{abs} relates to A^c_j in the same manner as Q_{abs} relates to $Q^{(1)}_{abs}$, $f_0(\varepsilon)$ is maxwellian distribution function, A is normalization constant.

Equation (12) gives simple description of combined action of two effects, (a) "enlightenment" of the media due to QL-reduction of absorption, the term $Q^{(1)}_{abs}$, and (b) depletion[6] of electron velocity distribution "tail" due to radiation losses, the term Q_{em}.

Substitution of eq.(12) into eqs.(1)-(3) leads, with allowing for eq.(10), to a complex multi-dimensional integro- differrential equation (in variables r,ω,n,ζ and p).

Numerical and analytic analysis shows that solving this problem may be essentially simplified within the framework of a rapidly-convergent computation scheme based on the formula (12) and analytic description for the intensity J obtained by the GEP method.

REFERENCES

1. G.B.Rybicki, In: Methods in Radiative Transfer , Ed. W.Kalkofen (Cambridge Univ.press, Cambridge, U.K., 1984), p.21.
2. V.A.Abramov, V.I.Kogan, V.S.Lisitsa, In: Reviews of Plasma Physics, Eds. M.A.Leontovich andB.B.Kadomtsev, (Consultants Bureau, N.Y., 1987) v.12, p.151.
3. S.Tamor, Fusion Technol. 3 293 (1983); Nucl.Instr. and Meth.Phys.Res. A271 37 (1988).
4. B.A.Trubnikov, In: Reviews of Plasma Physics, Ed. M.A.Leontovich (Consultants Bureau, N.Y., 1979), v.7, p.345.
5. S.Atzeni, B.Coppi, G.Rubinacci, Report PTP-81/7, MIT (1981).
6. Kudryavtsev, V.S., In: Plasma Physics and the Problems of Thermonuclear Reactions, Ed. M.A.Leontovich, (Pergamon Press, London, 1959), v.3.

SOURCE TO DETECTOR SPECTRUM TRANSFORMATION AND ITS INVERSE FOR THE PEGASUS Z-PINCH

Walter Matuska, Huan Lee, Robert Hockaday and Darrell Peterson
Los Alamos National Laboratory, Los Alamos, NM 87545

ABSTRACT

We have developed a ray-tracing code which enables us to calculate the spectrum from a 2-D source simulation and compare directly with experimental data. This code also allows us to study the various spectral components which can potentially be used to determine the source from measured data.

INTRODUCTION

Pegasus, the Los Alamos pulsed power Z-pinch facility, generates x-rays by imploding a thin cylinderical foil. Peterson has modeled the implosion dynamics with a 2-D, 3-temperature radiation magnetohydrodynamic (RMHD) code. These Eulerian calculations show good agreement with various experimental measurements, as reported in a separate paper at this conference.[1] Recently we have developed a ray-tracing code which calculates side-on spectra from 2-D Eulerian configurations at specified times using temperature and density dependent, multi-group opacities. Calculated spectra agree reasonably well with measured spectrometer data (Pegasus I, shot 41 is emphasized in this paper). We also are developing schemes, coupling the ray-tracing code with a parameter optimization technique, to determine the temperature and density distribution which gives the best fit to measured spectral data.

PEGASUS EXPERIMENT

The Pegasus foils are 2.0 cm in length and 5.0 cm in radius and are centered in a 56 cm radius vacuum chamber. The x-ray diagnostics are at the end of 1 to 3 meter long line-of-sight (LOS) pipes mounted normal to the vacuum chamber. The x-ray diagnostics considered here consist of a transmission grating spectrometer (TGS), x-ray diodes (XRD), a pinhole camera and a framing camera. This particular TGS is relatively new and is still being developed. Its early time signal is questionable and will not be included. The TGS measures the time dependent spectrum between photon energies of about 70 eV and 2.5 keV. The data are resolved into 49 bins of increasing width for increasing photon energies. A typical late time spectrum from the TGS, along with the error envelope, is shown in Fig. 1. The large error envelope at low photon energies is due to uncertainties in calibration; however, we believe the bottom values to be more accurate. The large error envelope around 400 eV is inherent since the spectrum does not monotonically decrease with increasing photon energy. Consequently, higher order corrections to the spectrum for the second peak have the effect of subtracting two large numbers to get a small number, hence a large error envelope. The second peak around 800 eV is sometimes questioned, but we believe it to be real, based on the calculations below.

The XRDs produce time dependent currents from the Z-pinch spectrum. The detector responses of different XRD channels are shown in Fig. 2. We mention that we plan to measure the second peak of the spectrum around 800eV, with an XRD channel having a single lobed response past 400 eV photon energy, since its existence is not firmly established.

The pinhole camera makes time-integrated x-ray images for four different responses. One such image as displayed in Fig. 3 shows two bright spots. Even though they are not on axis, the intensity contours are somewhat symmetric about the line passing through these two spots, implying an approximate symmetry about a tilted axis. (Calculations, of course, are about an axis at r=0.)

The framing camera, which is under development, gives multiple 2-D x-ray images at several different times. The different images at one time are for different photon energy responses. The framing camera will be calibrated and should allow us to do the axially symmetric 2-D fit discussed below.

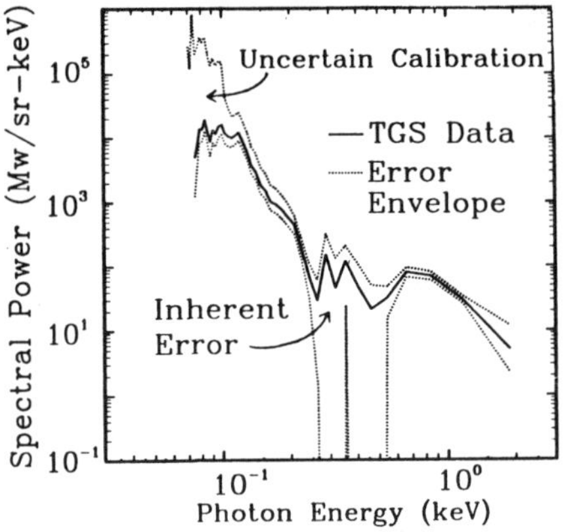

Fig. 1. TGS data at 19.15 μsec.

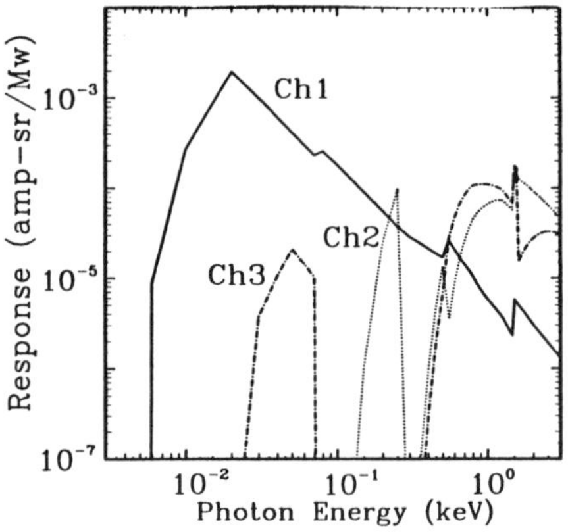

Fig. 2. Response functions for three different XRD detector channels.

SIMULATION OF THE Z-PINCH

The actual imploded foil was 7500Å Al backed with 1000Å parylene. The 2-D simulation, in which all material was taken to be Al, is discussed in detail by Peterson[1]. The simulation assumes the pinch to have a reflection symmetry about z = 0.0 cm. The simulations have high density hot spots on axis and density decreasing with radius. Densities are approximately 10^{-1}, 10^{-3}, 10^{-5} and 10^{-6} g/cc at radii of 0.0, 0.4, 2.0 and 2.2 cm, respectively, at 19.25 μsec. The three temperatures decrease with radius, see Fig. 4, out to a radius where the density is low, and then increase. This temperature increase in the very low

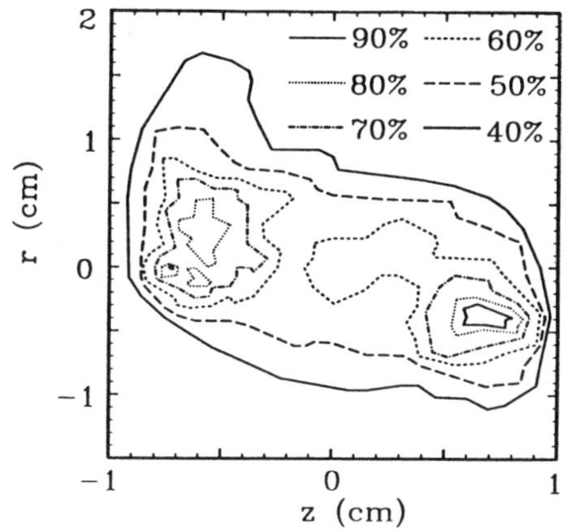

Fig. 3. Film density contours from time-integrated pinhole camera.

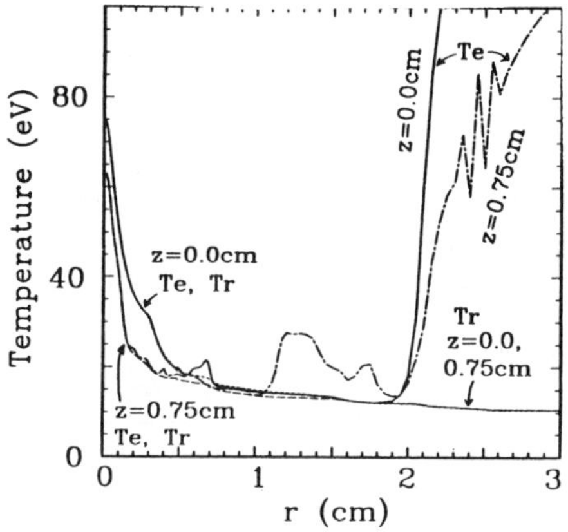

Fig. 4. Electron (Te) and radiation (Tr) temperatures versus radius for two values of z in the 2-D Eulerian calculation at 19.25 μsec.

density region is believed to be a code artifact. The three temperatures are close together (except in this low density region). This is important since our method of computing the spectrum assumes local thermal equilibrium (LTE).

SPECTRA CALCULATIONS

Spectra are calculated along the rays traced through the 2-D Eulerian calculational grid as illustrated in Fig. 5. One ray passes through the center of each rectangular calculational zone in a plane perpendicular to the line-of-sight (LOS). Since the sizes of the x-ray source and the detector are much smaller than the distance between them, we can consider all rays connecting them to be parallel. Let $I_\nu(x,t)$ be the specific intensity of the x-rays along the direction toward the detector, at a source point x and time t. The ray-tracing code simply integrates the equation

$$\frac{d}{dx}I_\nu(x,t) + (\kappa_\nu + \sigma_\nu)I_\nu(x,t) = \kappa_\nu B_\nu(T(x,t))$$

over x, where x is the traced distance, $B_\nu(T)$ is the specific intensity for black-body radiation at temperature T, κ is the absorption coefficient, σ the scattering coefficient, and ν the photon frequency. We note that κ and σ depend on x through their dependence on density and temperature. This approach can provide us a realistic spectrum if LTE is approximately satisfied. Our model includes out-scattering but no in-scattering. This is acceptable since the absorption opacity is always much larger than the scattering opacity in the applications considered. The spectrum from a given ray is in units of brightness. The brightness spectra along all rays, multiplied by the cross sectional area of the associated zone, can be summed to get a total output spectrum (spectral power). Also partial spectra can be calculated by starting photons only in zones within a specified radial interval while including all attenuation.

Column densities along rays (Fig. 6) and hν dependent opacities (Fig. 7) are helpful in understanding the features in a spectrum. In Fig. 6 the sum is only over half the traced distance, and r is the distance between the ray and the center line, see Fig. 5.

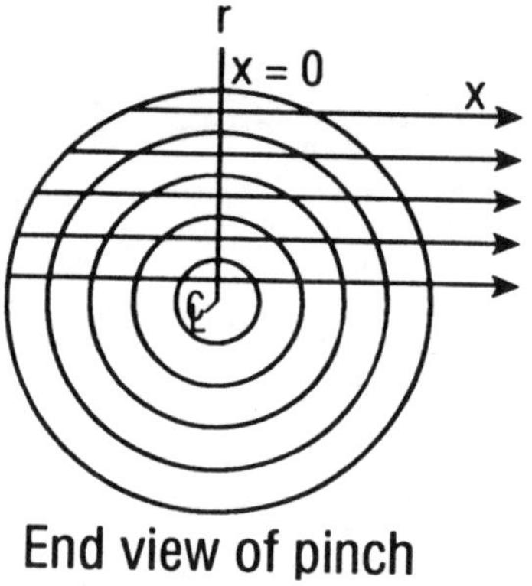

Fig. 5. Schematic of ray-tracing through 2-D grid.

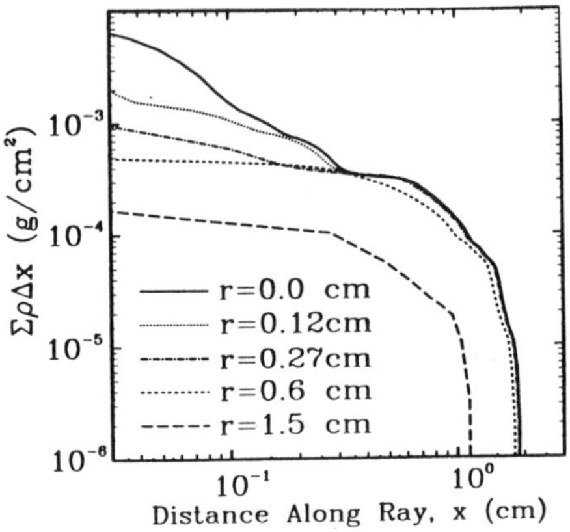

Fig. 6. Column densities at 19.25 μsec tracing along rays from a point x to the outside, where x=0 is defined in Fig. 5.

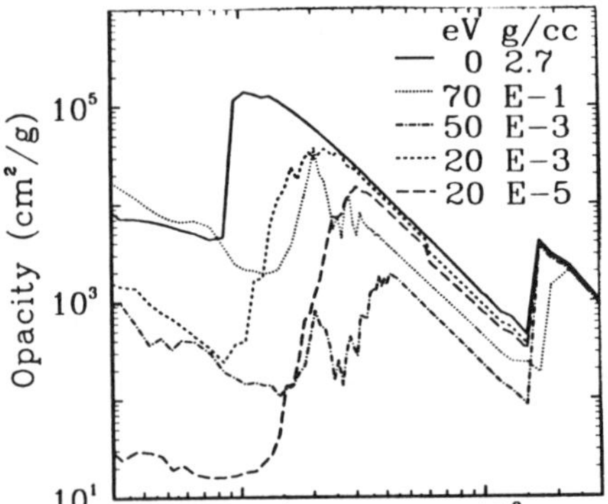

Fig. 7. Al opacities for five temperature and density pairs as indicated.

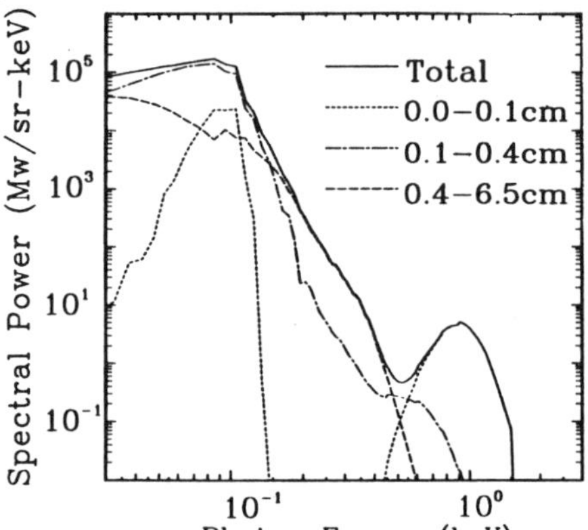

Fig. 8. Calculated total spectrum at 19.25 μsec which is the sum of partial spectra from three radial intervals.

Insight can be gained by studying how a total output spectrum is composed of partial spectra and how the spectrum brightness shape changes over (r,z), see Figs. 8 and 9. The peak at 800 eV comes from the pinch center, the lowest photon energies come from farther out and the spectrum in the intermediate range come from the outside. These features are consistent with the column densities and hν dependent opacities. The 60 eV Planckian from the pinch center contributes mainly to the spectrum below the K-edge of Al where the opacity is low, forming the peak at 800 eV. After this peak the spectrum decreases like a 60 eV Planckian and then drops off suddenly at the K-edge. The center of the pinch emits approximately as a Planckian since it is optically thick to all photon energies, see Figs. 6 and 7. Likewise, the outer radii of the pinch dominate the spectrum in the 200 to 400 eV range where the opacity is high, and nothing from the center can get out.

The brightness spectra vary over (r,z), with the spectra varying little along the axis but varying greatly off axis, see Fig. 9. The high energy peak is only found at small values of r. With increasing r, and decreasing density, the low energy part of the spectrum decreases while the spectrum in the 200 to 400 eV range does not change. This is consistant with the column densities and opacities.

The question of calculational sensitivity to various approximations is addressed in Fig. 10. Curve 1 is a spectrum from a fine zoned Eulerian calcula-

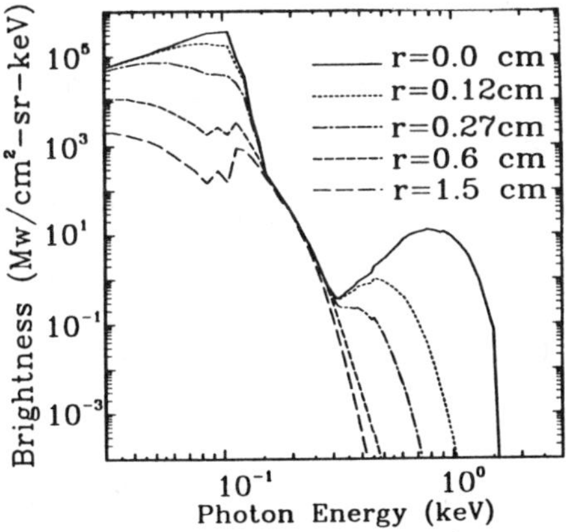

Fig. 9. Calculated spectral brightness of rays at 19.25 μsec. All rays are at z=0.25 cm but lie at different distances r from the center line.

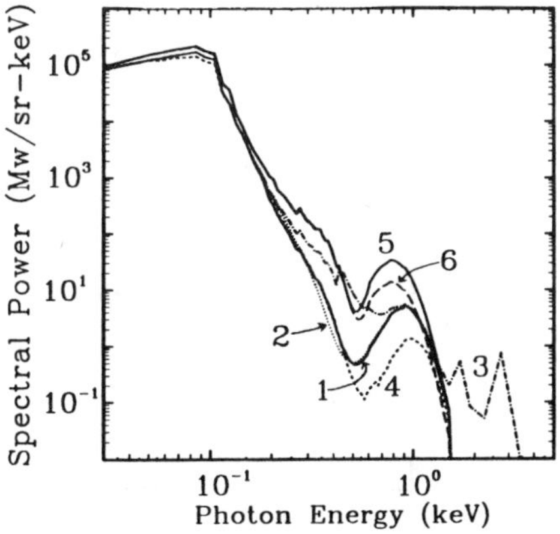

Fig. 10. Comparison of spectra calculated with various approximations at 19.25 μsec.

tion, Δr=0.0125 cm and Δz=0.025 cm, with only densities above 10^{-5} g/cc contributing to the spectrum. The high temperatures past a radius of 1.9 cm are not included in this spectrum calculation, curve 3 being the only curve shown which includes these high temperatures at large radii. Curve 2 only includes densities above 10^{-4} g/cc. Curves 1 and 2 are almost alike, and we believe these two to be the best calculations. Curve 3 includes the high temperatures at large radii and all densities, 10^{-8} g/cc being the lowest. In all spectra except curve 4, pure Al opacities are used. Curve 4 opacities assume the Al and parylene backing to be uniformly mixed. Pure Al spectrum calculations assume the parylene to be on axis of the pinch and not contributing to the spectrum. Curve 5 is for a coarse zoned Eulerian calculation, Δr=0.05 cm and Δz=0.05 cm, with temperature and density cut-offs as in curve 1. For curve 6, the coarse zoned Eulerian calculation is subzoned into finer zones in the radial direction, with interpolation done on temperature, before calculating a spectrum. Subzoning the fine zoned calculation makes little change in curve 1. Even though there is some variation among the curves at high photon energies, they all have mostly the same features.

COMPARISON WITH EXPERIMENTAL RESULTS

We find the calculated and TGS spectra to agree as well as one could hope at this early stage of the project. Peak time and late time comparisons are shown in Figs. 11 and 12, respectively. The comparisons are on an absolute

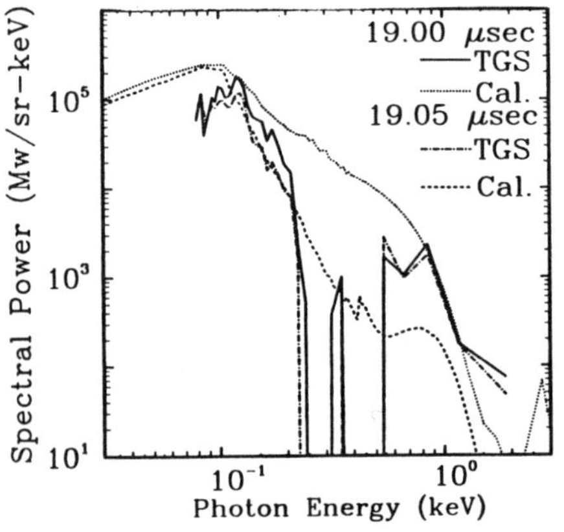

Fig. 11. Calculated spectra and TGS data compared at two times.

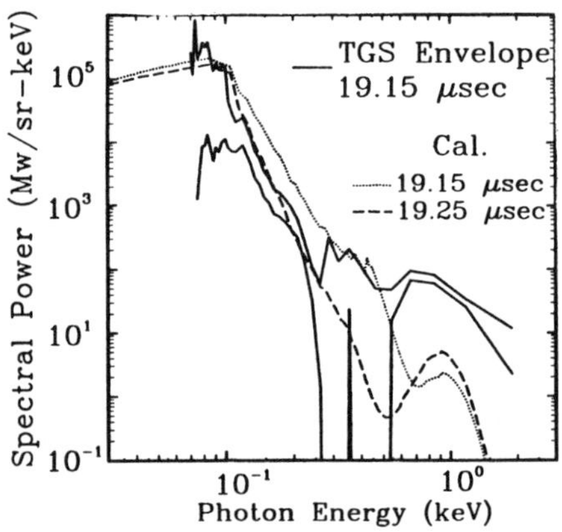

Fig. 12. Calculated spectra at two times compared with TGS error envelope from Fig. 1.

scale. Both have the large peak at low photon energies, a dip at 300 eV and a second peak at 800 eV. The calculated low photon energy peak is much higher at late times than the lower curve of the TGS error envelope, which we currently believe to be closer to the actual spectrum. The optimization fitting scheme below addresses this difference.

Spectra from calculations and the TGS are folded with the XRD responses in Fig. 2 to calculate currents which are directly comparable to the measured XRD currents, see Fig. 13. Calculation and XRDs are on an absolute scale. The relative magnitudes of the TGS currents are correct , but the TGS is normalized to the XRD (channel 2) at peak time (19.0 μsec). The major features of these three sets of currents agree well, but differences exist in detail. These differences are being addressed.

OPTIMIZED FITTING SCHEME

Figures 11-13 show some differences between calculation and experiment. We can alter the temperature and/or density distribution at a given time of an Eulerian calculation, with the use of a parameter optimization scheme [2], to make the calculated spectrum better fit the data. The optimization scheme does least square fitting for an arbitrary non-linear system. The variables to be determined by the fit are temperature and/or density multipliers at specified radii. Multipliers are defined between specified radii by interpolation. These multipliers are applied to an Eulerian grid as a function of radius, independent of z, and a spectrum is calculated for this new temperature and density distribution. Also different importance weights can be assigned to the different parts of the spectrum when finding the sum of squares. Calculated spectra at 19.25 μsec, with and without optimization, are compared with data at 19.15 μsec in Fig. 14. We use 19.25 μsec since it is close in time and better approximates the data, see Fig. 12. Also the coarse zoned Eulerian grid was used with sub-

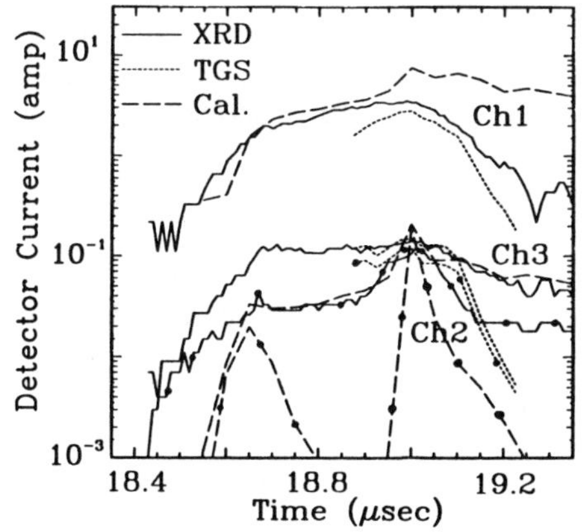

Fig. 13. Comparison of XRD currents. Channel 2 curve has dots.

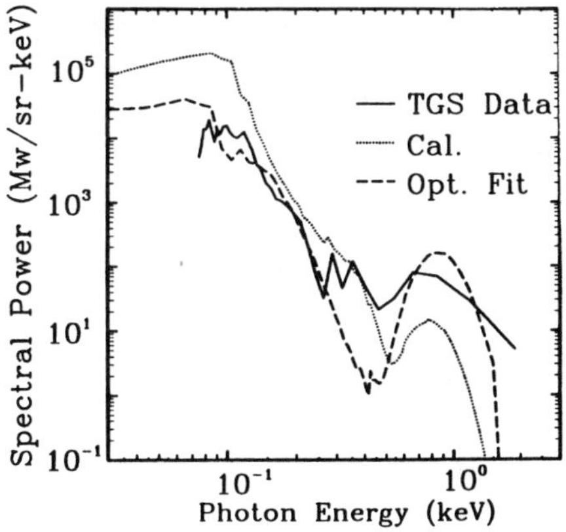

Fig. 14. Spectrum from optimized fit compared with data at 19.15μsec.

zoning in the optimization process to keep computation time reasonable. The importance weights were set low between 280 and 500 eV. The optimized spectrum fits the data better, especially the two peaks. The temperature multipliers here are 1.4, 0.7, 0.3, 1.6, 0.4 and 0.4 at radii of 0.0, 0.1, 0.2, 1.0, 1.7 and 6.5 cm, respectively. The density multipliers were fixed at 1.0 inside 0.2 cm and optimized to 0.04 outside 1.0 cm, interpolation in between. Since this fit is underconstrained, the fit is not unique and details of the fit cannot be believed, but higher temperatures on axis and lower temperatures at small radii is probably correct.

We are also developing a fitting scheme, based on the framing camera described above, which will completely constrain the fit. The fit does not require an existing 2-D calculation. At a given time, the framing camera records several images, each having a response of the type in Fig. 2. The pinhole camera shows in Fig. 3 that an assumption of symmetry about a slanted axis is probably acceptable. The data from the several images, at a given spacial point (r,z), can be used to fit a spectrum along a ray through that point. See Fig. 5. For a constant z and starting at the outermost r, the fitting scheme determines the temperature and density distribution that produces a spectrum which best matches the framing camera data at the given point, after being folded with the response functions. Next we move to a slightly smaller radius and solve for a temperature and density in this next ring, keeping fixed the temperatures and densities already determined at the outer radii. This process is repeated for small steps in r to the center of the pinch. All sweeps at given z are independent of the others, except one can make a check on the total mass and compare the total spectrum to the TGS data.

ACKNOWLEDGEMENTS

We acknowledge George Idzorek and Hal Kruse for the XRDs and pinhole camera, respectively; John Benage and Dave Scudder for the framing camera; Richard Bowers, Jack Brownell, and Art Greene for theoretical discussions; and Joe Mack, who years ago introduced one of the authors to non-linear parameter optimization.

REFERENCES

1. D.L. Peterson et al., these proceedings.
2. D.J. Wilde and C. S. Beightler, Foundations of Optimizations (Prentice-Hall, N.J., 1967), page 298.

HIGH-POWER Z PINCHES AS X-RAY LASER PHOTOPUMPS

T.J. Nash, R.B. Spielman, L. Ruggles, and M. Vargas
Sandia National Laboratories, Albuquerque, NM, 87185
USA

ABSTRACT

Using Saturn as a driver, we are pursuing both photoresonantly pumped and photoionization/recombination lasers. Our lasing targets are gas cells with thin windows that are pumped by a z pinch 2 cm away radiating 10 TW. In both schemes the lasant and gas fill is neon. To increase our chances of measuring the resonantly photopumped lasing transition we have introduced potassium into a sodium z pinch and have eliminated oxygen from the gas cell windows. We have measured the spatial dependence of ionization balance across the gas cell, and this measurement is consistent with propagation of a shock front across the gas cell target. We have measured blue-shifted satellites to several Li-like neon transitions that may indicate return-current driven jetting a high 1.5e8 cm/sec velocity. Using a gold z-pinch we have shown that keV radiation is necessary to excite the He-like lines of neon. An attempt at a single shot gain measurement also indicates that radiation is not the only source of gas cell heating.

INTRODUCTION

X-ray lasers demonstrated to date have been pumped by collisional excitation and recombination. [1-10] Although resonant photopumping could potentially produce an efficient x-ray laser [11-20], the shortest wavelength at which photopumped gain has been measured is 2163 Å in Be-like C. [21]

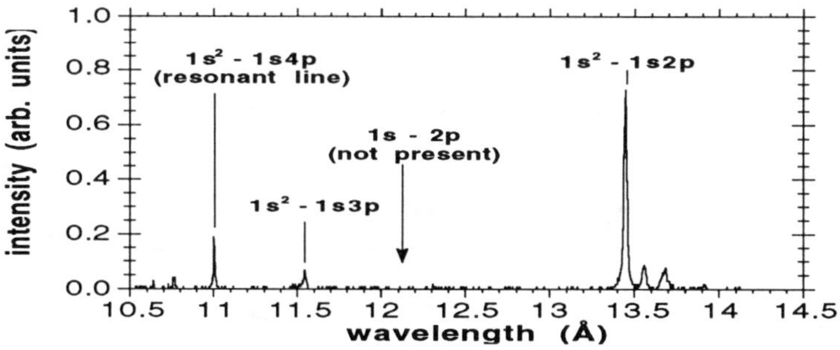

Figure 1. The 4 to 1 line of He-like neon is measured to be twice the intensity of the 3 to 1 line indicating a 4 to 3 inversion.

Using elliptical crystal spectroscopy, resonant photopumping of the n=4 level of He-like neon by a sodium z pinch on the pulsed power driver

Saturn has been demonstrated.[22] The measured neon K shell spectrum showing the 4 to 3 inversion in He-like neon is reviewed in figure 1. The 200 GW pump is sufficient to produce a measureable 4 to 3 inversion in He-like neon. Details are presented in Reference 22. The next logical step in this x-ray laser research is to measure the strongest predicted laser transition, the 4f-3d singlet line at 231 Å.

The atomic energy level diagram including the coincident lines for the sodium/neon x-ray laser scheme is shown in figure 2a. The resonant photopump of typically 200 GW in the He-like sodium 2 to 1 transition is provided by the z pinch load of the Saturn accelerator. Saturn also emits up to 10 TW total radiation which photoionizes the neon L shell in the neon lasant target, and is predicted to drive a shock wave from the side window of the gas cell into the neon.[23] The 4p singlet level of He-like neon is resonantly photopumped. Electron and ion collisions distribute the population and hence the 4 to 3 inversion over the n=4 sublevels and the largest gain is predicted for the 4f to 3d singlet transition at 231 Å.[13,19]

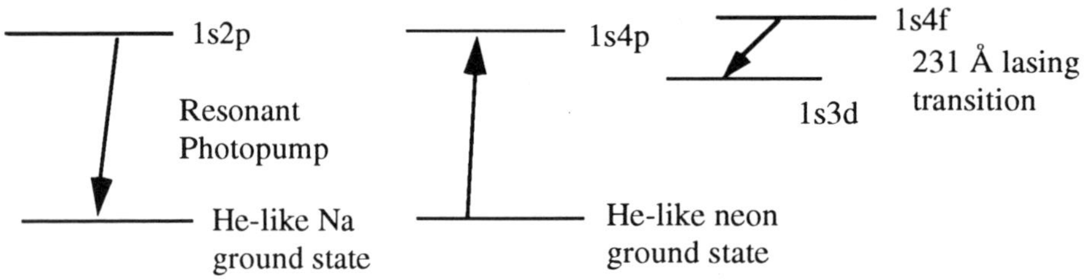

Figure 2a. Atomic energy level diagram for the resonanatly photopumped sodium/neon x-ray laser scheme.

Figure 2b shows the grotian diagram for the photoionization/recombination x-ray laser scheme. This scheme is more versatile in that is does not require a resonant pump and thus several combinations of pump and lasant are possible. It also uses a more significant fraction of the pinch radiation in the pumping.

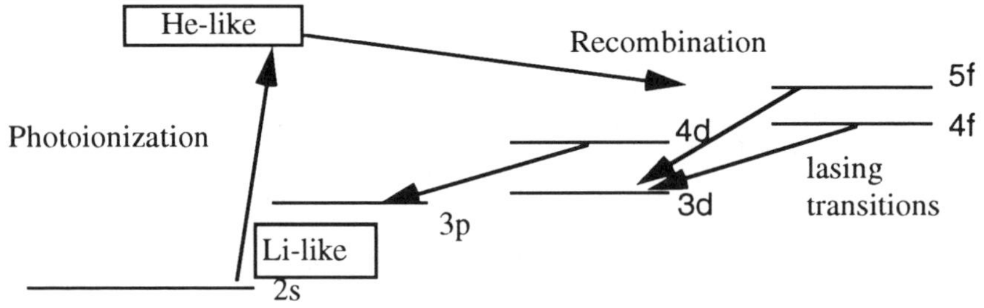

Figure 2b. Atomic Energy Level Diagram for the Photoionization/Recombination laser schemes.

The experimental arrangement is shown in figure 3. Saturn delivers approximately 10 MA with a 60 nsec rise time to a sodium z pinch load. The pinch radiates up to 10 TW with 200 GW in the resonant photopump. The pulsewidth of the resonant photopump is 20 to 30 nsec. A neon gas cell is placed with its side window 18 mm from the pinch axis, just outside of the pinch current return posts. A typical neon fill pressure of 10 torr is confined by a 4000 Å Lexan side window and an end window that is either 2000 Å Lexan or 5000 Å polystyrene. The pumping radiation is incident on and ionizes the side window, and also heats the neon gas target. XUV radiation is measured along an axial line of sight passing through the end window with a spatially resolved 10 nsec gated MCPIGS (microchannel plate (MCP) intensified one meter grazing incidence spectrometer).[24] The XUV radiation has also been measured by a flat field variable line spaced grazing incidence grating spectrometer that has spectral and spatial resolution superior to the MCPIGS.

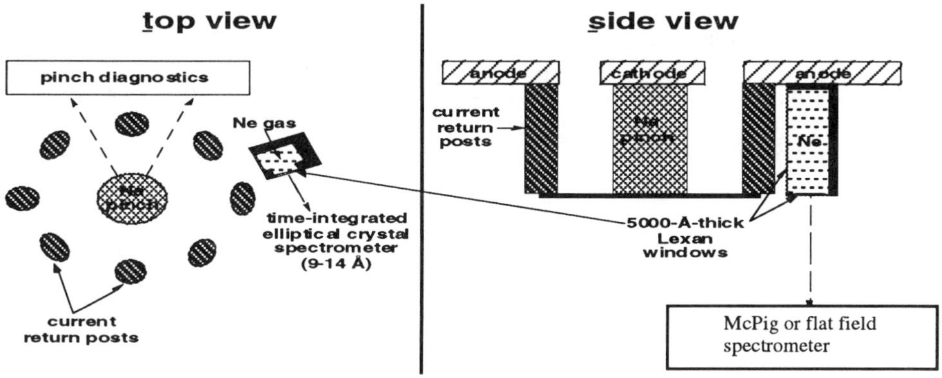

Figure 3. Experimental Arrangement

INSTRUMENTATION

We measure the XUV spectra from the neon gas cell with a spatially resolved single time gated MCPIGS. The MCP is curved to the one meter diameter of the Rowland circle. The instrumental acceptance is f/40 up to a 3 meter radius 2 degree grazing incidence spherical gathering mirror.

A 6 mm wide by 150 mm long stripline on the MCP is gated by a 10 nsec 900 volt pulse to provide a single 10 nsec resolved time frame. Typically the gate is centered on the peak of the z pinch x-ray emission. On the short 6 mm width of the strip we spatially resolve with an imaging cylindrical mirror at 5 degrees grazing incidence. With spatial magnification of 1/2.7 the spatial resolution at the target is better than 1 mm. However due to the large 150 mm length of the spectral focus, the spatial focus degrades rapidly away from the spatial focus wavelength. At spectral locations 15 Å from the spatial focus, the spatial resolution degrades to worse than 1 mm. On the MCPIGS instrument we use two vacuum spacers, one for spatial focus at 230 Å and one for spatial focus at 80 Å, assuming the use of a typical 1200 l/mm grating.

For the spatial focus at 230 Å we use a 1200 l/mm grating blazed at 5 degrees, and a 1000 Å Al filter to remove higher orders. For spatial focus at 80 Å we use no filter and a 1200 l/mm grating blazed at 2 degrees. The entrance slit width is typically 20 microns and the grating is half masked.

A spherical spectral gathering mirror sets the field of view as 1 mm wide at the gas cell, the other dimension of the gas cell being imaged by the cylindrical mirror. A closed to closed fast valve with 3 mm by 3 cm aperture open for 500 microseconds protects the mirrors from shot debris.

The microchannel plate is backed by a fiber optic faceplate coated with P11 phosphor. The spectral data is recorded on Kodak type 2484 film, then digitized and computer processed.

We have addressed the limitation in the spatial imaging of the MCPIGS by using a flat field variable line spaced grating spectrometer that maintains spatial focus along the entire spectral range. This instrument also has spectral resolution superior to the MCPIGS that has revealed useful information on line shapes to be discussed in the results section. The flat field instrument uses no spectral gathering mirror as we deployed it. Its field of view, unlike the MCPIG, includes the entire 6 mm diameter of the gas cell end window. We have found that because of the increased throughput of the flat field spectrometer it was necessary to use a 2000 Å thick aluminum filter to remove higher orders from the vicinity of candidate laser transitions.

RESULTS

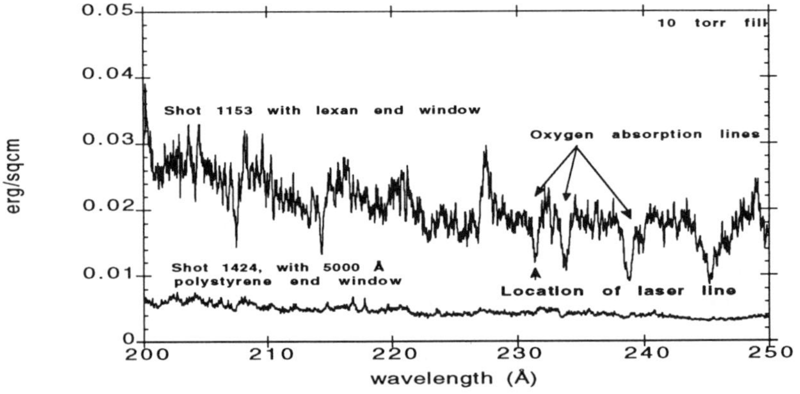

Figure 4. A polystyrene end window eliminates oxygen absorption.

An attempt at measuring the 231 Å line is depicted in figure 4. When viewed through a 2000 Å thick lexan end window the spectrum is plagued with oxygen absorption lines, one of which falls precisely on the location of the resonantly photopumped laser line. We eliminated these oxygen lines by using a 5000 Å oxygen-free polystyrene end window. With this end window the oxygen absorption lines disappear and the background level, likely due to oxygen lines as well as continuum emission, is reduced. Estimates of instrumental sensitivity and fluorescent yield in the laser line indicate that we are unlikely to detect the 231 Å transition unless it is lasing with a gain greater than 1 cm^{-1}.

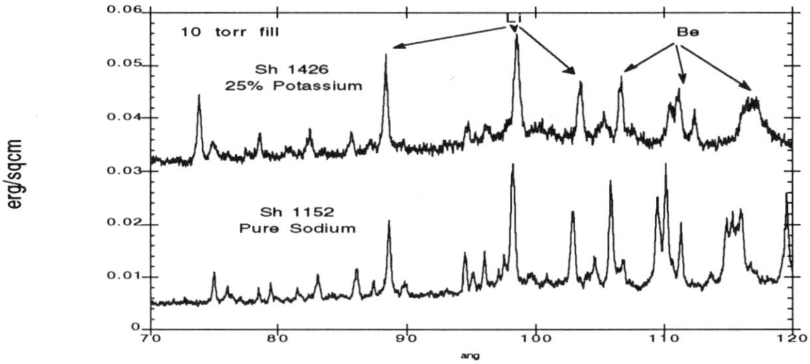

Figure 5. 25% potassium in the sodium pinch increases the neon ionization state.

Measurements of ionization balance are made by recording spectra in the 70 to 120 Å window. This region includes bright 3 to 2 transitions in Be-like, Li-like, and He-like neon. With a pure sodium z-pinch pump the lines of Be-like and Li-like neon are about equal as depicted in the bottom trace of figure 5. One reason we are not detecting the laser line could be a paucity of He-like neon. Complete photoionization of the neon L shell requires copious radiation at 240 to 450 eV energies, and a sodium plasma is lacking in transitions near these energies. In order to increase the photoionization pump we introduce potassium into the z pinch with the idea that potassium L radiation could help photoionize the neon L shell and produce more He-like neon. With 25% potassium in the driving pinch the ratio of Li-like to Be-like neon line intensities increases a factor of 2 indicating a higher ionization level and thus more He-like neon is present when potassium is used in the pinch.

The effect of different amounts of potassium introduced into the sodium pinch was scanned on a shot to shot basis. The results are shown in figure 6. The total number of wires in the z pinch load array is 16. The power in the 250 to 450 eV window is measured by a titanium-filtered carbon x-ray diode. This diode in figure 6 shows the power in this window increasing a factor of 3 in going from 0 to 25% potassium in the load. The 11 Å resonant photopump line power and total sodium K shell power, measured by x-ray diodes, bolometers, and photoconductive detectors, do not significantly decrease up to a 25% fraction (4 wires) of potassium. However at 50% potassium (8 wires) the sodium K shell and resonant photopump power fall off dramatically. The data of figure 6 shows that 25%

potassium in the sodium load optimizes the ability of the z pinch to both photoionize the neon L shell and resonantly photopump the He-like neon.

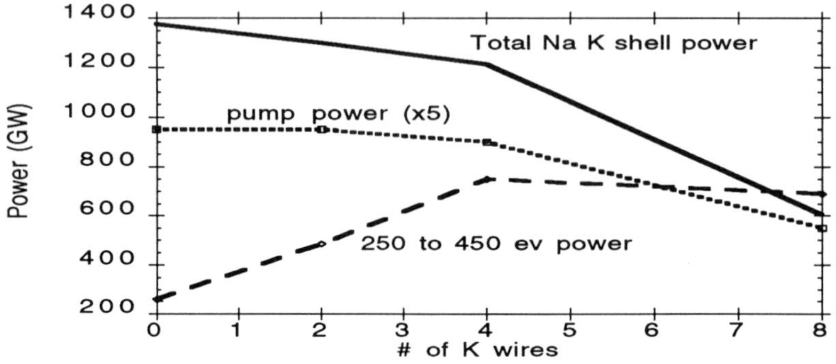

Figure 6. The optimum amount of potassium in the sodium pinch is 4 wires or 25% of the total mass.

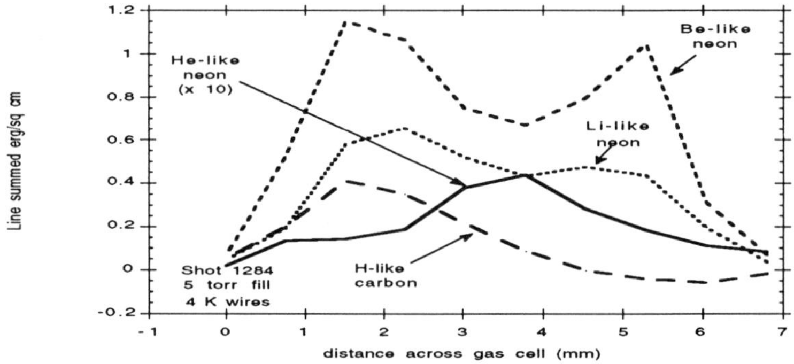

Figure 7. Spatial Profiles of lines of various ionization states

The spatial distributions of lines of He-like, Li-like, and Be-like neon, and H-like carbon, recorded with a 10 nsec gate centered on the radiation power pulse are shown in figure 7. The lines are the 78.3 Å 3d-2p triplet transition for He-like neon, the 98.2 Å 3d-2p transition for Li-like neon, the cluster of 3d-2p triplet transitions around 110 Å for Be-like neon, and Balmer alpha at 182 Å for H-like carbon. The side window is located at position 0 in figure 7. Be-like and Li-like neon are seen to burn out at the center of the gas cell where He-like emissions peak. The neon is likely heated by a radiatively driven shock wave that moves from the side window across the gas cell. The H-like carbon in figure 7 comes from the side window and its location of intensity fall-off indicates the position of the shock. The shock is estimated to travel 1 mm during the 10 nsec data acquisition gate. The shock evidently heats a localized region of the neon, burning through the L shell and enhancing He-like neon in concert with direct photoionization.

As mentioned above, the He-like neon line measured is the 78.3 Å 3d to 2p triplet transition array. Modelling indicates that this line should be stronger than other He-like neon XUV lines because the lower 2p triplet level of this transition has a slow decay time and a large population. The small measured intensity of the 3d-2p He-like line points out the difficulty of measuring the candidate laser transition which is predicted to be about 10 times weaker in fluorescence.

Radiation may not be the only mechanism by which the neon gas cell is heated. In figure 8 we show a time integrated spectrum from the neon gas cell recorded with the flat field spectrometer. The lines of Li-like neon are seen to have a "satellite" blue shifted by an amount proportional to the wavelength of the line. This is suggestive of a doppler shift, which would indicate jetting out of the neon gas cell at a very high velocity of 1.5e8 cm/sec. A possible mechanism for this acceleration is a J x B force, which would imply that the neon gas cell draws significant return current. The gas bag would need to draw tens of kiloamperes in radial current to accelerate to neon to these high velocities.

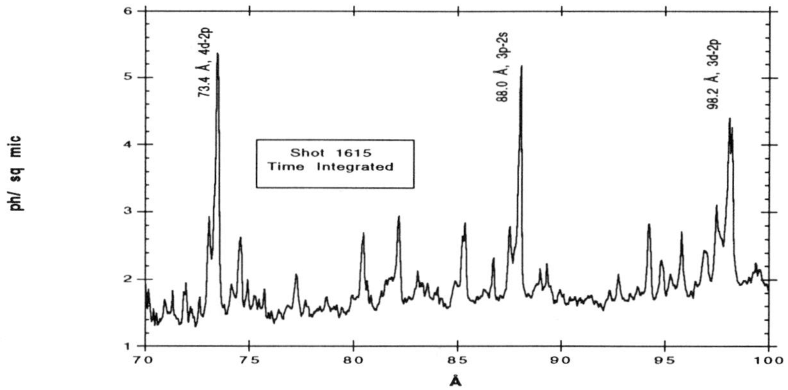

Figure 8. Li-like neon lines display blue shifted satellites that indicate plasma jetting likely driven by return current.

We have identified the 4d to 2p triplet transition in He-like neon due to the better sensitivity of the flat field instrument. The detected line is shown in figure 9. The relative intensity of this line with respect to the 3d to 2p triplet would indicate a 4 to 3 inversion[25] and corroborates the results of ref 22.

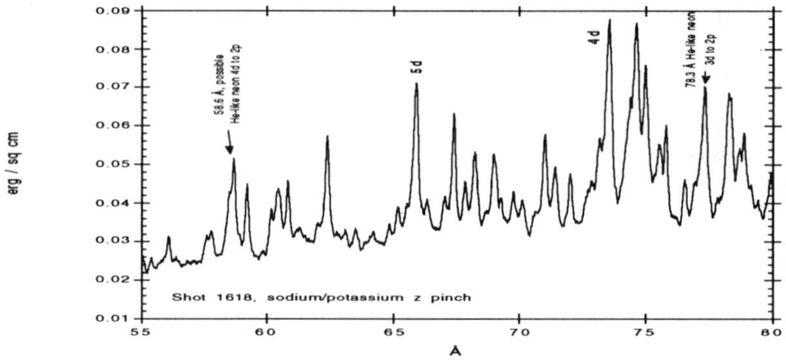

Figure 9. We may have detected the 4d to 2p triplet transition in He-like neon, using a sodium-potassium pinch.

The He-like lines are modelled to be due to keV pump radiation. In figure 10 we display the neon spectrum pumped by a gold spectrum which has comparitively less keV pump than a sodium/potassium pinch. With the gold pinch there are no He-like emissions. This particular shot which is time resolved during the radiation pump also shows an extreme amount of plasma jetting, again likely driven by current. The current, therefore, is incapable of exciting the He-like states. Evidently He-like lines can only be pumped by keV radiation.

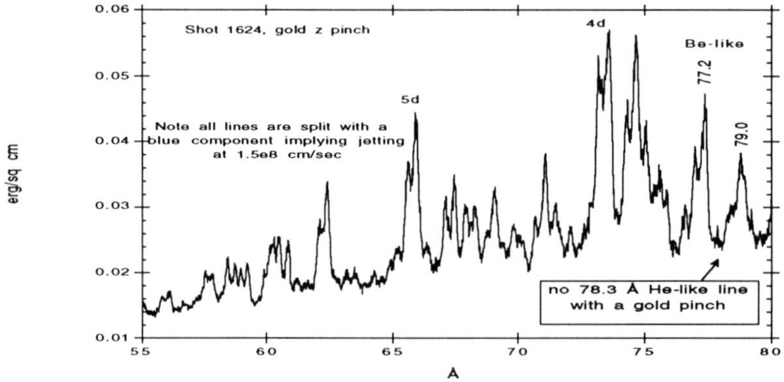

Figure 10. With the reduced keV pump of a gold pinch we see no emissions from He-like neon.

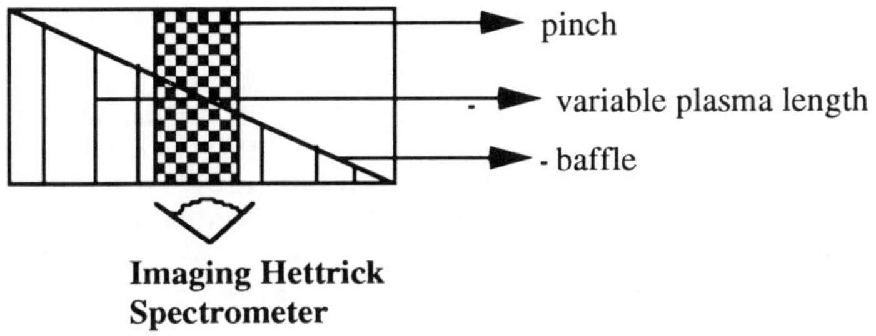

Figure 11. Geometry of single shot gain measurement attempt.

We have attempted a single shot gain length measurement. The geometry of the technique is shown in figure 11. The neon gas cell is diagonally baffled to only allow radiation to heat the plasma along a triangular portion of the neon. Viewed from below the flat field spectrometer images along a direction of increasing plasma length.

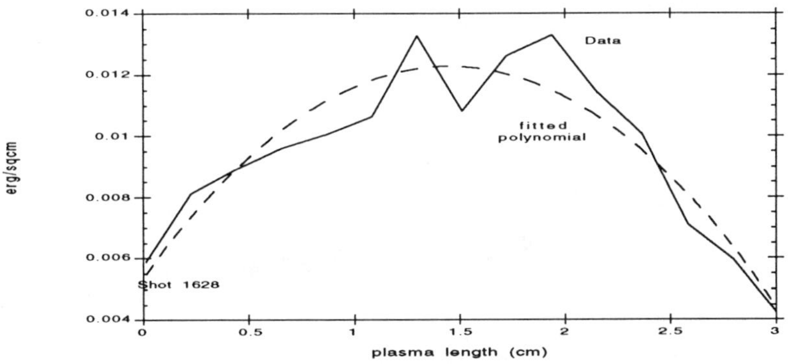

Figure 12. Intensity of Li-like neon 4d-3p line at 282.4 Å as a function of baffled plasma length.

The results from this experiment are rather discouraging as depicted in figure 12 which shows the intensity of a candidate recombination laser line as a function of plasma length. The line basically peaks along the center of the gas bag and falls off to either side. The profile suggest that the neon gas cell is not affected by the baffle, perhaps because it is heated by current, or because collisional processes or a diverging shock equilibrate the entire neon plasma even though only half of it is illuminated. The fall off of the Li-like neon line on either side of figure 12 is likely due to the cooling effect of the gas cell side walls. That the emissions peak along the center could also be suggestive of radiative self-pumping since most radiation source functions generally peak in the geometric center of the medium.

Besides having unwanted extra heating mechanisms the neon gas cell target is riddled with absorption problems due to the side walls and end window. The pinch side of the end window is shadowed from radiation and likely is not as hot as the rest of the end window. In figure 13 we show a time-gated neon spectrum revealing several candidate recombination laser lines as recorded from the side window and gas cell center. Along the side window the 282.4 Å line is considerably absorbed with respect to harder 5 to 3 transitions below 200 Å. We could therefore expect lasing measurements above 200 Å along the side window to be unwise. Evidently the 282.4 Å line is too soft to transmit through the cool shadowed portion of the end window. However along the gas cell center, where the end window is well illuminated the 282.4 Å line is well transmitted. Here the 5 to 3 lines are weaker, likely because the recombinations peak near the higher densities of the side window. One final feature of figure 13 are very wide absorption lines from Li-like carbon. These have been indentified as coming from the gas cell target side wall. Indeed spatially resolved spectra show cool plama from these side walls limits the hot region of the neon gas cell to the central 2 mm, and it is possible that hot neon plasma measured to be localized to the gas cell center is more due to cooling by surrounding walls than to any localized heating effect.

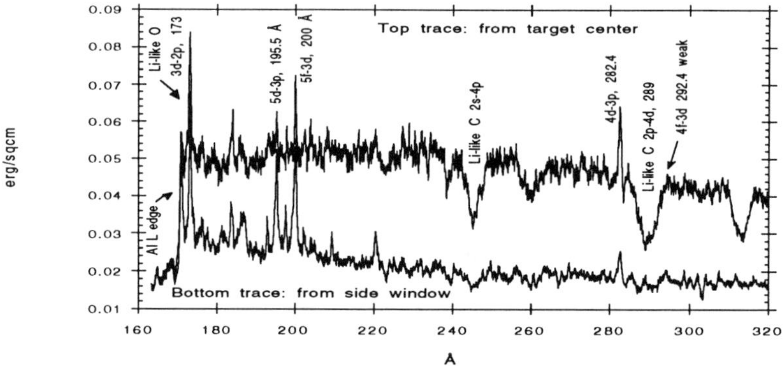

Figure 13. The 4d-3p line is absorbed by the shadowed portion of the end window. Fat C absorption lines indicate dense target side wall plasma.

CONCLUSIONS

We have measured the ionization balance across a neon gas cell irradiated by a 10 TW z-pinch. The measurements are consistent with a radiatively driven shock wave propagating through the neon. We have increased the overall level of ionization and the amount of He-like neon in the gas cell by introducing 25% potassium into the z pinch load. This increases the ionizing soft x-ray radiation a factor of 3. Oxygen absorption lines have also been eliminated from the spectrum by using an oxygen-free 5000 Å polystyrene end window. From the measurement of fast plasma jetting and the lack of response of the neon plasma to a radiation blocking baffle we conclude that return current may be flowing in the gas cell. Current does not appear to affect the excitation of the He-like neon states because of the higher excitation energy required.

This work performed at Sandia National Laboratories supported by the U. S. Department of Energy under contract DE-AC04_76DP00789.

REFERENCES

1. D.L. Matthews, et al., Phys. Rev. Lett 54, 110 (1985)
2. M.D. Rosen, et al., Phys. Rev. Lett. 54, 106 (1985)
3. B.J. MacGowan, et al., Phys. Rev. Lett. 59, 2157 (1987)
4. T.N. Lee, E.A. McLean, and R.C. Elton, Phys. Rev. Lett. 59, 1185 (1987)
5. S. Suckewer, et al., Phys. Rev. Lett. 55,1753 (1985)
6. C. Chenais-Popovics, et al., Phys. Rev. Lett. 59, 2161 (1987)
7. P. Jaegle, et al., J. Opt. Soc. Am. B 4, 503 (1987)
8. D. Kim, et al., J. Opt. Soc. Am. B 6, 115 (1989)
9. J.C. Moreno, et al., Phys. Rev. A 39, 6033 (1989)
10. G. Jamelot, et al., Appl. Phys. B 50, 239 (1990)
11. V.A. Bhagavatula, J. Appl. Phys. 47, 4535 (1976)
12. J.P. Apruzese, et al., J. Appl. Phys. 53, 4020 (1982)
13. J.P. Apruzese and J. Davis, Phys. Rev. A 31, 2976 (1985)
14. J.P. Apruzese, et al., Phys. Rev. A 35, 4896 (1987)
15. F.C.Young, et al., Appl. Phys. Lett. 50, 1053 (1987)
16. J. Nilsen, Phys. Rev. 40, 5440 (1989)
17. B.N. Chichkov and E.E. Fill, Phys. Rev. A 42, 599 (1990)
18. Y.T. Lee, et al., J. Quant. Spectros. Rad. Transfer 43, 335 (1990)
19. J. Nilsen and E.A. Chandler, Phys. Rev. A 44, 4591 (1991)
20. J. Nilsen, J. Quant. Spectrosc. Radiat. Transfer 46, 547 (1991)
21. N. Qi and M. Krishnan, Phys. Rev. Lett. 59, 2051 (1987)
22. J.L. Porter, et al., Phys Rev. Lett. 68, 796 (1992)
23. J.P. Apruzese, et al., in Proceedings of the Second International Colloquium on X-ray Lasers, York, England, edited by G.J. Tallents (Institute of Physics, Bristol, 1991), p. 39
24. T.J. Nash, et al., Rev. Sci. Instrum. 61 2810 (1990)
25. J.P. Apruzese, private communication (1993)

IMAGING X-RAY SPECTROSCOPY IN Z-PINCH EXPERIMENTS

S.A.Pikuz*, A.I.Erko**, A.Ya.Faenov***
*P.N.Lebedev Physical Institute, Moscow, Russia
**Institute for Microelectronics Technology, Chernogolovka, Moscow region, Russia
***VNIIFTRI, Mendeleevo, Moscow region, Russia

ABSTRACT

X-ray spectroscopy with high spectral (up to $\Delta\lambda/\lambda = 10^{-4}$) and spatial resolution (up to microns) is discussed. Devices based on crystals, difraction and Bragg-Fresnel elements and there application in Z- and X-pinches experiments are observed.

INTRODUCTION

X-ray spectroscopic techniques enable measuring the most important plasma parameters[1]. The range of wavelengths from 1 to 100 Å is of main interest in the development of spectroscopic methods of plasma diagnostics. In particular, the plasma parameters vary between very wide limits during compression of the plasma filament in studies on fast Z-pinches[2]. High spatial and spectral resolution are required to obtain adequate information about plasma parameters. Here we discuss the methods of imaging spectroscopy by using crystals and artificial X-ray optical elements.

DEVICES BASED ON CRYSTALS

Spectrographs with slits. Crystals themselves due to Bragg equation possess spatial resolution in direction of dispersion and may be used to obtain two dimensional plasma images in light of separate spectral line. Spatial resolution in these case depends on width of spectral lines and may be higher than 100 μm. This possibility is most fruitful in experiments with large plasma objects in powerful machines like Angara-5[3]. Figs.1,2 demonstrate experimental results obtained by spestrographs with convex mica crystals and slit or diaphragm in front of crystal for spatial resolution in perpendicular direction. Similar devices with CsAP crystals we widely use in X-pinch experiments[4] on BIN machine (Fig.3).

For wavelengths and shapes measurements with high spectral resolution used Johann-type spectrographs. On Figs.4,5 spectra of Ne-like Ni ions presented. Accuracy of wave lengths measurement achieved 10^{-3} Å.

Spherically curved crystals. High spectral and spatial resolution can be attained by using spectrographs with spherically curved crystals[1]. Instruments of these kind were first used for diagnostics of laser plasmas but have not found road application, mainly because it has not jet been possible to prepare crystals with a bending radius of less than 300 mm. This limit the detection range and luminosity of the instrument and increas its size. In our study we have constructed

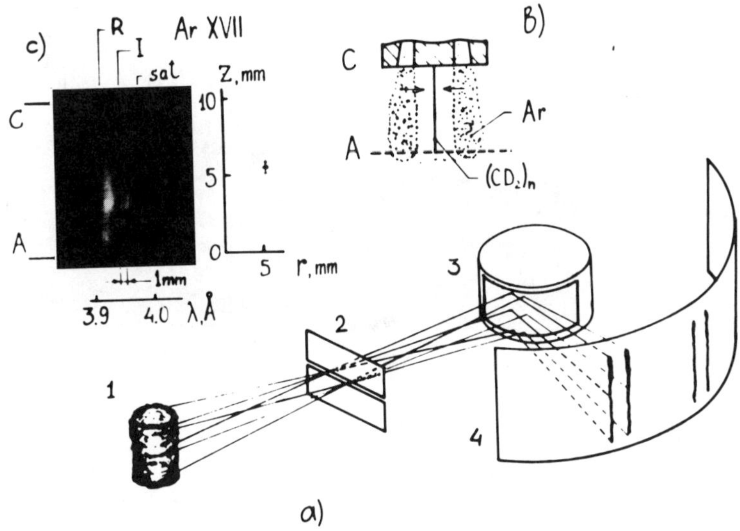

Fig.1: Convex mica crystal spectrograph with axial spatial resolution in Angara-5 experiment: (*a*) spectrograph scheme (1 — pinch, 2 — slit, 3 — mica crystal, 4 — photofilm), (*b*) diode arrangement, (*c*) spectrogram of Ar-plasma.

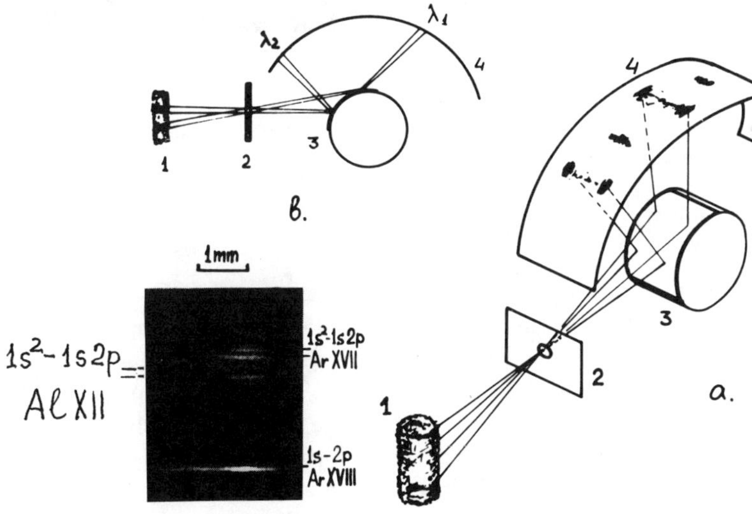

Fig.2: Convex mica crystal spectrograph with radial spatial resolution in Angara-5 experiment: (*a*) scheme of spatial resolution in dispertion direction forming, (*b*) spectrograph scheme (1 — pinch, 2 — pin-hole, 3 — mica crystal, 4 — photofilm), (*c*) spectrogram of Al/Ar-plasma.

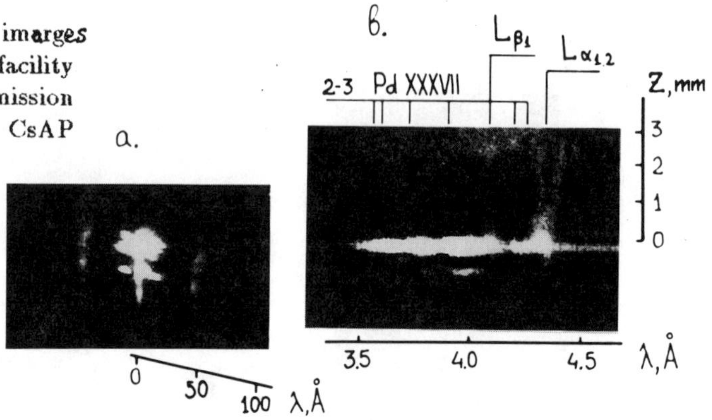

Fig.3: X-pinch plasma imarges obtained on BIN facility by using of (*a*) transmission grating and (*b*) convex CsAP crystal.

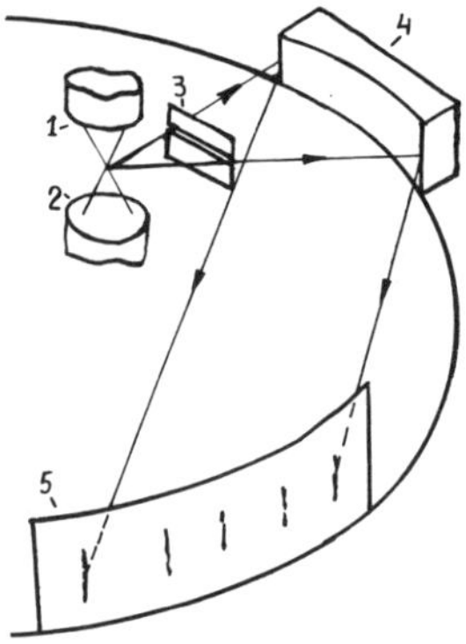

Fig.4: Scheme of Johann-type spectrograph with spatial resolution: 1 — anode, 2 — cathode, 3 — slit, 4 — curved crystal, 5 — photofilm

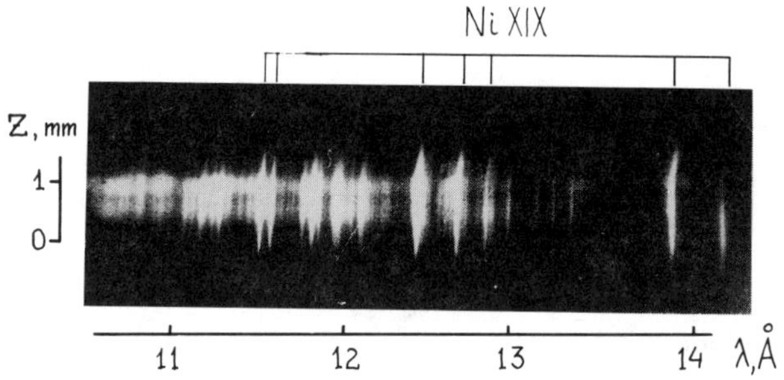

Fig.5: Spectrogram of X-pinch Ni-plasma.

spherical analysers based on mica crystals with bending radius $R = 100 - 200$ mm and diameter D up to 30 mm (or sizes 15 mm ×35 mm). The high reflection of mica in different orders, combined with its large field of view, makes it possible to cover a large spherical range, from 3 to 19 Å. This crystals may be used in an X-microscope scheme[5] (Fig.6). Fig.7 shows the setup for using the spherical crystal in the focusing spectrograph mode with spatial resolution. In this case the spectrum is formed on the Rowland circle in the sagittal plane and the spatial resolution is ensured by focusing in the meridional direction. The magnification in this case is given by $K = b/a = \cos 2\varphi = 2m^2(\lambda/2d)^2 - 1$, were φ is the angle of incidence of radiation on the crystal, $a = R\cos\varphi/\cos 2\varphi$ and $b = R\cos\varphi$ are the distances from the source to crystal and from the crystal to the photographic film, respectively, d — interplanar distance of crystal. On Fig.8 spectra obtained by spherically curved crystal is compared with one obtain by convex CsAP crystal spectrograph.

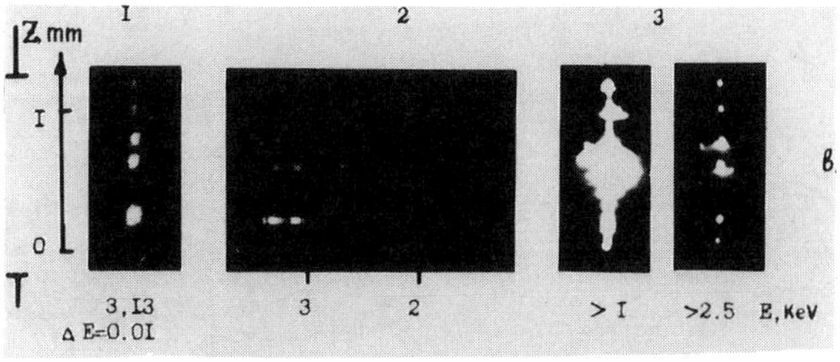

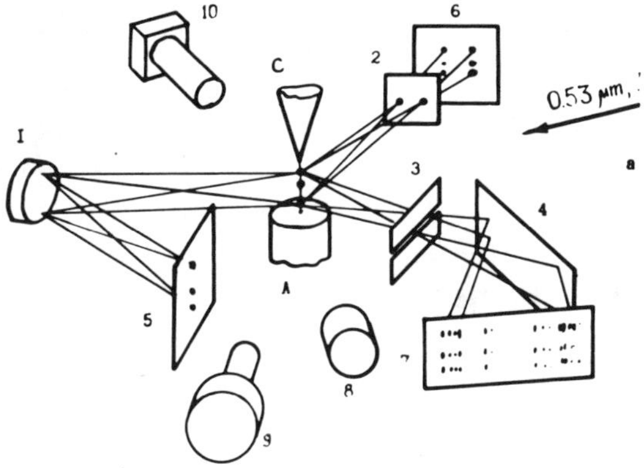

Fig.6: (*a*) Experimental arrangement for X-ray diagnostic on BIN facility: 1 — spherically bent mica crystal in the microscope mode, 2 — pin-hole cameras, 3 — spatial resolution slit, 4 — CsAP crystal, 5-7 — photofilms, 8 — calorimeter, 9 — photomultiplier with scintilator, 10 — optical frame cameras; (*b*) results obtained with exploding Pd-wire: 1 — monochromatic image of plasma in 5-th number reflection of mica crystal, 2 — plasma spectrum obtaned using spectrograph with a plane crystal and slit, 3 — images at various X-ray hardness obtaned using pin-hole cameras.

ARTIFICIAL X-RAY OPTICAL ELEMENTS

Transmission diffraction grating. By using of this type dispersive elements it possible to investigate pinch radiation in very wide spectral range (5–100 Å) with high spatial resolution but very rough spectral resolution. In our study we used a spectrographs with a transmission grating[6], consisting of a system ofunsupported strips with period 1 μm, arranged inside an aperture of diameter 25 μm (Fig.9). Since the membrane in which the grating was formed was comparatively large (dia ≈ 1,5 mm), we placed an auxiliary screen of 50-μm-thick tantalum foil with a 150 μm diameter hole in front of the grating to prevent direct exposure to hard X-rays. The images obtained indicate that the plasma filament had a fairly complex structure (Fig.10).

Bragg-Fresnel multilayer lens (BFML). Recent achievements in X-ray optics domain significantly expand the experimental possibilities in spectral diagnostics of an high-temperature plasma sources. Development of an artificial X-ray optical elements with volume Fresnel zones[7], BFML allows not only to overcome the choice restrictions of nature and synthetic crystals, which can be used in X-ray devices. BFML also able considerably improve some characteristics of such devices, for example, relative aperture and spatial resolution.

The BFML with straight structures which provide linear focusing were used in our experiments. These lens was made by using electron beam lithography,

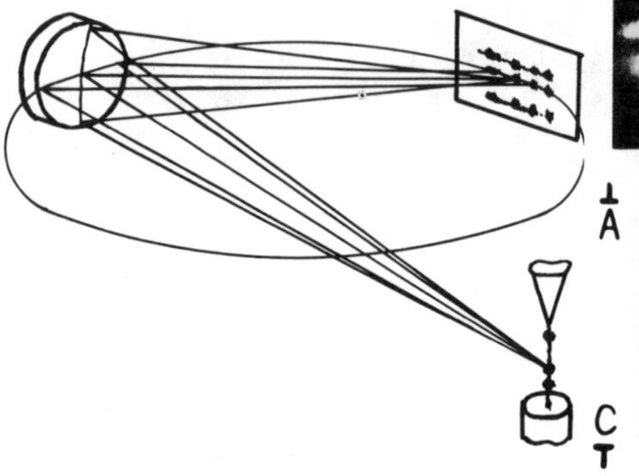

Fig.7: Setup of spherically curved crystal in the spectrograph mode.

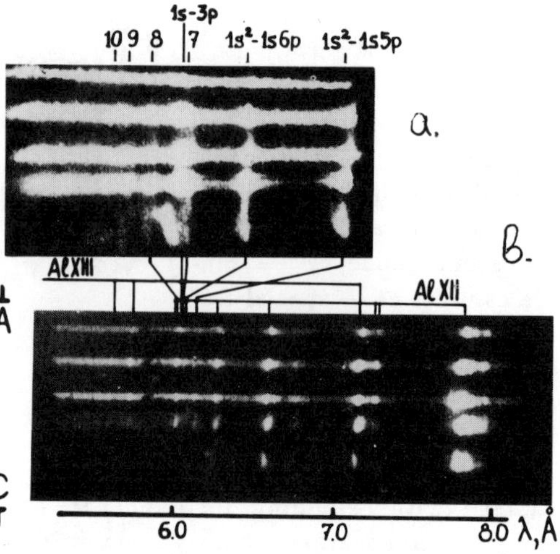

Fig.8: Al-plasma spectra obtained by using of spherically curved mica crystal (a) and (b) convex CsAP crystal.

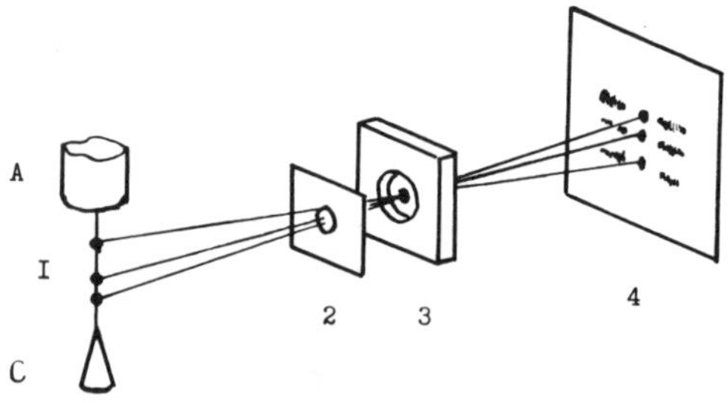

Fig.9: Setup of transmission grating: A — anode, C — cathode, 1 — plasma, 2 — screen, 3 — grating, 4 — photofilm.

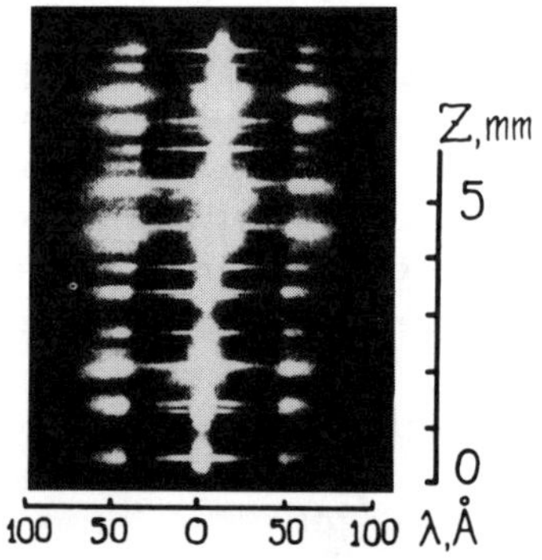

Fig.10: Image of the exploded W-wire of 8 μm in diameter obtained by pin-hole transmission grating using.

optical lithography, and ion beam etching processes. The aperture of lens of 1 mm ×5.6 mm and the external minimum zone width of 3.5 μm was done. The BFML were prepared by ion etching a 141 layer of 32 Å period W/Si multilayer mirror coated by magnetron. This lens had the focus distance of 7 cm at the wavelength of 1.54 Å.

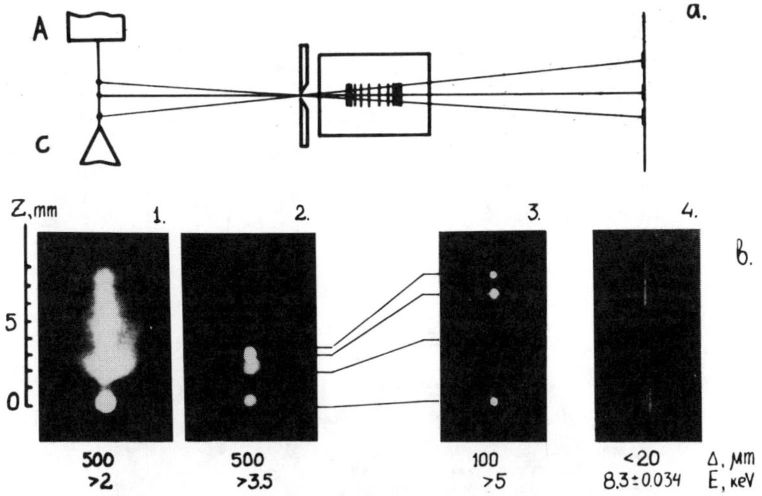

Fig.11: (a) The scheme of lens operation in emission of λ = 1.5 Å and (b) the experimental results: 1–3 — pinhole camera images, 4 — lens obtained images (E — filter cut-off energy in pin-hole camera, D — real spatial resolution of pin-hole camera).

Working at the wavelength of 1.5 Å the slit of $\approx$ 100 μm was placed in front of the lens (see Fig.11a). In Fig.11b one can see pinhole camera images: 1, 2, 3 obtained after the filters with different cut-off energy and plasma image obtained by Bragg-Fresnel lens in emission with wave length of 1.5 Å. The dimensions of plasma hot point image is about 20–30 μm, that is why we say that there are the extremely compact high gradient plasma regions which seem to be illuminated by the electron beam generated by current break in pinch on the second compression stage.

Zonal plate structure on crystal. The traditional methods which use the slit parallel to dispersion direction have not-high relative aperture and can be really used if the width slit (and accordingly spatial resolution) is not more than $\approx 20 - 30$ μm. We proposed to obtain the high-temperature plasma source spectrum with high spatial resolution in wide enaugh spectral interval by using the Bragg-Fresnel lens.

For these purposes the linear Bragg-Fresnel lens was calculated and fabricated in IMT Russian Academy of Sciences. Calculated parameters of this lens are: focal distance f_o = 7 cm at the wavelength λ_o = 7.75 Å; the last zone width 0.5 μm; total aperture of $\approx$ 1 cm ×108.5 μm. Mica crystal with lattice period $2d = 19,884$ Å as a substrate was used. Bragg-Fresnel lens was formed by electron beam lithography and wet etching in the metall layer, deposited on the cryslal surface. The thickness of the layer was disigned to provide π phase shift in the reflected X-ray beam coming through the layer.

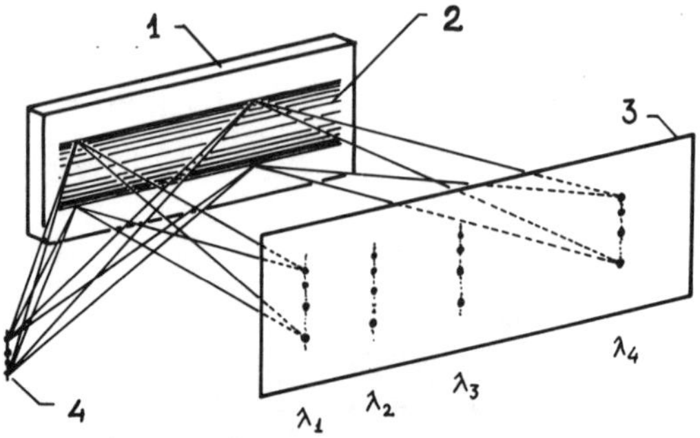

Fig.12:The scheme of using the linear Bragg-Fresnel lens for plasma spectra obtaining with spatial resolution: 1 — mica crystal, 2 — linear Bragg-Fresnel lens with $f_o = 7$ cm for $\lambda_o = 7,75$ Å, 3 — photofilm, 4 — pinch.

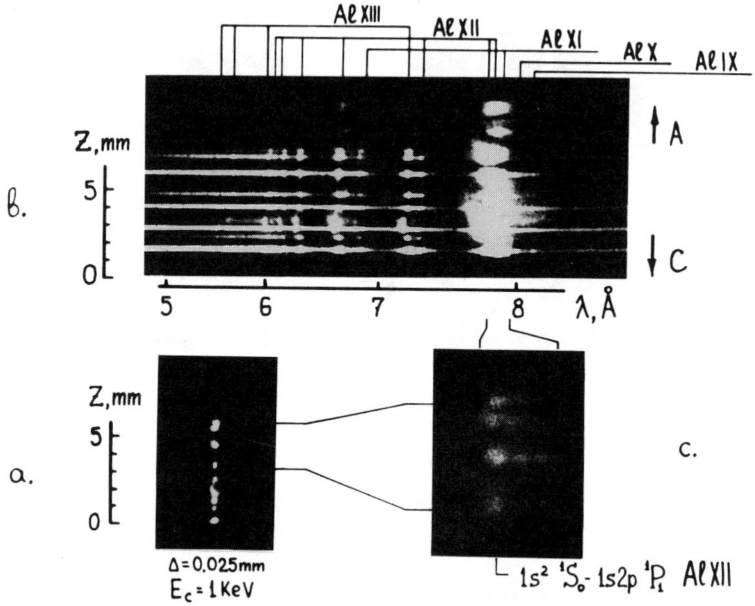

Fig.13: Al-wire explosion in Z-pinch geometry: (a) Al-plasma (pin-hole diameter — 25 μm, cut-off filter energy — 1 kev; (b) plasma spectrum obtained by using the plane crystal CsAP spectrograph; (c) plasma image in radiation of resonance line He-like AlXII ion obtained by using Bragg-Fresnel lens (z — coordinate in cathode - anode direction).

The scheme of the experiment with Bragg-Fresnel lens is shown in Fig.12. First Bragg-Fresnel lens was used for obtaining Z-pinch plasma image in radiation close by the resonance line of He-like ion AlXII. In Fig.13 you can see plasma spectrum obtained on the plane crystal spectrograph and the image obtained by using Bragg-Fresnel lens. In this case Bragg-Fresnel lens was set in such way that the image on the wave length $\lambda_o = 7,75$ Å was with magnification $\gamma_o = 1,5$.

CONCLUSION

By using devices described it is possible to investigate plasma parameters with high spatial and spectral resolution in wide spectral range. Table gives comparison of devises parameters.

Table.

	1	2	3	4	5
Spectrographs with slits:					
flat crystal	1–20	0.1–0.2	10^{-3}	100	20
convex crystal	1–20	1	10^{-3}	100	20
Johann scheme	1–20	0.05–0.2	10^{-4}	—	20
Spectrograph with spherically	1–20	0.05–0.2	10^{-4}	—	20
curved crystal					
X-ray microskope	1–20	0.01	—	5–10	5-10
Devices with artifical elements:					
transmission grating	5–100	1	10^{-2}	—	25
Bragg-Fresnel lens	1–15	0.01	10^{-2}	0.5–1	20
zonal plate structure on crystal	3–20	0.1–0.2	10^{-3}	100	0.5–1

1 — spectral range, Å;
2 — relative spectral range $(\lambda_{max} + \lambda_{min})/2(\lambda_{max} - \lambda_{min})$;
3 — spectral resolution $\Delta\lambda/\lambda$;
4 — spatial resolution in direction of dispersion, μm;
5 — spatial resolution in direction perpendicular to dispersion, μm.

REFERENCES

1. V.A.Boiko, A.V.Vinogradov, S.A.Pikuz, I.Ju.Skobelev, A.Ja.Faenov. J.Sov. Laser Reas. **6**, No 2, 85 (1985).

2. S.M.Zakharov, G.V.Ivanenkov, A.A.Kolomensky, S.A.Pikuz, A.I.Samokhin. Fiz.Plazmy **9**, 469 (1983) [Sov.J.Plasma Phys. **9**, 271 (1983)].

3. See the paper of Angara-5 group in this book.

4. B.A. Bryunetkin, G.V. Ivanenkov, S.Ya. Khakhalin, A.R. Mingaleev, S.A. Pikuz, V.M. Romanova, I.Yu. Skobelev, A.Ya. Faenov, T.A.Shelkovenko. Kvant. Elektronika **20**, No 3 (1993).

5. L.Belyaev, A.B.Gil'varg, Yu.A.Mikhailov, S.A.Pikuz, G.V.Sklizkov, A.Ya. Faenov, S.I.Fedotov. Sov.J.Quant.Electron. **6**, 1121 (1976).

6. B.A.Bryunetkin, G.V.Ivanenkov, S.A.Pikuz, A.Ja.Faenov, T.A.Shelkovenko. Sov.Tech.Phys.Lett. **17**, 689 (1991).

7. A.I.Erko. J. X-ray Sci.Technol. **2**, 297 (1990).

SOFT X-RAY SPECTRA ANALYSIS IN A HIGH-CURRENT Z-PINCH

F.B. Rosmej, O.N. Rosmej
Institute of Spectroscopy, 142092 Troitzk, Russia
present address: Ruhr-Universität Bochum, Institut für Experimentalphysik V, W-4630 Bochum, Germany

S.A. Komarov, V.O. Mishensky, J.G. Utjugov
TRINITI (ANGARA-V-1), 142092 Troitzk, Russia

ABSTRACT

Experiments at the high current (up to 2.5 MA) Z-pinch facility "ANGARA-V-1" were performed with Al, Cu, W and CD_2 wires and outer Ar, Xe and D_2 gas shells (p=0-1.8 atm). Single shot soft X-ray spectra were recorded by means of a curved X-ray cristal spectrograph in a wide spectral wavelength interval. The characteristic X-ray emission lines of H-, He- and Li-like ions were analysed with the elaborated spectra simulation codes "SAT" and "META". The electron temperature T_e, density n_e and ionisation temperature T_z were determined with respect to the non-equilibrium plasma state, optical thickness and electron beam excitation. Additional information about the plasma developement was obtained from the emission lines of Na, Cl and Si, which were added to the wire material for diagnostic purposes. For Ar we obtained electron temperatures up to $T_e \approx 1.2$ keV and electron densities of about $n_e \approx 10^{21}$ cm^{-3}, for Al $T_e \approx 600$ eV, $n_e \approx 3 \cdot 10^{20}$ cm^{-3}. A low ionisation temperature in practically all discharges indicates non-equilibrium states of overheated plasmas. Information about the plasma geometrie was obtained from pinhole framings in different spectral ranges. Discharges with copper wires showed a strong decrease of the hot spot creation with outer gas shells. Pinhole framings were analysed with a high resolution by means of a topographical method.

I. INTRODUCTION

Investigations of high current Z-Pinches for the thermonuclear fusion concepts are reported in many articles (e.g. Refs. 1-3). The optimisation of the fusion processes are directly connected with the enhancement of the pinch stability, lifetime, current coupling and neutron production.

In the frame of an international collaboration[4] experiments for the investigations of the physics of transient dense high current Z-Pinch plasmas were carried out at the "ANGARA-V-1" facility (see also the contributions of Smirnov, Choi, Etlicher, Larour, Wessel et al. in this volume). Here, we focus on the detailled analysis[5] of soft X-ray emission lines from the wire material, diagnostic elements and the outer gas shells, determine the plasma parameters in various ionisation stages, characterise the non-equilibium plasma state and report on a detailed analysis of pinhole framings by means of a topographical method.

Soft x-ray spectra were recorded by means of a convex mica cristal with curvature radius 2.5 cm in the wide spectral range from $3\text{-}18{\cdot}10^{-10}$ m. Further experimental details are described elsewhere.[1,2,4,5]

II. EXPERIMENTAL OBSERVATIONS AND SPECTRA SIMULATIONS

Soft X-ray emission lines originated from single and double excited states of H-, He- and Li-like Na, Al, Si, Cl and Ar ions are investigated. For spectra simulation analyses we employed the codes "SAT"[6] and "META"[7]. These codes involve full collisional radiative modellings of non-equilibrium and optically thick plasmas, include innershell processes, electron beam excitation and are based upon the simultaneous simulation of many emission lines from single and double excited states. In addition to the single excited levels in H-, He- and Li-like ions, the simulations include as well the complete 2l2l´- and 1s2l2l´-satellite spectra involving all double excited levels in a full collisional radiative modelling. Details of the codes are described elsewhere.[6,7]

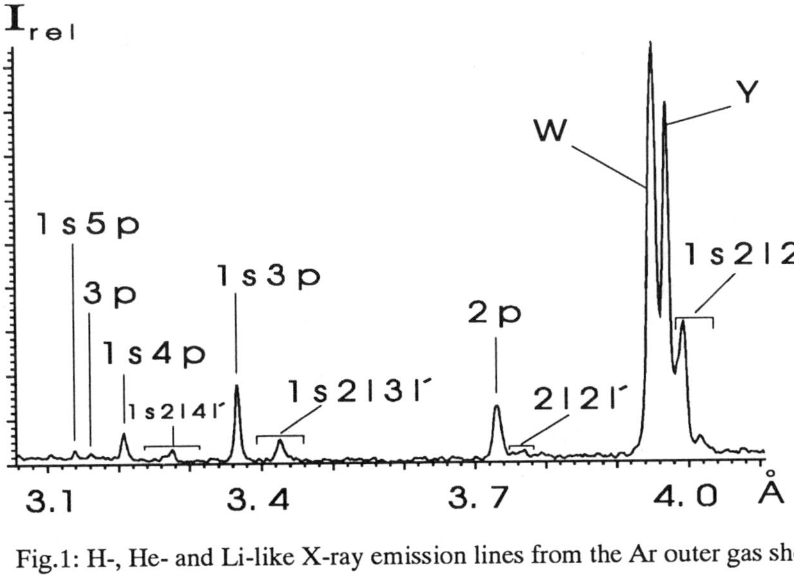

Fig.1: H-, He- and Li-like X-ray emission lines from the Ar outer gas shell

By using time integrated spectra for the analysis, we investigated not only the temperature and density parameters but payed also attention to the duration of the emission of characteristic X-ray lines. Because of the transient nature of the plasma, we determine the plasma parameters from emission lines originated from one ionisation stage only. To characterise the non-equilibrium situation (overheated or overcooled plasmas) the concept of the ionisation temperature T_z[8] was used. The wire material was doted with different Z_n diagnostic elements (solutions of NaCl, SiO_2 coatings) to investigate the ionisation processes by means of different elements simultaneously in single shots and to estimate possible influences from the emitting ion itself on the plasma parameter determination.

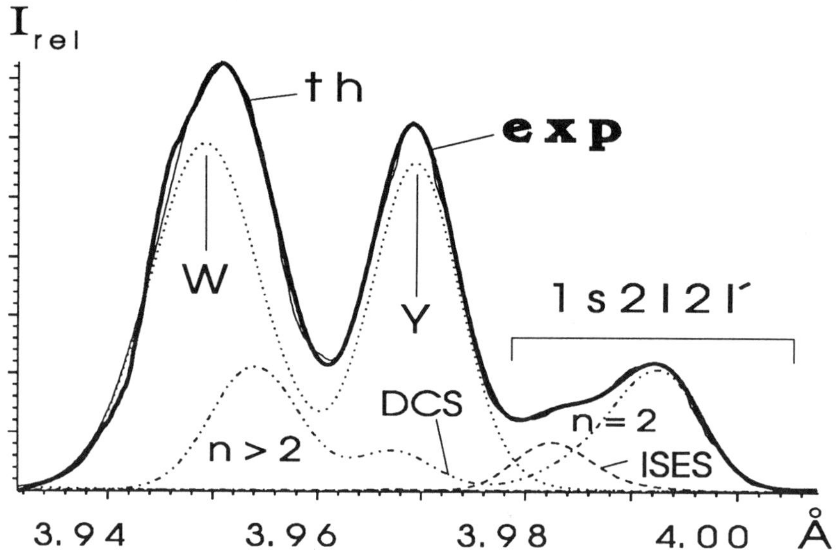

Fig. 2: Spectra simulation of the W-,Y-lines, DC and IS excited satellites

Fig. 1 shows the single shot X-ray emission lines of H-, He- and Li-like ions from the Ar outer gas shell. Fig. 2 shows the spectra simulation in the He- and Li-like ionisation stage depicting the experimental spectrum (exp) and the different contributions of the theoretical one (DCS=dielectronic captured satellites, ISES=innershell excited satellites, W and Y contribution, th=sum of all contributions). As can be seen from fig. 2, the visible line profile of the resonance line $W=1s^2\text{-}1s2p\ ^1P_1$ has a considerable higher full width at half maximum (FWHW) than the intercombination line $Y=1s^2\text{-}1s2p\ ^3P_1$. This is partly caused by unresolved higher order Li-like ($1s^2nl\text{-}1s2l'nl''$)-satellites (indicated by $n > 2$ in fig. 2) and by photoabsorption. If we "clean" the resonance transition W and intercombination transition Y from satellite contributions, we obtain a FWHM-ratio of about R_{FWHM} := FWHM(W)/FWHM(Y) ~ 1.3 which is caused by photoabsorption and an intensity ratio R_I := I(Y)/I(W) ~ 0.65. From investigations of the line profiles with increased resolution in the fourth reflection order, we obtain a W-line width of about $5 \cdot 10^{-13}$ m, indicating that the plasma velocity is not greater than about $1.5 \cdot 10^7$ cm/s. Slight asymmetries on the blue wing of the W-line profile indicate differential plasma motion in connection with photoabsorption (expansion).

Photoabsorption and electron beams cause serious difficulties for the electron density determination. Optical thickness drives the density sensitive intensity ratio of the Y- and W-line into an insensitive one[9] giving rise to underestimations. Moreover, despite of the high discharging voltage we expect highly energetic electrons penetrating into the plasma and causing additional excitation of X-ray lines appart the thermal ones.[10] In consequence the diagnostic ratio of the Y- and W-line is considerable changed because of the different asymptotics for direct ($\sigma \sim \ln\mathcal{E}/\mathcal{E}$) and exchange transitions ($\sigma \sim \mathcal{E}^{-3}$). Fig. 3 shows the influence on the diagnostic ratio for different fractions of highly energetic electrons. Because enough detailed information about the penetration of the beam into the dense pinch plasma and it´s fractional contribution are not known we use the concept of the second temperature with T_{Beam}=30 keV.[7] As can be clearly seen even a small beam fraction of only 1% can result in density overestimations by an order of magnitude!

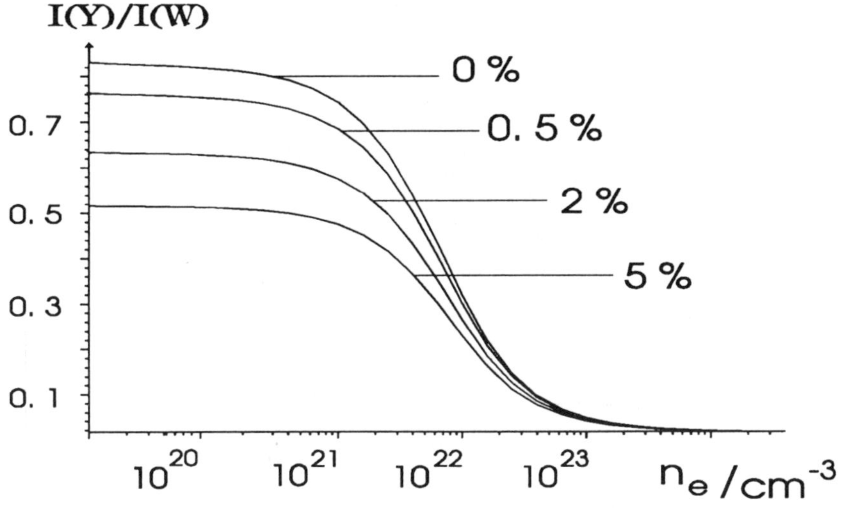

Fig. 3: Intensity ratio of the Y- and W-line in dependence of different beam fractions and electron density n_e

Neglection of electron beams gives an electron density of about $n_e \approx 2 \cdot 10^{21}$ cm^{-3} for the experimental spectrum from fig. 2. The Li-like 1s2l2l´-satellites are still in the density insensitve region[6], giving only an upper limit of $n_e < 10^{22}$ cm^{-3}. Although photoabsorption normaly increases the difficulties in the parameter evaluation, the broadening of the W-line can be used in a little bit tricky way to obtain a lower limit for the electron density in the presence of highly energetic electrons. This is an absolut necessary requirement because the diagnostic ratio can be located in the insensitive region even for rather low ratios (see fig. 3). With a maximum plasma diameter of about 1000 μm, we obtain for the observed width ratio R_{FWHM} (taken into account the apparatus profile) a lower limit of about $n_e > 2 \cdot 10^{20}$ cm^{-3}. This value in turn can be used to obtain an upper estimation of the beam fraction f in the He-like ionisation stage. From fig. 3 we estimate f < 2%. Further results of the simulations are: $T_e(He) \sim 900$ eV, $T_z(Li,He) \sim 400$ eV indicating an overheated plasma. Similar considerations carried out for the Ly-α line and it's adjacent 2l2l'-satellites give: $T_e(H) \sim 1200$ eV, $T_z(He,H) \sim 1000$ eV, $n_e(H) < 10^{21}$ cm^{-3}. The higher temperatures in the H-like ionisation stage indicate increasing electron temperatures.

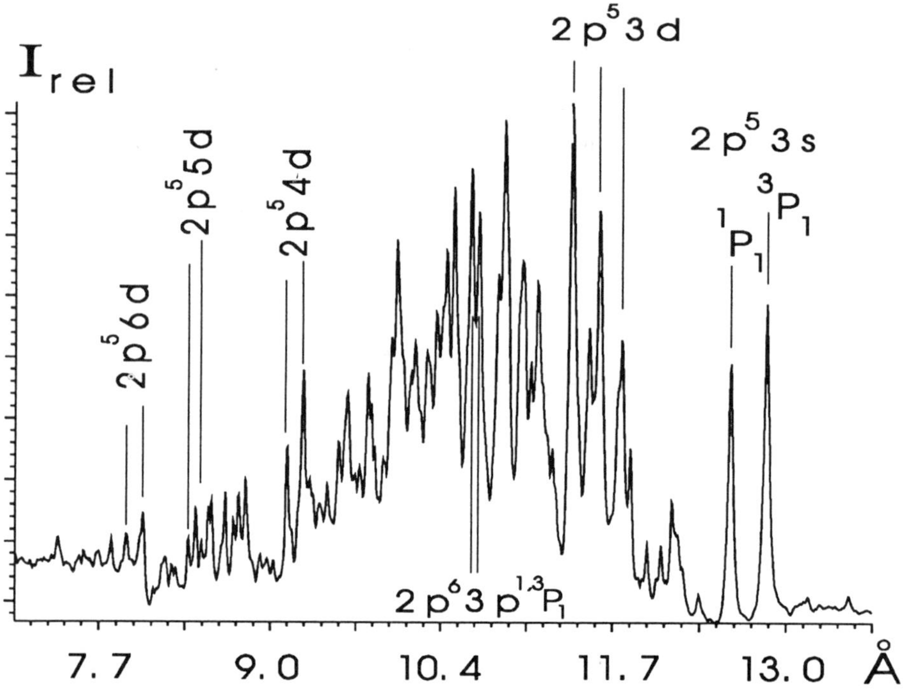

Fig. 4: Ne-like spectrum from a 20 μm copper wire

Discharges with Cu-wires were performed with and without outer Xe gas shells. Fig. 4 shows the emission lines from the copper L-shell. The line identification showed, that the single excited emission lines originate mainly from Ne-like ions, Li-like transitions were not observed, so the electron temperature seems not higher than some hundreds of eV. For further diagnostic purposes, the wire was covered with SiO_2. Fig. 5 shows the X-ray emission lines from H-, He- and Li-like Si: $T_e \sim 500$-700 eV, $n_e \sim 3 \cdot 10^{20}$ cm^{-3}. These parameters are in agreement with the absence of Li-like transitions in Cu. Figs. 6a/b show the topographical analysis of pinhole framings together with the real scale from X-ray quanta with an energy greater than about 1 keV.

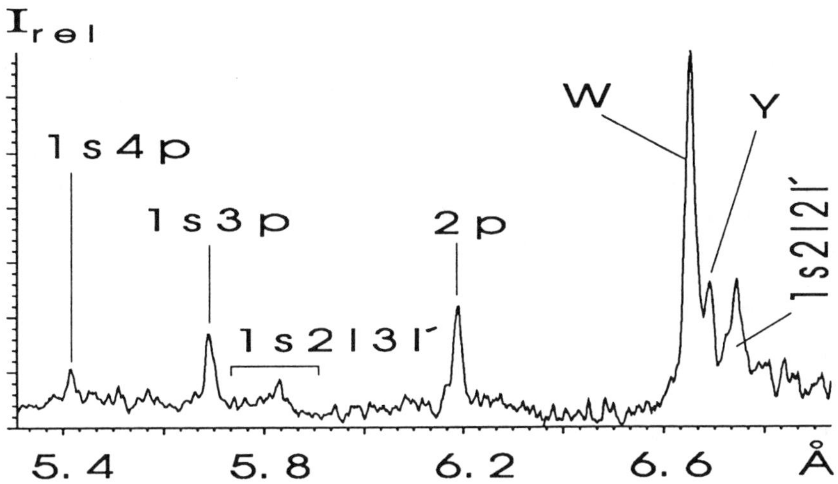

Fig. 5: Soft X-ray spectrum from H-, He- and Li-like ions, SiO_2-coating

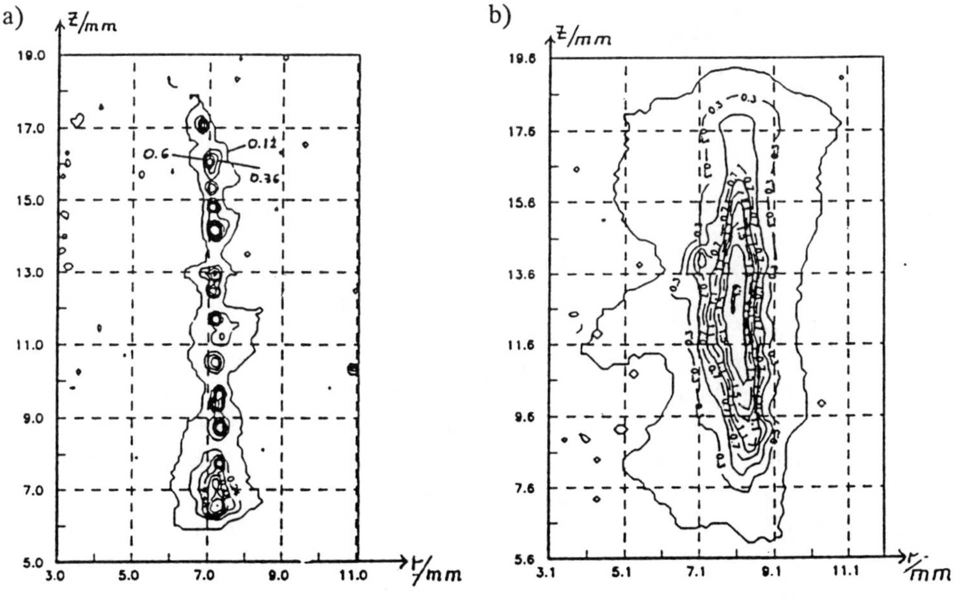

Fig. 6: Topographical framings, E > 1 keV, a) 20 µm Cu, b) 20 µm Cu + 0.4 atm Xe

Fig. 6a corresponds to discharges without Xe gas shell, fig. 6b was done with an outer Xe shell (p=0.4 atm). As can be clearly seen the outer gas shell prevents the creation of the "hot spots". The diameter of the hot spots are about 200-500 μm. With outer gases, we notice only one hot core from the framings (see fig. 6b), its length is about 4 mm, its width about 0.5 mm. Space resolved Si-spectra[11] from the coating showed, that this hot core is the origin of strong emissions from the H-like Si-ions.

Fig. 7 shows the emission spectrum from a discharge driven with pure Xe. The wavelength interval indicates emission lines from Ni-like ions. Without reliable reference lines a detailed line identification is very difficult. The intensity is rather weak and about a factor of 100 lower than for the characteristic He-like Ar lines.

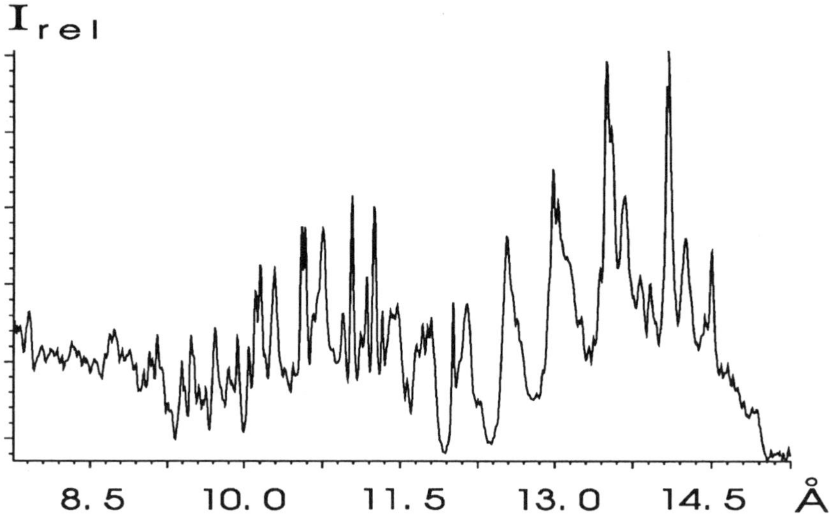

Fig. 7: Unidentified Xe spectrum in the Ni-like wavelength interval

The hot spot formation in the case of Al wires (20 μm) was considerable less, with outer gas shells similar framings like fig. 6b were observed. Fig. 8 shows the X-ray spectrum from H-, He- and Li-like ions. For diagnostic purposes, the wire was covered with a solution of NaCl. Corresponding n=2 resonance lines from H- and He-like ions were observed, their intensity was about a factor of 20 less than the corresponding Al lines. Although the He-like resonance line W from Na and Cl showed up with a similar intensity, the H-like Lyman-α line from Cl was much weaker. This indicates, that the temperature and the $n_e\tau$-parameter of the plasma was not high enough to cause a higher emission from the Cl ions. The parameter analysis for Al gave: $T_e(He) \approx 350\text{-}450$ eV, $T_z(Li,He) \approx 250$ eV, $n_e(He) \approx 3\text{-}10 \cdot 10^{20}$ cm^{-3}. In the H-like ionisation stage: $T_e(H) \approx 500\text{-}600$ eV, $T_z(He,H) \approx 500$ eV, $n_e(H) < 10^{21}$ cm^{-3}. The parameters obtained from the H-like Al ions are in agreement with those obtained from the Na and Cl resonance lines. The W-line width indicates plasma velocities not greater than about $3 \cdot 10^7$ cm/s.

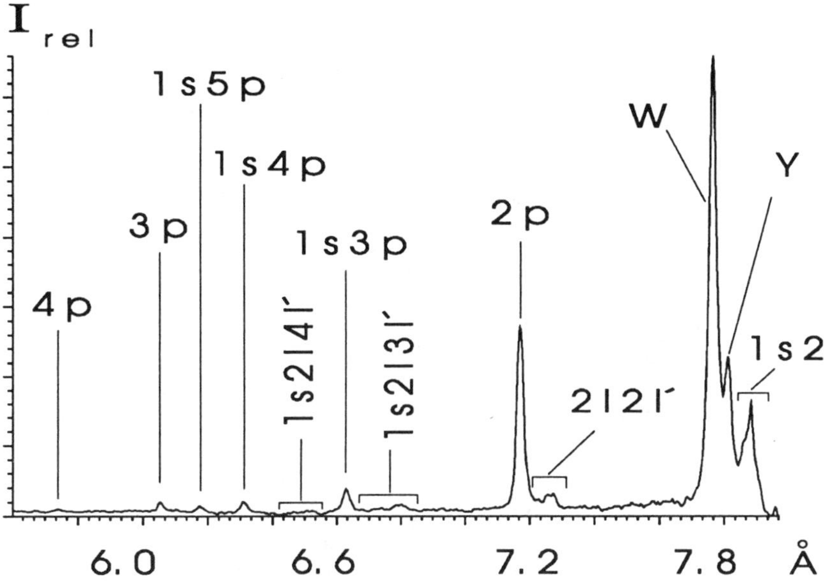

Fig. 8: H-, He- and Li-like soft X-ray emission lines from a 20 μm Al wire

One of us (F.B.Rosmej) greatly acknowledges the financial support from the Alexander von Humboldt-Stiftung.

References

1. A.V. Batyunin, A.N. Bulatov et al., Sov. J. Plasma Phys. **16**, 597 (1990)
2. V.D. Vikharev, G.S. Volkov et al. Sov. J. Plasma Phys. **16**, 217 (1990)
3 W. Kies, G. Decker, M.Mälzig, C. van Calker et al., J. Appl. Phys. **70**, 7261 (1991)
4. V. Smirnov, P.Choi, B. Etlicher, J. Larour, F. Wessel et al.: "An international collaboration in the study of the physics of transient dense magnetized plasmas in a large facility in the FSU", APS 34, Seatle, USA 1992, continued 1993
5. F.B. Rosmej, S.A. Komarov, O.N. Rosmej, V.O. Mishensky: "Plasma parameter evaluation at the ANGARA-5 device by means of soft x-ray spectra analysis", Report JEX 92, TRINITI, Troitzk, Russia
6. F.B. Rosmej, to be published in JQSRT
7. O.N. Rosmej, F.B. Rosmej, this volume
8. A. Gabriel, Mon. Not. R. astr. Soc. **160**, 99 (1972)
9. D. Duston, J. Davis: Phys. Rev. A. **21**, 1664 (1980)
10. C.A. Jones, D.R. Kania, Phys. Rev. Lett. **55**, 1993 (1985)
11. S.A Danko, private communication 93

EFFECTS OF HIGHLY ENERGETIC ELECTRONS AND NONSTATIONARITY ON HE-LIKE DIELECTRONIC SATELLITES IN DENSE PLASMA

O.N. Rosmej, F.B. Rosmej
Institute of Spectroscopy, Troitzk, Russia
present address: Institut für Exp. Phys. V, Ruhr-Universität Bochum, Germany

ABSTRACT

The influence of highly energetic electrons and nonstationarity on the H-like resonance line and its adjacent He-like satellites 1s2l-2l2l´ are investigated. Both of this effects result in an additional population (besides the dielectronic capture channel) of the autoionising 2l2l´-levels through the excitation from the metastable levels of single excited He-like ions. This channel can be rather strong especially for double excited levels with low rates of dielectronic capture, which are widely used for density diagnostics. It is shown, that the neglection of the non-equilibrium plasma state and the excitation by highly energetic electrons can result in serious overestimations of the electron density and temperature. Fully time dependent simulations are carried out for an experimental aluminium spectrum from a low inductance vacuum spark. We show that effects of highly energetic electrons and nonstationarity provide electron temperatures two times lower and electron densities an order of magnitude lower than simulations based upon thermal-equilibrium interpretations.

I. INTRODUCTION

Direct measurements of energetic electrons in the plasma at the time of pinch and investigations of the influence of this nonthermal electron distribution on X-ray spectra from an argon gas-puff plasma were carried out by Kania and Jones.[1,2] They have observed a short intense 20 keV electron beam produced at the time of pinch. It was shown that the actual plasma electron density and temperature were more modest than previously thought[3], i.e. $N_e=2\cdot10^{20}$ cm^{-3} and temperature T = 400 eV rather than $N_e=10^{21}$ cm^{-3} and T=1000 eV based on a thermal-equilibrium interpretation of the data. The measurement of the energetic distribution of fast electrons emitted from the PF-150 device by means of a magnetic spectrometer[4] showed that more than 70% of fast electrons have a mean energy 47 keV and FWHM of about 20-30 keV.

Unfortunately we have not enough information about the energy distribution of beams penetrating the pinch and its fractional contribution in vacuum spark plasma. But taking into account the fact, that the distribution can be rather broad[4,5,6] we simulated nonthermal X-ray spectra using a two-temperature approach: all electron-collisional processes were governed by plasma electrons with the temperature T_e, density N_e and by energetic electrons with the second temperature T_{beam}=10-50 keV and the density N_{beam}.

II. NON-EQUILIBRIUM SIMULATIONS OF X-RAY SPECTRA: PROGRAM "META"

The program calculates in dependence on the plasma parameters $T_e(t)$ and $N_e(t)$ the time dependent populations $n_i(t)$ of the following levels: in B-like ions $1s^22s^22p$, $1s^22s2p^2$, in Be-like ions $1s^22s^2$, $1s^22s2p$, in Li-like ions $1s^22s$, $1s^22p$, $1s^23$s,p,d, $1s^24$s,p,d,f, in He-like ions $1s^2$, 1s2s $^1S_0$, 1s2s $^3S_1$, 1s2p $^3P_0$, 1s2p $^3P_1$, 1s2p $^3P_2$, 1s2p $^1P_1$, 1s3s,p,d ^1L, 1s3s,p,d ^3L, 1s4s,p,d,f ^1L, 1s4s,p,d,f ^3L, all autoionising levels $2s^2$ $^1S_0$, 2s2p $^1P_1$, 2s2p $^3P_0$, 2s2p $^3P_1$, 2s2p $^3P_2$, $2p^2$ $^3P_0$, $2p^2$ $^3P_1$, $2p^2$ $^3P_2$, $2p^2$ $^1D_2$, $2p^2$ $^1S_0$, in H-like ions 1s, 2s, 2p, 3p, 4p, nucleus.

Simulations of X-ray spectra of higly ionised nonthermal transient plasma are carried out in the frame of a full collisional radiative modeling. For calculations of collisional and radiative transitions between the levels the following atomic data were used: collisional transitions between double excited levels[7]; collisional excitation, ionization and photorecombination from ground and single excited states[8,9]; values of energies, rates of autoionization and radiative decay.[10]

III. INFLUENCE OF ENERGETIC ELECTRONS ON HE-LIKE DIELECTRONIC SATELLITES

A. The ratio of J-satellite to H-like resonance line.

The ratio of J-satellite (transition 1s2p $^1P_1$-$2p^2$ $^1D_2$) to the H-like resonance line 1s-2p is widely used in the X-y ray spectroscopy for the determination of the electron temperature.

In thermal-equilibrium plasma this ratio is proportional to the ratio of the rates of dielectronic capture and electron-collisional excitation. Fig.1 demonstrates the behavior of the J-sat/Res-H ratio under the influence of energetic electrons with T_{beam}=10 keV and different fractions $f(\%)_{beam}=N_{beam}/N_e$ in aluminium plasma. One can see, that for rather low plasma temperatures in the presence of energetic electrons in plasma the ratio of J-satellite to the resonance line is very small. Due to the exponential dependence of dielectronic capture and electron-collisional excitation in this temperature region all collisional processes are governed by fast electrons, the J-sat/Res-H ratio is proportional $1/T_{beam}$ and is, therefore, very low.

The set of curves (fig. 1) with different fractions f_{beam} of energetic electrons gives more than one variant for the interpretation of the experimental ratio. For example, if J-sat/Res-H=0.02 the electron temperature obtained from the thermal-equilibrium model will be T_e=1000 eV, for plasma containing 10% of energetic electrons T_e=250 eV and 1000 eV and for f_{beam}=1% T_e=130 eV and 1000 eV. As can be seen from fig. 1, with increasing electron temperature the influence of highly energetic electrons decreases and in consequence the ambiguity of the plasma temperature determination decreases as well.

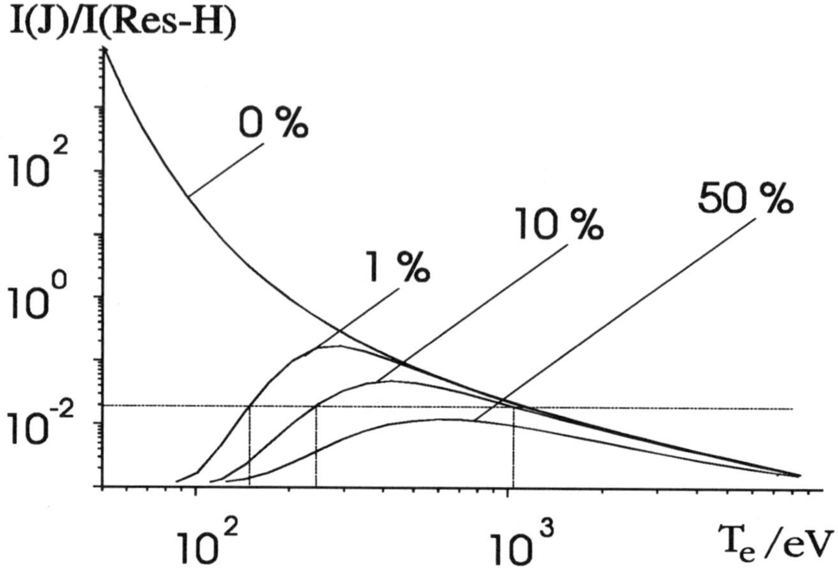

Fig. 1: The ratio of J-satellite to H-like resonance line of Al depending on the plasma temperature T_e and the fraction of energetic electrons f_{beam} for $N_e=10^{20}$ cm^{-3}, T_{beam}=10 keV

B. The total ''triplet''- to ''J''-satellite intensity ratio.

The idea to use the total ''triplet''- to ''J''-satellite ratio, i.e. (1s2s $^3S_1$-2s2p ^3P+1s2p ^3P-2p^2 ^3P) / (1s2p $^1P_1$-2p^2 $^1D_2$) for the determination of the electron density is well known.[11,12] Some of the triplet double excited levels (2p^2 ^3P) are populated rather weak by dielectronic capture and therefore the intensities of radiative transitions emanating from this levels are low. As the electron density increases, the collisional exchange between the double excited levels starts to become important, the populations of the 2p^2 ^3P states grow and the ''triplet''- to ''J''-satellite ratio starts to be sensitive to the electron concentration. Fig. 2 shows this effect for aluminium plasma with T_e=300 eV. The energetic electrons with the second temperature T_{beam}=10 keV provide an additional channel for the population of the autoionising levels: excitation from single excited metastable levels of He-like ions. This channel can be rather strong especially for double excited levels with low dielectronic capture like 2p^2 ^3P and the intensity of the ''triplet'' group grows in comparison with the intensity of ''J''-satellite. Therefore, the presence of energetic electrons in plasma can lead to an ambiguity in the determination of the plasma density. Fig. 2 shows, that the experimental ''triplet''- to ''J''-satellite ratio of 1.2 can be interpreted by a set of electron densities and fractions of energetic electrons: $N_e=10^{23}$ cm^{-3} in thermal-equilibrium plasma and $5\cdot10^{21}$ cm^{-3} for plasma with 10% of fast electrons.

The behaviour of the ''triplet''-satellite group depending on the electron density is shown in fig. 3. With increasing electron density the satellite intensities increase as well and the ´´center of heavity´´ of this group moves from (1s2s $^3S_1$-2s2p ^3P)-satellites, designated on the picture as ´´3S´´, to (1s2p ^3P-2p^2 ^3P)-satellites, designated as ´´3P´´.

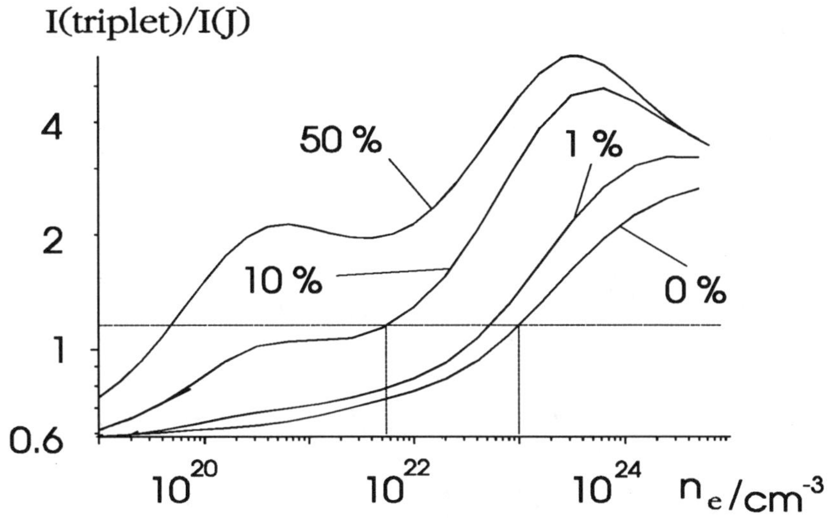

Fig. 2: The total ´´triplet´´- to ´´J´´-satellite ratio for different N_e and different fractions f_{beam} for Al plasma: T_e=300 eV, T_{beam}=10 keV

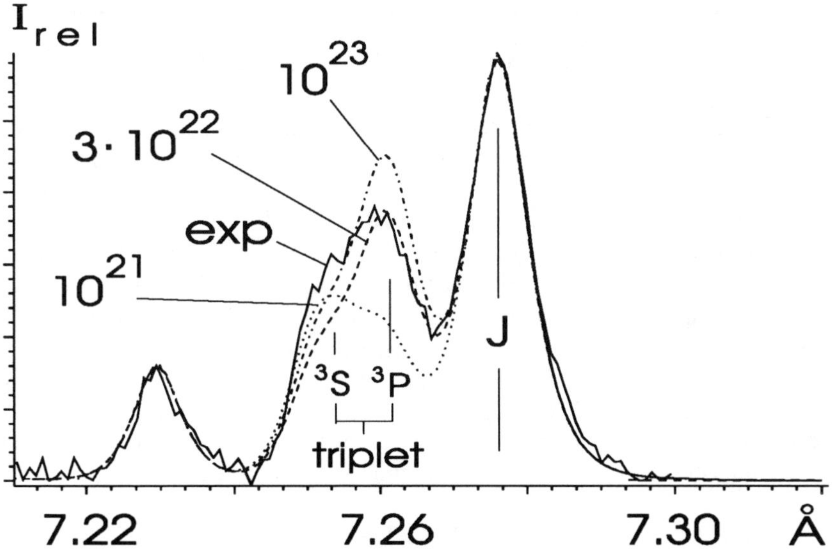

Fig. 3: The density dependence of the intensity of ´´triplet´´ satellites (1s2l^{3}L-2l2l´^{3}L´) for aluminium plasma with T_e=500 eV

IV. DIELECTRONIC SATELLITES IN TRANSIENT PLASMA

Spectra simulations taking into account the nonstationarity of radiation are very important for investigations of dense plasma with lifetimes less than about 10 ns. The development of fast X-ray techniques for the registration of time-resolved spectra can provide interesting information about the evolution of plasma parameters.

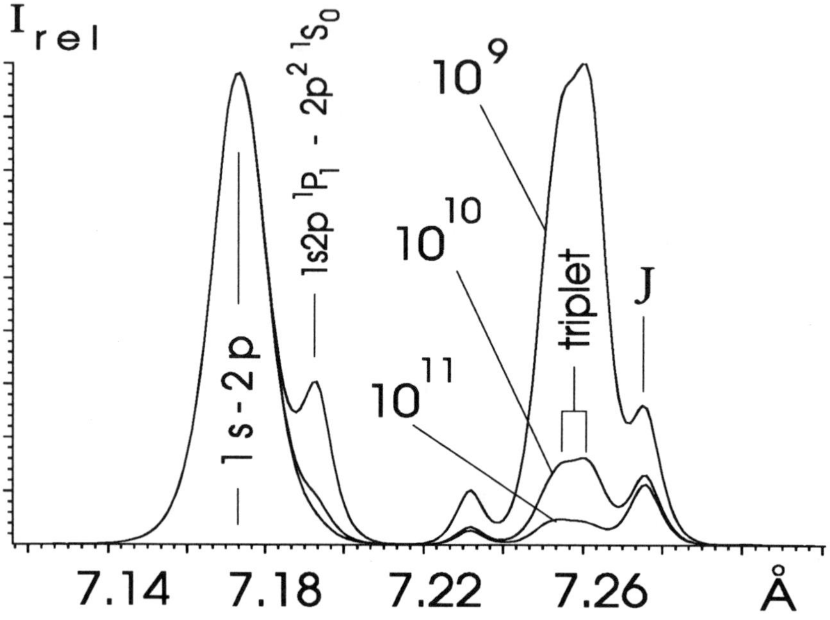

Fig. 4: Intensities of He-like dielectronic satellites depending on $N_e \cdot \tau$ ($cm^{-3} \cdot s$) in transient aluminium plasma with T_e=500 eV and $N_e=10^{21}$ cm^{-3}

Fig. 4 demonstrates the behaviour of He-like dielectronic satellites adjacent to the H-like resonance line depending on the product of the electron density and time $N_e \cdot \tau$ for aluminium plasma with T_e=500 eV and $N_e=10^{21}$ cm^{-3}. The effect of nonstationarity on the dielectronic satellites is similar to the effect of highly energetic electrons. If the lifetime of the plasma is less than the time needed for ionisation equilibrium the population of H-like ions can be very low in comparison to the population of He-like ions. In consequence the excitation from single excited levels of the He-like ions started to be more important for the population of the autoionising levels than dielectronic capture from the ground state of H-like ions.

V. SIMULATIONS OF EXPERIMENTAL SPECTRA

The experimental spectrum from aluminium vacuum spark plasma is shown in fig.5. Details of the vacuum spark device are described elsewhere.[13] The spectra simulation uses the H-like resonance line and ´´J´´-satellite for the determination of the electron temperature T_e and the ´´triplet´´-satellites for the determination of the electron density N_e. The comparison of the experimental spectrum with numerical simulations (fig. 3) shows, that the redistribution of intensities inside the ´´triplet´´ group with the increasing concentration does not permit ´´to fulfill´´ the left part of this satellite group. The numerical simulations including effects of highly energetic electrons and nonstationarity are shown in fig. 6. Because of the same mechanism of the both effects the results of simulations look very similar but the electron temperature differs in two times. One can not distinguish nonthermal and transient plasma looking at the H-like resonance line and adjacent dielectronic satellites. Using additionaly the ratio of He- and H-like resonance lines we can avoid this ambiguity. Indeed, if to compare the experimental ratio of the resonance lines (≈0.1) with nonthermal (=0.08) and transient (=0.02) calculations one can conclude, that the intensity distribution inside the satellite group is caused mainly by nonthermal electrons. The thermal-equilibrium interpretation (fig. 3) with the electron temperature T_e=500 eV and the electron density $N_e=3\cdot10^{22}$ cm^{-3} provides the ratio 0.8 which is too far from the experimental one.

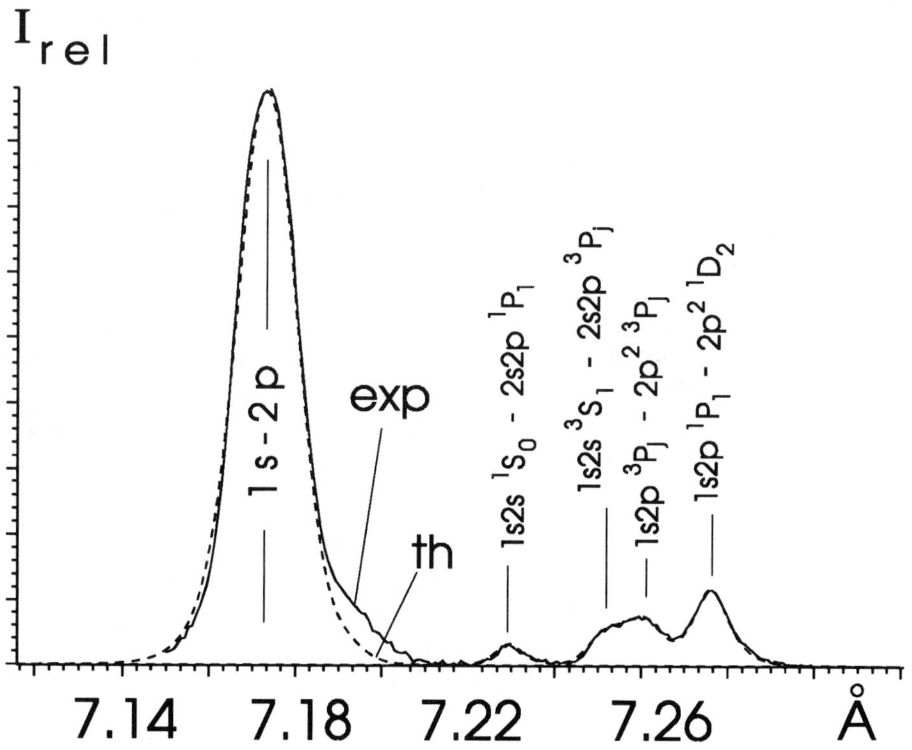

Fig. 5: The experimental spectrum of the aluminium vacuum spark plasma (exp) and the simulation of this spectra using the code ´´META´´ (th)

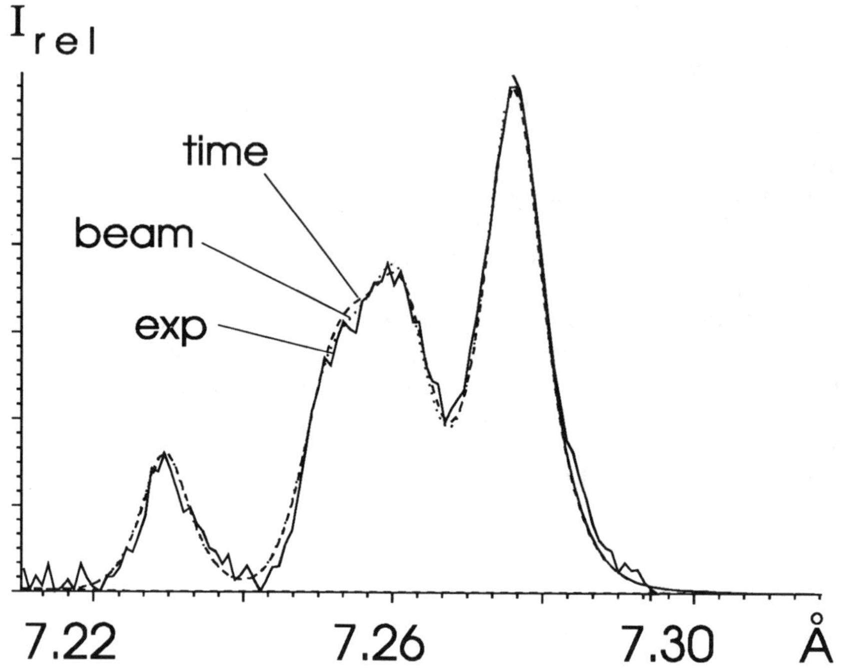

Fig. 6: The comparison of different numerical simulations of the aluminium spectra. The curve (.......): T_e=270 eV, N_e=1-2·10^{21} cm^{-3}, T_{beam}=20 keV, f_{beam}=2-3%; the curve (-----): T_e=530 eV, N_e=2·10^{21} cm^{-3}, N_e·τ=4·10^{10} cm^{-3}·sec

VI. CONCLUSION

We have shown that the effects of higly energetic electrons and nonstationarity on He-like dielectronic satellites can be rather strong and has to be taken into account to exclude serious overestimations of the electron density and temperature in X-ray spectra simulations.

One of us (F.B. Rosmej) greatly acknowledges the financial support from the Alexander von Humboldt-Stiftung.

REFERENCES

1. D.R. Kania, L.A. Jones. Phys. Rev. Lett. **53**, 166 (1984)
2. D.R. Kania, L.A. Jones. Phys. Rev. Lett. **55**, 1993 (1985)
3. P.G. Burkhalter, J. Shilton, A. Fisher, R.D. Cowan, J. Appl. Phys. **50**, 4532 (1979)

4. S. Czekaj, S. Denus, A. Szydlovski, S. Sledzinski in Proceedings of the 4-th international Workshop on Plasma Focus & Z-pinch research, Warsaw, 1985
5. R. Burhenn, B.S Harn, S. Gossling, H.-J. Kunze, D. Mielzarski, J. Phys. D **17**, 1665 (1984)
6. R. Beier, Zeitschrift für Physik A **292**, 292 (1979)
7. S.J. Goett, D.H. Sampson, R. E.H. Clark, At. Data Nucl. Data Tab. **28**, 279 (1983)
8. I.I. Sobelman, L.A. Vainshtein, E.A. Yukov: "Excitation of Atoms and Broadening of Spectral Lines", Springer, New York 1981
9. A.V. Vinogradov, I.Yu. Skobelev, E.A. Yukov, Sov. J. Quant. Electronic **5,** 630 (1975)
10. L.A. Vainshtein, U.I. Safronova, At. Data Nucl. Data Tables **21**, 49 (1978)
11. V.I. Bayanov, V.A. Boiko, A.V. Vinogradov et al., JETP Lett. **24**, No.6, 352 (1976)
12. D. Duston, J. Rogerson, J. Davis, M. Blaha, Phys. Rev. A **28**, 2968 (1983)
13. R. Beier and H.-J. Kunze, Zeitschrift für Physik A **285,** 347 (1978)

X-PINCHES, VACUUM SPARKS, AND OTHER PINCHES

MICROSECOND GAS-PUFF PLASMA IMPLOSION AT GPX DEVICE

A.Bartnik*, G.V.Ivanenkov**, L.Karpinski*, S.A.Pikuz**, T.A.Shelkovenko**
*Institute of Plasma Physics and Laser Microfusion, Warsaw, Poland
**P.N.Lebedev Physical Institute, Moscow, Russia

ABSTRACT

There are presented hollow cylindrical gas-puff plasmas of electrical discharges at GPX device (200 kA, 40 kV, 1 μs, Ne/Ar-mixture) experimental results. Investigation of plasma with time and spatial resolution in wide spectral range was enabled by the set of diagnostics. Plasma processes were investigated in visible and IR light by frame and streak cameras, in XUV by transmission grating spectrograph and XRD, in sort X-rays by convex and spherically bent crystal spectrograph, X-ray crystal microscope, pin-holes, *pin*-diodes. Plasma parameters ($n_e = 10^{19}$–10^{22} cm^{-3}, $T_e = 0.2$–1.3 keV) were controlled by spectroscopic methods using H- and He-like Ne and Ar spectra. Kink instability and hot spots were observed simultaneously in maximum X-ray emission regimes. Also argon K_α line emission from plasma was observed. Plasma stabilization methods are discussed.

INTRODUCTION

There exist many methods of creation of hot plasmas of high density. The use of the capacitor bank discharge with the current rise time of about 1 μs is the simplest one. The first gas-puff Z-pinch experiments were carried out using such devices. Then gas-puff Z-pinch experiment was repeated with the help of high power machines for soft X-ray, neutron and high magnetic field generation. Simpler microsecond devices may be used for new diagnostics development and testing.

These experiments continue our previous work [1]. The investigations were carried out by GPX device based on 3-μF-capacitor being charged up to 40 kV. The current amplitude was 200 kA with rise time about 1 μs. Plasma load was formed by electrical breakdown of hollow gas shell produced in vacuum anode-cathode gap. Working gas (the Ar/Ne 3:1 mix) was injected through an annular nozzle in cathode via fast opening valve. The diode had gap width 1.5–2.5 cm, outer nozzle diameter was 1.8–4.5 cm. Its entrance slit was 0.5 mm and the exit one — 2 mm. Gas velocity corresponded to 3–5 Mach number while average shell thickness was about 0.5 cm.

EXPERIMENTAL RESULTS

The experimental setup included a number of diagnostics, which enabled plasma radiation investigation in various wavelength ranges from visible light to soft X-rays. Plasma dynamics was observed by streak and frame cameras in visible light (fig.1). Streak camera photographs showed the variations of visible radiation distribution along the plasma radius within time interval 0.5–2 μs. Frame camera

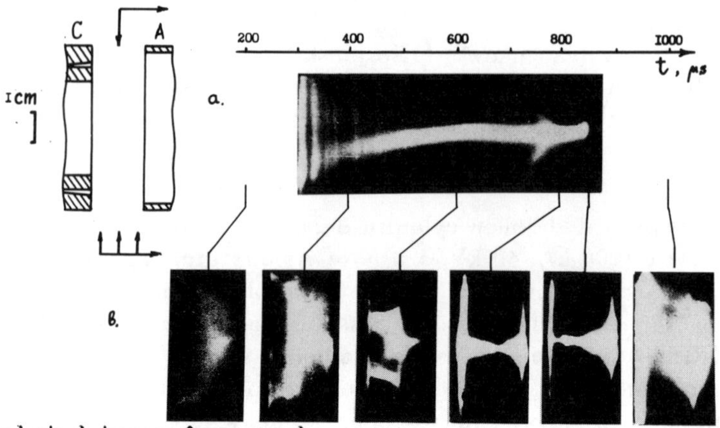

Fig.1: Typical pinch images from streak (*a*) and frame (*b*) cameras obtained on GPX device.

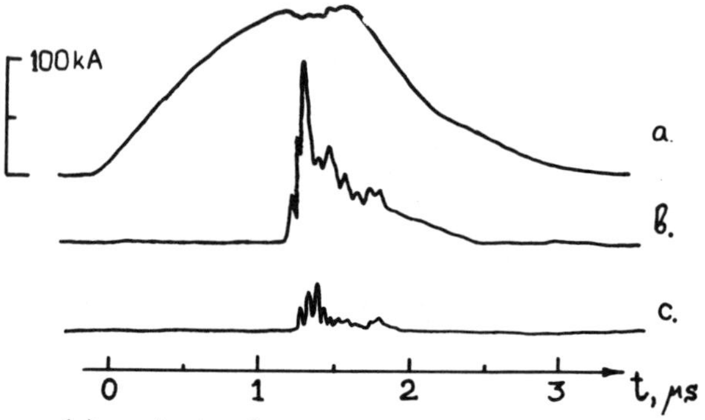

Fig.2: Current traces (*a*) and signals from *pin*-diodes placed behind aluminum filters 6,6 μm (*b*) and 13,2 μm (*c*) for optimum mass density.

with the exposition time 15 ns enabled to obtain plasma photographs in fixed moments of discharge.

Two *pin*-diodes with time resolution about few nanoseconds placed in 37.5 cm from pinch axis registered X-rays signals with quantum energy above 3.5 keV ($\lambda > 4$ Å) (fig.2). Time evolution of XUV radiation was measured by XRD with Al cathode. None filter was placed between XRD and plasma, so radiation was measured in wide range 2–100 Å with time resolution of the same order as *pin*-diodes. Spectrometer with two convex CsAP crystals enabled to register He- and H-like Ne-lines (10–14 Å) and He- and H-like Ar lines (3–4 Å). Space distribution of soft X-rays was registered time integrated by 100 and 10 μm pin-hole with 6.6 and 13.2 Al filters (fig.3).

The number of diagnostic instruments was applied in addition to those used in previous experiments [1]. There were soft X-ray microscope on the basis of spherical crystal mirror (fig.4), Johann spectrograph with spherically bent mica

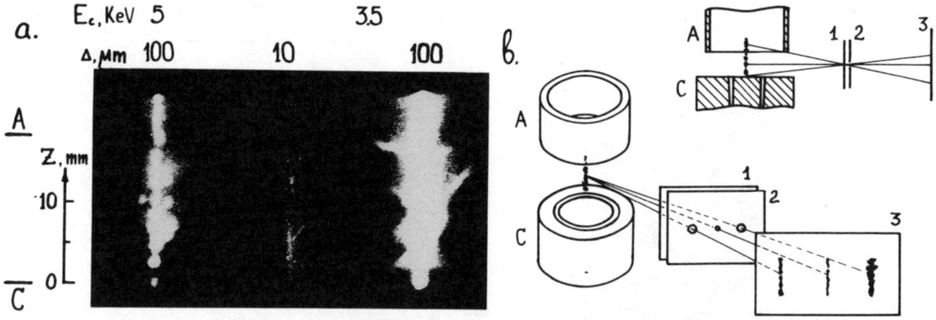

Fig.3: Pinch obscurograms: *a*) plasma image; *b*) scheme of inner-tube plasma and anode imaging (A — anode, C — cathode; 1 -- filter, 2 — pin-holes plane, 3 — photofilm).

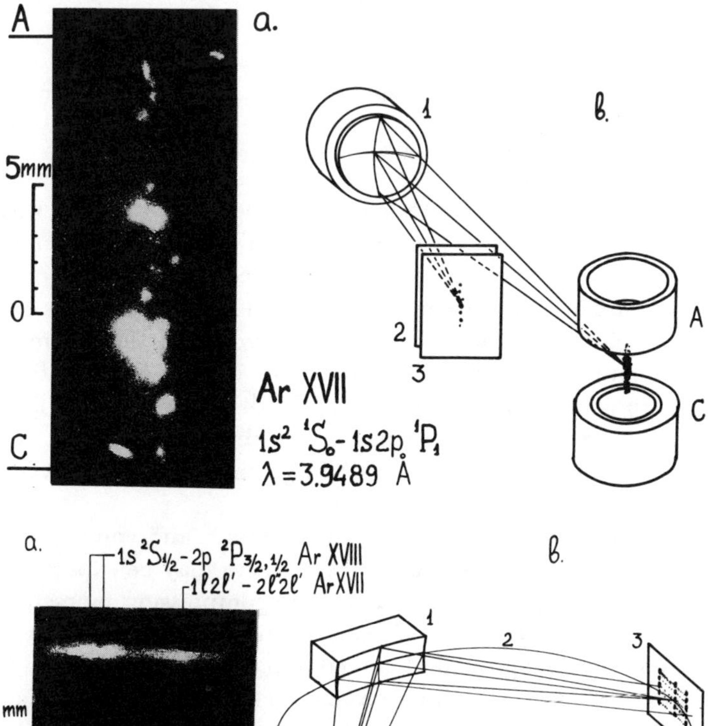

Fig.4: X-ray microscope: *a*) 6-shot series plasma image in He-like Ar ion resonance line; *b*) microscope scheme (A – anode, C – cathode; 1 – mount with spherical mica crystal, 2 – filter, 3 – photographic film).

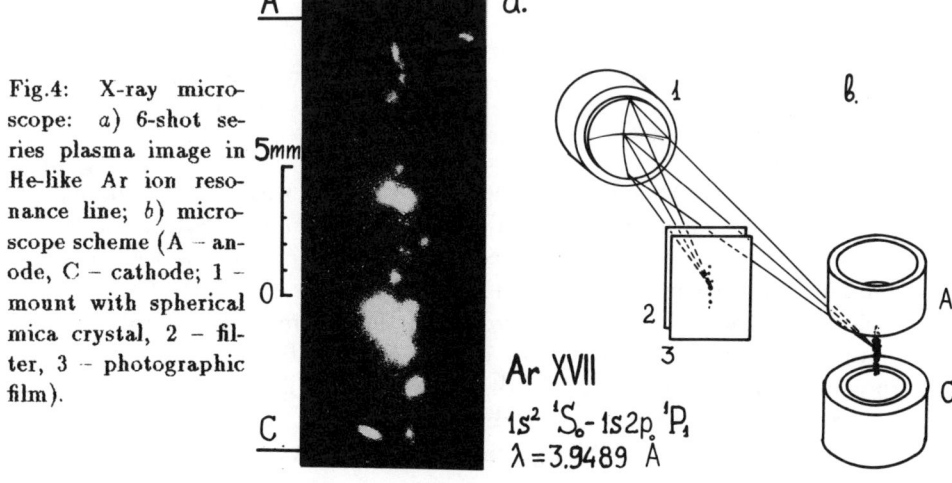

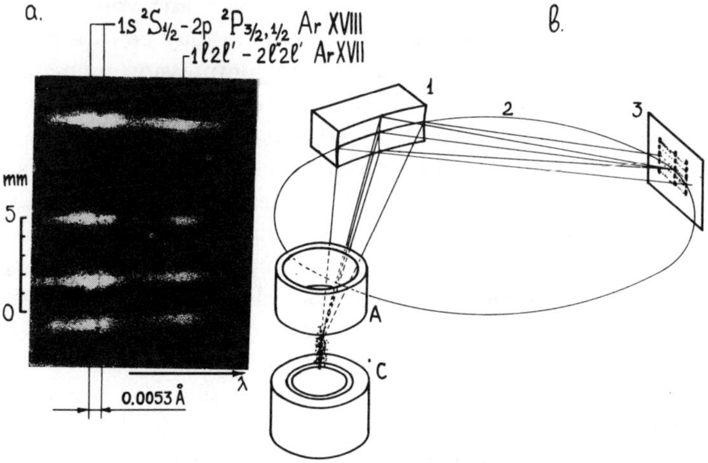

Fig.5: Johann spectrograph with spherical crystal: *a*) spectrum of Ar close to H-like ion resonance line; *b*) schematic diagram of space resolved spectrum imaging (A – anode, C – cathode; 1 – mount with spherically bent crystal, 2 – Rowland circle, 3 – photographic film).

crystal (fig.5) and the broad bound spectrograph with transmission grating (fig.6).

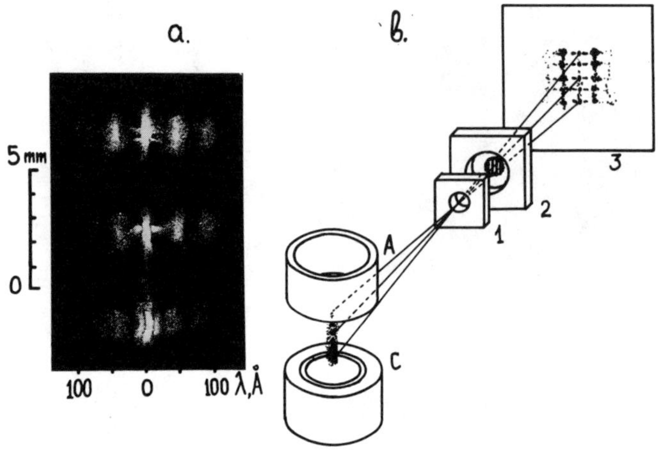

Fig.6: Transmission grating spectrograph: *a*) one shot plasma spectrum; *b*) schematic diagram of spectrograph (A — anode, C — cathode; 1 — tantalum diaphragm, 2 — mount with grating, 3 — photographic film).

X-ray microscope with spherically bent mica crystal with 100 mm radius enabled to obtain plasma images in He-like Ar resonance line. Perfect correspondence between line wavelength and lattice constant mica crystal enabled to obtain high luminous intensity in 5-th reflection number (Bragg's angle = 84.5°) with spatial resolution of the order of tens microns. In the scheme without entrance slit [2], Johann spectrograph with the same crystal with spatial resolution registered spectrum of argon in the area of H-like ion resonance line and $1s^2\ {}^1S_0 - 1s3p\ {}^1P_1$ He-like ion line. Transmission grating spectrograph [2] enabled to expand the registration range up to 100 Å. The grating was made in tungsten diaphragm with 1 μm period and 25 μm diameter.

According to the earlier experiments [1] and some estimations, simultaneous existence of kink and sausage instabilities may be expected. The most developed kink instability occurs in cases when compression moment converges with maximum of the current. As a result a number of bright hot spots (100 μm) may be seen at the background of curved and much cooler plasma column. So as the energy reserve is used for plasma column movement, further increase of the radiation output may be obtained during the decaying of such instability.

The most intense radiation was observed when plasma compressed near the current maximum (fig.2). Its XUV component was registered by XRD (fig.7). Without kink instability, hot spots radiation was much higher. In these cases the distributions of hot spots in anode-cathode gap was more uniform (fig.3), and H-like Ar ions were registered even in one shot.

A lot of experimental facts confirms the existence of electron beams in the diode. Obscurograms from series of shots (fig.3) show the edge of the anode tube (one can notice the existence of plasma inside the tube). This radiation may

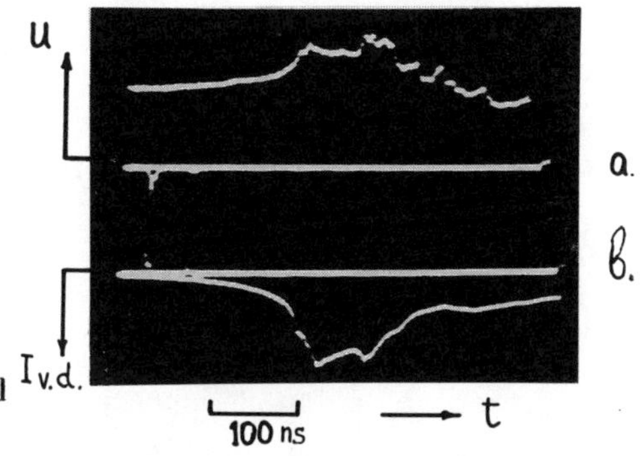

Fig.7: Voltage (a) and XRD (b) signal traces for optimum mass of the gas.

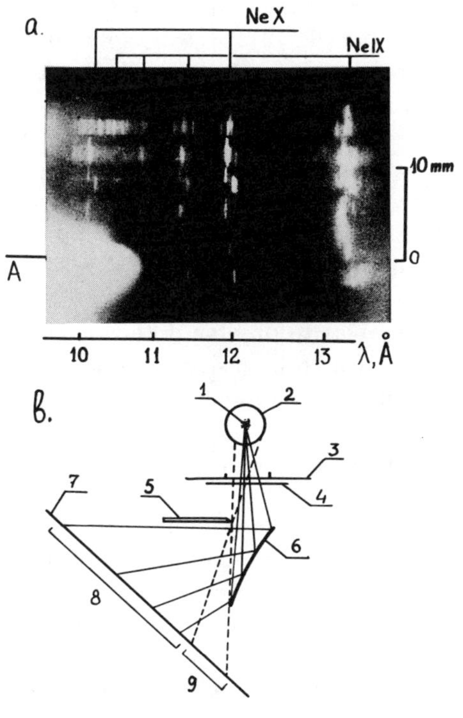

Fig.8: Convex crystal spectrograph: a) 6-shot series neon spectrum; b) schematic diagram of spectrum and the anode image forming (1 — plasma, 2 — anode tube, 3 — horizontal slit, 4 — filter, 5 — vertical knife, 6 — crystal, 7 — photographic film, 8 — spectrum image zone, 9 — anode image zone).

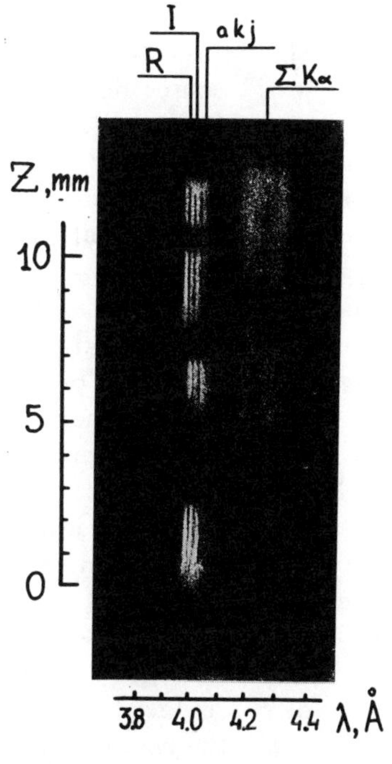

Fig.9: Spectrum of argon close to He-like resonance line from convex crystal spectrograph.

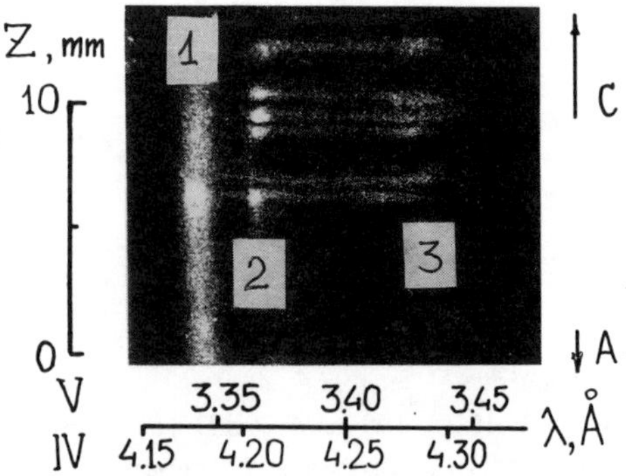

Fig.10: Spectrum of argon close to $1s^2 - 1s3p$ He-like ions in 5-th number of reflection and K_α weakly ionized argon in 4-th number.

be seen on the Ne spectrogram as well (fig.8). The image was formed here by "obscura" between the shield and the edge of crystal (see 5 and 6 on fig.8).

But the most interesting is the registration of K_α line of low ionized argon. It was obtained using convex crystal (fig.9) and Johann (fig.10) spectrographs. In the first case the instrument works as "X-ray photocamera" creating two-dimensional image in individual spectral line. The radiation may be seen as a wide strip corresponding to the source diameter (2 mm). Image analyses on fig.10 confirm that the line is really K_α one. Because of high dispersion only this line may occur with sufficient precision $\Delta\lambda/\lambda = 3 \cdot 10^{-3}$ in the range of registration at the fourth number of mica reflection. Taking into account poor reflection at this number one can expect that this line is about 10 time more intensive than $1s^2 - 1s3p$ one obtained at 5-th number.

The transversal distribution of hot spots enables to observe different configurations of plasma column registered in stages of high intensive radiation. The distribution as the dual points on obscurograms of the straight plasma column expects two following compression of hot spots. In this rather typical case two singularities on current trace appear. The first of them, as a rule, is connected with higher X-ray peak (fig.2).

Because of higher radiation intensity from straight plasma column, 3 cm in diameter tube was inserted inside the anode to stabilize the kink mode of instability. The mass of gas in this case corresponded with plasma collapse before current maximum. It was that very case when X-ray output was maximal for all shots with 18 mm nozzle. It seems that the inner tube acting as an anode in anode-cathode gap reduced the effective length of plasma column. Fraction of the gas injected into the space between the tubes did not take part in discharge process. It is important that the transversal spread of hot spots in different pulses was much smaller than 0.5 mm. The brightest spectra of H-like ions of Ar (the hottest plasma) was registered in this very case.

While increasing the nozzle diameter the axial and radial sizes ratio of plasma column decreased because of plasma contact with the surface of anode tube. The low thermal spectra intensity for earlier collapse and the existence of electron beams are the evidence of this. At the same time the decrease of nozzle diameter leads to increase of nonuniformity of gas mass along the axis (the reason of nonuniformity is the impact of shell inner surface when it reaches the axis while gas jet flowing towards the anode). Without the additional stabilization of plasmas, the highest radiation output took place in case of nozzle diameter of 35 mm.

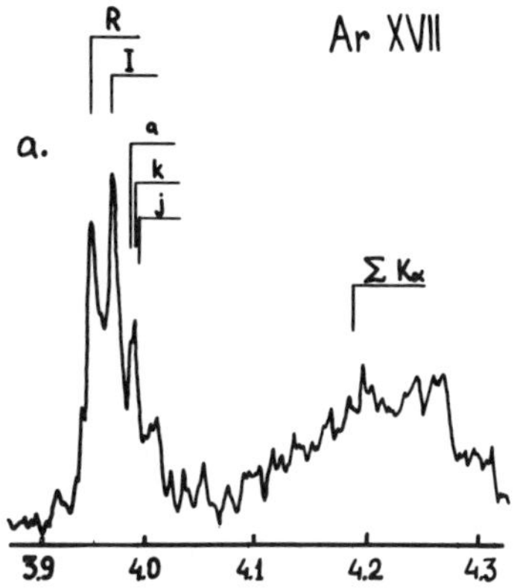

Fig.11: Spectrum densitogram of He-like ions of argon close to the resonance line.

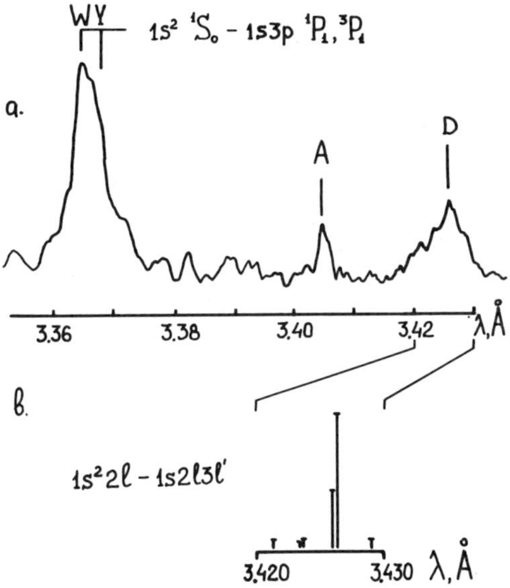

Fig.12: Spectrum of He-like Ar ions close to $1s^2 - 1s3p$ line (a) and relative intensities of satellites for the case of dielectronic mechanism of level populating (b).

H- and He-like Ar ions radiation spectra registered in experiments enabled to measure maximum temperature T_e and electron density n_e. The electron temperature in hot spots was obtained from intensity ratio of dielectronic satellites of $1s^2\ {}^1S_0 - 1s2p\ {}^1P_1$ (fig.11) and $1s^2\ {}^1S_0 - 1s3p\ {}^1P_1$ (fig.12) lines of He-like Ar ions and the corresponding line intensities using the method from [6,7]. $1s2p\ {}^2P_{3/2} - 1s2p\ {}^2D_{5/2}$ transition was taken for T_e estimation using satellites of $1s^2\ {}^1S_0 - 1s2p\ {}^1P_1$ line. This is the most intense well separated satellite line which intensity is determined only by dielectronic recombination.

Electron temperature T_e measured in brightest hot spots was about 1.2 keV. The estimation of T_e obtained from satellites of $1s^2\ {}^1S_0 - 1s3p\ {}^1P_1$ gave the value in the range 1.2–1.3 keV. However this estimation is less reliable because of the absence of satellites with the intensity connected only with the dielectronic recombination only. The brightest 16 and 23 satellites the intensity of which may be expected to be determined first of all by dielectronic recombination were used to calculate T_e. The estimations gave the values in the range 1.2–1.3 keV. The intensity ratio of satellites to resonance line of H-like ions of Ar gave for different

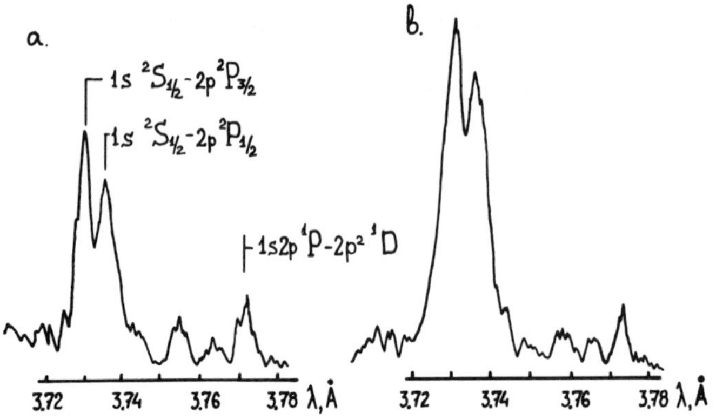

Fig.13: H-like Ar ions densitograms.

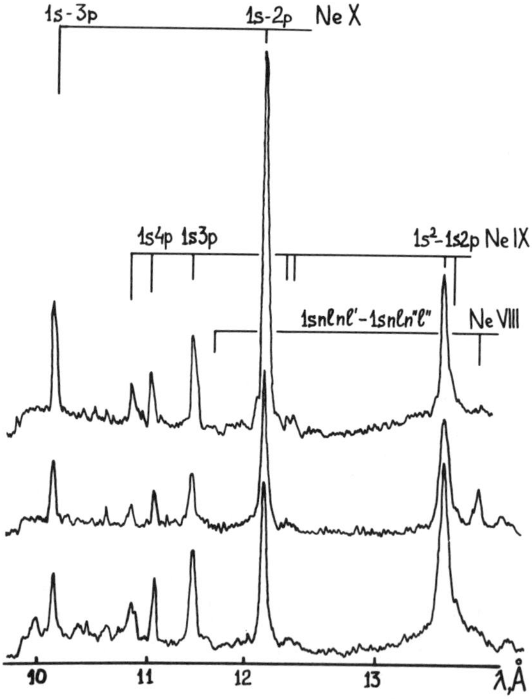

Fig.14: Neon spectrum densitograms for 3 different hot spots.

hot spots T_e = 1–1.2 keV (fig.13).

The electron density n_e was obtained from the ratio of intensities of resonance to intercombination line in He-like Ar (fig.14) and from relation of the dielectronic satellites intensities of H-like ions of Ar [6]. Both estimation procedures gave n_e =(0.5–1)$\cdot 10^{22}$ cm^{-3}. It is important to notice that the temperature $T_e > 1$ keV was obtained in individual particularly successful shots, in which stabilization tube was introduced. Gas shell mass and geometry were precisely chosen in the last cases.

Plasma parameters measured from H- and He-like Ne spectra (fig.3,14) were obviously much lower (T_e = 300–400 eV, T_e = 10^{20} cm^{-3}) because of those ions existence in such temperatures. Size of emitting aria may be estimated using half width of these lines. One can obtain the value about 100 μm. These lines occurred more frequently than high ionized Ar-ions lines. However Ar-ions existed in all areas being seen on obscurugrams as hot-spots. Emitting zones were, of course, sufficiently smaller than ones for Ne lines.

The registrations by transmission grating gave similar results. The areas of emission in $\lambda < 20$ Å range were less than 100 μm in size, more cooler areas radiating in $\lambda > 40$ Å were larger, about 1–2 mm and for the case with stabilization even 0.2–0.5 mm.

Transmission grating spectrograph enabled to estimate absolute XUY output from one hot spot. The value is about 0.05–0.2 J/Å in the range λ = 50–100 Å. It well coincides with spectral integrated XRD registrations and time integrated calorimetric measurements. Integral short wavelength ($\lambda < 200$ Å) radiation output from Z-pinch exceeds 100 J.

REFERENCES

1. A.Bartnik, G.V.Ivanenkov, L.Karpinski, S.A.Pikuz, T.A.Shelkovenko. Preprint No 56. M. FIAN. 1991; Trudy VNIIFTRI "Metody issledovanija radiacionno-stolknovitelnykh processov v lazernoj plazme", 50. M. 1991.
2. B.A.Bryunetkin, G.V.Ivanenkov, S.A.Pikuz, A.Ja. Faenov, T.A. Shelkovenko. Pis'ma v ZhTF **17**, No 19, 24 (1991).
3. B.A.Bryunetkin, G.V.Ivanenkov, S.A.Pikuz, V.G. Roman, I.Ju.Skobelev, A.Ja.Faenov, T.A.Shelkovenko. Ibid. 16.
4. A.Bartnik, G.V.Ivanenkov, L.Karpinski, S.A.Pikuz, T.A.Shelkovenko. Preprint No 30. M. FIAN. 1991; Fiz. plazmy **16**, 1482 (1990).
5. R.B.Baksht, A.V.Luchinskii, A.V.Fedjunin. Preprint No 30. Tomsk's Sci. Centre. 1990.
6. V.A.Boiko, A.V.Vinogradov, S.A.Pikuz, I.Ju.Skobelev, A.Ja.Faenov. J.Sov. Laser Reas. **6**, No 2, 85 (1985).
7. V.A.Boiko, S.A.Pikuz, U.I.Safronova, A.Ja.Faenov. Preprint No 147. M. FIAN. 1971.
8. V.Chaker, V.Bareau et.al. Rev.Sci.Instr. **60**, 3386 (1989).

INHOMOGENEOUS Z-PINCH INVESTIGATION ON "ANGARA-5-1"

A.N.Batunin, A.V.Branitsky, I.N.Frolov, E.V.Grabovsky, V.A.Kornilo, D.V.Kuznetsov, A.G.Lisitsyn, S.F.Medovschikov, V.O.Mishensky, A.R.Mingaleev[+], S.L.Nedoseev, L.B.Nikandrov, V.M.Romanova[+], T.A.Shelkovenko[+], V.P.Smirnov, A.N.Starostin, S.V.Trofimov, G.M.Olejnik, G.S.Volkov, E.G.Utjugov, S.V.Zakharov.
Troitsk Institute of Innovation and Thermonuclear Investigations, Troitsk, Moscow reg. Russia. +-Lebedev Institute, Moscow, Russia

Introduction

There are three different kinds of physical loads, coupled conventionally with pulsed power generators : high Density Liners or Z-Pinches (DLZP), Plasma Opening Switches (POS) and Electron or Ion high current Diodes (EID). Most specific feature of each one is a density of charged particles, greatest for DLZP, smaller for POS and EID , plasma quasineutrality being broken for the latter.

An intention to increase particles energy and hardness of X-ray radiation, produced in DZP, forces using of smaller density and linear mass of the load due to limitation of total energy , delivering to the pinch.

We made some experiments on "Angara-5-1", using small linear mass loads ($\mu\sim 1$-40 mkg/cm) as a part of double-zone inhomogeneous Z-pinches. In series double-zone Z-pinch the low mass zone was situated under anode upon high mass zone . In parallel double-zone pinch the low mass zone surrounded high mass zone coaxially. An instability and poor compression of the low mass zones was specific feature of these inhomogeneous pinches.

Series double-zone Z-pinch [1,2] (Fig.1).

Fast gas valve produced D_2 cloud at the axis of concentrator. To the moment of power pulse a linear density of deuterium near cathode was of $N\sim 2\cdot 10^{18}$ 1/cm , an axial gradient of linear density here was $dN/dz \sim (5\div 10)\cdot 10^{17} cm^{-2}$. It means, that gas filling of the gap was rather inhomogeneous and gas density near anode can be ~10 times smaller then near cathode.

The maximal neutron yield ($2*10^{12}$n/pulse) occurred at this inhomogeneous gas filling of the gap. Streak photographs of the pinch in visible light through the slits oriented perpendicular to the pinch axis at 5 mm down anode (a) and at 5 mm up cathode (b) are shown in Fig.2. Visible pinch diameter near cathode is ≈ 3 mm, implosion velocity ≈ $4 \cdot 10^7$ cm/sec. Maximal pinch compression in anode zone is delayed by ≈ 35 ns with respect to cathode zone. The minimal diameter in a visible light of plasma column in anode zone is ≈ 12 mm. Soft X-ray pinhole photographs in Fig.3 confirm the presence of uncompressed rare plasma near anode and presence of hot dense pinch near cathode of discharge gap. Maximum of energy spectrum of radial neutron flux is at 2.45 MeV. Spectral maximum of axial flux, directed opposite to ions acceleration vector, is shifted to lower energy, 1.8÷2.0 MeV. As well as energy spectrum, the total neutron yield is also different along these two directions. The anisotropy increases with total neutron yield growth, and typically is 2-3.

Neutrons energy spectrum and total neutron yield anisotropy can be explained by interaction of ions, accelerated to energy 200÷500 keV in anode rare plasma , like in a virtual ion diode, with the hot dense plasma of pinch-target (n_e $=10^{20-21}$ cm^{-3}, T_e~0.8-1.0 keV) near cathode of discharge gap.

Fig.4 shows the calculated trajectories for deuterons with different starting radiuses in the case of initial values of pinch voltage and current 400 kV and 2.0 MA respectively. Ions, being generated at anode region , are drifting radially in crossed axial electric and azimuthal magnetic fields of virtual EID, till they achieve axial zone with small magnetic field and are accelerated by electric field of virtual diode along pinch axis. Therefore, the area of ions collection zone can be greater then Z-pinch crossection because low density plasma near anode isn't compressed during pinch process. Self focusing of ion beam to the pinch - target is inherent for successive axial positioning of these two zones. This effect could be useful to increase neutron production efficiency of Z-pinches.

Parallel double-zone Z-pinch

Schematic of target unit of parallel double - zone pinch is shown at Fig.5 . It was proposed and investigated as a double liner scheme previously in [3]. Z-pinch in low mass solid gas , surrounding thin fiber, used to carry the bulk of current while the current is rising and to allow it to switch over to flow on the fiber at the moment when peak current is reached, was investigated in joint experiment "JEX'92".

In our experiments cathode nozzles could produce low mass annular or solid flux of Ar or Xe with $\mu \cong 10-40$ mkg/cm, 2-3 cm diameter. Foam loads (agar-agar, doped 20% KCl and 20% Mo) 200-300 mkg/cm of 1-3 mm diameter were situated at the axis of the nozzle between solid cathode and anode net, cathode - anode gap being of 1 cm.

Visible light streak-camera was used with two slits- radial at the middle of anode - cathode gap and axial, directed along pinch axis. X-ray streak - camera with radial slit and collimated X-ray filtered diodes (XRD) were used for soft X-ray time-resolved measurements. Laser shadowgraphy of gas current shell gave 3 frames with intervals 14 and 9 ns. Time integrated X-ray spectrometers (with and without spatial resolution), X-ray pinhole cameras and calorimeter were used too.

In our experiments low mass current shells became unstable at rather small compression ratios. It was happened as for solid so for annular gas fluxes, and was independent , in general, on parameters of axial load.

Visible light streak pictures show radial and azimuthal splitting of lighting layers of current shell. It begins as earlier as mass is lower. The instability is especially strong for annular Xe shell (Fig.6) . It occurs at the beginning of current shell compression. Shell stops its radial movement and axial foam load bursts into shining in 2-3 ns at this moment, axial slit indicates simultaneous lighting of whole length of foam cylinder. Some nanosecond later another wave of radiation passes in radial (to the axis) or axial (from cathode to anode) direction through the foam load and axial inhomogeneity of its lighting occurs. Disturbed shell begins its compression again and next phases of Z-pinch of total mass occur the ordinary successive compression and expansion.

X-ray streak picture of foam load (Fig. 7a) looks like visible light picture in most compressed phase (Fig.7b,c), being different only in some details .

XRD signals have two oscillating maximums: the former coincides with the beginning of foam load flash , the latter related to the total compression of Z-pinch (Fig.8).

For the case of annular Xe shell , time integrated pinhole camera shows nicely shaped, nonexpanded foam load and intensive X-ray radiation at the axis of the foam. This radiation has an axial inhomogeneity, and a tubular structure in some cases (Fig.9a,b,c).

Time- and space-integrated X-ray spectrum shows He- and H-like K and Cl lines as for 20%KCL doping, so for 20%KCl+20%Mo doping of the foam load. It allows to conclude that some part of foam plasma has an electron temperature of ~ 1keV (Fig.10).

A fine structure of current shell changes dramatically during initial phase of compression. Shadow pictures of Ar plasma shell (solid flux, I~ 1MA, V~ 0.4 MV) are shown at Fig.11a,b,c. Small scale ($\lesssim$ 0.1 cm) plasma density perturbations cover all surface of the shell (a,b). A front of electron density increasing propagates from cathode to anode after the perturbations (c). Visible light break shows that the shell lighting flash moves from cathode to anode with phase velocity of $\sim 10^8$ cm/s (Fig.11d). Taking into account that intensity of visible light from Ar (Z_1>1) plasma is proportional to $\sim n_A^2 Z_1^3$, it is possible to suggest, that this "hot breakdown" of current shell is related to sharp increasing of ionization degree of Ar plasma from $Z_1 \geqslant 1$ to $Z_1 >> 1$. It could be due to anomalous resistivity effects in low density plasma with high current density, consequently the magnetic flux could penetrate into the volume, surrounded by the current shell, to the load of small impedance or filling gas. High intensity oscillations of V(t) and dI/dt signals and hard X-ray pulsed radiation in this time confirm the idea.

Discussion

It is possible to suggest, that instability of current shell originates the POS process. The current of 1-2 MA could switch from shell to foam load during some nanosecond in that case, giving more then 10^{14}A/s of dI/dt value. But now we have not direct measurements of a current in the axial load. Indirect witness of current effects in the foam load is an inhomogeneous radiation of foam plasma at the axis, "hot spots", which are usual for high density Z-pinches. Axial inhomogeneity occurs as well in the case of nonsymmetric position of curved foam load of 1 mm diameter inside of annular gas flux.

Theory of anomalous resistivity of low density high current plasma, based on the Electron MHD [4], proposes a criterion of its validity as $(\omega_{pi}.a/c)\leq 1$, where a - specific size of plasma, c - light velocity, ω_{pi}- plasma ion frequency. This criterion is fulfilled for parameters of Z-pinches plasma in our experiments. Consequently, the anomalous conductivity $\sigma_{eff}\sim\omega_{pe}^{2}/\omega_{He}$ can be used to explain the POS process in our case [4]. The same theory can explain the poor compression of low density plasma of inhomogeneous D_2 Z-pinch.

Different proposals, based on MHD simulations, explain the experimental data as a result of $\gtrsim$ 50% current penetration inside Xe shell with ~10% of its mass [7], or as a result of fast shock wave generation in Xe shell, which penetrates to foam being responsible for the former maximum of X-ray radiation at least [5-6].

As concerns the processes in a foam plasma, we have not a theory and have a few experimental results now. We can suggest only that the foam plasma is very inhomogeneous at the initial moment of current flow through it. Low density and small mass plasma component fills most part of the foam bulk. This plasma can define the main electrical and some thermodynamical and radiative features of the foam load. Anyway, it is worth to investigate it in more details.

References

1.Физика плазмы, I99I,вып.9,стр.I

2.A.V.Batunin, G,S,Volkov, et. al. Beam-plasma interaction in axially inhomogeneous Z-pinch. "BEAMS'92, PH-15.

3.V.P.Smirnov, T.V.Grabovsky, et. al. "BEAMS'90, I-07, p.61-78.

4.П.В.Сасоров, "К теории плазменных размыкателей", ПЖЭТФ,I992,56,6I4-6I7.

5.В.А.Гасилов, С.Ф.Григорьев, С.В.Захаров. "Динамика магнитного сжатия газовых струй". Препринт ИАЭ, 4767/6, I989.

6. А.В.Браницкий, В.Д.Вихарев, и др. Физика плазмы, т.I7, вып.5,I99I, 53I-54I.

7.Л.Б. Никандров и др."Образование плазменного предвестника при схлопывании многопроволочных лайнеров". ПЖЭТФ,I987,т.45,стр.23-25.

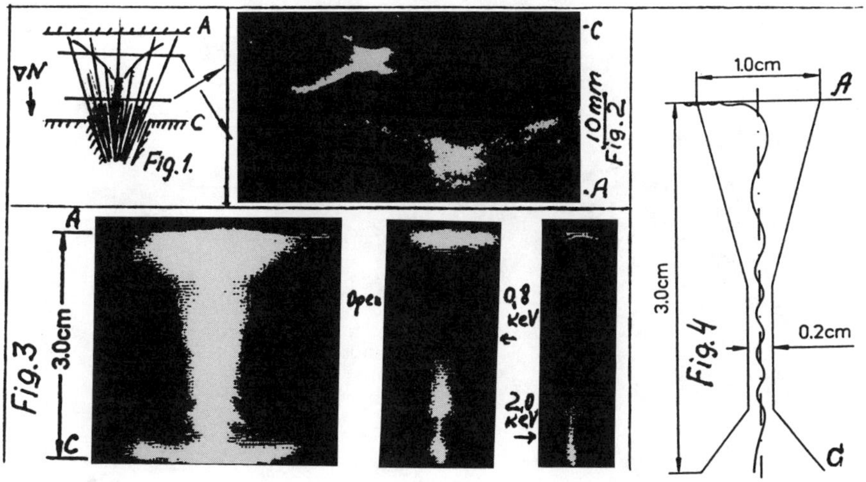

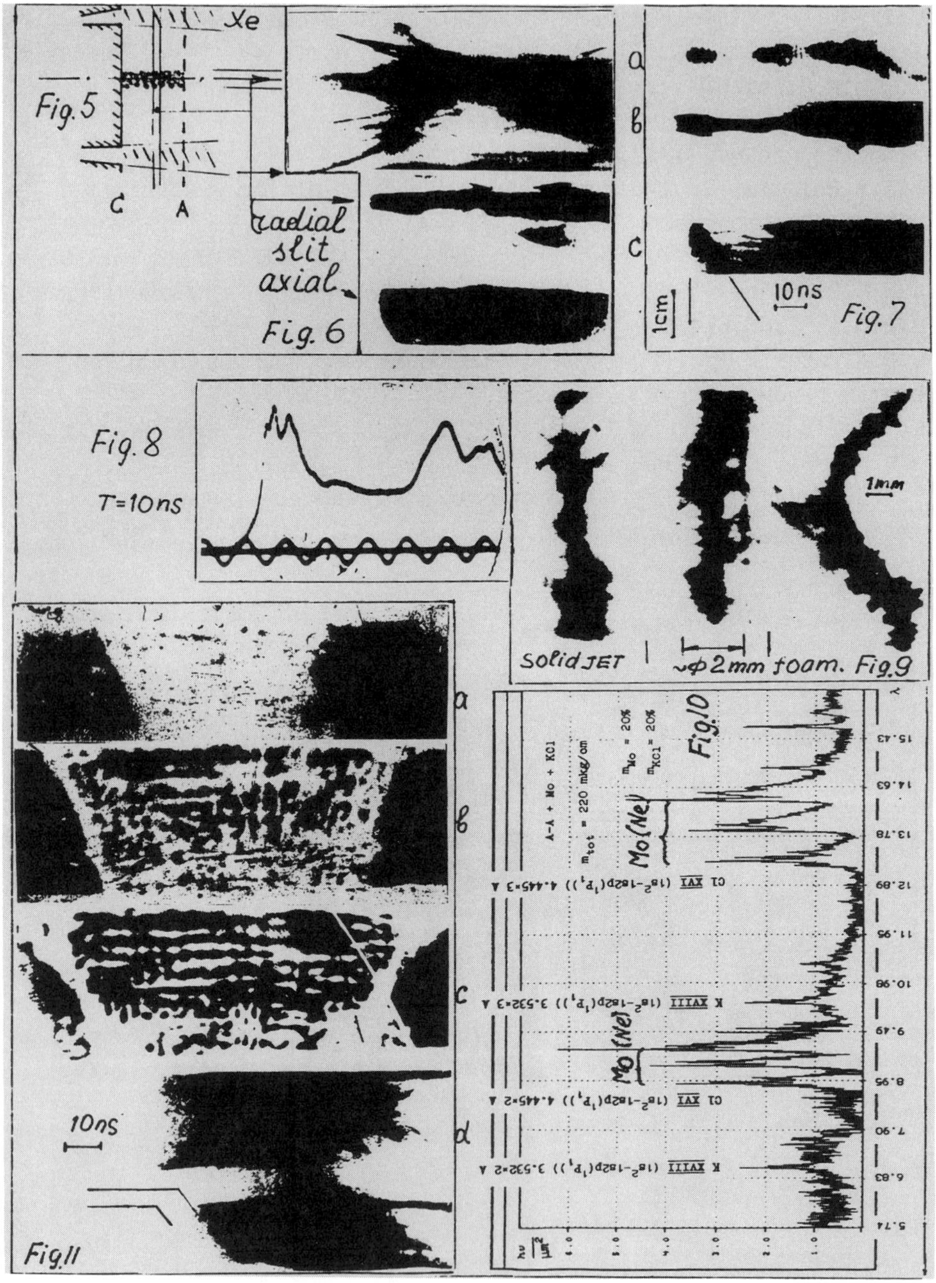
Xe
Fig.5
C A
radial
slit
axial
Fig. 6
a
b
c
1cm
10ns
Fig.7
Fig. 8
T=10ns
1mm
Solid JET
~Φ2mm foam.
Fig.9
Fig.10
A-A + Mo + KCl
m_tot = 220 mkg/cm
m_Mo = 20%
m_KCl = 20%
Mo(Ne)
Cl XVI (1s²-1s2p(¹P₁)) 4.445=3 Å
K XVIII (1s²-1s2p(¹P₁)) 3.532=3 Å
Cl XVI (1s²-1s2p(¹P₁)) 4.445=2 Å
K XVIII (1s²-1s2p(¹P₁)) 3.532=2 Å
a
b
c
d
10ns
Fig.11

X-PINCH IN HIGH-CURRENT DIODE

B.A.Bryunetkin*, A.Ya.Faenov*, G.V.Ivanenkov**,
S.Ya.Khakhalin*, A.R.Mingaleev**, S.A.Pikuz**,
V.M.Romanova**, T.A.Shelkovenko**, I.Yu.Skobelev*
*NPO VNIIFTRI, Mendeleevo, Moscow region, Russia
**P.N.Lebedev Physical Institute, Moscow, Russia

ABSTRACT

The review of X-pinch investigations in high current diode of BIN facility (250 kA, 100 ns) is presented. The main purposes were to investigate pinch forming processes and hot dense plasma properties. X-pinch is also considered as a source for multiple charged ions spectroscopy and for X-ray optics testing. The set of diagnostics applied in these experiments allowed us to investigate the pinch forming processes in different configurations of crossed wires loads. High spectral and space resolved measurements of plasma radiation in 1–200 Å range, absolute energy measurements and electron beam registration were provided. Plasma parameters were obtained from relative intensities and shapes of multiple charged ions spectral lines. Electron density of plasma with the temperature $T_e = 0.2$–1 keV variated from 10^{23} cm^{-3} in hot spot to 10^{18} cm^{-3} during plasma expansion. In recombining plasma, an inversion of Al He-like ions levels population was registrated. Total radiation output of 0.5 mm pinch reached hundreds Joules in 2–100 Å range during 100 ns.

INTRODUCTION

The development of methods for the creation and diagnostics of hot dense plasma is a great interest to many fields of plasma physics: ICF, interaction of intense ultrashort laser pulses with matter, X-ray lasers. On the other hand, for many problems, for example microradiography, creation of inversion in the schemes with resonance photopumping, medical-biological investigations, compact powerfull sources of X-ray are needed. To this point of view, the fast Z-pinch in high-current diode[1,2] is seems very prospective physical object. For example, it was shown[3] that the spectral intensity of the hot point emission is 10^5 – 10^6 W/Å·sr in the region 10 – 100 Å. It allows to consider this X-ray source as a possible one for pumping the active medium of shortwavelength lasers[4] and for other applied problems. An additional interest is due to the fact that many details of the hot point creation in fast Z-pinch is not clear now.

To achive the ultrahigh plasma parameters and X-ray radiation characteristics, the arrangement named "X-pinch" [5] was investigated in this work. It is necessary to remark, that some diagnostics tools (spectrometer with spherically bent mica crystal, Bragg-Fresnel lens) have a unique combination of efficiency, spectral and space resolution.

EXPERIMENTAL AND DIAGNOSTIC FACILITY

A fast pulse storage — facility (BIN) [6] is assembled as follows: Marx generator — intermediate capacitor — single forming line — vacuum diode. Storage energy of forming line at a charge voltage of 600 kV was 3.25 kJ. The output current in the short-circuit exceeded 270 kA. In experiments with a plasma load the diode current reached 250 kA with pulse duration of about 100 ns.

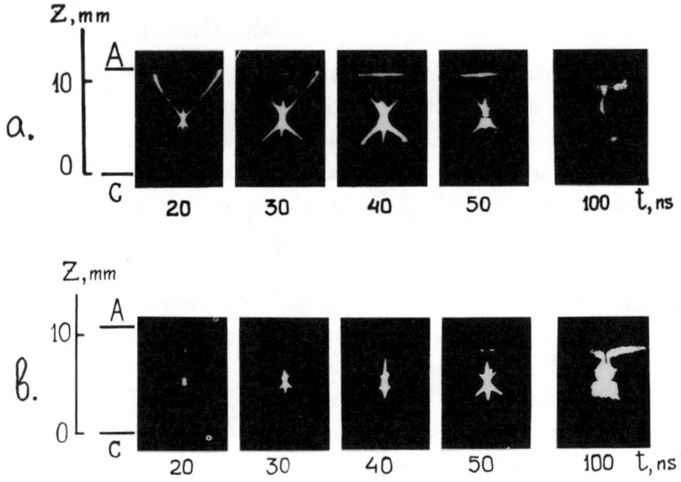

Fig.1: Visible framing-camera pictures of the X-pinch evolution: (*a*) 2-wire load, (*b*) 4-wires load (tungsten, diameter 8 μm).

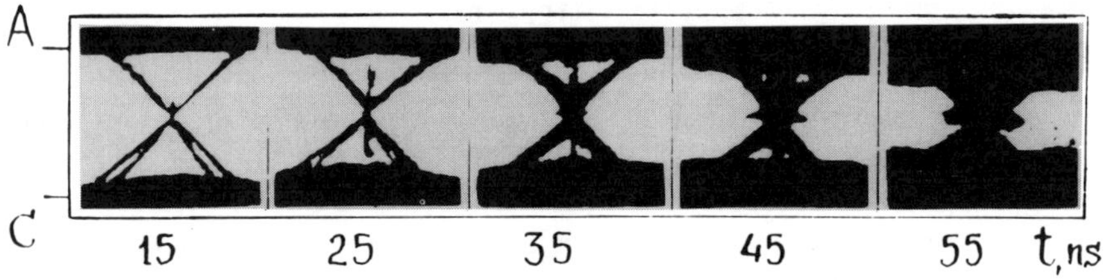

Fig.2: Shadowgrams of plasma in the case of four-wires load.

Thin metal wires and dielectric fibres crossed in the 10–15 mm gap were used as the load of the vacuum diode. In this case plasma pinching took place in a fixed space point in contrast to the case of single wire or liner. Due to the possibility of plasma flow from the hot spot of X-pinch, the higher specific plasma parameters can be achieved than in case of single wire. Plasma processes were investigated with the help of wide number of diagnostics in different spectral ranges. The pinch forming processes were observed in visible and infrared light with the help of 5-frames fast camera (Fig.1) and a YAG-laser (Fig.2). Vacuum UV radiation was recorded by vacuum X-ray diode and spectrometers with transmission grating (Fig.3) (for details see [7]). Soft X-ray radiation was investigated by pin-hole cameras, scintillation detectors, various types of crystal spectrometers (with plane and curved crystals according to Johann scheme and with spherically bent crystal) and Bragg-Fresnel lens[8]. Because of its very high efficiency, a spectrometer with a spherically bent crystal[9] allowed the observation of multicharged ions spectra with high spectral and spatial resolution not only in the hot point area but also at some distance (up to several mm) from it.

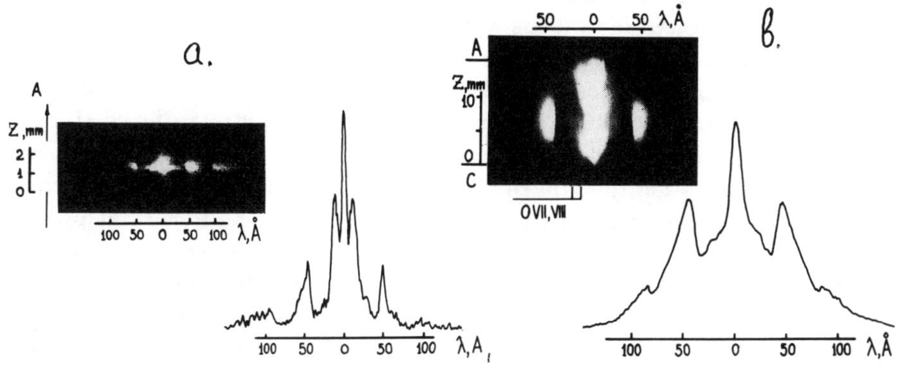

Fig.3: Transmission grating XUV spectra of the X-pinch plasma in the spectral region 5–100 Å: (*a*) tungsten grating, period 1 μm, diameter 25 μm; (*b*) gold grating, period 1 μm, sizes 0.1 mm × 1 mm.

X-PINCH PLASMA FORMING PROCESSES

The main stages of the pinch forming process are shown on the photos obtained by multi-frames camera. On Fig.1a, the evolution of 2-wires X-pinch (W, diameter 6 μm) is shown. The expansion velocity of the luminous area reaches 10 cm/μs along axis and 2 cm/μs transversely after 30 ns from current beginning. Up to 50 ns, cold external plasma absorb the radiation from the central core. Photos of the 4-wires X-pinch are shown on Fig.1b. Here maximal luminous area expansion velocities reach $1.5 \cdot 10^7$ cm/s in axial direction and $0.3 \cdot 10^7$ cm/s in transversal one. Cold wires cores were seen up to 100 ns. The similar picture was observed by shadowgraphy (Fig.2).

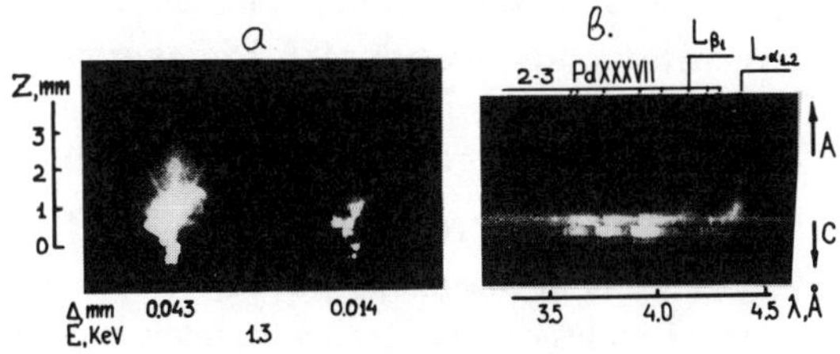

Fig.4: Pin-hole image (*a*) and X-ray spectra (*b*) of the X-pinch (Pd-wires, diameter 20 μm).

Local hot spots are well seen on the pin-hole imagings (Fig.4). The asymmetry in hard X-ray can be explained by electron beam generated in pinch-point. This substantiates availability of cold plasma stream characteristical radiation lines on spectrograms (Fig.4).

So, it is possible to create in X-pinch localized point-like X-ray sources with pulse duration of about 10 ns. The wire material determines the plasma size (from 5 to 200 μm) and the emission spectral range.

Estimations of the plasma compression were carried out in the basis of radiating hot point model[2]. It originates in the radiation collapse idea that has been put forward before for plasma focus and micro-pinch [10]. According to this idea, the evolution of a X-pinch single hot point is as follows: metal electrical explosion plasma expansion and luminosity of skin-layer been observed on it's background have instability near cross point as a result. Hot plasma stream flows along the system axis during the first stage ($\approx$ 10 ns) and increases the magnetic compression. Its result is cumulation of shock wave on the axis and its subsequent pinch surface exposure changes the balance of magnetic and thermal pressure. But heating now not only favours expansion but also increases repeated ionization and radiation (in continious spectrum — bremsstrahlung and recombinating ones). While current exceedes some critical value (Pease-Braginskii current in correspondence with ionization multiplicity) the second compression appeares in which radiation losses are of great significance.

Here radiation from the most inner layers can unlock that sufficiently increase expansion velocity ($\approx$ 0.1 ns). Calculations for aluminium wire with diameter 10 μm show that in our case the conditions for the appearance of He- and H-like ions exist. The computed values for T_e and n_e were $T_e \simeq 300$ eV, $n_e \simeq 6 \cdot 10^{22}$ cm^{-3}, and the experimental ones were $T_e \simeq 400$ eV, $n_e \simeq 5 \cdot 10^{22}$ cm^{-3} [2,11]. The calculated pinch radius is about 10 μm, the resistance is about 1 Ohm. The estimate of the applied pinch voltage ($\simeq$ 100 kV) is consistent with the data from bremsstrahlung radiation from the anode where the quantum energy reaches 60 keV (an electron beam is generated during pinch decay). Resistance was observed experimentally as well as beam generation, just analogous process are observed in case of X-pinch.

Energy estimations showed the K-line radiation output can reach about 0.1–10 J ($\simeq$ 1 GW of power). Kinetic energy of the pinch and magnetic field power have values of the same magnitude. The plasma stream carries about 0.1 μg of material that corresponds to about 0.5 J in energy. Photorecombination radiation output of pinch exceedes 1 J, but its power is on the level of 1 GW. The total radiation energy in wavelength range from 5 to 100 Å is about 1 kJ.

THE DIAGNOSTICS OF X-PINCH PLASMA

For the measurement of plasma parameters we used X-ray spectroscopic methods. The plasma emission spectra have been recorded on film, and the spectrographs used allowed spatial resolution for one coordinate z, i.e. the anode-cathode direction. All spectrograms are integrals of plasma radiation over both time and two spatial coordinates normal to z-axis. The integral character of spectra is most crucial for plasma regions close to the X-pinch hot point, where plasma parameters gradients are very large. In this case, the spectrograph sums as well the plasma radiation at the moments of both hot point creation and decay as it integrates over the transverse expansion of plasma source with very small ($\leq 10\ \mu$m) initial dimension. This effect is less crucial for plasma regions far from the hot point, because i) here the plasma is more homogeneous in transverse direction and ii) in this case the spatial coordinate z is approximately equivalent to some temporal stage of plasma expansion.

For this reason, to diagnize a hot point plasma one must use spectral lines which are weakly excited in the decaying plasma, i.e., when the plasma temperature is sharply decreased (for example, the transitions from autoionizing levels of multiply charged ions may be used). For such lines the intensities observed correspond mainly to the moment when the hot point plasma has maximal of temperature and density.

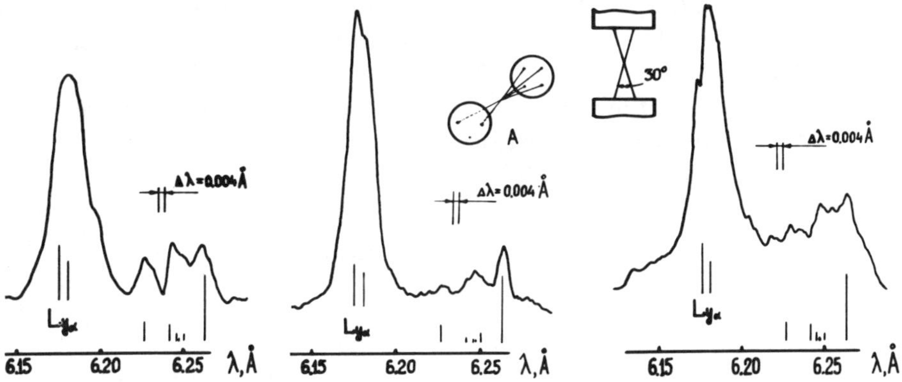

Fig.5: Densitograms of plasma spectra near resonance line Ly_α of Si*XIV* H-like ion, recodied at the different conditions of explosion a glass wires in X-pinch configuration.

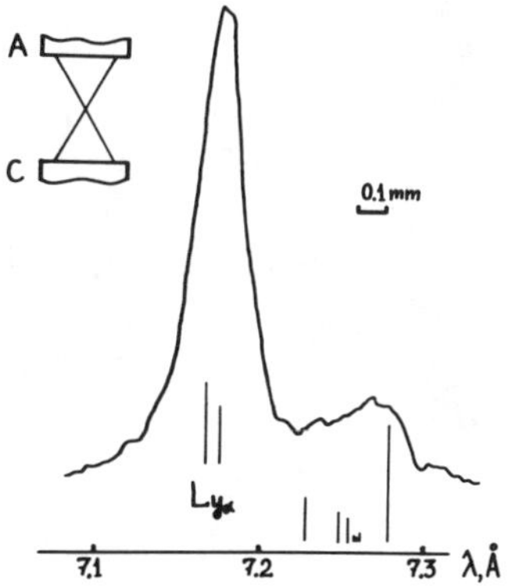

Fig.6: Densitogram of the emission spectra of the plasma near resonance line Ly_α of Al*XIII* H-like ion, obtained by means of CsAP flat-crystal spectrometer.

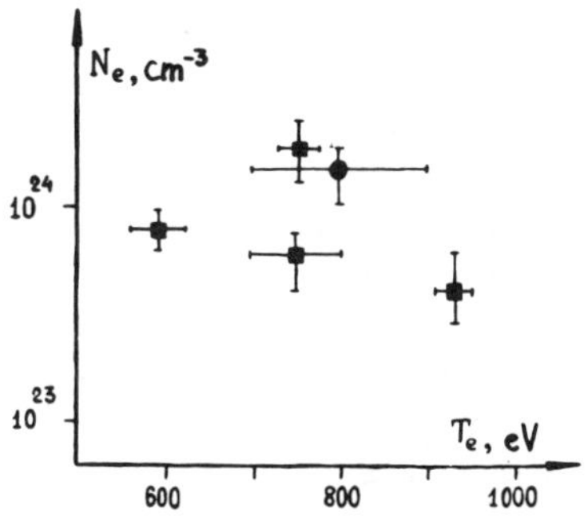

Fig.7: The parameters of X-pinch hot point plasma measured in the experiments on glass and aluminium thin wires.

X-pinch hot point parameters determination. The plasma electron density measurements were carried out by means of relative intensities of dielectronic satellites to resonance lines of the H-like ions Al*XIII* and Si*XIV*. The ratio of the sum of triplet satellites $2p^2\ ^3P \rightarrow 1s2p\ ^3P$ and $2s2p\ ^3P \rightarrow 1s2s\ ^3S$ intensities to the singlet satellite $2p^2\ ^1D_2 \rightarrow 1s2p\ ^1P_1$ intensity was used. It follows from the calculations [12] (see also [13]), that this ratio depends only very slightly on plasma temperature and is determined mainly by the electron density. Figs.5,6 show the densitograms of plasma emission spectra obtained with the help of spherical mica and flat CsAP crystal spectrographs. Fig.5 corresponds to wire explosions of glass fibres and Fig.6 to that of aluminium wires. The density was derived by comparison of the experimental intensity ratio value with calculations [12]; the results are shown in Fig.7.

The plasma electron temperature measurements were carried out by means of intensity ratio of the dielectronic satellite $2p^2\ ^1D_2 - 1s2p\ ^1P_1$ to the H-like resonance line (see, for example, [14]). The values of T_e obtained are also presented in Fig.7. It can be seen from Fig.7 that although the hot point plasma parameters in different spots are not the same, the typical values are in the ranges $T_e \simeq 550 - 950$ eV and $n_e \simeq 3 \cdot 10^{23} - 2 \cdot 10^{24}$ cm^{-3}.

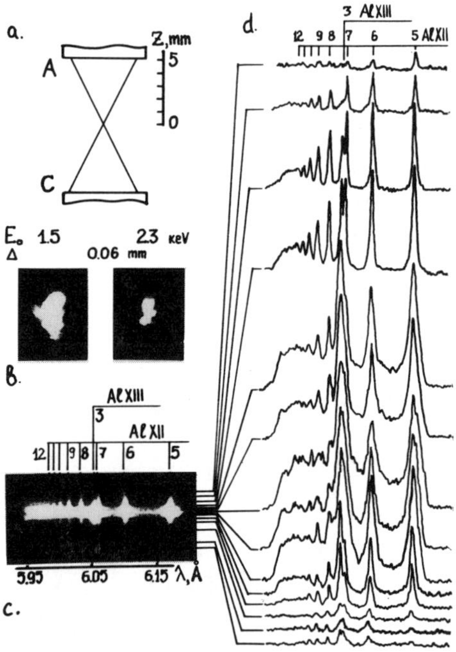

Fig.8: (*a*) Diode setup; (*b*) pin-hole images obtained by an explosion of a Al-wire at the different cut-off energies filters; (*c*) emission spectra of the Al-plasma, obtained by means of spherically curved mica crystal; (*d*) densitograms of Al plasma spectra on the different distances from then hot spot.

Plasma parameters for the decay stage of the hot points. In order to investigate the spectra radiated by plasma regions at some distance from the hot point, we employed a flat CsAP crystal spectrograph. This spectrograph covers a range sufficient to observe the series of the $1snp\ ^1P_1 \rightarrow 1s^2\ ^1S_0$ transitions ($n =$ 5–9) in the He-like aluminium ion. Both the spectra obtained in our experiments and the X-ray pin-hole camera results are presented in Fig.8.

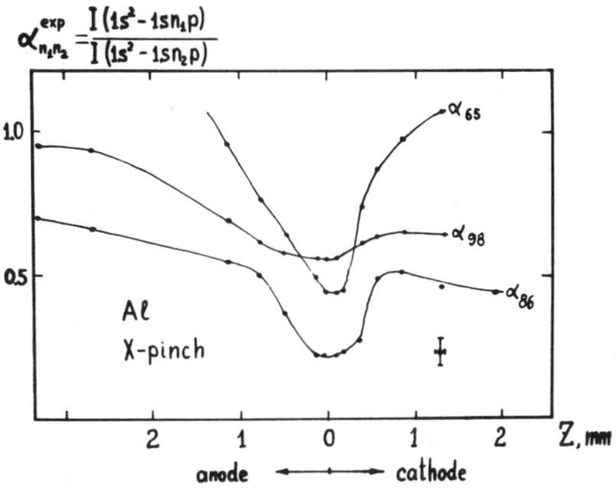

Fig.9: The spatial distribution along anode-cathode axis of some He-like Al*XII* ion spectral lines relative intensities.

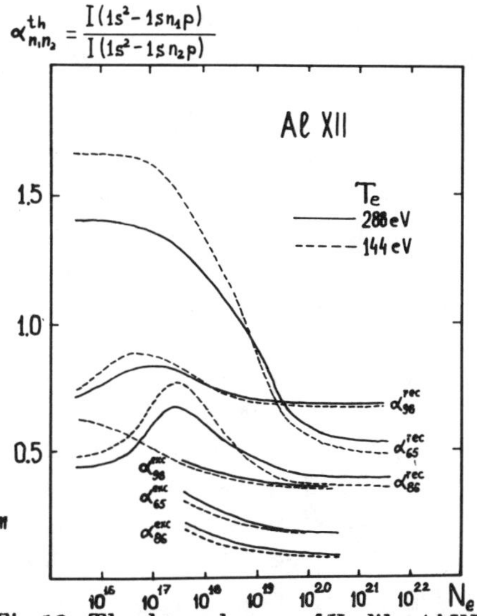

Fig.10: The dependences of He-like Al*XII* ion spectral lines relative intensities on plasma electron density calculated for T_e = 288 eV (solid lines) and T_e = 144 eV (dashed lines). Upper indexes "rec" and "exc" correspond to the cases of level populating by only recombination or excitation processes, respectively.

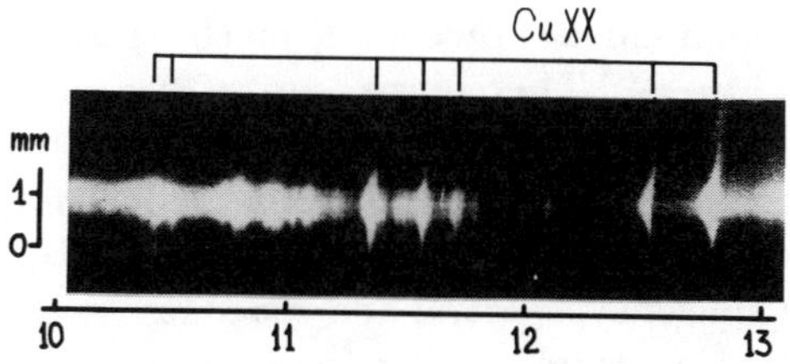

Fig.11: The spectrograms of Cu-plasma.

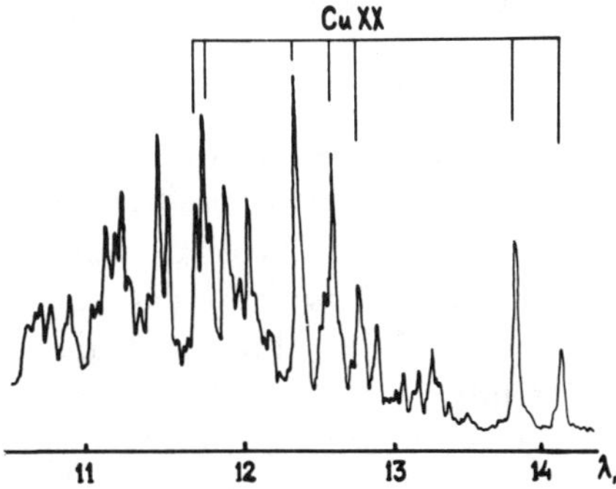

Fig.12: The densitograms of Cu plasma spectra, emitted by spatial regions placed at the different distances from hot point.

The spatial distribution along z-axis of some He-like Al*XII* spectral lines is shown in Fig.9. To derive the plasma parameters, intensity ratios were compared with numeric results calculated in quasistationary radiative-collisional plasma kinetic model from [15]. The theoretical results are presented in Fig.10 for two limiting cases: i) when the excited levels of Al*XII* ion are populated only by excitation from its ground state (ionizing or steady-state plasma) and ii) when these levels are populated only by recombination of H-like Al*XIII* ion (recombining plasma).

We can conclude from the comparison of experimental and theoretical intensities ratios that at the distances $|z| \geq 300\ \mu m$ from the hot point ionization state is recombining (supercold plasma). In this case, the intensity ratios mentioned depend only slightly on plasma temperature and can be used for plasma density determination (Fig.10). The values of n_e obtained by this method are shown in Table for the region $|z| \geq 300\ \mu m$. It should be noted that the recombining character of the plasma state in this spatial region was also confirmed by the inversion of some Al*XII* levels (for example, the $1s6p\ ^1P_1$ and $1s7p\ ^1P_1$ levels populations are greater than $1s5p\ ^1P_1$ one), which are observed surely on spectrograms presented in Fig.8.

Table

The electron density far from hot point placed at $z = 0$

z, mm	-2	-1.7	-1.3	-1	-0.7	0.3	0.5	0.83
n_e, 10^{18} cm^{-3}	0.3	1	5.4	16	25	26	17	10

The observation of L-spectra of heavy ions. In our experiments we also observed L-spectra of multiply charged Ni and Cu ions excited in the X-pinch plasma. The spectrogram and densitogram of Cu plasma emission spectra in region $\lambda \simeq 12.5$–12.9 Å are shown in Figs.11,12. It can be seen from Fig.12 that spectral lines emitted by hot point plasma a strongly asymmetric. In our opinion, this effect is caused by self-absorption of spectral lines in the dense plasma.

SUMMARY

In conclusion, we reviewed the main results of our present investigations. The explosion of the thin wires in the gap of high-current diod in X-pinch geometry results in a X-ray source of unique brightness. The comparison of numerical simulation results with the experimental observations shows, that the radiating hot point model, based on the radiation collapse conception, describes sufficiently well the process of the hot point evolution. The spectroscopic arrangement that has been used in the experiment, has high efficiency, spectral and space resolution. The ultrahigh values of plasma parameters measured in our experiments allow statement that this object has a unique energy density.

REFERENCES

1. S.M.Zakharov, G.V.Ivanenkov, A.A.Kolomensky et al. Fiz. Plazmy **9**, 469 (1983) (in Russian); A. Bartnik et al. Ibid. **16**, 1482 (1990) (in Russian).

2. S.M.Zakharov, G.V.Ivanenkov, A.A.Kolomensky et al. Ibid. **13**, 206 (1987) (in Russian).
3. R.C.Elton. "X-ray Lasers", Academic, Boston, MA, (1990).
4. E.Fill et al. Izv. Acad. Sci. USSR **55**, 794 (1991); J.Nilsen. Ibid. **55**, 782 (1991); Opt.Lett. **15**, 798 (1990); B.A.Bryunetkin, A.Ya.Faenov, S.A.Pikuz, I.Yu.Skobelev. Laser and Particle Beams. **10** (1992) (in press).
5. S.M.Zakharov, G.V.Ivanenkov, A.A.Kolomensky et al. Pis'ma v ZhTF **8**, 1060 (1982) (in Russian); G.V.Ivanenkov, A.N.Lebedev, S.A.Pikuz. P.N. Lebedev Phys.Inst.Preprint No 210, M. (1989); N.Qi, D.A.Hammer, D.H. Kalantar et al. JQSRT **44**, 519 (1990).
6. G.V.Ivanenkov, A.R.Mingaleev, S.A.Pikuz et al. P.N.Lebedev Phys. Inst. Preprint N 50, M., (1992) (in Russian).
7. B.A.Bryunetkin, G.V.Ivanenkov, S.A.Pikuz et al. Pis'ma v ZhTF **17**, 16 (1991) (in Russian).
8. Yu.A.Agafonov, B.A.Bryunetkin, A.I.Erko et al. Ibid. **18**, 56 (1992) (in Russian).
9. B.A.Bryunetkin, G.V.Ivanenkov, S.A.Pikuz et al. Ibid. **17**, 24 (1991) (in Russian).
10. V.V.Vihrev, K.G.Gureev. ZhTF **48**, 2264 (1978) (in Russian); V.V. Vihrev, V.V.Ivanov, K.N.Koshelev. Fiz. Plasmy **8**, 1211 (1982) (in Russian).
11. A.Ya.Faenov, S.Ya.Khakhalin, A.A.Kolomensky et al. J.Phys. **D18**, 1347 (1985).
12. A.V.Vinogradov, I.Yu.Skobelev. Pis'ma v ZhETF **27**, 97 (1978) (in Russian).
13. J.G.Lunney, J.F.Seely. Phys.Rev.Lett. **16**, 342 (1981); J.F.Seely, R.H. Dixon, R.C.Elton. Phys.Rev. **A 23**, 1437 (1981); I.Yu.Skobelev, S.Ya.Khakhalin. Opt. i Spektrosk. **59**, 22 (1985) (in Russian); P.S.Antsiferov, K.N.Koshelev, V.I.Krauz et al. Fiz. Plasmy **16**, 1319 (1990) (in Russian).
14. V.A.Boiko, A.Ya.Faenov, S.A.Pikuz, U.I.Safronova. Mon.Not.R.Astr. Soc. **181**, 107 (1977).
15. I.Yu. Skobelev, S.Ya. Khakhalin, S.I. Yakovlenko. Proc. VNIIFTRI 4, M., (1986) (in Russian).

OBSERVATIONS OF THE VACUUM SPARK UNDER DIFFERENT CONDITIONS OF OPERATION

H.Chuaqui, M.Favre, L.Soto, E.Wyndham.
Facultad de Física, Pontificia Universidad Católica de Chile,
Casilla 306, Santiago-22, Chile.

ABSTRACT

Observations of a Vacuum Spark discharge plasma are presented when driven from a 1.5Ω 120ns Coaxial line at peak current up to 180kA. The line may be operated with the line gap shorted allowing the Hybrid Vacuum Spark mode, or with the line gap operational giving a rectangular voltage pulse. In both cases the rate of rise of the current is 1-2x10^{12}A/s, however the initial voltage conditions are different. A Nd:YAG laser may be focused onto either the Anode or onto the Cathode to initiate the discharge. With the Laser focused onto the Anode intense electron beams are generated, with the formation of a low density gap plasma, except near the anode where the density is much higher and where filamentary density structures are observed. On applying the laser to the Cathode in the Hybrid mode, the additional formation of a very dense narrow pinch near the cathode at peak current is seen. For currents in excess of 90kA, Hot Spots form in this channel and are associated with the generation of energetic electron beams. The observations presented here differ with comparable work in the initial conditions required for the formation of dense Hot Spots, in their position and in their reproducibility. The electron density in the channel is measured at 10^{19} cm^{-3} over 0.5mm, whereas the Hot Spots have dimensions of 50 μm and temperatures of approximately 700eV.

INTRODUCTION

The Vacuum Spark has been the object of considerable interest for many years and several extensive reviews[1,2] have been published. It is apparent, however, that as the specific energy density deposited in the plasma has increased dramatically in the last decade or so with the development of pulse power techniques, the Vacuum Spark has not received the same attention as other Z pinch discharges. One of the few such experiments which has applied this new technology used a switched coaxial water line, the Don accelerator[3], with a rather short 50ns, 100kA, 2Ω pulse. Other short pulse work has been at much lower energies, either using small fast capacitors[4], or coaxial cables with stored energies from 1 to 10J [5,6]. Under these conditions, the gap plasma densities are low and the X-ray emission is from electron beam interaction with the anode. To obtain the keV temperatures and electron density above 10^{19} cm^{-3} currents of the order 100kA are required. In general these are usually associated with low inductance capacitor banks with over 1kJ stored energy and 1μs quarter period.

In this work we present observations of vacuum spark formation with the characteristic densities and temperatures of larger machines but using a 1.5Ω, 120ns coaxial line with a line gas gap and short transfer section operating with between 250 and 500J stored in the Marx bank. In previous work[7], a new mode of operation of the Vacuum Spark discharge was introduced in which the high dI/dt value associated with a switched coaxial water line was combined with an applied voltage at the moment of the preionising discharge. In this "hybrid" regime the fast transition to conduction of the discharge itself switches the line, whereas in the case of normal operation of a switched coaxial line gap the large value of dI/dt occurs when this breaks. In the case of what we define as "normal" operation the initial plasma is formed by a pulsed Nd:YAG focused onto one of the electrodes and as there is no applied voltage expands freely. The initial plasma parameters when the application of the external voltage is applied, depend on the laser energy deposited, the electrode material, and the relative delay between the laser pulse and the applied voltage. The situation is different in the hybrid mode as the laser is applied when there is already a voltage ramp on the centre electrode. The time of the preionization is chosen so that breakdown occurs when the line is fully charged, the voltage ramp coming from the charging of the centre line by the Marx. Thus the initial conditions are significantly different and differences in observations are found. For example, X-ray emission from the anode is observed for 150ns until the main current associated with breakdown starts. Although the initial electron densities in both schemes is insufficient to register using interferometry, a significantly different density evolution is observed during the current rise phase of the discharge. However, the occurrence of dense hot spots at the time of peak conduction occurs in the same region for both cases, that is, close to the cathode. These are observed to occur only when the cathode is illuminated. In contrast, when the anode is illuminated by the laser only strong X-ray emission is observed from, or close to, the anode. The inter electrode densities are lower than the former case and no dense pinches are seen.

In the following sections the comparative behaviour will be detailed and discussed for four different schemes of operation: two for normal operation and two for hybrid operation. The two cases in each scheme correspond to preionization on the cathode or on the anode respectively. Results are presented for peak discharge currents of between 90 and 180kA with values of dI/dt up to 2×10^{12} A/s in pulses of up to 250ns.

EXPERIMENTAL DETAILS AND RESULTS

The details of the transmission line for use in the hybrid mode have been published in ref. 7, where discharges of up to 90kA are described. In this paper the transmission line was also operated with the line gap operational in order to allow a full study of comparative behaviour. The pulse obtained in the normal mode of operation has a duration of 120ns, whereas in the hybrid mode, in which all the line is charged, the pulse is about 250ns. The diagnostics are the same as in ref. 7 except that a second laser with a shorter pulse of 8ns was used for the holographic interferometry giving better interference fringes. The initiating preionising laser has an energy of 0.25J. In the case of the formation of the preionizing plasma on the cathode, the laser was focused through the anode onto the cathode, as described previously; whereas in the case of anode illumination, the laser was focused through a 2.5mm cathode axial hole onto the solid conical anode. In all the series of shots the cathode was made of a Tungsten-Copper alloy (W78/Cu22) to reduce erosion to a minimum, whereas the anode was made of this alloy for anode illumination or pure copper for cathode illumination.

Observations for hybrid operation at peak currents of up to 180kA are presented first. In Figure 1 observations are shown for the case of anode preionization. The voltage is seen to ramp up without current in the load until approximately 100ns before peak voltage at which time the laser preionization occurs. Hard X-ray emission from the anode is observed immediately with the Cu filtered PIN diode, but there is no appreciable current for some 100ns after the laser pulse at which time conduction starts in the gap with a 40ns rise time. The broad band emission is observed with the Al filtered Quantrad PIN diode and coincides in form with the hard X-ray (>5keV) detector. The emission effectively ceases when the applied voltage is 10% or less of the initial value. In this figure two pinhole broadband images are shown; the upper photograph is for one shot and the lower photograph shows the integrated emission from five shots. The bulk of the emission is beam target in origen with a small amount from a low density gap plasma. The electron density was measured using holographic interferometry. The observations for this case may be described briefly. The only measurable density occurs within two millimetres of the anode and is concentrated near the point where the initiating laser is focussed. The peak current is approximately 100kA whereas for the same applied voltage for laser preionization on the cathode, the peak current are 50% or more higher.

Observations for operation in the hybrid mode with cathode preionization are shown in Figures 2 and 3. In the first of these the X-ray emission is shown with both spatial and temporal resolution. The emission from the whole volume is shown using two BPX 65 PIN diodes, one of which is filtered with 10μm Ag with 10% relative cut off at 7keV and the other filtered with Cu and mylar for a 2keV cut off. The Quantrad PIN diode is masked so as to see the emission within 5mm of the cathode, the mask, however, is transparent for energies over 40keV which explains the first part of the signal at low current. The long pulse seen at peak current from the whole volume coincides with two 15ns pulses observed from the cathode side of the volume. This may be compared with the four pinhole image of this shot, each image is filtered as indicated. The image of the hardest X-rays (>13keV) at 12 o'clock shows only an emission from the anode immediately opposite where the laser focuses, whereas the broadband emission seen with the Al filter at 3 o'clock clearly shows two hot spots, one contiguous with the cathode and the other 1.5mm from the cathode. The upper set of images corresponds to the traces at the left, while the lower set are from four consecutive shots. Comparison with the 2μm Cu filtered image at 9 o'clock do not show the hot spots but do show emission from an anode blow off plasma which in turn is not seen in the 10μm Ag filtered image at 6 o'clock. The sharp spatial cut off of the Al image of the hot spots indicate a fifty micron diameter, while the relative intensities indicate a temperature af approximately 700eV. It is notable from the multiple exposure that the two hot spots are spatially very reproducible, which is a characteristic already noted of this mode of operation.

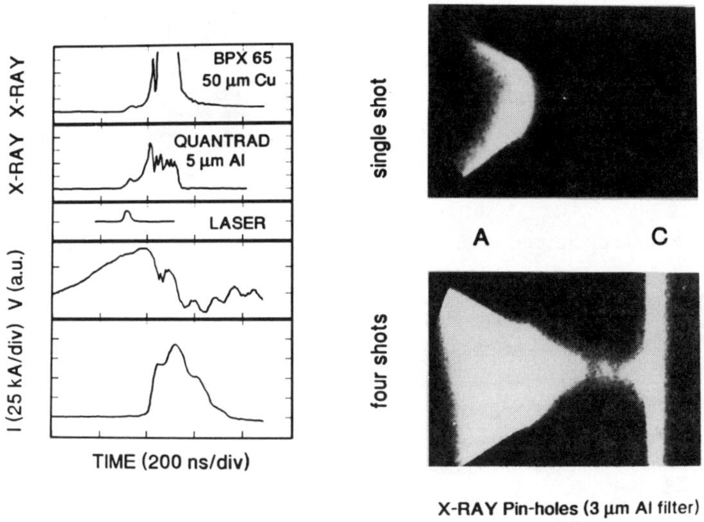

Figure 1. Hybrid mode operation, Laser incident on Anode. Pinhole Filters are 3μm Al.

In Figure 3 the electron plasma density is shown in the moment of formation of a hot spot. In the shadowgram it is easier to estimate the spatial dimensions of the hot spot, the dark blob, at less than 100μm. This in turn is embedded in a narrow 0.5mm diameter pinch some 3mm long stemming from the cathode. A value of 10^{19} cm^{-3} for the average pinch density is obtained.

In Figure 4 observations are shown of operation with the line gas gap operational giving the characteristic rectangular voltage and current waveforms of normal operation of the vacuum spark. The preionization is on the anode surface. The current pulse lasts 230ns FWHM with a peak current of 100kA. The interferograms correspond to a current of 50kA taken on the falling edge of the current. The broad band X-ray emission from the cathode half of the volume.is shown by the PIN diode trace. A short pulse is characteristic on the rising edge of the current occurring some 60ns into the current pulse when the applied voltage is at its maximum value.

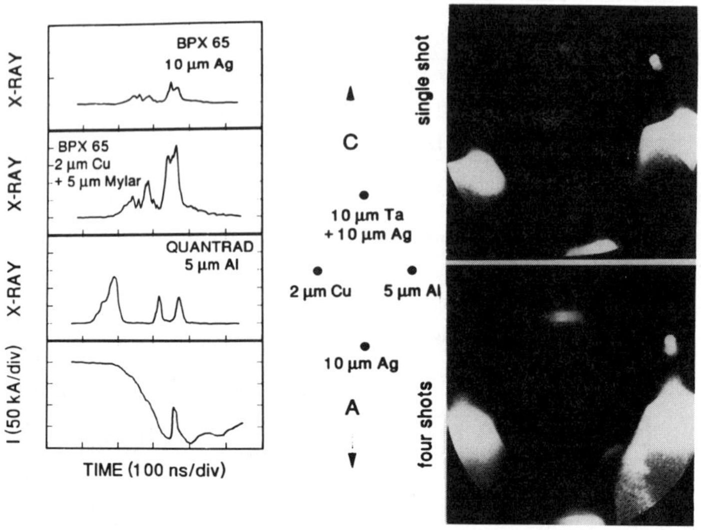

Figure 2. Hybrid mode operation, Laser incident on Cathode. Multi filtered Pinhole images.

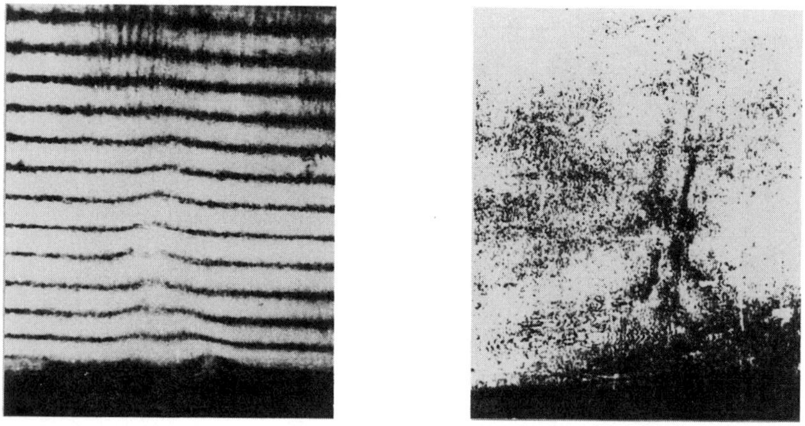

Figure 3. Interferogram and Shadowgram during the formation of a Hot Spot, which is embedded in the dense pinch close to the Cathode (lower dark border).

Interferograms taken at this time show a mm scale length plume emanating from the laser impact point but no other features. At the later time of Figure 4 the density distribution is confined to a column of less than 5mm diameter near the electrodes and 3mm elsewhere. Filamentary structures are clearly observeable near the anode. Near the cathode a dense relatively cold plasma is observed to form opposite the laser focus point.

Such filamentary structures are also seen in the anode plasma for the case of laser prionization at the cathode as may be seen in Figure 5. In this shadowgram and interferogram the time into the discharge and the maximum current are very close to the conditions of Figure 4. There are marked differences, especially at the cathode end, where, instead of an extensive plasma formation, there is a submillimetre diameter pinch a few mm long stemming from the cathode in which a hot spot forms. This pinch has a life time of no more than 40ns. The anode plasma is more extensive than in the previous case and is not seen to emit X-rays in an Al filtered pin hole image so a low temperature is inferred. The plasma is optically dense and probably of higher density than the previous case.

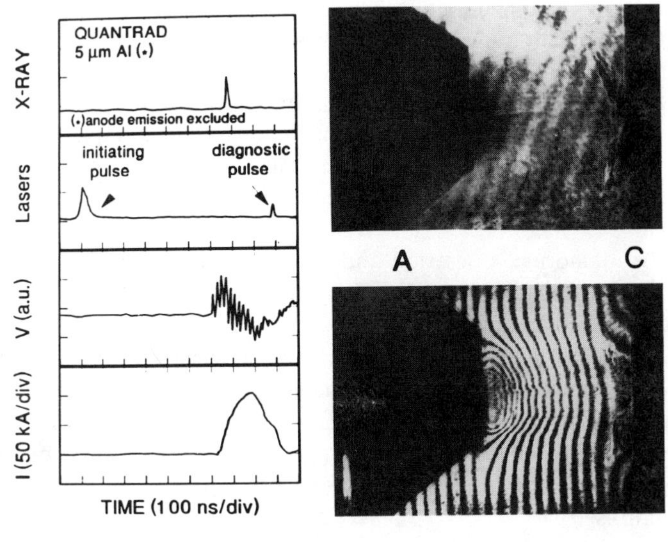

Figure 4. Normal mode operation. Laser incident on Anode 900ns before Voltage is applied.

DISCUSSION

It is interesting to compare these results with those of Zakharov et al. in which a 5J laser was focused onto the cathode in a very similar geometry. They found a dense pinch forming from the cathode after only 40ns, with a dense anode plasma observable after approximately 80ns. They also find a string of hot spots forming in the middle of the inter electrode gap, these being embedded in a dense pinch. They infer that the laser initiated plasma has a line density of 10^{15} to 10^{16} cm^{-1}, whereas

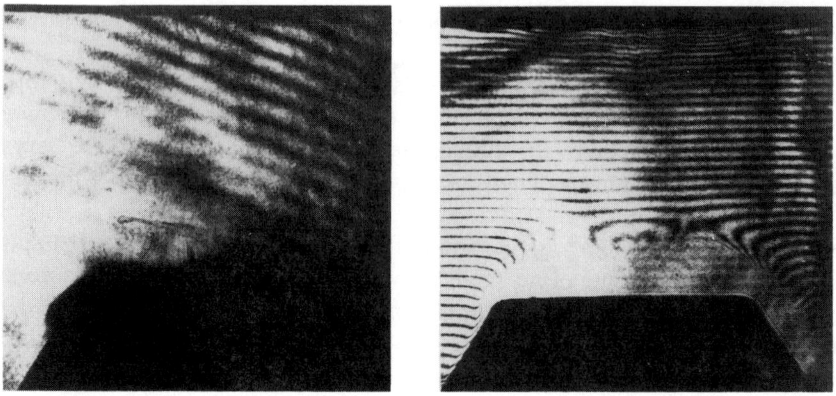

Figure 5. Normal mode operation. Laser incident on Cathode. Shadowgram and Interferogram taken 200ns after start of the current. Tight dense pinch forms at Cathode (top).

with the laser energy available here at least an order of magnitude lower line density is expected. This is also consistent with the much longer time required here for the formation of an observable inter electrode plasma using the same diagnostic. The means of increasing the interelectrode material is through electron beam evaporation of anode material.

In the case of normal operation and anode preionization the X-ray pulse on the rising current edge coincides with the Tungsten ion time of flight for the observed applied voltage. An ion beam may also be the cathode evaporation mechanism required to explain the form and the high density of the plasma at the cathode. In both cases filamentary structures are observed in the anode plasma. The contrast in density and temperature between the interelectrode plasma with its strong nonthermal electron distribution associated with the current and the relatively cold and dense anode plasma may be considered for electron thermal instabilities to explain the filamentary structures[8].

It may be inferred that the better results from the point of view of hot spot formation conditions obtained in the hybrid mode are due to the higher density which occurs when the laser plasma expands in a strong electric field. The mass of electrode material released in the field free case with the laser energy of 0.25J is estimated at 4×10^{-8}g, which satisfies the line density criterion for a successful pinch given in ref. 3, although there may well be room for improvement on the theoretical model in use at that time, so the limits are, in reality, not well defined. While retaining the condition of ref. 3, a successful pinch may be defined as one in which the value of the current is just sufficient for radiative collapse[9], which for the conditions of the experiment require of the order of 60-80kA.

CONCLUSIONS

A comparative study of the Vacuum Spark has been presented for the first time in which both normal and hybrid modes have been compared under identical geometry. The low laser preionizing energy of this experiment have produced considerable differences with previous work. It is hoped to extend the present results to peak currents in excess of 200kA in all modes of operation with laser energies up to a few joules. In addition Faraday rotation as a diagnostic will attempt to observe in a qualitative way the current distribution in the cathode pinch. Multi-frame holographic interferometry and shadowgrams will greatly enhance observational diagnostics in each shot.

ACKNOWDGEMENTS

We would like to thank professor M. G. Haines and Dr. P. Choi of Imperial College for stimulating conversations of these results. We thank FONDECYT, Fundación ANDES for their various financial contributions to the research programme. L. Soto is a scholarship student of Fundación ANDES. We thank the British Council for their support under the LINK scheme.

REFERENCES

1. E.D.Korop, B.E.Meierovich et al., Sov. Phys. Usp., **22**, 727, (1979)
2. K.N.Koshelev, N.R.Pereira, J. Appl. Phys. **69** (10), R21 (1991)
3. S.M.Zakharov, G.V.Ivanenkov, A.A.Kolomenskii, S.A.Pikuz, A.I.Samokhin, Sov. J.Plasma Phys. **10**(3), 303, (1984)
4. C.S.Wong, S.Lee, C.X.Ong, O.H.Chin, Jap. J. Appl. Phys. **28**, 1264, (1989)
5. S.P.Bugaev, R.B.Gaksht, E.A.Litvinov, V.P.Stas'ev, High Temp. **14**, 1027, (1976)
6. M.Skowronek, P.Romeas, IEEE Trans. Plasma Science, **PS-15** (5), 589, (1987)
7. E.Wyndham, H.Chuaqui, M.Favre, L.Soto P.Choi, J. Appl. Phys. **71**, 4164, (1992)
8. M.G.Haines, J. Plasma Phys. **12**, 1-14, (1974)
9. P.S.Antsiferov, K.N.Koshelev, A.E.Kramida, A.M.Panin, J.Phys. D: Appl. Phys. **22**,1073, (1989)

DYNAMICS OF AN X-PINCH PLASMA FROM TIME-RESOLVED DIAGNOSTICS

D. H. Kalantar and D. A. Hammer
Laboratory of Plasma Studies, Cornell University, Ithaca, NY 14853

A. E. Dangor, J. M. Bayley, and F. N. Beg
Blackett Laboratory, Imperial College, London, UK

ABSTRACT

Aluminum wire x-pinch and z-pinch experiments were conducted using the XP pulsed power generator at Cornell University. Time resolved measurements of plasma expansion and sausage instability growth along the wires were made using a sub-nanosecond pulsed nitrogen laser and x-ray backlighting. Rapid (0.9 to 2.9 cm/μs) unstable expansion of a coronal plasma formed around both x-pinch and z-pinch wires was observed, while a dense core plasma expanded more slowly (0.1 to 0.8 cm/μs) and uniformly. The x-pinch cross point is an intense, compact L-shell and K-shell radiation source. The cross region imploded to a minimum radius just before the intense emission of soft x-rays. Time resolved characteristics of the K-shell radiation implied electron densities as high as 10^{22} cm^{-3} and temperatures of a few hundred eV for 12 to 75 μm wires and 185 to 360 kA peak current. Z-pinches showed pinch points along the full wire length.

INTRODUCTION AND SUMMARY

Aluminum (Al) x-pinch[1] and z-pinch experiments were conducted using a low inductance pulser to deliver a peak current of 300 to 400 kA in a 100 ns (fwhm) pulse to the load. Two wire x-pinch and single wire z-pinch plasmas were generated both alone and in parallel with a separate x-pinch used for x-ray backlighting. The pinch dynamics may be summarized as follows: Early in the pulse, a plasma is formed on the surface of the wires. This coronal plasma expands rapidly, showing sausage instability growth. A dense core plasma expands more slowly and uniformly within the corona. At the cross point of x-pinch loads, the coronal plasma implodes in its self-magnetic field, and intense bursts of soft x-rays are emitted. Backlit images of the x-pinch show the implosion and disruption of the dense core at the cross point near the time soft x-rays are emitted. The x-ray streak of a spatially resolved x-pinch showed intense emission in bursts of ≤1 ns over several 10s of ns from separate bright spots. Streaked x-ray spectra imply an electron temperature of 500 eV and a density of 10^{22} cm^{-3} at the implosion points. For a single wire z-pinch, the outer envelope of plasma compresses and x-rays are emitted from points randomly located along the full length of the load.

EXPERIMENTAL PROCEDURE

The experiments described here were carried out with the XP pulser, the first stage of which is a Marx generator which pulse charges an intermediate storage capacitor in approximately 700 ns. The intermediate storage capacitor, consisting of four coaxial cylinders, discharges through a self-break gas spark gap switch into the output capacitor in approximately 200 ns. The output capacitor consists of three coaxial cylinders that discharge through an array of 8 parallel self-triggering water switches, delivering up to 630 kA in a 100 ns full width at half maximum (fwhm) current pulse to the load via a low inductance vacuum power feed. The experiments described here were carried out with a peak current which varied with the load configuration from 300 to 400 kA. A sample current and voltage trace is shown in Figure 1.

Time resolved data were obtained with up to four filtered photoconducting diodes (PCDs)[2], a pulsed nitrogen laser, x-ray backlighting, and visible light and x-ray streak imaging. Additional diagnostics included time integrated pinhole cameras, and a KAP curved crystal spectrograph. The diagnostics were placed radially around the pulser load region (Figure 2).

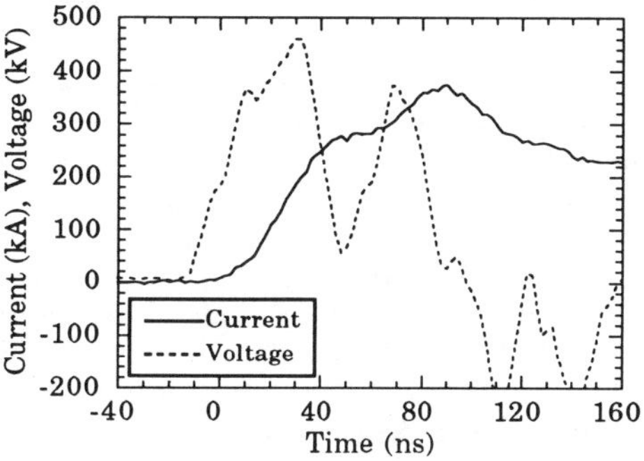

Figure 1: Current and voltage for a typical XP pulse (obtained with two x-pinches in parallel).

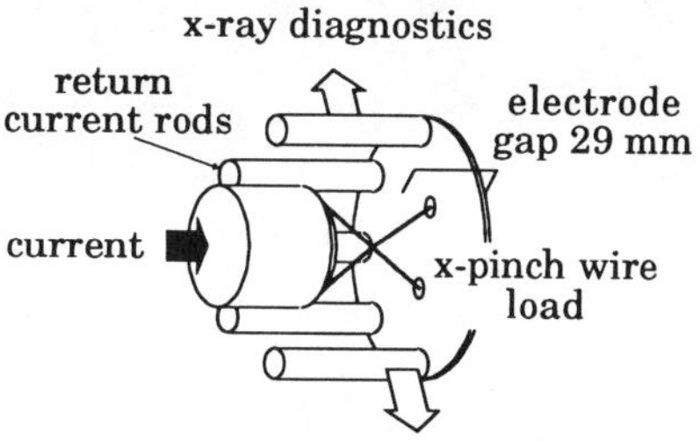

Figure 2: Schematic diagram of the x-pinch.

The PCDs were filtered with 25 μm beryllium (Be), 15 μm titanium (Ti) or 10 μm Al, and 25 μm Be with 20 μm Mylar. The Be filtered PCDs recorded primarily soft x-ray K-shell radiation, and the Ti or Al filtered PCD recorded harder radiation in the range 3 to 7 keV. Up to 3 PCDs viewed the pinch from the same direction.

An array of four 25 to 100 μm pinholes was used with different filters to record time integrated images of the x-pinch and z-pinch loads through different filters from the same direction. Images of intense UV and L-shell emission were recorded through thin (6 μm) aluminized Mylar. Soft x-ray emission was recorded through 25 μm Be and images of the harder (few keV) radiation were recorded through 10 μm Al. Two pieces of Kodak Exploration GWL[3] film were placed in the camera. The first film provided the images described, and the second film provided an image of the harder radiation only. X-ray streak pinhole images were obtained for several x-pinches using a Kentech streak camera with a photocathode consisting of CsI deposited on 25 μm Be.

Calibrated Kodak DEF x-ray film[4] was used in the curved crystal spectrograph to record spectra of the Al K-shell emission. Time integrated line intensity ratios were used to estimate plasma temperature and density. X-ray spectra from the spectrograph were recorded with the streak camera for several x-pinches, from which time resolved plasma parameters were obtained.

The short pulse nitrogen laser[5] consisted of an atmospheric pressure oscillator and a low pressure (100 torr) amplifier. The output laser pulse was a 1 cm diameter beam at 337 nm with a pulse duration of <1 ns (fwhm) and an energy of about 0.3 mJ. The optical system permitted simultaneous schlieren and interferometry imaging of the plasma. Images were recorded using 400 speed Kodak TMAX black and white film. The time of the laser pulse with respect to the machine current pulse was varied over a sequence of nominally identical pulses.

X-ray backlit images of x-pinch and z-pinch wire loads were made by using an x-pinch in parallel with the load to be imaged. A backlit image of the pinch was recorded using the harder (≥3 keV) radiation emitted at small spots at the cross position of the source x-pinch, with the time an image was obtained changed by varying the x-pinch backlighter wire diameter.

X-PINCH X-RAY SOURCE

The x-pinch is a compact source of intense soft x-ray radiation.[1,6] Time integrated pinhole images of a 25 μm Al x-pinch are shown in Figure 3. The peak current through the load was 320 kA. Ultraviolet (UV) and L-shell radiation is emitted from the full wire length (Figure 3a). Soft x-rays in the form of K-shell radiation is emitted from the cross region (Figures 3b and 3c). Co-located with the source of soft x-ray emission are spot sources of harder radiation that are <10 μm in size (Figure 3d). The effective photon energy of this radiation was determined from its attenuation through a series of 20 μm Mylar and 5 μm Al foils. For images filtered with 10 μm Al, the effective energy is approximately 7 keV. For images filtered with 15 μm titanium (Ti), it is 4-5 keV because of the absorption edge of the Ti at 4.9 keV[7].

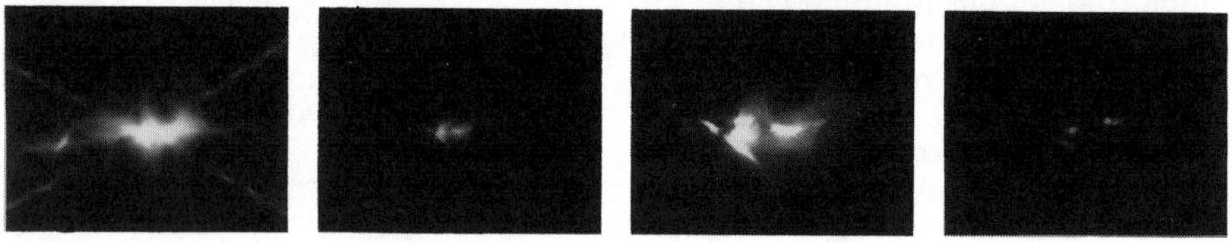

3 mm ▭ 1 mm ▭

a) aluminized Mylar b) 25 μm Be c) 25 μm Be d) second film

Figure 3: Pinhole images from an x-pinch made from two 25 μm Al wires at 320 kA.

The characteristics of the x-pinch as a soft x-ray source depend on the wire diameter and current that is applied. For 100 μm Al x-pinches with a 360 kA peak current, no soft x-rays are emitted. Smaller wire loads, such as a 37 μm Al x-pinch, give intense x-ray emission. The PCDs recorded the time of x-ray emission. Soft x-rays were emitted in bursts that were as short as the 2 ns resolution of our data acquisition system, and as long as several ns. The minimum duration of spots recorded with the streak camera using a 15 μm Ti foil indicated harder radiation bursts that were ≤1 ns, again limited by the resolution of the diagnostic system. Large diameter wire loads first emitted x-rays late in the pulse. The time at which an x-pinch first emits x-rays was reproducible for a given current and load configuration. Subsequent bursts were random in time. The time of the first burst varied with the wire diameter as shown in Figure 4.

A sample spectrum of the x-ray emission from an Al x-pinch is shown in Figure 5, showing both K-shell line emission and free-bound continuum. There is also the hard radiation component consisting of 3 to 7 keV radiation. The overall yield of harder radiation was only a small fraction (<1%) of the total x-ray yield.

K-shell intensities recorded on x-ray film were calculated for several H- and He- like Al lines[8], including the intercombination transition satellite of the He-like resonance line. The Ration codes[9] were used to graph line intensity ratios for a range of temperatures and densities assuming a source size of 0.1 to 0.5 mm including opacity effects in the CRE model. The electron temperature was determined for several x-pinches using these graphs. Temperatures were in the range 400 to 800 eV. From the slope of the free-bound continuum, temperatures were determined to be 600 to 1600 eV. Electron densities in the range $1\text{x}10^{19}$ to $1\text{x}10^{20}$ cm^{-3} were inferred for these x-pinches. There was no significant trend with the x-pinch wire size.

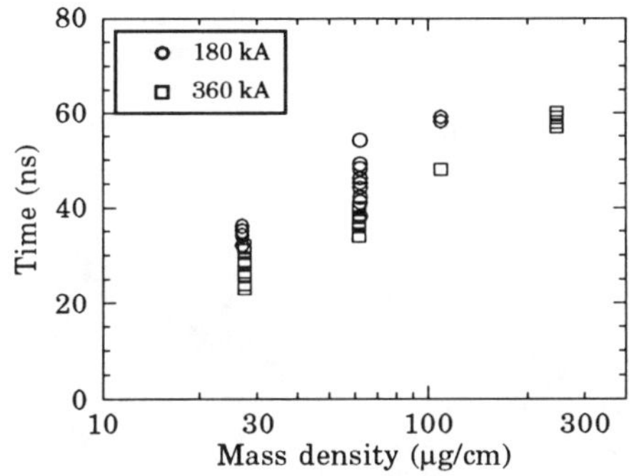

Figure 4: Time of x-ray emission as a function of the mass per unit length of the load.

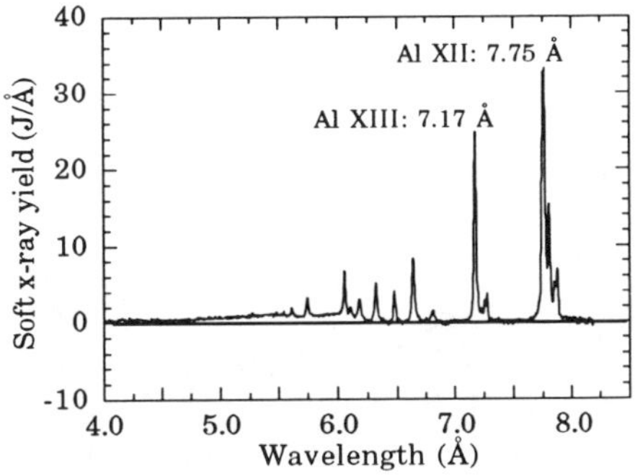

Figure 5: Typical Al K-shell spectrum.

Electron density and temperature measurements made from time and space integrated spectra can be expected to be only estimates since the lines and the free-bound continuum involve emission from several bright regions that were formed at different times. Time resolved K-shell spectra were obtained for several x-pinches by attaching the Kentech x-ray streak camera to the crystal spectrograph box. The Al spectrum from 5.6 to 7.3 Å was imaged on the photocathode slit, and the spectrum swept in time. A sample time dependent spectrum is shown in Figure 6. There were four bursts of x-ray emission during the pinch, each resulting in a spectrum on the film. The Ration codes were again used to generate spectra for an Al plasma with a range of parameters, and the results were compared with the experimental data shown in Figure 6. Linewidths of the He-like series lines indicated the electron density was $1x10^{21}$ to $1x10^{22}$ cm^{-3}. The ratio of the higher order H and He-like transitions suggested the temperature was approximately 400 to 600 eV. The spectrum from the four bursts of soft x-rays provided similar plasma parameter estimates, except line intensity ratios for the second of the two middle bursts implied a slightly higher temperature than the others, but still within the range specified.

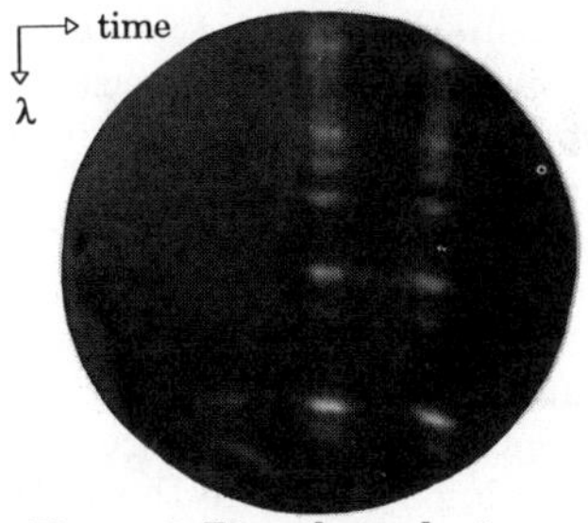

Figure 6: Time dependent x-ray spectrum.

Even these time dependent temperature and density measurements are based on the equilibrium calculation from the Ration codes. Individual bursts of x-ray emission last 1 to 5 ns, times that are long compared to the atomic transition times for x-ray transitions, but comparable to the time it takes for the source region to implode from 30 μm to 10 μm. Therefore, with time resolution of ~1 ns, a range of densities and temperatures may still be in view at any instant in time.

PLASMA DYNAMICS

A sequence of schlieren images of 100 μm x-pinches at 300 kA is shown in Figure 7. The maximum radius of the coronal plasma along the x-pinch limbs expands at 2.3 cm/μs starting 17 ns into the pulse, and undergoes sausage instability. For comparison, a sequence of schlieren images of a 50 μm x-pinch at 185 kA is shown in Figure 8. The plasma expands with a radial velocity of 0.9 cm/μs starting 17 ns into the current pulse. At the cross position, the plasma envelope pinches and jets of plasma stream along the axis away from the cross. The 100 μm x-pinch did not emit x-rays, whereas the 50 μm x-pinch emitted intense soft x-rays from the cross region. The time of the first burst of x-ray emission occurred approximately 60 ns into the current pulse (Figure 1), between the images 8b and 8c in the schlieren sequence.

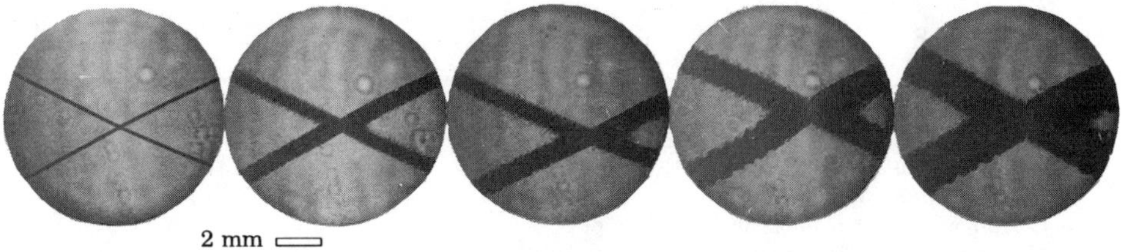

Figure 7: Schlieren image sequence for a 100 μm Al x-pinch at 300kA.

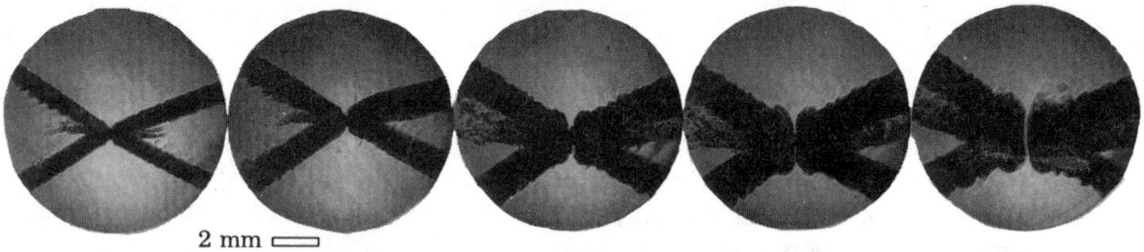

Figure 8: Schlieren image sequence for a 50 μm Al x-pinch at 185 kA.

The self-magnetic compression of the cross point was observed in schlieren images for several x-pinch loads that emitted bursts of soft x-rays, including 75 μm and 37 μm Al x-pinches at 340 and 360 kA. The laser beam was split into its S and P-polarization components using a glan-laser calcite crystal polarizer. The S-polarization was delayed 6.6 ns and reinserted into the optical path before encountering the x-pinch plasma. After the schlieren collection optics, a second polarizer separated the two polarized images so that they could be separately recorded. Using this technique, we obtained two images of the compression of the cross point for a single 37 μm x-pinch (Figure 9). The coronal plasma pinches at the cross point just prior to soft x-ray emission. The radial compression velocity measured from pairs of schlieren images is 1.1 cm/μs, while the x-pinch limb regions are expanding at 1.8 cm/μs. At the moment the first burst of soft x-rays is emitted, the plasma is re-expanding with a radial expansion velocity of approximately 6 cm/μs.

A sequence of schlieren images of a 37 μm Al z-pinch is shown in Figure 10. The coronal plasma expands with sausage instability starting 14 ns into the pulse. The expansion velocity of the maximum plasma radius is 3.8 cm/μs, and the amplitude of the sausage instability, $(r_{max}-r_{min})/2$, increases at 3.6 cm/μs during the current pulse. As the amplitude increases, the dominant wavelength of the sausage instability increases such that the amplitude of the instability remains about 2.2 times the wavelength. For a 100 μm Al z-pinch at 330 kA, the amplitude growth rate is only

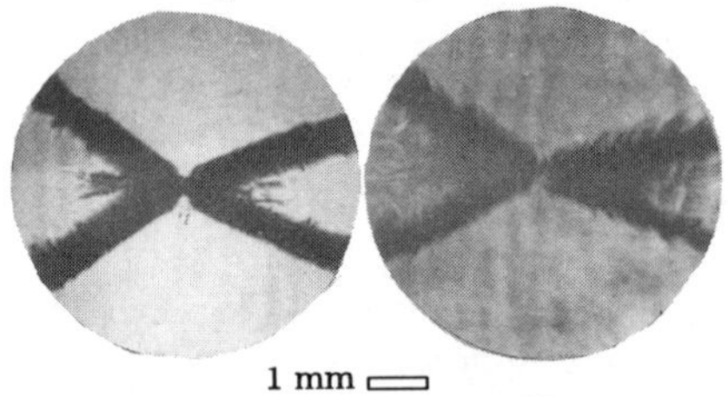

Figure 9: Schlieren images of the compression of the cross point for 37 μm Al wires.

1.8 cm/μs, while the radial expansion is 2.9 cm/μs.

The expansion rate for x-pinch limb plasmas and z-pinch plasmas varied with wire diameter and peak current. Figure 11 shows the maximum limb coronal plasma radial expansion for the x-pinches. The rate of expansion ranged from 0.9 cm/μs to 2.6 cm/μs. The maximum radius as a functionof time for z-pinches is shown in Figure 12. The expansion rate ranges from 1.1 to 2.9 cm/μs.

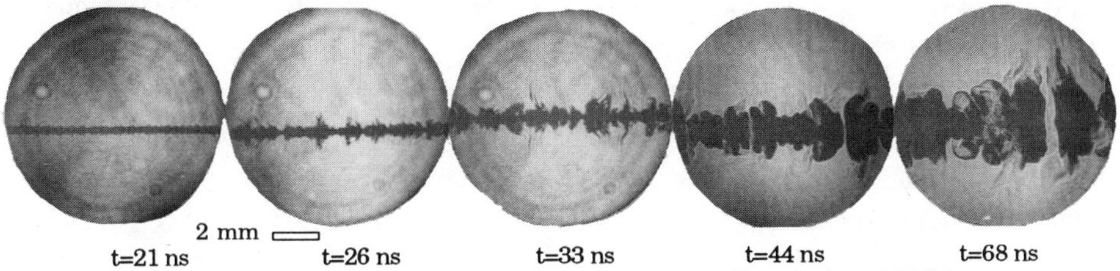

Figure 10: Schlieren image sequence for a 37 μm Al z-pinch at 330 kA.

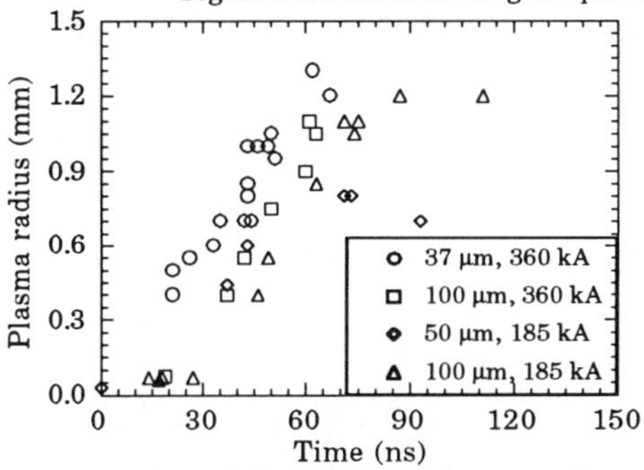

Figure 11: X-pinch limb coronal plasma radius as a function of time.

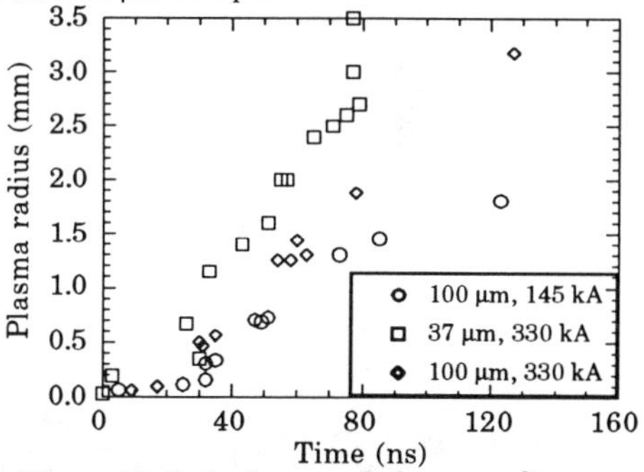

Figure 12: Z-pinch coronal plasma radius as a function of time.

Schlieren sequences from different diameter wire x-pinches and z-pinches at different current levels show the time at which the plasma starts to expand varies with the load. For a fixed current, the plasma from smaller diameter wire loads expands earlier than larger diameter loads. The time this expansion begins is later for a lower peak current. Figure 13 shows the time expansion begins for the different x-pinch and z-pinch loads. This time was determined from a linear fit of the maximum x-pinch limb or z-pinch plasma diameter as a function of time.

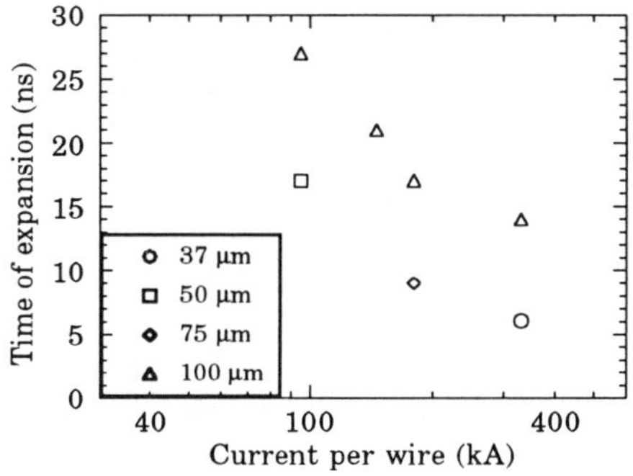

Figure 13: Time of coronal plasma expansion.

The schlieren images show a sharp boundary for the coronal plasma early in time. Refraction of the laser beam limits the usefulness of the interferometer to late in the current pulse when features such as the expanding jets imaged in Figure 8 become spread out in space. Early in time, interferograms of the pinch plasmas show fringes that end abruptly at the envelope of the corona. Late in time, interference fringes can be traced through the lower density regions, indicating electron densities of up to a few 10^{19} cm^{-3} before refraction bends the scene beam away from the collection optics. The numerical aperture of the collection optics was 0.004, limiting the line-integrated electron density gradient for the scene beam to be collected to 8×10^{19} cm^{-3}. In addition, for x-pinches and z-pinches that show a separation of the bulk coronal plasma late in time, the presence of the interference fringes in the gap region allows us to establish an upper bound for the plasma density in the gap. A single fringe shift for the nitrogen laser corresponds to a line-integrated density of 6.6×10^{18} cm^{-3}mm. For a 25 μm Al z-pinch that shows a gap in the schlieren images late in time, the plasma electron density in the gap is less than 10^{17} cm^{-3} (Figure 14).

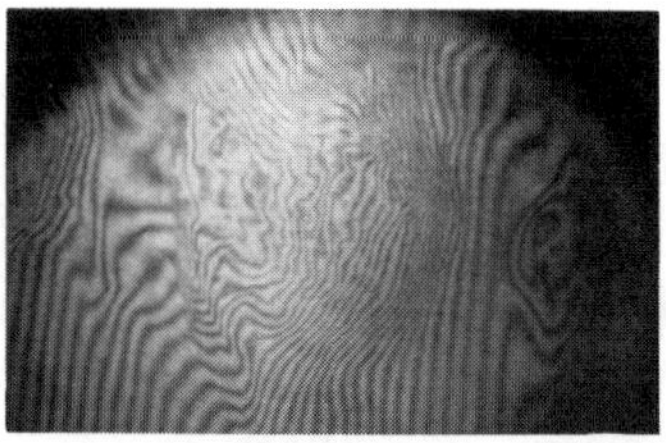

1 mm

Figure 14: Interferogram showing the gap in a z-pinch plasma.

X-ray backlit images of 100 μm and 50 μm x-pinches at 185 kA peak current are shown in Figures 15 and 16. These images were obtained using a second x-pinch placed in parallel with the x-pinch of interest. Al and Ti wires from 25 to 50 μm in diameter were used for the x-pinch backlighter source to provide time resolution. The radiation was filtered with 10 μm Al or 15 μm Ti before reaching the film. The <10 μm harder radiation spots from the backlighter x-pinch provided the spatial resolution in the backlit images. Since the x-ray emission occurs in several short bursts during the current pulse, several images may be recorded at different times from the same pulse as shown, for example, with the image in Figure 15c.

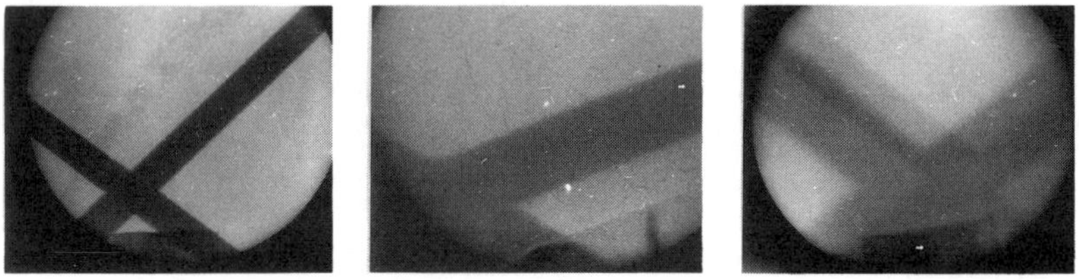

t=35 ns 0.25 mm t=42 ns t=43, 50 ns

Figure 15: Sequence of x-ray backlit images of a 100 μm Al x-pinch at 185 kA.

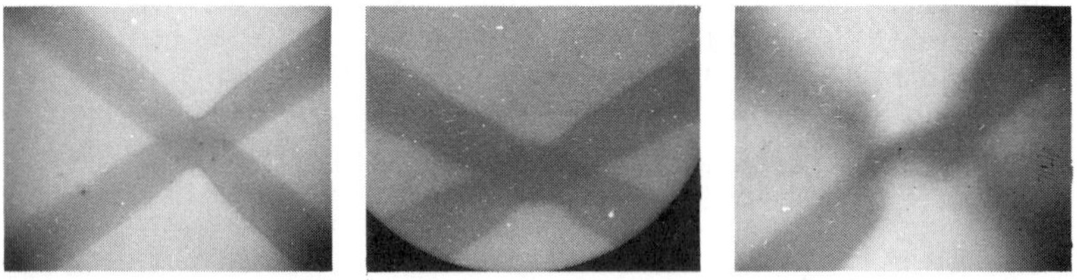

t=33 ns 0.25 mm t=36 ns t=57 ns

Figure 16: Sequence of x-ray backlit images of a 50 μm Al x-pinch at 185 kA.

For the 100 μm x-pinch, the dense core plasma expands uniformly along the full length of the wires with a velocity of 0.8 cm/μs. The backlit images from the 50 μm Al x-pinch in Figure 16 show expansion followed by compression of the dense core at the cross position. The images in Figure 16c showing the disruption of the dense core at the cross point was obtained with a 50 μm Al x-pinch as the backlighter. Since the two loads were identical, we expect the first burst of x-ray emission from both the backlighter and the x-pinch we were imaging to occur at the same time within the 2 ns time resolution of the data acquisition system. Therefore, this image represents the disruption in the dense core plasma at or near the time of x-ray emission.

The dense core plasma in both x-pinches and z-pinches expanded much slower than the lower density coronal plasma. For a 100 μm x-pinch, the dense core expands uniformly. At the same time, the coronal plasma expands with sausage instability. The smaller wire loads show distortion of the dense core at the cross point as the coronal plasma pinches and the x-pinch emits soft x-rays. Single wire 75 μm and 100 μm z-pinches showed a uniform dense core that expanded at 145 kA peak current. The dense core for a 25 μm Al z-pinch at the same current showed non-uniform expansion. This z-pinch load also pinched, emitting bursts of intense soft x-rays.

Attenuation of the hard x-rays by the dense core plasma was used to determine the fraction of material remaining in the dense core. We assumed the dense core was composed of neutral or low ionization states of Al. The attenuation of 7 keV radiation (i.e. using the 10 μm Al filter) in several images indicated that >50% of the initial wire mass remained in the dense core.

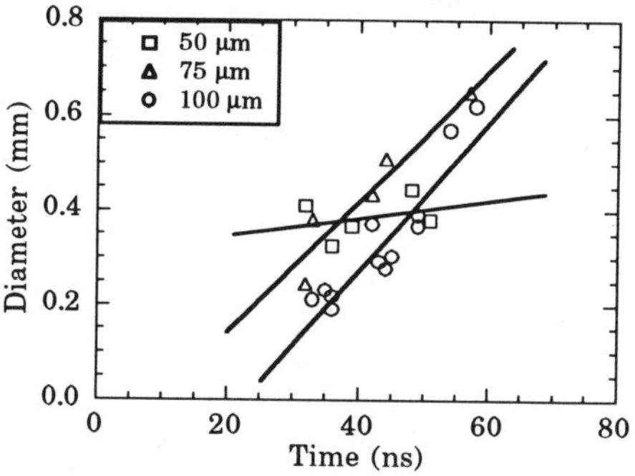

Figure 17: Dense core plasma diameter as a function of time.

ACKNOWLEDGMENTS

Dr. R. B.. Spielman of the Sandia National Laboratory provided the calibrated PCDs. The XP pulser was constructed with funding from New York State Science and Technology Foundation Contract No. RDG 90179, and the nitrogen laser was assembled with support from NSF under Grant No. 89-19960. The work was supported by NSF Grant No ECS-9113695 and by NATO Grant CRG 910936.

REFERENCES

1 S. M. Zakharov, G. V. Ivanenkov, A. A. Kolomenskii, S. A. Pikuz, A.I. Samokhin, and J. Ullschmied, Sov. Tech. Phys. Lett. **8**, 456 (1982).

2 R. B. Spielman, W. W. Hsing, and D. L. Hanson, Rev. Sci. Instrum. **59**, 1804 (1988).

3 Kodak Corporation, private communication (1991); B. L. Henke, F. G. Fujiwara, M. A. Tester, C. H. Dittmore, and M. A. Palmer, J. Opt. Soc. Am. B **1**, 828 (1984).

4 P. D. Rockett, C. R. Bird, C. J. Hailey, D. Sullivan, D. B. Brown, and P. G. Burkhalter, Applied Optics **24**, 2536 (1985).

5 D. H. Kalantar, Ph.D. dissertation (in preparation).

6 D. H. Kalantar, D. A. Hammer, K. C. Mittal, N. Qi, J. R. Maldonado, and Y. Vladimirsky, J. Vac. Sci. Technol. B **9**, 3245 (1991).

7 B. L. Henke, P. Lee, T. J. Tanaka, R. L. Shimabukuro, and B. K. Fujikawa, At. Data Nucl. Data Tables **27**, 1 (1982).

8 J. P. Apruzese, D. Duston, and J. Davis, J. Quant. Spectrosc. Radiat. Transfer **36**, 339 (1986).

9 R. W. Lee, private communication (1990).

THE STUDY OF COMPACT PLASMA SOURCE OF SXR OF VACUUM SPARK TYPE WITH CAPILLARY CONCENTRATOR AND IT´S APPLICATION

V.L.Kantsyrev, K.I.Kopytok, A.S.Shlyaptseva
Scientific Research Institute of Technical Glass
Moscow, 117218, Russia

ABSTRACT

The results are presented dealing with the working out and study of the plasma source of soft X-ray (SXR) of the new type.

Experimental set up included compact low-inductance vacuum spark (LIVS) with initial energy supply equal up to 2.5 kJ and glass-capillary concentrator (GCC) of SXR. The characteristics of SXR of vacuum spark and properties of SXR were studied using diagnostic complex.The coefficient of conversion of initial energy supply into SXR (η) amounted to 0.01 in range 1.2nm. Value η had peak dependence on atomic number of anode Za. The spectra were recorded belonging to Ne-like, F-like ions of Fe, Cu ions and He-like, H-like ions of Al, Ti, Fe.

Glass capillary concentrator consists of about several hundreds glass capillaries. Flux density of SXR in focusing spot was up to 10^5-10^6 Wt/cm^2, density of energy is up to 20-30 mJ/cm^2 at diameter of SXR focusing spot equal to about 2-3mm in the range 0.7-1.0 nm.

The plasma source of the new type is intended for X-ray microscopy, study of influence of SXR on the surface of solid state. It allows to carry out experiments making only on electron synchrotronic sources of SXR.

INTRODUCTION

The study and the working out of plasma sources of SXR based on the dense Z-pinches (for example, low-inductance vacuum spark, gas - puff) have besides the pure scientific purposes the applied meaning. By that the basic problem was to produce the maximal values of energy density E and flux density q of SXR impulse on the surface of pattern emitted. As the source of SXR we choose the low-inductance vacuum spark (LIVS), while this source had high value of the coefficient of conversion of energy flux of the source into SXR η - up to 0.01, the small effective size of emitted region (averaged during the series of impulses d) - few millimeters. This source is a compact one with rather simple construction. However, characteristics of all compact plasma source of SXR aren't sufficient good for practical applications. On practice the coefficient of the use of this emission ξ doesn't exceed $\xi = 10^{-3}$. It is necessary to apply non-selective focusing systems for the spectral range λ = 1.0- 1.5 nm., as the plasma sources have wide spectrum. Among the recent X-ray optics, optics of grazing incidence is progressive in this spectral range.

The possibility of transferring and focusing of SXR from laser plasma with glass capillaries were experimentally set in papers[2,3,4], later glass-capillary concentrator (GCC) was constructed from few hundreds capillaries. The value $q=10^5$ Wt/cm^2 (λ <1.2nm) was reached in spot with 2 mm diameter. However, the density of SXR energy didn't exceed $E=(2-3)10^{-3}$ J/cm^2, what didn't differ from the value of E for laser plasma at the distance 3...5cm without focusing. The better results must be obtained using LIVS. The purpose of our investigations was to work out the new type of SXR source, which posses the better parameters in comparison with traditional LIVS.

EXPERIMENTAL SET UP

The investigations were carried out on experimental set up shown on Fig.1. LIVS was constructed in scheme using trigger ignition. The initial energy supply was up to 2.5kJ, maximal discharge current was up to 250 kA, discharge half-period - 2mcs. X-ray complex allowed to study SXR in the range 1.8 nm. The complex consists of the following: sensors with X-ray filters (Al, Be, mailar) and X-ray film UF-VR, UF-SHC; thermo-luminescence detectors (TLD); X-raypin-diodes, three-channel pin-hole camera, survey spectrograph with convex mica crystal.

THE DEVELOPMENT OF LIVS CHARACTERISTICS

The striking peak dependence of η on Za was experimentally observed in the spectral SXR range λ<1.5 nm (Fig.2). The similar dependence η(Za) was observed for laser plasma in papers[7,8]. As Te increasing the peaks were shifting to great Za.

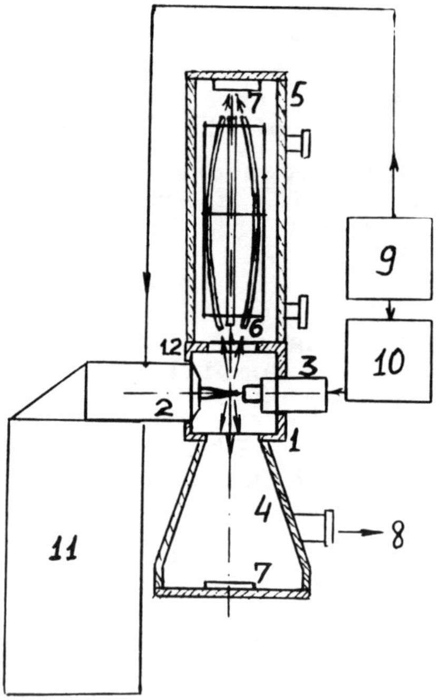

1. Vacuum chamber.
2. Anode.
3. Cathode knot.
4. Camera with diagnostic complex.
5. GCC camera.
6. GCC.
7. Emitted patterns and detectors.
8. vacuum system.
9, 10. Blocks for control.
11. Low-inductance capacity.
12. Block of GCC protection.

Fig.1. The scheme of the LIVS source with GCC.

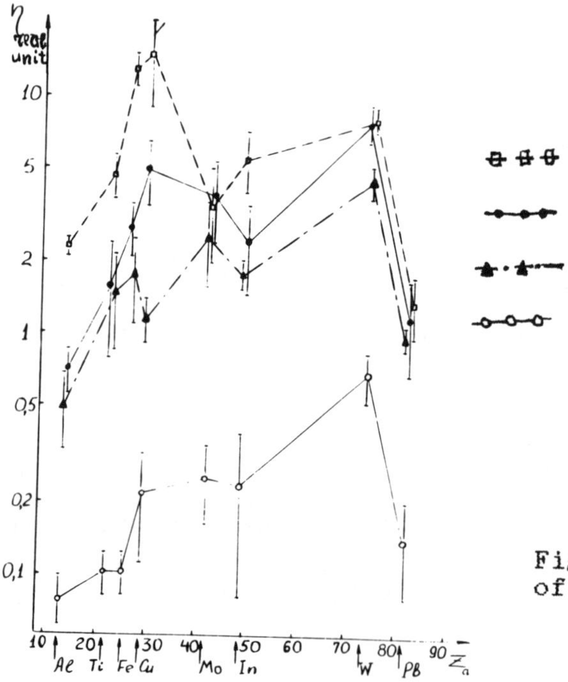

Fig.2.The dependence of η (Za) for LIVS.

The analysis of spectra of multiply-charged ions (Fig. 3) and imagines of discharge spacing made using pin-hole camera has shown that the basic contribution into formation of dependence η (Za) and the total intensity of SXR was caused by linear radiation of multiply-charged ions of plasma points. The mechanism of formation of dependence η (Za) was analyzed in paper[9] and it is the same for all plasma sources.

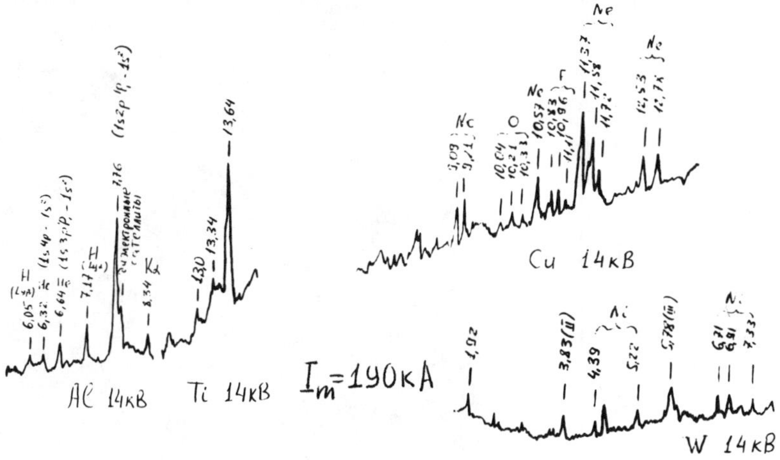

Fig. 3. Spectra of multiply-charged ions of LIVS. Im - maximal discharge current.

The spectra of LIVS contain the lines of multiply-charged ions with nuclear charge up to Z=74 (WXXXXVII) and ionization potential up to 8 keV (Fe XXV). The maximal intensity falls in spectral region (λ=1.2nm). The maximal value of η = 0.01, maximal number of plasma points (up to 4-5) and minimal size d (0.5 - 1.5 mm) were reached at the value of current Im = 150-230kA and spacing between electrodes L=6-8mm (for anodes from Fe and W).

GLASS-CAPILLARY CONCENTRATOR (GCC).

In the present paper we constructed GCC on the basis of GCC made in paper[5]. It consists from 450 glass capillaries with inner diameter - 400 mcm, outer diameter - 600mcm, length - 500mm (Figs.1,4). In GCC the total reflection of SXR from the inner capillary walls by the grazing incidence of radiation. The focus of GCC was equal to 60 mm. The diameter of the smallest focusing spot was equal to 1-2mm.

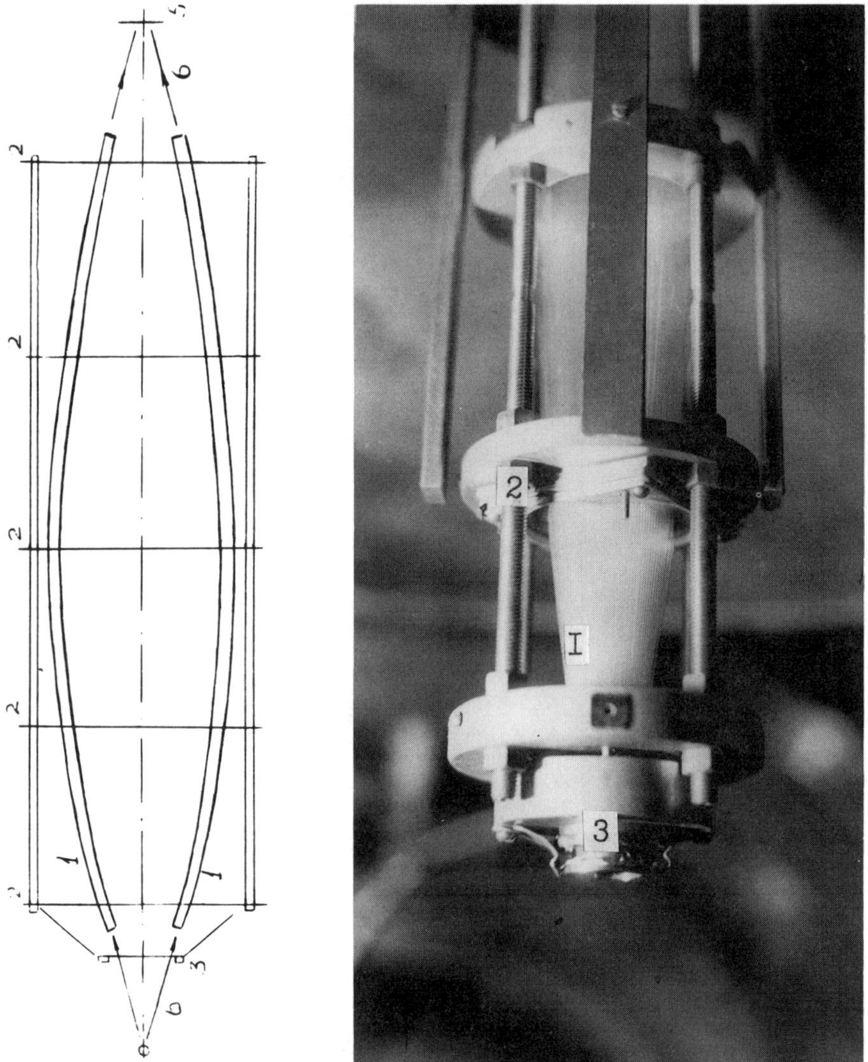

1. Glass capillary.
2. Supporting perforated membranes and bearing bars.
3. Protective block
4. Hot plasma.
5. Detector or pattern emitted.
6. SXR.

Fig. 4. The outside diameter and scheme of GCC.

It appears to be possible to use GCC only in the spectral range 0.67-1.5 nm where protecting filters existed (Be-foils and mailar films). Using such filters the device was constructed for protection of entrance window of GCC. Than the combination of improved LIVS and GCC was achieved: the spectral transmition range of GCC corresponds to the maximum of spectral intensity of LIVS with Fe anode (coinciding to the maximum of η (Za)); d is close to diameter of focusing spot of GCC.

EXPERIMENTAL RESULTS OF THE STUDY OF NEW COMPACT SOURCE OF SXR AND IT'S APPLICATIONS

The size of focusing spot was dependent on the spectral region of registration (at λ lim= 0.9 nm to 1-3 mm). The value of E, measured by TLD, varied from 15...30 up to 0.3 mJ/cm^2.The value of the coefficient of concentration - K (the ratio of energy density E in focusing spot to E at the same distance from plasma without focusing) reached to K=30. The measuring of SXR impulse duration allowed to estimate the value of q=10^6 Wt/cm^2 in comparison with 10^5 Wt/cm^2 in paper[5]. This coincides with estimations of calculations of radiation loading on the mailar film with width equal to 3 mcm which caused it's partial melting in focusing spot what was really observed in experiment. The values q and E obtained in the present paper are typical for experiments with synchrotrone sources.

The studied source of SXR was used in high-resolution biological microscopy of non-colored histological shears of liver and lung fabric with registration of the images on the X-ray resist. The image resolution was achieved equal to 0.2 mcm.

The new source of SXR was also used in radiative physics of dielectric and semiconductors surfaces. At doze of SXR greater than 0.1...0.2J/cm^2 it was observed the effects of the changing of coloring of electrochromic alkaline tungsten and niobium phosphate glasses and the changing of stecheometry of surface layers of chalcogenide glass-like semiconductors. The effects found could be used in technology of microcircuit and integral-optical circuits production.

About the progress in GCC development. The flexibility of the glass-capillary focus system form gives an opportunity to concentrate SXR from not only point SXR sources but from ones with extended emitted region (for example, gas-puff). It is possible to change the form of cross-section of SXR beam. In the last case the system works as glass-capillary converter which can be applied for investigating of temporal behavior of space distribution of SXR of plasma. Such device is shown on Fig.5. It transforms SXR in visible emission at the entrance window of IC (image converter) and has the changing cross section. The inner diameter of capillaries is equal to 80 mcm. It appears to be rather real to create GCC with capillaries having inner diameter equal to 10-20 mcm what will provide the space resolution close to 20-30 mcm.

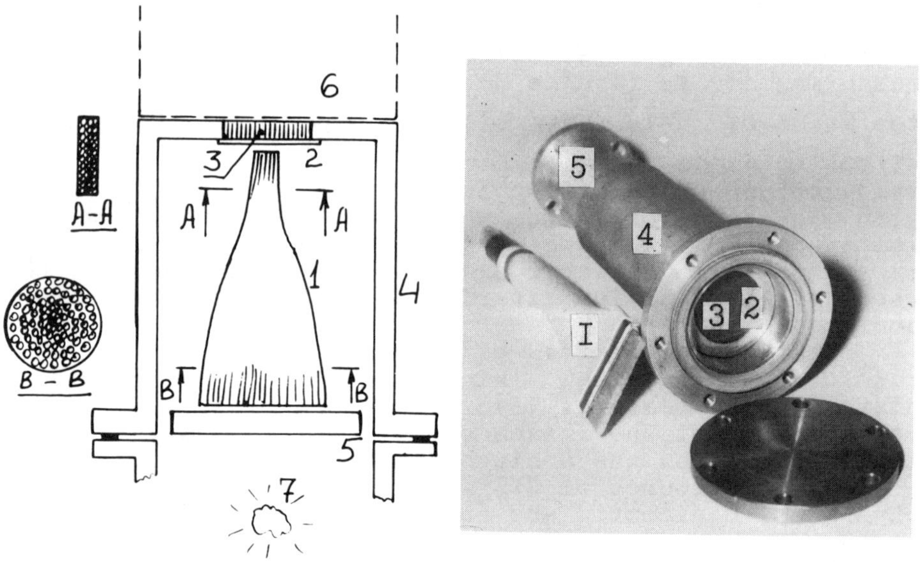

1. GCC.
2. Scintillate layer.
3. Fiber-optical disk.
4. Vacuum shell.
5. Protective block.
6. IC
7. Plasma

Fig. 5. The outside diameter of GCC and photo of transforming module made on it´s basis.

CONCLUSION

The new type of compact plasma source of SXR is worked out and created - the low-inductance vacuum spark with glass-capillary concentrator. It is shown experimentally the possibility of it's application in biological microroentgenography and radiative physics of dielectrics and semiconductors surfaces. The perspectives were discussed dealing with the development of glass-capillary concentrators and converters.

ACKNOWLEDGEMENTS

The authors thanks M.Ya. Yakovlev for attention to the work; Tolstichin O.I., Petruchin O.G.,Korotkova T.I., Shashin V.I. for help in calculation of models, production of some experimental equipment; Bykovsky Yu.A., Anayin O.B., Faenov A.Ya. for helpful discussions .

REFERENCES

1. N.G.Basov, Yu.A.Bykovsky, A.V.Vinogradov, V.L.Kantsyrev, Surface N 9, 5 (1985).
2. P.J.Mallozzi et all, J.Appl.Phys.,45, 1891 (1974).
3. O.B.Ananyin, Yu.A.Bykovsky, V.L.Kantsyrev, Yu.P.Kozyrev. Patent USSR N 520863. Bul.izobret. i otkr., N 11 , 229 (1979), priority from 15.10.1974.
4. Yu.A.Bykovsky, V.L.Kantsyrev,Yu.P.Kozyrev, Reports of XIY Intern.Congress of Photography, 630 (1980).
5. O.B.Ananyin, Yu.A.Bykovsky, A.A.Zhuravlev, V.L.Kantsyrev, Lett.J.Techn.Phys., 16, 55 (1990).
6. A.K.Zhverkov, A.A.Krivtsov, V.L.Kantsyrev, A.S.Shlyaptseva, J.Techn.Phys.,56, 975 (1987).
7. O.B.Ananyin, Yu.A.Bykovsky, V.L.Kantsyrev, Yu.P.Kozyrev, A.M.Raspopin. J.Quant.Elect., 4, 965 (1977).
8. Yu.A.Bykovsky, V.L.Kantsyrev,Yu.P.Kozyrev, J.Quant.Elect., 6, 414 (1979).
9. A.V.Vinogradov, B.N.Chichkov, J.Quant.Elect., 10, 741 (1983).

OBSERVATION OF THE JETS IN THE GAS EMBEDDED INTERRUPTED z-PINCH

P. Kubeš, J. Kravárik, J. Hakr, J. Píchal, P. Kulhánek
Czech Technical University, Faculty of Electrical Engineering
Technická 2, 166 27 Prague 6, Czech Republic

ABSTRACT

There are presented the implications of the interferometric, schlieren and spectroscopic diagnostics in the diode pulse discharge in nitrogen between two cone-shaped electrodes (discharge energy 0.5 kJ, max. current 50 kA, 1 st discharge period 4 μs , gas pressure 5 kPa).

In this discharge, there are formed two jets with the filamentary structure due to the inertial force of the ions of the current layer. These jets accelerate the ions evaporated from the electrodes and the ions from the breakdown current layer into the central part of the discharge. The filaments and the shape of both jets are relatively stable during the first halfperiod of the discharge current.

The space and time development of electron density, density gradient and temperature were measured and results were used for the calculation of the jets' and filaments' stationary regime conditions.

EXPERIMENTS AND RESULTS

The z-pinch had been formed in the discharge between two cone-shaped copper electrodes (top angle 90° and diameter 10 mm each, interelectrode distance 10 mm) placed in the glass vacuum chamber filled with air under the pressure 5 kPa. The used energy source - 3 μF condenser battery - was charged to a voltage of 18 kV and the discharge current reached its maximum 50 kA at time 1 μs after the breakdown.

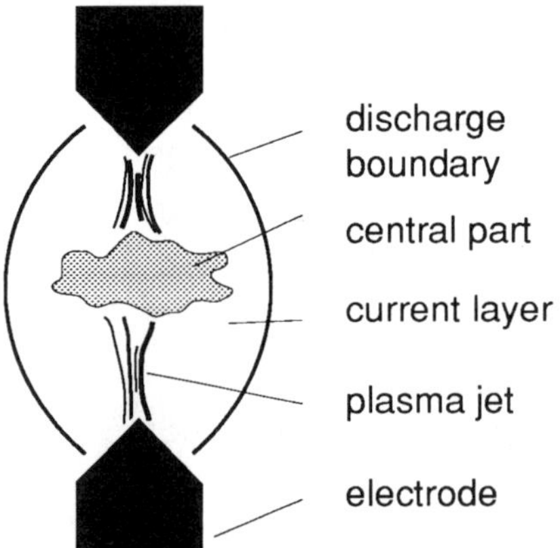

Fig. 1.: Discharge scheme

The plasma creates a current layer in the interelectrode region and the dense z-pinch is formed along the layer's axis. The discharge cylindrical shape is interrupted due to the magnetic and electric field's configuration and two comparatively stable plasma jets with the filamentary structure are formed. The jets throw the plasma towards the central region (where the plasma density is high) with very interesting filamentary structure and inhomogeneity's configuration. The discharge scheme is presented on Fig. 1.

Assuming a cylindrical geometry and slow changes of

fields in the plasma layer surrounding the z-pinch, the results of the Michelson interferometry enabled the discharge evolution study as well as the electron density calculations.

The Michelson interferometry results were not applicable for calculations of the filaments' electron density average value in the small inhomogeneity regions characterized by large gradient changes, therefore the schlieren method with the variable diameter of the focal point disk screen was used [1].

Both methods, the Michelson interferometry and the schlieren method, comprise a ruby laser switched in the Q-regime as the light source. The laser light pulse length was 0.04 μs, but due to the synchronization and the delay of the light pulse from the discharge ignition, there was a possibility to get a series of experiments. The experiments were carried out every 0.1 μs in the 1 μs time interval after the discharge ignition, the results being registered with a camera. During every discharge there was taken one snap only, but the sequence of measurements was very well acceptable for the discharge evolution study due to the discharge good repeatability (25 % variance about the mean).

The electron temperature was determined from the measurement of the intensities of the spectral lines and continuum for the visible wave band.

Plasma in the current layer is generated by the Joule heating and the outside boundary of this layer is given by the position of the shock resp. ionization wave. These waves are propagated from the z-axis, where the breakdown was initiated, in the radial direction with velocity 10^4 ms^{-1}. The electron temperature $T_e \sim 2$ eV, the electron density $n_e \sim 5\times10^{23}$ m^{-3}, magnetic induction $B \sim 1$ T and current density $j \sim 5\times10^8$ Am^{-2} are known in this outside layer from our experiments. The magnetic pressure $p_B = B^2/2\mu \sim 6\times10^6$ Pa is higher than the plasma pressure $p_T = 2n_ekT_e \sim 3\times10^5$ Pa.

The magnetic energy is dominant and the electrons are accelerated by the force density $\mathbf{j} \times \mathbf{B}$. This force is transferred to the ions and the plasma accelerates towards the z-axis of the discharge. The average thermal ion velocity (6×10^3 ms^{-1}) is lower than the velocity of the ion acoustic waves (8×10^3 ms^{-1}). The plasma seems to be stable considering the ion acoustic turbulence because the average electron velocity in the field of the discharge (3×10^2 ms^{-1}) is lower than both the velocities mentioned above [2]. The flow of the plasma to the axis is laminar. The velocity 10^4 ms^{-1} of the accelerated plasma on the boundary of the jets was estimated from interferometric measurements and calculated from the one dimensional MHD model [3]. This compressed plasma has parameters: electron density $n_e \sim 5\times10^{24}$ m^{-3}, electron temperature $T_e \sim 2$ eV, plasma pressure $p_T = 2n_ekT_e \sim 3\times10^6$ Pa, dynamic pressure $p_D = \rho v^2/2 \sim 6\times10^6$ Pa. The radial velocity 10^4 ms^{-1} of the ions is higher than the velocity of the Alfven and ion acoustic waves and it is actual to suppose the existence of the turbulence in this area.

The densest part of the z-pinch is created in the volume around the z axis. This part is interrupted due to the barrel shape of the magnetic and electric field's configuration and two comparatively stable plasma jets connected with the tops of the electrodes are formed at time 0.4 μs after the breakdown. The disintegration of the jets coincides with the end of the first half period of the discharge (≈ 2 μs). The diameter (0.3÷1) mm of both jets with the length 3 mm is increasing during their life-time. The anode jet is abruptly spread during time interval (0.6÷0.8) μs after the breakdown in a consequence of the ions' evaporation from the anode. The filaments presented in the jets have the direction of the streamlines, diameter 0.1 mm and electron density (5÷8) $\times 10^{24}$ m^{-3} during the life-time of the jets (0.5÷2) μs. The life-time of these filaments is not known. The electron density $n_e \sim 5\times 10^{24}$ m^{-3} of the plasma in the jets was estimated from the schlieren experiments.

Both jets accelerate the plasma into central part with velocity 10^4 ms^{-1} and flux 5×10^{16} ions and electrons per 0.1 μs. The flux from the cathode jet seems to be twice lower than from the anode one. The plasma accelerated from the opposite directions is cumulated in the central part with structure similar to the cauliflower shape. The filaments remind the whirlwinds.

CONCLUSION

The double jets' plasma configuration is formed in the discharge between two cone-shaped electrodes. Ions from the current layer and from the electrodes penetrate into the jets and are accelerated into the central part. These phenomena are controlled by the *z*-pinch effect in the barrel shape configuration of the electric and magnetic fields. The jets are working during the first half-period of the discharge. The filamentary structure is presented on the schlieren pictures.

REFERENCES

1. P. Kubeš et al.: Czech J. Phys. B35 (1985), 155
2. B. B. Kadomcev: Kolektyvnye yavlenya v plazme, Moskva, Nauka 1988, 210
3. P. Kubeš et al.: in Proc. Intern. Conf. on Plasma Physics, Innsbruck 1992, Vol. 16C, Part 1, p. 667

SMALL RADIATIVE Z-PINCH WITH LOW-Z PLASMA

J. Rauš, A. Krejčí
Institute of Plasma Physics, Czech Academy of Sciences,
P.O. Box 17, 182 11 Prague 8, Czech Republic

ABSTRACT

Gas-puff Z-pinch research at the IPP Prague is briefly reviewed. The XUV and soft x-ray measurements, performed with elements from carbon to argon, are summarized. Coupling of driver and load parameters for optimum generation of K-shell radiation of the mentioned elements is discussed using simple considerations.

Z-PINCH RESEARCH AT THE IPP PRAGUE

The world-wide growing interest in dense imploding plasmas during the 80's simulated construction and operation of new experimental facilities, equipped with both large machines for thermonuclear or other high-power studies, and small devices, serving mostly as x-ray sources for some applications or basic research [1]. Especially the last reason was the case of small (5 - 10 kJ) gaseous Z-pinch at the Institute of Plasma Physics in Prague.

Since 1988 we worked with argon and neon gas-puffs because these gases are widely used in pinch experiments and enable to compare the results with other laboratories. Time-resolved x-ray diagnostics with filtered XRD and semiconductor detectors (surface-barrier, PIN diode) as well as simple snowplow calculations allowed us to make a first approach to implosion dynamics and x-ray emission of the Z-pinch [2]. Time-integrated x-ray measurements with flat crystal spectrograph and pinhole cameras have followed [3]. As the investigation of small pinch plasma in keV-region concern only the hot spots, the "temporal resolution" of these diagnostics is done by the lifetime of the spots in order of several nanoseconds. Mentioned series of measurements made possible to determine the size of hot spots (~ 30 - 40 μm), but especially provided the spectral data about important K-shell transitions of both gases. From the ratio of He-like resonance line and its dielectronic satellites the T_e of argon hot spot (~ 1 keV) was computed. The electron density of this spot (~ 3 x 10^{21} cm^{-3}) was derived from the intensities of resonance and intercombination lines. These methods are standardly used however they are sensitive to the opacity effect [4].

The mentioned parameters make from Ar and Ne hot spots rather unique physical phenomena. Of course, it is not necessary to overestimate the importance of the spots in small Z-pinches, as they represent only small fraction of total amount of the mass, energy, and radiation of plasma. Because the T_e of bulk plasma is much lower than T_e

of the spots, the bulk does not emit in the keV-region. In a similar sense, the results given by visible light diagnostics, both spectrometry and interferometry, are limited, because the main information about the bulk plasma behaviour (e.g. radiation maximum in blackbody approximation) coincide with VUV and XUV ranges (10 - 100 eV and 100 - 1000 eV, respectively).

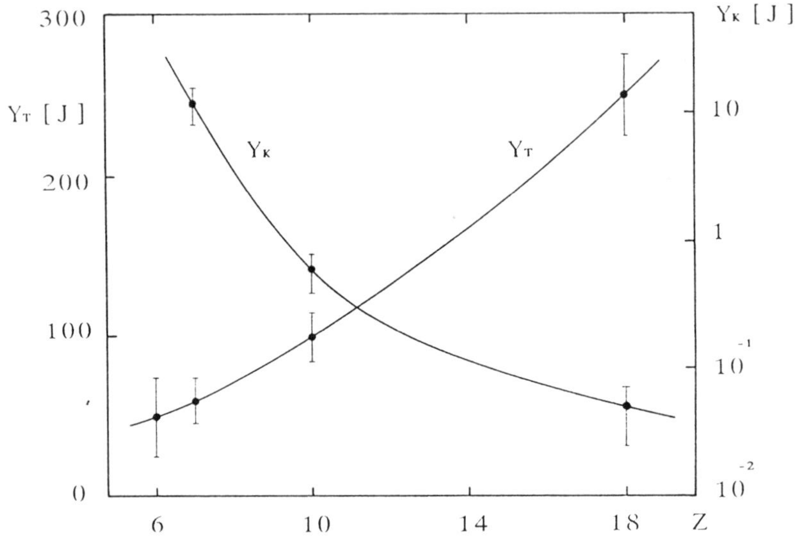

Fig. 1. Dependence of the total x-ray yield Y_T and the K-shell yield Y_K on atomic number in pinch with current 200 kA.

In further phase of our Z-pinch experiment (since 1990) we improved the coupling of electrical circuit to the pinch, so we have reached higher current (200 kA/1.2 μs) with less input energy (4.3 kJ). For the study of total x-ray yield from bulk plasma we used bare XRD. We tried to scale the K-shell yield Y_K and the total yield Y_T with current between 100 and 200 kA for argon, neon, and nitrogen [5]. When we changed the working media to the elements with lower atomic number (nitrogen) the value of Y_T decreased, as expected. On the contrary, the K-shell radiation (both lines and recombination continuum) increased remarkably (fig. 1); due to lower ionization and excitation energies, this radiation is emitted by the whole plasma fibre (fig. 2). The dimension of bright spots is several hundreds of μm.

Unlike argon or neon, the K-shell emission of nitrogen is laying in the XUV region (400 - 700 eV). Analysis of such radiation is more complicated than the soft x-ray diagnostics above 1 keV. Of course, an array of XRD's with submicrometer filters, e.g. nitrocellulose combined with aluminium, can be used for a rough spectral analysis [6]. For detailed time-resolved x-ray study an advanced diagnostics have to be exploited. We utilized multichannel polychromator with W-Si multilayer mirrors as dispersive elements ($\Delta\lambda/\lambda \sim 0.05$), submicrometer light-tight metallic filters, and

PIN diodes for detection+). The x-ray signals in selected spectral windows allowed us to reconstruct a time evolution of T_e in nitrogen pinch, reported in [7]. Peak temperatures have reached 60 - 100 eV in the bulk plasma and 200 - 400 eV in the hot spots.

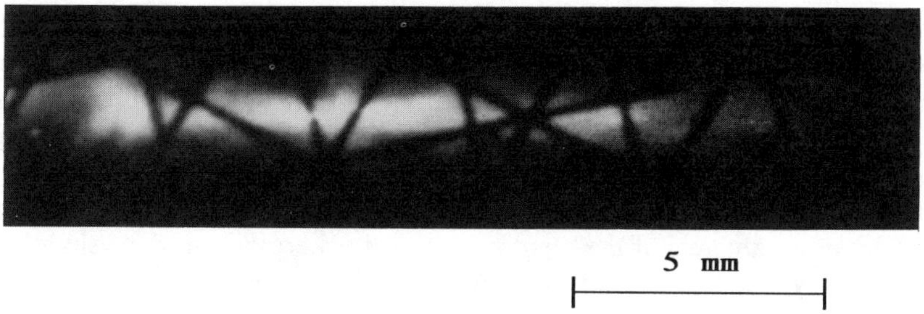

Fig. 2. XUV pinhole photograph of nitrogen pinch taken behind 0.35 μm Sn filter, i.e. radiation below 500 eV: no spots but all plasma column is visible. (The black lines mark a supporting mesh of the filter).

In order to extend our scalings to lower Z, measurements with carbon plasma have been done. For practical reasons a pure carbon could not be used, but propane (C_3H_8) and acetylene (C_2H_2) could. Due to low binding energies, the molecules of these gases are decomposed in the starting phase of pinch discharge and the system then implodes to axis as two plasma layers, the hydrogen first. This was observed also in the mixtures of gases before [8]. The x-ray yields of carbon confirm the expected tendency (see fig. 1), the portion of the hydrogen continuum is negligible (several %).

ENERGY INPUT AND K-SHELL OUTPUT COUPLING

The simple equation of motion of the hollow plasma cylinder, caused by the magnetic forces generated when a current flows through it, is often used for rough description of Z-pinch implosion. The basic conclusions, following from it,are:

a) For optimum energy transport to the plasma the implosion must occur near the current maximum; to reach it for sinusoidal currents achievable at Z-pinches driven by condenser banks, the circuit and load parameters have to fulfill the condition (cf. [9])

$$I^2 t^2 / m r_0^2 = 4 \times 10^7 \qquad \text{[in SI units]}, \qquad (1)$$

where I is the maximum current, t its quarter-period, and m the mass per unit length of the load with initial radius r_0.

+) The apparatus was prepared and measurements performed in cooperation with the Ioffe Physical-Technical Institute in St. Petersburg, Russia.

b) Conserving the above condition, the final kinetic energy per unit length of the plasma shell is near to

$$W = (\mu_0/4\pi)\ ln(r_0/r_f)\ I^2 \qquad (2)$$

with r_f the final plasma radius (just before striking and thermalization of the plasma shell on the pinch axis). This energy is generally assumed to be the dominant energy input to the plasma.

Let us to use these results for comparing the possibilities of small and large machines for investigation of K-shell emission using various working media. Experimental observations (e.g. [10]) indicate that maximum K-shell emission occurs, at given driver parameters, if the mentioned kinetic energy is equal or slightly higher than the energy needed for ionization of the whole plasma volume up to 50 % He-like and 50 % H-like ionization state plus thermal energy, equivalent to plasma temperature corresponding to this ionization state. The appropriate energy per ion can be approximated by [11]

$$E_i = 2.4 \times 10^{-19} Z^{3.5} \qquad [\mathrm{J}] \qquad (3)$$

with Z the atomic number of the working medium. (If the energy per ion is lower, the K-shell emission is dominated by hot spots.)

In the range of interest here, $Z = 6 \div 18$, the atomic mass can be approximated as

$$m_i = 3.35 \times 10^{-27} Z \qquad [\mathrm{kg}]\ . \qquad (4)$$

Finally, if N is the number of ions per unit length, combining the optimum driver--load coupling with the maximum K-shell emission, we have

$$(\mu_0/4\pi)\ ln(r_0/r_f)\ I^2 = 2.4 \times 10^{-19} N Z^{3.5} \quad , \qquad (5)$$

$$I^2 t^2 = 1.34 \times 10^{-19} N Z r_0^2 \quad . \qquad (6)$$

Assuming that the value r_0/r_f does not change with Z significantly, which seems to be reasonable if the final ionization state is the same in all cases, we obtain

$$I^2/N \sim Z^{3.5} \quad \text{and} \quad t^2/r_0^2 \sim Z^{-2.5}\ . \qquad (7)$$

As an example, let us consider our experiments with nitrogen, referred above. They were carried out at 1 cm initial radius and the current of 200 kA with 1.2 μs quarter-period. From absolute values of the calculated parameters (see fig. 3) it seems that the current risetime is too long to reach the mentioned optimum (note that the used equations describe only roughly the relevant processes). On the other hand, from the above relations follows that the conditions for generation of K-shell radiation in the 200 kA/1.2 μs machine with nitrogen should be similar as those for argon at 1 MA with 400 ns risetime, with the same number of atoms and initial load geometry. We have in mind especially ionization state and electron temperature corresponding to efficient excitation of K-shell transitions, and ion density; electron density should scale as $n_e = (Z - 1.5)\, n_i$. Of course, spectral region of the K-shell emission differs for different working media (the photon energy increases strongly with Z).

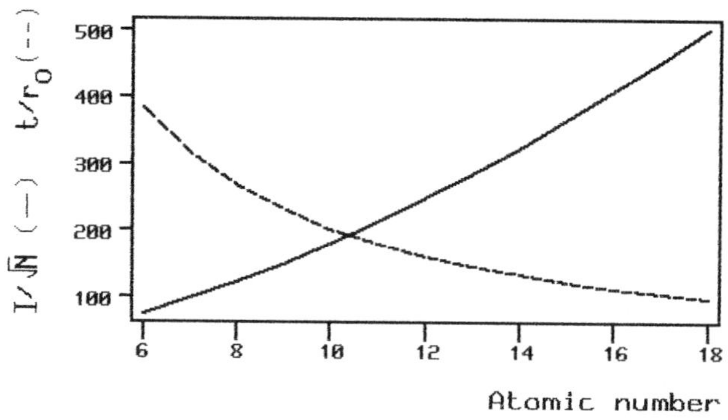

Fig. 3. Coupling of driver and load parameters for optimum generation of K-shell radiation, following eqs. (5) and (6). The ratio of the initial and final radii is assumed to be $r_0/r_f = 10$. Driver: I is the maximum current in kA, t is the quarter-period in ns. Load: N (in 10^{17} cm^{-1}) is the number of atoms per unit length, r_0 (in cm) is the initial radius.

Though these considerations include by far not all processes important for K-shell emission, it is seen that the "small" Z-pinches should be studied not only as radiation sources for individual applications; commonly with large devices they are suitable, using lower-Z working media, for basic research of processes connected with generation of intense K-shell radiation.

REFERENCES

[1] Proc. 2nd Int. Conf. on Dense Z-Pinches (eds. N.R.Pereira, J.Davis, N.Rostoker) Laguna Beach, 1989 (AIP Conf. Proc. 195, New York, 1989).
[2] A. Krejčí, Czech. J. Phys. 40, 182 (1990).
[3] A. Krejčí, E. Krouský, O. Renner, Czech. J. Phys. 40,1244 (1990); also in [1], p. 474.
[4] D. Duston, J. Davis, Phys. Rev. A 21, 932 (1980).
J.P. Apruzese et al., Phys. Rev. A 24, 1001 (1981).
[5] A. Krejčí, Plasma Sources Sci. Technol. 1, 271 (1992).

[6] J. Rauš, A. Krejčí, V. Piffl, Plasma Sources Sci. Technol. 1, 202 (1992).
[7] A. Krejčí et al., Proc. 9th Int. Conf. on High-Power Particle Beams, Washington DC, 1992, in print.
[8] J. Bailey et al., Appl. Phys. Lett. 40, 460 (1982).
[9] J. Katzenstein, J. Appl. Phys. 52, 676 (1981).
[10] C. Deeney et al., Phys. Rev. A 44, 6762 (1991).
[11] K.G. Whitney et al., J. Appl. Phys. 67, 1725 (1990).

RADIATIVE COLLAPSE IN VACUUM SPARK PLASMAS OF INTERMEDIATE AND HIGH Z ELEMENTS

D. Stutman, M. Finkenthal and J.L. Schwob
Racah Institute of Physics, Hebrew University, Jerusalem 91904, Israel

ABSTRACT

The soft X-ray emission of intermediate and high Z elements from a low inductance vacuum spark, has been analyzed by space-resolved spectroscopy and modeling of the emission. In the case of intermediate Z elements, the observed spatial extent and the estimated plasma parameters of the regions emitting L-shell ions transitions are consistent with the predictions of the radiative collapse model of the spark. For high Z elements, the estimated characteristics of the plasma emitting radiation of the M-shell ions seem to indicate that in this case too a radiative collapse may take place, triggered by the intense line emission of Ni-like ions.

INTRODUCTION

The radiative collapse (RC) model of the spark has been developed in order to explain the formation of micro-pinches in the vacuum spark plasmas of intermediate Z (iron group) elements.[1] According to this model, in a first stage of the discharge the plasma radius decreases and the electron temperature and density increase gradually, with the ohmic heating power per unit length (W_{ohm}) balancing the power losses associated with optically-thick, low energy M-shell radiation (W_{rad}) and particle outflow (W_{flow}). However, the onset of optically-thin, energetic L-shell line emission drastically increases W_{rad} and allows a fast compression of the plasma (a radiative collapse), which leads finally to the formation of the hot, dense micro-pinch.

In order to test the validity of such a scenario, it has been observed that time-integrated, but space-resolved single shot soft X-ray spectra may be of relevance: according to the model, the L-shell emission should arise from a very fast contracting plasma and therefore large differences should be observed between the spatial extent of the regions emitting lines of the first and the last ionization stages of the L-shell. As regards higher Z (rare-earths group) elements, it has been reasoned that if the RC model also applies, the radiative collapse could be triggered by the onset of M-shell emission and therefore similar effects should be observed in the space-resolved M-shell spectra of high Z plasmas.

EXPERIMENTAL

A low inductance vacuum spark of the three-electrode type has been used in the experiments.[2] The discharge parameters have been: anode-cathode gap 4 mm, capacitor bank of 24 μF, external inductance of ≈ 80 nH and high voltage of 17 kV.

Single-shot, space-resolved spectra in the range 7-30 A have been obtained using a high efficiency flat crystal spectrometer equipped with RbAP organic crystals

(2d=26 A) or multilayer W/C mirrors (2d=44 A) and photographic detection.[3] A spectral range of 8 to 12 A could be covered in a single shot with this spectrometer, with a spectral resolution ranging from 0.05 A to 0.2 A and space resolution of ≈ 50 μm along the discharge axis. In order to perform accurate line intensity comparisons and to measure the time integrated intensity of the space-resolved emission, the elements composing the spectrometer have been photometrically calibrated in the 8-25 A spectral range. In the analysis of the flat crystal data have been also used single or multiple shot, high resolution (0.01-0.02 A) soft X-ray spectra obtained with a grazing incidence 2m Schwob-Fraenkel spectrograph.

RESULTS AND DISCUSSION

I Intermediate Z elements. The space-resolved spectra of elements with Z in the range 22-28 have revealed large differences between the spatial extent and time integrated brightness of the plasma emitting resonance lines of the first L-shell ions (Ne and F-like) and that emitting lines of the last L-shell ions (B to Li-like). Thus:

- the line emission of the B-to Li-like ions is highly localized (figure 1a). The axial extent of the emitting plasma is less than ≈ 20 μm, since almost no penumbra may be observed in the image obtained with a 50 μm slit; its radial extent has an upper limit of the same order, since the observed line width coincides with the estimated crystal broadening. This emission is contained within a sharp trace of continuum, for which the lack of penumbra indicates that it also originates from a micron sized plasma .
- on the other hand, the emission of the Ne-like and F-like ionization stages arises from a much more extended plasma column, having axial and radial dimensions of the order of some hundreds of microns (figure 1b and figure 4). A core of more intense emission may be observed in this column, centered on the continuum trace. The change in the spatial extent occurs quite abruptly, after the F-like stage or even after the Ne-like stage, when strong pinches occur.
- the photographic density produced by the line emission of the constricted plasma (e.g. 2p-3d lines of Li-like ions) is nearly equal to that produced by the most intense lines emitted by the extended plasma (2p-3d transitions of the Ne-like ions). Correcting for the changes in instrumental response and unfolding the effect of spatial integration over sources having the above estimated dimensions, one obtains that the time integrated brightness of the line emission from the constricted plasma is by at least an order of magnitude higher than that from the extended plasma. If the dimensions predicted in reference 1 for the constricted plasma (about 2 μm x 10 μm) are assumed, its time integrated brightness would by two orders of magnitude higher.

In view of its characteristics, the extended plasma having a dominant population of Ne-like and F-like ions has been called the "pre-pinch" stage of the discharge. The average electron density in the pre-pinch has been estimated using the comparison between the experimental intensity ratios in the Ne-like Ti resonance spectrum and those predicted by a collisional radiative (CR) model for the level populations of this ion at T_e=200 eV.[4] The population mechanisms considered have been excitation from the ground state of the Ne-like ion together with variable contributions of either inner-shell ionization from the Na-like stage or recombination from the F-like stage. The model indicated that the above ratios depend quite sensibly on the electron density (figure 2a) and only weakly on the temperature. Self-absorption has been accounted for in the "escape-factor" approximation and assuming an uniform plasma. However, for the density estimate have been used only transitions of very similar (and relatively small) oscillator strength (i.e. 2p-3s and 2s-3p): for such lines the intensity ratios may be expected to be only marginally affected by self-absorption, up to quite high plasma densities.

The experimental intensities have been corrected for instrumental effects and normalized to that of a 2p-3s transition (figure 2b). A least-squares fit of the CR model predicted intensity ratios to the experimental ones indicated that the pre-pinch Ne-like Ti spectrum is formed dominantly by excitation from the ground state of the ion, at an electron density of about 8×10^{19} cm^{-3}. Assuming that over a small Z range, the observed pre-pinch parameters may be scaled using the Bennet relation at fixed current ($N_{Z1}\ T_{Z1}\ d_{Z1}{}^2 \approx N_{Z2}\ T_{Z2}\ d_{Z2}{}^2 \approx$ const.) and taking $T_e \sim \chi_{\text{Ne-like ion}}$, gives then for the Fe pre-pinch ($d_{Ti}\approx 300\ \mu m$, $d_{Fe}\approx 100\ \mu m$), an electron temperature $T_e \approx 350$ eV and density $N_e \approx 3 \times 10^{20}$ cm^{-3}. At such densities the 2p-3s and 2s-3p transitions are indeed estimated to be only weakly affected by self-absorption.[4, 8]

Using such density estimates and the experimental time integrated brightness of an optically thin (e.g. 2p-3s) transition, an evaluation of the typical duration of the Ne-like emission can be made. Since the observed spectra are consistent with excitation from the ground, the time integrated brightness may be approximated by:

$$B_{2p\text{-}3s} \approx d/4\pi\ \Delta t\ A_{3s\rightarrow 2p}\ (N_e/Z_{avg})\ (N_{gr}/N_{ion})\ (N_{3s}/N_{gr})\ \ [\text{photons/cm}^2\ \text{sr}]$$

where : d is plasma diameter, Δt the duration of the emission, $A_{3s\rightarrow 2p}$ the radiative transition probability, N_{3s}/N_{gr} the fractional level population, N_{gr}/N_{ion} the abundance of the ground state Ne-like ion relative to the total ionic density N_{ion} and Z_{avg} the average charge state in the plasma. Among these quantities, only the fractional level population has a strong dependence on the electron density in the range 10^{18}-10^{21} cm^{-3}, defining thus with good approximation a unique correspondence between the time integrated brightness $B_{2p\text{-}3s}$, the duration of the emission Δt and the electron density N_e. It follows that for a Fe discharge the pre-pinch emission occurs only for about 0.8 ns (figure 3). Due to the quite large uncertainties involved in the evaluation of the brightness and of the density, the above value is not better than an order of magnitude estimate. However, this still leaves the duration of the pre-pinch emission in the range from fractions of ns to a few ns; this is a very short interval on the time scale of the discharge (of the order of hundreds of ns).

Using the experimental pre-pinch parameters and the above estimated duration of the emission and performing a weighted sum over all the resonance transitions in the experimental Ne-like Fe XVII spectrum, one obtains that from this contribution alone, one has for the Fe pre-pinch: $W_{rad}/W_{ohm} \approx 70$ and $W_{rad}/W_{flow} \approx 150$ (with W_{ohm} and W_{flow} computed as in reference 1).

If it is accepted that the observed differences in the spatial extent of the L-shell emission reflect a shrinkage of the plasma, the above results (see also figure 4) seem

to be qualitatively - and to a large extent quantitatively - consistent with the predictions of the RC model for the evolution of an iron spark plasma:
- the stage prior to the collapse (the "pre-pinch") is predicted to be a plasma column of about 100 μm diameter and about 500 μm height, with average Ne ≈ 4 10^{20} cm^{-3}, Te ≈ 400 eV and having a dominant Ne-like and F-like ion population. Its predicted lifetime is ≈ 1 ns since W_{rad}/W_{ohm} ≈ 100.
- the thermally unstable pre-pinch plasma collapses quite fast (in about 0.2 ns), with the effect that the last ions with an open L-shell (B-like to Li-like) emit already from a point like (about 2 μm diameter, 5-10 μm height) very dense (Ne ≈ 10^{23} cm^{-3}) and relatively cold (T_e ≈ 800 eV) plasma. Since in these conditions the plasma becomes a surface radiator (and taking also into account its predicted lifetime of ≈ 0.1 ns), the time integrated brightness of its emission should be by about two orders of magnitude higher than that of the pre-pinch, consistent with the experiment.
- finally, due to the appearance of anomalous resistivity the collapsed plasma is heated to very high temperatures (several keV) forming the hot, dense and short-lived micro-pinch, which emits lines of He-like and H-like Fe ions. The predictions concerning the micro-pinch have been in good measure confirmed by previous research.[5]

II High Z elements. Space-resolved soft X-ray spectra of the of rare-earth elements (Z=58-65) have been recorded[6] using the same experimental setup and methods as in the above. The space-resolved spectra (figure 1c) have indicated that:
- the spark spectrum is dominated by intense Ni-like 3d-4f, 3d-4p and 3p-4s transitions, with weaker Cu-like and Zn-like unresolved arrays (originating from superimposed 3d-4f transitions having a 4l (l=s,p,d) spectator electron). This emission arises from an extended plasma region having an axial extent around 300 μm and a radial extent of about 100 μm. The image of the extended plasma is superimposed on a sharp trace of intense continuum originating from a micron sized plasma .
- the emission of the ionization stages higher than about Co-like consists mostly of intense unresolved transition arrays and is contained within the continuum trace, indicating thus that the emitting plasma has an axial extent of no more than a few tens of microns. Considering as in I, the effect of spatial integration on the observed intensities, the plate exposure recorded in these arrays seems to indicate a sensibly higher brightness for this emission.

Collisional radiative modeling of the Ni-like Ce (Z=58) ion emission indicated that in this case also, the intensity ratios of the resonance transitions are density sensitive in the range 10^{18}-10^{22} cm^{-3} and quite insensible to variations in the electron temperature.[6] T_e=500 eV has been assumed for the extended plasma, based on the comparison between the observed charge state distribution and the computed temperature dependence[7] of the fractional abundance of Ce ions at N_e=5 x 10^{20} cm^{-3}.

The opacity effects have been incorporated in the model using again the escape factor approximation and as above, only transitions of comparable and relatively small oscillator strength (3d-4p and 3p-4s) have been used in the density estimate. The observed Ni-like Ce spectrum was thus found to be consistent with an electron density in the emitting plasma of about 2 x 10^{20} cm^{-3}. Using as above, the experimental time integrated brightness of an optically thin transition (3d-4p), the density estimate and the ratios N_{4p}/N_{gr} and N_{gr}/N_{ion} computed in reference 6 and 7 respectively, a duration of the Ni-like emission of about 0.6 ns has been estimated.

A straight comparison with the intermediate Z case (see figure 4), seems to indicate that the Ni-like spark emission of high Z elements is also consistent with a pre-pinching phase of the constriction. Estimating as above the radiative losses due to Ni-like emission only and the ohmic heating, one obtains again that the former are by nearly two orders of magnitude larger than the latter, i.e. this emission alone may cause the radiative collapse of the plasma.

CONCLUSIONS

Admitting that the differences observed in the spatial extent of the emission of the various ionization stages reflect a shrinkage of the plasma, the intermediate Z results are in good measure consistent with the main qualitative and quantitative predictions of the RC model for the stages of the spark plasma prior to the "hot-point". Moreover, the high Z results seem to indicate that the RC model may be also valid for high Z vacuum spark plasmas - at least in a qualitative way and for the stages prior to the "hot-point" investigated here. The radiative collapse could be triggered in this case by the intense resonance emission of the Ni-like ions.

There may be however processes unaccounted for in the above models. Thus, by decreasing the external inductance from 80 nH to about 40 nH we have obtained Ne-like spectra in which one of the 2p-3s transitions (line G for Ti and F for Fe) is anomalously enhanced (figure 5 versus 2b), increasing the Ne-like radiative losses by about 30%. As may be seen from figure 2a, for a wide range of densities and independently of the excitation mechanism considered, the CR model predicts nearly equal intensities for the lines F and G. We do not have presently an explanation for this effect. It may be interesting to note however, that for the 40 nH discharge the pre-pinch density and temperature should reach the range where maximal gain is predicted for the 3s-3p transitions [8]. Thus, in the event of amplified spontaneous emission on a 3s-3p transition, the 2p-3s radiative power losses could be enhanced and moreover, could depend critically on the spatial extent of the pre-pinch.

REFERENCES

1. V.V. Vikhrev, V.V. Ivanov and K.N. Koshelev, Sov. J. Plasma Phys. **8**, 688 (1983)
2. J.L. Schwob and B.S. Fraenkel, Phys. Lett. **40A**, 81 (1972)
3. M. Finkenthal, D. Stutman, J.L. Schwob and J.H. Underwood, J. de Physique, **49**, C1-87, (1988)
4. M. Finkenthal, D. Stutman, P. Mandelbaum, A.L. Osterheld, W.H. Goldstein and M. Chen, J. Phys. B: At. Mol. Opt. Phys. **22**, 1133, (1989)
5. K.N. Koshelev, V.I. Krauz, N.G. Reshetniak, R.G. Salukvadze, Yu.V. Sidel'nikov and E. Ya. Khautiev, J. Phys. D: Appl. Phys. **21**, 1827, (1989)
6. M. Finkenthal, D. Stutman and A. Bar-Shalom, J. Appl. Phys. **65**, 3786, (1989)
7. Y.T. Lee, private communication
8. U. Feldman, J.F. Seely and A.K. Bhatia, J. Appl. Phys. **56**, 2475, (1984)

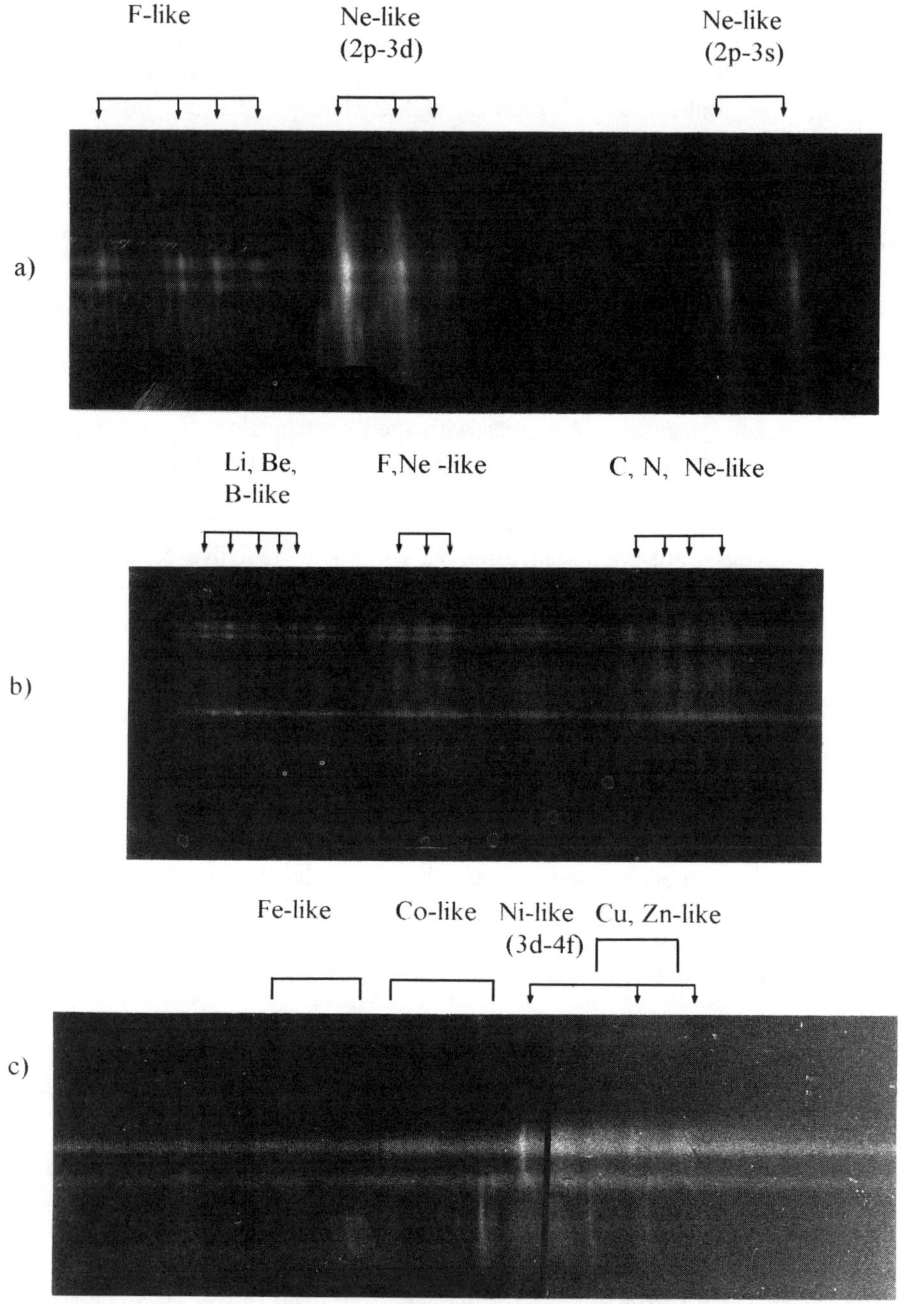

Fig. 1 Space-resolved soft X-ray spectra of intermediate and high Z spark discharges
a) intermediate Z, pre-pinch emission (Co anode)
b) intermediate Z, constricted plasma emission (V anode)
c) high Z, M-shell ions emission (Ce anode)

Figure 2a. Predicted density dependence of Ne-like Ti line intensities at Te=200 eV.[4] Intensities are normalized to that of the $2p^6$-$2p^5 3d\ ^1P_1$ transition (line C). The following excitation mechanisms have been considered:

a) inner-shell ionization from the $2s^2 2p^6 3l$, l=s,p,d levels of Na-like Ti ions

b) excitation from the $2p^6\ ^1S_0$ ground state of Ne-like ions

c) recombination from the $2s^2 2p^5$ and $2s2p^6$ levels of F-like Ti ions

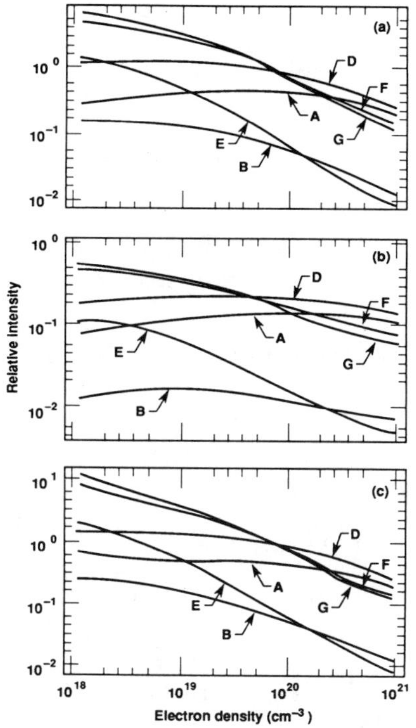

key	upper level	
A	$2s^2 2p^6 3p$	$^1P_1$
B		$^3P_1$
C	$2p^5 3d$	$^1P_1$
D		$^3D_1$
E		$^3P_1$
F	$2p^5 3s$	$^1P_1$
G		$^3P_1$

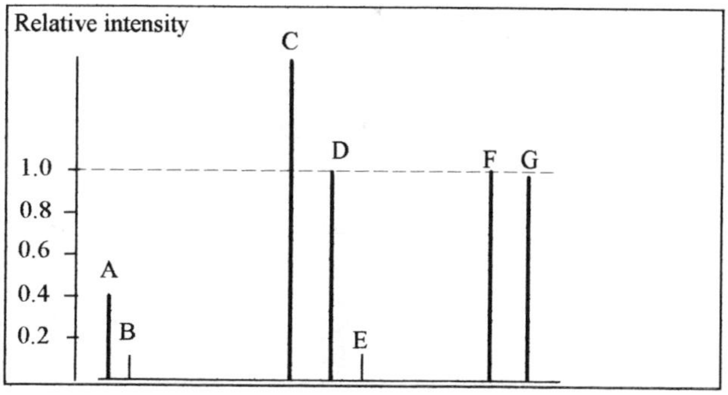

Figure 2b. Experimental intensity ratios in the Ne-like Ti pre-pinch spectrum, after correction for instrumental effects. Intensities are normalized to that of the optically thin line G. The estimated uncertainity in the intensity ratios is around 20 %.

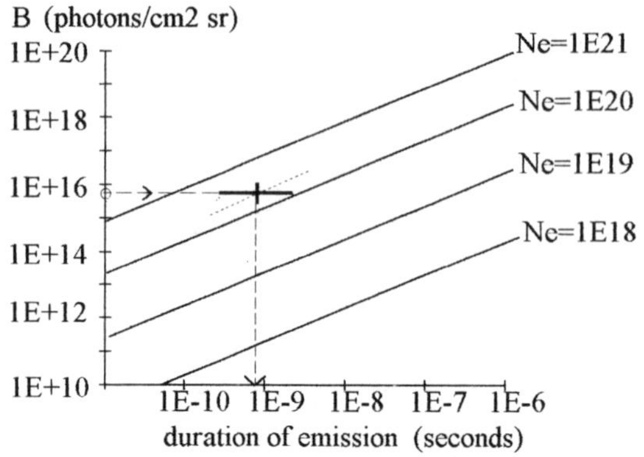

Figure 3. Estimate of the duration of emission of the Fe pre-pinch, using the predicted and the experimental value of the time integrated brightness of the 2s-3p $^1S_0$-$^3P_1$ transition and the density estimate. The fractional level population computed in [8] has been used and a 100 μm diameter, optically thin plasma has been considered.

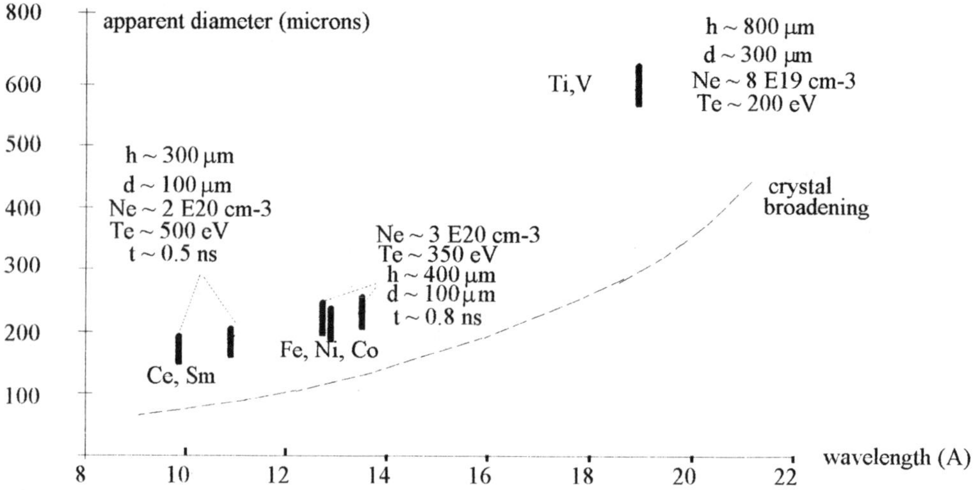

Figure 4. Summary of the intermediate and high Z pre-pinch estimated properties.

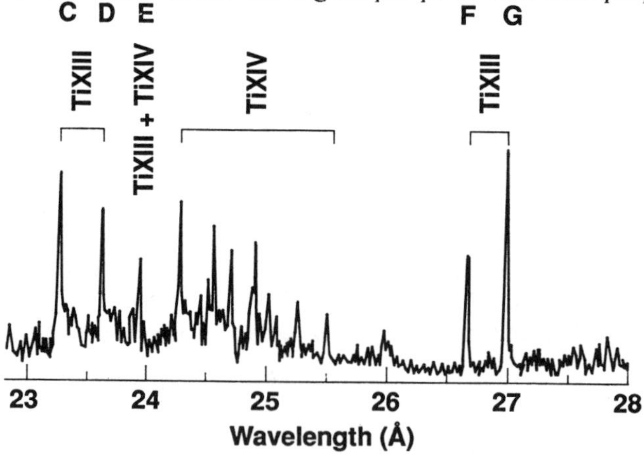

Figure 5. Anomalous intensity ratios in the Ne-like Ti emission. Single-shot, grazing incidence spectrum (2400 l/mm grating) obtained with 40 nH external inductance.

EFFECT OF OPERATING PRESSURE ON PLASMA FORMATION IN THE TRANSIENT HOLLOW CATHODE DISCHARGE

C.S. Wong, C.X. Ong, S.P. Moo
Plasma Research Laboratory, Physics Department
and
Pulse Technology and Instrumentation Laboratory, IPT
University of Malaya
59100 Kuala Lumpur, MALAYSIA

and

P. Choi
The Blackett Laboratory, Imperial College
London SW7 2BZ, U.K.

ABSTRACT

The discharge development of a triggered pulsed hollow cathode discharge is investigated with respect to two regimes of operating pressure: the high pressure regime at ~ 10^{-2} mbar and the low pressure regime at ~ 10^{-5} mbar. The hollow cathode discharge studied here is performed with the UMFX device powered by either eight 2700 pF door-knob capacitors connected in parallel and discharged at 40 kV or a single 1.85 μF capacitor discharged at 20 kV. The characteristics of the discharge is monitored by using magnetic probe for dI/dt measurement; PIN diodes and XRD for time-resolved X-ray signals; and pinhole camera for time-integrated X-ray imaging. At the high operating pressure regime, X-rays are observed to be emitted from the tip of the anode only. At the low operating pressure regime, however, the heating of the plasma by the main capacitor discharge is significantly enhanced. In the case of low energy discharge (with 22 nF capacitor bank), a plume-like plasma emitting in the soft X-ray region is observed to be formed attached to the tip of the anode. The pre-breakdown electron beam, which is a well-known phenomenon in this type of discharge, is also observed to be significantly enhanced. For the case of discharges performed with the 1.85 μF capacitor, dense plasma spots are observed at low operating pressure of 10^{-4} mbar or lower. These dense plasma spots are not observed for discharges operated at pressures higher than 10^{-4} mbar

INTRODUCTION

The production of a pre-breakdown electron beam in the transient hollow cathode discharge and its associated properties have been the subject of many investigations[1,2,3,4]. It has been observed experimentally that[4] the intensity of this electron beam is strongly affected by the operating pressure of the discharge. Specifically, the electron beam intensity is greatly enhanced by the lowering of the operating pressure from as high as ~ 10^{-1} mbar to as low as ~ 10^{-4} mbar. Such an enhanced electron beam has been proposed as a possible method of triggering the vacuum spark discharge instead of the conventional sliding spark method.

In the operation of a vacuum spark triggered by the enhanced electron beam produced by the transient hollow cathode effect[5], the sequence of events may be described as follow: On charging up the capacitor the voltage is held directly across the electrodes but electrical breakdown is prohibited by the low pressure in the inter-electrode space. The inter-electrode separation used is typically 1 to 3 mm so that even at an operating pressure of 10^{-1} mbar, the discharge is operating at the left-hand arm of the Paschen curve. To initiate the discharge, a high voltage low energy spark is produced between a third electrode and the cathode just behind the cathode opening in the hollow cathode region. This triggering spark acts to seed the space in the hollow cathode region with some initial electrons which will be accelerated towards the anode tip following the electric field lines penetrated into the hollow cathode region. These initial electrons will multiply when they collide with atoms or ions to produce collisional ionizations, leading to the formation of the enhanced pre-breakdown electron beam. On hitting the anode tip, this electron beam will vaporize some anode material and in the process also gives rise to the emission of x-ray from the anode tip. Breakdown of the main discharge gap will occur when the vapor of the anode material expands to fill the inter-electrode gap. The subsequent passage of the discharge current through the vapor of the anode material heats it to hot and dense condition. The present paper is mainly concerned with the plasma heating effect during the main discharge of the vacuum spark.

EXPERIMENTAL SETUP

The transient hollow cathode discharge chamber and the electrode system is shown schematically in Fig. 1. The cylindrical chamber is made of copper and has a diameter of 4 cm

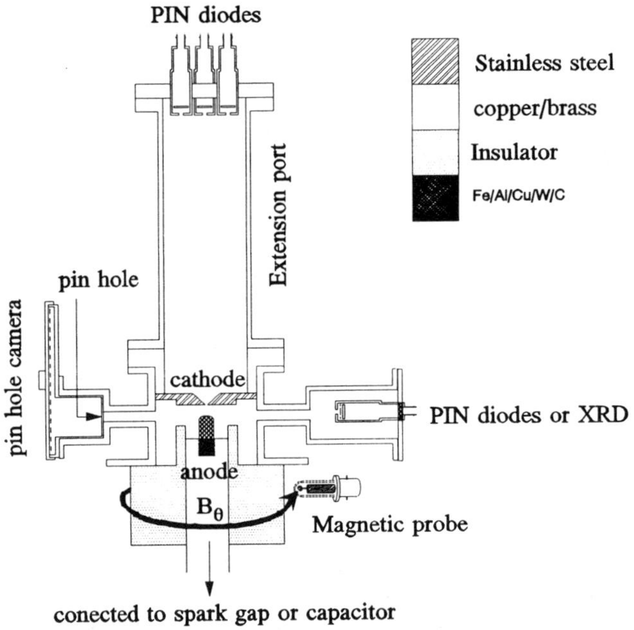

Fig. 1 Schematic of the transient hollow cathode device.

and a length of 4 cm. The cathode is made of stainless steel and has a 6 mm diameter hole on axis to provide the hollow cathode configuration. The anode is made of iron in the present setup and has a diameter of 7 mm. The chamber is pumped to a pressure of 10^{-5} mbar using a small diffusion pump backed by a rotary pump. For the purpose of investigating the effect of operating pressure on the discharge, the pressure is adjusted by regulating the pumping rate over the range from 10^{-1} to 10^{-5} mbar.

The transient hollow cathode discharge is powered either by eight 2700 pF door-knob capacitors connected in parallel(the UMFX-III), or by a single 1.85 μF capacitor(UMVS-III). For the case of UMFX-III, the anode is connected to the positive plate of the capacitor bank via an atmospheric air spark gap. The spark gap helps to hold off a portion of the voltage to enable discharges at 40 kV to be performed. On the other hand, the anode of the UMVS-III is mounted directly onto the positive terminal of the 1.85 μF capacitor so that on charging up the capacitor, the full voltage will be held across the electrodes. At a base pressure of below 10^{-3} mbar, the UMVS-III system is capable of holding off 20 kV with an inter-electrode separation of 3 mm. In both cases, the discharge is initiated by applying an external 40 kV low energy spark across an auxiliary trigger electrode and the cathode. The auxiliary trigger electrode is located inside the hollow cathode as illustrate in Fig. 1.

The diagnostics used in this study include the measurement of the rate of change of the discharge current by the magnetic pickup coil. The x-ray emission from the anode tip as well as the plasma are monitored using two channels of PIN diode (Quantrad 100-PIN-250) and one channel of XRD. One PIN channel is mounted end-on at a distance of 67.5 cm from the anode tip; whereas the second channel is mounted side-on at a distance of 31.5 cm from the anode tip. Both PIN diodes are filtered by a combination of 24 μm of aluminized mylar and 5 μm of iron foil. The XRD has aluminium cathode with a brass wire mesh as anode and is mounted side-on at a distance of roughly 28 cm from the anode tip of the discharge system. It has no absorption foil and is biased at 400 V. Time-integrated x-ray imaging of the discharge is obtained by using a pinhole camera with a pinhole size of 300 μm diameter filtered by 12 μm of aluminized mylar to exclude visible light. Kodak X-Omat film is used for recording the x-ray image.

RESULTS AND DISCUSSION

For the present study, the inter-electrode separation of the system is fixed at 3 mm. The operating pressure is varied from 10^{-2} mbar to 10^{-5} mbar. It is particularly interesting to compare the results obtained at the two extreme ends of this range of operating pressure as illustrated in Fig. 2 and Fig. 3. In Fig. 2, the signals of dI/dt, end-on PIN, side-on PIN and side-on XRD are

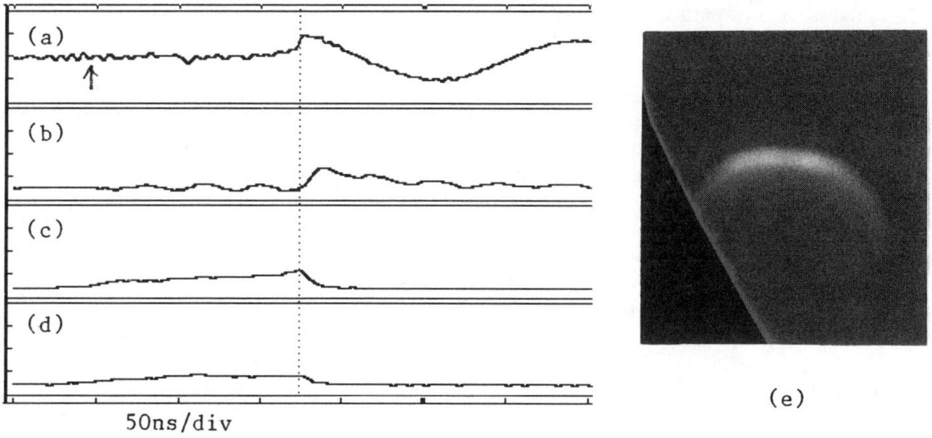

Fig. 2 Results obtained for a 40 kV discharge of UMFX-III at P = 5 x 10^{-2} mbar.
(a) dI/dt; 10V/div. (b) XRD; 200mV/div. (c) side-on PIN; 15V/div. (d) end-on PIN; 10V/div.
(e) X-ray pinhole image.

obtained simultaneously for a discharge of UMFX-III at 40 kV and at an operating pressure of 5 x 10^{-2} mbar. In this discharge, both end-on and side-on PIN signals show the slowly building up of a pre-breakdown electron beam over a duration of about 120 ns. This electron beam is

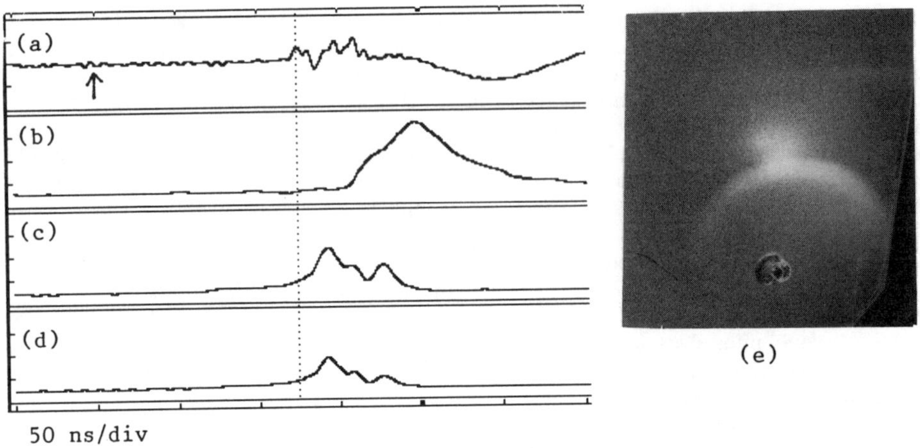

Fig. 3 Results obtained for a 40 kV discharge of UMFX-III at P = 5 x 10^{-5} mbar. (a) dI/dt; 10V/div. (b) XRD; 200mV/div. (c) Side-on PIN; 15V/div. (d) End-on PIN; 10V/div. (e) X-ray pinhole image.

similar to those observed previously[2] and it will disappear once the inter-electrode spacing become fully conducting and breakdown occurs. The conductivity of the inter-electrode space may be increased due to collisional ionization of the gas medium as well as some anode material thrown into the space due to bombardment of the anode tip by the electron beam. The conductivity seems to build up gradually so that the super-exponential electron avalanche is not observed in this particular discharge. Upon breaking down, the gas medium and vapor of anode material present in the inter-electrode gap is heated up rapidly as indicated by the sharp rising XRD signal. The temperature reached, however, is not sufficiently high to emit in the x-ray region which explains the absent of the PIN diode signal, and the x-ray pinhole image shows that x-rays are emitted from the tip of the anode only.

In contrast, the discharge of UMFX-III at low operating pressure seems to show a different pre-breakdown electron beam activity as can be seen from Fig. 3. The electron beam has a much lower intensity initially as compare to the high pressure case. It starts to rise exponentially at about 80 ns after it has been initiated (Note: The instant of electron beam initiation as deduced from the dI/dt signal has been marked by an arrow in Fig. 2 and Fig. 3.) followed by an almost super-exponential rise indicating the occurrence of electron avalanche process. This results in an electron beam of high intensity which collapses rapidly as soon as breakdown of the main gap occurs. In this case, the medium that is being heated up by the discharge current consists of predominantly the vapor of the anode material, which is iron for the present setup, due to the low operating pressure of this discharge. Several events are observed to be taking place during the plasma heating process in this particular discharge. At around 12 ns after the discharge has started, a dip in the dI/dt signal and correspondingly a sharp rise in the XRD signal are observed. A small spike is also observed on the falling edge of the PIN diode signal. After another 20 ns or so, a second dI/dt dip which is more severe than the first dip, occurs giving rise to a further increase in the XRD signal and a third sharp spike in the PIN diode signals. The fact that the intensity of the pre-breakdown electron beam is much higher at low operating pressure suggests that a much denser vapor of the anode material will be formed. Thus the passage of discharge current through the inter-electrode space becomes more difficult and it is possible that intermittent discruption of the discharge current may occur, resulting in the appearence of the dI/dt dip, which is believed to be the cause of the first dI/dt dip. This abrupt change in the dI/dt may also induce localised electric field that accelerate electrons to form beam. This electron beam will be less energetic than the pre-breakdown electron beam and the x-ray emitted from the anode tip due to bombardment by this beam will be soft enough to be detected by the XRD. On the other hand, during the second dI/dt dip which is much more severe than the first dip, it is reasonable to believe that an iron plasma has been sufficiently heated to emit soft x-ray which gives rise to the "plume-like" x-ray image above the tip of the anode as shown in Fig. 3(e).

Similar experiments have been carried out on the UMVS-III of which the same electrode system is mounted directly onto a single 1.85 μF capacitor. The pressure has also been varied over the range of 10^{-2} to 10^{-5} mbar. For operating pressure of higher than 10^{-4} mbar, no hot spot formation has been observed. Time-integrated x-ray pinhole image shows that x-ray is only emitted from the tip of the anode. However, at operating pressure of 10^{-4} mbar or lower, dense plasma spots emitting intense soft x-ray are observed. An example of a hot-spot producing discharge of the UMVS-III vacuum spark device is shown in Fig. 4. In this discharge, severe dip

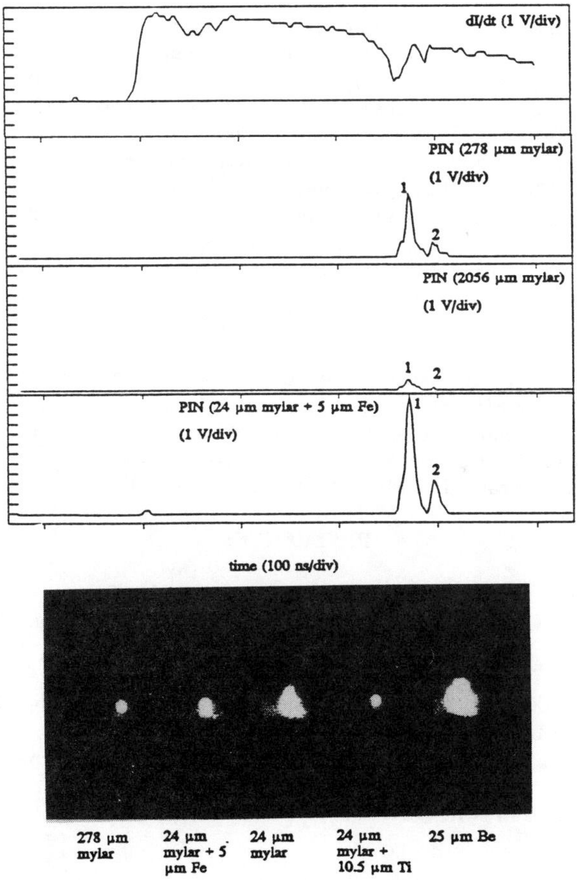

Fig. 4 Results obtained for a 20 kV UMVS-III vacuum spark discharge showing the simultaneous signals of dI/dt and three channels of PIN diodes with different filters. A five pinholes x-ray pinhole camera was used in this case.

in the dI/dt signal is observed at about 250 ns after breakdown, where the discharge current is near to its first maximum. As can be seen from the PIN diode signals and the x-ray pinhole images, the plasma has been heated strongly to produce two dense plasma spots. The electron temperatures of these plasma spots have been estimated by the foil absorption technique to be about 1 to 3 keV.

CONCLUSIONS

It is clear from the above experiments performed on the UMFX-III and UMVS-III devices that at low operating pressure of 10^{-4} mbar or lower, the transient hollow cathode discharge is capable of heating the vapor of the anode material effectively to extreme plasma condition. This is not the case at high operating pressure. The pre-breakdown electron beam production is also greatly enhanced at low operating pressure. With a more intense pre-breakdown electron beam at low operating pressure, more anode material will be vaporised to fill the inter-electrode gap and hence a denser channel of current conduction is expected. This implies that the initial phase of the discharge after breakdown will be resistive and Joule heating effect will play an important role. Subsequently, at least in the case of UMVS-III, pinching of a plasma column takes place which may lead eventually to constriction of the column and radiative collapse which has been postulated as a possible mechanism of hot spot formation[6,7]. For the case of UMFX-III, it has been shown that despite the low electrical input energy employed, the heating of the vapor of the anode material can be significantly enhanced by operating the device at low pressure. For a 40 kV, 17 J discharge at an operating pressure of 5 x 10^{-5} mbar, x-ray emitting plume-like plasma cloud has been produced.

ACKNOWLEDGEMENTS

The UMFX and UMVS were developed with the support of two research grants from the Malaysian Ministry of Science, Technology and the Environment under IRPA Programmes 04-40 and 02-033. The attendence of CSW at the Third International Dense Z-pinch Conference was supported by IRPA Programme 04-40.

The authors are grateful to Mr. J. Singh and Mr. T.S. Toh for technical support.

REFERENCES

1. J. Christiansen and Ch. Schultheiss, Z. Phys. **A290**, 35(1979).
2. P. Choi, H.H. Chuaqui, M. Favre and E.S. Wyndham, IEEE Trans. Plasma Sci. **PS-15**, 428(1987).
3. K. Mittag, P. Choi and Y. Kaufman, Nucl. Instrum. & Methods **A292**, 465(1990).
4. C.S. Wong, C.X. Ong, S. Lee and P. Choi, IEEE Trans. Plasma Sci. **PS-20**, 405(1992).
5. C.X. Ong, Ph.D. thesis, University of Malaya, Malaysia (1992).
6. P.S. Antsiferov, K.N. Koshelev, A.E. Kramida, A.M. Panin, J. Phys. D: Appl. Phys. **22**, 1073(1989).
7. K.N. Koshelev and N.R. Pereira, J. Appl. Phys. **69**, R21(1991).

Evidence for a Multiphase Discharge Channel in Single Al wire Z-Pinch Experiment.

E.J. Yadlowsky, R.C. Hazelton, J.J. Moschella, and T.B. Settersten
HY-Tech Research
104 Centre Court
Radford, VA 24141

ABSTRACT

The vapor phase separating the initial solid state from the final plasma state of wire loads in pulse power devices is often assumed to be very short lived and of no consequence to the subsequent load implosion for fast current pulses. However, researchers have attributed Cu K_α radiation and load stability to a persistent solid core in experiments with Cu wires or frozen D_2 fibers, respectively. The spatial structure of Al Z-Pinch loads has been studied with streaked absorptiongrams, holographic interferometry, gated multiple pinhole cameras, and debris collectors to address this issue. A coronal plasma surrounding a neutral rich inner region is seen 310 ns into a current pulse having a quarter period of 1.5 μs. The observations made of the coronal plasma separating from the neutral inner region in two wire experiments has ramifications for multiple wire arrays. This phenomenon may account for pre-cursor plasma formation, softened implosion, a small fraction of the mass radiating in the K shell and straggling in the load assembly on the axis. The results indicate that the initial plasma formation process on individual wires of an array is very important to the subsequent load implosion and must be modeled realistically.

I. INTRODUCTION

The question of how a solid wire or fiber in a high voltage diode is converted to an ionized plasma by the current/voltage pulse has been addressed by numerous researchers.[1-4] A rapidly rising current pulse is assumed to uniformly vaporize and ionize the wires in an array with the vapor phase so short on generators with large $\dot{I}$ that it can be ignored.[1] At the other end of the spectrum, where $\dot{I}$ is low, the surface of the wire is assumed to vaporize and ionize first; therefore, forming a non-uniform current channel.[2-5] Ablation and ionization of the remaining core proceeds as a result of energy deposition by electrons, ions, and radiation from the coronal plasma.

The initial start up phase is assumed to have little consequence in the shorter rise time regime, whereas numerous physical phenomena have been attributed to the coexistence of different material phases in the current channel for longer rise times. Aranchuk *et al.*[5], attributed the Cu K_α radiation they observed to energetic electrons from the tenuous current-carrying coronal plasma striking a solid Cu core containing most of the wire mass. Similarly, the hydrodynamic stability of frozen D_2 fiber loads has been linked to a solid core in the pinch by Lindemuth *et al.* [4]. Figura *et al.*[6], and Yadlowsky *et al.*[7], have observed a solid core well into the current pulse in experiments on optical fibers and on carbon fiber loads, respectively.

This work reports studies on 25μm Al and 12.7μm Ni wire loads exploded by current pulses with $\dot{I}$~0.1-1.0x10^{12} A/sec. Diagnostic techniques included streaked absorptiongrams, holographic interferograms, gated multiple pinhole cameras, filtered x-ray diodes, and particle debris collectors. The results show that a multi-phase current channel is formed , which appears to consist of a neutral rich core surrounded by a coronal plasma. These results explain many of the effects observed on wire array experiments, and indicates that the sudden transition approximation often assumed for convenience may not be appropriate.

II. EXPERIMENTAL SYSTEM

The experiments were carried out on a capacitor discharge system delivering 250kA to a 25μm Al load in 1.5μs. A 0.5 J discharge system was used to administer a preconditioning current pulse with a peak amplitude of 0.5 kA and rise time of 100ns in an attempt to vaporize the load prior to firing the main capacitor. A plasma opening switch (POS) was used to shorten the current rise time to 80-100ns while delivering 80-100kA to the Al load.

Magnetic pick-up coils located in the shadow of the return current posts are used to monitor the load current. The x-ray emission was studied using GaAs photoconducting detectors which are filtered with aluminized polymer foils to pass a range of x-ray energies while excluding visible and uv radiation. Aluminized parylene foils 0.2μm thick are used for energies less than 0.3keV, while single and double layers of Kimfoil and 8μm Kapton foils are used as high pass filters for energies exceeding 1keV. The diodes are apertured and aligned to view the central 4 mm of the load. A gated multiple pinhole framing camera, with 7 pinholes aligned with 7 micro channel plate strip line detectors, is used to produce a sequence of 2-D images of the load emission. The pinholes are filtered as described above to study the plasma volume emitting x-rays in these energy ranges. The striplines can be gated independently or simultaneously.

An image intensified streak camera is used to observe the absorption effects of the load using a 900ns collimated dye laser pulse to back light the load. The laser is tuned to 585.2nm and the plasma emission is filtered with a narrow band filter (2 nm) centered on the laser frequency. A lens at the exit port of the vacuum chamber corrects refractive effects of the plasma and magnifies the image 12X to allow microscopic probing of the plasma spatial structure.

A double exposure holographic ruby interferometer is used to measure the distribution of electrons and neutrals. A f/2 lens at the exit port of the vacuum chamber magnifies the image 3X and corrects for refractive effects as described above. Early in the current pulse, the plasma density is not very high and the fringe pattern is orderly, making analysis straightforward. Later in the pulse the number of fringe shifts is large, and the pattern is turbulent, making it difficult to follow individual fringes. Single exposure holograms representing gated absorptiongrams of the entire length of the pinch, have proved useful in studying the late time behavior of the pinch.

Debris produced during the Z-Pinch discharge was collected on 1cm x 1cm quartz slides and analyzed using a SEM with EDAX to determine the chemical composition of the particles and their size distribution. Particles less than 1μm in diameter could not be chemically identified. A collimator consisting of two apertures was also used in an attempt to collect debris from the load region while excluding particles from the electrodes. A complementary collector consisting of a Lucite cylinder concentric with the load and located inside the return current cage was used as a large solid angle detector (2π steradians). The collected debris was dissolved in acid and spectroscopically analyzed to determine its composition.

III. RESULTS

The initial expansion phase of the wire, exploded by the capacitor bank without the opening switch, is shown in the streaked absorptiongram in Fig. 1. The expansion phase is followed by an oscillatory phase (as in Fig. 1) or a pinch. Figure 1 shows considerable spatial structure in the plasma not seen in typical emission streaks.

The first significant pulses of soft x-rays are emitted about 300-400ns into the pulse. These low energy pulses (E < 300eV) occur at about the same time that the plasma column becomes transparent to the 585.2 nm laser radiation. More energetic x-rays (E < 1KeV) are emitted later in the current pulse (600-800ns), and are also accompanied by rapid changes in plasma opacity. The multiple pinhole framing camera indicated that the low energy x-rays are emitted from a thin cylindrical shell 1-3mm in diameter. Small localized regions appear to be the sources of the higher energy radiation.

The Abel inverted density profile of the plasma column, inferred from an interferogram obtained 310 ns into the current pulse for a 35 kV charge is shown in Fig. 2. The figure clearly shows a neutral Al vapor core surrounded by a coronal plasma. Since the index of refraction for Al vapor was not known to the authors, an average value of the index for He and Ar[8] was used in this analysis.

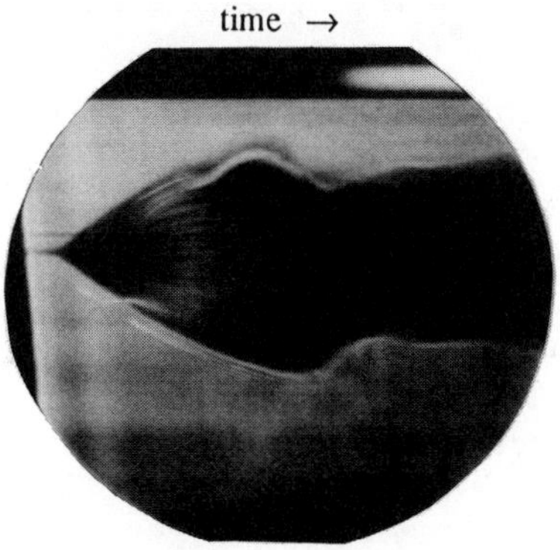

Fig.1. Streaked absorptiongram obtained with main capacitor bank.(300ns duration)

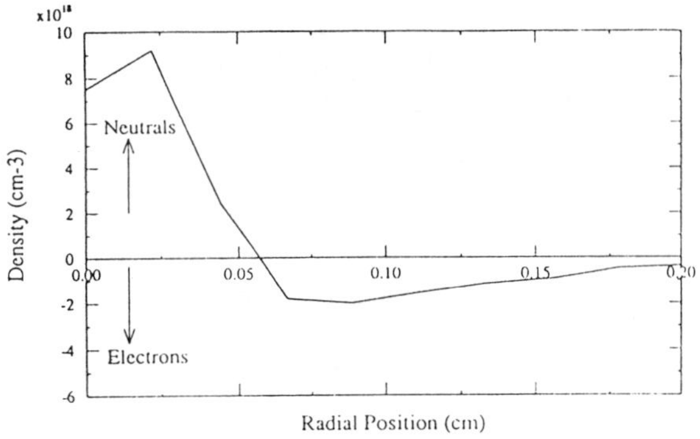

Fig. 2. Abel inferred density profile 310ns into current pulse from main bank.

The SEM micrograph of the particle debris in Fig. 3 reveals typical Al particulates in the form of spherical particles. The density of particles on a substrate utilizing a collimator to view the load region is considerably less than that on an unobstructed substrate. Ni particles, characteristic of the load and Fe and Al particles, characteristic of the electrodes and return current structure were observed. The chemical composition of the debris on the Lucite collector is presented in Table I. The results are for two experiments consisting of 10 shots each with Al and Ni loads.

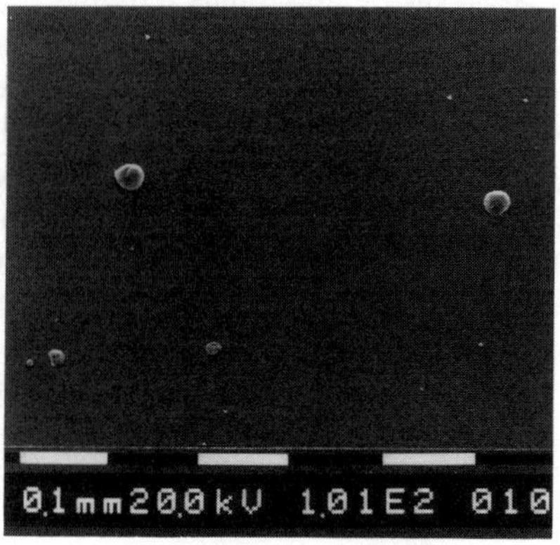

Fig. 3. Micrograph of particulate debris obtained with SEM.

The preconditioner was introduced in an attempt to vaporize the Al wire prior to the application of the main current pulse. Interferograms of the wire exploded by the fast preconditioner pulse show bending of the fringes which is associated with neutrals. Optical emission spectra also show neutral Al lines at 394.4 and 396.15nm, which are strongest 500ns after the start of the pulse. Although the interferometer measurements indicate that some vaporization of the Al wire occurs, the SEM analysis of the particulate debris indicates the presence of 1-3μm diameter spheres which can account for 50% of the total initial wire mass. The average size of the particles decreased form 15-20μm at a 15kV charge on the preconditioner to 0.5-2μm at 40 kV.

TABLE I. Chemical Composition of Debris

Element	Minimum Detection Limit (μg)	Detected Mass (μg) Al loads (113μg Al)	Detected Mass (μg) Ni loads (96μg Ni)
Al	0.5	32	31
Fe	1.5	58	126
Ni	2	4	14.5
Cu	0.5	8.5	5.5

Slight increases in x-ray yield were observed when the preconditioner was fired 400 to 600ns before the main current pulse. Absorptiongrams taken during the preconditioning phase reveal a thin refractive shell expanding radially with a velocity of ~ 0.27cm/μm. A streaked absorptiongram obtained later in the current pulse (Fig.4) shows an imploding refractive discontinuity which appears to bounce or reflect from the opaque core. An interferogram taken 137ns into the current pulse (Fig.5) shows a thin plasma shell surrounding a predominately neutral core. It is logical to associate this plasma shell with the imploding refractive discontinuity observed in Fig.4, and attribute it to gases liberated early in the preconditioner pulse, which are subsequently ionized and imploded by the main current pulse.

One effect of the opening switch on the plasma dynamics obtained with the preconditioner is the reduction in radial expansion velocity seen with the switch (~0.02 cm/μs) relative to that observed without the switch (0.3 to 0.7 cm/μs). The range of values obtained with the switch is comparable to those observed with the preconditioner (0.01 to 0.2 cm/μs).

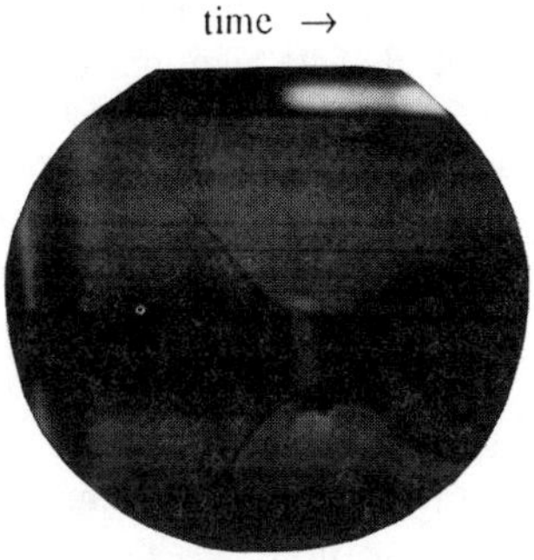

Fig.4. Streaked absorptiongram and main capacitor bank. The 300ns streaks begin 75ns after the main bank

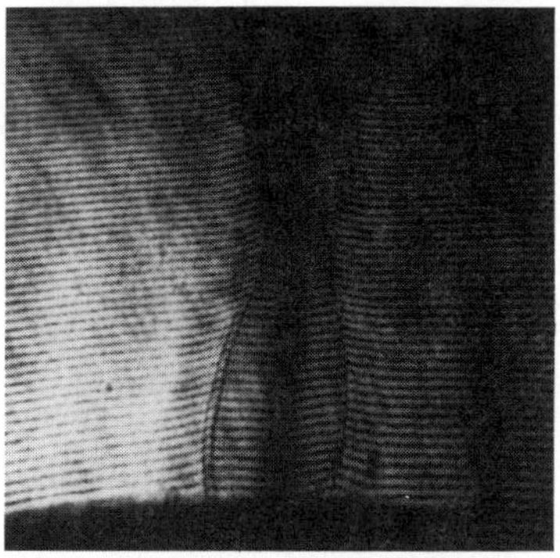

Fig. 5. Interferogram obtained 137ns into current pulse from main bank when preconditioner fired 500ns before main bank.

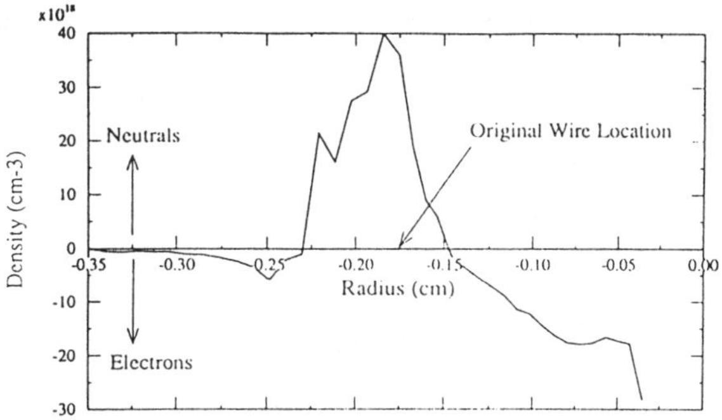

Fig.6. Abel inverted density profile of a two wire load. Origin of coordinates at midplane and original position of one wire at 0.15cm.

The observation of multiphase channels in a single wire discharge raises the question of how these components may separate when magnetic forces between individual wires in an array implode the load. The density profile inferred from an interferogram of a two wire load 270ns into the current pulse is shown in Figure 6. Plasma shells surrounding neutral cores are clearly evident in the density distribution with significant plasma present midway between the original locations of the wires is indicated in the figure.

IV. DISCUSSION OF RESULTS

The measurements indicate that a plasma and neutral vapors, possibly can coexist in these Z-Pinches to various degrees at different times during the discharge. It is important to establish a level of confidence in these measurements before definitive statements can be made about the load composition. This task is complicated by the fact that the index of neutral Al vapor is not known to the authors, and two wavelength interferometry is not available to distinguish between electrons and neutrals. With the Gladstone-Dale value[8] of the index of refraction coefficient (10% of the electron value), neutral rich or electron rich regions imply an average ionization less than or more than 10% respectively. If the density profile in Fig.2 is integrated over the discharge cross section, linear densities of 0.36×10^{17}/cm and 1.06×10^{17}/cm neutrals and electrons, respectively, are obtained. If the coronal plasma is assumed singly ionized, this would account for 45% of the wire mass. Higher degrees of ionization in the coronal plasma would yield a correspondingly smaller degree of load vaporization.

The debris collected after the discharge can also indicate the fraction of the load not vaporized if the load and electrode debris can be distinguished. Ni loads were chosen for this purpose since the electrodes contain both stainless steel and Al. The pure Ni particles observed with the collimated detector implies a load origin for the debris. Although the simultaneous presence of Al and Fe particles characteristic of the electrodes raises the question of whether the Ni particles came directly from the load or were first deposited on the electrodes, it is not clear how they could have coalesced from a vapor coating on the electrodes. The chemical composition of the debris in Table I does not clarify the situation. Both the Al and Ni experiments indicate that less than 5% of the load mass is collected by the Lucite detector even though it should collect about half the load mass if the latter was vaporized into 4π steradians. If the absence of load mass on the collector is not due to plasma flow fields which prevent the vaporized load from depositing on the collector, it is an indication that the load was not totally vaporized.

The plasma dynamics provide additional evidence that the wire is not vaporized and ionized early in the discharge. An upper limit on the linear plasma density can be inferred from the value of the pinch current where the magnetic pressure equals the plasma pressure as give by the Bennett relation, $I^2 = 3.2 \times 10^{-10} N_i(Z+1)T$. A lower limit on the plasma temperature can be obtained from the initial expansion velocity if the plasma is assumed to expand freely with the ion acoustic velocity, $C_s = [\gamma T(Z+1)/M]^{1/2}$. Assuming $\gamma = 3/2$, a value of 10.8 is obtained for $T(Z+1)$. The Bennett relation then yields a value for $N_i = 0.22 N_w$, where N_w is the linear density of atoms in the wire load. A Saha equilibrium model can then be used to infer the average ion charge, $Z \sim 2.05$ and plasma temperature $T = 3.5$ eV. The equilibrium assumption was checked by comparing the average ionization time expected for the temperatures and densities inferred with the duration of the experiment. The value 3.2 ns, obtained using the ionization rate of McWhirter [9], is sufficiently less than the 90 ns expansion phase to justify the assumption.

A self consistency check of these estimates can be made by comparing the plasma transmission at the dye laser wavelength that is expected for these values of ion density and temperature with that observed experimentally. A value for $\eta = e^{-\alpha d}$ of 0.4 is obtained on axis using the absorption coefficient α calculated by J. Apruzese[10]. This value for η is larger than the observed value of 0.15, but is of the correct order of magnitude.

The observation of a discharge channel consisting of a coronal plasma surrounding a core containing neutral Al atoms is consistent with the observation or speculation of others. Studies of x-ray emission from a 20μm diameter Cu wire exploded on a 0.8 TW accelerator led Aranchuk *et al.*[5] to conclude that only 2-7% of the wire mass carried current and took part in the radiation processes. They concluded that the CuK_{α} radiation observed in the experiment, was emitted when plasma electrons struck the remaining mass contained in a solid core. Recently, interferometric techniques were used by Figura *et al.* [6] to study optical fiber loads of various diameters. Their results revealed a coronal plasma surrounding a solid core 147 ns into a current pulse having a rise time of 50ns. Lindemuth[4] attributed the stability of frozen D_2 fiber loads to a solid core which persisted well into the current pulse.

An important consequence of a multiple phase current channel in wire array loads is the potential for the magnetic forces to implode the low mass coronal plasma before the remainder is ionized. This results in a precursor plasma assembling on axis first which softens the implosion of the subsequent load. This is seen in Fig. 6 where part of the coronal plasma has reached the midpoint of the two current carrying wires before neutral vapor around the wires has ionized. These results are consistent with those of Aivozov *et al.*[11], who used a laser technique to detect a precursor plasma arriving on axis prior to the main plasma "liner".

V. CONCLUSIONS

The results show that the current channel that forms around a 25μm Al wire consists of a coronal plasma surrounding a neutral rich core. The initial current flow can liberate absorbed atoms on the wire surface and/or vaporize the wire surface. The neutral cloud is quickly vaporized by the rapidly rising current pulse forming a coronal plasma which shields the core from additional current buildup. Subsequent vaporization and ionization of the core results from electrons, ions and radiation ablating the wire surface. These results indicate that the cold start situation is more complicated than previously considered and that theories must model the vaporization and ionization phases realistically. In particular, sheath formation, which shields the interior core from additional current build up, must be included. Implosion theories must also include azimuthally asymmetric plasma motion that can take into account separation of the coronal plasma from the core.

The composite phase discharge channel can account for many features presently observed in multiple wire load experiments, and suggests that tailoring the current pulse to vaporize and ionize the load prior to implosion may result in better implosion characteristics.

REFERENCES

[1] H.W. Bloomberg, M. Lampe, and D.G. Colombant, J. Appl. Phys. 51, 5277

(1980).
[2] S.M. Zakharov, G.V. Pikuz, and A.I. Samokhin, Sov. J. Plasma Phys. 9, 271 (1983).
[3] T.J. Tucker and R.P. Toth, EBW1: A Computer Code for the Prediction of the Behavior of Electrical Circuits Containing Exploding Wire Elements, Sandia Report SAND-75-0041, 1975.
[4] I.R. Lindemuth, G.H. McCall, and R.A. Nebel, Phys. Rev. Lett 62, 269 (1989).
[5] L.E. Aranchuk, G.S. Bogolyubskii, G.S. Volkov, V.D. Korolev, Yu. V. Kova, V.I. Liksonov, A.A. Lukin, L.B. Nikandrov, O.V. Tel'kovslaya, M.V. Tilupov, A.S. Cherenko, V.Ya. Tsarfin, and V.V. Yankov, Sov. J. Plasma Phys. 12, 765 (1986).
[6] E.S. Figura, G.H. McCall, and A.E. Dangor, Phys. Fluids B3, 2835 (1991).
[7] E.J. Yadlowsky, R.C. Hazelton, J.J. Moschella, and T.B. Settersten,in 9th. Inter. Conf. on High Power Part. Beams, Washington, D.C., May 1992.
[8] C.M. Vest, Holographic Interferometry (John Wiley & Sons, N.Y., 1979), p. 355.
[9] R.W.P. McWhirter, in Plasma Diagnostics Techniques, R.H. Huddlestone and S.L. Leonard, Eds. (Academic Press, N.Y., 1965).
[10] J. Apruzese, Private Communications.
[11] I.K. Aivozov, V.D. Vikarev, G.S. Volkov, L.B. Nikandrov, V.P. Smirnov, and V.Ya.Tsarfin, Sov. J. Plasma Phys. 14, 110 (1988).
[12] P. Sheehey, I. Lindemuth, R. Lovberg, and R. Riley,Jr., in 9th. Inter. Conf. on High Power Part. Beams, Washington D.C., May 1992.

A WIDER VIEW

GENERATION OF LARGE MAGNETIC FIELDS IN A Z-PINCH

R.K.Appartaim* and A.E.Dangor
Blackett Laboratory, Imperial College
London, U.K.

ABSTRACT

Experiments to generate large magnetic fields by flux compression have been performed in a z-pinch. Two z-pinch plasma configurations were investigated: (i) a novel plasma liner developed from a thin cylindrical soap film and (ii) a conventional gas pinch formed in a uniform cylinder of gas. The apparatus, experiments and the results obtained are described in this paper. A comprehensive investigation of the gas pinch shows that an optimum set of conditions exists, for given discharge parameters, for obtaining the highest magnetic field. Using a current of 500 kA with $\tau_{1/4} = 2$ μs, an initial magnetic field of 0.3 T has been compressed to 38 T, a compression ratio of over 120. The scaling of the attainable peak magnetic field with discharge parameters has been studied and compared with simplified models of the compression.

INTRODUCTION

Large magnetic fields (>10 T) are of interest in the laboratory due to applications in applied physics. These include solid state physics research[1], the investigation of the properties of solids when subjected simultaneously to ultra-high pressures in the Mbar range and magnetic fields as high as 500 T[2] and the acceleration of macro-particles to velocities in excess of 10^4 m/s[3]. Certain schemes have also been proposed which may use megagauss magnetic fields (1 gauss=10^{-4} T) for single shot, cyclic, high energy particle acceleration[4,5].

Large fields are usually more conveniently produced in a pulsed fashion due to their associated high electromagnetic energy density. Two approaches are common: (1) High electrical currents may be discharged into an inductive device such as a solenoid[6,7] or a thin wire z-pinch load[8]. Megagauss fields have been produced by high energy capacitor discharges of MA currents into single turn coils. (2) Magnetic flux may be compressed within a conducting shell when the shell is collapsed by the application of an external force. Fields of up to 1500-2500 T were reported by Fowler *et al.*[8] and Smirnov *et al.*[9] using stainless steel or copper cylindrical shells driven by chemical explosives. In an electromagnetic drive scheme the conductor is imploded by the JxB force due to a large current passed through it. Felber *et al.*[9] report a peak field of 4200(±1100) T, achieved with a neon gas-puff driven by a 7.4 MA, 60 ns pulse.

In this paper, we describe some flux compression experiments in a z-pinch and a characterization of the plasma and magnetic fields in terms of the pinch discharge parameters.

*Lab. of Plasma Studies, Cornell Univ., N.Y.

EXPERIMENTAL APPARATUS

The experimental apparatus may be divided into three main sections.

1. The capacitor bank which drives the z-pinch current.
2. The apparatus which generates the seed axial magnetic field.
3. The z-pinch assembly where the z-pinch is housed.

1. Z-Pinch Capacitor Bank

Figure 1 shows a circuit diagram of the capacitor bank which consists of four 12.8 µF, 50 kV capacitors switched by gas-insulated switches. The capacitors are connected in a 2x2 series/parallel configuration to form a double, two-stage Marx generator. The total electrical energy stored is 64 kJ but due to age considerations charging was limited to 40 kV, which corresponds to about 40 kJ. The maximum current is then 850 kA, with a rise time of 2 µs.

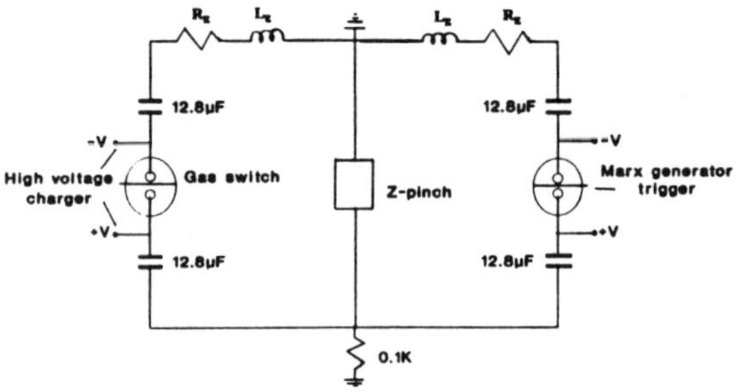

Figure 1 Z-pinch capacitor bank circuit

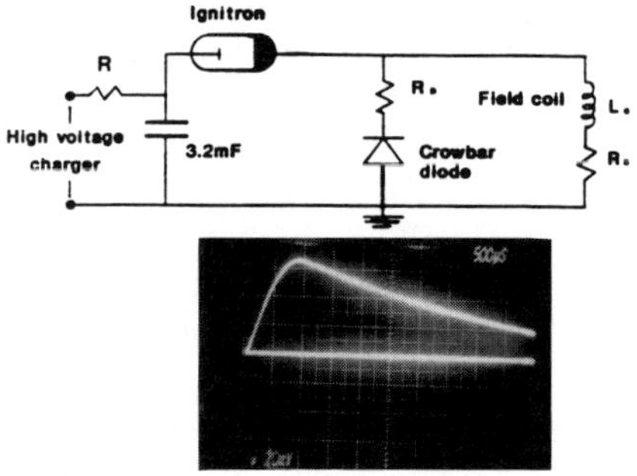

Figure 2 Field generating circuit circuit and oscillogram

2. Axial Magnetic Field Apparatus

The axial magnetic field was produced by energizing two 151/2 turn coils connected in a Helmholtz configuration, with a 0.7 kV, 3.2 mF capacitor bank discharge. A 'crowbar' switch which isolates the coil from the rest of the circuit at the first current maximum gives a longer field duration than is otherwise obtainable with an ignitron bank switch. The field generating circuit and an oscillogram of the field are shown in Figure 2. A uniform field distribution was measured in the z-pinch region by means of a small pick-up coil. The z-pinch was activated at a predetermined time during the field decay. In this way the seed magnetic field was varied.

3. Z-Pinch Assembly

Figures 3 and 4 show respectively, the pinch apparatus developed for the soap film experiments and the compressional gas pinch experiments. Both pinch assemblies were constructed with non-magnetic stainless steel.

DIAGNOSTICS

Three kinds of diagnostics were used to observe the z-pinch and the axial magnetic field. These were as follows.

1. Magnetic induction probes for measuring the plasma current, the capacitor bank current and the axial magnetic field.
2. Time-resolved optical photography which consisted of a visible streak camera and a 1 ns gated image intensifier (Kentech Instr. U.K.) for both end-on and side-on observations.
3. Visible and near UV spectroscopic observations using a Hilger 'medium' quartz spectrograph and a Bentham SMD-3B spectrometer. Time-resolved observations were achieved by coupling the Bentham spectrometer to the streak camera.

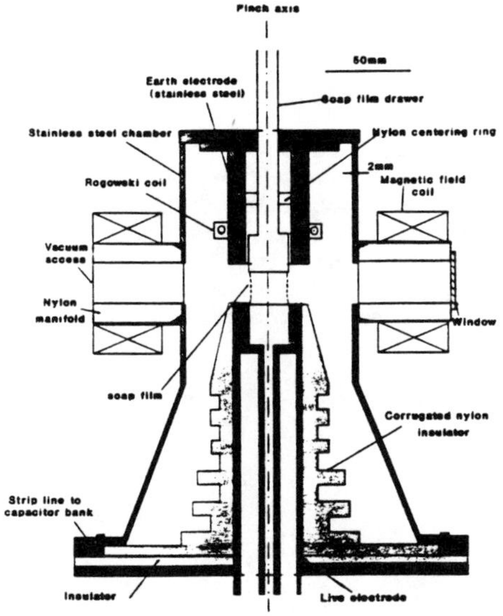

Figure 3 Soap film pinch assembly

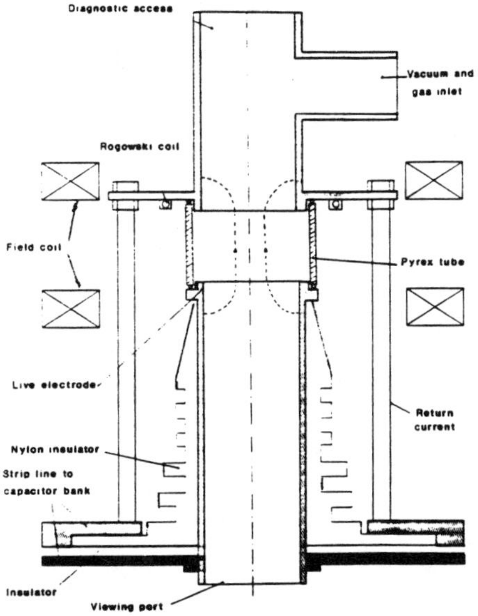

Figure 4 Gas pinch apparatus

MODELLING

The two z-pinch configurations used in the experiments, namely a hollow cylindrical plasma liner and a pinch formed in a uniform cylinder of gas, were investigated with the aid of simplified models of plasma implosion and field compression. The models were based on the equation of motion of a current sheath with a trapped axial magnetic field. This allowed a comparison between the two compression schemes and served as a guide in determining the discharge parameters required to obtain the highest compressions. The liner was assumed to be an annular shell of negligible thickness and fixed mass/length in which the seed flux is trapped. The equation of motion of such a liner, driven by the self magnetic field of the axial current is

$$\frac{d^2y}{dx^2} = \beta^2\left(\frac{1}{y^3} - y\right) - \frac{i^2}{y} \tag{1}$$

where, in normalized units, y is the radius, x is time, i is the pinch current and β is the normalized applied field. For the gas pinch, the mass of the plasma shell increases during the implosion as more particles are swept up. The implosion is modelled with either the snowplough [10] or the shock model [11,12]. The snowplough model, modified to include a trapped magnetic field was particularly useful in predicting the collpase time and gave some idea of the final compression ratio. The shock model supplied the spatial structure of the current sheath. The equation of motion of the pinch using the snowplough model is

$$\frac{d}{dx}\left\{(y^2-1)\frac{dy}{dx}\right\} = \frac{i^2}{y} + \beta^2 y - \frac{\beta^2}{y^3} \qquad (2)$$

Eq. (1) and (2) were solved in conjunction with the equation for the current obtained from a capacitor discharge circuit, written in normalized units as

$$(1-\alpha \ln y)\frac{di}{dx} - i\frac{\lambda}{y}\frac{dy}{dx} - q\alpha^2 = 0 \qquad q = q_o - \int i dx$$

The parameters α and γ describe respectively, the matching of the implosion to the time of peak current and the relative magnitude of the pinch inductance to the total external inductance prior to the collapse. These models were used to investigate the parameter space relevant to the proposed experiments and provided insight into the implosion and compression characteristics. The main result was that a higher field compression was predicted for the liner than for the gas pinch for similar discharge parameters such as mass/length, seed magnetic field and peak discharge current. The variation of the normalized radius at peak compression with a is shown for the two cases in Figure 5.

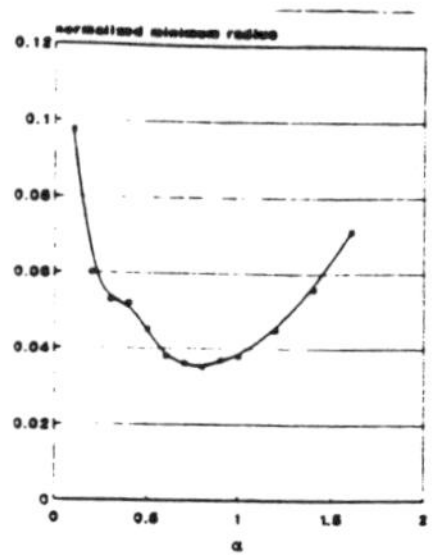

Figure 5(a) Final radius versus a for plasma liner

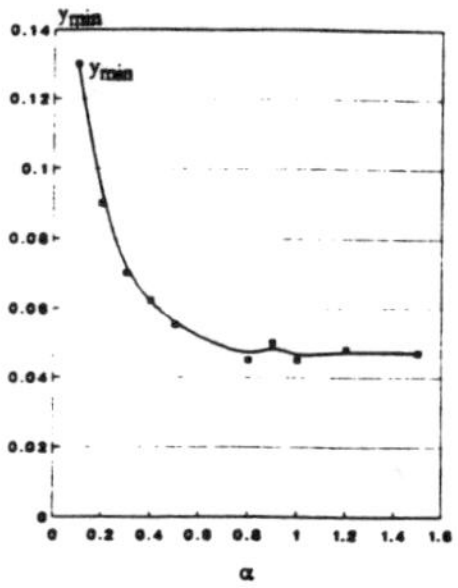

Figure 5(b) Final radius versus a for gas pinch

EXPERIMENTS

1. Soap Film Implosions

The ability of thin soap films to adopt simple geometric shapes when bounded within a fixed framework was exploited as a starting point in the production of thin plasma liners. Such films have good axial and azimuthal uniformity when drawn between the two electrodes shown in Figure 3. The use of soap films for pinch implosions was first proposed by Linhart[13]. A mass per unit length of about 0.5 mg/cm required to match the implosion with the current pulse as predicted by the model discussed above, can be obtained with a film of 30 mm diameter with a thickness of 5000 Å. Starting with such a film usually 12-15 mm high, cylindrical quasi-membrane 'foils' were produced by evacuating the discharge vessel to less than 10^{-3} torr. A peak discharge current of 640 kA was then applied.

The initial results indicated non-uniform breakdown and current conduction, especially in shots in which the applied voltage was not very high, i.e. ≤40 kV. Three side-on 1 ns single frame photographs of the film plasma taken at various times during the discharge are shown in Figure 6(a). Introduction of ultraviolet flash preionisation significantly improved the breakdown and a compression of the film was achieved. A streak photograph of the plasma discharge with the preioniser is shown in Figure 6(b). This shows a well defined plasma boundary and a somewhat uniform collapse. The compression velocity reaches 1.0 x 10^6 cm/s. The observed compression ratios were rather low and thus did not allow successful flux compression experiments. Nevertheless, the scheme of generating plasma liners using soap films and modest equipment has been shown to be feasible. Better preionisation appears to be necessary before these plasma liners can be used for magnetic flux compression.

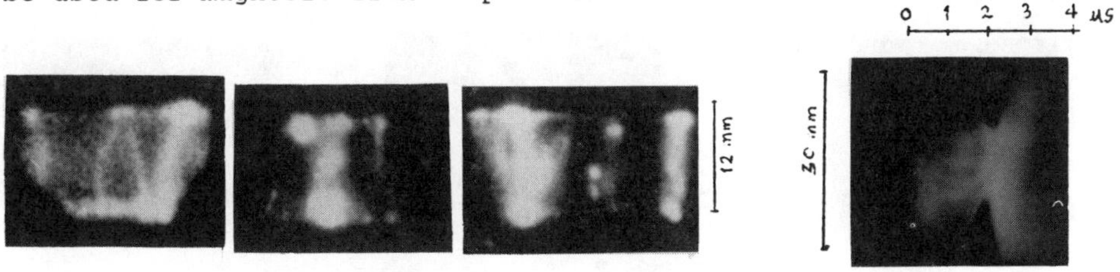

(a) (b)

Figure 6 **(a)** 1 ns frames of soap film plasma
(b) streak photograph of plasma

2. Gas Pinch Experiments

The gas pinch experiments were conducted using the apparatus shown in Figure 4. Two insulating tubes were used: a quartz tube of 30 mm inner diameter (I.D.) and a pyrex tube of 53.5 mm I.D. The gases investigated were hydrogen, deuterium, helium, neon, argon and krypton. The seed magnetic field was varied between 0 and 1 T and a maximum current of 530 kA was used, corresponding to a capacitor energy of 16 kJ. The main observations are summarized as follows.

2.1 Photographic Observations

a. A delay always occurs between current initiation and the time the discharge detaches from the insulating wall. This wall hang-up time, was observed to increase with the atomic mass of gas and filling pressure but decreased with increasing dI/dt. The measured times were a factor of 4-6 lower than the prediction of the neutral depletion theory [14,15] but over an order of magnitude more than predicted by the pressure balance theory[16]. A dependence on the seed field was not established.

b. A well defined piston and shock structure was observed during the implosion (see Figure 7).

c. The plasma was uniform both axially and radially (Figure 8).

d. The time to pinch increased with filling pressure and atomic mass of gas but decreased with current.

e. The seed magnetic field did not significantly influence the early stages of the pinch but increased the pinch radius (Figure 9).

f. The luminosity of the region between the central pinch column and the wall increased after maximum compression.

g. The plasma remained pinched until well after peak current. The seed field increased the time in which the plasma remained pinched (Figure 10).

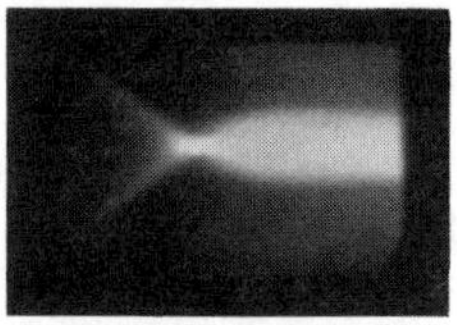

Figure 7

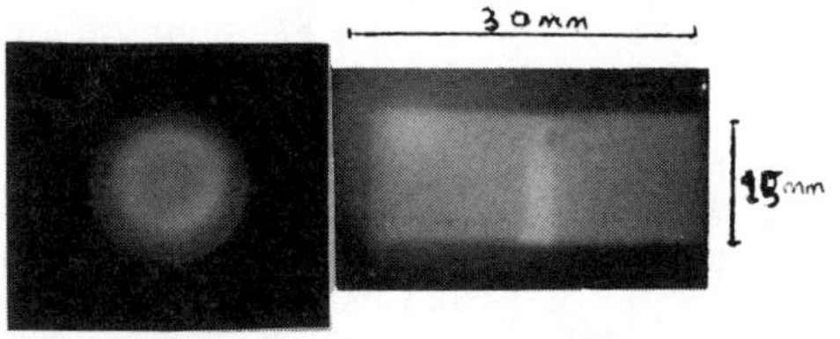

Figure 8

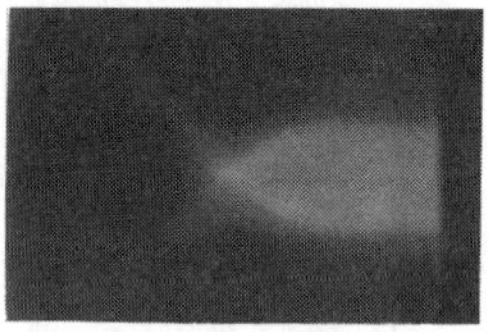

Figure 9

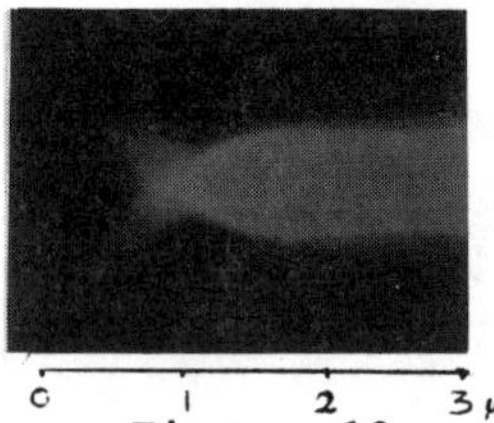

Figure 10

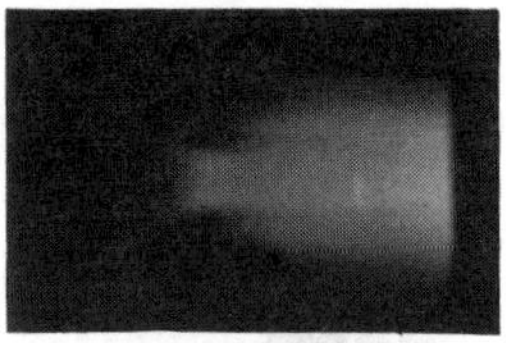

Figure 11

A comparison of the observed implosion history with the models discussed earlier suggested that the discharge current may not be fully confined to a thin sheath behind the dense luminous plasma as usually assumed but more probably distributed over a substantial part of the insulating tube. Infact an interesting plasma structure consisting of a central plasma column and a less bright outer plasma of (Figure 11) may well be due to a tenuous plasma in the "vacuum" region stagnating on the central pinched plasma. Good agreement between the observations and the models could only be achieved when the plasma current was modelled as

$$i(x) = \frac{Z_o(x)}{Z_o(x) + Z_p(x)} i_R(x)$$

where $i_R(x)$ is the normalized current measured by the Rogowski coil, $Z_o(x)$ represents a variable impedance at the insulating tube wall and $Z_p(x)$, the impedance of the plasma assumed to be of the form $Z_p(x)=Z_o(x)(1-y_p)$. y_p and x refer to the normalized pinch radius and time, respectively.

2.2 Magnetic Field Measurements

Using estimates based on flux conservation within the observed plasma radius, field compression was investigated in all the gases used. The best results were obtained in argon and krypton. The axial magnetic field in these gases was then measured in two ways: (1) with a calibrated four-turn, 1 mm diameter pick-up coil wound with 0.25 mm Cu wire and mounted on the pinch axis (2) a single turn diamagnetic loop 73 mm in diameter wound around the pinch tube. Typical oscillograms of the magnetic pick-up coil and the diamagnetic loop are shown in Figure 12. Some features of the field measurements are as follows:

a. The diamagnetic loop and the pick-up coil showed that maximum magnetic field occured at minimum radius, in agreement with streak photograph observations and flux conservation.

b. As the seed field was increased, the width of the magnetic field versus time curve, Bz(t), showed a corresponding increase, in keeping with the plasma radius observed in the corressponding streak photographs.

c. The pick-up coil in general, indicated lower field values. This was attributed to a plasma sheath which was observed to form on the glass jacket of the probe as shown in Figure 13.

The dependence of the peak magnetic field on filling pressure and plasma current is shown in Figures 14 and 15, using the field estimates based on flux conservation. The figures show an optimum set of conditions for the highest field compression. The highest magnetic field, using a peak current of 500 kA was measured at 38 T in 2 torr argon with a seed field of 0.3 T, indicating a compression ratio of 126.

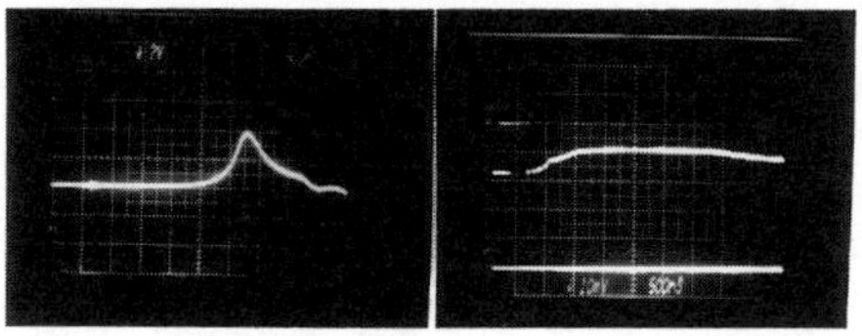

Figure 12

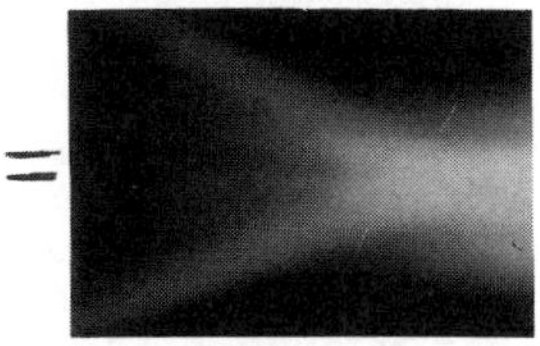

Figure 13

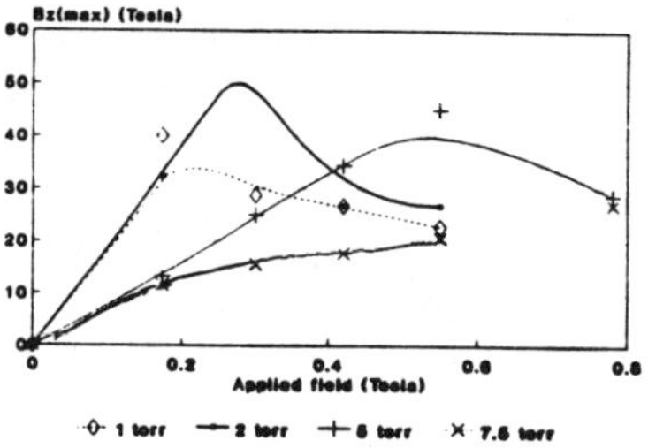

Figure 14

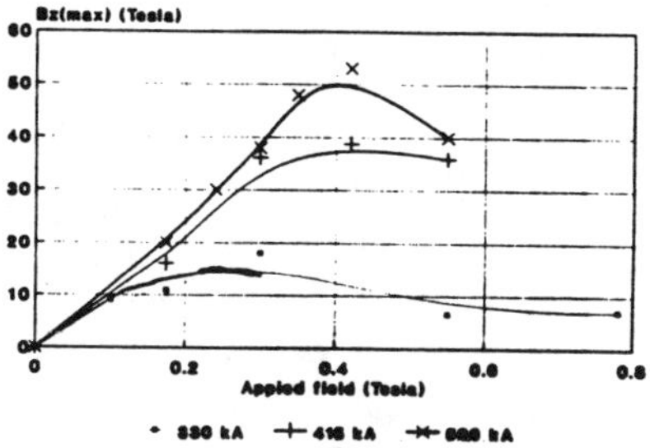

Figure 15

2.3 Spectroscopic Observations

The object of the spectroscopic observations was to investigate the level of ionisation of the plasma and the electron number density during the implosion. Both time-integrated and time-resolved observations were performed. The main results were as follows:

a. Time-integrated spectra were dominated by impurity lines from the wall and the z-pinch electrodes and showed a high level of continuum emission.

b. In the time-resolved observation, the continuum emission was observed to set in before maximum compression and persisted for about 200 ns or more.

c. Spectral line profiles were found to be predominantly Stark broadened.
d. Using the Stark broadening measurements, the electron density was measured to be a few times $10^{18}cm^{-3}$ in the imploding luminous sheath for a 5 torr argon plasma discharge.
e. The electron temperature of the fastest imploding sheath, observed in 1 torr hydrogen, was measured to be 1.2 (±0.2) eV from the line-reversal observed on the H_{α} emission.

CONCLUSION

Large magnetic fields have been produced by flux conservation in a z-pinch formed in a uniform cylinder of gas. A comprehensive investigation of the compression parameters showed that an optimum set of conditions can be found which gives the highest compressed fields. A maximum field of 38 T was achieved using a current of 500 kA. The z-pinch plasma was characterized by optical photography and spectroscopic observations. The observed scaling of the compressed field with applied current indicated that fields in excess of 100 T were possible at full capacitor energy.

ACKNOWLEDGEMENTS

The experiments with the soap film plasma liner were performed in collaboration with Prof. J.G. Linhart, Univ. of Ferrara, Italy. The technical assistance of R. Swain and his team and also of J. Beckwith is greatly appreciated.

REFERENCES

1. Miura, N. High Magnetic Fields in Semiconductors (Ed. Landwehr G.)Springer Series in Solid State Physics, 618, (1989)
2. Pavlovski, A.I. Megagauss Tech. and Pulsed Power Appl., Proc.4th Int. Conference on Megagauss Magnetic Fields, Plenum Press, N.Y. (1986)
3. Degnan, J. H. Megagauss Fields and Power Systems, **65**(Ed. Titov V.M. and Shetsov G.A.) MG-V, Nova Science, N.Y. (1990)
4. Terletskii, La Sov. Phys. J.E.T.P., **32**, 301, (1975)
5. Panasyuk A. I. Megagauss Fields and Power Systems, p169 (Ed. Titov V.M. and Shetsov G.A.) MG-V, Nova Science, N.Y. (1990)
6. Bocharov Yu. N. Megagauss Fields III(Ed. Titov V.M. and Shetsov G.A.) 1984
7. Shearer J.W. J. Appl. Phys. **40**, 4490 (1969)
8. Fowler C.M. *et al* J. Appl. Phys. **31** 588 (1960)
9. Felber F.S. *et al* J. Appl. Phys. **64**(8) 3831 (1988)
10. Rosenbluth M.N. Los Alamos Report, LA - 1850 (1954)
11. Allen J.E. Proc. of Phys. Soc. B **70**, 24 (1957)
12. Potter D. Nucl. Fusion **18** (6) 813 (1978)
13. Linhart, G. J., IEEE Transactions on Plasma Sc. **16**(4) (1988)
14. Braginsky S.I. Plasma Phys. and C.T.R. **2**, 28 (1959)
15. Osovets S.M. Plasma Phys. and C.T.R. **3**, 193 (1959)
16. Wheeler C.B. J. Phys. D. Appl. Phys. **7**, 363 (1974)

CREATION OF AN INTERNATIONAL CENTRE FOR DENSE MAGNETIZED PLASMAS

A. Bernard *

THREE MOTIVES FOR AN INTERNATIONAL CENTRE

It is recognized that instabilities play a large—even predominant—role in plasmas and it is the aim of the Centre to make significant progress in the understanding of the general situation in which current flows in the z-pinch configuration.

From previous experience obtained from several researchers in the past one knows that plasma Focus installations offer the possibility of studying a number of phenomena without the need of the very fast capacitor banks that are required for purely cylindrical z-pinch plasmas. This turns into a financial advantage when one considers the creation of a new Centre. The geometry of the device offers also very good possibilities for the simultaneous use of many diagnostic techniques which (as will appear later) are essential for the understanding of the plasma phenomena. The community of physicists that have joined their efforts to start common activities in the subject have been convinced of the wealth of results that could be investigated in plasma Focus research. But from what has been done in the past twenty years it is also apparent that the methodology of research has to be changed.

Indeed a number of information obtained though refined methods of measurements of a given laboratory unfortunately cannot be referred to and compared with another. So that the main phenomena are found over and over by all laboratories while the understanding of the "fine" structure of the physics has made smaller progress than anticipated.

An essential idea which is at the core of the creation of a new Centre is to establish a common facility that will serve as a reference one: all types of measurements will be able to be installed on an experiment of the Centre so that information obtained from them will be correlated. This lead to the second essential point which is the need to have a collaboration between many countries. None at the present time is able by itself to carry the effort that is required. On the contrary if it is shared into several laboratories there is no difficulty of principle inasmuch as every country—even in an early state of scientific development—can provide the

* Alain Bernard is the Secretary of the Working Group for the creation of the ICDMP. All information concerning the project can be obtained from him at the following address: Direction des Relations Internationales, CEA, 31, rue de la Fédération, 75752 Paris Cedex 15.

study of partial aspects of the physics (for instance of specific x-ray features) and development for instance of one piece of equipment that would in a second phase be moved to the Centre facility. The study can be done even in a small laboratory because the phenomena are essentially the same so that the first studies can be done locally. In the project that will be described in more details in the following sections a group of countries would collaborate on an Equal Status basis so that the total effort both in national laboratories and in the ICDMP would lead to an organization of the research programmes in the field.

At the present time the structure is not formalized and the participation of all interested countries is welcome. Such participation could take the form of having a member in the Board of Direction of the Centre. The Centre would serve as other research Centres for the formation of students and their preparation to PhD diplomas. But to take also into account the specific needs of developing countries a training programme will be organized along the line of an already existing formation created by the Asian African Association for Plasma Training (AAAPT). It is done to give students of physics who have received theoretical knowledge in their country the practical training in technology and general fields of physics which is sometimes lacking.

The third motivation of the Centre is applications and technology development. Plasma Focus experiments are known to be powerful sources of electron and ion beams and correspondingly of x-rays and neutrons (when deuterium is used as filling gas) which have already been used for various applications. Much more can be done along these lines and particularly in developing countries in which it may be advantageous to go direct to a newer and more advanced technology when an older one is already protected by patents and industrial property.

PROJECT DEVELOPMENT

The first idea on a common Centre originated during discussions between H. Bruzzone and V. Gribkov at a topical meeting held in Buenos Aires in January 1990. A Workshop was organized by V. Gribkov in Zvenigorod, near Moscow at the end of the year which decided to set up a Working Group with the purpose of studying the various aspects of the new Centre and of writing the basis document that describes the aims and the characteristics of the proposed Centre. Since the first meeting in Paris in June 1991 the group has been convening regularly every 3-4 months with all its members actually contributing to discussions and text drafts.

The group has worked as an open structure which has regularly increased from the beginning. At this time the following countries are represented: Argentina, Chile, Czech republic, Germany, France, Italy, Japan, Malaysia, Poland, P. R. of China, Romania, Russia, Singapore, United Kingdom, United States.

The contribution of the group will be a self-contained document in which the three main objectives are described and also the means to achieve the goals. A large effort has been spent in the description of what the Centre should be: facilities, diagnostic techniques and the relative weight of training, formation, and research. From this latter consideration one has decided to aim at the following figures for the personnel present in the Centre at the same time:

- 12 permanent physicists and engineers
- 10 technicians
- 6 PhD students
- 6 exchange scientists
- 6 trainees

The main facilities would be:

- one 1 MJ, 60 kV
- two 100 kJ, 40 kV
- several other (fast risetime, Marx, low energy...)
- laser system for interaction
- all relevant diagnostic techniques

Finally a chart of the likely schedule of construction and operation will be added to the proposal. Although some aspects will clearly depend on the siting general features are common. In order that the international cooperation starts at the earliest possible date it is considered during the construction phase that interested laboratories begin a coordinated programme of research in existing facilities.

At this stage some elements are still missing to give an evaluation of the capital costs. Yearly, running operational costs would be in the range of 2.5—3.5 million US dollars.

During the present phase of the project the Group has had the support of several national institutions and of UNESCO. Their help is gratefully acknowledged.

Z-PINCH TRIGGER OF AN AXIAL NUCLEAR DETONATION

J.G. Linhart
Department of Physics, University of Ferrara,
Via Paradiso 12, 44100 Ferrara, Italy

ABSTRACT

An HD implosion of a pellet in spherical geometry cannot ignite advanced fuels. In cylindrical geometry, where the region of the spark is accessible to other means of energy delivery, a detonation can be started and propagated even in pure deuterium. A mechanism which could achieve this, based on the compression of a fibre by a Z-pinch driven by a liner implosion is described.

INTRODUCTION

Experiments have demonstrated that the ignition of inertially confined DT pellets can be achieved[1]. It is also known that explosive thermonuclear devices burn advanced fuels (D_2, Li, Be and their mixtures).

No such certainty applies to magnetic confinement fusion devices. The only fact which is certain so far is that (provided they work) they will be technologically complicated and expensive.

The prospects that ICF in DT pellets will do much better as far as technological feasibility and cost is concerned are not very bright either. This is due to the difficulties of both T breeding in explosive systems and the conversion of neutron energy into heat and electricity. The extrapolation of ICF to advanced fuels appears, therefore, highly desirable.

However, it can be shown that a spherical implosion of a D-pellet (or another advanced fuel pellet) cannot provide the required spark energy[2]. This would be overcome if the implosion provides sufficient energy to spark ignite a central DT core, which in turn ignites the surrounding D mantel[3,4]. However, it can be shown that the trigger energies required are acceptable only if volume compressions in excess of 10^4 can be envisaged. This implies impossibly small tolerances on the spherical symmetry of the target and on the time dependence of the driver.

These constraints are much weaker in cylindrical systems, where the region of the DT trigger (or another easy to ignite trigger[5]) is accessible to heating and compression, independently of the D-fuel (Fig. 1).

The feasibility of such a system depends on the properties of propagation of nuclear detonations in tamped, narrow cylindrical channels.

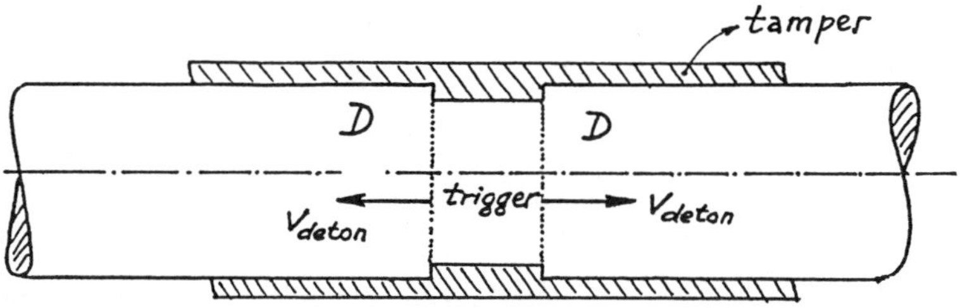

Fig. 1

We shall start by mentioning briefly the ignition and detonation criteria in DT and D in plane geometry, apply them to detonation waves in an inertially confined cylinder of DT and D and finally suggest that a Z-pinch could provide suitable parameters for such a system.

1. DETONATION IN DT AND D

It has been shown[6,7,8] that a plane shock wave in DT becomes a detonation wave provided the surface energy density U is

$$U \geq 200 b^{-1} \qquad (\mathrm{MJ/cm^2}) \tag{1}$$

where $b = 1$ if only the energy of α's is deposited within the shocked region. The mean thickness x of the shock is given by

$$x \cdot \frac{n}{n_s} \geq 1.26 b^{-1} \tag{2}$$

and the optimum mean temperature $T_{op} \sim 2 \cdot 10^8 (°\mathrm{K})$.

As the detonation develops, the energy content of the wave increases, the x and T increase as well as the speed of propagation. The structure of a fully developed detonation in DT corresponds to $x_d \sim 20 \frac{n_s}{n}$, $T \sim 3 \cdot 10^9 (°\mathrm{K})$ if $b = 1$ [9].

Similar considerations for D give

$$U = 31.5 b^{-1} \quad (\mathrm{GJ/cm^2}) \tag{3}$$

$$x \frac{n}{n_s} \geq 110 b^{-1} \tag{4}$$

at $T_{op} \sim 3.5 \cdot 10^8 (°K)$. Such a detonation in DT resp. D can be propagated in a rigid cylinder absorbing no energy from the plasma. The total trigger energy necessary will then be

$$U = \pi a^2 \cdot u \tag{5}$$

where a is the radius of the cylinder. The U_{min} will depend on what is the narrowest channel (i.e., a_{min}) in which a detonation wave can still propagate. There are two main mechanisms which will determine the a_{min}

1. The heat loss from the detonation to the walls of the cylinder.
2. The expansion of the cylinder.

The situation corresponds to a two-dimensional MHD plasma flow coupled with the liberation of nuclear energy and should be treated by two-dimensional numerical simulation. We shall limit ourselves here to only an approximate one-dimensional treatment.

2. HEAT LOSS TO A LINER

In order that a detonation wave in DT can propagate in a cylindrical channel (radius a) the heat loss Q must be inferior to the nuclear output W_F. Considering only the conduction of heat we have for $T = \text{const}$ at $r < a - \delta$, $T = 0$ at $r = a$ and heat conductivity $\kappa = 2 \cdot 10^{-5} T^{5/2}$

$$\dot{Q} = \frac{2\pi}{\lg \frac{a}{a-\delta}} \int_0^T k\,dT \cong 3.6 \frac{a}{\delta}\left(1 + \frac{1}{2}\frac{\delta}{a}\right)^{-1} 10^{-5} T^{7/2} \tag{6}$$

$$\dot{W}_F = 4.4 \cdot 10^{-6} (a - \phi\delta)^2 n^2 < \sigma v > \cdot b \tag{7}$$

where $0 < \phi < 1$. We shall consider only α-energy deposition and consequently $b = 1$. From $\dot{W}_F > \dot{Q}$ follows

$$a > \underbrace{\frac{2.86}{n_s}\left[\xi\left(1 + \frac{1}{2}\xi\right)\right]^{-1/2} (1 - \phi\xi)^{-1}}_{F} \frac{T^{7/4}}{\sqrt{< \sigma v >_{DT}}} \cdot \frac{n_s}{n} \tag{8}$$

where $\xi = \delta / a$. The a_{min} obtains for $\xi \cong (4\phi + 1)^{-1}$, $T \sim 7 \cdot 10^7 (°K)$. A likely $\phi \cong 1/2$ and therefore

$$a_{min} \cong 1.1 \frac{n_s}{n} \tag{9}$$

Using the same model for a D-detonation we obtain

$$\alpha \cong 2.54 \cdot 10^{-23} F \frac{T^{7/4}}{\sqrt{<\sigma v>_{DD}}} \frac{n_s}{n} \qquad (10)$$

where the minimum of $\sqrt{T^{7/2}/<\sigma V>_{DD}}$ is 10^{23} at $T \sim 4 \cdot 10^7 (°K)$ and therefore,

$$\alpha_{\min} \cong 4.8 \frac{n_s}{n} \qquad (11)$$

The $\alpha_{\min}$ corresponds to a point on the $<\sigma v>$ curve tangential to the line $T^{7/2}$ (on the log log scale, Fig. 2). Such a situation is physically unstable. If $\alpha > \alpha_{\min}$ one obtains two points of intersection $A_1 A_2$ where A_2 corresponds to a stable state.

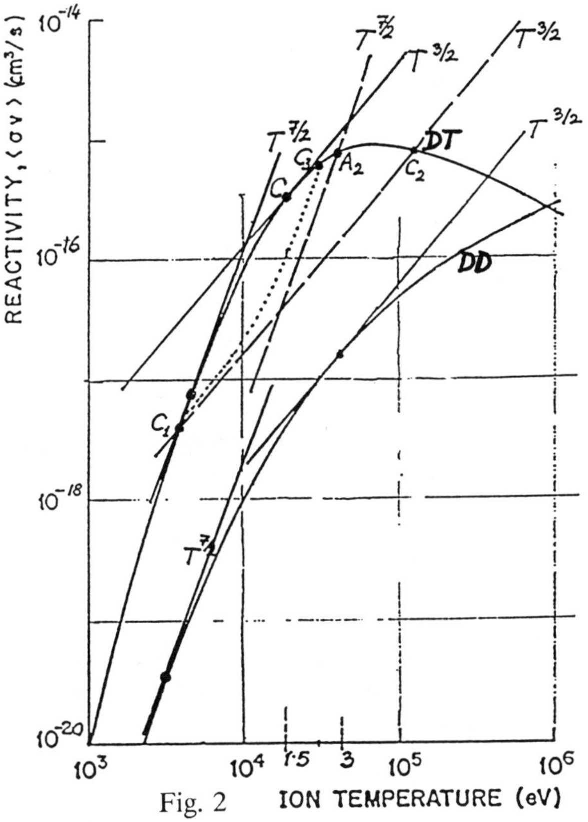

Fig. 2

The propagation at speed v_p of the nuclear burn represents an additional energy-loss

$$\dot{W}_s = 2\pi a^2 \cdot 3nkT \cdot v_p \tag{12}$$

This must be covered by the difference

$$\Delta W = 2\int_0^{x_d} (\dot{W}_F - \dot{Q})dx \tag{13}$$

where x_d is the mean length of the burning nuclear zone. However, x_d is limited by the requirement that the energy ΔW must reach the detonation wave front in time short enough to be transformed into $\dot{W}_s$. The speed of energy transfer in the burning fuel will be the sound speed $v_s = \sqrt{2kT/M}$. We shall take as x_d a length between x (Eqs. 2,4) and the fully developed x_d (putting $x_d - \gamma x$).

In our simplified "slug" model we shall use directly Eqs. (2), (6) and (7) and obtain for DT

$$\pi a^2 3nkTv_p \cong 1.26\frac{n_s}{n}\gamma(a^2 - a_m^2)4.4\cdot 10^{-6}(1-\phi\xi)^2 n^2 <\sigma v>$$

obtaining for $\phi = 1/2$

$$v_p = 0.15\cdot 10^{33}\gamma\frac{<\sigma v>}{T}\left(1 - \frac{a_{\min}^2}{a^2}\right) \tag{14}$$

But v_p must be a shock speed corresponding to the pressure in the burning plasma and, therefore,

$$v_p = \sqrt{\frac{32}{3}\frac{kT}{M}}$$

Substituting this into Eq. (14) we get an equation (for γ, a, T) of the same form as that from which criteria 1 and 2 were derived, only the nuclear input is diminished by a factor $(1-(a_{\min}^2/a^2))\cdot(1-\phi\xi)^2$. It can be written in the form

$$1.3\cdot 10^{-28}\cdot\left[\gamma\cdot\left(1 - \frac{a_{\min}^2}{a^2}\right)\right]^{-1}\cdot T^{3/2} = <\sigma v> \tag{14a}$$

which can be plotted on the log T scale (Fig. 2) giving again two intersection points C_1 and C_2, of which C_2 corresponds to a stable propagation. From the tangential point C we get $T_{opt} \cong 2 \cdot 10^8 (^\circ K)$ and $\gamma_{min}(1-(a^2_{min}/a^2)) = 1.23$. It follows that in case of $a = 2a_{min}$ we get $\gamma_{min} = 1.63$. This is also the measure by which the original trigger (spark) energy (Eq. 1) must be increased. However, as we move from C_1 towards C_2 we encounter again the line $T^{7/2}$ corresponding to the radial heat loss which brings us to the A_2 working point. Obviously the true stable working point will occur at a lower temperature (such as C_3).

The a_{min} (for DT and D_2) can be made smaller by at least an order of magnitude if an insulating magnetic field is present. In order that $\kappa_\perp < \kappa$ one must satisfy

$$\frac{eB}{mc} \cdot \tau_e \gg 1$$

where τ_e is the electron collision time. This can be written as

$$B \gg B_0 = 0.2 \frac{n}{n_s} \cdot \left(\frac{T_0}{T}\right)^{3/2} \qquad \text{(MGauss)} \qquad (15)$$

where $T_0 = 4 \cdot 10^7 (^\circ K)$.

Provided that Eq. (15) is satisfied the critical a will be reduced according to

$$a_B = \frac{\alpha_0 B_0}{B} \cdot a \qquad (16)$$

where $\alpha_0 \sim 10^{8}$.

Ex: $n/n_s = 1000$, $T_s/T = 0.1$ then $B_0 \cong 6.4 (\text{MGauss})$ and in order that $a/a_B = 10$ we must have $B = 640 (\text{MGauss})$.

The pressure p corresponding to a density $n = 1000 n_s$ and $T = 4 \cdot 10^8 (^\circ K)$ is about $5.6 \cdot 10^{12}$ bar, which corresponds to a magnetic pressure of a field $B \sim 10^{10}$ (Gauss), a much higher field than that needed for magnetic thermal insulation. Consequently, if a process exists capable of compressing the D and T to pressures of the order of 10^{12} bar then it will also be able to cope with the compression of the same D-T plasma plus a frozen B field for thermal insulation.

Taking the values of a for $T = 4 \cdot 10^8 (°K)$ in deuterium and assuming $a = a_{min}$ and a ten-fold reduction in a we get $a_B = 81 (\mu m)$ and the U_{min} required for a detonation propagation is approx. $3.4 (MJ)$. This must be provided by the D-T trigger.

We should also consider, in the case of the D-T reaction, the α-particle loss to the walls. This loss will be also cut down by the magnetic field. The criterion for this to be effective is

$$B_a a_B \gg 2.7 \cdot 10^5 \tag{17}$$

If, e.g., $a_B = 30 \mu m$ then $B_a \gg 3.4 \cdot 10^7 (Gauss)$, a field which is of the same order of magnitude as B appearing in Eq. (17). Consequently, if the magnetic field is big enough to cut down the heat conduction, it will also be effective in preventing the α loss to the walls.

3. ENERGY LOSS DUE TO EXPANSION OF THE CHANNEL

This is a complex situation represented in a single form in Fig. 3. The value of a_{min} may correspond to two regimes:

1) The a_{min} must be large enough so that a detonation can propagate.
2) One wishes to have a good nuclear burn in the channel.

Let us consider the first situation.

In order that some energy is left for the propagation of burn in DT we require that

$$\pi a^2 \frac{1}{4} n^2 <\sigma v> Q dt \quad > \quad 2\pi a \cdot p \cdot da \tag{18}$$

where $p = 2nkT$ and obtain

$$a n <\sigma v> Q \quad > \quad 16 kT \cdot \dot{a}$$

The maximum $\dot{a} \sim v_s \cdot \kappa^{-1}$ and, therefore, a safe estimate is

$$a > \frac{22.6 k^{3/2}}{n_s \sqrt{MQ}} \frac{T^{3/2}}{<\sigma v>} \cdot \frac{n_s}{n} \kappa^{-1} \tag{19}$$

The minimum of $T^{3/2} / <\sigma v> \sim 0.6 \cdot 10^{28}$ at $T \sim 2 \cdot 10^8 °K$. The minimum a will be, therefore,

$$a_{\min} > 0.43\kappa^{-1}\frac{n_s}{n} \tag{19a}$$

Similarly, for a burn propagation in D we have

$$a_{\min} > 13.4\kappa^{-1}\frac{n_s}{n}$$

at $T = 3.6 \cdot 10^{8}{}^\circ\mathrm{K}$.

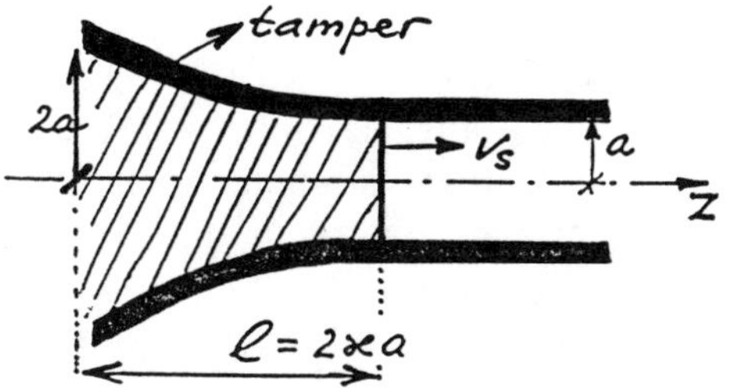

Fig. 3

We can make again the same approximation as we have made for the heat loss and let the shock loss be compensated by $(a^2 - a^2_{\min})$ × nuclear output and find the effective γ giving the increase of the lengths x (Eqs. 2 and 4).

The mere possibility of nuclear burn propagation does not, in the case of a non-rigid tamper, assure a good nuclear burnup. It is clear that an expansion from a to $2a$ will result in an energy loss as well as four-fold decrease in reactivity. In order that this should not affect seriously the nuclear burnup we shall require that

$$\ell \geq \ell_{\mathrm{burn}} \tag{21}$$

where $\ell = 2\kappa \cdot a$, $\kappa = \sqrt{M_L/M_p} \cdot F$, the coefficient F taking into account possible magnetic confinement and liner thickness[10]. Also $\ell_{\mathrm{burn}} = \tau_{\mathrm{burn}} \cdot v_s$, $\tau_{\mathrm{burn}} = 1/n < \sigma v >$ which gives

$$a\kappa > \frac{v_\ell}{n < \sigma v >} = \sqrt{\frac{2k}{M}}\, n_s^{-1} \frac{\sqrt{T}}{< \sigma v >} \cdot \left(\frac{n_s}{n}\right) \tag{22}$$

For DT the $f(T) = \sqrt{T}/ < \sigma v >$ has a minimum at $T \sim 40(\mathrm{keV})$ of about $(1/3)10^{20}$ giving

$$a > 5.3 \cdot \left(\frac{n_s}{n}\right) \kappa^{-1} \qquad (23a)$$

whereas for DD the function $f(T)$ has a shallow minimum at $T > 1000(\mathrm{keV})$. At $100\,\mathrm{kV}$ we have

$$a > 123 \cdot \left(\frac{n_s}{n}\right) \kappa^{-1} \qquad (23b)$$

which decreases by only a factor of 1.6 when $T = 1000(\mathrm{keV})$.

It should be possible to slow down the expansion of the burning plasma by interposing between the liner L and the DT (or DD) plasma a layer of ablator. If this ablator absorbs a sizeable fraction of the heat loss from the burning plasma it will expand and for a time comparable to l/v_t will block the expansion. This is known as the bootstrap compression[11] or an inverse ablation. We shall not discuss this in detail at this point; the influence of this effect l can be included in the coefficient F in κ.

From the comparison of Eqs. (9, 11, 19a, 20, 23a and 23b) it is clear that the largest a_{min} correspond to the requirement of good nuclear burnup. The heat loss (with magnetic insulation) and the expansion loss give smaller a_{min}'s and of about equal order of magnitude.

4. Z-PINCH COMPRESSED BY LINER IMPLOSION

The cylindrical spark and axial detonation wave maybe provided by a liner implosion on a Z-pinch compressing a cylindrical DT and D channel. As the pressure cushion between the Z-pinch and the liner is a B_ϕ field it is possible to convert the kinetic energy of a spherical liner-collapse into a uniform cylindrical compression[5]. Compressing the central channel via the B_ϕ field has a further advantage of softening the liner impact and effecting an almost adiabatic compression. On the other hand, the central portion can be separately heated by Joule's and shock heat deposition. The liner can be driven by a heavy ion beam or perhaps by a plasma drive. It would be difficult and premature to propose a design of a pellet corresponding to this idea. However, some general description of the axial DT spark and the D channel can be attempted, assuming that one aims at $n/n_s = 1000$. Let us consider the situation at a moment before maximum compression (Fig. 4).

The liner L is almost stopped by the B_ϕ field which is squeezed almost entirely into the small central space Ω. In this way the D channel is compressed by a pressure $p_0 \sim \rho_L v_L^2$, whereas the DT section is exposed to $p = p_0(r_L/a)^2$ and is also heated by the current I_e.

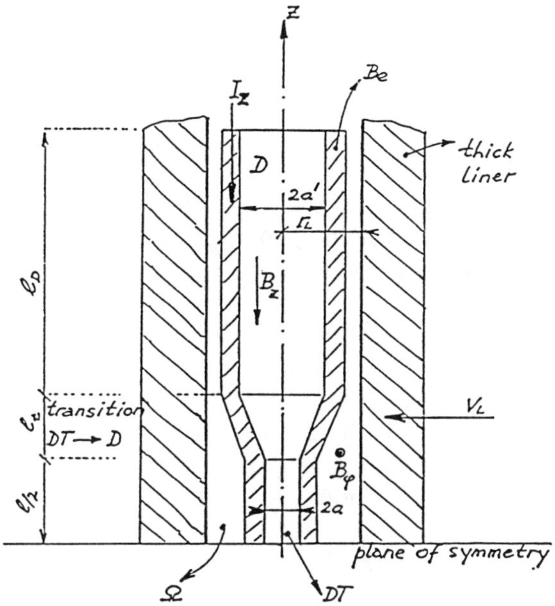

Fig. 4

If the compression of D is adiabatic and $n_D = 1000 n_s$, then $p_0 \cong 1.6 \cdot 10^{15} (\mathrm{dyn/cm})^2$. If $\rho_L = 50 \mathrm{g/cm}^3$ then $v_L = 1.8 \cdot 10^7 (\mathrm{cm/sec})$.

The central pressure p must be high enough to counter the pressure of the heated DT. If the temperature of the DT is, i.e., equal to $1.5 \cdot 10^8 (°\mathrm{K})$ at $n = 1000 n_s$ then $p = 2 \cdot 10^{17} \mathrm{dyn/cm}^2$. It follows that $r_L / a \cong 10$.

From Eq. (23) and for a feasible $\kappa = 5$ we get $a = 11 (\mu\mathrm{m})$. The length l is then approx. $1 (\mathrm{mm})$. The trigger energy will be
$W_0 + \sim 70(1 + 4/3 \ln(r_L/a)) \sim 187 (\mathrm{kJ})$. The nuclear output from the DT trigger (burn 30%) will be about 1(MJ), one half of which streaming into the transition section l_t (where the % of T decreases from 50 to 0). From Eq. (23b) the pure D section will have $a' > 0.12/\kappa$. For a thick liner we can take $\kappa = 20$ and get $a' > 60 (\mu\mathrm{m})$. From our previous considerations (Eqs. 3 and 5), the DT output appears to be enough to propagate a detonation wave in the D channel. The total output of the system $(l_D = 2\mathrm{mm})$ will be about 1 (GJ).

The total kinetic energy of the liner (assuming 20% compression efficiency) will be about 2 MJ, the energy of the driver 5 MJ. The gain is, therefore, approx. 200.

CONCLUSIONS

Axial triggers (sparks) appear to offer several advantages as compared to spherical ones, especially when one intends to ignite advanced fuels. Considering that the use of cylindrical implosions will, most likely, involve the compression of multi-MGauss fields, the associated physical problems are new and complicated and consequently it is impossible, as yet, to make calculations as precise as have been done for the spherical pellets. It is clear that the next step should be numerical simulations of the problems set out in this paper.

REFERENCES

1. Halite-Centurion (press 1988).
2. M.M. Basko, Nucl. Fusion 30, 2443 (1990).
3. S. Skupsky, Nucl. Fusion 18, 843 (1978).
4. G.H. Miley, Laser Interaction and Related Phenomena, Plenum Press, NY (1981).
5. J.G. Linhart, Czech. J. Phys. 42, 985 (1992).
6. J.G. Linhart, Nucl. Fusion 10, 211 (1970).
7. M.S. Chu. Phys. Fluids 15, 413 (1972).
8. M.A. Liberman and A.L. Velikovich, J. Plasma Phys. 31, 381 (1984).
9. R.A. Gross, Phys. of High Energy Density, Proc. E. Fermi School, Varenna 1969, p.245.
10. J.G. Linhart, Proc. DZ Pinch Workshop, Nice 1988 (Acad. Press).
11. J.G. Linhart, Nuovo Cimento D9, 541 (1987).

INERTIAL-ELECTROSTATIC CONFINEMENT NEUTRON/PROTON SOURCE[I]

G.H. Miley, J. Javedani, Y. Yamamoto[*], R. Nebel[†], J. Nadler[‡],
Y. Gu, A. Satsangi, and P. Heck
Fusion Studies Laboratory
University of Illinois at Urbana-Champaign
103 South Goodwin
Urbana, Illinois 61801 USA

ABSTRACT

There is considerable demand in the scientific community for a neutron generator with an output of 10^6-10^8 neutrons/second (n/s) that can be switched on or off, emit fusion neutrons, be self-calibrating, and offer portable operation. An Inertial Electrostatic Confinement (IEC)-based neutron generator is proposed to meet these needs.

In an IEC device, ion beams are injected into a spherical vacuum vessel containing one or more sets of spherical wire grids. A high fusion rate is generated in a dense plasma region created in the center of the innermost grid by intersecting ion beams and the associated potential well structure. Here, we describe a unique configuration, termed the IECGD, where a gaseous discharge in a single-gridded device serves as the ion source. Operation in the newly discovered "Star" discharge mode maximizes the effective grid transmission factor for ions. This configuration, then, provides a simple, rugged, low-cost fusion neutron source, operating in the 10^6 D-D n/s or 10^8 D-T n/s range. The extension to an intense MeV proton source is also possible.

Experimental results for the IECGD are presented, with a discussion of corresponding theoretical studies using a Vlasov-Poisson solver (IXL) and an electrostatic particle-in-cell code (PDS1).

INTRODUCTION

Previous experimental IEC device studies employing ion-gun injectors demonstrated the ability to generate ~10^9 D-T n/s at maximum currents and voltages set by grid-cooling requirements and voltage breakdown limits.[1] However, the goal of present experiments is to develop a simpler, rugged IEC that can dependably generate steady-state yields of ~10^6-10^7 D-D n/s. Neutron sources in this strength range are commonly used to calibrate neutron detectors and diagnostics.[2] For example, the conventional 1-Ci PuBe neutron source emits ≈2×10^6n/s. Such sources are used to calibrate BF_3 proportional neutron counters and neutron-field monitors for personnel safety (e.g., the "snoopy" detector). Sources of this type are also frequently used for

---

[I]Work sponsored by U.S. Department of Energy Contract 9-XG2-Y5958-1
[*]Institute of Atomic Energy, Kyoto University, Uji, Kyoto 611, Japan
[†]Los Alamos National Laboratory, Los Alamos, NM, 87545
[‡]U.S. Department of Energy, Idaho Falls, ID 83401

classroom or student laboratory use. However, there are several disadvantages associated with radioisotope neutron sources: 1) the radioisotopes have relatively short half-lives (^{252}Cf, for example, has a 2.6 year half-life); 2) the neutrons emitted have a broad energy spectrum; and 3) the source must be stored in protective shielding when not in use. The IEC generator represents a potentially attractive alternative that overcomes these problems.

An IEC generator using the D-D reaction does not involve a hazardous radionuclide, thus avoiding licensing problems associated with controlled materials. (Some radioactivity is generated through neutron-activation of structures and by tritium produced by the D-D reaction. However, the total inventory of radioactivity is very low, representing minimal radiological health concerns.) It also appears that an IEC-type generator can be produced for about one-fourth the capital cost of a comparable neutron generator which uses a conventional accelerator-solid target design.

The feature which distinguishes the present IEC design from earlier devices of this type is the use of a single grid to produce a gaseous discharge for ion production. The grid simultaneously serves to extract high energy ions from the discharge. To distinguish this unique design from others, we term it an IECGD (IEC - Gaseous Discharge). This is important, in that IEC designs using other ion sources, such as external ion guns (see Hirsch's design[1]) have the potential for much higher neutron production and have even been proposed for scale-up to a fusion power reactor. The IECGD, on the other hand, avoids the complication of the ion guns and has the goal of using the simplest and cheapest design for neutron production in the 10^6-10^7n/s range.

DESCRIPTION OF DEVICE

Two different IECGDs, designated "A" and "B," were employed in studying the physics of IECs. As stressed earlier, both use a unique grid-discharge design, as opposed to the ion gun-injected IEC used by Hirsch, et al.[1,3,4] (See Fig. 1.) IEC-A uses a 30-cm dia. vacuum vessel made out of 0.48-cm 304 stainless steel. The other vessel, IEC-B, has a 61-cm dia. (Note: While IEC-A and -B were operated as IECGDs, the "GD" was not included in the device designation, since they could also function with other ion sources if desired.) IEC-A is vacuum-pumped with an 80-liters/second (l/s) turbo-pump, backed by a mechanical roughing pump. IEC-B is initially vacuum-pumped by a sorption pump and then switched to a 1000 l/s cryopump. Vacuum base pressures of 10^{-7}Torr are achievable in both devices without baking the chambers. The results reported in this work are for IEC-A, which incorporates nearly the optimum size and design elements for a portable neutron/proton source. IEC-B is used primarily for studying IEC scaling laws and IEC diagnostics.

Different-sized cathode grids were installed to study the effects of grid size. Typically, cathode grids of 7.6-cm and 3.75-cm dia. were used in IEC-A. The grids were made from various sizes of T302/304 stainless steel wire: 0.80 mm, 1.04 mm, and 1.30 mm in diameter. All of the grids had a geometric transparency of ~80-97%, with an estimated <3% deviation from exact sphericity.

Prior to operation, the IECGD is conditioned to remove absorbed gas impurities, using extended glow discharge operation. Then the vessel is pumped down, back-filled to ~5-20 mTorr, and a 10- to 80-kV electric potential is applied to the cathode to initiate the glow discharge. The voltage and pressure are generally related by the traditional Paschen voltage breakdown relation,[5] where the voltage is a function of a pressure-length product. For the IECGD, the "length" in this relation is identified with the distance from the grid to the vessel wall (vs. the grid diameter).[6]

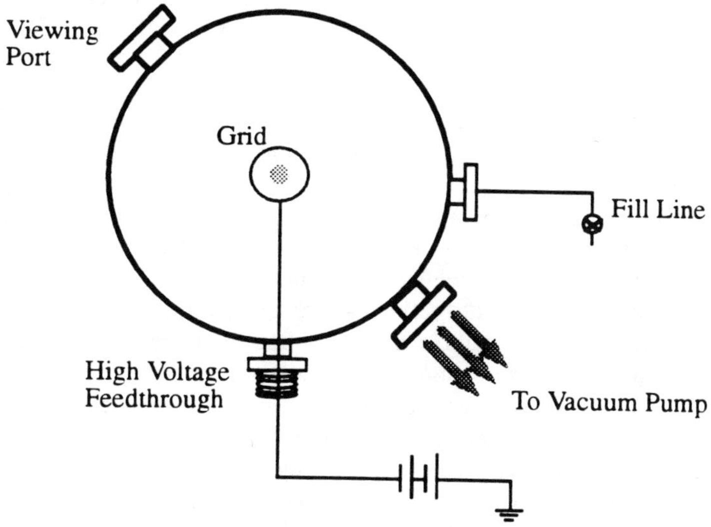

(a)

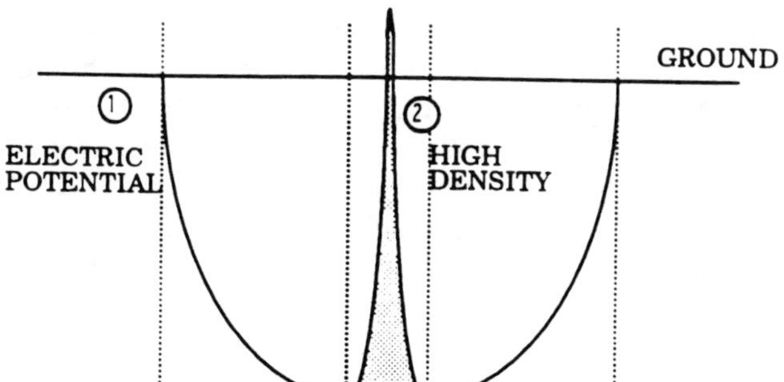

(b)

Figure 1. a) Basic components of an IECGD
b) The associated electric potential

The diagnostics employed included pressure sensors and current and voltage meters on the cathode power supply. A BF_3 proportional counter, placed 40 cm from the IEC chamber and surrounded by a 9 cm-thick polyethylene cylinder (for thermalization of neutrons), was used to measure neutron source strength. This neutron-counting system allows neutron detection over a wide range of fluxes: from high counting rates ($\sim 10^5$ cpm) to low counting rates, ultimately limited by the background rate of ~10 cpm. Interference by induced electronic noise was minimized by electronically shielding the preamp and cables. The detector was calibrated in situ with a 1-Ci PuBe source.

RESULTS

We have categorized glow discharge operation of the IECGD according to three discharge "modes": Star, Central Spot, and Halo mode. These names are quite descriptive of the visual appearances of the light emitted from the discharges, as depicted in Fig. 2. All three modes were found to be easily reproducible and stable. Each is associated with a different potential well structure, hence neutron production rate, for given operating parameters, i.e. cathode current and voltage. As noted in the following description of these modes, each requires a unique combination of operating parameters, i.e. voltage, current, pressure, and grid parameters.

The Star mode was used extensively in the present experiments, and unless otherwise noted, all of the neutron measurements were taken in this mode. It is distinguished by microchannels or "spokes" radiating outward from a center spot. As verified by magnetic deflection experiments, the spokes are primarily composed of ion beams, aligned so that they pass through the center of the openings delineated by the grid-wires. At the center of the cathode grid, where the spokes intersect, a bright spot is formed. This mode is very efficient for neutron production, since the large effective grid transparency allows numerous passes of ions through the center spot before being intercepted by the grid. The Star mode is typically obtained in both IEC-A and IEC-B at lower operating pressures (<10 mTorr) and higher voltages (>30 kV), using a carefully formed grid with good sphericity and high transparency (>95%).

The Central Spot mode is initiated in the same manner as the Star mode, but requires the addition of a larger grid mounted concentric to the cathode-grid. This grid is biased slightly negative (~25% of the cathode bias).[6] In this mode, the spokes vanish and only a central bright spot (hence its name) remains. However, for similar operating voltages and currents, this mode gives only one-third of the neutron output produced by the Star mode. Because of this lower neutron emission rate and the complication of the added grid, this mode was not normally used in neutron source studies.

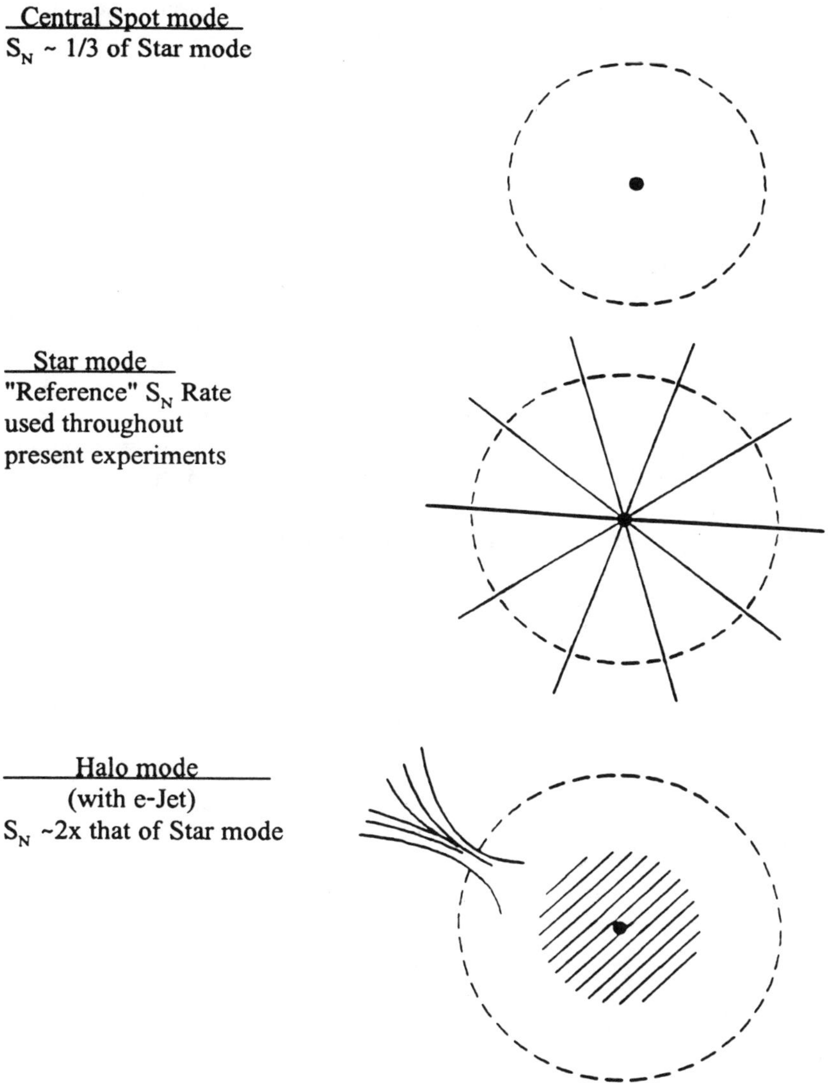

Figure 2. Characterization of the different glow discharge modes. The dashed circle represents the cathode grid; S_N is the neutron source strength.

The final, quite unique mode, the Halo, is initiated in the same manner as the Star mode, but usually at higher pressures, and hence, lower cathode voltages. However, to achieve the Halo, a physical modification of the grid structure is necessary. The size of one of the grid openings is approximately doubled, causing an intense jet of electrons to flow through this larger gap. A bright white, spherical halo is formed concentric to the cathode grid with a bright spot at the center. The Halo has always been accompanied by the electron jet, noted above, which we conclude is a fundamental characteristic of the mode. The mechanism responsible for the Halo mode is under continuing investigation, but it is thought that the Halo structure represents a virtual cathode, resulting from the formation of a double potential well.[7] Because of the asymmetry involved in this mode, we have not proposed it for neutron applications, although it generally offers higher rates of neutron emission than the Star mode. Rather, we are searching for a way to enter this mode while avoiding the jet. This appears to require a double-gridded structure, and thus does not lend itself to as simple a configuration as the Star mode.

Plots of measured neutron source strength versus cathode current in IEC-A are shown in Fig. 3 for different cathode voltages and currents. The neutron yield increases linearly with current, and scales strongly with voltage, in agreement with the modified beam-background model described in Ref. 8. The scaling with voltage roughly corresponds to the variation of the fusion cross section with energy, i.e. approaches an experimental increase in this range of voltages. For the practical applications of interest here, we have limited operation to ~70 kV, so that more complex high-voltage handling equipment is avoided. With this voltage limit, the maximum current is then set by the power heating limits for the grid, as described later. It is seen from Fig. 3 that the target source rate of 10^6 D-D n/s is achieved at 70 kV with a current of ~15 mA.

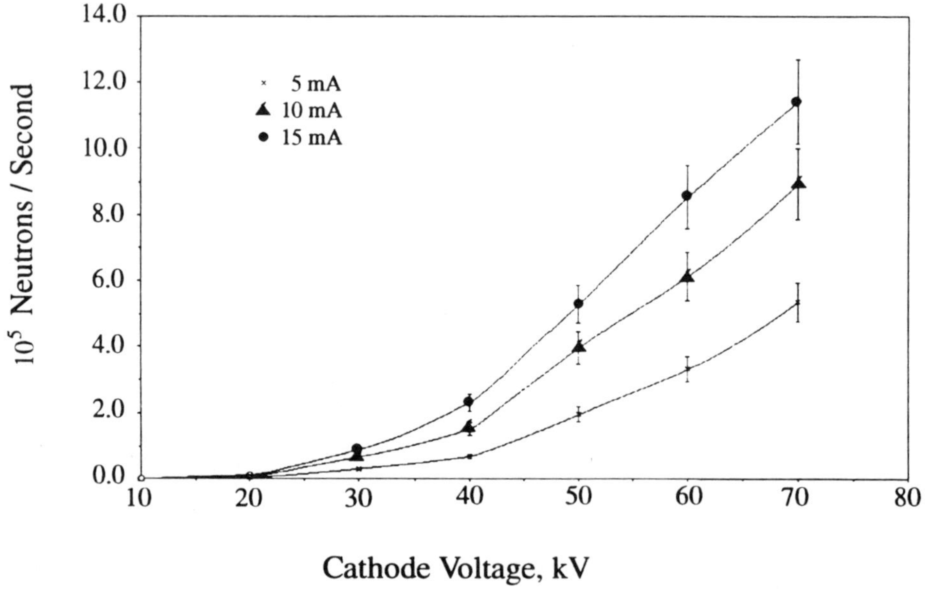

Figure 3. Observed variation of neutron output for IEC-A with voltage for several ion current levels.

The variation of neutron production with grid design is significant, but represents a complicated issue which is not addressed here. All experiments shown in Fig. 3 used the standard IEC-A grid described earlier. Results from experiments with five different grids were presented in Ref. 8. In general, the best grid design should not only maximize the transparency without disturbing microchannel (Star) formation, but also have sufficient structural strength to maintain a near-spherical shape under high-power conditions.

Based on this data, parameters for an optimized version of our device, designed for production of ~10^6 D-D n/s, are summarized in Table I.

Table I. Operating parameters for a steady-state $10^6 \mathrm{sec}^{-1}$ neutron source

PARAMETER	VALUE
Cathode Voltage	70 kV
Cathode Current	15 mA
IEC Diameter	30 cm
Grid Cathode Diameter	3.75 cm
Geometric Grid Transparency	95%
Background D_2 Pressure	8 mTorr

The device shown in the table is restricted to 70 kV for convenience. If higher operating voltages were possible and desirable for a particular application, higher neutron yields would be obtained. Such an extrapolation is not straightforward, however. In the Star mode, the operating voltage is an inverse function of the background pressure.[6,8] Thus, an increase in the operating voltage requires a decrease in the background pressure. However, the neutron yield still increases, since the increase in cross section due to higher ion energies at the higher voltage more than offsets the decrease in background pressure.

The predicted neutron yields for the device in Table I with both D-D and D-T are shown in Fig. 4, as a function of current. The maximum current allowed is determined by the cathode-grid melting limit. If a larger power supply were used, operation to currents over 35 mA should be possible, without changing the grid design or using active cooling of the grid structure. Above that current, active cooling becomes necessary to protect the grid. At yet higher currents, in order to minimize the cooling problem and also to maintain reasonable power supply costs, it appears desirable to switch to a pulsed, capacitor-type power unit. D-T operation, as projected in Fig. 4, would produce higher neutron yields than the D-D reaction, roughly proportional to the ratio of the respective reactions' cross sections. Such operation of the device would be quite similar to the present deuterium case, but a second gas-handling system for the tritium would be necessary. This was not attempted in present experiments, however, because of the restrictive licensing requirements associated with tritium-handling under current Nuclear Regulatory Commission regulations. Some confidence that the predicted results could be obtained is provided, however, from the results reported earlier by Hirsch,[1] who demonstrated this D-T to D-D yield ratio experimentally in an ion gun-injected IEC.

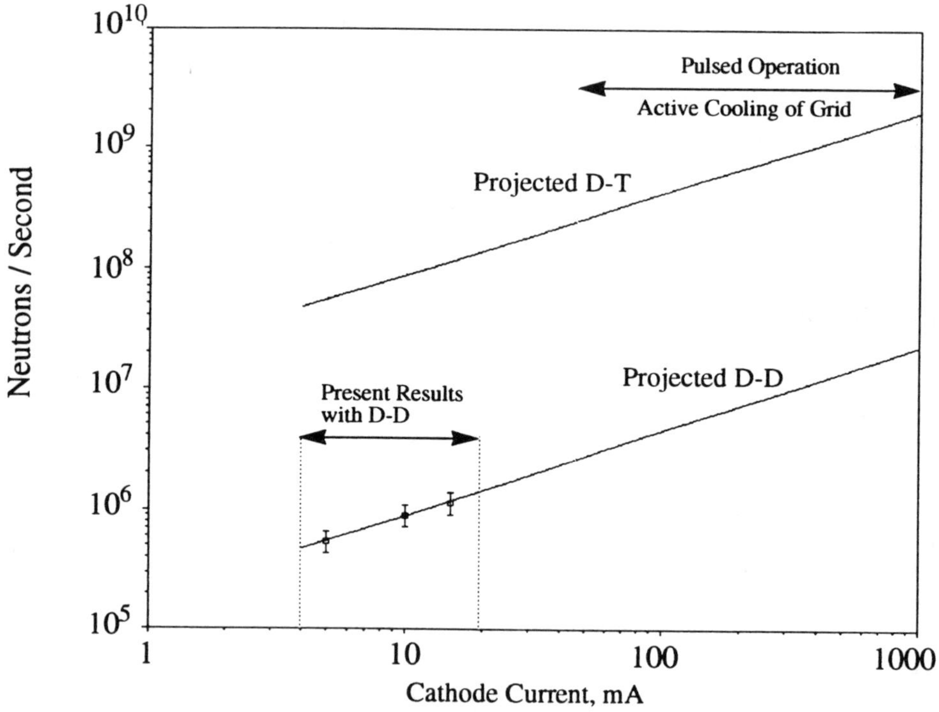

Figure 4. Current scaling projection for D-D and D-T operation of IEC-A, with parameters cf Table I.

Present grids used stainless steel wire construction (vs. high-melting-point materials such as tungsten or molybdenum, which pose more difficult fabrication problems). The current limit can be determined by assuming that, as an upper limit, all of the power from the power supply is deposited into the grid. (This represents a conservative estimate, since only the ions strike the cathode. The electrons, which constitute at least one-half of the charge-carriers, strike the anode or vessel wall, depositing their energy there.) Steel melts at 1800 K. If the maximum temperature before deformation occurs is assumed to be 1500 K, in the present grid design the current is limited to ~35 mA at a 70-kV cathode potential. For safety, the actual current used in present devices was about one-half of this theoretical limit.

If tungsten grids were used, they could operate at twice the temperature limit for steel grids. The radiated power varies as absolute temperature to the fourth power, so the grid could operate at sixteen times the current limit for steel, or ~0.4 amps. Thus, tungsten grids hold promise as a means to further increase the neutron output in future devices.

In summary, operation with a simple device such as studied here (i.e. one using a stainless steel grid employing only radiative cooling and a modest 70-kV power supply) provides a steady-state source strength of ~1.2 x 10^6 D-D n/s or ~1.2 x 10^8 D-T n/s. Much higher yields could be obtained if tungsten grids were used or if the grids were actively cooled, e.g. via water cooling in a tubular design. A reasonable target for such designs would be 10^7 D-D n/s. Still further modifications to achieve higher currents (e.g. a pulsed design with capacitors) could, in principle, achieve even higher neutron yields.

COMPUTER SIMULATIONS

Two computer codes are employed to analyze the IECGD experiments. One, the IXL code, uses a one-dimensional Vlasov description coupled to Poisson's equation.[9,10] While the many assumptions used in this treatment make its accuracy questionable, IXL runs on a personal computer (PC). Thus, its main utility has been for survey studies designed to identify trends. For example, IXL has been used to study ion and electron current requirements for initiating a double potential well. Results from some typical calculations are shown in Fig. 5. As these results indicate, the threshold current predictions for double potential well formation obtained from IXL are higher than those obtained experimentally. Still, the trends predicted appear to be consistent with our experimental results.

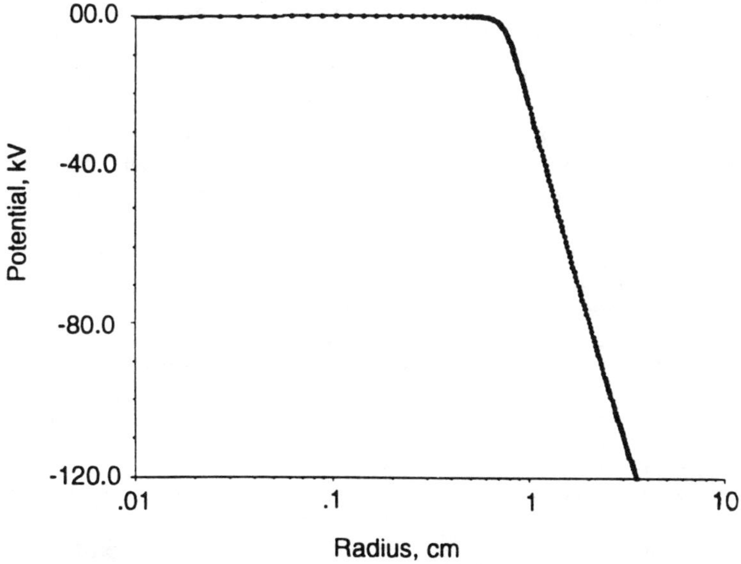

Figure 5. IXL study of threshold for double potential well. Cathode grid located at 3.75-cm radius is held at 120 kV. Ion injection current is 0.6 Amp, and electron current is 1 Amp. Recirculation ratio is 1. Fusion rate is 4.33×10^6 per second. While not clearly visible on this scale, a small drop in potential of ~1 kV occurs in the core region (r < 0.1 cm), signaling the onset of the double potential.

The PDS1 particle-in-cell code allows detailed one-dimensional phase space and three-dimensional velocity space studies, including a variety of atomic physics and secondary electron interactions. This code is a modification of the original PD code, developed at UC-Berkeley for use in plasma chemistry studies.[11] It was recently modified by R. Nebel et al.[12] at Los Alamos National Laboratory (LANL) for use in IECGD studies by adding grid- and fusion-relevant atomic physics effects. The code has several advantages over the IXL code: 1) the PDS1 code also includes the physics of electrons and ions in the outer core region; and 2) the input parameters to the code are based on known experimental parameters, such as pressure, voltage, etc. The code also simulates self-consistent discharge processes, including external electrical circuits. However, the PDS1 run-time, even on a CRAY computer, is protracted. As a result, its main use thus far has been restricted to studies of local conditions, e.g. plasma conditions around grid wires.

Our most recent application of PDS1 has been to study beam flow through the wire grids and the associated effect of secondary electron production. The input parameters for a typical problem of this type are given in Table II, and the corresponding output is illustrated in Fig. 6. In this example, poor ion penetration through the grid is observed. Consequently, as seen from the ion density profile in Fig. 6b and the corresponding ion velocity phase in the core region plotted in Fig. 6c, (compared to the plot of the grid location shown in 6d), relatively few ions reach the core. Further calculations confirm that this is due to space charge effects associated with too few electrons being generated via ion-grid collisions. This result suggests that the neutron production could be increased even further in present devices, if a method of introducing more electrons into the grid region is found. One approach under study is to use coated grids, designed to have a higher secondary electron emission. An alternate approach would be to use a separate electron emitter (e.g. a hot filament) in the outer region.

Table II. Typical conditions for input to PDS1 code shown for steady-state operation of IEC-A with deuterium

PARAMETER	VALUE
Chamber Diameter	30 cm
Grid Cathode Diameter	3.75 cm
Cathode Voltage	35 kV
Cathode Current	25 mA
Geometric Grid Transparency	95%
Background Pressure	9 mTorr

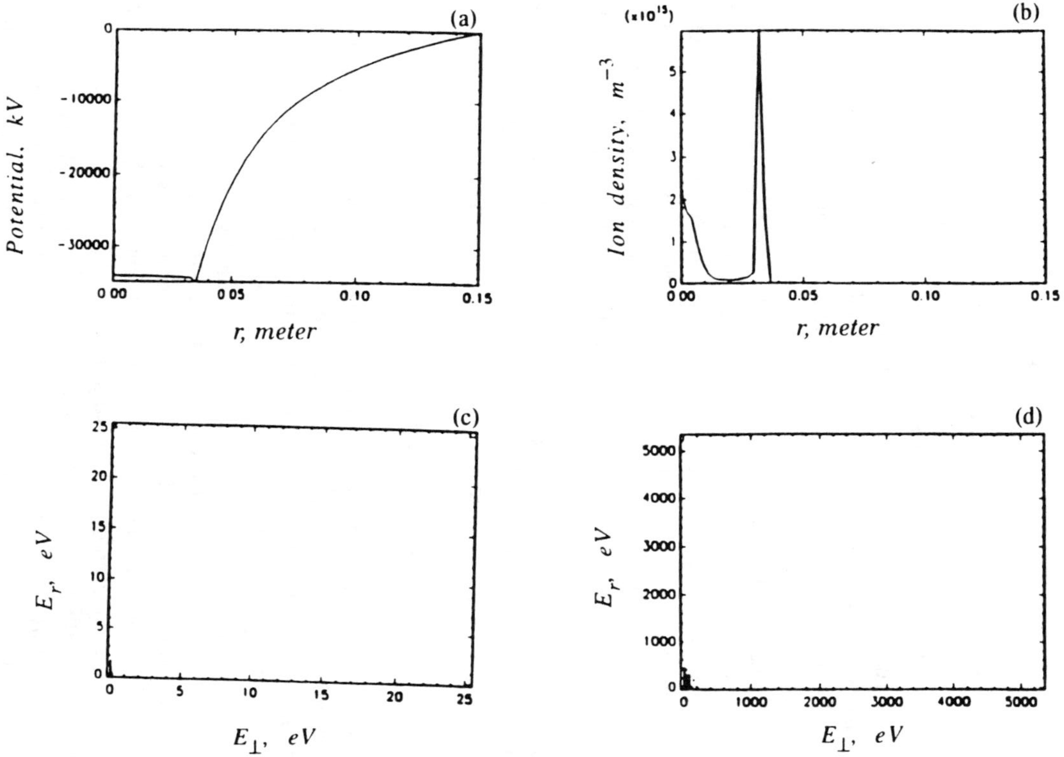

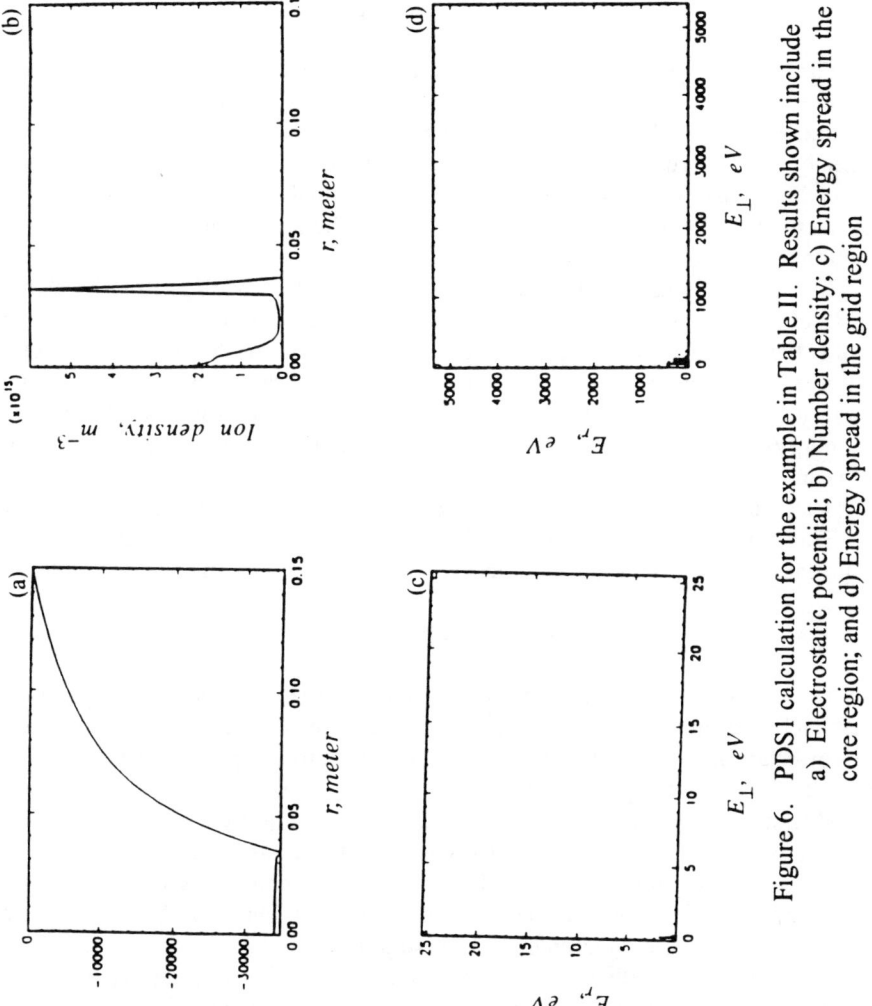

Figure 6. PDS1 calculation for the example in Table II. Results shown include a) Electrostatic potential; b) Number density; c) Energy spread in the core region; and d) Energy spread in the grid region

PROTON SOURCE

The IECGD also potentially represents an important MeV proton source. Indeed, one of the diagnostics, used in the D-D device studies described here for neutron production, was a solid-state detector for measurement of the 3-MeV proton also produced by the D-D reaction. Since the reaction branches for D-D are roughly equally probable, the neutron and proton source strengths are equal. The energetic protons easily escape from the kV confinement fields, making their use for external irradiation applications feasible.

The IECGD is also ideally suited for production of 14-MeV protons by using the D-^{3}He reaction instead of the D-D reaction. This energetic proton exceeds the energy threshold for production of key positron emitters via proton-induced reactions. Operation at ~80 kV would produce a D-^{3}He proton source rate equal to the D-D neutron emission rate at that voltage, since the two reactions have roughly equal cross sections in that energy (i.e. voltage) range. A variety of important applications for such energetic protons can be envisioned, e.g. isotope production for Positron Emission Tomography (P.E.T. scan) facilities. However, many proton applications require quite high proton yields. Thus, for these uses, one of the advanced designs discussed earlier (e.g. pulsed, high-current operation with tungsten grids and active cooling) appears desirable.

FIGURES OF MERIT

The IECGD neutron generator must compete with other modest-strength neutron sources, such as beam-solid target neutron generators (a small accelerator that directs a deuterium beam onto a tritiated solid target[2]) and radioisotopes, such as ^{252}Cf. One significant advantage of an IECGD generator is its modest capital cost. This will be discussed first, followed by a discussion of the other "qualitative" figures of merit.

The cost estimate for a generator based on the design given in Table I totals ~$20k for the basic equipment. If a profit margin and assembly costs are added, a total of ~$35k appears reasonable. By comparison, a typical beam-solid target generator, capable of producing a comparable neutron source strength, sells for $70k. Further, tritiated solid targets typically last only ~100 hours, after which they require replacement at a cost of ~$16k. Grid wires must be replaced periodically in the IECGD, but their cost is negligible in comparison to tritiated targets. [Note: The useful life of the cathode-grids has not been determined; however, none were damaged, under normal operation, during the course of this study (~100 hours operation). Unless the softening temperature is exceeded, the grid should remain intact until weakened by sputtering. However, the build-up of sputtered metallic impurities on insulator surfaces may cause voltage-hold problems well before the grid itself is damaged. The impurity layer can be removed, but the device must be shut down and partly disassembled for this maintenance.]

The present device consumes approximately one liter of deuterium gas per hour of operation, since the vacuum pump is allowed to run continuously and a leak valve is used to admit deuterium to the chamber. At a cost of ~$2.00/l, the D_2 expense for 100 hours' operation is ~$200. However, for the proposed portable device, we envision constructing the vacuum vessel, evacuating it, backfilling it with D_2 to ~8 mTorr, and then hermetically sealing the chamber, eliminating the need for continuous gas consumption. A hermetically sealed IECGD would significantly lower the total cost of the device and also essentially eliminate the cost of D_2. Major savings (~\$6K) would be realized by eliminating the vacuum pump on the portable unit, although a central vacuum fill-station would be required. This station would be used to periodically refill, recondition, and reseal a number of portable devices from various locations. Much additional research needs to be carried out to determine grid useful life and optimum operating parameters under sealed conditions, however, before this design can become a commercial reality.

Another operating cost for the IEC is purchased electricity. Full-power operation of the cathode grid (15 mA at 70 kV), generating 1.2×10^6 n/s, would consume ~1.05 kW. Operation of the device for 100 hours requires 105 kWh of input electricity for the cathode grid, with an additional 140 kWh required to operate the vacuum pumps (if run continuously). At ~\$0.05 per kWh, this represents a negligible expense. A beam-solid target device would have comparable electricity costs. A radioisotope source would not require electricity, but as noted earlier, the replacement cost after decay represents a major recurring expense.

Table III displays additional figures of merit of the IECGD, compared to alternate sources. The solid targets used in the accelerator-type neutron generators pose several problems unique to these devices. Since the targets are tritiated, they fall under radioisotope handling regulations. And as the generator is used, the target's effectiveness typically decreases with time, until it is ultimately exhausted and must be replaced.

Table III. Advantages of an IECGD-Generator

OVER A BEAM-SOLID TARGET GENERATOR:
- ♦ Steady state (vs. pulsed generators)
- ♦ Longer run time without maintenance
- ♦ No tritiated targets

OVER A RADIOISOTOPE:
- ♦ D-D and D-T neutrons possible
- ♦ On/Off capability
- ♦ No significant radioactive inventory
- ♦ Variable source strength
- ♦ Self-calibration with proton detector
- ♦ Constant intensity
- ♦ No storage shielding required
- ♦ Longer life

The IECGD should be quite competitive for applications currently using radioisotopes, e.g. for neutron detector calibration and/or source checks. Numerous laboratories and radiation facilities use neutron diagnostics that require calibration. In some instances, a simple daily neutron source check is required, and in others an accurate calibration is needed. An IECGD could supply the former, while an IECGD coupled with an internal proton detector (to simultaneously measure this D-D source rate by proton detection) would provide a method of easily recalibrating the source with each use.

One unique characteristic of the IECGD is that it provides a fusion neutron source with isotropic emission from virtually a point. By contrast, accelerator-solid target devices give neutron emission strongly biased toward the forward direction. ^{252}Cf sources are isotropic, but the energy distribution differs from that provided by fusion. Consequently, the IECGD appears to be particularly well-suited for special applications requiring an isotropic fusion neutron source. One example is the calibration of neutron tomography diagnostics on a tokamak fusion device, such as TFTR or JET. In that case, the IECGD could be moved about inside the torus to map the responses of the neutron detector array to a "point" source at various locations in the plasma volume.

CONCLUSION

Experiments described here, using a glow discharge version of the IEC, namely the IECGD, have demonstrated its potential for use as a low-level 2.54-MeV neutron source, and eventually as a 14-MeV proton source. A unique feature that enhances the neutron yield is the newly-discovered Star mode. Employing a single-gridded design, this device offers a simple, rugged unit with a relatively low capital cost. In addition to offering steady-state operation, the IECGD is ideally suited for applications requiring a single-point source with isotropic fusion neutron emission.

Ultimately, more advanced versions of this device, operating at lower pressures (non-glow discharge mode), offer the possibility of yet better ion confinement via a double potential well structure in the central core region. Such extensions could provide not only a very high-intensity neutron or proton source, but also the basis for a fusion reactor design.[13,14] Several important extensions of the physics and engineering of the devices described here must be carried out, however, before the feasibility of such advanced IEC devices can be established.

REFERENCES

1. R. Hirsch, J. Appl. Phys., 38, 4522 (1967).
2. C.A. Barnes, et al., Rev. Sci. Instrum., 61, 3151 (1990).
3. P. Farnsworth, U.S. Patent #3,258,402, June 28, 1966; U.S. Patent #3,386,883, June 4, 1968.
4. R. Hirsch, U.S. Patent #3,664,920, May 23, 1972; U.S. Patent #3,530,497, September 22, 1970.
5. A. VonEngel, Electric Plasmas: Their Nature and Uses (Taylor & Francis, Ltd, New York, 1983), pp. 121-127.
6. T.A. Hochberg, "Characterization and Modeling of the Gas Discharge in a SFID Neutron Generator," M.S. Thesis, Department of Nuclear Engineering, Univ. of Illinois, Urbana, IL (1992).

7. Y. Gu, M.S. "Self-Generating Electron-Beam Method for Nuclear Fusion in Spherical Inertial Electrostatic Confinement (SEIC)," Thesis, Department of Nuclear Engineering, Univ. of Illinois, Urbana, IL (1993).
8. J.H. Nadler, G.H. Miley, Y. Gu, and T. Hochberg, Fusion Tech., 21, 1639 (1992).
9. D. Smithe, "XL Reference Manual," Mission Research Corporation Report, MRC/WDC-R-226, Mission Research Corporation, Washington, DC (1990).
10. K. King and R.W. Bussard, "EKXL: A Dynamic Poisson-Solver for Spherically-Convergent Inertial-Electrostatic Confinement Systems," Energy/Matter Conversion Corporation Report, EMC2-1191-03, EMC2, Manassas, VA (1991).
11. C.K. Birdsall, et al., "PDP1, PDC1, PDS1 Plasma Device 1 Dimensional Bounded Electrostatic Codes," Reference Manual Version 2.0, Plasma Theory and Simulation Group, Electronics Research Laboratory, Univ. of California, Berkeley, CA (1990).
12. R.A. Nebel, L. Turner, R.W. Bussard, J. Bates, H.R. Lewis, and G.H. Miley, Bult. APS, 37, 1582 (1992).
13. G.H. Miley, R. Burton, J. Javedani, Y. Yamamoto, et al., "Inertial Electrostatic Confinement as a Power Source for Electric Propulsion," Proceedings of the NASA Vision 21 Conference, NASA Lewis Research Center, Cleveland, OH (March 30-31, 1993).
14. S. Dean, R. Gross, R. Krakowski, et al., "Report to EPRI on the Inertial Electrostatic Fusion Project," Electric Power Research Institute, Palo Alto, CA (Nov. 2, 1992).

A POWERFUL CAPACITOR BANK FOR DENSE Z-PINCH INVESTIGATIONS

A.N.Mokeev, V.V.Prut

Kurchatov Institute, Moscow, Russia

ABSTRACT

A powerful capacitor bank was designed and built to investigate the possibility of increasing the neutron yield up to the 10^{13}-10^{14} DD neutrons per pulse and to generate ultrahigh magnetic fields up to 1GG. The bank consists of 60 independent 20 kJ modules. The module contains four 5 kJ, 40 kV capacitors and a gaseous spark gap. Each module is connected to the load by means of four low inductive coaxial cables, each about 11 m long. The inductance of module and output cables is 100 nH. The spark gaps are fired by a generator wich consists of a capacitor, a water pulse-forming line and an untriggered gaseous spark gap. The bank has a stored energy of 1.2MJ, capacitance of 1.5 mF, inductance of 3 nH, operates at voltages up to 40kV, and provides I ~ 13 MA, 5 mks risetime current pulse to a 4 nH load.

INTRODUCTION

The achievement of ultrahigh magnetic fields is possible either in a plasma or metal Z-pinch. Dense, high-temperature plasmas and compressed substances are important in megagauss physics, both in the experiments and in their application. It has been shown that the neutron yield scales as I^4 and the maximum attainable magnetic field scales as I^5 [1,2]. Therefore a high voltage capacitor bank with low internal inductance and resistance is most suitable and convenient for this purposes.

CAPACITOR BANK DESIGN

Layout

The capacitor bank consists of two racks 10 meters long, 1.2 meter wide, 3 meters high. The ceiling height is 4.5 meters. Each rack contains 30 modules arranged in three storeys. The module [3] contains four capacitors connected by two buses 1.2 meter long and 0.5 meter wide. The pressurised spark gap switch is placed on the high voltage bus and connected to the load by four cables.

Electrical characteristics of the capacitor bank

Maximum stored energy	1.2 MJ
Operating voltage	10 - 40 kV
Total capacitance	1.5 mF
Resistance	1 mΩ
Number of capacitors	240
Number of modules	60
Bank inductance	3 nH
Minimum load inductance	4 nH
Maximum current (calculated)	13 MA
Current risetime	5 mks

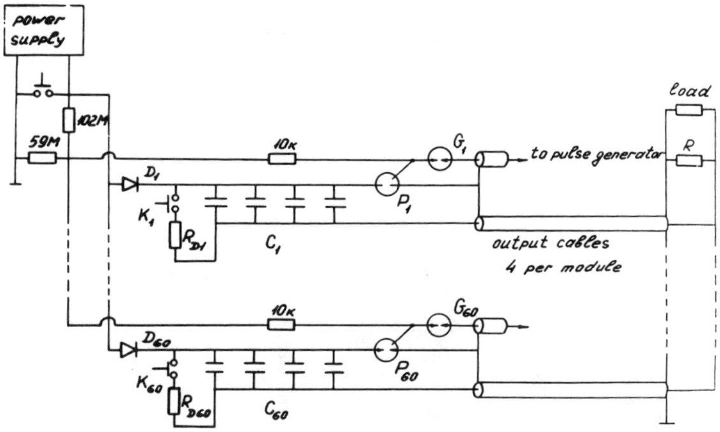

Components

D_1-D_{60}	90 kV, 0.5 A
K_1-K_{60}	grounding relay
R_{D1}-R_{D60}	dump resistors
P_1-P_{60}	spark gaps
G_1-G_{60}	isolation spark gaps
R	10 Ω

Fig.1 Circuit diagram of the capacitor bank

The bank is designed to withstand capacitor, or cable faults without damage to neighboring components. Each module is isolated with a charging diode. The diodes prevent flow of current from one module into another through charging leads, in the event of a capacitor or cable short.

The capacitors were selected for their reliability. These capacitors use paper for insulation and castor oil as impregnant. At 40 kV charge voltage, 500 kA peak current, and 70 percent reversal, these capacitors have expected life exceeding 10^5 discharges (at 90 percent survival).

The cable has inductance of 40 nH per meter and nominal characteristic impedance of 10 Ω. The modules were tested at charge voltage of 45 kV and discharge current of 500 kA[3].

Spark gap design

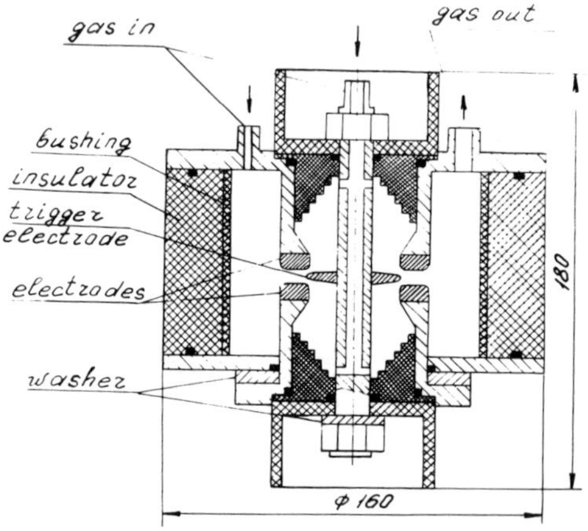

Fig.2 Drawing of the spark gap

Fig.2 shows the pressurised spark gap working at pressures up to 6 atm [4]. Working gas is dry air. The electrodes is made of tungsten alloy. The insulator is made of nylon and the bushing is made of teflon. This switch has been operated without cleaning at current amplitude up to 320 kA for an excess of 10^4shots. Com mutation delay is 15 ns.

The spark gaps are fired by 50 kV, 90 ns long, 10 ns rise time pulse from pulse generator [5]. It consists of 1.5 meter water line charged by 2 mkF, 45 kV capacitor. The line has a charachteristic impedance of 1 Ω. The output cable has a charactheristic impedance of 75 Ω. When the line is charged the gas switch is fired and voltage pulse is trasmitted to the spark gaps.

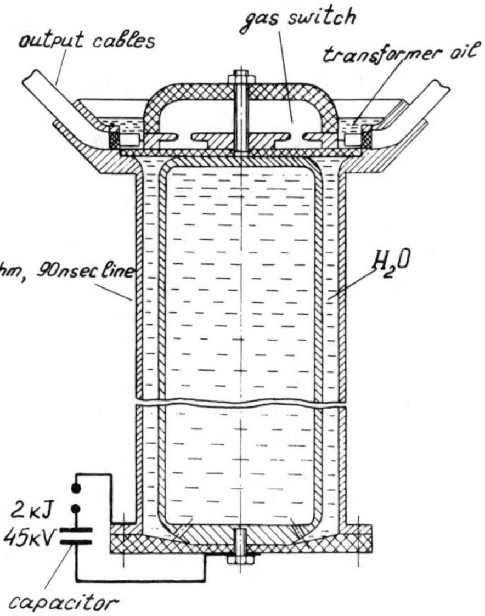

Fig.3. Drawing of the pulse generator.

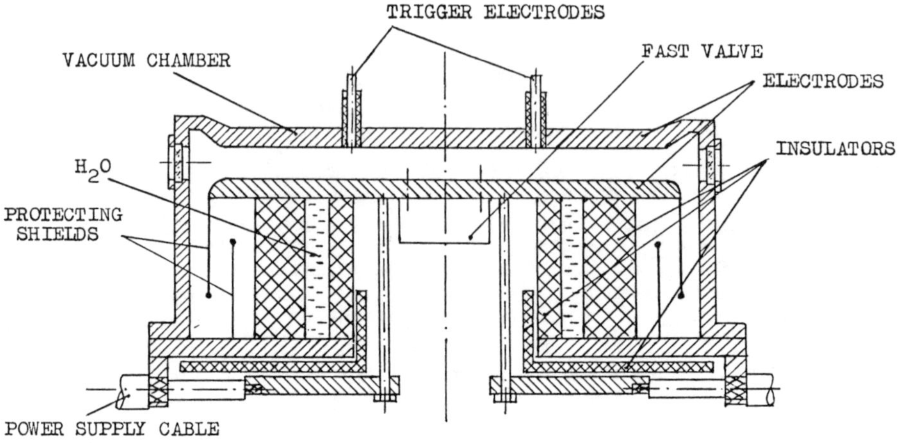

Fig.4 Drawing of the discharge chamber

DISCHARGE CHAMBER

The discharge chamber (Fig.4) includes two steel electrodes: cathode (the lower one), 90 cm in diameter, anode (the upper one), 98 cm in diameter,separated by cylindrical polyethylene insulator, protected with two steel shields against radiation. The electrodes are connected to the bank by 240 coaxial cables.

A ring-line electrodynamic walve is installed at the cathode from which the gas entered the chamber through the inlet orifices at the diameter of 20 cm. The gas pressure in the walve is 10 atm. 48 trigger electrodes are mounted into the anode at the diameter of 38 cm. When the gaseous sheath reaches the trigger electrodes they are fired by 30 kV pulse generator, similar to the one described above. It noticeably improves the discharge characteristics: stability, reproducibility and symmetry.

PERFOMANCE

Bank inductance and resistance have been measured. To model the Z-pinch discharge [6] the capacitor bank has been loaded on the discharge chamber of a gas puff Z-pinch (Fig.4). The elecrodes were connected to each other at the center of the chamber by means of 20 cm diameter copper gasket. The chamber inductance is 55 nH. At 40 kV charge voltage and 5.7 MA peak current the logarithmic decrement is 0.5.

REFERENCES

1.V.V.Prut, in: Proceeding of the Firth International Conference on Megagauss Field Generation and Related Topics", (Ed. V.M.Titov and G.A.Shvetsov), New York: Nova Science Publishers, 1989, p. 73.
2.R.B.Spielman, T.W.Hussey, D.L.Hanson, and S.F.Lopez, ibid, p.43.
3.A.N.Mokeev, V.V.Prut, PTE, No.6, 1986, p.153, (in Russian).
4.A.I.Zemskov, A.N.Mokeev, V.V.Prut, PTE, No.1, 1984, p.133. (in Russian).
5.V.V.Matveev,V.V.Prut,P.A.Suslov,S.A.Shibaev, PTE,No.1, 1983, p.90, (in Russian).
6.V.V.Matveev,A.N.Mokeev, and V.V.Prut, ref. [1], p.73.

INERTIAL CONFINEMENT FUSION IN A Z-θ PINCH

H. U. Rahman and P. Ney
Institute of Geophysics and Planetary Physics
University of California, Riverside, California 92521

F. J. Wessel and N. Rostoker
Department of Physics
University of California, Irvine, California 92717

ABSTRACT

The implosion of a dense θ pinch plasma driven by an annular Z-pinch is considered. A cryogenic fiber is coaxially located within an annular gas jet Z-pinch. The imploding Z-pinch traps an applied axial magnetic field and conserves flux. The axial magnetic field increases due to compression of the outer pinch and attains values in excess of tens of megagauss with field risetimes of order, $dB/dt = 10MG/ns$, with an order of magnitude shorter rise time than the Z-pinch current. An azimuthal θ-current is induced on the fiber surface with a similar rise time; which could not otherwise be achieved in a simple fiber Z-pinch due to the large inductance initially presented by the small diameter axial plasma. The implosion transfers Z-pinch kinetic energy to the magnetic field and then to the θ-pinch. The final plasma pressure of the θ- pinch exceeds the magnetic pressure so that the θ-pinch then expands–the only confinement is inertial. To reach break-even and beyond, with reasonable Z-pinch currents, we increase the radiative losses by seeding the discharge with high Z impurities. Seeding a cryogenic Hydrogen-like fiber with high Z-impurities would substantially reduce the preheat in the θ-pinch, over a non-seeded discharge, and increase the final density by radiative collapse. Thus $n\tau \sim 10^{14}$ cm^{-3}-s could be reached with a confinement time of 0.01 ns and density of 10^{25}cm^{-3} predicted in this paper.

Radiative collapse was first proposed by Pease and Braginskii[1] for hydrogenic Z-pinches in a Bennett type equilibrium. Ohmic heating decreases with current I like $(1/I)$ and Bremsstrahlung increases like I. If the current exceeds the value $I_c = 1.66$ MA, the radiation dominates. If the current and magnetic pressure are maintained the plasma would be radiation cooled enhancing collapse. The collapse would cease when the absorption length becomes comparable to the plasma radius. For higher Z-plasmas, such as krypton[2] or for hydrogen seeded with impurities,[3] the required current can be much less.

There is experimental evidence for radiative collapse in Z-pinches,[4] vacuum sparks[5], and a plasma focus[6] seeded with high Z-impurities. In the latter case there are indications that in regions of micron scale dimensions densities of the order of $10^{24} cm^{-3}$, and perhaps higher, were attained comparable to those achieved in high energy laser driven implosions.

Recently there have been Z-pinch experiments of a solid fiber.[7–11] The initial load inductance of such discharges is large and the current rise time is at least 10's of ns. The plasma expands until the magnetic pressure is sufficiently large to contain the material pressure and the plasma eventually goes unstable without achieving high density.

To address these limitations we consider the Z-θ pinch[12] illustrated in Fig. 1. In this hybrid scheme an annular Z-pinch plasma implodes onto an axial B_z-magnetic field compressing it to multi- megagauss fields with an order of magnitude shorter rise time than the Z-current. Coaxial with the Z-pinch is a solid fiber or straw which breaks down and ionizes, due to the inductive electric field resulting from the large dB/dt, and forms a θ-pinch. The rise time of the current in the θ-pinch is a few ns and the combined Z-θ configuration is more stable than a simple Z-pinch. Key features of the Z-θ pinch have been studied on a small-scale pulsed power source at UC-Irvine[12–15] and on higher power machines in the USSR[16] and at Sandia National Laboratories.[17] The measured axial fields were 1.6 MG (UCI), 2.5 MG (USSR) and 40 MG (Sandia).

The simplest model for the coupled Z-θ pinch is developed assuming a thin shell annular implosion of a normal Z-pinch with an entrained axial magnetic field. On axis we assume a uniform column of high-density plasma, which thereby avoids the computational problem of "cold start".

The dynamic equation for the outer Z pinch is,

$$\frac{d^2R}{dt^2} = -\frac{{I_{max}}^2}{100M{r_0}^2}\sin^2\left(\frac{\pi t}{2t_0}\right)\frac{1}{R} - \frac{{B_0}^2}{4M}R\left(1 - \frac{1}{R^4}\right), \tag{1}$$

where R is the normalized radius of the annulus, $R = r/r_0$ and r_0 is the initial radius, M is the mass per unit length in μgm/cm, I_{max} is the maximum current in MA, B_0 is the applied axial field in MG, t and t_0 are the time and quarter period risetime of the current in ns.

The nonlinear differential equation (1) was numerically solved as a function of time for a sinusoidal current profile $I = I_{max}\sin(\pi t/2t_0)$ with the following initial conditions: $I_{max} = 10$ MA, $B_0 = 20$ kG, $t_0 = 50$ ns, $R_0 = 4$ cm , $M = 38$ μgm/cm. The results are displayed in Figure 2 for the normalized Z-current, outer radius, and compressed magnetic field as a function of time. These initial values provide the optimum compression for a given current profile. We see that the rise time of $B_z(t)$ is about a factor of 100 less than the rise time of the current $I_z(t)$ and the maximum value could be above 100 MG. A magnetic field of such large value and fast risetime will most certainly initiate a θ- pinch on the surface of the coaxial fiber, justifying the cold start assumption included above.

The purpose of the magnetic field is to accelerate plasma and compress it to high density and temperature. Magnetic confinement involving pressure balance equilibrium for a sufficient time for break-even is not a possibility. To illustrate this we consider some simple analytic estimates, based upon Ohmic heating using Spitzer resistivity. The induced θ-current is determined from Ampere's law where the profile for B_z is obtained by solving the magnetic diffusion equation. For cylindrical geometry the current profile is thus,

$$J_\theta = (\frac{cB_{z0}}{2\,\pi\,a})\,\{\frac{J_1(\xi_1\tau)}{J_1(\xi_1)}\}\,e^{-\xi_1^2\,\tau}, \tag{2}$$

where $\tau = Dt/a^2$, $D = c^2\eta/4\pi$, J_1 is the first order Bessel function, $\xi_1^2 = 5.76$, and η is the resistivity of the newly formed.θ-pinch plasma. Knowing the value of J_θ, one can estimate the ohmic heating as,

$$W = \int \eta\,J_\theta{}^2\,2\,\pi\,r\,dr\,dt = \frac{a^2\,B_{z0}^2}{2\,\xi_1^2} \cong \frac{B_{z0}^2}{8\,\pi}\,\text{area}\,(\text{erg cm}^{-1}). \tag{3}$$

If we ignore radiative energy losses for the moment, then all the available magnetic energy will be utilized for heating purposes. The final temperature then depends on the final value of the magnetic field at the surface B_{zo}, provided there is enough time available for field diffusion into the fiber plasma. The diffusion time can be estimated as,

$$\tau = (1/1.56)\,(a^2/D) = (4\,\pi\,a^2)/(56\,c^2\,\eta), \tag{4}$$

which is inversely proportional to the resistivity. For classical Spitzer resistivity the time is on the order of a few ns that is comparable to the early stages of the rise time of the magnetic field. During this phase, when Ohmic heating is effective, a relationship between temperature, density and magnetic field can be obtained from the pressure balance equation,

$$T(eV) = 2.5\,[\frac{B^2(MG)}{n(10^{22}\,cm^{-3})}] \tag{5}$$

If the fiber is maintained at a constant radius then a fusion temperature of 10 keV by Ohmic heating alone will require a compressed magnetic field of 100 MG maintained for at least 4 ns in order to satisfy the Lawson criterion. These conditions are impossible to achieve with present pulse-power technology.

Indeed the concept of pressure balance as in a Bennett equilibrium is quite inappropriate. It is apparent from the calculations of B_z illustrated in Fig. 2 that the peak magnetic field is maintained for much less than a nanosecond. The inner θ-pinch must therefore be treated dynamically and not as a quasi-equilibrium as is customary in the usual treatment of magnetic confinement for pinches. There is an essential physical difference in that compressional heating is much more important than Ohmic heating.

The dynamic equation for the (inner) θ-pinch is,

$$\frac{d^2a}{dt^2} = \frac{B_0^2}{a_0^2 n_0}\left[-4\frac{a}{R^4} + 800\left(\frac{n_0 T_0}{B_0^2}\right)\frac{T}{a} - 2a\left(1 - \frac{1}{a^4}\right)\right], \tag{6}$$

and energy equation is,

$$\frac{dT}{dt} = -2(\gamma - 1)\frac{T}{a}\frac{da}{dt} + \frac{(\gamma - 1)\times 10^{-22}}{n_0 T_0}a^2\left[\frac{dW_{ohm}}{dt} - \frac{dW_{rad}}{dt} + \frac{dW_\alpha}{dt}\right], \tag{7}$$

where the radius a and temperature T are normalized to their respective initial values a_0 and T_0, and n_0 is the initial density in units of 10^{22} cm^{-3}. Heating is achieved by a combination of Ohmic, adiabatic compression, and α- particle interaction. Energy losses arise from bremsstrahlung and cyclotron radiation which are included in dW_{rad}/dt. The heating and energy loss rates can be written as:

$$\frac{dW_{ohm}}{dt} = 2\times 10^7\left(\frac{B_0^2}{a_0^2 T_0^{1.5}}\right)\frac{1}{a^2 R^4 T^{1.5}}, \tag{8}$$

$$\frac{dW_{rad}}{dt} = 2.37\times 10^{20}(n_0^2\sqrt{T_0})\frac{\sqrt{T}}{a^4} + 2.8\times 10^{16}(n_0 T_0 B_0^2)\frac{T}{a^2 R^4}, \tag{9}$$

$$\frac{dW_\alpha}{dt} = 3.22\times 10^{24}\left(\frac{n_0^2}{T_0^{2/3}}\right)\left(\frac{1}{a^2} - 2n_\alpha\right)^2 \exp\left[-\left(\frac{1.99}{T_0^{1/3}}\right)\frac{1}{T^{1/3}}\right], \tag{10}$$

where

$$\frac{dn_\alpha}{dt} = 9.2\left(\frac{n_0}{T_0^{2/3}}\right)\frac{1}{a^4 T^{2/3}}\exp\left[-\left(\frac{1.99}{T_0^{1/3}}\right)\frac{1}{T^{1/3}}\right] - 2\frac{n_\alpha}{a}\frac{da}{dt}, \tag{11}$$

γ=5/3 or 2 for collisional or collisionless plasma, respectively. The initial conditions for the inner pinch are $a_0 = 200$ μm radius, $n_0 = 10^{22}$ cm^{-3}, and temperature $T_0 = 20$ eV.

Figure 3 displays the results for the plasma radius, a, density, n, temperature, T, burn fraction, f and gain G as a function of time during the final stages of collapse for a D-T θ-pinch plasma when ohmic heating, adiabatic heating, α-particle heating, and radiative losses are included. The collapse of theta pinch plasma occurs on a time scale of 10-20 pico seconds (density rise time) as compared to the magnetic field compression time of few hundred pico seconds. The maximum parameters of the plasma are, $n = 2.5 \times 10^{25}$ cm^{-3}, $T = 40$ keV, $f \approx 10\%$, $G > 3$. This correspond to a neutron yield of 10^{18} cm^{-1}. Figure 4 displays different heating and radiation losses included in our model. This figure shows that Ohmic heating starts quite early and the radiation losses (mainly Bremsstrahlung) are not enough to keep the plasma cold. The main mechanism for heating and compression leading to the fusion conditions is adiabatic compression that appears at later stages of the collapse and is more effective if the plasma temperature at early stages is low. Ohmic heating in the early stages is a pre-heating that limits the final density that can be achieved by compression. Seeding the plasma with high Z-impurities will increase the radiative loss in this early stage of Ohmic heating to keep the plasma at low temperature ie., to avoid pre-heating at the early stages of compression. The present calculations do not include this possibility. Once high density and high temperature are reached that allows fusion to take place the α-particles can add further energy due to their trapping in the strong compressed magnetic field. This assumption may be valid because the gyroradius for 3.5 MeV α-particles is about 20 μm. Even if these α-particles are not trapped one can get a gain of more than 2 with a burn fraction of about 8%.

Considering stability, the Z-θ pinch is the most stable of a general class of pinch configurations. Rayleigh-Taylor instabilities usually occur during the acceleration phase of the Z-pinch. Stability of Z-θ pinch during the acceleration phase depends on the details of the magnetic field distribution which involves magnetic diffusion. In this configuration of the Z-θ pinch, part of the B_θ magnetic field that accelerates the plasma, and part of the axial B_Z field is trapped in the annular plasma. This combination of B_θ and B_Z components produces a sheared magnetic field that can stabilize the Rayleigh-Taylor instability. The UCI experiments have already shown this improved stability during the acceleration and final compression phase of a Z-pinch onto a quartz fiber optic by means of Mach-Zender interferometry and Schlieren imaging[4,12,15].

After the acceleration phase, peak compression is attained which may be followed by one or more bounces. The duration of peak compression is of the order of 10 picoseconds. Conven-tional MHD stability analysis assumes an equilibrium characterized by distributions $B_z(r)$, $B_\theta(r)$ and $P(r)$ where P is pressure. Then perturbations about this stationary state are investigated. In the present investigation the duration of peak compression is of the same order of magnitude as the characteristic MHD growth rate so that the usual MHD instabilities would not have time to grow. The only instabilities that are significant are dynamic instabilities, such as the Rayleigh-Taylor instability. This instability is ubiquitous in inertial confinement fusion of all types. It is generally controlled by accelerating a shell over a small number of shell thicknesses and design- ing the shell so that initial perturbations are small. In the case of the $Z - \theta$ pinch there is the additional possibility of stabilization by magnetic shear.

The present calculations show that a $Z - \theta$ pinch with a current of 10 MA can achieve $n\tau \sim 2.5 \times 10^{14}$ and a temperature of 40 keV. We speculate that higher values of nt can be obtained by reducing the ohmic preheating so that the inner plasma can be compressed to higher density. This involves seeding the plasma so that radiation cooling keeps the temperature from rising (radiation cooling dominates over the increased ohmic heating from high Z-impurities). This effect has been called radiative collapse and has been demonstrated experimentally.[4,5,6] The present calculations have been repeated without ohmic heating, in which case a much higher value of density was obtained. A proper calculation must incorporate atomic physics of the impurity radiation, as well as the modification of the Spitzer resistivity.

1. S. I. Braginskii in Plasma Physics and the Problem of Controlled Thermonuclear Fusion, Pergamon Press (1961), p. 135; R. S. Pease, Proc. Phys. Soc. London B **70**, 11 (1957).

2. J. P. Apruzese and P. C. Kepple, Proc. 2nd Int. Conf. on High Density Pinches, Laguna Beach, April 26-29 (1989).

3. J. W. Shearer, Phys. Fluids **19**, 1426 (1976).

4. J. Bailey, Y. Ettinger, A. Fisher and N. Rostoker, Appl. Phys. Lett. **40**, 460 (1982); J. Appl. Phys. **60**, 1939 (1986).

5. E. Ya Gol'ts et al., Phys. Lett. A **115**, 114 (1986); Phys. Lett. A **119**, 359 (1987).

6. K. N. Koshelev et al., J. Phys. D - Appl. Phys. **21**, 1827 (1988) (printed in U.K.).

7. J. D. Sethian, A. E. Robsen, K. A. Gerber and A. W. DeSilva, proc. 2nd Int. Conf. on High Density Pinches; AIP Conf. Proc. (1989), N. R. Pereria, J. Davis, and N. Rostoker, Editors, p. 308.

8. Op. Cit, J. Hammel, p. 303. and J. E. Hammel, D. W. Scudder, Proc. 14th European Conf. on Controlled Fusion and Plasma Physics (EPS, Petit- Lancy, 1987) Part 2, 450.

9. E. S. Figura, G. H. McCall, and A. E. Dangor, Phys. Fluids **B3**, 2835(1991).

10. F. C. Young, S. J. Stephanakis, and D. Mosher, J. Appl. Phys. **48**, 3642(1977).

11. W. Kies, G. Decker, M. Malzig, and C. Van Calker, et. al., J. Appl. Phys. **70**, 7261(1991).

12. H. U. Rahman, P. Ney, F. J. Wessel, A. Fisher and N. Rostoker, Proc. 2nd Int. Conf. on High Density Pinches, Laguna Beach, April 26-29 (1989), AIP Conf. Proc., p. 195.

13. F. J. Wessel, F. S. Felber, N. C. Wild, H. U. Rahman, E. Ruden and A. Fisher, Appl. Phys. Lett. **48**, 1119(1986).

14. F. J. Wessel, N. C. Wild, A.Fisher, H. U. Rahman, A. Ron and F. S. Felber, Rev. Sci. Instrum. **57**, 2247(1986)

15. F. S. Felber, F. J. Wessel, N. C. Wild, H. U. Rahman, A. Fisher, C. M. Fowler, M. A. Liberman, A. L. Velikovich, J. Appl. Phys., **64**, 3831-3845(1988).

16. N. A. Ratakhin, S. A. Sorokin, S. A. Chaikovsky, Proc. Seventh Int. Conf. on High Power Particle Beams,Vol. II, 1204, (1988), and S. A. Sorokin and S. A. Chaikovsky, AIP Conf. Proc. (1989), N. R. Pereria, J. Davis, and N. Rostoker, Editors, p. 345.

17. F. S. Felber, M. M. Malley, F. J. Wessel, M. K. Matzen, M. A. Palmer, R. B. Spielman, M. A. Liberman and A. L. Velikovich, Phys. Fluids, **31**, 2053, (1988).

18. B. B. Kadomtsev, in Reviews of plasma physics, p. 153-199, Consultant Bureau, New York (1966).

19. M. N. Rosenbluth, Report LA 2030, Los Alamos, New Mexico (1959).

20. M. N. Rosenbluth, Proc. 2nd Int. Conf. on Peaceful Uses of Atomic Energy, Geneva (1958).

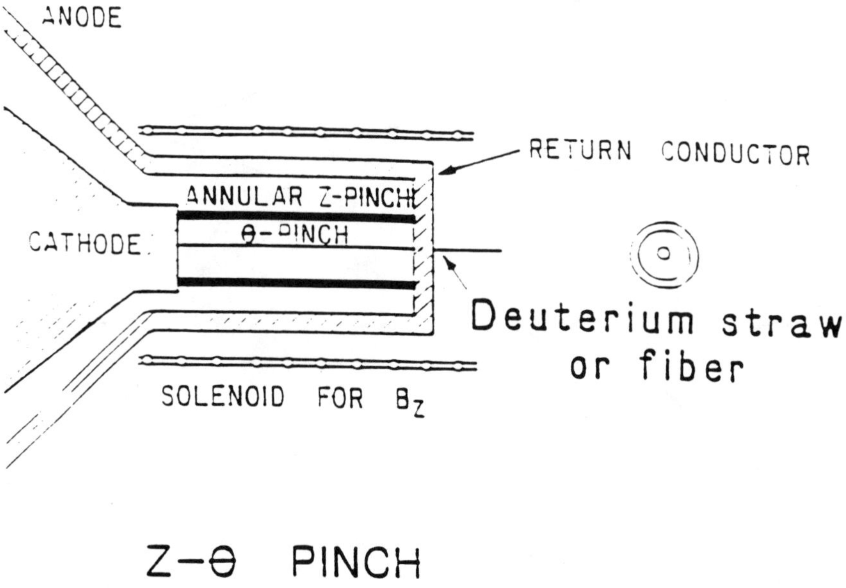

Fig.1. Schematic configuration of Z-θ pinch.

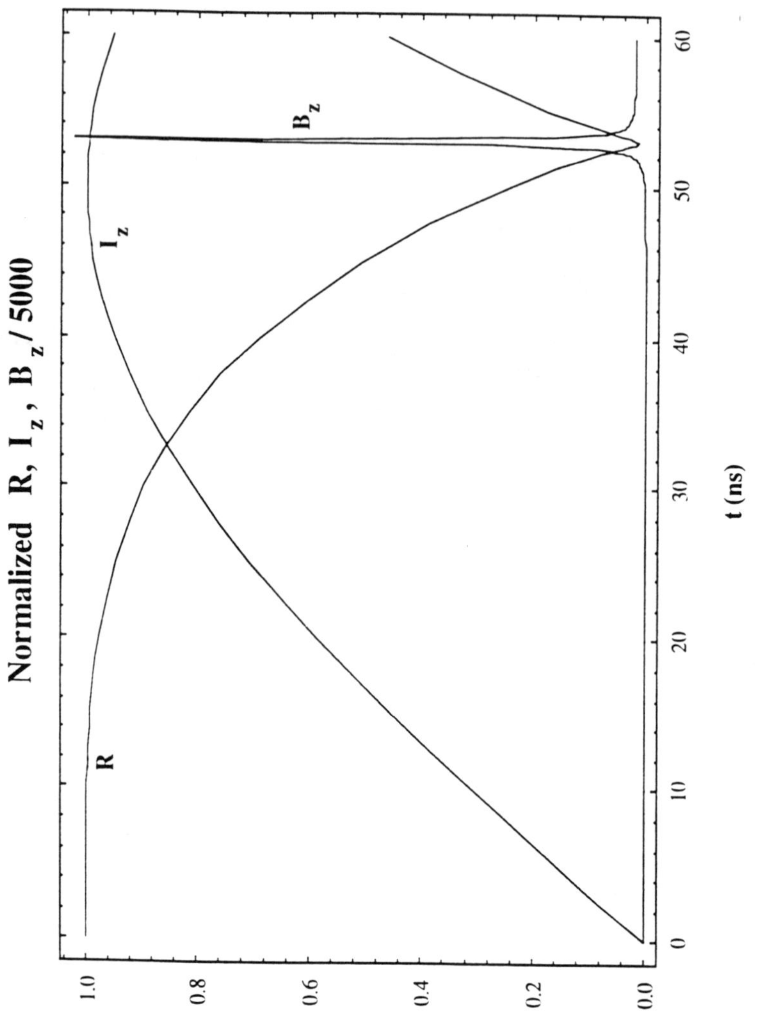

Fig.2 Normalized plots of axial current, I_z, outer radius of Z-pinch and the compressed magnetic field , B_z that is further divided by 5000.

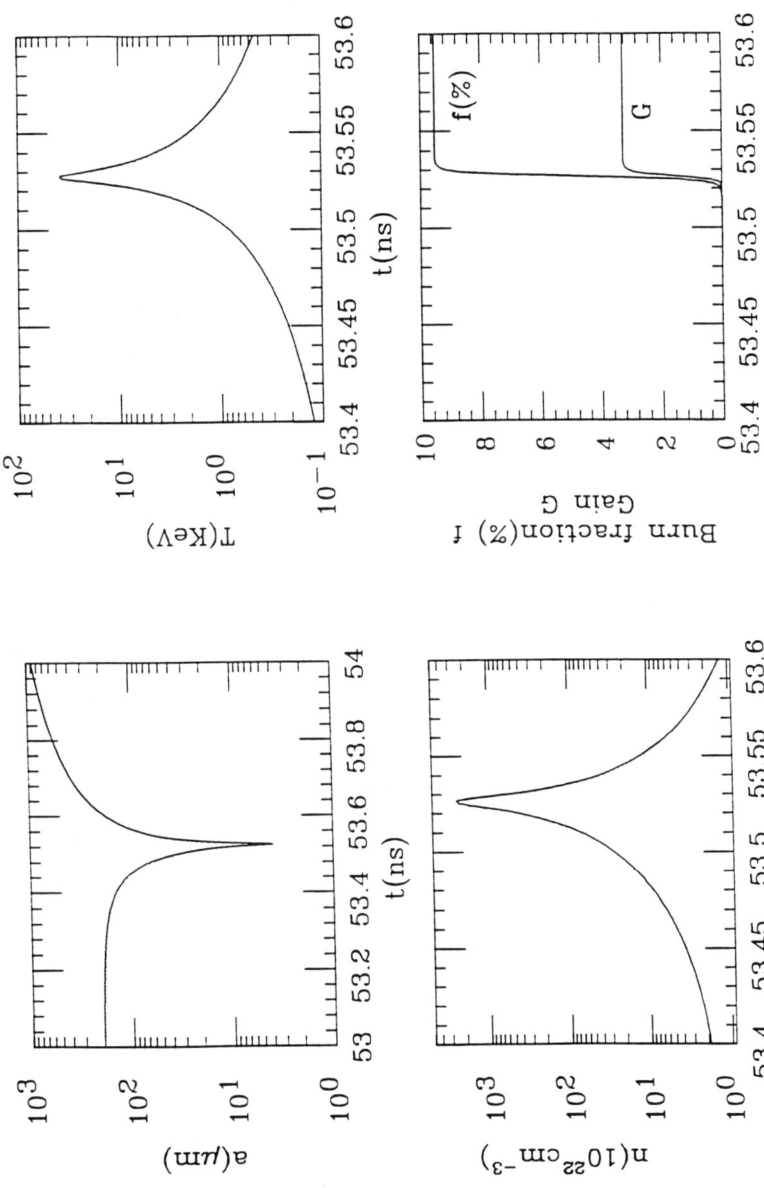

Fig.3 Dynamics of the θ-pinch displaying the plots of the radius, a, temperature, T, density n, burn fraction f nad the gain, G versus time during last stages of the compression.

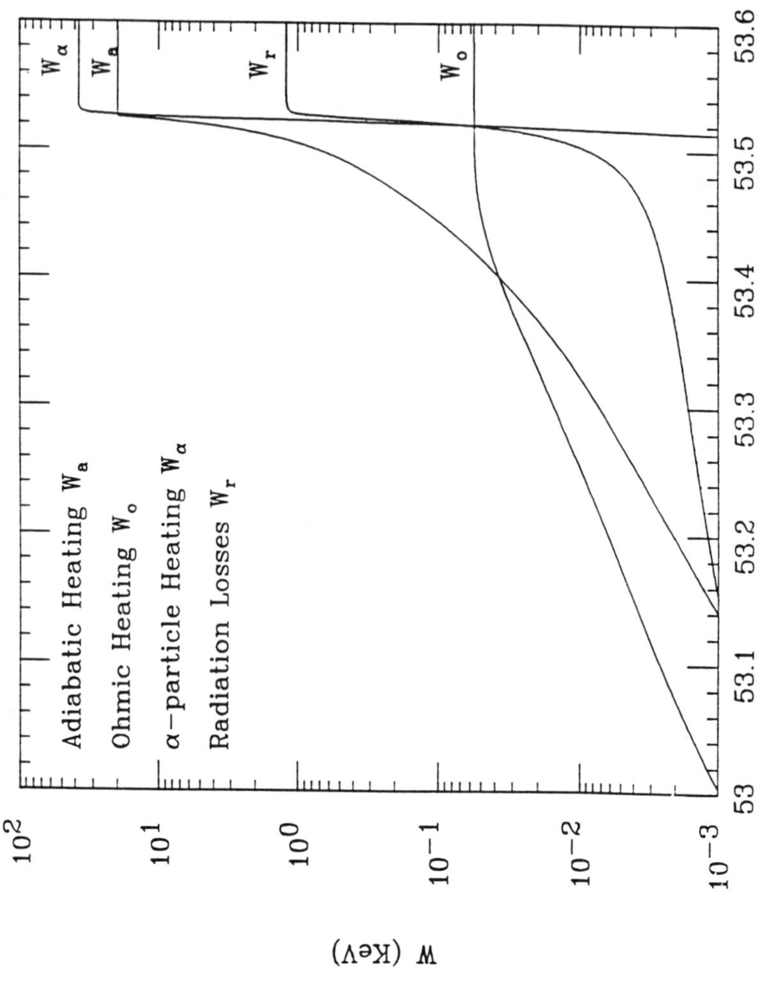

Fig.4 Plots of various heating and loss components as function of time also during last phase of compression.

PROSPECTS FOR FUSION WITH DENSE Z-PINCHES

A.E. Robson
Plasma Physics Division, Naval Research Laboratory
Washington DC 20375, USA

ABSTRACT

Assuming that a quasi-static z-pinch formed from a frozen DT fiber is stable as long as the current is rising, and that the magnetic energy of the pinch can be recovered with > 90% efficiency, it is shown that a pulsed energy-producing cycle is possible yielding about 7.5MJ per pulse from a 30 cm long fiber. Practical details are given of a reactor that would operate at a pulse repetition frequency of 40 Hz, and have an output of about 100 MW(e).

INTRODUCTION

Since a z-pinch is a rather simple way of creating very high temperature and density, albeit for a very short time, it is reasonable to ask whether, and under what circumstances, it might form the basis of a fusion reactor. The question may be resolved into two parts: first, whether a pulsed cycle can be devised in which, at least in principle, a z-pinch would produce significantly more energy in fusion than was used in creating it, and, second, whether a practical means could then be developed for continually repeating the cycle so that an effectively continuous power output could be obtained.

Z-pinches fall into two classes, according to their method of formation. Dynamic pinches, formed by the rapid implosion of a gas shell or a wire array, have been successfully used to create plasma of high energy density for pulsed radiation sources[1]. Work done by pinch forces provides the energy input to the plasma, but the magnetic pressure of the current is usually much less than the stagnation pressure of the pinch, so the only confinement after collapse is inertial. The requirements for fusion in such a system have been discussed by Linhart[2], and are quite formidable. In the quasi-static pinch, on the other hand, the plasma pressure is always balanced by the magnetic pressure and the energy input is by a combination of ohmic heating and adiabatic compression. This means that longer containment times are possible, at least in principle, but the main concern for such a pinch is whether it would remain stable long enough for significant fusion to occur; on the basis of ideal MHD theory, it would not. Experiments on quasi-static pinches formed from frozen deuterium fibers[3,4] appeared to be stable for much longer than predicted by ideal MHD, at least as long as the current was rising[4]. In spite of considerable theoretical effort[5], this result is still unexplained, and it is not clear whether it would obtain at the pinch currents necessary for a fusion reactor. For the purpose of this paper, it will be assumed that it would.

Apart from stability, the principal problem for a fusion pinch is the large amount of magnetic energy in the region surrounding the plasma. It follows simply from the Bennett relation that, for a highly constricted pinch, the magnetic energy in the vacuum is about ten times the thermal energy in the plasma. Unless this energy is recovered, the $n\tau$ product required for breakeven is about ten times the usual Lawson value; efficient recovery of the magnetic energy will thus be an essential feature of any z-pinch reactor.

The first thoughtful study of the dense z-pinch (DZP) as a fusion power reactor was performed by Hagenson et al.[6] who used zero- and one-dimensional models to make preliminary calculations of scaling and systems energy balance. They derived a point design in which the pinch current rose to 1.45 MA in 320 ns and was crowbarred for approximately 2 μs. The burnup fraction was 0.8 and the Q (fusion energy/initial electrical energy) was 33, without any magnetic energy recovery. These remarkable results depended upon three very optimistic assumptions: that the pinch was completely stable throughout the current pulse; that there was no coupling of the α-particle energy into the pinch; and that the return conductor was only 10 cm in diameter. More recent studies by Robson[7,8] have assumed that the pinch is stable only while the current is rising, have included the effect of α-particle heating and have used more realistic reactor dimensions. This paper is a continuation of these studies. The results, though considerably less optimistic than those of Hagenson et al, do not preclude the possibility of a viable fusion system based on the DZP

ZERO-DIMENSIONAL MODELING

In a zero-dimensional pinch model the profiles of density, current density and temperature are kept constant while the current I, the pinch radius a, and the line density N are treated as variables. Pressure equilibrium is assumed at all times, and the temperature is determined from I and N through the Bennett relation. Such modeling is quite appropriate for a quasi-static DZP such as a fiber pinch surrounded by vacuum; it has been shown[6] that it leads to results closely similar to those obtained from one-dimensional codes, and of course allows much faster scanning of parameter space. A detailed description of the model used here has been published elsewhere.[7,8]

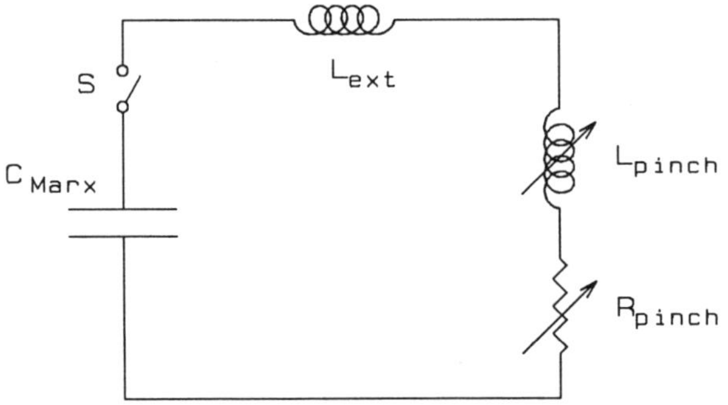

Figure 1. Equivalent circuit of DZP

The pinch is driven by the circuit shown in Fig. 1. This consists of a Marx generator, represented by its erected capacitance C_{Marx}, and an external inductance L_{ext}, associated with the electrodes and the feeds to them; the pinch is represented by a variable resistance R_{pinch} and a variable inductance, L_{pinch}, whose values are derived from the zero-D model. With C charged, the switch S is closed to initiate a half-cycle of current; when the current falls to zero, S is opened, leaving C charged in the opposite polarity. If the final energy in the capacitor is less than the initial energy, the difference represents the net energy E_{net} supplied to the pinch, and is used to derive $Q = E_{fusion}/E_{net}$.

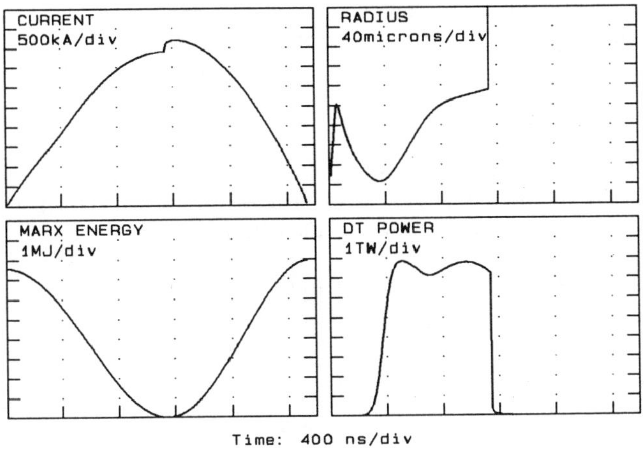

Figure 2. Power-producing cycle of a DZP in DT

TABLE I. Parameters of point design

Fiber diameter	46 μm
Fiber length	30 cm
Return current radius	100 cm
Erected Marx capacitance, C_{Marx}	0.5 μF
External inductance, L_{ext}	0.5 μH
Erected Marx voltage	5.5 MV
Initial energy in Marx	7.56 MJ
Final energy in Marx	7.93 MJ
Fusion energy (17.6 MeV/reaction)	5.84 MJ
Fusion energy (22.4 MeV/reaction[a])	7.43 MJ
Burnup:	0.14

[a] includes energy of neutron capture in Li

Figure 2 shows the results of a typical cycle, whose main parameters are given in Table I. The current rises initially at 4.5 $kA.ns^{-1}$, which is the rate at which fiber pinch experiments have shown enhanced stability. The stages of the pinch are evident in the graph of radius vs time. There is an initial expansion as the solid fiber is converted into plasma (this is simulated by a separate, two-zone model), followed by a contraction as the pinch pressure rises faster than the kinetic pressure. This contraction is assisted by radiation losses. At 400 ns the fusion reaction rate becomes appreciable and the deposition of α-particle energy into the pinch causes a steady expansion, even though the current is still rising. During this pressure-driven expansion, work is done on the magnetic field and some of the α-particle energy is converted into electrical energy. The onset of instability at peak current cannot be modeled in zero dimensions and so is simulated by a large increase in the resistivity of the pinch, leading to rapid expansion and a current fall time that is somewhat shorter than the current rise time. The external circuit in the model is lossless and the final energy in the capacitor is 5% greater than the initial energy, the excess resulting from the direct conversion of α-particle energy. The fractional burnup of the DT fiber is 0.14.

If the magnetic energy of the pinch could be recovered with 95% efficiency, that is, if the circuit losses were no greater than 5% in one half-cycle (which translates to a circuit "Q" of about 60), the sequence described above could be repeated indefinitely with no further external energy input (fusion Q = ∞). If the circuit losses were any greater, they would have to be made up by the diversion of a fraction of the electrical output derived from the fusion thermal energy. It is generally considered uneconomic to divert more than 20% of the output power for this purpose and, assuming a thermal-to-electrical conversion efficiency of 0.33, the required energy recovery efficiency would then be reduced to about 89%.

TABLE II. Sensitivity study for 46 μm fiber

C μF	E_M MJ	E_f MJ	I_{max} MA	f_{bu}	η_0 %	η_{20} %
1.0	15.1	20.4	5.8	0.38	>100	92.4
0.9	13.6	17.4	5.6	0.33	99.8	91.4
0.8	12.1	14.8	5.3	0.28	98.5	90.5
0.7	10.6	12.2	4.9	0.23	97.3	89.7
0.6	9.1	9.7	4.6	0.18	96.2	89.1
0.5	7.6	7.4	4.2	0.14	95.1	88.6
0.4	6.1	5.5	3.8	0.10	94.1	88.2
0.3	4.5	3.7	3.3	0.07	93.8	88.4
0.2	3.0	2.1	2.6	0.04	96.3	91.7

TABLE III. Sensitivity study for 40 μm fiber

C μF	E_M MJ	E_f MJ	I_{max} MA	f_{bu}	η_0 %	η_{20} %
1.0	15.1	22.0	5.7	0.54	>100	97.2
0.9	13.6	19.0	5.5	0.47	>100	95.1
0.8	12.1	15.9	5.2	0.39	>100	93.5
0.7	10.6	13.1	4.9	0.32	>100	92.1
0.6	9.1	10.4	4.5	0.26	98.5	90.9
0.5	7.6	7.9	4.2	0.20	97.1	90.2
0.4	6.1	5.7	3.7	0.14	95.6	89.4
0.3	4.5	3.7	3.2	0.09	94.5	89.1
0.2	3.0	2.0	2.5	0.05	94.8	90.6

The sensitivity of these conclusions to the choice of initial conditions is illustrated in Table II. With all other parameters unchanged, the value of the capacitor C is varied; this varies the initial stored energy and the peak current, but the initial dI/dt is unchanged. The quantities tabulated are: the initial Marx energy, E_M; the total fusion output (22.4 MeV/reaction), E_f; the peak current, I_{max}; the fractional burnup, f_{bu}; the energy recovery efficiency for a self-sustaining cycle, η_0; and the corresponding efficiency for 20% recirculating power.

It can be seen that increasing the peak current increases the fusion output and the fractional burnup, but also requires greater efficiency of energy recovery. Table III gives a corresponding set of figures with the fiber diameter reduced from

46 μm to 40 μm. This leads to somewhat increased fusion output on account of greater fractional burnup, but the requirements for energy recovery are even more stringent. Note that in the absence of energy recovery the value of Q is only about 1, that is, barely breakeven.

If, in an attempt to increase the fusion output, the fiber diameter is increased above 60 μm a new phenomenon is observed: the decrease in pinch radius that follows the initial expansion is not halted by α-particle heating but continues until the pinch radius is < 1 μm. This is the process of radiative collapse[9], which causes most of the pinch energy to be converted into radiation and the fusion yield to be drastically reduced. With fibers of smaller diameter, which reach fusion temperatures at lower currents, the onset of α-particle heating forestalls radiative collapse and allows the Pease-Braginskii current to be exceeded by a large margin.

Surveys of parameter space, such as illustrated in Tables II and III, do not lead to an obviously optimum working point; rather, it is necessary to balance the desirability of maximum fractional burnup with the necessity for efficient energy recovery. It seems that energy recovery efficiencies of 90% to 95% will be essential, and that burnup fractions of 0.1 to 0.2 will have to be accepted. The cycle illustrated by Fig 1 and Table I represents a compromise and has been used as the basis for a conceptual reactor design.

A DZP REACTOR

Because of the incomplete burnup of the already rather small amount of DT fuel in a single fiber, the fusion output per pulse is only 7.43 MJ, including the energy of neutron capture, and so rapid repetition of the cycle is necessary to produce significant power output. The motivation of this work has been the possibility of a naval reactor, for which an output of 100 MW(e), or about 130,000 shp, would be adequate. To generate this output with a thermal energy conversion efficiency of 0.33, the repetition rate must be about 40 Hz, at which frequency the thermal power output is essentially continuous. The technical challenge is to extrapolate the present single-shot z-pinch experiments to a system that will operate continuously at 40 Hz.

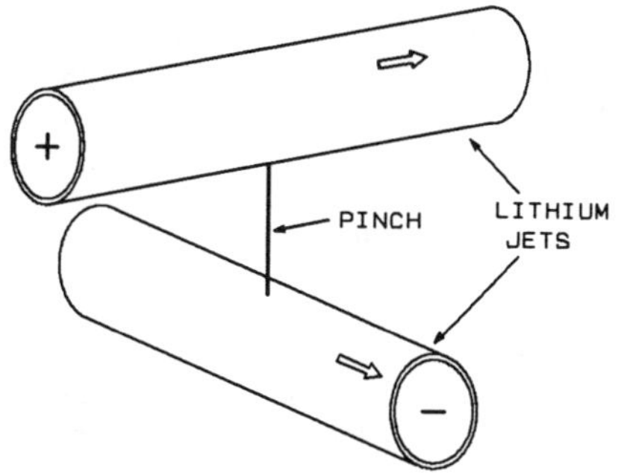

Figure 3. Electrode arrangement consisting of orthogonal lithium jets

Heating and erosion of the electrodes under continuous operation dictate that they should be formed from flowing liquid metal. Since it is virtually impossible to make both electrodes of liquid metal in co-axial geometry an alternative arrangement is proposed that consists of two orthogonal, non-intersecting, hollow jets of liquid lithium in which the pinch is formed across the shortest distance between them (Fig. 3). Each electrode is fed electrically from both ends, which minimizes the inductance of the feeds. An adequate jet velocity is 25 $m.s^{-1}$, which is not difficult to achieve. With this arrangement, the hot spots formed on the electrodes by the pinch (which will continue to evaporate long after the pulse) will be carried away from the pinch region and should not compromise the electrical strength of the system by the time of the next pulse.

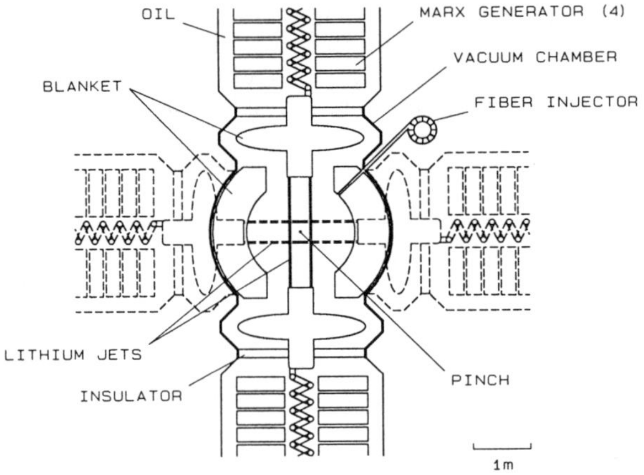

Figure 4. DZP reactor chamber

A conceptual reactor layout that incorporates this electrode arrangement is shown in Fig. 4. Each electrode jet is projected from a large terminal on one side of the reactor vessel and caught in a corresponding terminal on the opposite side. The lithium is carried to and from the terminals through helical pipes whose inductance provides a high impedance to ground. Additional pumps may be situated in or close to the terminals. Each jet is connected via the terminals to two Marx generators of the same polarity, but the electrodes have opposite polarities with respect to ground. The terminals form part of the reactor blanket and shield the insulators from the neutron flux emitted by the pinch. The open circuit voltage is ± 2.75 MV and the size of the terminals and electrodes is arranged so that the maximum electric field in the system does not exceed 100 $kV.cm^{-1}$. The total inductance of this arrangement is estimated as 500 nH, which was the value used in the calculations given above.

The power supply consists of four identical Marx generators arranged as shown in Fig. 5 and Fig 6. Each generator has an erected capacitance of 0.5 μF at 2.75 MV and stores 1.9 MJ. The dimensions of the generators shown here are based on capacitors with a specific energy density of 30 $kJ.m^{-3}$, which is a factor of 20 less than that of modern energy storage capacitors, but a factor of two greater than that of power-factor correction capacitors. To achieve efficient recovery of the magnetic energy the switches must be of very low-loss design, and it is anticipated that the capacitors would be connected in series by light-activated semiconductor switches (LASS) which conduct in either direction on command. Although at present only demonstrated on a small scale, LASS technology[10] has the potential for large scale application and, coupled with expected developments in capacitor technology, could form the basis of very compact and efficient power supplies for this and other applications.

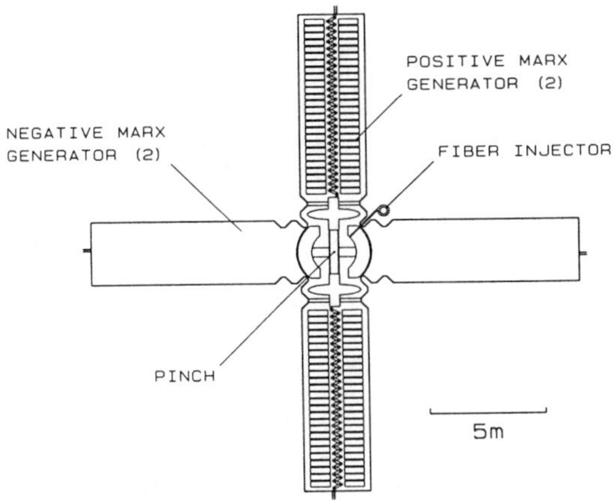

Figure 5. DZP reactor: plan view

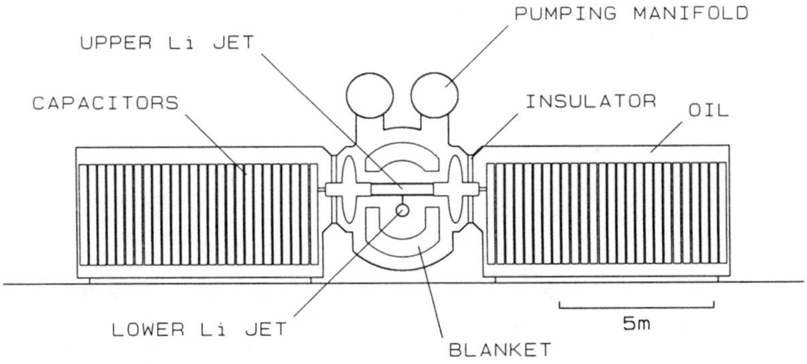

Figure 6. DZP reactor: side view

The fiber injector produces 30-cm long fibers at the rate of 40 per second, and projects them sideways into the gap between the electrodes. In order that there should be only one fiber in the reactor at the time of the pulse, the velocity of projection must be at least 60 $m.s^{-1}$. The fibers are continuously extruded, then cut to length and projected into the chamber by means of a centrifugal device operating at cryogenic temperature. The injection path contains a rotary shutter that closes at the time of the pinch to shield the injector cryogenics from the neutron pulse.

Assuming 14% burnup of the DT fuel, the gas load at 40 Hz operation will be 17 $Torr\text{-}l.s^{-1}$ (DT) and 2.8 $Torr\text{-}l.s^{-1}$ (He). The pumping of DT is accomplished automatically by the exposed liquid lithium surfaces of the electrode jets, which act as continuous getter pumps with a pumping speed of about 3×10^6 $l.s^{-1}$ for DT, resulting in a DT partial pressure of about 6×10^{-6} Torr. The helium is pumped externally by means of large turbomolecular pumps; the pumping speed at the chamber is 5×10^4 l.s^{-1} resulting in a He partial pressure of 6×10^{-5} Torr. This is comparable to the base pressure encountered in most pulsed power experiments. The liquid lithium in the electrodes is injected at the reactor inlet temperature of 350 °C, at which the vapor pressure of lithium is only 8.5×10^{-6} Torr. It is ultimately the pumping of the helium "ash" that limits the repetition rate, and hence the power output, of a reactor with a single pinch chamber.

The blanket structure, including the high-voltage terminals, will be subjected to 14 MeV neutron flux and will require periodic replacement, which should not be a difficult task because of the simple geometry of the system. The closest solid material (the "first wall") is 95 cm from the pinch; the mean neutron wall loading is about 12 MW.m^{-2}, which is comparable to the wall loading in other high-power-density fusion concepts[11].

As well as forming the electrodes, liquid lithium is also used in the blanket for heat removal and tritium breeding. On account of the relatively small burnup, removal of bred tritium from the lithium stream is less of a problem than removal of unburned DT that has been pumped by the electrode jets. It seems that molten salt contactors[12] would be suitable for this task: the number required would be determined by the concentration of tritium in the lithium stream, which in turn would be determined by the allowed tritium inventory. Because the lithium inventory in a DZP reactor will be much smaller than in, say, a tokamak, a higher concentration of tritium may be allowed, resulting in a correspondingly smaller extraction system.

TABLE IV Reactor power densities

System	MPD kW(e)/t	VPD MW(e)/m^3
ARIES tokamak	100	0.6
DZP reactor	5000	3.3
PWR (fission)	1000	5.0

In comparing different fusion reactor concepts, and in comparing fusion reactors with fission reactors, two figures of merit are often used[13]. These are the fusion power core (FPC) mass power density (MPD) expressed in kW(electric)/tonne, and the FPC volume power density (VPD) expressed in MW(electric)/m^3. The larger these figures, the more economically attractive the reactor. In Table IV the DZP reactor is compared with a typical tokamak reactor (ARIES), and the pressurized water fission reactor (PWR).

CONCLUSIONS

Although at the present stage of understanding of DZP physics it may seem premature to discuss a DZP reactor, this study has been undertaken in the belief that proponents of alternative fusion concepts should produce at least zero-order reactor designs based on reasonable, even optimistic, physics assumptions so that the obvious technical problems may be identified at an early stage, and the likelihood of finding solutions may be assessed, before too much effort is expended on a physics program. Most of the technical problems of the DZP seem capable of solution, although the high-voltage power supplies with 95% efficient energy recovery will require the most development; here the technical problem is complicated by the uncertainty of the plasma behavior after the pinch has gone unstable. If the instabilities are axially symmetrical (m = 0), it seems likely that most of the magnetic energy could be recovered. In the event of m = 1 instability, in which magnetic energy is converted into energy of transverse plasma motion, the outlook is less optimistic, but it should be pointed out that, so far, only the m = 0 instability has been seen in DZP experiments.

Apart from the obvious question of stability, there is some concern that deviations from the idealized model used here might reduce the already small region of practicability; for example, any increase in the pinch resistance over the Spitzer value will lower the plasma density, and hence the reaction rate, and reduce the burnup from its already low value. There are obviously serious questions of physics and technology that must be answered before the fusion potential of the DZP can be realistically assessed, but the possibility that it could lead to a small, compact fusion reactor, economically competitive with the PWR, should provide encouragement for further research.

ACKNOWLEDGMENT

Some of the ideas presented here were developed while the author was on sabbatical at Imperial College, and discussions of reactor matters with members of the College are gratefully acknowledged. This work was supported by the Office of Naval Research.

REFERENCES

1. N.R. Pereira and J. Davis, J. Appl. Phys. 63, R1 (1988).
2. J.G. Linhart, Nucl. Fusion 10, 211 (1970).
3. J.E. Hammel and D.W. Scudder, in Proceedings of the 14th European Conference on Controlled Fusion and Plasma Physics, Madrid, 1987 (EPS, Petit-Lancy, Switzerland 1987), Pt. 2, p.450.
4. J.D. Sethian, A.E. Robson, K.A. Gerber and A.W. DeSilva, Phys. Rev. Lett. 59, 892 (1987); ibid. p. 1790.
5. J. Scheffel and M. Coppins, Nucl. Fusion 33, 101 (1993).
6. R.L. Hagenson, A.S. Tai, R.A. Krakowski and R.W. Moses, Nucl. Fusion 21, 1351 (1981).
7. A.E. Robson, Nucl. Fusion 28, 2171 (1988).
8. A.E. Robson, in Proceedings of the Second International Conference on Dense Z-pinches, Laguna Beach, CA, 1989 (Conference Series #195, AIP, NY 1989), p. 362.
9. See, for example, A.E. Robson, Phys. Fluids B 1, 1834 (1989).
10. See, for example, SPIE Proceedings Vol. 1632 Optically Activated Switching II (1992).
11. R.L. Hagenson, R.A. Krakowski and H. Dreicer, Nucl. Instrum. Methods 207, 241 (1983).
12. V.A. Moroni, R.D. Wolson and G.E. Staahl, Nucl. Technology 25, 83 (1975).
13. R.A. Krakowski, Fusion Tech. 20, 121 (1991).

Author Index

A

B

C

D

E

F

G

H

I

J

K

L

M

N

O

P

R

S

AIP Conference Proceedings

		L.C. Number	ISBN
No. 84	Physics in the Steel Industry (APS/AISI, Lehigh University, 1981)	82-72033	0-88318-183-5
No. 85	Proton-Antiproton Collider Physics – 1981 (Madison, WI)	82-72141	0-88318-184-3
No. 86	Momentum Wave Functions – 1982 (Adelaide, Australia)	82-72375	0-88318-185-1
No. 87	Physics of High Energy Particle Accelerators (Fermilab Summer School, 1981)	82-72421	0-88318-186-X
No. 88	Mathematical Methods in Hydrodynamics and Integrability in Dynamical Systems (La Jolla Institute, 1981)	82-72462	0-88318-187-8
No. 89	Neutron Scattering – 1981 (Argonne National Laboratory)	82-73094	0-88318-188-6
No. 90	Laser Techniques for Extreme Ultraviolet Spectroscopy (Boulder, CO, 1982)	82-73205	0-88318-189-4
No. 91	Laser Acceleration of Particles (Los Alamos, NM, 1982)	82-73361	0-88318-190-8
No. 92	The State of Particle Accelerators and High Energy Physics (Fermilab, 1981)	82-73861	0-88318-191-6
No. 93	Novel Results in Particle Physics (Vanderbilt, 1982)	82-73954	0-88318-192-4
No. 94	X-Ray and Atomic Inner-Shell Physics – 1982 (International Conference, U. of Oregon)	82-74075	0-88318-193-2
No. 95	High Energy Spin Physics – 1982 (Brookhaven National Laboratory)	83-70154	0-88318-194-0
No. 96	Science Underground (Los Alamos, NM, 1982)	83-70377	0-88318-195-9
No. 97	The Interaction Between Medium Energy Nucleons in Nuclei – 1982 (Indiana University)	83-70649	0-88318-196-7
No. 98	Particles and Fields – 1982 (APS/DPF University of Maryland)	83-70807	0-88318-197-5
No. 99	Neutrino Mass and Gauge Structure of Weak Interactions (Telemark, 1982)	83-71072	0-88318-198-3
No. 100	Excimer Lasers – 1983 (OSA, Lake Tahoe, NV)	83-71437	0-88318-199-1
No. 101	Positron-Electron Pairs in Astrophysics (Goddard Space Flight Center, 1983)	83-71926	0-88318-200-9

No. 102	Intense Medium Energy Sources of Strangeness (UC-Santa Cruz, CA, 1983)	83-72261	0-88318-201-7
No. 103	Quantum Fluids and Solids – 1983 (Sanibel Island, FL)	83-72440	0-88318-202-5
No. 104	Physics, Technology and the Nuclear Arms Race (APS, Baltimore, MD, 1983)	83-72533	0-88318-203-3
No. 105	Physics of High Energy Particle Accelerators (SLAC Summer School, 1982)	83-72986	0-88318-304-8
No. 106	Predictability of Fluid Motions (La Jolla Institute, 1983)	83-73641	0-88318-305-6
No. 107	Physics and Chemistry of Porous Media (Schlumberger-Doll Research, 1983)	83-73640	0-88318-306-4
No. 108	The Time Projection Chamber (TRIUMF, Vancouver, 1983)	83-83445	0-88318-307-2
No. 109	Random Walks and Their Applications in the Physical and Biological Sciences (NBS/La Jolla Institute, 1982)	84-70208	0-88318-308-0
No. 110	Hadron Substructure in Nuclear Physics (Indiana University, 1983)	84-70165	0-88318-309-9
No. 111	Production and Neutralization of Negative Ions and Beams (3rd Int'l Symposium) (Brookhaven, NY, 1983)	84-70379	0-88318-310-2
No. 112	Particles and Fields – 1983 (APS/DPF, Blacksburg, VA)	84-70378	0-88318-311-0
No. 113	Experimental Meson Spectroscopy – 1983 (7th International Conference, Brookhaven, NY)	84-70910	0-88318-312-9
No. 114	Low Energy Tests of Conservation Laws in Particle Physics (Blacksburg, VA, 1983)	84-71157	0-88318-313-7
No. 115	High Energy Transients in Astrophysics (Santa Cruz, CA, 1983)	84-71205	0-88318-314-5
No. 116	Problems in Unification and Supergravity (La Jolla Institute, 1983)	84-71246	0-88318-315-3
No. 117	Polarized Proton Ion Sources (TRIUMF, Vancouver, 1983)	84-71235	0-88318-316-1
No. 118	Free Electron Generation of Extreme Ultraviolet Coherent Radiation (Brookhaven/OSA, 1983)	84-71539	0-88318-317-X
No. 119	Laser Techniques in the Extreme Ultraviolet (OSA, Boulder, CO, 1984)	84-72128	0-88318-318-8

No. 120	Optical Effects in Amorphous Semiconductors (Snowbird, UT, 1984)	84-72419	0-88318-319-6
No. 121	High Energy e^+e^- Interactions (Vanderbilt, 1984)	84-72632	0-88318-320-X
No. 122	The Physics of VLSI (Xerox, Palo Alto, CA, 1984)	84-72729	0-88318-321-8
No. 123	Intersections Between Particle and Nuclear Physics (Steamboat Springs, CO, 1984)	84-72790	0-88318-322-6
No. 124	Neutron-Nucleus Collisions: A Probe of Nuclear Structure (Burr Oak State Park, 1984)	84-73216	0-88318-323-4
No. 125	Capture Gamma-Ray Spectroscopy and Related Topics – 1984 (Int'l Symposium, Knoxville, TN)	84-73303	0-88318-324-2
No. 126	Solar Neutrinos and Neutrino Astronomy (Homestake, 1984)	84-63143	0-88318-325-0
No. 127	Physics of High Energy Particle Accelerators (BNL/SUNY Summer School, 1983)	85-70057	0-88318-326-9
No. 128	Nuclear Physics with Stored, Cooled Beams (McCormick's Creek State Park, IN, 1984)	85-71167	0-88318-327-7
No. 129	Radiofrequency Plasma Heating (Sixth Topical Conference) (Callaway Gardens, GA, 1985)	85-48027	0-88318-328-5
No. 130	Laser Acceleration of Particles (Malibu, CA, 1985)	85-48028	0-88318-329-3
No. 131	Workshop on Polarized ^{3}He Beams and Targets (Princeton, NJ, 1984)	85-48026	0-88318-330-7
No. 132	Hadron Spectroscopy – 1985 (International Conference, Univ. of Maryland)	85-72537	0-88318-331-5
No. 133	Hadronic Probes and Nuclear Interactions (Arizona State University, 1985)	85-72638	0-88318-332-3
No. 134	The State of High Energy Physics (BNL/SUNY Summer School, 1983)	85-73170	0-88318-333-1
No. 135	Energy Sources: Conservation and Renewables (APS, Washington, DC, 1985)	85-73019	0-88318-334-X
No. 136	Atomic Theory Workshop on Relativistic and QED Effects in Heavy Atoms (Gaithersburg, MD, 1985)	85-73790	0-88318-335-8
No. 137	Polymer-Flow Interaction (La Jolla Institute, 1985)	85-73915	0-88318-336-6
No. 138	Frontiers in Electronic Materials and Processing (Houston, TX, 1985)	86-70108	0-88318-337-4

No. 139	High-Current, High-Brightness, and High-Duty Factor Ion Injectors (La Jolla Institute, 1985)	86-70245	0-88318-338-2
No. 140	Boron-Rich Solids (Albuquerque, NM, 1985)	86-70246	0-88318-339-0
No. 141	Gamma-Ray Bursts (Stanford, CA, 1984)	86-70761	0-88318-340-4
No. 142	Nuclear Structure at High Spin, Excitation, and Momentum Transfer (Indiana University, 1985)	86-70837	0-88318-341-2
No. 143	Mexican School of Particles and Fields (Oaxtepec, México, 1984)	86-81187	0-88318-342-0
No. 144	Magnetospheric Phenomena in Astrophysics (Los Alamos, NM, 1984)	86-71149	0-88318-343-9
No. 145	Polarized Beams at SSC & Polarized Antiprotons (Ann Arbor, MI & Bodega Bay, CA, 1985)	86-71343	0-88318-344-7
No. 146	Advances in Laser Science—I (Dallas, TX, 1985)	86-71536	0-88318-345-5
No. 147	Short Wavelength Coherent Radiation: Generation and Applications (Monterey, CA, 1986)	86-71674	0-88318-346-3
No. 148	Space Colonization: Technology and The Liberal Arts (Geneva, NY, 1985)	86-71675	0-88318-347-1
No. 149	Physics and Chemistry of Protective Coatings (Universal City, CA, 1985)	86-72019	0-88318-348-X
No. 150	Intersections Between Particle and Nuclear Physics (Lake Louise, Canada, 1986)	86-72018	0-88318-349-8
No. 151	Neural Networks for Computing (Snowbird, UT, 1986)	86-72481	0-88318-351-X
No. 152	Heavy Ion Inertial Fusion (Washington, DC, 1986)	86-73185	0-88318-352-8
No. 153	Physics of Particle Accelerators (SLAC Summer School, 1985) (Fermilab Summer School, 1984)	87-70103	0-88318-353-6
No. 154	Physics and Chemistry of Porous Media—II (Ridgefield, CT, 1986)	83-73640	0-88318-354-4
No. 155	The Galactic Center: Proceedings of the Symposium Honoring C. H. Townes (Berkeley, CA, 1986)	86-73186	0-88318-355-2
No. 156	Advanced Accelerator Concepts (Madison, WI, 1986)	87-70635	0-88318-358-0
No. 157	Stability of Amorphous Silicon Alloy Materials and Devices (Palo Alto, CA, 1987)	87-70990	0-88318-359-9
No. 158	Production and Neutralization of Negative Ions and Beams (Brookhaven, NY, 1986)	87-71695	0-88318-358-7

No. 159	Applications of Radio-Frequency Power to Plasma: Seventh Topical Conference (Kissimmee, FL, 1987)	87-71812	0-88318-359-5
No. 160	Advances in Laser Science—II (Seattle, WA, 1986)	87-71962	0-88318-360-9
No. 161	Electron Scattering in Nuclear and Particle Science: In Commemoration of the 35th Anniversary of the Lyman-Hanson-Scott Experiment (Urbana, IL, 1986)	87-72403	0-88318-361-7
No. 162	Few-Body Systems and Multiparticle Dynamics (Crystal City, VA, 1987)	87-72594	0-88318-362-5
No. 163	Pion–Nucleus Physics: Future Directions and New Facilities at LAMPF (Los Alamos, NM, 1987)	87-72961	0-88318-363-3
No. 164	Nuclei Far from Stability: Fifth International Conference (Rosseau Lake, ON, 1987)	87-73214	0-88318-364-1
No. 165	Thin Film Processing and Characterization of High-Temperature Superconductors (Anaheim, CA, 1987)	87-73420	0-88318-365-X
No. 166	Photovoltaic Safety (Denver, CO, 1988)	88-42854	0-88318-366-8
No. 167	Deposition and Growth: Limits for Microelectronics (Anaheim, CA, 1987)	88-71432	0-88318-367-6
No. 168	Atomic Processes in Plasmas (Santa Fe, NM, 1987)	88-71273	0-88318-368-4
No. 169	Modern Physics in America: A Michelson-Morley Centennial Symposium (Cleveland, OH, 1987)	88-71348	0-88318-369-2
No. 170	Nuclear Spectroscopy of Astrophysical Sources (Washington, DC, 1987)	88-71625	0-88318-370-6
No. 171	Vacuum Design of Advanced and Compact Synchrotron Light Sources (Upton, NY, 1988)	88-71824	0-88318-371-4
No. 172	Advances in Laser Science—III: Proceedings of the International Laser Science Conference (Atlantic City, NJ, 1987)	88-71879	0-88318-372-2
No. 173	Cooperative Networks in Physics Education (Oaxtepec, Mexico, 1987)	88-72091	0-88318-373-0
No. 174	Radio Wave Scattering in the Interstellar Medium (San Diego, CA, 1988)	88-72092	0-88318-374-9
No. 175	Non-neutral Plasma Physics (Washington, DC, 1988)	88-72275	0-88318-375-7
No. 176	Intersections Between Particle and Nuclear Physics (Third International Conference) (Rockport, ME, 1988)	88-62535	0-88318-376-5

No. 177	Linear Accelerator and Beam Optics Codes (La Jolla, CA, 1988)	88-46074	0-88318-377-3
No. 178	Nuclear Arms Technologies in the 1990s (Washington, DC, 1988)	88-83262	0-88318-378-1
No. 179	The Michelson Era in American Science: 1870–1930 (Cleveland, OH, 1987)	88-83369	0-88318-379-X
No. 180	Frontiers in Science: International Symposium (Urbana, IL, 1987)	88-83526	0-88318-380-3
No. 181	Muon-Catalyzed Fusion (Sanibel Island, FL, 1988)	88-83636	0-88318-381-1
No. 182	High T_c Superconducting Thin Films, Devices, and Applications (Atlanta, GA, 1988)	88-03947	0-88318-382-X
No. 183	Cosmic Abundances of Matter (Minneapolis, MN, 1988)	89-80147	0-88318-383-8
No. 184	Physics of Particle Accelerators (Ithaca, NY, 1988)	89-83575	0-88318-384-6
No. 185	Glueballs, Hybrids, and Exotic Hadrons (Upton, NY, 1988)	89-83513	0-88318-385-4
No. 186	High-Energy Radiation Background in Space (Sanibel Island, FL, 1987)	89-83833	0-88318-386-2
No. 187	High-Energy Spin Physics (Minneapolis, MN, 1988)	89-83948	0-88318-387-0
No. 188	International Symposium on Electron Beam Ion Sources and their Applications (Upton, NY, 1988)	89-84343	0-88318-388-9
No. 189	Relativistic, Quantum Electrodynamic, and Weak Interaction Effects in Atoms (Santa Barbara, CA, 1988)	89-84431	0-88318-389-7
No. 190	Radio-frequency Power in Plasmas (Irvine, CA, 1989)	89-45805	0-88318-397-8
No. 191	Advances in Laser Science—IV (Atlanta, GA, 1988)	89-85595	0-88318-391-9
No. 192	Vacuum Mechatronics (First International Workshop) (Santa Barbara, CA, 1989)	89-45905	0-88318-394-3
No. 193	Advanced Accelerator Concepts (Lake Arrowhead, CA, 1989)	89-45914	0-88318-393-5
No. 194	Quantum Fluids and Solids—1989 (Gainesville, FL, 1989)	89-81079	0-88318-395-1
No. 195	Dense *Z*-Pinches (Laguna Beach, CA, 1989)	89-46212	0-88318-396-X
No. 196	Heavy Quark Physics (Ithaca, NY, 1989)	89-81583	0-88318-644-6

No. 197	Drops and Bubbles (Monterey, CA, 1988)	89-46360	0-88318-392-7
No. 198	Astrophysics in Antarctica (Newark, DE, 1989)	89-46421	0-88318-398-6
No. 199	Surface Conditioning of Vacuum Systems (Los Angeles, CA, 1989)	89-82542	0-88318-756-6
No. 200	High T_c Superconducting Thin Films: Processing, Characterization, and Applications (Boston, MA, 1989)	90-80006	0-88318-759-0
No. 201	QED Structure Functions (Ann Arbor, MI, 1989)	90-80229	0-88318-671-3
No. 202	NASA Workshop on Physics From a Lunar Base (Stanford, CA, 1989)	90-55073	0-88318-646-2
No. 203	Particle Astrophysics: The NASA Cosmic Ray Program for the 1990s and Beyond (Greenbelt, MD, 1989)	90-55077	0-88318-763-9
No. 204	Aspects of Electron-Molecule Scattering and Photoionization (New Haven, CT, 1989)	90-55175	0-88318-764-7
No. 205	The Physics of Electronic and Atomic Collisions (XVI International Conference) (New York, NY, 1989)	90-53183	0-88318-390-0
No. 206	Atomic Processes in Plasmas (Gaithersburg, MD, 1989)	90-55265	0-88318-769-8
No. 207	Astrophysics from the Moon (Annapolis, MD, 1990)	90-55582	0-88318-770-1
No. 208	Current Topics in Shock Waves (Bethlehem, PA, 1989)	90-55617	0-88318-776-0
No. 209	Computing for High Luminosity and High Intensity Facilities (Santa Fe, NM, 1990)	90-55634	0-88318-786-8
No. 210	Production and Neutralization of Negative Ions and Beams (Brookhaven, NY, 1990)	90-55316	0-88318-786-8
No. 211	High-Energy Astrophysics in the 21st Century (Taos, NM, 1989)	90-55644	0-88318-803-1
No. 212	Accelerator Instrumentation (Brookhaven, NY, 1989)	90-55838	0-88318-645-4
No. 213	Frontiers in Condensed Matter Theory (New York, NY, 1989)	90-6421	0-88318-771-X 0-88318-772-8 (pbk.)
No. 214	Beam Dynamics Issues of High-Luminosity Asymmetric Collider Rings (Berkeley, CA, 1990)	90-55857	0-88318-767-1
No. 215	X-Ray and Inner-Shell Processes (Knoxville, TN, 1990)	90-84700	0-88318-790-6

No. 216	Spectral Line Shapes, Vol. 6 (Austin, TX, 1990)	90-06278	0-88318-791-4
No. 217	Space Nuclear Power Systems (Albuquerque, NM, 1991)	90-56220	0-88318-838-4
No. 218	Positron Beams for Solids and Surfaces (London, Canada, 1990)	90-56407	0-88318-842-2
No. 219	Superconductivity and Its Applications (Buffalo, NY, 1990)	91-55020	0-88318-835-X
No. 220	High Energy Gamma-Ray Astronomy (Ann Arbor, MI, 1990)	91-70876	0-88318-812-0
No. 221	Particle Production Near Threshold (Nashville, IN, 1990)	91-55134	0-88318-829-5
No. 222	After the First Three Minutes (College Park, MD, 1990)	91-55214	0-88318-828-7
No. 223	Polarized Collider Workshop (University Park, PA, 1990)	91-71303	0-88318-826-0
No. 224	LAMPF Workshop on (π, K) Physics (Los Alamos, NM, 1990)	91-71304	0-88318-825-2
No. 225	Half Collision Resonance Phenomena in Molecules (Caracas, Venezuela, 1990)	91-55210	0-88318-840-6
No. 226	The Living Cell in Four Dimensions (Gif sur Yvette, France, 1990)	91-55209	0-88318-794-9
No. 227	Advanced Processing and Characterization Technologies (Clearwater, FL, 1991)	91-55194	0-88318-910-0
No. 228	Anomalous Nuclear Effects in Deuterium/Solid Systems (Provo, UT, 1990)	91-55245	0-88318-833-3
No. 229	Accelerator Instrumentation (Batavia, IL, 1990)	91-55347	0-88318-832-1
No. 230	Nonlinear Dynamics and Particle Acceleration (Tsukuba, Japan, 1990)	91-55348	0-88318-824-4
No. 231	Boron-Rich Solids (Albuquerque, NM, 1990)	91-53024	0-88318-793-4
No. 232	Gamma-Ray Line Astrophysics (Paris-Saclay, France, 1990)	91-55492	0-88318-875-9
No. 233	Atomic Physics 12 (Ann Arbor, MI, 1990)	91-55595	088318-811-2
No. 234	Amorphous Silicon Materials and Solar Cells (Denver, CO, 1991)	91-55575	088318-831-7

No. 235	Physics and Chemistry of MCT and Novel IR Detector Materials (San Francisco, CA, 1990)	91-55493	0-88318-931-3
No. 236	Vacuum Design of Synchrotron Light Sources (Argonne, IL, 1990)	91-55527	0-88318-873-2
No. 237	Kent M. Terwilliger Memorial Symposium (Ann Arbor, MI, 1989)	91-55576	0-88318-788-4
No. 238	Capture Gamma-Ray Spectroscopy (Pacific Grove, CA, 1990)	91-57923	0-88318-830-9
No. 239	Advances in Biomolecular Simulations (Obernai, France, 1991)	91-58106	0-88318-940-2
No. 240	Joint Soviet-American Workshop on the Physics of Semiconductor Lasers (Leningrad, USSR, 1991)	91-58537	0-88318-936-4
No. 241	Scanned Probe Microscopy (Santa Barbara, CA, 1991)	91-76758	0-88318-816-3
No. 242	Strong, Weak, and Electromagnetic Interactions in Nuclei, Atoms, and Astrophysics: A Workshop in Honor of Stewart D. Bloom's Retirement (Livermore, CA, 1991)	91-76876	0-88318-943-7
No. 243	Intersections Between Particle and Nuclear Physics (Tucson, AZ, 1991)	91-77580	0-88318-950-X
No. 244	Radio Frequency Power in Plasmas (Charleston, SC, 1991)	91-77853	0-88318-937-2
No. 245	Basic Space Science (Bangalore, India, 1991)	91-78379	0-88318-951-8
No. 246	Space Nuclear Power Systems (Albuquerque, NM, 1992)	91-58793	1-56396-027-3 1-56396-026-5 (pbk.)
No. 247	Global Warming: Physics and Facts (Washington, DC, 1991)	91-78423	0-88318-932-1
No. 248	Computer-Aided Statistical Physics (Taipei, Taiwan, 1991)	91-78378	0-88318-942-9
No. 249	The Physics of Particle Accelerators (Upton, NY, 1989, 1990)	92-52843	0-88318-789-2
No. 250	Towards a Unified Picture of Nuclear Dynamics (Nikko, Japan, 1991)	92-70143	0-88318-951-8
No. 251	Superconductivity and its Applications (Buffalo, NY, 1991)	92-52726	1-56396-016-8

No. 252	Accelerator Instrumentation (Newport News, VA, 1991)	92-70356	0-88318-934-8
No. 253	High-Brightness Beams for Advanced Accelerator Applications (College Park, MD, 1991)	92-52705	0-88318-947-X
No. 254	Testing the AGN Paradigm (College Park, MD, 1991)	92-52780	1-56396-009-5
No. 255	Advanced Beam Dynamics Workshop on Effects of Errors in Accelerators, Their Diagnosis and Corrections (Corpus Christi, TX, 1991)	92-52842	1-56396-006-0
No. 256	Slow Dynamics in Condensed Matter (Fukuoka, Japan, 1991)	92-53120	0-88318-938-0
No. 257	Atomic Processes in Plasmas (Portland, ME, 1991)	91-08105	0-88318-939-9
No. 258	Synchrotron Radiation and Dynamic Phenomena (Grenoble, France, 1991)	92-53790	1-56396-008-7
No. 259	Future Directions in Nuclear Physics with 4π Gamma Detection Systems of the New Generation (Strasbourg, France, 1991)	92-53222	0-88318-952-6
No. 260	Computational Quantum Physics (Nashville, TN, 1991)	92-71777	0-88318-933-X
No. 261	Rare and Exclusive B&K Decays and Novel Flavor Factories (Santa Monica, CA, 1991)	92-71873	1-56396-055-9
No. 262	Molecular Electronics—Science and Technology (St. Thomas, Virgin Islands, 1991)	92-72210	1-56396-041-9
No. 263	Stress-Induced Phenomena in Metallization: First International Workshop (Ithaca, NY, 1991)	92-72292	1-56396-082-6
No. 264	Particle Acceleration in Cosmic Plasmas (Newark, DE, 1991)	92-73316	0-88318-948-8
No. 265	Gamma-Ray Bursts (Huntsville, AL, 1991)	92-73456	1-56396-018-4
No. 266	Group Theory in Physics (Cocoyoc, Morelos, Mexico, 1991)	92-73457	1-56396-101-6
No. 267	Electromechanical Coupling of the Solar Atmosphere (Capri, Italy, 1991)	92-82717	1-56396-110-5
No. 268	Photovoltaic Advanced Research & Development Project (Denver, CO, 1992)	92-74159	1-56396-056-7

No. 269	CEBAF 1992 Summer Workshop (Newport News, VA, 1992)	92-75403	1-56396-067-2
No. 270	Time Reversal—The Arthur Rich Memorial Symposium (Ann Arbor, MI, 1991)	92-83852	1-56396-105-9
No. 271	Tenth Symposium Space Nuclear Power and Propulsion (Vols. I–III) (Albuquerque, NM, 1993)	92-75162	1-56396-137-7 (set)
No. 272	Proceedings of the XXVI International Conference on High Energy Physics (Vols. I and II) (Dallas, TX, 1992)	93-70412	1-56396-127-X (set)
No. 273	Superconductivity and Its Applications (Buffalo, NY, 1992)	93-70502	1-56396-189-X
No. 274	VIth International Conference on the Physics of Highly Charged Ions (Manhattan, KS, 1992)	93-70577	1-56396-102-4
No. 275	Atomic Physics 13 (Munich, Germany, 1992)	93-70826	1-56396-057-5
No. 276	Very High Energy Cosmic-Ray Interactions: VIIth International Symposium (Ann Arbor, MI, 1992)	93-71342	1-56396-038-9
No. 277	The World at Risk: Natural Hazards and Climate Change (Cambridge, MA, 1992)	93-71333	1-56396-066-4
No. 278	Back to the Galaxy (College Park, MD, 1992)	93-71543	1-56396-227-6
No. 279	Advanced Accelerator Concepts (Port Jefferson, NY, 1992)	93-71773	1-56396-191-1
No. 280	Compton Gamma-Ray Observatory (St. Louis, MO, 1992)	93-71830	1-56396-104-0
No. 281	Accelerator Instrumentation Fourth Annual Workshop (Berkeley, CA, 1992)	93-072110	1-56396-190-3
No. 282	Quantum 1/f Noise & Other Low Frequency Fluctuations in Electronic Devices (St. Louis, MO, 1992)	93-072366	1-56396-252-7
No. 283	Earth and Space Science Information Systems (Pasadena, CA, 1992)	93-072360	1-56396-094-X